Lecture Notes in Artificial Intel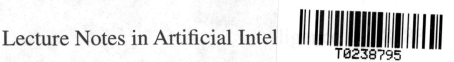

Subseries of Lecture Notes in Computer Science

LNAI Series Editors

Randy Goebel
  University of Alberta, Edmonton, Canada
Yuzuru Tanaka
  Hokkaido University, Sapporo, Japan
Wolfgang Wahlster
  DFKI and Saarland University, Saarbrücken, Germany

LNAI Founding Series Editor

Joerg Siekmann
  DFKI and Saarland University, Saarbrücken, Germany

Hendrik Blockeel   Kristian Kersting
Siegfried Nijssen   Filip Železný (Eds.)

Machine Learning and Knowledge Discovery in Databases

European Conference, ECML PKDD 2013
Prague, Czech Republic, September 23-27, 2013
Proceedings, Part I

 Springer

Volume Editors

Hendrik Blockeel
KU Leuven, Department of Computer Science
Celestijnenlaan 200A, 3001 Leuven, Belgium
E-mail: hendrik.blockeel@cs.kuleuven.be

Kristian Kersting
University of Bonn, Fraunhofer IAIS, Department of Knowledge Discovery
Schloss Birlinghoven, 53754 Sankt Augustin, Germany
E-mail: kristian.kersting@iais.fraunhofer.de

Siegfried Nijssen
Universiteit Leiden, LIACS, Niels Bohrweg 1, 2333 CA Leiden, The Netherlands
and KU Leuven, Department of Computer Science, 3001 Leuven, Belgium
E-mail: snijssen@liacs.nl

Filip Železný
Czech Technical University, Department of Computer Science and Engineering
Technicka 2, 16627 Prague 6, Czech Republic
E-mail: zelezny@fel.cvut.cz

Cover image: © eleephotography

ISSN 0302-9743   e-ISSN 1611-3349
ISBN 978-3-642-40987-5   e-ISBN 978-3-642-40988-2
DOI 10.1007/978-3-642-40988-2
Springer Heidelberg New York Dordrecht London

Library of Congress Control Number: 2013948101

CR Subject Classification (1998): I.2.6, H.2.8, I.5.2, G.2.2, G.3, I.2.4, I.2.7, H.3.4-5, I.2.9, F.2

LNCS Sublibrary: SL 7 – Artificial Intelligence

Typesetting: Camera-ready by author, data conversion by Scientific Publishing Services, Chennai, India

Printed on acid-free paper

Springer is part of Springer Science+Business Media (www.springer.com)

Preface

These are the proceedings of the 2013 edition of the European Conference on Machine Learning and Principles and Practice of Knowledge Discovery in Databases, or ECML PKDD for short. This conference series has grown out of the former ECML and PKDD conferences, which were Europe's premier conferences on, respectively, Machine Learning and Knowledge Discovery in Databases. Organized jointly for the first time in 2001, these conferences have become increasingly integrated, and became one in 2008. Today, ECML PKDD is a world–leading conference in these areas, well–known in particular for having a highly diverse program that aims at exploiting the synergies between these two different, yet related, scientific fields.

ECML PKDD 2013 was held in Prague, Czech Republic, during September 23–27. Continuing the series' tradition, the conference combined an extensive technical program with a variety of workshops and tutorials, a demo track for system demonstrations, an industrial track, a nectar track focusing on particularly interesting results from neighboring areas, a discovery challenge, two poster sessions, and a rich social program.

The main technical program included five plenary talks by invited speakers (Rayid Ghani, Thorsten Joachims, Ulrike von Luxburg, Christopher Re and John Shawe-Taylor) and a record–breaking 138 technical talks, for which further discussion opportunities were provided during two poster sessions. The industrial track had four invited speakers: Andreas Antrup (Zalando), Ralf Herbrich (Amazon Berlin), Jean-Paul Schmetz (Hubert Burda Media), and Hugo Zaragoza (Websays). The demo track featured 11 software demonstrations, and the nectar track 5 talks. The discovery challenge, this year, focused on the task of recommending given names for children to soon–to–be–parents. Twelve workshops were held: Scalable Decision Making; Music and Machine Learning; Reinforcement Learning with Generalized Feedback; Languages for Data Mining and Machine Learning; Data Mining on Linked Data; Mining Ubiquitous and Social Environments; Tensor Methods in Machine Learning; Solving Complex Machine Learning Problems with Ensemble Methods; Sports Analytics; New Frontiers in Mining Complex Pattern; Data Analytics for Renewable Energy Integration; and Real–World Challenges for Data Stream Mining. Eight tutorials completed the program: Multi–Agent Reinforcement Learning; Second Order Learning; Algorithmic Techniques for Modeling and Mining Large Graphs; Web Scale Information Extraction; Mining and Learning with Network–Structured Data; Performance Evaluation of Machine Learning Algorithms; Discovering Roles and Anomalies in Graphs: Theory and Applications; and Statistically Sound Pattern Discovery.

The conference offered awards for distinguished papers, for the paper from ECML / PKDD 2003 with the highest impact after a decade, and for the best demonstration. In addition, there was the novel Open Science Award. This award was installed in order to promote reusability of software, data, and experimental setups, with the aim of improving reproducibility of research and facilitating research that builds on other authors' work.

For the first time, the conference used a mixed submission model: work could be submitted as a journal article to Machine Learning or Data Mining and Knowledge Discovery, or it could be submitted for publication in the conference proceedings. A total of 182 original manuscripts were submitted to the journal track, and 447 to the proceedings track. Of the journal submissions, 14 have been published in the journal, as part of a special issue on ECML PKDD 2013, and 14 have been redirected to the proceedings track. Among the latter, 13 were accepted for publication in the proceedings. Finally, of the 447 submissions to the proceedings track, 111 have been accepted. Overall, this gives a record number of 629 submissions, of which 138 have been scheduled for presentation at the conference, making the overall acceptance rate 21.9%.

The mixed submission model was introduced in an attempt to improve the efficiency and reliability of the reviewing process. Reacting to criticism on the conference–based publication model that is so typical for computer science, several conferences have started experimenting with multiple reviewing rounds, continuous submission, and publishing contributions in a journal instead of the conference proceedings. The ECML PKDD model has been designed to maximally exploit the already existing infrastructure for journal reviewing. For an overview of the motivation and expected benefits of this new model, we refer to *A Revised Publication Model for ECML PKDD*, available at arXiv:1207.6324.

These proceedings of the 2013 European Conference on Machine Learning and Principles and Practice of Knowledge Discovery in Databases contain full papers of work presented at the main technical track, abstracts of the journal articles and invited talks presented there, and short papers describing the demonstrations and nectar papers. We thank the chairs of the demo track (Andreas Hotho and Joaquin Vanschoren), the nectar track (Rosa Meo and Michèle Sebag), and the industrial track (Ulf Brefeld), as well as the proceedings chairs Yamuna Krishnamurthy and Nico Piatkowski, for their help with putting these proceedings together. Most importantly, of course, we thank the authors for their contributions, and the area chairs and reviewers for their substantial efforts to guarantee and sometimes even improve the quality of these proceedings. We wish the reader an enjoyable experience exploring the many exciting research results presented here.

July 2013

Hendrik Blockeel
Kristian Kersting
Siegfried Nijssen
Filip Železný

Organization

ECML PKDD 2013 was organized by the Intelligent Data Analysis Research Lab, Department of Computer Science and Engineering, of the Czech Technical University in Prague.

Conference Chair

Filip Železný   Czech Technical University in Prague,
                              Czech Republic

Program Chairs

Hendrik Blockeel   KU Leuven, Belgium & Leiden University,
                              The Netherlands
Kristian Kersting   University of Bonn & Fraunhofer IAIS,
                              Germany
Siegfried Nijssen   Leiden University, The Netherlands & KU
                              Leuven, Belgium
Filip Železný   Czech Technical University in Prague,
                              Czech Republic

Local Chair

Jiří Kléma   Czech Technical University in Prague,
                              Czech Republic

Publicity Chair

Élisa Fromont   Université Jean Monnet, France

Proceedings Chairs

Yamuna Krishnamurthy   TU Dortmund, Germany
Nico Piatkowski   TU Dortmund, Germany

Workshop Chairs

Niels Landwehr   University of Potsdam, Germany
Andrea Passerini   University of Trento, Italy

Tutorial Chairs

Kurt Driessens   Maastricht University, The Netherlands
Sofus A. Macskassy   University of Southern California, USA

Demo Track Chairs

Andreas Hotho   University of Würzburg, Germany
Joaquin Vanschoren   Leiden University, The Netherlands

Nectar Track Chairs

Rosa Meo   University of Turin, Italy
Michèle Sebag   Université Paris-Sud, France

Industrial Track Chair

Ulf Brefeld   TU Darmstadt, Germany

Discovery Challenge Chair

Taneli Mielikäinen   Nokia, USA

Discovery Challenge Organizers

Stephan Doerfel   University of Kassel, Germany
Andreas Hotho   University of Würzburg, Germany
Robert Jäschke   University of Kassel, Germany
Folke Mitzlaff   University of Kassel, Germany
Jürgen Müller   L3S Research Center, Germany

Sponsorship Chairs

Peter van der Putten   Leiden University & Pegasystems,
                               The Netherlands
Albert Bifet   University of Waikato & Yahoo, New Zealand

Awards Chairs

Bart Goethals   University of Antwerp, Belgium
Peter Flach   University of Bristol, UK
Geoff Webb   Monash University, Australia

Open-Science Award Committee

Tias Guns   KU Leuven, Belgium
Christian Borgelt   European Center for Soft Computing, Spain
Geoff Holmes   University of Waikato, New Zealand
Luis Torgo   University of Porto, Portugal

Web Team

Matěj Holec, Webmaster   Czech Technical University in Prague,
                                  Czech Republic
Radomír Černoch   Czech Technical University in Prague,
                                  Czech Republic
Filip Blažek   Designiq, Czech Republic

Software Development

Radomír Černoch   Czech Technical University in Prague,
                                  Czech Republic
Fabian Hadiji   University of Bonn, Germany
Thanh Le Van   KU Leuven, Belgium

ECML PKDD Steering Committee

Fosca Gianotti, Chair   Universitá di Pisa, Italy
Jose Balcazar   Universitat Polytécnica de Catalunya, Spain
Francesco Bonchi   Yahoo! Research Barcelona, Spain
Nello Cristianini   University of Bristol, UK
Tijl De Bie   University of Bristol, UK
Peter Flach   University of Bristol, UK
Dimitrios Gunopulos   University of Athens, Greece
Donato Malerba   Universitá degli Studi di Bari, Italy
Michèle Sebag   Université Paris-Sud, France
Michalis Vazirgiannis   Athens University of Economics and Business,
                                  Greece

Area Chairs

Henrik Boström   Stockholm University, Sweden
Jean-François Boulicaut   University of Lyon, France
Carla Brodley   Tuft University, USA
Ian Davidson   University of California, Davis, USA
Jesse Davis   KU Leuven, Belgium
Tijl De Bie   University of Bristol, UK
Janez Demšar   University of Ljubljana, Slovenia
Luc De Raedt   KU Leuven, Belgium

Pierre Dupont   UC Louvain, Belgium
Charles Elkan   University of California, San Diego, USA
Alan Fern   Oregon State University, USA
Johannes Fürnkranz   TU Darmstadt, Germany
Joao Gama   University of Porto, Portugal
Thomas Gärtner   University of Bonn and Fraunhofer IAIS,
                           Germany
Aristides Gionis   Aalto University, Finland
Bart Goethals   University of Antwerp, Belgium
Geoff Holmes   Waikato University, New Zealand
Andreas Hotho   University of Würzburg, Germany
Eyke Hüllermeier   Philipps-Universität Marburg, Germany
Manfred Jaeger   Aalborg University, Denmark
Thorsten Joachims   Cornell University, USA
George Karypis   University of Minnesota, USA
Stefan Kramer   University of Mainz, Germany
Donato Malerba   University of Bari, Italy
Dunja Mladenic   Jožef Stefan Institute, Slovenia
Marie-Francine Moens   KU Leuven, Belgium
Bernhard Pfahringer   University of Waikato, New Zealand
Myra Spiliopoulou   Magdeburg University, Germany
Hannu Toivonen   University of Helsinki, Finland
Marco Wiering   University of Groningen, The Netherlands
Stefan Wrobel   University of Bonn and Fraunhofer IAIS,
                           Germany

Program Committee

Leman Akoglu
Mohammad Al Hasan
Aris Anagnostopoulos
Gennady Andrienko
Annalisa Appice
Cedric Archambeau
Marta Arias
Hiroki Arimura
Ira Assent
Martin Atzmüller
Chloe-Agathe Azencott
Antonio Bahamonde
James Bailey
Jose Balcazar
Christian Bauckhage
Roberto Bayardo
Aurelien Bellet

Andras Benczur
Bettina Berendt
Michael Berthold
Indrajit Bhattacharya
Albert Bifet
Mario Boley
Francesco Bonchi
Gianluca Bontempi
Christian Borgelt
Zoran Bosnic
Abdeslam Boularias
Kendrick Boyd
Pavel Brazdil
Ulf Brefeld
Björn Bringmann
Wray Buntine
Robert Busa-Fekete

Toon Calders
Andre Carvalho
Francisco Casacuberta
Michelangelo Ceci
Loic Cerf
Duen Horng Chau
Sanjay Chawla
Weiwei Cheng
Fabrizio Costa
Sheldon Cooper
Vitor Costa
Bruno Cremilleux
Tom Croonenborghs
Boris Cule
Tomaz Curk
James Cussens
Martine De Cock
Colin de la Higuera
Juan del Coz
Francois Denis
Jana Diesner
Wei Ding
Janardhan Doppa
Devdatt Dubhashi
Ines Dutra
Sašo Džeroski
Tina Eliassi-Rad
Tapio Elomaa
Seyda Ertekin
Floriana Esposito
Ines Faerber
Fazel Famili
Hadi Fanaee Tork
Elaine Faria
Ad Feelders
Stefano Ferilli
Carlos Ferreira
Jordi Fonollosa
Antonino Freno
Elisa Fromont
Fabio Fumarola
Patrick Gallinari
Roman Garnett
Eric Gaussier
Ricard Gavalda

Pierre Geurts
Rayid Ghani
Fosca Giannotti
David Gleich
Vibhav Gogate
Michael Granitzer
Dimitrios Gunopulos
Tias Guns
Jiawei Han
Daniel Hernandez Lobato
Frank Hoeppner
Thomas Hofmann
Jaako Hollmen
Arjen Hommersom
Vasant Honavar
Tamás Horváth
Dino Ienco
Elena Ikonomovska
Robert Jäschke
Frederik Janssen
Szymon Jaroszewicz
Ulf Johansson
Alipio Jorge
Kshitij Judah
Hachem Kadri
Alexandros Kalousis
U Kang
Panagiotis Karras
Andreas Karwath
Hisashi Kashima
Samuel Kaski
Latifur Khan
Angelika Kimmig
Arno Knobbe
Levente Kocsis
Yun Sing Koh
Alek Kolcz
Andrey Kolobov
Igor Kononenko
Kleanthis-Nikolaos Kontonasios
Nitish Korula
Petr Kosina
Walter Kosters
Georg Krempl
Sergei Kuznetsov

Helge Langseth

Pedro Larranaga

Silvio Lattanzi

Niklas Lavesson

Nada Lavrač

Gregor Leban

Chris Leckie

Sangkyun Lee

Ping Li

Juanzi Li

Edo Liberty

Jefrey Lijffijt

Jessica Lin

Francesca Lisi

Corrado Loglisci

Eneldo Loza Mencia

Peter Lucas

Francis Maes

Michael Mampaey

Giuseppe Manco

Stan Matwin

Michael May

Mike Mayo

Wannes Meert

Ernestina Menasalvas

Rosa Meo

Pauli Miettinen

Bamshad Mobasher

Joao Moreira

Emmanuel Müller

Mohamed Nadif

Alex Nanopoulos

Balakrishnan Narayanaswamy

Sriraam Natarajan

Aniruddh Nath

Thomas Nielsen

Mathias Niepert

Xia Ning

Niklas Noren

Eirini Ntoutsi

Andreas Nürnberger

Gerhard Paass

David Page

Rasmus Pagh

Spiros Papadimitriou

Panagiotis Papapetrou

Andrea Passerini

Mykola Pechenizkiy

Dino Pedreschi

Jian Pei

Nikos Pelekis

Ruggero Pensa

Marc Plantevit

Pascal Poncelet

Aditya Prakash

Kai Puolamaki

Buyue Qian

Chedy Raïssi

Liva Ralaivola

Karthik Raman

Jan Ramon

Huzefa Rangwala

Umaa Rebbapragada

Jean-Michel Renders

Steffen Rendle

Achim Rettinger

Fabrizio Riguzzi

Celine Robardet

Marko Robnik Sikonja

Pedro Rodrigues

Juan Rodriguez

Irene Rodriguez-Lujan

Ulrich Rückert

Stefan Rüping

Jan Rupnik

Yvan Saeys

Alan Said

Lorenza Saitta

Antonio Salmeron

Scott Sanner

Raul Santos-Rodriguez

Sam Sarjant

Claudio Sartori

Taisuke Sato

Lars Schmidt-Thieme

Christoph Schommer

Michèle Sebag

Marc Sebban

Thomas Seidl

Giovanni Semeraro

Junming Shao
Pannaga Shivaswamy
Jonathan Silva
Kevin Small
Koen Smets
Padhraic Smyth
Carlos Soares
Mauro Sozio
Eirini Spyropoulou
Ashwin Srinivasan
Jerzy Stefanowski
Benno Stein
Markus Strohmaier
Mahito Sugiyama
Einoshin Suzuki
Sandor Szedmak
Andrea Tagarelli
Nima Taghipour
Nikolaj Tatti
Matthew Taylor
Maguelonne Teisseire
Evimaria Terzi
Ljupco Todorovski
Luis Torgo
Panagiotis Tsaparas
Vincent Tseng
Grigorios Tsoumakas
Antti Ukkonen
Athina Vakali

Guy Van den Broeck
Matthijs van Leeuwen
Joaquin Vanschoren
Michalis Vazirgiannis
Shankar Vembu
Celine Vens
Sicco Verwer
Enrique Vidal
Herna Viktor
Christel Vrain
Jilles Vreeken
Byron Wallace
Fei Wang
Xiang Wang
Takashi Washio
Jörg Wicker
Gerhard Widmer
Aaron Wilson
Chun-Nam Yu
Jure Zabkar
Gerson Zaverucha
Bernard Zenko
Min-Ling Zhang
Elena Zheleva
Arthur Zimek
Albrecht Zimmermann
Indre Zliobaite
Blaz Zupan

Demo Track Program Committee

Alan Said
Albert Bifet
Andreas Nürnberger
Bettina Berendt
Christian Borgelt
Daniela Stojanova
Gabor Melli
Geoff Holmes
Gerard de Melo
Grigorios Tsoumakas
Jaako Hollmen

Lars Schmidt-Thieme
Michael Mampaey
Mikio Braun
Mykola Pechenizkiy
Omar Alonso
Peter Reutemann
Peter van der Putten
Robert Jäschke
Stephan Doerfel
Themis Palpanas

Nectar Track Program Committee

Maria Florina Balcan
Christian Böhm
Toon Calders
Luc De Raedt
George Karypis

Hugo Larochelle
Donato Malerba
Myra Spiliopoulou
Vicenc Torra

Additional Reviewers

Rohit Babbar
Aubrey Barnard
Christian Beecks
Alejandro Bellogin
Daniel Bengs
Souhaib Ben Taieb
Mansurul Bhuiyan
Sam Blasiak
Patrice Boizumault
Teresa Bracamonte
Janez Brank
George Brova
David C. Anastasiu
Cécile Capponi
Annalina Caputo
Jeffrey Chan
Anveshi Charuvaka
Claudia d'Amato
Xuan-Hong Dang
Ninh Dang Pham
Lucas Drumond
Wouter Duivesteijn
François-Xavier Dupé
Ritabrata Dutta
Pavel Efros
Dora Erdos
Pasqua Fabiana Lanotte
Antonio Fernandez
Georg Fette
Manoel França
Sergej Fries
Atsushi Fujii
Patrick Gabrielsson
Esther Galbrun
Michael Geilke

Christos Giatsidis
Robby Goetschalckx
Boqing Gong
Michele Gorgoglione
Tatiana Gossen
Maarten Grachten
Xin Guan
Massimo Guarascio
Huan Gui
Amaury Habrard
Ahsanul Haque
Marwan Hassani
Kohei Hayashi
Elad Hazan
Andreas Henelius
Shohei Hido
Patricia Iglesias Sanchez
Roberto Interdonato
Baptiste Jeudy
Hiroshi Kajino
Yoshitaka Kameya
Margarita Karkali
Mehdi Kaytoue
Fabian Keller
Mikaela Keller
Eamonn Keogh
Umer Khan
Tushar Khot
Benjamin Kille
Dragi Kocev
Jussi Korpela
Domen Kosir
Hardy Kremer
Tanay K. Saha
Gautam Kunapuli

Martijn Lappenschaar
Yann-ael Le Borgne
Florian Lemmerich
Fabio Leuzzi
Jurica Levatic
Tomer Levinboim
Juanzi Li
Jialu Liu
Xu-Ying Liu
Tuve Löfström
Miguel Lopes
Thomas Low
Ana Luisa Duboc
Lucrezia Macchia
Abdun Mahmood
Fragkiskos Malliaros
Elio Masciari
Peiman M. Barnaghi
Daniil Mirylenka
Anastasia Mochalova
Tsuyoshi Murata
Benjamin Negrevergne
Mohammad Nozari Zarmehri
Uros Ocepek
Rasaq O. Otunba
Aline Paes
Pance Panov
Roberto Paredes
Brandon Parker
Ioannis Partalas
Darko Pevec
Jean-Philippe Peyrache
Matej Piculin
Anja Pilz
Fábio Pinto
Gianvito Pio
Cristiano Pitangui
Yoann Pitarch
Mahmudur Rahman

Sutharshan Rajasegarar
Domenico Redavid
Xiang Ren
Joris Renkens
Francois Rousseuau
Delia Rusu
Justin Sahs
Germán Sanchis-Trilles
Jan Schlüter
Christoph Scholz
Sohan Seth
Fan Shi
Yuan Shi
Ivica Slavkov
Tadej Stajner
Daniela Stojanova
Erik Strumbelj
Katerina Tashkova
Stefano Teso
Hoang Thanh Lam
David T.J. Huang
Erik Tromp
Yuta Tsuboi
Emmanouil Tzouridis
Ugo Vespier
Seppo Virtanen
Jonas Vlasselaer
Petar Vracar
Jun Wang
Xing Wang
Christian Wirth
Frank W. Takes
Lan Zagar
Jure Zbontar
Christos Zigkolis
Anca Zimmer
Marinka Zitnik
Kaja Zupanc

Guest Editorial Board (Journal Track)

Luc De Raedt   KU Leuven, Belgium
Luis Torgo   University of Porto, Portugal
Marie-Francine Moens   KU Leuven, Belgium
Matthijs van Leeuwen   KU Leuven, Belgium
Michael May   Fraunhofer IAIS, Germany
Michael R. Berthold   Universität Konstanz, Germany
Nada Lavrač   Jožef Stefan Institute, Slovenia
Nikolaj Tatti   University of Antwerp, Belgium
Pascal Poupart   University of Waterloo, Canada
Pierre Dupont   UC Louvain, Belgium
Prasad Tadepalli   Oregon State University, USA
Roberto Bayardo   Google Research, USA
Soumya Ray   Case Western Reserve University, USA
Stefan Wrobel   University of Bonn and Fraunhofer IAIS,
                                Germany
Stefan Kramer   University of Mainz, Germany
Takashi Washio   Osaka University, Japan
Tamás Horváth   Fraunhofer IAIS, Germany
Tapio Elomaa   Tampere University of Technology, Finland
Thomas Gärtner   University of Bonn and Fraunhofer IAIS,
                                Germany
Tijl De Bie   University of Bristol, UK
Toon Calders   Eindhoven University of Technology,
                                The Netherlands
Willem Waegeman   Ghent University, Belgium
Wray Buntine   NICTA, Australia

Additional Reviewers (Journal Track)

Babak Ahmadi   Emma Brunskill
Amr Ahmed   Michelangelo Ceci
Leman Akoglu   Sanjay Chawla
Mohammad Al Hasan   Weiwei Cheng
Massih-Reza Amini   KyungHyun Cho
Bart Baesens   Tom Croonenborghs
Andrew Bagnell   Florence d'Alché-Buc
Arindam Banerjee   Bhavana Dalvi
Christian Bauckhage   Kurt De Grave
Yoshua Bengio   Bolin Ding
Albert Bifet   Chris Ding
Andrew Bolstad   Jennifer Dy
Byron Boots   Sašo Džeroski
Karsten Borgwardt   Alan Fern
Kendrick Boyd   Luis Ferre
Ulf Brefeld   Daan Fierens

Marcel van Gerven
Martijn van Otterlo
Lieven Vandenberghe
Joaquin Vanschoren
Michalis Vazirgiannis
Aki Vehtari
Byron Wallace
Thomas J. Walsh
Chao Wang
Pu Wang
Shaojun Wang
Randy Wilson

Han-Ming Wu
Huan Xu
Zhao Xu
Jieping Ye
Yi-Ren Yeh
Shipeng Yu
Dengyong Zhou
Shenghuo Zhu
Arthur Zimek
Albrecht Zimmermann
Indre Zliobaite

Sponsors

Gold Sponsor
Winton Capital   http://wintoncapital.com

Silver Sponsors
Cisco Systems, Inc.   http://www.cisco.com
Deloitte Analytics   http://www.deloitte.com
KNIME   http://www.knime.com
Yahoo! Labs   http://www.yahoo.com

Bronze Sponsors
CSKI   http://www.cski.cz
Definity Systems   http://www.definity.cz
DIKW Academy   http://dikw-academy.nl
Google   http://research.google.com
Xerox Research Centre Europe   http://www.xrce.xerox.com
Zalando   http://www.zalando.de

Prize Sponsors
Data Mining and Knowledge Discovery   http://link.springer.com/
                                        journal/10618
Deloitte Analytics   http://www.deloitte.com
Google   http://research.google.com
Machine Learning   http://link.springer.com/
                                        journal/10994
Yahoo! Labs   http://www.knime.com

Abstracts of Invited Talks

Using Machine Learning Powers for Good

Rayid Ghani

The past few years have seen increasing demand for machine learning and data mining—both for tools as well as experts. This has been mostly motivated by a variety of factors including better and cheaper data collection, realization that using data is a good thing, and the ability for a lot of organizations to take action based on data analysis. Despite this flood of demand, most applications we hear about in machine learning involve search, advertising, and financial areas. This talk will talk about examples on how the same approaches can be used to help governments and non-prpofits make social impact. I'll talk about a summer fellowship program we ran at University of Chicago on social good and show examples from projects in areas such as education, healthcare, energy, transportation and public safety done in conjunction with governments and non-profits.

Biography

Rayid Ghani was the Chief Scientist at the Obama for America 2012 campaign focusing on analytics, technology, and data. His work focused on improving different functions of the campaign including fundraising, volunteer, and voter mobilization using analytics, social media, and machine learning; his innovative use of machine learning and data mining in Obama's reelection campaign received broad attention in the media such as the New York Times, CNN, and others. Before joining the campaign, Rayid was a Senior Research Scientist and Director of Analytics research at Accenture Labs where he led a technology research team focused on applied R&D in analytics, machine learning, and data mining for large-scale & emerging business problems in various industries including healthcare, retail & CPG, manufacturing, intelligence, and financial services. In addition, Rayid serves as an adviser to several start-ups in Analytics, is an active organizer of and participant in academic and industry analytics conferences, and publishes regularly in machine learning and data mining conferences and journals.

Learning with Humans in the Loop

Thorsten Joachims

Machine Learning is increasingly becoming a technology that directly interacts with human users. Search engines, recommender systems, and electronic commerce already heavily rely on adapting the user experience through machine learning, and other applications are likely to follow in the near future (e.g., autonomous robotics, smart homes, gaming). In this talk, I argue that learning with humans in the loop requires learning algorithms that explicitly account for human behavior, their motivations, and their judgment of performance. Towards this goal, the talk explores how integrating microeconomic models of human behavior into the learning process leads to new learning models that no longer reduce the user to a "labeling subroutine". This motivates an interesting area for theoretical, algorithmic, and applied machine learning research with connections to rational choice theory, econometrics, and behavioral economics.

Biography

Thorsten Joachims is a Professor of Computer Science at Cornell University. His research interests center on a synthesis of theory and system building in machine learning, with applications in language technology, information retrieval, and recommendation. His past research focused on support vector machines, text classification, structured output prediction, convex optimization, learning to rank, learning with preferences, and learning from implicit feedback. In 2001, he finished his dissertation advised by Prof. Katharina Morik at the University of Dortmund. From there he also received his Diplom in Computer Science in 1997. Between 2000 and 2001 he worked as a PostDoc at the GMD Institute for Autonomous Intelligent Systems. From 1994 to 1996 he was a visiting scholar with Prof. Tom Mitchell at Carnegie Mellon University.

Unsupervised Learning with Graphs:
A Theoretical Perspective

Ulrike von Luxburg

Applying a graph–based learning algorithm usually requires a large amount of data preprocessing. As always, such preprocessing can be harmful or helpful. In my talk I am going to discuss statistical and theoretical properties of various preprocessing steps. We consider questions such as: Given data that does not have the form of a graph yet, what do we loose when transforming it to a graph? Given a graph, what might be a meaningful distance function? We will also see that graph–based techniques can lead to surprising solutions to preprocessing problems that a priori don't involve graphs at all.

Biography

Ulrike von Luxburg is a professor for computer science/machine learning at the University of Hamburg. Her research focus is the theoretical analysis of machine learning algorithms, in particular for unsupervised learning and graph algorithms. She is (co)–winner of several best student paper awards (NIPS 2004 and 2008, COLT 2003, 2005 and 2006, ALT 2007). She did her PhD in the Max Planck Institute for Biological Cybernetics in 2004, then moved to Fraunhofer IPSI in Darmstadt, before returning to the Max Planck Institute in 2007 as a research group leader for learning theory. Since 2012 she is a professor for computer science at the University of Hamburg.

Making Systems That Use Statistical Reasoning Easier to Build and Maintain over Time

Christopher Re

The question driving my work is, how should one deploy statistical data–analysis tools to enhance data–driven systems? Even partial answers to this question may have a large impact on science, government, and industry—each of whom are increasingly turning to statistical techniques to get value from their data.

To understand this question, my group has built or contributed to a diverse set of data–processing systems: a system, called GeoDeepDive, that reads and helps answer questions about the geology literature; a muon filter that is used in the IceCube neutrino telescope to process over 250 million events each day in the hunt for the origins of the universe; and enterprise applications with Oracle and Pivotal. This talk will give an overview of the lessons that we learned in these systems, will argue that data systems research may play a larger role in the next generation of these systems, and will speculate on the future challenges that such systems may face.

Biography

Christopher Re is an assistant professor in the department of Computer Sciences at the University of Wisconsin-Madison. The goal of his work is to enable users and developers to build applications that more deeply understand and exploit data. Chris received his PhD from the University of Washington, Seattle under the supervision of Dan Suciu. For his PhD work in the area of probabilistic data management, Chris received the SIGMOD 2010 Jim Gray Dissertation Award. Chris's papers have received four best papers or best–of–conference citations (best paper in PODS 2012 and best–of–conference in PODS 2010, twice, and one in ICDE 2009). Chris received an NSF CAREER Award in 2011.

Deep–er Kernels

John Shawe-Taylor

Kernels can be viewed as shallow in that learning is only applied in a single (output) layer. Recent successes with deep learning highlight the need to consider learning richer function classes. The talk will review and discuss methods that have been developed to enable richer kernel classes to be learned. While some of these methods rely on greedy procedures many are supported by statistical learning analyses and/or convergence bounds. The talk will highlight the trade–offs involved and the potential for further research on this topic.

Biography

John Shawe-Taylor obtained a PhD in Mathematics at Royal Holloway, University of London in 1986 and joined the Department of Computer Science in the same year. He was promoted to Professor of Computing Science in 1996. He moved to the University of Southampton in 2003 to lead the ISIS research group. He was Director of the Centre for Computational Statistics and Machine Learning at University College, London between July 2006 and September 2010. He has coordinated a number of European wide projects investigating the theory and practice of Machine Learning, including the PASCAL projects. He has published over 300 research papers with more than 25000 citations. He has co-authored with Nello Cristianini two books on kernel approaches to machine learning: "An Introduction to Support Vector Machines" and "Kernel Methods for Pattern Analysis".

Abstracts of Industrial
Track Invited Talks

ML and Business: A Love–Hate Relationship

Andreas Antrup

Based on real world examples. the talk explores common gaps in the mutual understanding of the business and the analytical side; particular focus shall be on misconceptions of the needs and expectations of business people and the resulting problems. It also touches on some approaches to bridge these gaps and build trust. At the end we shall discuss possibly under–researched areas that may open the doors to a yet wider usage of ML principles and thus unlock more of its value and beauty.

Bayesian Learning in Online Service: Statistics Meets Systems

Ralf Herbrich

Over the past few years, we have entered the world of big and structured data—a trend largely driven by the exponential growth of Internet–based online services such as Search, eCommerce and Social Networking as well as the ubiquity of smart devices with sensors in everyday life. This poses new challenges for statistical inference and decision–making as some of the basic assumptions are shifting:

- The ability to optimize both the likelihood and loss functions
- The ability to store the parameters of (data) models
- The level of granularity and 'building blocks' in the data modeling phase
- The interplay of computation, storage, communication and inference and decision–making techniques

In this talk, I will discuss the implications of big and structured data for Statistics and the convergence of statistical model and distributed systems. I will present one of the most versatile modeling techniques that combines systems and statistical properties—factor graphs—and review a series of approximate inference techniques such as distributed message passing. The talk will be concluded with an overview of real–world problems at Amazon.

Machine Learning in a Large diversified Internet Group

Jean-Paul Schmetz

I will present a wide survey of the use of machine learning techniques across a large number of subsidiaries (40+) of an Internet group (Burda Digital) with special attention to issues regarding (1) personnel training in state of the art techniques, (2) management buy–in of complex non interpretable results and (3) practical and measurable bottom line results/solutions.

Some of the Problems and Applications of Opinion Analysis

Hugo Zaragoza

Websays strives to provide the best possible analysis of online conversation to marketing and social media analysts. One of the obsessions of Websays is to provide "near–man–made" data quality at marginal costs. I will discuss how we approach this problem using innovative machine learning and UI approaches.

Abstracts of Journal Track Articles

The full articles have been published in *Machine Learning* or *Data Mining and Knowledge Discovery*.

Fast sequence segmentation using log–linear models
Nikolaj Tatti
Data Mining and Knowledge Discovery
DOI 10.1007/s10618-012-0301-y

Sequence segmentation is a well–studied problem, where given a sequence of elements, an integer K, and some measure of homogeneity, the task is to split the sequence into K contiguous segments that are maximally homogeneous. A classic approach to find the optimal solution is by using a dynamic program. Unfortunately, the execution time of this program is quadratic with respect to the length of the input sequence. This makes the algorithm slow for a sequence of non–trivial length. In this paper we study segmentations whose measure of goodness is based on log–linear models, a rich family that contains many of the standard distributions. We present a theoretical result allowing us to prune many suboptimal segmentations. Using this result, we modify the standard dynamic program for 1D log–linear models, and by doing so reduce the computational time. We demonstrate empirically, that this approach can significantly reduce the computational burden of finding the optimal segmentation.

ROC curves in cost space
Cesar Ferri, Jose Hernandez-Orallo and Peter Flach
Machine Learning
DOI 10.1007/s10994-013-5328-9

ROC curves and cost curves are two popular ways of visualising classifier performance, finding appropriate thresholds according to the operating condition, and deriving useful aggregated measures such as the area under the ROC curve (AUC) or the area under the optimal cost curve. In this paper we present new findings and connections between ROC space and cost space. In particular, we show that ROC curves can be transferred to cost space by means of a very natural threshold choice method, which sets the decision threshold such that the proportion of positive predictions equals the operating condition. We call these new curves rate–driven curves, and we demonstrate that the expected loss as measured by the area under these curves is linearly related to AUC. We show that the rate–driven curves are the genuine equivalent of ROC curves in cost space, establishing a point–point rather than a point–line correspondence. Furthermore, a decomposition of the rate–driven curves is introduced which separates the loss due to the threshold choice method from the ranking loss (Kendall

τ distance). We also derive the corresponding curve to the ROC convex hull in cost space: this curve is different from the lower envelope of the cost lines, as the latter assumes only optimal thresholds are chosen.

A framework for semi–supervised and unsupervised optimal extraction of clusters from hierarchies

Ricardo J.G.B. Campello, Davoud Moulavi, Arthur Zimek and Jörg Sander
Data Mining and Knowledge Discovery
DOI 10.1007/s10618-013-0311-4

We introduce a framework for the optimal extraction of flat clusterings from local cuts through cluster hierarchies. The extraction of a flat clustering from a cluster tree is formulated as an optimization problem and a linear complexity algorithm is presented that provides the globally optimal solution to this problem in semi–supervised as well as in unsupervised scenarios. A collection of experiments is presented involving clustering hierarchies of different natures, a variety of real data sets, and comparisons with specialized methods from the literature.

Pairwise meta–rules for better meta–learning–based algorithm ranking

Quan Sun and Bernhard Pfahringer
Machine Learning
DOI 10.1007/s10994-013-5387-y

In this paper, we present a novel meta–feature generation method in the context of meta–learning, which is based on rules that compare the performance of individual base learners in a one–against–one manner. In addition to these new meta–features, we also introduce a new meta–learner called Approximate Ranking Tree Forests (ART Forests) that performs very competitively when compared with several state–of–the–art meta–learners. Our experimental results are based on a large collection of datasets and show that the proposed new techniques can improve the overall performance of meta–learning for algorithm ranking significantly. A key point in our approach is that each performance figure of any base learner for any specific dataset is generated by optimising the parameters of the base learner separately for each dataset.

Block coordinate descent algorithms for large–scale sparse multiclass classification

Mathieu Blondel, Kazuhiro Seki and Kuniaki Uehara

Machine Learning

DOI 10.1007/s10994-013-5367-2

Over the past decade, ℓ_1 regularization has emerged as a powerful way to learn classifiers with implicit feature selection. More recently, mixed–norm (e.g., ℓ_1/ℓ_2) regularization has been utilized as a way to select entire groups of features. In this paper, we propose a novel direct multiclass formulation specifically designed for large–scale and high–dimensional problems such as document classification. Based on a multiclass extension of the squared hinge loss, our formulation employs ℓ_1/ℓ_2 regularization so as to force weights corresponding to the same features to be zero across all classes, resulting in compact and fast–to–evaluate multiclass models. For optimization, we employ two globally–convergent variants of block coordinate descent, one with line search (Tseng and Yun in Math. Program. 117:387423, 2009) and the other without (Richtrik and Tak in Math. Program. 138, 2012a, Tech. Rep. arXiv:1212.0873, 2012b). We present the two variants in a unified manner and develop the core components needed to efficiently solve our formulation. The end result is a couple of block coordinate descent algorithms specifically tailored to our multiclass formulation. Experimentally, we show that block coordinate descent performs favorably compared to other solvers such as FOBOS, FISTA and SpaRSA. Furthermore, we show that our formulation obtains very compact multiclass models and outperforms ℓ_1/ℓ_2–regularized multiclass logistic regression in terms of training speed, while achieving comparable test accuracy.

A comparative evaluation of stochastic–based inference methods for Gaussian process models

Maurizio Filippone, Mingjun Zhong and Mark Girolami

Machine Learning

DOI 10.1007/s10994-013-5388-x

Gaussian process (GP) models are extensively used in data analysis given their flexible modeling capabilities and interpretability. The fully Bayesian treatment of GP models is analytically intractable, and therefore it is necessary to resort to either deterministic or stochastic approximations. This paper focuses on stochastic–based inference techniques. After discussing the challenges associated with the fully Bayesian treatment of GP models, a number of inference strategies based on Markov chain Monte Carlo methods are presented and rigorously assessed. In particular, strategies based on efficient parameterizations and efficient proposal mechanisms are extensively compared on simulated and real data on the basis of convergence speed, sampling efficiency, and computational cost.

Probabilistic topic models for sequence data

Nicola Barbieri, Antonio Bevacqua, Marco Carnuccio, Giuseppe Manco and Ettore Ritacco
Machine Learning
DOI 10.1007/s10994-013-5391-2

Probabilistic topic models are widely used in different contexts to uncover the hidden structure in large text corpora. One of the main (and perhaps strong) assumptions of these models is that the generative process follows a bag–of–words assumption, i.e. each token is independent from the previous one. We extend the popular Latent Dirichlet Allocation model by exploiting three different conditional Markovian assumptions: (i) the token generation depends on the current topic and on the previous token; (ii) the topic associated with each observation depends on topic associated with the previous one; (iii) the token generation depends on the current and previous topic. For each of these modeling assumptions we present a Gibbs Sampling procedure for parameter estimation. Experimental evaluation over real–word data shows the performance advantages, in terms of recall and precision, of the sequence–modeling approaches.

The flip–the–state transition operator for restricted Boltzmann machines

Kai Brügge, Asja Fischer and Christian Igel
Machine Learning
DOI 10.1007/s10994-013-5390-3

Most learning and sampling algorithms for restricted Boltzmann machines (RBMs) rely on Markov chain Monte Carlo (MCMC) methods using Gibbs sampling. The most prominent examples are Contrastive Divergence learning (CD) and its variants as well as Parallel Tempering (PT). The performance of these methods strongly depends on the mixing properties of the Gibbs chain. We propose a Metropolis–type MCMC algorithm relying on a transition operator maximizing the probability of state changes. It is shown that the operator induces an irreducible, aperiodic, and hence properly converging Markov chain, also for the typically used periodic update schemes. The transition operator can replace Gibbs sampling in RBM learning algorithms without producing computational overhead. It is shown empirically that this leads to faster mixing and in turn to more accurate learning.

Differential privacy based on importance weighting

Zhanglong Ji and Charles Elkan
Machine Learning
DOI 10.1007/s10994-013-5396-x

This paper analyzes a novel method for publishing data while still protecting privacy. The method is based on computing weights that make an existing dataset, for which there are no confidentiality issues, analogous to the dataset that must be kept private. The existing dataset may be genuine but public already, or it may be synthetic. The weights are importance sampling weights, but to protect privacy, they are regularized and have noise added. The weights allow statistical queries to be answered approximately while provably guaranteeing differential privacy. We derive an expression for the asymptotic variance of the approximate answers. Experiments show that the new mechanism performs well even when the privacy budget is small, and when the public and private datasets are drawn from different populations.

Activity preserving graph simplification

Francesco Bonchi, Gianmarco De Francisci Morales, Aristides Gionis and Antti Ukkonen
Data Mining and Knowledge Discovery
DOI 10.1007/s10618-013-0328-8

We study the problem of simplifying a given directed graph by keeping a small subset of its arcs. Our goal is to maintain the connectivity required to explain a set of observed traces of information propagation across the graph. Unlike previous work, we do not make any assumption about an underlying model of information propagation. Instead, we approach the task as a combinatorial problem.

We prove that the resulting optimization problem is **NP**–hard. We show that a standard greedy algorithm performs very well in practice, even though it does not have theoretical guarantees. Additionally, if the activity traces have a tree struc-ture, we show that the objective function is supermodular, and experimentally verify that the approach for size–constrained submodular minimization recently proposed by Nagano et al (2011) produces very good results. Moreover, when applied to the task of reconstructing an unobserved graph, our methods perform comparably to a state–of–the–art algorithm devised specifically for this task.

ABACUS: frequent pattern mining based community discovery in multidimensional networks

Michele Berlingerio, Fabio Pinelli and Francesco Calabrese
Data Mining and Knowledge Discovery
DOI 10.1007/s10618-013-0331-0

Community Discovery in complex networks is the problem of detecting, for each node of the network, its membership to one of more groups of nodes, the com-munities, that are densely connected, or highly interactive, or, more in general, similar, according to a similarity function. So far, the problem has been widely studied in monodimensional networks, i.e. networks where only one connection between two entities may exist. However, real networks are often multidimen-sional, i.e., multiple connections between any two nodes may exist, either re-flecting different kinds of relationships, or representing different values of the same type of tie. In this context, the problem of Community Discovery has to be redefined, taking into account multidimensional structure of the graph. We define a new concept of community that groups together nodes sharing memberships to the same monodimensional communities in the different single dimensions. As we show, such communities are meaningful and able to group nodes even if they might not be connected in any of the monodimensional networks. We devise ABACUS (frequent pAttern mining–BAsed Community discoverer in mUltidimensional networkS), an algorithm that is able to extract multidimensional communities based on the extraction of frequent closed itemsets from monodimensional community memberships. Experiments on two different real multidimensional networks confirm the meaningfulness of the introduced concepts, and open the way for a new class of algorithms for community discovery that do not rely on the dense connections among nodes.

Growing a list

Benjamin Letham, Cynthia Rudin and Katherine A. Heller
Data Mining and Knowledge Discovery
DOI 10.1007/s10618-013-0329-7

It is easy to find expert knowledge on the Internet on almost any topic, but obtaining a complete overview of a given topic is not always easy: Information can be scattered across many sources and must be aggregated to be useful. We introduce a method for intelligently growing a list of relevant items, starting from a small seed of examples. Our algorithm takes advantage of the wisdom of the crowd, in the sense that there are many experts who post lists of things on the Internet. We use a collection of simple machine learning components to find these experts and aggregate their lists to produce a single complete and meaningful list. We use experiments with gold standards and open–ended experiments without gold standards to show that our method significantly outperforms the state of the art. Our method uses the ranking algorithm Bayesian Sets even when its underlying independence assumption is violated, and we provide a theoretical generalization bound to motivate its use.

What distinguish one from its peers in social networks?

Yi-Chen Lo, Jhao-Yin Li, Mi-Yen Yeh, Shou-De Lin and Jian Pei
Data Mining and Knowledge Discovery
DOI 10.1007/s10618-013-0330-1

Being able to discover the uniqueness of an individual is a meaningful task in social network analysis. This paper proposes two novel problems in social network analysis: how to identify the uniqueness of a given query vertex, and how to identify a group of vertices that can mutually identify each other. We further propose intuitive yet effective methods to identify the uniqueness identification sets and the mutual identification groups of different properties. We further conduct an extensive experiment on both real and synthetic datasets to demonstrate the effectiveness of our model.

Spatio–temporal random fields: compressible representation and distributed estimation

Nico Piatkowski, Sangkyun Lee and Katharina Morik
Machine Learning
DOI 10.1007/s10994-013-5399-7

Modern sensing technology allows us enhanced monitoring of dynamic activities in business, traffic, and home, just to name a few. The increasing amount of sensor measurements, however, brings us the challenge for efficient data analysis. This is especially true when sensing targets can interoperate—in such cases we need learning models that can capture the relations of sensors, possibly without collecting or exchanging all data. Generative graphical models namely the Markov random fields (MRF) fit this purpose, which can represent complex spatial and temporal relations among sensors, producing interpretable answers in terms of probability. The only drawback will be the cost for inference, storing and optimizing a very large number of parameters—not uncommon when we apply them for real–world applications.

In this paper, we investigate how we can make discrete probabilistic graphical models practical for predicting sensor states in a spatio–temporal setting. A set of new ideas allows keeping the advantages of such models while achieving scalability. We first introduce a novel alternative to represent model parameters, which enables us to compress the parameter storage by removing uninformative parameters in a systematic way. For finding the best parameters via maximum likelihood estimation, we provide a separable optimization algorithm that can be performed independently in parallel in each graph node. We illustrate that the prediction quality of our suggested method is comparable to those of the standard MRF and a spatio–temporal k–nearest neighbor method, while using much less computational resources.

Table of Contents – Part I

Reinforcement Learning

A Cascaded Supervised Learning Approach to Inverse Reinforcement
Learning 1
  Edouard Klein, Bilal Piot, Matthieu Geist, and Olivier Pietquin

Learning from Demonstrations: Is It Worth Estimating a Reward
Function? 17
  Bilal Piot, Matthieu Geist, and Olivier Pietquin

Recognition of Agents Based on Observation of Their Sequential
Behavior 33
  Qifeng Qiao and Peter A. Beling

Learning Throttle Valve Control Using Policy Search 49
  Bastian Bischoff, Duy Nguyen-Tuong, Torsten Koller,
  Heiner Markert, and Alois Knoll

Model-Selection for Non-parametric Function Approximation in
Continuous Control Problems: A Case Study in a Smart Energy
System 65
  Daniel Urieli and Peter Stone

Learning Graph-Based Representations for Continuous Reinforcement
Learning Domains 81
  Jan Hendrik Metzen

Regret Bounds for Reinforcement Learning with Policy Advice 97
  Mohammad Gheshlaghi Azar, Alessandro Lazaric, and
  Emma Brunskill

Exploiting Multi-step Sample Trajectories for Approximate Value
Iteration 113
  Robert Wright, Steven Loscalzo, Philip Dexter, and Lei Yu

Markov Decision Processes

Expectation Maximization for Average Reward Decentralized
POMDPs 129
  Joni Pajarinen and Jaakko Peltonen

Properly Acting under Partial Observability with Action Feasibility
Constraints 145
  Caroline P. Carvalho Chanel and Florent Teichteil-Königsbuch

Iterative Model Refinement of Recommender MDPs Based on Expert
Feedback 162
    Omar Zia Khan, Pascal Poupart, and John Mark Agosta

Solving Relational MDPs with Exogenous Events and Additive
Rewards 178
    Saket Joshi, Roni Khardon, Prasad Tadepalli, Aswin Raghavan, and
    Alan Fern

Continuous Upper Confidence Trees with Polynomial Exploration –
Consistency 194
    David Auger, Adrien Couëtoux, and Olivier Teytaud

Active Learning and Optimization

A Lipschitz Exploration-Exploitation Scheme for Bayesian
Optimization 210
    Ali Jalali, Javad Azimi, Xiaoli Fern, and Ruofei Zhang

Parallel Gaussian Process Optimization with Upper Confidence Bound
and Pure Exploration 225
    Emile Contal, David Buffoni, Alexandre Robicquet, and
    Nicolas Vayatis

Greedy Confidence Pursuit: A Pragmatic Approach to Multi-bandit
Optimization 241
    Philip Bachman and Doina Precup

A Time and Space Efficient Algorithm for Contextual Linear Bandits... 257
    José Bento, Stratis Ioannidis, S. Muthukrishnan, and Jinyun Yan

Knowledge Transfer for Multi-labeler Active Learning 273
    Meng Fang, Jie Yin, and Xingquan Zhu

Learning from Sequences

Spectral Learning of Sequence Taggers over Continuous Sequences 289
    Adrià Recasens and Ariadna Quattoni

Fast Variational Bayesian Linear State-Space Model 305
    Jaakko Luttinen

Inhomogeneous Parsimonious Markov Models 321
    Ralf Eggeling, André Gohr, Pierre-Yves Bourguignon,
    Edgar Wingender, and Ivo Grosse

Explaining Interval Sequences by Randomization 337
    Andreas Henelius, Jussi Korpela, and Kai Puolamäki

Itemset Based Sequence Classification 353
    Cheng Zhou, Boris Cule, and Bart Goethals

A Relevance Criterion for Sequential Patterns 369
    Henrik Grosskreutz, Bastian Lang, and Daniel Trabold

A Fast and Simple Method for Mining Subsequences with Surprising
Event Counts.. 385
    Jefrey Lijffijt

Relevant Subsequence Detection with Sparse Dictionary Learning 401
    Sam Blasiak, Huzefa Rangwala, and Kathryn B. Laskey

Time Series and Spatio-temporal Data

Future Locations Prediction with Uncertain Data 417
    Disheng Qiu, Paolo Papotti, and Lorenzo Blanco

Modeling Short-Term Energy Load with Continuous Conditional
Random Fields 433
    Hongyu Guo

Data Streams

Fault Tolerant Regression for Sensor Data.......................... 449
    Indrė Žliobaitė and Jaakko Hollmén

Pitfalls in Benchmarking Data Stream Classification and How to Avoid
Them 465
    Albert Bifet, Jesse Read, Indrė Žliobaitė, Bernhard Pfahringer, and
    Geoff Holmes

Adaptive Model Rules from Data Streams............................ 480
    Ezilda Almeida, Carlos Ferreira, and João Gama

Fast and Exact Mining of Probabilistic Data Streams................. 493
    Reza Akbarinia and Florent Masseglia

Graphs and Networks

Detecting Bicliques in GF[q] 509
    Jan Ramon, Pauli Miettinen, and Jilles Vreeken

As Strong as the Weakest Link: Mining Diverse Cliques in Weighted
Graphs 525
    Petko Bogdanov, Ben Baumer, Prithwish Basu,
    Amotz Bar-Noy, and Ambuj K. Singh

How Robust Is the Core of a Network? 541
  Abhijin Adiga and Anil Kumar S. Vullikanti

Community Distribution Outlier Detection in Heterogeneous
Information Networks 557
  Manish Gupta, Jing Gao, and Jiawei Han

Protein Function Prediction Using Dependence Maximization.......... 574
  *Guoxian Yu, Carlotta Domeniconi, Huzefa Rangwala, and
  Guoji Zhang*

Improving Relational Classification Using Link Prediction
Techniques 590
  Cristina Pérez-Solà and Jordi Herrera-Joancomartí

A Fast Approximation of the Weisfeiler-Lehman Graph Kernel for RDF
Data 606
  Gerben K.D. de Vries

Efficient Frequent Connected Induced Subgraph Mining in Graphs of
Bounded Tree-Width 622
  Tamás Horváth, Keisuke Otaki, and Jan Ramon

Continuous Similarity Computation over Streaming Graphs 638
  Elena Valari and Apostolos N. Papadopoulos

Trend Mining in Dynamic Attributed Graphs...................... 654
  *Elise Desmier, Marc Plantevit, Céline Robardet, and
  Jean-François Boulicaut*

Sparse Relational Topic Models for Document Networks 670
  Aonan Zhang, Jun Zhu, and Bo Zhang

Author Index 687

Table of Contents – Part II

Social Network Analysis

Incremental Local Evolutionary Outlier Detection for Dynamic Social
Networks 1
    Tengfei Ji, Dongqing Yang, and Jun Gao

How Long Will She Call Me? Distribution, Social Theory and Duration
Prediction 16
    *Yuxiao Dong, Jie Tang, Tiancheng Lou, Bin Wu, and
    Nitesh V. Chawla*

Discovering Nested Communities 32
    Nikolaj Tatti and Aristides Gionis

CSI: Community-Level Social Influence Analysis 48
    *Yasir Mehmood, Nicola Barbieri, Francesco Bonchi, and
    Antti Ukkonen*

Natural Language Processing and Information Extraction

Supervised Learning of Syntactic Contexts for Uncovering Definitions
and Extracting Hypernym Relations in Text Databases 64
    Guido Boella and Luigi Di Caro

Error Prediction with Partial Feedback 80
    *William Darling, Cédric Archambeau, Shachar Mirkin, and
    Guillaume Bouchard*

Boot-Strapping Language Identifiers for Short Colloquial Postings 95
    Moises Goldszmidt, Marc Najork, and Stelios Paparizos

Ranking and Recommender Systems

A Pairwise Label Ranking Method with Imprecise Scores and Partial
Predictions 112
    Sebastien Destercke

Learning Socially Optimal Information Systems from Egoistic Users 128
    Karthik Raman and Thorsten Joachims

Socially Enabled Preference Learning from Implicit Feedback Data 145
    *Julien Delporte, Alexandros Karatzoglou, Tomasz Matuszczyk, and
    Stéphane Canu*

Cross-Domain Recommendation via Cluster-Level Latent Factor
Model 161
    *Sheng Gao, Hao Luo, Da Chen, Shantao Li, Patrick Gallinari, and
    Jun Guo*

Minimal Shrinkage for Noisy Data Recovery Using Schatten-p Norm
Objective 177
    Deguang Kong, Miao Zhang, and Chris Ding

Matrix and Tensor Analysis

Noisy Matrix Completion Using Alternating Minimization 194
    Suriya Gunasekar, Ayan Acharya, Neeraj Gaur, and Joydeep Ghosh

A Nearly Unbiased Matrix Completion Approach 210
    Dehua Liu, Tengfei Zhou, Hui Qian, Congfu Xu, and Zhihua Zhang

A Counterexample for the Validity of Using Nuclear Norm as a Convex
Surrogate of Rank 226
    Hongyang Zhang, Zhouchen Lin, and Chao Zhang

Efficient Rank-one Residue Approximation Method for Graph
Regularized Non-negative Matrix Factorization 242
    Qing Liao and Qian Zhang

Maximum Entropy Models for Iteratively Identifying Subjectively
Interesting Structure in Real-Valued Data 256
    Kleanthis-Nikolaos Kontonasios, Jilles Vreeken, and Tijl De Bie

An Analysis of Tensor Models for Learning on Structured Data 272
    Maximilian Nickel and Volker Tresp

Learning Modewise Independent Components from Tensor Data Using
Multilinear Mixing Model 288
    Haiping Lu

Structured Output Prediction, Multi-label and Multi-task Learning

Taxonomic Prediction with Tree-Structured Covariances 304
    Matthew B. Blaschko, Wojciech Zaremba, and Arthur Gretton

Position Preserving Multi-Output Prediction 320
    *Zubin Abraham, Pang-Ning Tan, Perdinan, Julie Winkler,
    Shiyuan Zhong, and Malgorzata Liszewska*

Structured Output Learning with Candidate Labels for Local Parts 336
    Chengtao Li, Jianwen Zhang, and Zheng Chen

Shared Structure Learning for Multiple Tasks with Multiple Views   353
  Xin Jin, Fuzhen Zhuang, Shuhui Wang, Qing He, and Zhongzhi Shi

Using Both Latent and Supervised Shared Topics for Multitask
Learning   369
  *Ayan Acharya, Aditya Rawal, Raymond J. Mooney, and
  Eduardo R. Hruschka*

Probabilistic Clustering for Hierarchical Multi-Label Classification of
Protein Functions   385
  *Rodrigo C. Barros, Ricardo Cerri, Alex A. Freitas, and
  André C.P.L.F. de Carvalho*

Multi-core Structural SVM Training   401
  Kai-Wei Chang, Vivek Srikumar, and Dan Roth

Multi-label Classification with Output Kernels   417
  Yuhong Guo and Dale Schuurmans

Transfer Learning

Boosting for Unsupervised Domain Adaptation   433
  Amaury Habrard, Jean-Philippe Peyrache, and Marc Sebban

Automatically Mapped Transfer between Reinforcement Learning
Tasks via Three-Way Restricted Boltzmann Machines   449
  *Haitham Bou Ammar, Decebal Constantin Mocanu,
  Matthew E. Taylor, Kurt Driessens, Karl Tuyls, and Gerhard Weiss*

Bayesian Learning

A Layered Dirichlet Process for Hierarchical Segmentation of Sequential
Grouped Data   465
  Adway Mitra, Ranganath B.N., and Indrajit Bhattacharya

A Bayesian Classifier for Learning from Tensorial Data   483
  *Wei Liu, Jeffrey Chan, James Bailey, Christopher Leckie,
  Fang Chen, and Kotagiri Ramamohanarao*

Prediction with Model-Based Neutrality   499
  Kazuto Fukuchi, Jun Sakuma, and Toshihiro Kamishima

Decision-Theoretic Sparsification for Gaussian Process Preference
Learning   515
  M. Ehsan Abbasnejad, Edwin V. Bonilla, and Scott Sanner

Variational Hidden Conditional Random Fields with Coupled Dirichlet
Process Mixtures 531
    Konstantinos Bousmalis, Stefanos Zafeiriou,
    Louis–Philippe Morency, Maja Pantic, and Zoubin Ghahramani

Sparsity in Bayesian Blind Source Separation and Deconvolution 548
    Václav Šmídl and Ondřej Tichý

Nested Hierarchical Dirichlet Process for Nonparametric Entity-Topic
Analysis 564
    Priyanka Agrawal, Lavanya Sita Tekumalla, and
    Indrajit Bhattacharya

Graphical Models

Knowledge Intensive Learning: Combining Qualitative Constraints with
Causal Independence for Parameter Learning in Probabilistic Models ... 580
    Shuo Yang and Sriraam Natarajan

Direct Learning of Sparse Changes in Markov Networks by Density
Ratio Estimation 596
    Song Liu, John A. Quinn, Michael U. Gutmann, and
    Masashi Sugiyama

Greedy Part-Wise Learning of Sum-Product Networks 612
    Robert Peharz, Bernhard C. Geiger, and Franz Pernkopf

From Topic Models to Semi-supervised Learning: Biasing Mixed-
Membership Models to Exploit Topic-Indicative Features in Entity
Clustering 628
    Ramnath Balasubramanyan, Bhavana Dalvi, and William W. Cohen

Nearest-Neighbor Methods

Hub Co-occurrence Modeling for Robust High-Dimensional kNN
Classification 643
    Nenad Tomašev and Dunja Mladenić

Fast kNN Graph Construction with Locality Sensitive Hashing 660
    Yan-Ming Zhang, Kaizhu Huang, Guanggang Geng, and
    Cheng-Lin Liu

Mixtures of Large Margin Nearest Neighbor Classifiers 675
    Murat Semerci and Ethem Alpaydın

Author Index 689

Table of Contents – Part III

Ensembles

AR-Boost: Reducing Overfitting by a Robust Data-Driven
Regularization Strategy 1
  Baidya Nath Saha, Gautam Kunapuli, Nilanjan Ray,
  Joseph A. Maldjian, and Sriraam Natarajan

Parallel Boosting with Momentum 17
  Indraneel Mukherjee, Kevin Canini, Rafael Frongillo, and
  Yoram Singer

Inner Ensembles: Using Ensemble Methods Inside the Learning
Algorithm 33
  Houman Abbasian, Chris Drummond, Nathalie Japkowicz, and
  Stan Matwin

Statistical Learning

Learning Discriminative Sufficient Statistics Score Space
for Classification 49
  Xiong Li, Bin Wang, Yuncai Liu, and Tai Sing Lee

The Stochastic Gradient Descent for the Primal L1-SVM Optimization
Revisited 65
  Constantinos Panagiotakopoulos and Petroula Tsampouka

Bundle CDN: A Highly Parallelized Approach for Large-Scale
ℓ_1-Regularized Logistic Regression 81
  Yatao Bian, Xiong Li, Mingqi Cao, and Yuncai Liu

MORD: Multi-class Classifier for Ordinal Regression 96
  Kostiantyn Antoniuk, Vojtěch Franc, and Václav Hlaváč

Identifiability of Model Properties in Over-Parameterized Model
Classes 112
  Manfred Jaeger

Semi-supervised Learning

Exploratory Learning 128
  Bhavana Dalvi, William W. Cohen, and Jamie Callan

Semi-supervised Gaussian Process Ordinal Regression 144
    P.K. Srijith, Shirish Shevade, and S. Sundararajan

Influence of Graph Construction on Semi-supervised Learning 160
    Celso André R. de Sousa, Solange O. Rezende, and
    Gustavo E.A.P.A. Batista

Tractable Semi-supervised Learning of Complex Structured Prediction
Models 176
    Kai-Wei Chang, S. Sundararajan, and S. Sathiya Keerthi

PSSDL: Probabilistic Semi-supervised Dictionary Learning 192
    Behnam Babagholami-Mohamadabadi, Ali Zarghami,
    Mohammadreza Zolfaghari, and Mahdieh Soleymani Baghshah

Unsupervised Learning

Embedding with Autoencoder Regularization 208
    Wenchao Yu, Guangxiang Zeng, Ping Luo, Fuzhen Zhuang,
    Qing He, and Zhongzhi Shi

Reduced-Rank Local Distance Metric Learning 224
    Yinjie Huang, Cong Li, Michael Georgiopoulos, and
    Georgios C. Anagnostopoulos

Learning Exemplar-Represented Manifolds in Latent Space
for Classification 240
    Shu Kong and Donghui Wang

Locally Linear Landmarks for Large-Scale Manifold Learning 256
    Max Vladymyrov and Miguel Á. Carreira-Perpiñán

Subgroup Discovery, Outlier Detection and Anomaly Detection

Discovering Skylines of Subgroup Sets 272
    Matthijs van Leeuwen and Antti Ukkonen

Difference-Based Estimates for Generalization-Aware Subgroup
Discovery 288
    Florian Lemmerich, Martin Becker, and Frank Puppe

Local Outlier Detection with Interpretation 304
    Xuan Hong Dang, Barbora Micenková, Ira Assent, and
    Raymond T. Ng

Anomaly Detection in Vertically Partitioned Data by Distributed Core
Vector Machines   321
    Marco Stolpe, Kanishka Bhaduri, Kamalika Das, and
    Katharina Morik

Mining Outlier Participants: Insights Using Directional Distributions
in Latent Models   337
    Didi Surian and Sanjay Chawla

Privacy and Security

Anonymizing Data with Relational and Transaction Attributes   353
    Giorgos Poulis, Grigorios Loukides, Aris Gkoulalas-Divanis, and
    Spiros Skiadopoulos

Privacy-Preserving Mobility Monitoring Using Sketches of Stationary
Sensor Readings   370
    Michael Kamp, Christine Kopp, Michael Mock, Mario Boley, and
    Michael May

Evasion Attacks against Machine Learning at Test Time   387
    Battista Biggio, Igino Corona, Davide Maiorca, Blaine Nelson,
    Nedim Šrndić, Pavel Laskov, Giorgio Giacinto, and Fabio Roli

Data Mining and Constraint Solving

The Top-k Frequent Closed Itemset Mining Using Top-k SAT
Problem   403
    Said Jabbour, Lakhdar Sais, and Yakoub Salhi

A Declarative Framework for Constrained Clustering   419
    Thi-Bich-Hanh Dao, Khanh-Chuong Duong, and Christel Vrain

SNNAP: Solver-Based Nearest Neighbor for Algorithm Portfolios   435
    Marco Collautti, Yuri Malitsky, Deepak Mehta, and Barry O'Sullivan

Evaluation

Area under the Precision-Recall Curve: Point Estimates and Confidence
Intervals   451
    Kendrick Boyd, Kevin H. Eng, and C. David Page

Applications

Incremental Sensor Placement Optimization on Water Network   467
    Xiaomin Xu, Yiqi Lu, Sheng Huang, Yanghua Xiao, and Wei Wang

Detecting Marionette Microblog Users for Improved Information
Credibility 483
    Xian Wu, Ziming Feng, Wei Fan, Jing Gao, and Yong Yu

Will My Question Be Answered? Predicting "Question Answerability"
in Community Question-Answering Sites 499
    Gideon Dror, Yoelle Maarek, and Idan Szpektor

Learning to Detect Patterns of Crime 515
    Tong Wang, Cynthia Rudin, Daniel Wagner, and Rich Sevieri

Space Allocation in the Retail Industry: A Decision Support System
Integrating Evolutionary Algorithms and Regression Models 531
    Fábio Pinto and Carlos Soares

Medical Applications

Forest-Based Point Process for Event Prediction from Electronic Health
Records 547
    Jeremy C. Weiss and C. David Page

On Discovering the Correlated Relationship between Static
and Dynamic Data in Clinical Gait Analysis 563
    Yin Song, Jian Zhang, Longbing Cao, and Morgan Sangeux

Computational Drug Repositioning by Ranking and Integrating
Multiple Data Sources 579
    Ping Zhang, Pankaj Agarwal, and Zoran Obradovic

Score As You Lift (SAYL): A Statistical Relational Learning Approach
to Uplift Modeling 595
    Houssam Nassif, Finn Kuusisto, Elizabeth S. Burnside,
    C. David Page, Jude Shavlik, and Vítor Santos Costa

Nectar Track

A Theoretical Framework for Exploratory Data Mining: Recent Insights
and Challenges Ahead 612
    Tijl De Bie and Eirini Spyropoulou

Tensor Factorization for Multi-relational Learning 617
    Maximilian Nickel and Volker Tresp

MONIC and Followups on Modeling and Monitoring Cluster
Transitions 622
    Myra Spiliopoulou, Eirini Ntoutsi, Yannis Theodoridis, and
    Rene Schult

Towards Robot Skill Learning: From Simple Skills to Table Tennis   627
 Jan Peters, Jens Kober, Katharina Mülling, Oliver Krömer, and
 Gerhard Neumann

Functional MRI Analysis with Sparse Models   632
 Irina Rish

Demo Track

Image Hub Explorer: Evaluating Representations and Metrics
for Content-Based Image Retrieval and Object Recognition   637
 Nenad Tomašev and Dunja Mladenić

Ipseity – A Laboratory for Synthesizing and Validating Artificial
Cognitive Systems in Multi-agent Systems   641
 Fabrice Lauri, Nicolas Gaud, Stéphane Galland, and Vincent Hilaire

OpenML: A Collaborative Science Platform   645
 Jan N. van Rijn, Bernd Bischl, Luis Torgo, Bo Gao,
 Venkatesh Umaashankar, Simon Fischer, Patrick Winter,
 Bernd Wiswedel, Michael R. Berthold, and Joaquin Vanschoren

ViperCharts: Visual Performance Evaluation Platform   650
 Borut Sluban and Nada Lavrač

Entityclassifier.eu: Real-Time Classification of Entities in Text
with Wikipedia   654
 Milan Dojchinovski and Tomáš Kliegr

Hermoupolis: A Trajectory Generator for Simulating Generalized
Mobility Patterns   659
 Nikos Pelekis, Christos Ntrigkogias, Panagiotis Tampakis,
 Stylianos Sideridis, and Yannis Theodoridis

AllAboard: A System for Exploring Urban Mobility and Optimizing
Public Transport Using Cellphone Data   663
 Michele Berlingerio, Francesco Calabrese, Giusy Di Lorenzo,
 Rahul Nair, Fabio Pinelli, and Marco Luca Sbodio

ScienScan – An Efficient Visualization and Browsing Tool for Academic
Search   667
 Daniil Mirylenka and Andrea Passerini

InVis: A Tool for Interactive Visual Data Analysis   672
 Daniel Paurat and Thomas Gärtner

Kanopy: Analysing the Semantic Network around Document Topics 677
    Ioana Hulpuş, Conor Hayes, Marcel Karnstedt, Derek Greene, and
    Marek Jozwowicz

SCCQL : A Constraint-Based Clustering System 681
    Antoine Adam, Hendrik Blockeel, Sander Govers, and Abram Aertsen

Author Index 685

A Cascaded Supervised Learning Approach to Inverse Reinforcement Learning

Edouard Klein[1,2], Bilal Piot[2,3], Matthieu Geist[2], and Olivier Pietquin[2,3,*]

[1] ABC Team LORIA-CNRS, France
[2] Supélec, IMS-MaLIS Research Group, France
firstname.lastname@supelec.fr
[3] UMI 2958 (GeorgiaTech-CNRS), France

Abstract. This paper considers the Inverse Reinforcement Learning (IRL) problem, that is inferring a reward function for which a demonstrated expert policy is optimal. We propose to break the IRL problem down into two generic Supervised Learning steps: this is the Cascaded Supervised IRL (CSI) approach. A classification step that defines a score function is followed by a regression step providing a reward function. A theoretical analysis shows that the demonstrated expert policy is near-optimal for the computed reward function. Not needing to repeatedly solve a Markov Decision Process (MDP) and the ability to leverage existing techniques for classification and regression are two important advantages of the CSI approach. It is furthermore empirically demonstrated to compare positively to state-of-the-art approaches when using only transitions sampled according to the expert policy, up to the use of some heuristics. This is exemplified on two classical benchmarks (the mountain car problem and a highway driving simulator).

1 Introduction

Sequential decision making consists in choosing the appropriate action given the available data in order to maximize a certain criterion. When framed in a Markov Decision Process (MDP) (see Sec. 2), (Approximate) Dynamic programming ((A)DP) or Reinforcement Learning (RL) are often used to solve the problem by maximizing the expected sum of discounted rewards. The Inverse Reinforcement Learning (IRL) [15] problem, which is addressed here, aims at inferring a reward function for which a demonstrated expert policy is optimal.

IRL is one of many ways to perform Apprenticeship Learning (AL): imitating a demonstrated expert policy, without necessarily explicitly looking for the reward function. The reward function nevertheless is of interest in its own right. As mentioned in [15], its semantics can be analyzed in biology or econometrics for instance. Practically, the reward can be seen as a succinct description of a task. Discovering it removes the coupling that exists in AL between understanding

---

* The research leading to these results has received funding from the European Union Seventh Framework Programme (FP7/2007-2013) under grant agreement n°270780.

H. Blockeel et al. (Eds.): ECML PKDD 2013, Part I, LNAI 8188, pp. 1–16, 2013.

the task and learning how to fulfill it. IRL allows the use of (A)DP or RL techniques to learn how to do the task from the computed reward function. A very straightforward non-IRL way to do AL is for example to use a multi-class classifier to directly learn the expert policy. We provide in the experiments (Sec. 6) a comparison between AL and IRL algorithms by using IRL as a way to do AL.

A lot of existing approaches in either IRL or IRL-based AL need to repeatedly solve the underlying MDP to find the optimal policies of intermediate reward functions. Thus, their performance depends strongly on the quality of the associated subroutine. Consequently, they suffer from the same challenges of scalability, data scarcity, etc., as RL and (A)DP. In order to avoid *repeatedly* solving such problems, we adopt a different point of view.

Having in mind that there is a one to one relation between a reward function and its associated optimal action-value function (via the Bellman equation, see Eq. (1)), it is worth thinking of a method able to output an action-value function for which the greedy policy is the demonstrated expert policy. Thus, the demonstrated expert policy will be optimal for the corresponding reward function. We propose to use a score function-based multi-class classification step (see Sec. 3) to infer a score function. Besides, in order to retrieve via the Bellman equation the reward associated with the score function computed by the classification step, we introduce a regression step (see Sec. 3). That is why the method is called the Cascaded Supervised Inverse reinforcement learning (CSI). This method is analyzed in Sec. 4, where it is shown that the demonstrated expert policy is near-optimal for the reward the regression step outputs.

This algorithm does not need to iteratively solve an MDP and requires only sampled transitions from expert and non-expert policies as inputs. Moreover, up to the use of some heuristics (see Sec. 6.1), the algorithm is able to be trained only with transitions sampled from the demonstrated expert policy. A specific instantiation of CSI (proposed in Sec. 6.1) is tested on the mountain car problem (Sec. 6.2) and on a highway driving simulator (Sec. 6.3) where we compare it with a pure classification algorithm [20] and with two recent successful IRL methods [5] as well as with a random baseline. Differences and similarities with existing AL or IRL approaches are succinctly discussed in Sec. 5.

2   Background and Notation

First, we introduce some general notation. Let E and F be two non-empty sets, E^F is the set of functions from F to E. We note Δ_X the set of distributions over X. Let $\alpha \in \mathbb{R}^X$ and $\beta \in \mathbb{R}^X$: $\alpha \geq \beta \Leftrightarrow \forall x \in X, \alpha(x) \geq \beta(x)$. We will often slightly abuse the notation and consider (where applicable) most objects as if they were matrices and vectors indexed by the set they operate upon.

We work with finite MDPs [10], that is tuples $\{S, A, P, R, \gamma\}$. The state space is noted S, A is a finite action space, $R \in \mathbb{R}^{S \times A}$ is a reward-function, $\gamma \in (0, 1)$ is a discount factor and $P \in \Delta_S^{S \times A}$ is the Markovian dynamics of the MDP. Thus, for each $(s, a) \in S \times A$, $P(.|s, a)$ is a distribution over S and $P(s'|s, a)$

is the probability to reach s' by choosing action a in state s. At each time step t, the agent uses the information encoded in the state $s_t \in S$ in order to choose an action $a_t \in A$ according to a (deterministic[1]) policy $\pi \in A^S$. The agent then steps to a new state $s_{t+1} \in S$ according to the Markovian transition probabilities $P(s_{t+1}|s_t, a_t)$. Given that $P_\pi = (P(s'|s, \pi(s)))_{s,s' \in S}$ is the transition probability matrix, the stationary distribution over the states ρ_π induced by a policy π satisfies $\rho_\pi^T P_\pi = \rho_\pi^T$, with X^T being the transpose of X. The stationary distribution relative to the expert policy π_E is ρ_E.

The reward function R is a local measure of the quality of the control. The global quality of the control induced by a policy π, with respect to a reward R, is assessed by the value function $V_R^\pi \in \mathbb{R}^S$ which associates to each state the expected discounted cumulative reward for following policy π from this state: $V_R^\pi(s) = \mathbf{E}[\sum_{t \geq 0} \gamma^t R(s_t, \pi(s_t))|s_0 = s, \pi]$. This long-term criterion is what is being optimized when solving an MDP. Therefore, an optimal policy π_R^* is a policy whose value function (the optimal value function V_R^*) is greater than that of any other policy, for all states: $\forall \pi, V_R^* \geq V_R^\pi$.

The Bellman evaluation operator $T_R^\pi : \mathbb{R}^S \to \mathbb{R}^S$ is defined by $T_R^\pi V = R_\pi + \gamma P_\pi V$ where $R_\pi = (R(s, \pi(s)))_{s \in S}$. The Bellman optimality operator follows naturally: $T_R^* V = \max_\pi T_R^\pi V$. Both operators are contractions. The fixed point of the Bellman evaluation operator T_R^π is the value function of π with respect to reward R: $V_R^\pi = T_R^\pi V_R^\pi \Leftrightarrow V_R^\pi = R_\pi + \gamma P_\pi V_R^\pi$. The Bellman optimality operator T_R^* also admits a fixed point, the optimal value function V_R^* with respect to reward R.

Another object of interest is the action-value function $Q_R^\pi \in \mathbb{R}^{S \times A}$ that adds a degree of freedom on the choice of the first action, formally defined by $Q_R^\pi(s, a) = T_R^a V_R^\pi(s)$, with a the policy that always returns action a ($T_R^a V = R_a + \gamma P_a V$ with $P_a = (P(s'|s, a))_{s,s' \in S}$ and $R_a = (R(s, a))_{s \in S}$). The value function V_R^π and the action-value function Q_R^π are quite directly related: $\forall s \in S, V_R^\pi(s) = Q_R^\pi(s, \pi(s))$. The Bellman evaluation equation for Q_R^π is therefore:

$$Q_R^\pi(s, a) = R(s, a) + \gamma \sum_{s' \in S} P(s'|s, a)Q(s', \pi(s')). \tag{1}$$

An optimal policy follows a greedy mechanism with respect to its optimal action-value function Q_R^*:

$$\pi_R^*(s) \in \underset{a}{\operatorname{argmax}} Q_R^*(s, a). \tag{2}$$

When the state space is too large to allow matrix representations or when the transition probabilities or even the reward function are unknown except through observations gained by interacting with the system, RL or ADP may be used to approximate the optimal control policy [16].

We recall that solving the MDP is the direct problem. This contribution aims at solving the inverse one. We observe trajectories drawn from an expert's deterministic[1] policy π_E, assuming that there exists some unknown reward R_E

---

[1] We restrict ourselves here to deterministic policies, but the loss of generality is minimal as there exists at least one optimal deterministic policy.

for which the expert is optimal. The suboptimality of the expert is an interesting setting that has been discussed for example in [7,19], but that we are not addressing here. We do not try to find this unknown reward R_E but rather a non trivial reward R for which the expert is at least near-optimal. The trivial reward 0 is a solution to this ill-posed problem (no reward means that every behavior is optimal). Because of its ill-posed nature, this expression of *Inverse Reinforcement Learning* (IRL) still has to find a satisfactory solution although a lot of progress has been made, see Sec. 5.

3   The Cascading Algorithm

Our first step towards a reward function solving the IRL problem is a classification step using a *score function-based multi-class classifier* (SFMC^2 for short). This classifier learns a score function $q \in \mathbb{R}^{S \times A}$ that rates the association of a given action^2 $a \in A$ with a certain input $s \in S$. The classification rule $\pi_C \in A^S$ simply selects (one of) the action(s) that achieves the highest score for the given inputs:

$$\pi_C(s) \in \underset{a}{\operatorname{argmax}}\, q(s,a). \tag{3}$$

For example, *Multi-class Support Vector Machines* [4] can be seen as SFMC^2 algorithms, the same can be said of the structured margin approach [20] both of which we consider in the experimental setting. Other algorithms may be envisioned (see Sec. 6.1).

Given a dataset $D_C = \{(s_i, a_i = \pi_E(s_i))_i\}$ of actions a_i (deterministically) chosen by the expert on states s_i, we train such a classifier. The classification policy π_C is not the end product we are looking for (that would be mere supervised imitation of the expert, not IRL). What is of particular interest to us is the score function q itself. One can easily notice the similarity between Eq. (3) and Eq. (2) that describes the relation between the optimal policy in an MDP and its optimal action-value function. The score function q of the classifier can thus be viewed as some kind of optimal action-value function for the classifier policy π_C. By inversing the Bellman equation (1) with q in lieu of Q_R^π, one gets R^C, the reward function relative to our score/action-value function q:

$$R^C(s,a) = q(s,a) - \gamma \sum_{s'} P(s'|s,a) q(s', \pi_C(s')). \tag{4}$$

As we wish to approximately solve the general IRL problem where the transition probabilities P are unknown, our reward function R^C will be approximated with the help of information gathered by interacting with the system. We assume that another dataset $D_R = \{(s_j, a_j, s'_j)_j\}$ is available where s'_j is the state an agent taking action a_j in state s_j transitioned to. Action a_j need not be chosen by any

---

^{2} Here, actions play the role of what is known as *labels* or *categories* when talking about classifiers.

particular policy. The dataset D_R brings us information about the dynamics of the system. From it, we construct datapoints

$$\{\hat{r}_j = q(s_j, a_j) - \gamma q(s'_j, \pi_C(s'_j))\}_j. \tag{5}$$

As s'_j is sampled according to $P(\cdot|s_j, a_j)$ the constructed datapoints help building a good approximation of $R^C(s_j, a_j)$. A regressor (a simple least-square approximator can do but other solutions could also be envisioned, see Sec. 6.1) is then fed the datapoints $((s_j, a_i), \hat{r}_j)$ to obtain $\hat{R}^C$, a generalization of $\{((s_j, a_j), \hat{r}_j)_j\}$ over the whole state-action space. The complete algorithm is given in Alg. 1.

There is no particular constraint on D_C and D_R. Clearly, there is a direct link between various qualities of those two sets (amount of data, statistical representativity, etc.) and the classification and regression errors. The exact nature of the relationship between these quantities depends on which classifier and regressor are chosen. The theoretical analysis of Sec. 4 abstracts itself from the choice of a regressor and a classifier and from the composition of D_C and D_R by reasoning with the classification and regression errors. In Sec. 6, the use of a single dataset to create both D_C and D_R is thoroughly explored.

Algorithm 1. CSI algorithm

Given a training set $D_C = \{(s_i, a_i = \pi_E(s_i))\}_{1 \leq i \leq D}$ and another training set $D_R = \{(s_j, a_j, s'_j)\}_{1 \leq j \leq D'}$
Train a score function-based classifier on D_C, obtaining decision rule π_C and score function $q : S \times A \to \mathbb{R}$
Learn a reward function $\hat{R}^C$ from the dataset $\{((s_j, a_j), \hat{r}_j)\}_{1 \leq j \leq D'}$, $\forall(s_j, a_j, s'_j) \in D_R, \hat{r}_j = q(s_j, a_j) - \gamma q(s'_j, \pi_C(s'_j))$
Output the reward function $\hat{R}^C$

Cascading two supervised approaches like we do is a way to inject the MDP structure into the resolution of the problem. Indeed, mere classification only takes into account information from the expert (i.e., which action goes with which state) whereas using the Bellman equation in the expression of $\hat{r}_j$ makes use of the information lying in the transitions (s_j, a_j, s'_j), namely information about the transition probabilities P. The final regression step is a way to generalize this information about P to the whole state-action space in order to have a well-behaved reward function. Being able to alleviate the ill effects of scalability or data scarcity by leveraging the wide range of techniques developed for the classification and regression problems is a strong advantage of the CSI approach.

4   Analysis

In this section, we prove that the deterministic expert policy π_E is near optimal for the reward $\hat{R}^C$ the regression step outputs. More formally, recalling from Sec. 2 that ρ_E is the stationary distribution of the expert policy, we prove that $\mathbf{E}_{s \sim \rho_E}[V^*_{\hat{R}^C}(s) - V^{\pi_E}_{\hat{R}^C}(s)]$ is bounded by a term that depends on:

- the classification error defined as $\epsilon_C = \mathbf{E}_{s \sim \rho_E}[\mathbb{1}_{\{\pi_C(s) \neq \pi_E(s)\}}]$;
- the regression error defined as $\epsilon_R = \max_{\pi \in A^S} \|\epsilon_\pi^R\|_{1,\rho_E}$, with:
  - the subscript notation already used for R_π and P_π in Sec. 2 meaning that, given an $X \in \mathbb{R}^{S \times A}$, $\pi \in A^S$, and $a \in A$, $X_\pi \in \mathbb{R}^S$ and $X_a \in \mathbb{R}^S$ are respectively such that: $\forall s \in S, X_\pi(s) = X(s, \pi(s))$ and $\forall s \in S, X_a(s) = X(s, a)$;
  - $\epsilon_\pi^R = R_\pi^C - \hat{R}_\pi^C$;
  - $\|.\|_{1,\mu}$ the μ-weighted L_1 norm: $\|f\|_{1,\mu} = \mathbf{E}_{x \sim \mu}[|f(x)|]$;
- the concentration coefficient $C_* = C_{\hat{\pi}_C}$ with:
  - $C_\pi = (1 - \gamma) \sum_{t \geq 0} \gamma^t c_\pi(t)$, with $c_\pi(t) = \max_{s \in S} \frac{(\rho_E^T P_\pi^t)(s)}{\rho_E(s)}$;
  - $\hat{\pi}_C$, the optimal policy for the reward $\hat{R}^C$ output by the algorithm ;
  The constant C_* can be estimated *a posteriori* (after $\hat{R}^C$ is computed). A *priori*, C_* can be upper-bounded by a more usual and general concentration coefficient but C_* gives a tighter final result: one can informally see C_* as a measure of the similarity between the distributions induced by $\hat{\pi}_C$ and π_E (roughly, if $\hat{\pi}_C \approx \pi_E$ then $C_* \approx 1$).
- $\Delta q = \max_{s \in S}(\max_{a \in A} q(s, a) - \min_{a \in A} q(s, a)) = \max_{s \in S}(q(s, \pi_C(s)) - \min_{a \in A} q(s, a))$, which could be normalized to 1 without loss of generality. The range of variation of q is tied to the one of R^C, $\hat{R}^C$ and $V_{\hat{R}^C}^{\pi_C}$. What matters with these objects is the relative values for different state action couples, not the objective range. They can be shifted and positively scaled without consequence.

Theorem 1. *Let π_E be the deterministic expert policy, ρ_E its stationary distribution and $\hat{R}^C$ the reward the cascading algorithm outputs. We have:*

$$0 \leq \mathbf{E}_{s \sim \rho_E}[V_{\hat{R}^C}^*(s) - V_{\hat{R}^C}^{\pi_E}(s)] \leq \frac{1}{1 - \gamma}(\epsilon_C \Delta q + \epsilon_R(1 + C_*)).$$

Proof. First let's recall some notation, $q \in \mathbb{R}^{S \times A}$ is the score function output by the classification step, π_C is a deterministic classifier policy so that $\forall s \in S, \pi_C(s) \in \operatorname{argmax}_{a \in A} q(s, a)$, $R^C \in \mathbb{R}^{S \times A}$ is so that $\forall (s, a) \in S \times A, R^C(s, a) = q(s, a) - \gamma \sum_{s' \in S} P(s'|s, a) q(s', \pi_C(s'))$, and $\hat{R}^C \in \mathbb{R}^{S \times A}$ is the reward function output by the regression step.

The difference between R^C and $\hat{R}^C$ is noted $\epsilon^R = R^C - \hat{R}^C \in \mathbb{R}^{S \times A}$. We also introduce the reward function $\mathfrak{R}^E \in \mathbb{R}^{S \times A}$ which will be useful in our proof, not to be confused with R_E the unknown reward function the expert optimizes:

$$\forall (s, a) \in S \times A, \mathfrak{R}^E(s, a) = q(s, a) - \gamma \sum_{s' \in S} P(s'|s, a) q(s', \pi_E(s')).$$

We now have the following vectorial equalities $R_a^C = q_a - \gamma P_a q_{\pi_C}$; $\mathfrak{R}_a^E = q_a - \gamma P_a q_{\pi_E}$; $\epsilon_a^R = R_a^C - \hat{R}_a^C$. Now, we are going to upper bound the term: $\mathbf{E}_{s \sim \rho_E}[V_{\hat{R}^C}^* - V_{\hat{R}^C}^{\pi_E}] \geq 0$ (the lower bound is obvious as V^* is optimal). Recall that $\hat{\pi}_C$ is a deterministic optimal policy of the reward $\hat{R}^C$. First, the term $V_{\hat{R}^C}^* - V_{\hat{R}^C}^{\pi_E}$ is decomposed:

$$V_{\hat{R}^C}^* - V_{\hat{R}^C}^{\pi_E} = (V_{\hat{R}^C}^{\hat{\pi}_C} - V_{R^C}^{\hat{\pi}_C}) + (V_{R^C}^{\hat{\pi}_C} - V_{R^C}^{\pi_E}) + (V_{R^C}^{\pi_E} - V_{\hat{R}^C}^{\pi_E}).$$

We are going to bound each of these three terms. First, let π be a given deterministic policy. We have, using $\epsilon^R = R^C - \hat{R}^C$: $V_{R^C}^\pi - V_{\hat{R}^C}^\pi = V_{\epsilon^R}^\pi = (I - \gamma P_\pi)^{-1} \epsilon_\pi^R$. If $\pi = \pi_E$, we have, thanks to the power series expression of $(I - \gamma P_{\pi_E})^{-1}$, the definition of ρ_E and the definition of the μ-weighted L_1 norm, one property of which is that $\forall X, \mu^T X \le \|X\|_{1,\mu}$:

$$\rho_E^T(V_{R^C}^\pi - V_{\hat{R}^C}^\pi) = \rho_E^T(I - \gamma P_{\pi_E})^{-1}\epsilon_{\pi_E}^R = \frac{1}{1-\gamma}\rho_E^T\epsilon_{\pi_E}^R \le \frac{1}{1-\gamma}\|\epsilon_{\pi_E}^R\|_{1,\rho_E}.$$

If $\pi \ne \pi_E$, we use the concentration coefficient C_π. We have then:

$$\rho_E^T(V_{R^C}^\pi - V_{\hat{R}^C}^\pi) \le \frac{C_\pi}{1-\gamma}\rho_E^T\epsilon_\pi^R \le \frac{C_\pi}{1-\gamma}\|\epsilon_\pi^R\|_{1,\rho_E}.$$

So, using the notation introduced before we stated the theorem, we are able to give an upper bound to the first and third terms (recall also the notation $C_{\hat{\pi}_C} = C_*$): $\rho_E^T((V_{R^C}^{\pi_E} - V_{\hat{R}^C}^{\pi_E}) + (V_{\hat{R}^C}^{\hat{\pi}_C} - V_{R^C}^{\hat{\pi}_C})) \le \frac{1+C_*}{1-\gamma}\epsilon_R$. Now, there is still an upper bound to find for the second term. It is possible to decompose it as follows:

$$V_{R^C}^{\hat{\pi}_C} - V_{R^C}^{\pi_E} = (V_{R^C}^{\hat{\pi}_C} - V_{R^C}^{\pi_C}) + (V_{R^C}^{\pi_C} - V_{\Re^E}^{\pi_E}) + (V_{\Re^E}^{\pi_E} - V_{R^C}^{\pi_E}).$$

By construction, π_C is optimal for R^C, so $V_{R^C}^{\hat{\pi}_C} - V_{R^C}^{\pi_C} \le 0$ which implies:

$$V_{R^C}^{\hat{\pi}_C} - V_{R^C}^{\pi_E} \le (V_{R^C}^{\pi_C} - V_{\Re^E}^{\pi_E}) + (V_{\Re^E}^{\pi_E} - V_{R^C}^{\pi_E}).$$

By construction, we have $V_{R^C}^{\pi_C} = q_{\pi_C}$ and $V_{\Re^E}^{\pi_E} = q_{\pi_E}$, thus:

$$\rho_E^T(V_{R^C}^{\pi_C} - V_{\Re^E}^{\pi_E}) = \rho_E^T(q_{\pi_C} - q_{\pi_E})$$

$$= \sum_{s \in S} \rho_E(s)(q(s,\pi_C(s)) - q(s,\pi_E(s)))[\mathbb{1}_{\{\pi_C(s) \ne \pi_E(s)\}}].$$

Using Δq, we have: $\rho_E^T(V_{R^C}^{\pi_C} - V_{\Re^E}^{\pi_E}) \le \Delta q \sum_{s \in S} \rho_E(s)[\mathbb{1}_{\{\pi_C(s) \ne \pi_E(s)\}}] = \Delta q \epsilon_C$. Finally, we also have:

$$\rho_E^T(V_{\Re^E}^{\pi_E} - V_{R^C}^{\pi_E}) = \rho_E^T(I - \gamma P_{\pi_E})^{-1}(\Re_{\pi_E}^E - R_{\pi_E}^C) = \rho_E^T(I - \gamma P_{\pi_E})^{-1}\gamma P_{\pi_E}(q_{\pi_C} - q_{\pi_E}),$$

$$= \frac{\gamma}{1-\gamma}\rho_E^T(q_{\pi_C} - q_{\pi_E}) \le \frac{\gamma}{1-\gamma}\Delta q \epsilon_C.$$

So the upper bound for the second term is: $\rho_E^T(V_{R^C}^{\hat{\pi}_C} - V_{R^C}^{\pi_E}) \le (\Delta q + \frac{\gamma}{1-\gamma}\Delta q)\epsilon_C = \frac{\Delta q}{1-\gamma}\epsilon_C$. If we combine all of the results, we obtain the final bound as stated in the theorem. $\qquad\square$

Readers familiar with the work presented in [5] will see some similarities between the theoretical analyses of SCIRL and CSI as both study error propagation in IRL algorithms. Another shared feature is the use of the score function q of the classifier as a proxy for the action-value function of the expert $Q_{R_E}^{\pi_E}$.

The attentive reader, however, will perceive that similarities stop there. The error terms occurring in the two bounds are not related to one another. As CSI makes no use of the feature expectation of the expert, what is known as $\bar{\epsilon}_Q$ in [5] does not appear in this analysis. Likewise, the regression error ϵ_R of this paper does not appear in the analysis of SCIRL, which does not use a regressor. Perhaps more subtly, the classification error and classification policy known in both papers as ϵ_C and π_C are not the same. The classification policy of SCIRL is not tantamount to what is called π_C here. For SCIRL, π_C is the greedy policy for an approximation of the value function of the expert with respect to the reward output by the algorithm. For CSI, π_C is the decision rule of the classifier, an object that is not aware of the structure of the MDP. We shall also mention that the error terms appearing in the CSI bound are more standard than the ones of SCIRL (*e.g.*, regression error vs feature expectation estimation error) thus they may be easier to control. A direct corollary of this theorem is that, given perfect classifier and regressor, CSI produces a reward function for which π_E is the *unique* optimal policy.

Corollary 1. *Assume that $\rho_E > 0$ and that the classifier and the regressor are perfect ($\epsilon_C = 0$ and $\epsilon_R = 0$). Then π_E is the* unique *optimal policy for $\hat{R}^C$.*

Proof. The function q is the optimal action-value function for π_C with respect to the reward R^C, by definition (see Eq. (4)). As $\epsilon_C = 0$, we have $\pi_C = \pi_E$. This means that $\forall s, \pi_E(s)$ is the only element of the set $\mathrm{argmax}_{a \in A} \, q(s, a)$. Therefore, $\pi_C = \pi_E$ is the unique optimal policy for R^C. As $\epsilon_R = 0$, we have $\hat{R}^C = R^C$, hence the result.

This corollary hints at the fact that we found a non-trivial reward (we recall that the null reward admits *every* policy as optimal). Therefore, obtaining $\hat{R}_C = 0$ (for which the bound is obviously true: the bounded term is 0, the bounding term is positive) is unlikely as long as the classifier and the regressor exhibit decent performance.

The only constraints the bound of Th. 1 implies on datasets D_R and D_C is that they provide enough information to the supervised algorithms to keep both error terms ϵ_C and ϵ_R low. In Sec. 6 we deal with a lack of data in dataset D_R. We address the problem with the use of heuristics (Sec. 6.1) in order to show the behavior of the CSI algorithm in somewhat more realistic (but difficult) conditions.

More generally, the error terms ϵ_C and ϵ_R can be reduced by a wise choice for the classification and regression algorithms. The literature is wide enough for methods accommodating most of use cases (lack of data, fast computation, bias/variance trade-off, etc.) to be found. Being able to leverage such common algorithms as multi-class classifiers and regressors is a big advantage of our cascading approach over existing IRL algorithms.

Other differences between existing IRL or apprenticeship learning approaches and the proposed cascading algorithm are further examined in Sec. 5.

5   Related Work

IRL was first introduced in [15] and then formalized in [9]. Approaches summarized in [8] can be seen as iteratively constructing a reward function, solving an MDP at each iteration. Some of these algorithms are IRL algorithms while others fall in the Apprenticeship Learning (AL) category, as for example the projection version of the algorithm in [1]. In both cases the need to solve an MDP at each step may be very demanding, both sample-wise and computationally. CSI being able to output a reward function without having to solve the MDP is thus a significant improvement.

AL via classification has been proposed for example in [12], with the help of a structured margin method. Using the non-trivial notion of metric in an MDP, the authors of [6] build a kernel which is used in a classification algorithm, showing improvements compared to a non-structured kernel.

Classification and IRL have met in the past in [13], but the labels were complete optimal policies rather than actions and the inputs were MDPs, which had to be solved. It may be unclear how SCIRL [5] relates to the proposed approach of his paper. Both algorithms use the score function of a classifier as a proxy to the action-value function of the expert with respect to the (unknown) true reward: $Q_R^{\pi_E}$. The way this proxy is constructed and used, however, fundamentally differs in the two algorithms. This difference will cause the theoretical analysis of both approaches (see Sec. 4) to be distinct. In SCIRL, the score function of the classifier is approximated *via* a linear parametrization that relies on the feature expectation of the expert $\mu^E(s) = E[\sum_{t \geq 0} \gamma^t \phi(s_t)|s_0 = s, \pi_E]$. This entails the use of a specific kind of classifier (namely linearly-parametrized-score-function-based classifiers) and of a method of approximation of μ^E. By contrast, almost any off-the-shelf classifier can be used in the first step of the cascading approach of this paper. The classification step of CSI is unaware of the structure of the MDP whereas SCIRL knows about it thanks to the use of μ^E. In CSI, the structure of the MDP is injected by reversing the Bellman equation prior to the regression step (Eq. 4 and (5)), a step that does not exist in SCIRL as SCIRL directly outputs the parameter vector found by its linearly-parametrized-score-function-based classifier. The regressor of CSI can be chosen off-the-shelf. One can argue that this and not having to approximate μ^E increases the ease-of-use of CSI over SCIRL and makes for a more versatile algorithm. In practice, as seen in Sec. 6, performance of SCIRL and CSI are very close to one another thus CSI may be a better choice as it is easier to deploy. Neither approach is a generalization of the other.

Few IRL or AL algorithms do not require solving an MDP. The approach of [17] requires knowing the transition probabilities of the MDP (which CSI does not need) and outputs a policy (and not a reward). The algorithm in [3] only applies to linearly-solvable MDPs whereas our approach does not place such restrictions. Closer to our use-case is the idea presented in [2] to use a subgradient ascent of a utility function based on the notion of relative entropy. Importance sampling is suggested as a way to avoid solving the MDP. This requires sampling trajectories according to a non-expert policy and the direct problem remains at the core of the approach (even if solving it is avoided).

6   Experiments

In this section, we empirically demonstrate the behavior of our approach. We begin by providing information pertaining to both benchmarks. An explanation about the amount and source of the available data, the rationale behind the heuristics we use to compensate for the dire data scarcity and a quick word about the contenders CSI is compared to are given Sec. 6.1. We supply quantitative results and comparisons of CSI with state-of-the art approaches on first a classical RL benchmark (the mountain car) in Sec. 6.2 and then on a highway driving simulator (Sec. 6.3).

6.1   Generalities

Data Scarcity. The CSI algorithm was designed to avoid repeatedly solving the RL problem. This feature makes it particularly well-suited to environments where sampling the MDP is difficult or costly. In the experiments, CSI is fed only with data sampled according to the expert policy. This corresponds for example to a situation where a costly system can only be controlled by a trained operator as a bad control sequence could lead to a system breakdown.

More precisely, the expert controls the system for M runs of lengths $\{L_i\}_{1 \le i \le M}$, giving samples $\{(s_k, a_k = \pi_E(a_k), s'_k)_k\} = D_E$. The dataset D_C fed to the classifier is straightforwardly constructed from D_E by dropping the s'_k terms: $D_C = \{(s_i = s_k, a_i = a_k)_i\}$.

Heuristics. It is not reasonable to construct the dataset $D_R = \{((s_k, a_k = \pi_E(s_k)), \hat{r}_k)_k\}$ only from expert transitions and expect a small regression error term ϵ_R. Indeed, the dataset D_E only samples the dynamics induced by the expert's policy and not the whole dynamics of the MDP. This means that for a certain state s_k we only know the corresponding expert action $a_k = \pi_E(s_k)$ and the following state s'_k sampled according to the MDP dynamics : $s'_k \sim P(\cdot|s_k, a_k)$. For the regression to be meaningful, we need samples associating the same state s_k and a different action $a \ne a_k$ with a datapoint $\hat{r} \ne \hat{r}_k$.

Recall that $\hat{r}_j = q(s_j, a_j) - \gamma q(s'_j, \pi_C(s'_j))$ (Eq. (5)); without knowing $s'_k \sim P(\cdot|s_k, a \ne a_k)$, we cannot provide the regressor with a datapoint to associate with $(s_k, a \ne a_k)$. We artificially augment the dataset D_R with samples $((s_j = s_k, a), r_{min})_{j;\forall a \ne \pi_E(s_j) = a_k}$ where $r_{min} = \min_k \hat{r}_k - 1$. This heuristics instructs the regressor to associate a state-action tuple disagreeing with the expert (*i.e.*, $(s_k, a \ne a_k)$) with a reward strictly inferior to any of those associated with expert state action tuples (*i.e.*, $(s_k, a_k = \pi_E(s_k))$). Semantically, we are asserting that disagreeing with the expert in states the expert visited is a bad idea. This heuristics says nothing about states absent from the expert dataset. For such states the generalization capabilities of the regressor and, later on, the exploration of the MDP by an agent optimizing the reward will solve the problem. Although this heuristics was not analyzed in Sec. 4 (where the availability of a more complete dataset D_R was assumed), the results shown in the next two subsections demonstrate its soundness.

Comparison with State-of-the-Art Approaches. The similar looking yet fundamentally different algorithm SCIRL [5] is an obvious choice as a contender to CSI as it advertises the same ability to work with very little data, without repeatedly solving the RL problem. In both experiments we give the exact same data to CSI and SCIRL.

The algorithm of [3] also advertises not having to solve the RL problem, but needs to deal with linearly solvable MDPs, therefore we do not include it in our tests. The Relative Entropy (RE) method of [2] has no such need, so we included it in our benchmarks. It could not, however, work with the small amount of data we provided SCIRL and CSI with, and so to allow for importance sampling, we created another dataset D_{random} that was used by RE but not by SCIRL nor CSI.

Finally, the classification policy π_C output by the classification step of CSI was evaluated as well. Comparing classification and IRL algorithms makes no sense if the object of interest is the reward itself as can be envisioned in a biological or economical context. It is however sound to do so in an imitation context where what matters is the performance of the agent with respect to some objective criterion. Both experiments use such a criterion. Classification algorithms don't have to optimize any reward since the classification policy can directly be used in the environment. IRL algorithms on the other hand output a reward that must then be plugged in an RL or DP algorithm to get a policy. In each benchmark we used the same (benchmark-dependent) algorithm to get a policy from each of the three rewards output by SCIRL, CSI and RE. It is these policies whose performance we show. Finding a policy from a reward is of course a non-trivial problem that should not be swept under the rug; nevertheless we choose not to concern ourselves with it here as we wish to focus on IRL algorithms, not RL or DP algorithms. In this regard, using a classifier that directly outputs a policy may seem a much simpler solution, but we hope that the reader will be convinced that the gap in performance between classification and IRL is worth the trouble of solving the RL problem (once and for all, and not repeatedly as a subroutine like some other IRL algorithms).

We do not compare CSI to other IRL algorithms requiring repeatedly solving the MDP. As we would need to provide them with enough data to do so, the comparison makes little sense.

Supervised Steps. The cascading algorithm can be instantiated with some standard classification algorithms and any regression algorithm. The choice of such subroutines may be dictated by the kind and amount of available data, by ease of use or by computational complexity, for example.

We referred in Sec.3 to *score-function based multi-class classifiers* and explained how the classification rule is similar to the greedy mechanism that exists between an optimal action-value function and an optimal policy in an MDP. Most classifications algorithms can be seen as such a classifier. In a simple k-nearest neighbor approach, for example, the score function $q(s, a)$ is the number of elements of class a among the k-nearest neighbors of s. The generic M-SVM model makes the score function explicit (see [4]) (we use a SVM in the mountain car experiment Sec. 6.2). In the highway experiment, we choose to use a structured margin classification approach [20]. We chose a SVR as a regressor in the mountain car experiment and a simple least-square regressor on the highway.

It is possible to get imaginative in the last step. For example, using a Gaussian process regressor [11] that outputs both expectation and variance can enable (notwithstanding a nontrivial amount of work) the use of reward-uncertain reinforcement learning [14]. Our complete instantiation of CSI is summed up in Alg. 2.

Algorithm 2. A CSI instantiation with heuristics

Given a dataset $D_E = (s_k, a_k = \pi_E(a_k), s'_k)_k$
Construct the dataset $D_C = \{(s_i = s_k, a_i = \pi^E(s_i)) = a_k\}$
Train a score function-based classifier on D_C, obtaining decision rule π_C and score function $q : S \times A \to \mathbb{R}$
Construct the dataset $\{((s_j = s_k, a_j = a_k), \hat{r}_j)_j\}$ with $\hat{r}_j = q(s_j, a_j) - \gamma q(s'_j = s'_k, \pi_C(s'_j = s'_k))$
Set $r_{min} = \min_j \hat{r}_j - 1$.
Construct the training set $D_R = \{((s_j = s_k, a_j = a_k), \hat{r}_j)_j\} \cup \{((s_j = s_k, a), r_{min})_{j; \forall a \neq \pi_E(s_j) = a_k}\}$
Learn a reward function $\hat{R}^C$ from the training set D_R
Output the reward function $\hat{R}^C : (s, a) \mapsto \omega^T \phi(s, a)$

6.2 Mountain Car

The mountain car is a classical toy problem in RL: an underpowered car is tasked with climbing a steep hill. In order to do so, it has to first move away from the target and climb the slope on its left, and then it moves right, gaining enough momentum to climb the hill on the right on top of which lies the target. We used standard parameters for this problem, as can for example be found in [16]. When training an RL agent, the reward is, for example, 1 if the car's position is greater than 0.5 and 0 anywhere else. The expert policy was a very simple hand crafted policy that uses the power of the car to go in the direction it already moves (*i.e.*, go left when the speed is negative, right when it is positive).

The initial position of the car was uniformly randomly picked in $[-1.2; -0.9]$ and its speed uniformly randomly picked in $[-0.07; 0]$. From this position, the hand-crafted policy was left to play until the car reached the objective (*i.e.*, a position greater than 0.5) at which point the episode ended. Enough episodes were played (and the last one was truncated) so that the dataset D_E contained exactly n samples, with n successively equal to 10, 30, 100 and 300. With these parameters, the expert is always able to reach the top on the first time it tries to climb the hill on the right. Therefore, a whole part of the state space (when the position is on the hill on the right and the speed is negative) is not visited by the expert. This hole about the state space in the data will be dealt with differently by the classifier and the IRL algorithms. The classifier will use its generalization power to find a default action in this part of the state space, while the IRL algorithms will devise a default reward; a (potentially untrained) RL agent finding itself in this part of the state space will use the reward signal to decide what to do, making use of new data available at that time.

In order to get a policy from the rewards given by SCIRL, CSI and RE, the RL problem was solved by LSPI fed with a dataset D_{random} of 1000 episodes of length 5 with a starting point uniformly and randomly chosen in the whole state space and actions picked at random. This dataset was also used by RE (and not by SCIRL nor CSI).

The classifier for CSI was an off-the-shelf SVM[3] which also was the classifier we evaluate, the regressor of CSI was an off-the-shelf SVR[4]. RE and SCIRL need features over the state space, we used the same evenly-spaced hand-tuned 7×7 RBF network for both algorithms.

The objective criterion for success is the number of steps needed to attain the goal when starting from a state picked at random ; the lesser the better. We can see Fig. 1 that the optimal policies for the rewards found by SCIRL and CSI very rapidly attain expert-level performance and outperform the optimal policy for the reward of RE and the classification policy. When very few samples are available, CSI does better than SCIRL (with such a low p-value for $n = 10$, see Tab. 1a, the hypothesis that the mean performance is equal can be rejected); SCIRL catches up when more samples are available. Furthermore, CSI required very little engineering as we cascaded two off-the-shelf implementations whereas SCIRL used hand-tuned features and a custom classifier.

Table 1. Student or Welch test of mean equality (depending on whether a Bartlett test of variance equality succeeds) p-values for CSI and SCIRL on the mountain car (1a) and the highway driving simulator (1b). High values ($> 1.0 \times 10^{-02}$) means that the hypothesis that the means are equal cannot be rejected.

<table>
<tr><td colspan="2" align="center">(a) Mountain Car</td><td colspan="2" align="center">(b) Highway Driving</td></tr>
<tr><td>Number of expert samples</td><td>p-value</td><td>Number of expert samples</td><td>p-value</td></tr>
<tr><td align="center">10</td><td align="center">**1.5e − 12**</td><td align="center">9</td><td align="center">3.0e − 01</td></tr>
<tr><td align="center">30</td><td align="center">3.8e − 01</td><td align="center">49</td><td align="center">**8.9e − 03**</td></tr>
<tr><td align="center">100</td><td align="center">1.3e − 02</td><td align="center">100</td><td align="center">**1.8e − 03**</td></tr>
<tr><td align="center">300</td><td align="center">7.4e − 01</td><td align="center">225</td><td align="center">**2.4e − 05**</td></tr>
<tr><td></td><td></td><td align="center">400</td><td align="center">**2.0e − 50**</td></tr>
</table>

6.3   Highway Driving Simulator

The setting of the experiment is a driving simulator inspired from a benchmark already used in [17,18]. The agent controls a car that can switch between the three lanes of the road, go off-road on either side and modulate between three speed levels. At all timesteps, there will be one car in one of the three lanes.

---

[3] http://scikit-learn.org/stable/modules/generated/sklearn.svm.SVC.html
[4] http://scikit-learn.org/stable/modules/generated/sklearn.svm.SVR.html

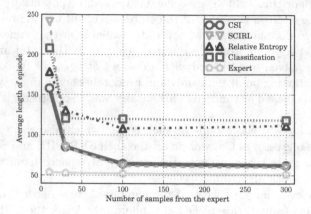

Fig. 1. Performance of various policies on the mountain car problem. This is the mean over 100 runs.

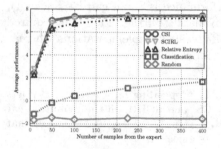 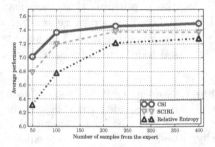

(a) Mean performance over 100 runs on the Highway driving problem.   **(b)** Zoom of Fig 2a showing the ranking of the three IRL algorithms.

Fig. 2. Results on the highway driving problem

Even at the lowest speed, the player's car moves faster than the others. When the other car disappears at the bottom of the screen, another one appears at the top in a randomly chosen lane. It takes two transitions to completely change lanes, as the player can move left or right for half a lane's length at a time. At the highest speed setting, if the other car appears in the lane the player is in, it is not possible to avoid the collision. The main difference between the original benchmark [17,18] and ours is that we made the problem more ergodic by allowing the player to change speed whenever he wishes to, not just during the first transition. If anything, by adding two actions, we enlarged the state-action space and thus made the problem tougher. The reward function R_E the expert is trained by a DP algorithm on makes it go as fast as possible (high reward) while avoiding collisions (harshly penalized) and avoiding going off-road (moderately penalised). Any other situation receives a null reward.

The performance criterion for a policy π is the mean (over the uniform distribution) value function with respect to R_E : $\mathbf{E}_{s\sim\mathcal{U}}[V_{R_E}^\pi(s)]$. Expert performance averages to 7.74 ; we also show the natural random baseline that consists in drawing a random reward vector (with a uniform law) and training an agent on it. The reward functions found by SCIRL, CSI and RE are then optimized using a DP algorithm. The dataset D_{random} needed by RE (and neither by CSI nor SCIRL) is made of 100 episodes of length 10 starting randomly in the state space and following a random policy. the dataset D_E is made of n episodes of length n, with $n \in \{3, 7, 10, 15, 20\}$.

Results are shown Fig. 2. We give the values of $\mathbf{E}_{s\sim\mathcal{U}}[V_{R_E}^\pi(s)]$ with π being in turn the optimal policy for the rewards given by SCIRL, CSI and RE, the policy π_C of the classifier (the very one the classification step of CSI outputs), and the optimal policy for a randomly drawn reward. Performance for CSI is slightly but definitively higher than for SCIRL (see the p-values for the mean equality test in Tab. 1b, from 49 samples on), slightly below the performance of the expert itself. Very few samples (100) are needed to reliably achieve expert-level performance.

It is very interesting to compare our algorithm to the behavior of a classifier alone (respectively red and green plots on Fig. 2a). With *the exact same data*, albeit the use of a very simple heuristics, the cascading approach demonstrates far better performance from the start. This is a clear illustration of the fact that using the Bellman equation to construct the data fed to the regressor and outputting not a policy, but a reward function that can be optimized on the MDP truly makes use of the information that the transitions (s, a, s') bear (we recall that the classifier only uses (s, a) couples). Furthermore, the classifier whose results are displayed here is the output of the first step of the algorithm. The classification performance is obviously not that good, which points to the fact that our algorithm may be empirically more forgiving of classification errors than our theoretical bound lets us expect.

7 Conclusion

We have introduced a new way to perform IRL by cascading two supervised approaches. The expert is theoretically shown to be near-optimal for the reward function the proposed algorithm outputs, given small classification and regression errors. Practical examples of classifiers and regressors have been given, and two combinations have been empirically (on two classic benchmarks) shown to be very resilient to dire lack of data on the input (only data from the expert was used to retrieve the reward function), with the help of simple heuristics. On both benchmarks, our algorithm is shown to outperform other state-of-the-art approaches although SCIRL catches up on the mountain car. We plan on deepening the analysis of the theoretical properties of our approach and on applying it to real world robotics problems.

References

1. Abbeel, P., Ng, A.: Apprenticeship learning via inverse reinforcement learning. In: Proc. ICML (2004)
2. Boularias, A., Kober, J.: Peters: Relative entropy inverse reinforcement learning. In: Proc. ICAPS, vol. 15, pp. 20–27 (2011)
3. Dvijotham, K., Todorov, E.: Inverse optimal control with linearly-solvable MDPs. In: Proc. ICML (2010)
4. Guermeur, Y.: A generic model of multi-class support vector machine. International Journal of Intelligent Information and Database Systems (2011)
5. Klein, E., Geist, M., Piot, B., Pietquin, O.: Inverse Reinforcement Learning through Structured Classification. In: Proc. NIPS, Lake Tahoe, NV, USA (December 2012)
6. Melo, F.S., Lopes, M.: Learning from demonstration using MDP induced metrics. In: Balcázar, J.L., Bonchi, F., Gionis, A., Sebag, M. (eds.) ECML PKDD 2010, Part II. LNCS, vol. 6322, pp. 385–401. Springer, Heidelberg (2010)
7. Melo, F., Lopes, M., Ferreira, R.: Analysis of inverse reinforcement learning with perturbed demonstrations. In: Proc. ECAI, pp. 349–354. IOS Press (2010)
8. Neu, G., Szepesvári, C.: Training parsers by inverse reinforcement learning. Machine Learning 77(2), 303–337 (2009)
9. Ng, A., Russell, S.: Algorithms for inverse reinforcement learning. In: Proc. ICML, pp. 663–670. Morgan Kaufmann Publishers Inc. (2000)
10. Puterman, M.: Markov decision processes: Discrete stochastic dynamic programming. John Wiley & Sons, Inc., New York (1994)
11. Rasmussen, C., Williams, C.: Gaussian processes for machine learning, vol. 1. MIT press, Cambridge (2006)
12. Ratliff, N., Bagnell, J., Srinivasa, S.: Imitation learning for locomotion and manipulation. In: International Conference on Humanoid Robots, pp. 392–397. IEEE (2007)
13. Ratliff, N., Bagnell, J., Zinkevich, M.: Maximum margin planning. In: Proc. ICML, p. 736. ACM (2006)
14. Regan, K., Boutilier, C.: Robust online optimization of reward-uncertain MDPs. In: Proc. IJCAI 2011 (2011)
15. Russell, S.: Learning agents for uncertain environments (extended abstract). In: Annual Conference on Computational Learning Theory, p. 103. ACM (1998)
16. Sutton, R., Barto, A.: Reinforcement learning. MIT Press (1998)
17. Syed, U., Bowling, M., Schapire, R.: Apprenticeship learning using linear programming. In: Proc. ICML, pp. 1032–1039. ACM (2008)
18. Syed, U., Schapire, R.: A game-theoretic approach to apprenticeship learning. In: Proc. NIPS, vol. 20, pp. 1449–1456 (2008)
19. Syed, U., Schapire, R.: A reduction from apprenticeship learning to classification. In: Proc. NIPS, vol. 24, pp. 2253–2261 (2010)
20. Taskar, B., Chatalbashev, V., Koller, D., Guestrin, C.: Learning structured prediction models: A large margin approach. In: Proc. ICML, p. 903. ACM (2005)

Learning from Demonstrations:
Is It Worth Estimating a Reward Function?

Bilal Piot[1,2], Matthieu Geist[1], and Olivier Pietquin[1,2]

[1] Supélec, IMS-MaLIS Research Group, France
{bilal.piot,matthieu.geist,olivier.pietquin}@supelec.fr
[2] GeorgiaTech-CNRS UMI 2958, France

Abstract. This paper provides a comparative study between Inverse Reinforcement Learning (IRL) and Apprenticeship Learning (AL). IRL and AL are two frameworks, using Markov Decision Processes (MDP), which are used for the imitation learning problem where an agent tries to learn from demonstrations of an expert. In the AL framework, the agent tries to learn the expert policy whereas in the IRL framework, the agent tries to learn a reward which can explain the behavior of the expert. This reward is then optimized to imitate the expert. One can wonder if it is worth estimating such a reward, or if estimating a policy is sufficient. This quite natural question has not really been addressed in the literature right now. We provide partial answers, both from a theoretical and empirical point of view.

1 Introduction

This paper provides a comparative study between two methods, using the Markov Decision Process (MDP) paradigm, that attempt to solve the imitation learning problem where an agent (called the apprentice) tries to learn from demonstrations of an expert. These two methods are Apprenticeship Learning (AL) [1] and Inverse Reinforcement Learning (IRL) [8]. In the AL framework, the agent tries to learn the expert policy or at least a policy which is as good as the expert policy (according to an unknown reward function). In the IRL framework, the agent tries to learn a reward which can explain the behavior of the expert and which is optimized to imitate it. AL can be reduced to classification [7,3,6,11] where the agent tries to mimic the expert policy via a Supervised Learning (SL) method such as classification. There exist also several AL algorithms inspired by IRL such as [1,10] but they need to solve recursively MDPs which is a difficult problem when the state space is large and the dynamics of the MDP is unknown.

The key idea behind IRL is that the reward is the most succinct representation of the task. However, as the outputs of IRL algorithms are rewards, it is still required to solve an MDP to obtain an optimal policy with respect to this reward. With AL algorithms, the output is a policy which can be directly used. However, this policy is fixed and cannot adapt to a perturbation of dynamics which could be done if one knew the true reward, as it is a representation of the

H. Blockeel et al. (Eds.): ECML PKDD 2013, Part I, LNAI 8188, pp. 17–32, 2013.
© Springer-Verlag Berlin Heidelberg 2013

task possibly independent of the dynamics. Thus, a natural question arises: in which circumstances is it interesting to use an IRL algorithm, knowing that it still needs to solve an MDP in order to obtain a policy?

First, we analyse the difference of value functions between the apprentice and the expert policies when a classifier is used as AL method (in the infinite horizon case). When compared to the sole (as far as we know) related result in IRL, quantifying the quality of an apprentice trained with the recently introduced SCIRL (Structured Classification based IRL) algorithm [5], this analysis tells us that estimating a reward only adds errors. Then, we perform an empirical study on the generic Garnet framework [2] to see if this first partial answer is confirmed. It turns out that it actually strongly depends on the (unknown) reward optimized by the expert: roughly, the less informative the reward is, the more IRL provides gains compared to AL. Finally, we push this empirical study even further by perturbing the dynamics of the MDP, which goes beyond the studied theory. In this case, the advantage of IRL is even clearer.

2   Background and Notations

2.1   General Notations

Let $\mathcal{X} = (x_i)_{\{1 \leq i \leq N_{\mathcal{X}}\}}$ be a finite set and $f \in \mathcal{R}^{\mathcal{X}}$ a function, f is identified to a column vector and f^T is the transposition of f. The powerset of $\mathcal{X}$ is noted $\mathbb{P}(X)$. The set of probability distributions over $\mathcal{X}$ is noted $\Delta_{\mathcal{X}}$. Let $\mathcal{Y}$ be a finite set, $\Delta_{\mathcal{X}}^{\mathcal{Y}}$ is the set of functions from $\mathcal{Y}$ to $\Delta_{\mathcal{X}}$. Let $\zeta \in \Delta_{\mathcal{X}}^{\mathcal{Y}}$ and $y \in \mathcal{Y}$, $\zeta(y) \in \Delta_{\mathcal{X}}$, which can be seen as the conditional distribution probability knowing y, is also noted $\zeta(.|y)$ and $\forall x \in \mathcal{X}, \zeta(x|y) = [\zeta(y)](x)$. Besides, let $A \in \mathbb{P}(\mathcal{X})$, then $\chi_A \in \mathbb{R}^{\mathcal{X}}$ is the indicator function on the subset $A \subset \mathcal{X}$. The support of f is noted $\mathrm{Supp}(f)$. Moreover, let $\mu \in \Delta_{\mathcal{X}}$, $\mathbb{E}_{\mu}[f]$ is the expectation of the function f with respect to the probability μ. Let $x \in \mathcal{X}$, $x \sim \mu$ means that x is sampled according to μ. Finally, we define also for $p \in \mathbb{N}^*$, the $\mathbb{L}_p$-norm of the function f: $\|f\|_p = (\sum_{x \in \mathcal{X}} (f(x)^p))^{\frac{1}{p}}$, and $\|f\|_{\infty} = \max_{x \in \mathcal{X}} f(x)$.

2.2   Markov Decision Process

A finite Markov Decision Process (MDP) is a tuple $\mathcal{M} = \{\mathcal{S}, \mathcal{A}, \mathcal{P}, \mathcal{R}, \gamma\}$ where $\mathcal{S} = (s_i)_{\{1 \leq i \leq N_{\mathcal{S}}\}}$ is the finite state space, $\mathcal{A} = (a_i)_{\{1 \leq i \leq N_{\mathcal{A}}\}}$ is the finite action space, $\mathcal{P} \in \Delta_{\mathcal{S}}^{\mathcal{S} \times \mathcal{A}}$ is the Markovian dynamics of the MDP, $\mathcal{R} \in \mathbb{R}^{\mathcal{S} \times \mathcal{A}}$ is the reward function and γ is the discount factor. A stationary and Markovian policy $\pi \in \Delta_{\mathcal{A}}^{\mathcal{S}}$ represents the behavior of an agent acting in the MDP $\mathcal{M}$. The set of all Markovian and stationary policies is noted $\Pi_{MS} = \Delta_{\mathcal{A}}^{\mathcal{S}}$. When the policy π is deterministic, it can also be seen as an element of $\mathcal{A}^{\mathcal{S}}$ and $\pi(s)$ is the action chosen by the policy π in state s. The quality of this behavior in the infinite horizon framework is quantified by the value function $v_{\mathcal{R}}^{\pi} \in \mathbb{R}^{\mathcal{S}}$ which maps to each state the expected and discounted cumulative reward for starting in this state and following the policy π afterwards: $\forall s \in \mathcal{S}, v_{\mathcal{R}}^{\pi}(s) = \mathbb{E}[\sum_{t \geq 0} \gamma^t \mathcal{R}(s_t, a_t)|s_0 = s, \pi]$.

A policy $\pi_{\mathcal{R}}^*$ (according to the reward $\mathcal{R}$) is said optimal if its value function $v_{\mathcal{R}}^*$ satisfies $v_{\mathcal{R}}^* \geq v_{\mathcal{R}}^\pi$ for any policy π and component wise.

Let P_π be the stochastic matrix $P_\pi = (\sum_{a \in A} \pi(a|s) \mathcal{P}(s'|s, a))_{\{(s,s') \in \mathcal{S}^2\}}$ and $\mathcal{R}_\pi \in \mathbb{R}^{\mathcal{S}}$ the function such that: $\forall s \in \mathcal{S}, \mathcal{R}_\pi(s) = \sum_{a \in A} \pi(a|s) \mathcal{R}(s, a)$. With a slight abuse of notation, we may write a the policy which associates the action a to each state s. The Bellman evaluation (resp. optimality) operators $T_{\mathcal{R}}^\pi$ (resp. $T_{\mathcal{R}}^*$) : $\mathcal{R}^{\mathcal{S}} \to \mathcal{R}^{\mathcal{S}}$ are defined as $T_{\mathcal{R}}^\pi v = \mathcal{R}_\pi + \gamma P_\pi v$ and $T_{\mathcal{R}}^* v = \max_\pi T_{\mathcal{R}}^\pi v$. These operators are contractions and $v_{\mathcal{R}}^\pi$ and $v_{\mathcal{R}}^*$ are their respective fixed-points: $v_{\mathcal{R}}^\pi = T_{\mathcal{R}}^\pi v_{\mathcal{R}}^\pi$ and $v_{\mathcal{R}}^* = T_{\mathcal{R}}^* v_{\mathcal{R}}^*$. The action-value function $\mathcal{Q}_{\mathcal{R}}^\pi \in \mathcal{S} \times \mathcal{A}$ adds a degree of freedom on the choice of the first action, it is formally defined as $\mathcal{Q}_{\mathcal{R}}^\pi(s, a) = [T_{\mathcal{R}}^a v_{\mathcal{R}}^\pi](s)$. We also write, when it exists, $\rho_\pi \in \mathbb{R}^{\mathcal{S}}$ the stationary distribution of the policy π (satisfying $\rho_\pi^T P_\pi = \rho_\pi^T$). The existence and uniqueness of ρ_π is guaranteed when the Markov chain induced by the matrix of finite size P_π is irreducible which will be supposed true in the remaining of the paper.

2.3   AL and IRL

AL and IRL are two methods that attempt to solve the imitation problem using the MDP paradigm. More precisely, in the AL framework, the apprentice, given some observations of the expert policy π_E, tries to learn a policy π_A which is as good as the expert policy according to the unknown reward $\mathcal{R}$ that the expert is trying to optimize (often the expert is considered optimal: $v_{\mathcal{R}}^{\pi_E} = v_{\mathcal{R}}^*$). This can be expressed numerically: the apprentice tries to find a policy π_A such that the quantity: $\mathbb{E}_\nu[v_{\mathcal{R}}^{\pi_E} - v_{\mathcal{R}}^{\pi_A}]$ is the lowest possible, where $\nu \in \Delta_S$. In general $\nu = \rho$ where ρ is the uniform distribution or $\nu = \rho_{\pi_E}$ (ρ_{π_E} is also noted ρ_E). In the IRL framework, the apprentice is trying to learn a reward $\hat{\mathcal{R}}$ which could explain the expert behavior. More precisely, given some observations of the expert policy π_E, the apprentice is trying to learn $\hat{\mathcal{R}}$ such that $\pi_E \approx \pi_{\hat{\mathcal{R}}}^*$. This can be expressed numerically, the apprentice is trying to learn a reward $\hat{\mathcal{R}}$ such that the quantities $\mathbb{E}_\nu[v_{\hat{\mathcal{R}}}^{\pi_{\hat{\mathcal{R}}}^*} - v_{\hat{\mathcal{R}}}^{\pi_E}]$ or $\mathbb{E}_\nu[v_{\mathcal{R}}^{\pi_E} - v_{\mathcal{R}}^{\pi_{\hat{\mathcal{R}}}^*}]$ are the lowest possible.

3   Theoretical Study

This section gives some theoretical insights into the question: Is it worth estimating a reward. First, we present a theoretical result for AL reduced to classification for the infinite horizon case. A proof of this result is given on the appendix 6. The result is an upper bound on the difference of the value functions of the expert and apprentice policies. As a previous bound for AL reduced to classification in the finite horizon case had been proposed in [11], we give an informal comparison of the two results. Besides, there is also a performance bound for an IRL algorithm [5] (SCIRL) which allows us to compare IRL and AL performances from a theoretical point of view. We choose to compare those bounds because the classification and the SCIRL algorithms does not need to resolve iteratively MDPs. Thus, there is no Approximate Dynamic programming error to deal with and to propagate to obtain the performance of the algorithm.

3.1   AL Reduced to Classification for the Infinite Horizon Case

A simple way to realize an AL method is by pure mimicry via an SL method such as classification. More precisely, we assume that some demonstrations examples $D_E = (s_i, a_i)_{\{1 \leq i \leq N\}}$ where $a_i \sim \pi_E(.|s_i)$ are available. Without loss of generality, we assume that the states s_i are sampled according to some probability distribution $\nu \in \Delta_\mathcal{S}$. So, the data (s_i, a_i) are sampled according to the distribution μ_E such that: $\mu_E(s, a) = \nu(s)\pi_E(a|s)$. Then, a classifier is learnt based on these examples (with discrete actions, it is a multi-class classification problem) thanks to an SL algorithm. This outputs a policy $\pi_C \in \mathcal{A}^\mathcal{S}$, which associates to each state an action. The quality of the classifier is quantified by the classification error: $\epsilon_C = \mathbb{E}_{\mu_E}[\chi_{\{(s,a) \in \mathcal{S} \times \mathcal{A}, \pi_C(s) \neq a\}}] = \sum_{s \in S} \sum_{a \in \mathcal{A}, a \neq \pi_C(s)} \nu(s)\pi_E(a|s)$. The quality of the expert (respectively to the unknown reward function $\mathcal{R}$) may be quantified with $v_\mathcal{R}^{\pi_E}$. Usually, it is assumed that the expert is optimal (that is, $v_\mathcal{R}^{\pi_E} = v_\mathcal{R}^*$), but it is not necessary for the following analysis (the expert may be sub-optimal respectively to $\mathcal{R}$). The quality of the policy π_C can also be quantified by its value function $v_\mathcal{R}^{\pi_C}$. In the following, we bound $\mathbb{E}_\nu[v_\mathcal{R}^{\pi_E} - v_\mathcal{R}^{\pi_C}]$ which represents the difference between the quality of the expert and the classifier policy. If this quantity is negative, that is fine, because (in mean), π_C is better than π_E. So, only an upper bound is computed. This upper-bound shows the soundness of the AL through classification method for the infinite horizon case.

Let define the following concentration coefficient: $C_\nu = (1 - \gamma) \sum_{t \geq 0} \gamma^t c_\nu(t)$ where $\forall t \in \mathbb{N}, c_\nu(t) = \max_{s \in \mathcal{S}} \frac{(\nu^T P_{\pi_E}^t)(s)}{\nu(s)}$. Notice that if $\nu = \rho_E$, which is a quite reasonable assumption, then $C_\nu = C_{\rho_E} = 1$.

Theorem 1. *Let π_C be the classifier policy (trained on the data set D_E to imitate the expert policy π_E). Let also ϵ_C be the classification error and C_ν the above defined concentration coefficient. Then $\forall \mathcal{R} \in \mathbb{R}^{\mathcal{S} \times \mathcal{A}}$:*

$$\mathbb{E}_\nu[v_\mathcal{R}^{\pi_E} - v_\mathcal{R}^{\pi_C}] \leq \frac{2C_\nu \|\mathcal{R}\|_\infty}{(1 - \gamma)^2}\epsilon_C.$$

The proof of Th. 1 is given on the appendix 6 and is based on the propagation of the classification error. In [11], the authors have established similar bounds in the finite horizon case. However, as most of AL and IRL algorithms considered so far the infinite horizon framework, we think that our result has its interest.

3.2   The Bound on the Finite-Horizon Case

In this section, we introduce specific notations to the finite horizon case and we interpret the results from [11]. Let consider a finite MDP $\mathcal{M} = \{\mathcal{S}, \mathcal{A}, \mathcal{P}, \mathcal{R}\}$ with horizon H and without discount factor γ. A Markovian and non-stationary policy is an element of the set Π_{MS}^H; if π is non-stationary, then π^t refers to the stationary policy that is equal to the tth component of π. Similarly to the infinite horizon case, we define the value function of the policy π at time t:

$$\forall s \in \mathcal{S}, v_{t,\mathcal{R}}^{\pi}(s) = \mathbb{E}[\sum_{t'=t}^{H} \mathcal{R}(s_{t'}, a_{t'})|s_t = s, \pi].$$

Let D_{π}^t be the distribution on state-action pairs at time t under policy π. In other words, a sample (s,a) is drawn from D_{π}^t by first drawing $s_1 \sim \nu \in \Delta_{\mathcal{S}}$, then following policy π for time steps 1 through t, which generates a trajectory $(s_1, a_1, \ldots, s_t, a_t)$, and then letting $(s,a) = (s_t, a_t)$. More formally, we have:

$$\forall 1 \leq t \leq H, \forall (s,a) \in \mathcal{S} \times \mathcal{A}, D_{\pi,\nu}^t(s,a) = (\nu^T(P_{\pi^1} \times \cdots \times P_{\pi^{t-1}}))(s)\pi^t(a|s).$$

In [11], the authors suppose the availability of the set of trajectories $D_E = (\omega_i)_{\{1 \leq i \leq N\}}$ where $\omega_i = (s_1^i, a_1^i, \ldots, s_H^i, a_H^i)$ with $s_1^i \sim \nu \in \Delta_{\mathcal{S}}$ and $(s_t^i, a_t^i) \sim D_{\pi_E}^t$ where $1 \leq t \leq H$ and π_E is the non-stationary and Markovian expert policy. In the finite horizon case, Apprenticeship Learning through classification will consists in learning an apprentice policy $\pi_C = (\pi_C^t)_{\{1 \leq t \leq H\}}$ thanks to H classifiers trained on the sets $D_E^t = (s_t^i, a_t^i)_{\{1 \leq i \leq N\}}$. Thus, for each set $D_E^t = (s_t^i, a_t^i)_{\{1 \leq i \leq N\}}$, we train a multi-class classifier and learn a deterministic policy π_C^t with classification error:

$$\epsilon_C^t = \mathbb{E}_{D_E^t}[\chi_{\{(s,a) \in \mathcal{S} \times \mathcal{A}, \pi_C^t(s) \neq a\}}].$$

We note $\epsilon_C = \max_{1 \leq t \leq H} \epsilon_C^t$. Then we have the following theorem:

Theorem 2. *Let π_E be the expert non-stationary and Markovian expert policy, D_E a set of N trajectories with $s_1^i \sim \nu \in \Delta_{\mathcal{S}}$ and π_C the policy learnt by the H classifiers, then:*

$$\mathbb{E}_{\nu}[v_{1,\mathcal{R}}^{\pi_E} - v_{1,\mathcal{R}}^{\pi_C}] \leq \min(2\sqrt{\epsilon_C}H^2, 4\epsilon_C H^3 + \delta_{\pi_E})\|\mathcal{R}\|_{\infty},$$

where $\delta_{\pi_E} = \frac{\mathbb{E}_{\nu}[v_{1,\mathcal{R}}^* - v_{1,\mathcal{R}}^{\pi_E}]}{\|\mathcal{R}\|_{\infty}}$ represents the sub-optimality of the expert.

It is possible to compare these results with our bound, even if one deals with the infinite horizon case and the other with the finite horizon case, by informally noticing that the introduction of the discount factor γ in the infinite horizon corresponds to an horizon of length $\frac{1}{1-\gamma}$: $\sum_{t \geq 0} \gamma^t = \frac{1}{1-\gamma}$. By replacing H by $\frac{1}{1-\gamma}$ in the the precedent bound, we obtain:

$$\mathbb{E}_{\nu}[v_{1,\mathcal{R}}^{\pi_E} - v_{1,\mathcal{R}}^{\pi_C}] \leq \min(\frac{2\sqrt{\epsilon_C}}{(1-\gamma)^2}, \frac{4\epsilon_C}{(1-\gamma)^3} + \delta_{\pi_E})\|\mathcal{R}\|_{\infty}.$$

So, if we informally identify the classification errors and the horizon H to $\frac{1}{1-\gamma}$, our bound is slightly better either by $\sqrt{\epsilon_C}$ or by $\frac{2}{1-\gamma}$. Moreover, as our bound is specific to the infinite horizon, it is more adapted to AL and IRL algorithms as most of them consider the infinite horizon case.

3.3   SCIRL and Its Performance Bound

[5] assume that the unknown reward is linearly parameterized by some feature vector. More precisely, let $\phi(s,a) = (\phi_1(s,a), \ldots, \phi_p(s,a))^T$ be a feature vector composed of $p \in \mathbb{N}^*$ basis functions $\phi_i \in \mathbb{R}^{S \times A}$, the parameterized reward function is $\mathcal{R}_\theta(s,a) = \theta^T \phi(s,a) = \sum_{1 \le i \le p} \theta_i \phi_i(s,a)$. Searching a good reward thus reduces to searching a good parameter vector $\theta \in \mathbb{R}^p$. The choice of features is done by the user. Moreover, SCRIL needs the estimation of the expert feature expectation ω_{π_E} [5] which is the expected discounted cumulative feature vector for starting in a given state, applying a given action and following the expert policy:

$$\omega_{\pi_E}(s,a) = \mathbb{E}[\sum_{t \ge 0} \gamma^t \phi(s_t, a_t) | s_0 = s, a_0 = a, \pi_E].$$

It can be seen that: $Q_{\mathcal{R}_\theta}^{\pi_E}(s,a) = \theta^T \omega_{\pi_E}(s,a)$. An estimation of the feature expectation $\hat{\omega}_{\pi_E}$ is done via the expert data set: D_E. The problem of estimating the expert feature is a policy evaluation problem. Then, SCIRL uses the estimation of the expert feature expectation $\hat{\omega}_{\pi_E}$ as the basis function of a linearly parameterized score-based multi-class classifier fed by the set D_E. The classification error is $\epsilon_C = \mathbb{E}_{\mu_E}[\chi_{\{(s,a) \in S \times A, \pi_C(s) \ne a\}}]$ with $\pi_C(s) = \text{argmax}_{a \in A} \theta_C^T \hat{\omega}_{\pi_E}(s,a)$ and θ_C the output of the score-based classifier. The reward outputted by the SCIRL algorithm is $\mathcal{R}_C = \theta_C^T \phi$. Then, the performance bound for this algorithm is:

$$0 \le \mathbb{E}_{\rho_E}[v_{\mathcal{R}_C}^* - v_{\mathcal{R}_C}^{\pi_E}] \le \frac{C_f}{(1-\gamma)}\left(\frac{2\|\mathcal{R}_C\|_\infty \epsilon_C}{1-\gamma} + \bar{\epsilon}_Q\right),$$

With $C_f = (1-\gamma)\sum_{t \ge 0} \gamma^t c_f(t)$ where $\forall t \in \mathbb{N}, c_f(t) = \max_{s \in S} \frac{(\rho_E^T P_{\pi_{\mathcal{R}_C}^*}^t)(s)}{\rho_E(s)}$. Moreover, $\bar{\epsilon}_Q = \mathbb{E}_{\rho_E}[\max_{a \in A} \epsilon_Q(.,a) - \min_{a \in A} \epsilon_Q(.,a)]$, where $\epsilon_Q(s,a) = \theta_C^T(\hat{\omega}_{\pi_E}(s,a) - \omega_{\pi_E}(s,a))$, is a measure of the error estimation of the feature expectation. This bound is specific to the reward $\mathcal{R}_C$ and the constant C_f is not equal to 1 when $\nu = \rho_E$, which makes this bound possibly quite worst than the pure classification bound, even when the expert feature expectation is perfectly estimated ($\bar{\epsilon}_Q = 0$). This seems to indicate that this IRL algorithm is less interesting than a simple classification algorithm in theory. However, in practice, we will see that for specific unknown rewards SCIRL can have much better performance than a classification algorithm (see Sec. 4).

4   Empirical Study

This section shows through experiments that the previous theoretical bounds does not tell everything about AL methods and IRL methods. Here, several experiments are conducted and show the interest of finding a reward thanks to a general framework of experiments called the Garnet framework. We choose a particular framework where all the problems are finite MDPs with a tabular representation. Even if those problems are not challenging, they allow comparing fairly the different approaches without the problem of bias induced by the choice of representation. The comparison is done between a pure classification algorithm and two recently published IRL algorithms which are SCIRL and Relative Entropy IRL (RE) [4], for which there is no known error analysis. The pure classification algorithm was chosen as a benchmark for the AL approach because it has a theoretical performance guarantee and does not need to resolve iteratively MDPs unlike most of the other algorithms. SCIRL and RE were chosen as benchmarks for the IRL approach because they also do not need to resolve iteratively MDPs which reduces the impact of Approximate Dynamic Programming (ADP) in the interpretation even if the outputted reward is optimized via the policy iteration algorithm. These experiments show that the choice of the underlying unknown reward, which is used in order to create the expert policy thanks to the policy iteration algorithm, is crucial. Indeed when the unknown reward is normally distributed on each state-action-couple the classification has quite good performance whereas it has quite low performance when the reward is sparse or state-only-dependent. The intuitive idea behind those results is: when the reward is too informative, the impact of the optimization horizon is reduced, which favors the classification approach.

4.1 AL and IRL Algorithms

The first algorithm is a pure classification algorithm. More precisely, it is multi-class classification algorithm fed by the set D_E using a structured large-margin approach [12] which consists in minimizing the following criterion with respect to $Q \in \mathbb{R}^{S \times A}$:

$$\mathfrak{L}_0(Q) = \frac{1}{N} \sum_{i=1}^{N} \max_{a \in A}[Q(s_i, a) + l(s_i, a)] - Q(s_i, a_i) + \lambda \|Q\|_2^2,$$

where $l(s, a) = 0$ when $\exists 1 \leq i \leq N, (s, a) = (s_i, a_i)$ and $l(s, a) = 1$ otherwise. The minimization is realized via a sub-gradient descent [9]. Then the policy obtained by the algorithm is a deterministic policy such that $\pi_C(s) \in \text{argmax}_{a \in A} Q^*(s, a)$ where Q^* is the output of the minimisation of the criterion $\mathfrak{L}_0$ via the sub-gradient descent. The two other algorithms are IRL algorithms. SCIRL (presented in Sec. 3.3) needs only the set D_E to be implemented and outputs a reward $\mathcal{R}_C$. The instantiation of SCIRL, in our experiments, is the one described in the original paper. In order to obtain a policy π_C, this reward is optimized by the policy iteration algorithm with respect to the reward $\mathcal{R}_C$. The policy iteration algorithm needs the knowledge of the whole dynamics of the MDP to be implemented but allows a comparison which does not depend on the choice of an ADP algorithm (we need solving an MDP to measure the efficiency of the estimate, but not to obtain the estimate). Like SCIRL, the RE algorithm supposes a linear parametrization of the reward. The principle of the Relative Entropy method is based on minimizing the relative entropy (KL divergence) between the empirical distribution of the state-action trajectories under a random policy and the distribution of the trajectories under a policy that matches the expert feature expectation [4]. The RE algorithm used in this paper is the one described in the original paper. It needs the set D_E and also requires a set D_P of sampled trajectories according to a non-expert policy. In the experiments, the random policy will be chosen in order to generate the set D_P (see Sec. 4.3). The output of the algorithm is a reward $\mathcal{R}_C$ and a policy iteration algorithm is also used to obtain the policy π_C relative to the outputted reward.

4.2 The Garnet Framework

The Garnet problems are a class of randomly constructed finite MDPs meant to be totally abstract while remaining representative of the kind of finite MDPs that might be encountered in practice [2]). The routine to create an instance of a stationary Garnet problem is characterized by 3 parameters and written as $Garnet(N_S, N_A, N_B)$. The parameters N_S and N_A are the number of states and actions respectively, and N_B is a branching factor specifying the number of next states for each state action pair. The next states are chosen at random from the state set without replacement. The probability of going to each next state is generated by partitioning the unit interval at $N_B - 1$ cut points selected randomly. The reward $\mathcal{R}(s, a)$ will be chosen depending on the experiments. For each Garnet problem, it is possible to compute an expert policy π_E thanks to the reward $\mathcal{R}$ via the policy iteration algorithm. Finally, the discount factor is fixed to 0.99.

4.3 Pure Classification Versus SCRIL and RE

The idea, in order to obtain a general result, is to run the same experiment on hundreds of MDPs and regroup the results at the end. All the algorithms are fed with data sets of the the following type: $D_E = (s_i, a_i)_{\{1 \leq i \leq N\}}$ where $a_i \sim \pi_E(.|s_i)$. More particularly, $D_E = (\omega_j)_{\{1 \leq i \leq K_E\}}$ where $\omega_j = (s_{i,j}, a_{i,j})_{\{1 \leq i \leq H_E\}}$ is a trajectory obtained by starting from a random state $s_{1,j}$ (chosen uniformly) and applying the policy π_E H_E times $(s_{i+1,j} \sim P(.|s_{i,j}, a_{i,j}))$. So, D_E is composed by K_E trajectories of π_E of length H_E and we have $K_E H_E = N$. We also fed the RE algorithms with a data set of sampled transitions $D_P = (s_i, a_i, s'_i)_{\{1 \leq i \leq N'\}}$ where $a_i \sim \pi_R(.|s_i)$ with π_R the random policy (uniform distribution over the actions for each state) and where $s'_i \sim P(.|s_i, a_i)$. Actually, D_P has the particular form $D_P = (\tau_j)_{\{1 \leq j \leq K_P\}}$ where $\tau_j = (s_{i,j}, a_{i,j}, s'_{i,j})_{\{1 \leq i \leq H_P\}}$ is a trajectory obtained by starting from a random state $s_{1,j}$ (chosen uniformly) and applying the policy π_R H_P times $(s'_{i,j} = s_{i+1,j} \sim P(.|s_{i,j}, a_{i,j}))$. So, D_P is composed by K_P trajectories of π_R of length H_P and we have $K_P H_P = N'$. Therefore, if we have for a given Garnet problem π_E and π_R, the set of parameters (K_E, H_E, K_P, H_P) is sufficient to instantiate sets of types D_E and D_P.

Our first experiment shows the performance of the algorithms when H_E is increasing and when the reward for each Garnet is chosen normally distributed for each state-action couple. The reward $\mathcal{R}(s, a)$ is selected randomly according to a normal distribution with mean 0 and with standard deviation 1. It consists in generating 100 Garnet problems of the type $Garnet(N_S, N_A, N_B)$, where N_S is uniformly chosen between 50 and 100, N_A uniformly chosen between 5 and 10 and

N_B uniformly chosen between 2 and 5 . This gives us the set of Garnet problems $\mathfrak{G} = (G_p)_{\{1 \leq p \leq 100\}}$. On each problem p of the set $\mathfrak{G}$, we compute π_E^p and π_R^p. The parameter H_E takes its values in the set $(H_E^k)_{\{1 \leq k \leq 11\}} = (50, 100, 150, .., 500)$, $K_E = 1$, $H_P = 10$, $K_P = 50$. Then, for each set of parameters (K_E, H_E^k, K_P, H_P) and each G_p, we compute 100 expert policy sets $(D_E^{i,p,k})_{\{1 \leq i \leq 100\}}$ and 100 random policy sets $(D_P^{i,p,k})_{\{1 \leq i \leq 100\}}$. Our criteria of performance for each couple $(D_E^{i,p,k}, D_P^{i,p,k})$ is the following: $T^{i,p,k} = \frac{\mathbb{E}_\rho[v_R^{\pi_E^P} - v_R^{\pi_C^{i,p,k}}]}{\mathbb{E}_\rho[v_R^{\pi_E}]}$, where π_E^p is the expert policy, $\pi_C^{i,p,k}$ is the policy induced by the algorithm fed by the couple $(D_E^{i,p,k}, D_P^{i,p,k})$ and ρ is the uniform distribution over the state space S. For the pure classifier, we have $\pi_C^{i,p,k}(s) \in \operatorname{argmax}_{a \in A} \hat{Q}^*(s, a)$ where $\hat{Q}^*$ is the minimizer of $\mathfrak{L}_0$. For the SCIRL and RE algorithms, $\pi_C^{i,p,k}$ is the policy obtained by optimizing the reward $\mathcal{R}_C$ outputted by the algorithm via the policy iteration algorithm. Our mean criterion of performance for each set of parameters (K_E, H_E^k, K_P, H_P) is: $T^k = \frac{1}{10000} \sum_{1 \leq p \leq 100, 1 \leq i \leq 100} T^{i,p,k}$. For each algorithm we plot $(H_E^k, T^k)_{\{1 \leq k \leq 15\}}$. Another criterion is also useful in order to interpret the results. For each Garnet problem and each set of parameters, we calculate the standard deviation $\text{std}^{p,k}$ for each algorithm:

$$\text{std}^{p,k} = \left\{ \frac{1}{100} \sum_{1 \leq i \leq 100} [T^{i,p,k} - \sum_{1 \leq j \leq 100} T^{j,p,k}]^2 \right\}^{\frac{1}{2}}.$$ Then we compute the mean standard deviation over the 100 Garnet problems for each set of parameters: $\text{std}^k = \frac{1}{100} \sum_{1 \leq p \leq 100} \text{std}^{p,k}$. For each algorithm we plot $(H_E^k, \text{std}^k)_{\{1 \leq k \leq 15\}}$. Results are reported on Fig. 1. Here, we see that the pure classification algorithm has a better performance over the IRL algorithms when the number of data is increasing. This can be explained by the particular shape of the reward which is particularly suited to make the pure classification algorithm work well and IRL algorithms work bad. Indeed, as there are rewards for each state-action couples

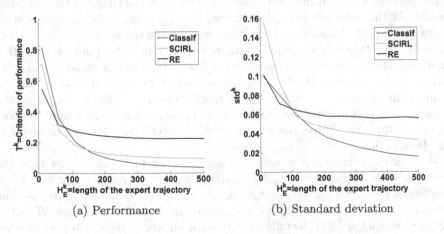

(a) Performance   (b) Standard deviation

Fig. 1. Garnets experiment: normally distributed reward

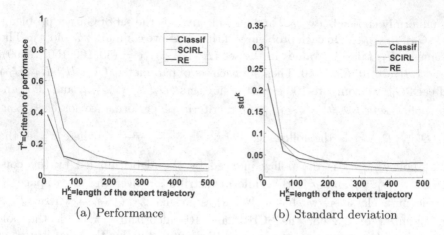

(a) Performance   (b) Standard deviation

Fig. 2. Garnets experiment: sparse reward

which are normally chosen, doing a misclassification is not so important as there will be rewards with the same form in the next states. However, as there are a lot of rewards everywhere, a lot of data is needed for an IRL algorithm to be able to estimate a meaningful reward. Another possible but complementary interpretation of those results is: as the reward is very informative, the choice of the action does not depend too much on the future states and the impact of the optimization horizon is strongly reduced.

The second experiment is exactly the same as the first one, except that the reward is no longer normally distributed. For each Garnet, we generate a reward with a small support: $\text{Supp}(\mathcal{R}) \leq \frac{N_S N_A}{50}$ by randomly choosing between 1 and $\frac{N_S N_A}{50}$ couples (s, a) such that $\mathcal{R}(s, a) \neq 0$ (reward randomly chosen between 0 and 1). For the other couples (s, a), $\mathcal{R}(s, a) = 0$. Results are reported on Fig. 2. Here, we see that the IRL algorithms work better than previously and the pure classification algorithms has its performance deteriorated a little bit compared to the previous experiment. This can be explained by the shape of the unknown reward. As the unknown reward is sparse, doing a misclassification on a state where the expert choose the action that gives a reward is important as there are only few state-action couples with rewards. Thus, the pure classification algorithm may have some problems with few data which is what we observe on Fig. 2(a). Moreover, the IRL algorithms have a better performance, maybe because the unknown reward has a simpler structure to learn. Again as the reward is less informative, the impact of the optimization horizon may be more important than for the previous reward which deteriorates the performance of the classification.

The third experiment is exactly the same as the first one, except that the reward is state-only-dependent. To construct a state-only-dependent reward, it is sufficient for each $s \in \mathcal{S}$ to select randomly a value $\mathcal{R}(s)$ according to a normal distribution with mean 0 and with standard deviation 1 and then $\forall (s, a) \in \mathcal{S} \times \mathcal{A} = \mathcal{R}(s, a) = \mathcal{R}(s)$. Results are reported on Fig 3. Here, the performance of the IRL algorithms is better than the second experiment and than the pure

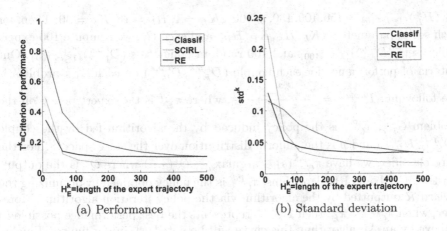

(a) Performance   (b) Standard deviation

Fig. 3. Garnets experiment: state-only-dependent reward

classification. This can be explained by the fact that the structure of the reward is even simpler. The pure classification see its performance deteriorated compared to the second experiment. As the unknown reward depends only on the state and not on the action, it is very important to follow the path of the expert to obtain a good performance. Thus, a misclassification on a given state which leads to a bad path can be very damageable and lead to bad performance.

5 Dynamics Perturbations

In this section, we want to show that it can be interesting to retrieve the reward in order to be more stable to dynamics perturbations. As the reward is seen as the most succinct hypothesis explaining the behavior of the expert, we can expect that the reward outputted by the IRL algorithms is such that its optimization will lead to a near-optimal behavior even if there is a dynamics perturbation. The dynamics perturbations considered are the ones which keep identical the structure of the MDP. The structure of the MDP is for a given state-action couple (s, a) the different states that could be reached by choosing the action a in state s, that is $\mathrm{Supp}(P_{s,a})$. The structure of the MDP is the set $(\mathrm{Supp}(P_{s,a}))_{\{(s,a)\in\mathcal{S}\times\mathcal{A}\}}$. Therefore a dynamic perturbation is the choice of a dynamics $\tilde{P}$ different from P such that: $(\mathrm{Supp}(P_{s,a}))_{\{(s,a)\in\mathcal{S}\times\mathcal{A}\}} = (\mathrm{Supp}(\tilde{P}_{s,a}))_{\{(s,a)\in\mathcal{S}\times\mathcal{A}\}}$.

The first experiment consists in in generating 100 Garnet problems of the type $Garnet(N_S, N_A, N_B)$, where N_S is chosen randomly between 50 and 100, N_A randomly chosen between 5 and 10 and N_B chosen randomly between 2 and 5 . This gives us the set of Garnet problems $\mathfrak{G} = (G_p)_{\{1\leq p\leq 100\}}$. Here, The reward $\mathcal{R}(s, a)$ is selected randomly according to a normal distribution with mean 0 and with standard deviation 1. Then for each G_p, we realize 50 dynamics perturbation and we obtain the set of Garnets problems $\tilde{\mathfrak{G}} = (G_{p,q})_{\{1\leq p\leq 100, 1\leq q\leq 50\}}$. On each problem p, q of the set $\tilde{\mathfrak{G}}$, we compute $\pi_E^{p,q}$ and $\pi_R^{p,q}$ and on each problem p of the set $\mathfrak{G}$, we compute π_E^p and π_R^p. The parameter H_E takes its values in the set $(H_E^k)_{\{1\leq k\leq 15\}} = (50, 100, 150, .., 500)$, $K_E = 1$, $H_P = 10$, $K_P = 50$. Then, for each set of parameters (K_E, H_E^k, K_P, H_P) and each G_p, we compute 100 expert policy sets $(D_E^{i,p,k})_{\{1\leq i\leq 100\}}$ and 100 random policy sets $(D_P^{i,p,k})_{\{1\leq i\leq 100\}}$. Our criteria of performance for each couple $(D_E^{i,p,k}, D_P^{i,p,k})$ on each $G_{p,q}$ problem is the following: $T^{i,p,q,k} = \frac{\mathbb{E}_\rho[v_R^{\pi_E^{p,q}} - v_R^{\pi_C^{i,p,k}}]}{\mathbb{E}_\rho[v_R^{\pi_E^{p,q}}]}$, where $\pi_E^{p,q}$ is the expert policy on the problem $G_{p,q}$, $\pi_C^{i,p,k}$ is the policy induced by the algorithm fed by the couple $(D_E^{i,p,k}, D_P^{i,p,k})$ and ρ is the uniform distribution over the state space S. For the pure classifier, we have $\pi_C^{i,p,k}(s) \in \text{argmax}_{a\in A} \hat{Q}^*(s,a)$ where $\hat{Q}^*$ is the output. For the SCIRL and RE algorithms, $\pi_C^{i,p,k}$ is the policy obtained by optimizing the reward $\mathcal{R}$ outputted by the algorithm via the policy iteration algorithm. Moreover, when $\pi_C^{i,p,k} = \pi_E^p$, then $T^{i,p,q,k}$ represents the best performance possible to achieve by an AL algorithm: this curve will be noted AL in our figures. Finally, when $\pi_C^{i,p,k} = \pi_R^p$, then $T^{i,p,q,k}$ represents the performance of the random policy and this curve will be noted Rand in our figures.

Our mean criterion of performance for each set of parameters (K_E, H_E^k, K_P, H_P) is: $T^k = \frac{1}{500000}\sum_{1\leq p\leq 100, 1\leq q\leq 50, 1\leq i\leq 100} T^{i,p,q,k}$. For each algorithm we plot $(H_E^k, T^k)_{\{1\leq k\leq 15\}}$. Another criterion is also useful in order to interpret the results. For each Garnet problem G_p and each set of parameters, we calculate the standard deviation $\text{std}^{p,k}$ for each algorithm:

$$\text{std}^{p,k} = \left\{\frac{1}{5000}\sum_{1\leq i\leq 100}^{1\leq q\leq 50}[T^{i,p,q,k} - \sum_{1\leq j\leq 100}^{1\leq q'\leq 50} T^{j,p,q',k}]^2\right\}^{\frac{1}{2}}.$$

Then we compute the mean standard deviation over the 100 Garnet problems for each set of parameters: $\text{std}^k = \frac{1}{100}\sum_{1\leq p\leq 100}\text{std}^{p,k}$. For each algorithm we plot $(H_E^k, \text{std}^k)_{\{1\leq k\leq 15\}}$. Results are reported on Fig. 4. Here, the reward is normally

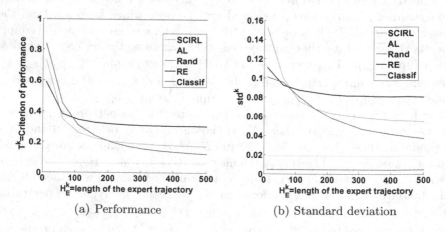

(a) Performance   (b) Standard deviation

Fig. 4. Perturbed dynamics: normally distributed reward

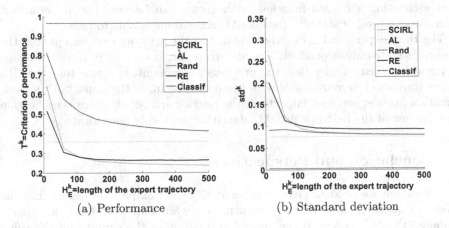

(a) Performance   (b) Standard deviation

Fig. 5. Perturbed dynamics: sparse reward

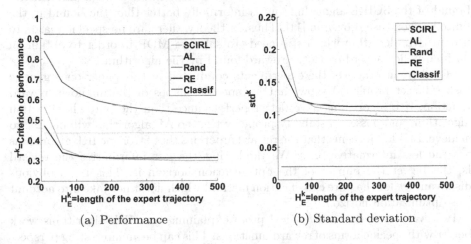

(a) Performance   (b) Standard deviation

Fig. 6. Perturbed dynamics: state-only-dependent reward

distributed so a dynamic perturbation may not deteriorate too much the expert policy. Indeed, as the reward is very informative, the impact of the optimization horizon must be very small and the perturbation of dynamics will not change too much the optimal policy. We can observe this on Fig. 4(a), where we see that the yellow curve noted AL is not so far away from 0. With this shape of reward, it is better to use a pure classification algorithm to have this stability property.

The second experiment is exactly the same as the previous one, except that the reward is sparse. Results are reported on Fig. 5. As the reward is sparse, we can expect that a dynamic perturbation leads to an important deterioration of the performance of the expert policy. Here, we see that IRL algorithms are under the yellow curve when the number of data is increasing, which means that no AL algorithms will be able to reach that level of stability. Thus, it seems that estimating a reward function in that case can be very useful because it guarantees a level of stability that no AL algorithms is able to provide.

The third experiment is exactly the same as the previous one, except that the reward is state-only-dependent. Results are reported on Fig. 6. Here the shape of reward is even simpler that the previous experiment. It seems that the IRL algorithms are even more stable with less data. Again, as the impact of the optimization horizon becomes important, the performance of the pure classification and the one of the best possible AL algorithm are really deteriorated.

6   Conclusion and Perspectives

In this paper, we tried to give some theoretical and empirical insights into the following question: is it worth estimating a reward function? First, we upper-bounded the difference between the value function of the expert and the value function of the apprentice policy, for AL reduced to classification in the infinite horizon case. This result gives a better bound than the theoretical performance bound of the SCIRL algorithm and is informally better than the bound in the finite horizon case proved in [11]. Thus, in theory, there are no specific reason to use an IRL algorithm which still needs to solve an MDP in order to obtain an optimal policy according to the reward found by the algorithm.

However, in practice, the experiments conducted in this paper on a generic task (Garnet problems) show that for specific shapes of the unknown reward function, IRL algorithms have better performance than the pure classification algorithms and possess a stability property that no AL algorithm will be able to achieve. Besides, it seems that the reward functions that favor the IRL algorithms are the less informative ones. We think that the less informative the reward is, the bigger the impact of the optimization horizon is. This is an obvious disadvantage for the pure classification method which doest not take into account this optimization horizon.

However, there is no theoretical proof explaining why IRL algorithms work better with specific forms of reward functions. This can be an interesting perspective to give more soundness to the experiments leaded in this paper. Moreover, it would be interesting to create an algorithm able to use data coming from different perturbed dynamics of the same MDP in order to learn a reward function which will be even less sensible to perturbed dynamics. This can be useful with applications where human are involved: in those kind of real-life applications, each human can be seen as a perturbed version of an MDP.

Acknowledgements. The research leading to these results has received partial funding from the European Union Seventh Framework Programme (FP7/2007-2013) under grant agreement n°270780.

References

1. Abbeel, P., Ng, A.Y.: Apprenticeship learning via inverse reinforcement learning. In: Proceedings of the 21st International Conference on Machine Learning, ICML (2004)
2. Archibald, T., McKinnon, K., Thomas, L.: On the generation of markov decision processes. Journal of the Operational Research Society (1995)
3. Atkeson, C.G., Schaal, S.: Robot learning from demonstration. In: Proceedings of the 14th International Conference on Machine Learning, ICML (1997)
4. Boularias, A., Kober, J., Peters, J.: Relative entropy inverse reinforcement learning. In: JMLR Workshop and Conference Proceedings, AISTATS 2011, vol. 15 (2011)
5. Klein, E., Geist, M., Piot, B., Pietquin, O.: Inverse reinforcement learning through structured classification. In: Advances in Neural Information Processing Systems 25 (NIPS) (2012)
6. Langford, J., Zadrozny, B.: Relating reinforcement learning performance to classification performance. In: Proceedings of the 22nd International Conference on Machine Learning, ICML (2005)
7. Pomerleau, D.: Alvinn: An autonomous land vehicle in a neural network. Tech. rep., DTIC Document (1989)
8. Russell, S.: Learning agents for uncertain environments. In: Proceedings of the 11th Annual Conference on Computational Learning Theory, COLT (1998)
9. Shor, N.Z., Kiwiel, K.C., Ruszcaynski, A.: Minimization methods for non-differentiable functions. Springer (1985)
10. Syed, U., Schapire, R.: A game-theoretic approach to apprenticeship learning. In: Advances in Neural Information Processing Systems 21 (NIPS) (2008)
11. Syed, U., Schapire, R.: A reduction from apprenticeship learning to classification. In: Advances in Neural Information Processing Systems 23 (NIPS) (2010)
12. Taskar, B., Chatalbashev, V., Koller, D., Guestrin, C.: Learning structured prediction models: A large margin approach. In: Proceedings of the 22nd International Conference on Machine Learning, ICML (2005)

Appendix: Proof of Th.1

We have:

$$v_{\mathcal{R}}^{\pi_E} - v_{\mathcal{R}}^{\pi_C} \overset{(a)}{=} T_{\mathcal{R}}^{\pi_E} v_{\mathcal{R}}^{\pi_E} - T_{\mathcal{R}}^{\pi_E} v_{\mathcal{R}}^{\pi_C} + T_{\mathcal{R}}^{\pi_E} v_{\mathcal{R}}^{\pi_C} - v_{\mathcal{R}}^{\pi_C}$$

$$\overset{(b)}{=} \gamma P_{\pi_E}(v_{\mathcal{R}}^{\pi_E} - v_{\mathcal{R}}^{\pi_C}) + T_{\mathcal{R}}^{\pi_E} v_{\mathcal{R}}^{\pi_C} - v_{\mathcal{R}}^{\pi_C},$$

$$\overset{(c)}{=} (I - \gamma P_{\pi_E})^{-1}(T_{\mathcal{R}}^{\pi_E} v_{\mathcal{R}}^{\pi_C} - v_{\mathcal{R}}^{\pi_C}),$$

Equality (a) holds because $T_{\mathcal{R}}^{\pi_E} v^{\pi_E} = v_{\mathcal{R}}^{\pi_E}$, Equality (b) is obtained by definition of $T_{\mathcal{R}}^{\pi_E}$ and Equality (c) is true by invertibility of $I - \gamma P_{\pi_E}$ where $I \in \mathbb{R}^{\mathcal{S} \times \mathcal{S}}$ is the identity matrix. The next step is to work on the term $T_{\mathcal{R}}^{\pi_E} v_{\mathcal{R}}^{\pi_C} - v_{\mathcal{R}}^{\pi_C}$. For any function $v \in \mathbb{R}^{\mathcal{S}}$, by definition of $T_{\mathcal{R}}^{\pi_E}$: $T_{\mathcal{R}}^{\pi_E} v = \mathcal{R}_{\pi_E} + \gamma P_{\pi_E} v$. Noticing that:

$$T_{\mathcal{R}}^{\pi_E} v_{\mathcal{R}}^{\pi_C}(s) - v_{\mathcal{R}}^{\pi_C}(s) = \sum_{s' \in \mathcal{S}} \sum_{a \in \mathcal{A}} \pi_E(s,a)[\mathcal{R}(s,a) + \gamma \mathcal{P}(s'|s,a) v_{\mathcal{R}}^{\pi_C}(s')] - v_{\mathcal{R}}^{\pi_C}(s),$$

and by definition of $Q_{\mathcal{R}}^{\pi_C}(s,a)$, we have:

$$\forall s \in S, T_{\mathcal{R}}^{\pi_E} v_{\mathcal{R}}^{\pi_C}(s) - v_{\mathcal{R}}^{\pi_C}(s) = \sum_{a \in \mathcal{A}} \pi_E(s,a) Q_{\mathcal{R}}^{\pi_C}(s,a) - v_{\mathcal{R}}^{\pi_C}(s),$$

$$= \sum_{a \in \mathcal{A}, a \neq \pi_C(s)} \pi_E(a|s)[Q_{\mathcal{R}}^{\pi_C}(s,a) - v_{\mathcal{R}}^{\pi_C}(s)].$$

So:

$$\nu^T(v_{\mathcal{R}}^{\pi_E} - v_{\mathcal{R}}^{\pi_C}) = \nu^T(I - \gamma P_{\pi_E})^{-1}[T_{\mathcal{R}}^{\pi_E} v_{\mathcal{R}}^{\pi_C} - v_{\mathcal{R}}^{\pi_C}],$$

$$= \sum_{s \in S} \sum_{t \geq 0} \gamma^t \frac{(\nu^T P_{\pi_E}^t)(s)}{\nu(s)} \nu(s)[T_{\mathcal{R}}^{\pi_E} v_{\mathcal{R}}^{\pi_C}(s) - v_{\mathcal{R}}^{\pi_C}(s)],$$

$$= \sum_{s \in S} \sum_{t \geq 0} \gamma^t \frac{(\nu^T P_{\pi_E}^t)(s)}{\nu(s)} \nu(s) \sum_{a \neq \pi_C(s)} \pi_E(a|s)[Q_{\mathcal{R}}^{\pi_C}(s,a) - v_{\mathcal{R}}^{\pi_C}(s)].$$

Thus by definition of C_ν:

$$\nu^T(v_{\mathcal{R}}^{\pi_E} - v_{\mathcal{R}}^{\pi_C}) \leq \frac{C_\nu}{1-\gamma} \sum_{s \in S} \sum_{a \in \mathcal{A}, a \neq \pi_C(s)} \nu(s)\pi_E(a|s)|Q_{\mathcal{R}}^{\pi_C}(s,a) - v_{\mathcal{R}}^{\pi_C}(s)|,$$

$$\overset{(d)}{\leq} \frac{C_\nu}{1-\gamma} \frac{2\|\mathcal{R}\|_\infty}{1-\gamma} \sum_{s \in S} \sum_{a \in \mathcal{A}, a \neq \pi_C(s)} \nu(s)\pi_E(a|s),$$

$$\overset{(e)}{=} \frac{2\|\mathcal{R}\|_\infty C_\nu \epsilon_C}{(1-\gamma)^2}.$$

Inequality (d) is true because $|Q_{\mathcal{R}}^{\pi_C}(s,a) - v_{\mathcal{R}}^{\pi_C}(s)| \leq \frac{2\|\mathcal{R}\|_\infty}{1-\gamma}$ and Equality (e) is true by definition of ϵ_C. This ends the proof.

Recognition of Agents Based on Observation of Their Sequential Behavior

Qifeng Qiao and Peter A. Beling

Department of Systems Engineering,
University of Virginia, VA, USA
{qq2r,pb3a}@virginia.edu

Abstract. We study the use of inverse reinforcement learning (IRL) as a tool for recognition of agents on the basis of observation of their sequential decision behavior. We model the problem faced by the agents as a Markov decision process (MDP) and model the observed behavior of an agent in terms of forward planning for the MDP. The reality of the agent's decision problem and process may not be expressed by the MDP and its policy, but we interpret the observation as optimal actions in the MDP. We use IRL to learn reward functions for the MDP and then use these reward functions as the basis for clustering or classification models. Experimental studies with *GridWorld*, a navigation problem, and the *secretary problem*, an optimal stopping problem, show algorithms' performance in different learning scenarios for agent recognition where the agents' underlying decision strategy may be expressed by the MDP policy or not. Empirical comparisons of our method with several existing IRL algorithms and with direct methods that use feature statistics observed in state-action space suggest it may be superior for agent recognition problems, particularly when the state space is large but the length of the observed decision trajectory is small.

1 Introduction

The availability of sensing technologies, such as digital cameras, global position system, infrared sensors, web technology and others, makes the computer easily access varieties of data recording the interaction between agents and the environment. As summarized in Figure 1, research in learning from the observed behavior has seen the development of approaches to activity recognition (It may be referred as different terms within the published literature, including plan recognition and goal recognition) and learning from demonstrations (It may be referred as imitation learning in other fields):

- Activity recognition: an activity can be described as a specific event or a combination of events, e.g. "go to bed", "cook a breakfast", "read a book" for the study of human activity recognition. The goal in activity recognition is a special event so that some optimal plan for a goal is compatible with the observations. A plan represents a mapping between state of a decision problem and action of an agent. The goal may change or it may consist of several

H. Blockeel et al. (Eds.): ECML PKDD 2013, Part I, LNAI 8188, pp. 33–48, 2013.

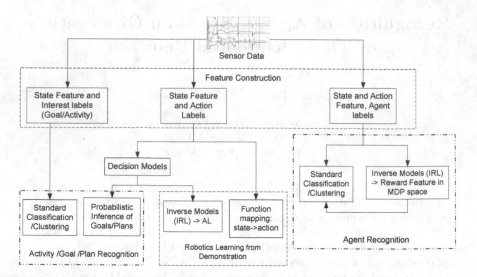

Fig. 1. Overview and Categorization of problems of learning from observation of decision-making behavior, including the widely studied problems that infer the goals of an agent and that learn how to make decisions, as well as our proposed new problem of recognizing agents based on their decision behavior

sub-activities or sub-goals. One may recognize an activity by applying classification/clustering algorithms directly to the feature vectors constructed from the observation data [22]. Alternatively, given a plan library or a set of goals as a prior, the entire trace of actions can be recognized and matched against a plan library or a set of possible goals [16]. Despite of the success of these methods, they assume that the plan library, a set of possible goals or some behavior model are known beforehand and provided as an input. Goal information is often completely unknown in practice, however, and so it is difficult to model the goal precisely.
- Learning from demonstrations: Much of work is focused on approximating the function mapping from observed experts' states to actions [2]. Alternatively, one may use demonstration data to inverse a decision model and a policy is then derived using this model, e.g. apprenticeship learning [4].

However, in practical applications we may not only be interested in reverse engineering of a decision-making process or imitating a behavior (identification), but also in determining whether two agents correspond to the same behavior pattern (clustering), or which decision-making pattern is being observed in an agent (classification). In this paper, we propose a new problem, termed *Behavior-based Agent Recognition*(BAR), that involves recognizing agents based on observation of their sequential behavior, instead of recognizing activities or actions.

This new problem is also motivated by varieties of applications in the real-world. E.g., we may find a way to train drivers by classifying the observed drivers into defensive driving and aggressive driving, even those drivers may have similar activities or goals, such as avoid driving over curbs or a collision. In the e-commerce market, if the web site can automatically categorize users based on observation of their on-line behavior, such as which buttons have been clicked by an user on which web page with what advertisements, a similar marketing strategy may work successfully on people in the same group. Another motivation comes from domains like high frequency trading of stocks and commodities, where there is considerable interest in identifying new market players and algorithms based on observations of trading actions, but little hope in learning the precise strategies employed by these agents [24][14].

A direct approach to BAR problem is to program some heuristic rules to recognize agents by decomposing complex behavior into a series of simple events and then evaluating them to reach a conclusion. However, programming the rules is hard. Alternatively, we can construct a feature vector to characterize agents based on the observed behavior, and then categorize the agents using these feature vectors. Consider image recognition as an illustration of this method. A computer learns to categorize images by representing every image as a multidimensional feature vector that consists of the components such as RGB color, texture, or other metrics. Here, an agent is an object like an image, and the feature vector needs to be constructed from the observed behavior. The key point in this process is how to find a high-level vector that can represent the underlying decision-making process. If the decision problem can be cast in the MDP framework, we propose to represent the agents with the reward functions of MDP models because they effectively influence the forward planning process.

IRL [13] addresses the task of finding a reward function that explains the observed behavior via the forward planning of a MDP. The observed behavior is assumed to maximize the long-term accumulative reward for that MDP. Most of recent work in IRL is focused on apprenticeship learning (AL), in which IRL is used as the core method for finding decision policies consistent with observed behavior [1] [21]. A number of IRL algorithms, being designed for apprenticeship learning or imitation learning, includes Max-margin planning [17], gradient tuning methods [12], linear solvable MDP [10], bootstrap learning [5], Gaussian process IRL [15] and Bayesian inference [6].

Our main contribution is to propose a new problem called BAR, and present a method based on IRL that models the problem faced by the agents as a MDP and assumes the reward function of the MDP model as a high-level abstraction of the decision behavior. The motivation is that even when the agent's behavior is not rational and we hardly learn the precise goals/plans, we still may categorize agents by using reward functions learned from the observed behavior.

On two well-known sequential decision-making problems, we compare our method with several existing IRL algorithms and with direct methods that use feature statistics observed in state-action space. The results show that our method using reward functions provides a formal way to solve the agent recognition problem and performs superior to other methods.

2   Preliminaries

We define the input of BAR problem as a tuple $\mathbf{B} = (D_1, D_2, \ldots D_N)$, where $D_n, n \in \{1, 2, \ldots, N\}$ is the observation of the n-th agent. For a classification problem, $D_n = (\mathcal{O}_n, y_n)$, where $\mathcal{O}_n$ is a set of observed decision trajectories and y_n is the class label for the n-th agent. The agents, who have the same decision-making pattern, are given the same class label. Similarly, for a clustering problem, D_n only consists of the observed decision trajectories $\mathcal{O}_n$.

We define the set of decision trajectories $\mathcal{O}_n = \{h_n^j\}$, $j = 1, 2, \ldots, |\mathcal{O}_n|$, where each trajectory h_n^j is defined as a series of state and action pairs: $\{(s, a)_n^t\}, t = 1, 2, \ldots, |h_n^j|$. Here, the s denotes the state for the decision problem and the a means the action selected by the agent at state s. Below are two definitions about the agent recognition problem.

Definition 1. *In general, an classification problem is: given a decision problem where the observed behavior $\mathcal{O}_n$ comes from, a universe $\mathcal{F}$ where the examples come from(the observed behavior is expressed in this space), a fixed set of classes $\mathcal{Y}$, and a training set X of labeled agents whose element $x \in \mathcal{F} \times \mathcal{Y}$, we use a learning algorithm to find a function $g : \mathcal{F} \to \mathcal{Y}$.*

Definition 2. *Given the observed behavior $\{\mathcal{O}_n\}_{n=1}^N$, a universe $\mathcal{F}$ where the examples come from, and a training set $X = \{f_1, f_2, \ldots, f_N\}$, where $f_n \in \mathcal{F}, n \in \{1, \ldots, N\}$, the clustering of N agents is the partitioning of X into K clusters $\{C_1, C_2, \ldots, C_K\}$ that satisfies (1) $\cup_{k=1}^K C_k = X$; (2) $C_k \neq \phi, k = 1, 2, \ldots, K$; (3) $C_k \cap C_{k'} = \phi, \forall k \neq k'$ and $k, k' \in \{1, 2, \ldots, K\}$.*

Next, we present the approaches to BAR problem in Section 3 and Section 4, and review the related IRL algorithms that have been used within our model in Section 5.

3   Two Direct Agent Recognition Models

In this section, we describe two approaches to agent recognition problem that construct feature vectors directly with the raw observation data.

The first method is called feature trajectory (FT). Consider a decision trajectory h_n^j. The vector to characterize the behavior in j-th decision trajectory is written as follows.

$$f(h_n^j) = [s_1, a_1, s_2, a_2, \ldots, s_{|h_n^j|}, a_{|h_n^j|}],$$

where $s_i, i \in \{1, 2, \ldots, |h_n^j|\}$ is a discrete random variable meaning the state index at i-th decision stage, and a_i represents the action selected at state s_i. E.g., we have a problem that can be defined by 3 states and 2 actions. Then $s_i \in \{1, 2, 3\}$ and $a_i \in \{1, 2\}$. In the observation, every trajectory starts from the same initial state. Given the observation set $\mathcal{O}_n$ for n-th agent, the feature vector f_n is obtained by computing this equation: $f_n = \frac{1}{|\mathcal{O}_n|} \sum_{j=1}^{|\mathcal{O}_n|} f(h_n^j)$, where the vector $f(h_n^j)$ is preprocessed by scale-normalization before averaging.

Then, the n-th agent is represented by a feature vector f_n. Consider a supervised learning problem. Given a real valued input vector $f_n \in \mathcal{F}$ and a category label $y_n \in \mathcal{Y}$, we aim to learn a function $g : \mathcal{F} \to \mathcal{Y}$.

The second method is called feature expectation (*FE*), which has been widely used by apprenticeship learning as a representation of the averaged long-term performance. Assume a basis function $\phi : \mathcal{S} \to [0,1]^d$, where $\mathcal{S}$ denotes the state space. The feature expectation $f_n = \frac{1}{|\mathcal{O}_n|} \sum_{j=1}^{|\mathcal{O}_n|} \sum_{s_t \in h_n^j} \gamma^t \phi(s_t)$, where $\gamma \in (0,1)$ is a discount factor. The associated apprenticeship learning algorithms aim to find a policy that performs as well as demonstrations by minimizing the distance between their feature expectations. Here, we only use the observed state sequence to compute the feature expectation vector for an agent, where the γ is manually defined constant, e.g. 0.95. Then, the n-th agent can be represented by the vector f_n that is built on $\mathcal{O}_n$.

4   A Behavior-Based Agent Recognition Model

First, we briefly review some background about MDP necessary for the next proposed method.

A finite MDP model $M = (\mathcal{S}, \mathcal{A}, R, \gamma, \mathcal{P})$ where $\mathcal{S}$ is the set of states, $\mathcal{A}$ is the set of actions, R is the reward function, γ is the discount factor, and $\mathcal{P} = \{\mathbf{P}_a\}_{a \in \mathcal{A}}$ is a set of transition probability matrices. The entries of $\mathbf{P}_a$, written as $\mathbf{P}_a(s, s')$, give the probability of transitioning to state $s' \in \mathcal{S}$ from state $s \in \mathcal{S}$ given the action is a. The rows of $\mathbf{P}_a$, denoted $\mathbf{P}_a(s,:)$, give a probability vector of transitioning from state s to all the states in $\mathcal{S}$. In a finite state space the reward function R may be considered as a vector, r, whose elements give the reward in each state.

In the MDP, a stationary policy is a function $\pi : \mathcal{S} \to A$. The value function for a policy π is $V^\pi(s_0) = E[\sum_{t=0}^{\infty} \gamma^t R(s_t) | p(s_0), \pi]$ where $p(s_0)$ is the distribution of the initial state and the action at state s_t is determined by policy π. Similarly, the Q function is defined as $Q(s,a) = R(s) + \gamma \sum_{s' \in \mathcal{S}} \mathbf{P}_a(s, s') V^\pi(s')$. At state s, an optimal action is selected by $a^* = \max_{a \in \mathcal{A}} Q(s,a)$.

Then, an instance of the IRL problem is written as a triplet $B = (M \setminus r, p(r), \mathcal{O})$, where $M \setminus r$ is a MDP model without the reward function and $p(r)$ is prior knowledge on the reward. The vector $p(r)$ can be a non-informative prior if we have no knowledge about the reward, or a Gaussian or other distribution if we model the reward as a specific stochastic process. Later in Section 5, we present the details for Bayesian IRL that has been used in our experiments.

Our behavior-based agent recognition method proceeds as follows.

1. Given the BAR problem with input $\mathbf{B}$, we use the set $\{\mathcal{O}_n\}, n \in \{1, 2, \ldots, N\}$ to construct the state space $\mathcal{S}$ and action space $\mathcal{A}$ for the decision-making problem. The $\mathcal{P}$ can be modelled using prior knowledge of the problem or estimated from the observed decision trajectories.
2. For every agent, we construct a MDP model, no matter whether the optimal policy of this MDP can match the observed behavior.

3. Apply IRL algorithms to learn the reward vector r_n for n-th agent.
4. Given the training set $\{r_1, r_2, \ldots\}$, where $r_n \in \mathcal{F}$ and the corresponding category label $y_n \in \mathcal{Y}$, we aim to train a classifier $g : \mathcal{F} \to \mathcal{Y}$.
5. Given a new agent, we repeat step 1-3 to get the reward vector for the agent and then predict the label for the behavior pattern using function $g : \mathcal{F} \to \mathcal{Y}$.

We use a MDP to model the decision problem faced by an agent under observation. The reality of the agent's decision problem and process may differ from the MDP model, but we interpret every observed decision of the agent as the choice of an action in the MDP. The dynamics of the environment in the MDP are described by the transition probabilities $\mathcal{P}$. These probabilities may be interpreted as being a prior, if known in advance, or as an estimation of the agent's beliefs of the dynamics. Next, we will show how to learn the reward functions by employing some exiting IRL algorithms.

5   Bayesian Framework for IRL

Most existing IRL algorithms have some assumption about the form of the reward function. Prominent examples include the model in [13], which we term linear IRL (*LIRL*) because of its linear nature, *WMAL* in [21], and *PROJ* in [1]. In these algorithms, the reward function is written linearly in terms of features as $R(s) = \sum_{i=1}^{d} \omega_i \phi_i(s) = \omega^T \phi(s)$, where $\phi : \mathcal{S} \to [0, 1]^d$ and $\omega^T = [\omega_1, \omega_2, \cdots, \omega_d]$.

Our computational framework uses Bayesian IRL to estimate the reward vectors in a MDP, which was initially proposed in [8]. The posterior over reward function for n-th agent is written as

$$p(r_n|\mathcal{O}_n) = p(\mathcal{O}_n|r_n)p(r_n) \propto \prod_{j=1}^{|\mathcal{O}_n|} \prod_{(s,a) \in h_n^j} p(a|s, r_n).$$

Then, the IRL problem is written as $\max_{r_n} \log p(\mathcal{O}_n|r_n) + \log p(r_n)$. For many problems, however, the computation of $p(r_n|\mathcal{O}_n)$ may be complicated and some algorithms use Markov chain Monte Carlo (MCMC) to sample the posterior probability. Considering the computation complexity to deal with a large number of IRL problems, we choose two IRL algorithms that have well defined likelihood function to reduce the computation cost, which are shown in the following subsections. The first algorithm in Section 5.1 has two assumptions on the reward functions: (1) it can be written linearly in terms of the state features; (2) it only depends on state. The second algorithm in Section 5.2 doesn't have these restrictions, and it not only can model the reward functions in nonlinear form but also consider the reward affected by both state and action.

5.1   IRL with Boltzmann Distribution

The IRL algorithm in [3], which we call maximum likelihood IRL (*MLIRL*), uses Boltzmann distribution to model likelihood function using $p(a|s, r_n) = \frac{e^{\beta Q(s,a)}}{\sum_{a \in \mathcal{A}} e^{\beta Q(s,a)}}$, where β denotes the degree of decision-making confidence.

The likelihood function is optimized via gradient ascent method as follows.

1. Initialize: Choose random set of reward weights ω_1.
2. Iterate for $t = 1$ to M do: Compute $Q(s,a)$ and $p(a|s, r_n)$ using ω_t; $L = \sum_{j=1}^{|\mathcal{O}_n|} \sum_{(s,a) \in h_n^j} \log p(a|s, r_n)$; $\omega_{t+1} \leftarrow \omega_t + \alpha_t \nabla_\omega L$.
3. Output reward $r_n = \omega_M^T \phi(s)$ for n-th agent.

The parameters β and M need to be defined as constants.

5.2   IRL with Gaussian Process

IRL algorithm, which is called $GPIRL$ in [15], uses preference relations to model the likelihood function $P(\mathcal{O}_n|r_n)$ and assumes the r_n is generated by Gaussian process for n-th observed agent.

Given a state, we assume an optimal action is selected according to Bellman optimality. At state s, $\forall \hat{a}, \breve{a} \in \mathcal{A}$, we define the *action preference relation* as:

- Action $\hat{a}$ is weakly preferred to $\breve{a}$, denoted as $\hat{a} \succeq_s \breve{a}$, if $Q(s, \hat{a}) \geq Q(s, \breve{a})$;
- Action $\hat{a}$ is strictly preferred to $\breve{a}$, denoted as $\hat{a} \succ_s \breve{a}$, if $Q(s, \hat{a}) > Q(s, \breve{a})$;
- Action $\hat{a}$ is equivalent to $\breve{a}$, denoted as $\hat{a} \sim_s \breve{a}$, if and only if $\hat{a} \succeq_s \breve{a}$ and $\breve{a} \succeq_s \hat{a}$.

Given the observation set $\mathcal{O}_n$, we have a group of preference relations at each state s, which is written as

$$\mathcal{E} \equiv \left\{ (\hat{a} \succ_s \breve{a}),\ \hat{a} \in \hat{\mathcal{A}},\ \breve{a} \in \mathcal{A} \setminus \hat{\mathcal{A}} \right\} \cup \left\{ (\hat{a} \sim_s \hat{a}'),\ \hat{a}, \hat{a}' \in \hat{\mathcal{A}} \right\},$$

where $\hat{\mathcal{A}}$ is the action subspace from observation $\mathcal{O}_n$.

Then, the likelihood function $p(\mathcal{O}_n|r_n) = \prod p(\hat{a} \succ_s \breve{a}) \prod p(\hat{a} \sim_s \hat{a}')$. The models of $p(\hat{a} \succ_s \breve{a})$ and $p(\hat{a} \sim_s \hat{a}')$ are defined in [15].

Let $\mathbf{r}$ be the vector of r_n containing the reward for $M = |\mathcal{A}|$ possible actions at T observed states. We have

$$\mathbf{r} = (\underbrace{\mathbf{r}_1(s_1), ..., \mathbf{r}_1(s_T)}, ..., \underbrace{\mathbf{r}_M(s_1), ..., \mathbf{r}_M(s_T)})$$
$$= (\qquad \mathbf{r}_1, \qquad \cdots, \qquad \mathbf{r}_M),$$

where $T = |\mathcal{S}|$ and $\mathbf{r}_m, \forall m \in \{1, 2, \ldots, M\}$, denotes the reward for action a_m.

Consider $\mathbf{r}_m$ as a Gaussian process. We denote by $k_m(s_i, s_j)$ the function generating the value of entry (i, j) for covariance matrix $\mathbf{K}_m$, which leads to $\mathbf{r}_m \sim N(0, \mathbf{K}_m)$. Then the joint prior probability of the reward is a product of multivariate Gaussian, namely $p(\mathbf{r}|\mathcal{S}) = \prod_{m=1}^{M} p(\mathbf{r}_m|\mathcal{S})$ and $\mathbf{r} \sim N(0, \mathbf{K})$. Note that $\mathbf{r}$ is completely specified by the positive definite covariance matrix $\mathbf{K}$, which is block diagonal in the covariance matrices $\{\mathbf{K}_1, \mathbf{K}_2..., \mathbf{K}_M\}$ based on the assumption that the reward latent processes are uncorrelated . In practice, we use a squared exponential kernel function, written as:

$$k_m(s_i, s_j) = e^{\frac{1}{2}(s_i - s_j)\mathbf{M}_m(s_i - s_j)} + \sigma_m^2 \delta(s_i, s_j),$$

where $\mathbf{M}_m = \kappa_m \mathbf{I}$ and $\mathbf{I}$ is an identity matrix. The function $\delta(s_i, s_j) = 1$, when $s_i = s_j$; otherwise $\delta(s_i, s_j) = 0$. Under this definition the covariance is almost unity between variables whose inputs are very close in the Euclidean space, and decreases as their distance increases.

Then, the *GPIRL* algorithm estimates the reward function by iteratively conducting the following two main steps:

1. Get estimation of $\mathbf{r}_{MAP}$ by maximizing the posterior $p(r_n|\mathcal{O}_n)$, which is equal to minimize $-\log p(\mathcal{O}_n|r_n) - \log p(r_n|\theta)$, where $\theta = (\kappa_m, \sigma_m)_{m=1}^M$ is the hyper-parameter controlling the Gaussian process. Above optimization problem has been proved to be convex programming in [15].
2. Optimize the hyper-parameters by using gradient decent method to maximize $\log p(\mathcal{O}_n|\theta, \mathbf{r}_{MAP})$, which is the Laplace approximation of $p(\theta|\mathcal{O}_n)$.

6 Experimentation

Our experiments simulate agent recognition problems and compare several IRL algorithms against the methods that construct feature vectors from raw observation data. We study two problems, *GridWorld* and the *secretary problem*. *GridWorld* sheds light on the task of recognizing agents whose underlying decision strategy can be matched by a policy of the MDP model. The secretary problem provides a more practical environment in which agents' true decision strategy may not be explained or expressed by any policy of the MDP that is used to model the decision-making problem. Agents in the secretary problem employ heuristic decision rules derived from experimental study of human behavior in psychology and economics.

To evaluate the recognition performance, we use the following algorithms: (1) Clustering: Kmeans [9]; (2) Classification: Support vector machine (SVM), K-nearest neighbours (KNN), Fisher discriminant analysis (FDA) and logistic regression (LR) [9]. We use clustering accuracy [23] and Normalized Mutual Information (NMI) [20] to compare clustering results.

6.1 *GridWorld* Problem

In the *GridWorld* problem, which is used as a benchmark experiment by Ng and Russell in [13], an agent starts from a given square and moves towards a destination square. The agent has five actions to take: moving in the four cardinal directions or staying put. With probability 0.65 the agent moves to its chosen location, with probability 0.15 it stays in the same location regardless of chosen action, and with probability 0.2 it moves in a random cardinal direction.

The small *GridWorld* has been widely used as a test domain by most of IRL algorithms. The observation data is collected when an agent is moving in the grid world. From the observation, the reward is learned to make the optimal policy of a MDP match the observed behavior. We investigate the agent recognition problem in terms of clustering and classification on a 10×10 *GridWorld* problem. Experiments are conducted according to the steps in Algorithm 1.

Algorithm 1. *GridWorld* experimentation steps

1: Input the variables $\mathcal{S}, \mathcal{A}, \mathcal{P}$, and two policies π_1 and π_2.
2: **for** $i = 1 \to 2$ **do**
3:     **for** $j = 1 \to 200$ **do**
4:         Model an agent's action selection using π_i + random Gaussian noise. With probability 0.65 the agent executes the selected action.
5:         Sample decision trajectories $\mathcal{O}_{ij}$, and make the ground truth label $y_{ij} = 0$, if $i = 1$; $y_{ij} = 1$, if $i = 2$.
6:         IRL has access to the problem $B = (\mathcal{S}, \mathcal{A}, \mathcal{P}, \gamma, \mathcal{O}_{ij})$ for this agent, and then infers the reward r_{ij}.
7:     **end for**
8: **end for**
9: Recognize these agents based on the reward r_{ij}.

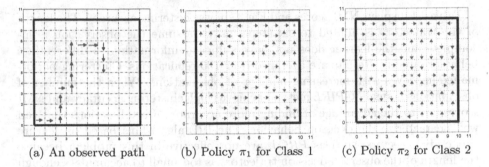

(a) An observed path   (b) Policy π_1 for Class 1   (c) Policy π_2 for Class 2

Fig. 2. The (a) shows an observed decision trajectory. The (b) and (c) illustrate underlying decision policy for two classes of agents. Colored arrow denotes the observed action.

For the input of the experimentation, we simulate two groups of agents, and make each group have 200 agents who adopt similar decision strategy moving in the grid world. Figure 2 (b) and (c) display the underlying policy used by two groups. Each policy represents a decision-making pattern, e.g. a group of agents may prefer the routes close to the border where the scenery is more attractive, while the other group may like passing by the center to avoid traffic. In each group, an agent's decision is simulated by adding Gaussian noise to his/her group's underlying policy. Here, agents may have multiple destinations to visit. Though these agents may have the same goal such as arriving at the destination in the shortest time, their decision patterns can still be different.

In the experiments, we find that a small number of short decision trajectories tends to present challenges to action feature methods, which is an observation of particular interest. Additionally, the length of trajectories may have a substantial impact on performance. If the length is so long that the observed agent reaches the destination in every trajectory, the problem can be easily solved based on observations. Thus, we evaluate and compare performance by making the length of decision trajectory small.

Table 1. NMI scores

$\mathcal{O}_n$	FE	FT	PROJ	GPIRL
4	0.0077	0.0012	0.0068	0.0078
8	0.0114	0.0016	0.0130	0.0932
16	0.0177	0.0014	0.0165	0.7751
20	0.0340	0.0573	0.0243	0.8113
30	0.0321	0.0273	0.0365	0.8119
40	0.0361	0.0459	0.0389	0.8123
60	0.0387	0.0467	0.0388	0.8149
80	0.0441	0.1079	0.0421	0.8095
100	0.0434	0.1277	0.0478	0.8149
200	0.0502	0.1649	0.0498	0.8149

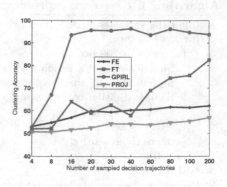

Fig. 3. Clustering accuracy

Table 6.1 displays NMI scores and Fig.3 shows clustering accuracy. The length of the trajectory is limited to six steps, as we assume the observation is incomplete and the learner does not have sufficient information to differentiate behavior directly. Results are averaged over 100 replications. Clustering performance improves with increasing number of observations. When the number of observations is small, *GPIRL* method achieves high clustering accuracy and NMI scores due to the advantage of finding more accurate reward functions that can well characterize the decision behavior. The IRL algorithms based on feature expectation vectors, such as *PROJ*, are not effective in this problem because the length of the observed decision trajectory is too small to accumulate enough observations that correctly approximate the long-term goal.

Considering the utilization of feature learning algorithms to improve the simple feature representations, we also run experiments with PCA-based features where the projection sub-space is spanned by those eigenvectors that correspond to the principal components $c = 10, 20, \ldots, 90$ for *FE* and $c = 2, 4, 6, 8, 10$ for *FT*. No significant changes in the clustering NMI scores and accuracy scores are observed. Therefore, we do not show the performance of PCA-based features in Table 6.1and Figure 3.

Fig.4 displays classification accuracy for a binary classification problem in which there are four hundred agents coming from two groups of decision strategies. The results are averaged over 100 replications with tenfold cross-validations. Four popular classifiers (SVM, KNN, FDA and LR) are employed to evaluate the classification performance. Results suggest that the classifiers based on IRL perform better than the simple methods, such as *FT* and *FE*, particularly when the number of observed trajectories and the length of the trajectory are small. The results support our hypothesis that recovered reward functions constitute an effective and robust feature space for clustering or classifying the agents abased on observation of their decision behavior.

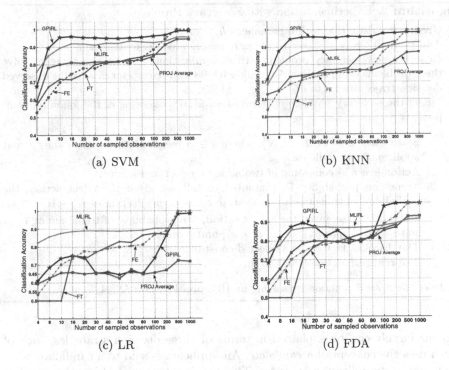

(a) SVM

(b) KNN

(c) LR

(d) FDA

Fig. 4. Classification results with respect to different classifier

6.2   Secretary Problem

The secretary problem is a sequential decision-making problem in which the binary decision to either stop or continue a search is made on the basis of objects already seen. As suggested by the name, the problem is usually cast in the context of interviewing applicants for a secretarial position. The decision maker interviews a randomly-ordered sequence of applicants one at a time. The applicant pool is such that the interviewer can unambiguously rank each applicant in terms of quality relative to the others seen up to that point. After each interview, the decision maker chooses either to move on to the next applicant, forgoing any opportunity to hire the current applicant, or to hire the current applicant, which terminates the process. If the process goes as far as the final applicant, he or she must be hired. Thus the decision maker chooses one and only one applicant. The objective is to maximize the probability that the accepted applicant is, in fact, the best in the pool.

To test our hypotheses on BAR, an ideal experiment would involve recognizing individual human decision makers on the basis of observations of hiring decisions that they make in secretary problem simulations. Experiments with human decision making for the secretary problem are reported on in [19][18], but raw data consisting of decision maker action trajectories is not available. However, a major conclusion of these studies is that the decisions made by the

---

Algorithm 2. Experimentation with Secretary Problem

1: Given a heuristic rule with a parameter h, k or ℓ.
2: Add random Gaussian noise to the parameter, which is written as $\hat{p}$.
3: Generate new secretary problem with N applications and let $n - th$ agent solve these problems using this heuristic rule with its own parameter $\hat{p}$. Save the observed decision trajectories into $\mathcal{O}_n$.
4: Model the secretary problem in terms of an MDP consisting of the following components:

    1. State space $\mathcal{S} = \{1, 2, \ldots, N\}$, where $s \in \mathcal{S}$ means that at time s the current applicant is a candidate.
    2. Action space $\mathcal{A}$ consisting of two actions: reject and accept.
    3. Transition probability $\mathcal{P}$, computed as follows: given the reject action, the probability of transitioning from state s_i to s_j, $p(s_j|s_i)$, is $\frac{s_i}{s_j(s_j-1)}$ if $s_j \geq s_i$, and 0 otherwise; given the accept action, the probability of transitioning from state s_i to s_j, $p(s_j|s_i)$, is 1 if $s_i = s_j$, and 0 otherwise.
    4. The discount factor γ is a selected constant.
    5. The reward function is unknown.

5: Infer the reward function by solving an IRL problem $B = (\mathcal{S}, \mathcal{A}, \mathcal{P}, \gamma, \mathcal{O}_n)$.

---

humans largely can be explained in terms of three decision strategies, each of which uses the concept of a *candidate*. An applicant is said to a candidate he or she is the best applicant seen so far. The decision strategies of interest are the:

1. *Cutoff rule (CR)* with cutoff value h, in which the agent will reject the first $h - 1$ applicants and accept the next candidate;
2. *Successive non-candidate counting rule (SNCCR)* with parameter value k, in which the agent will accept the first candidate who follows k successive non-candidate applicants since the last candidate; and
3. *Candidate counting rule (CCR)* with parameter value ℓ, in which the agent selects the next candidate once ℓ candidates have been seen.

The optimal decision strategy for the secretary problem is to use CR with a parameter that can be computed using dynamic programming for any value of N, the number of secretaries. As N grows, the optimal parameter converges to N/e and yields a probability of successfully choosing the best applicant that converges to $1/e$. Thus only one of the three decision strategies enumerated above can be viewed as optimal, and that only for a single parameter value out of the continuum of possible values. Human actions are usually suboptimal and tend to look like mixtures of CR (with a non-optimal parameter), SNCCR, and CCR [19]. As a surrogate for the action trajectories of humans, we use agents that we generate action trajectories for randomly sampled secretary problems using CR, SNCCR, and CCR. For a given decision rule (CR, SNCCR, CCR), we simulate a group of agents that adopt this rule, differentiating individuals in a group by adding Gaussian noise to the rule's parameter. The details of the process are given in Algorithm 2. We use IRL and observed actions to learn reward functions

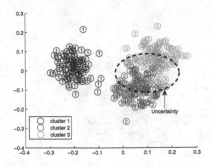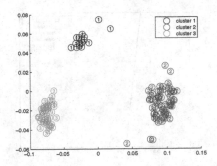

Fig. 5. Each circle denotes the feature vector for an agent, which is projected into 2D space by using PCA. The feature vectors provided by *FE* method are shown on the left. The reward vectors estimated by IRL are shown on the right.

for the MDP model given in Algorithm 2. It is critical to understand that the state space for this MDP model captures nothing of the history of candidates, and as a consequence is wholly inadequate for the purposes of modeling SNCCR and CCR. In other words, for general parameters, neither SNCCR nor CCR can be expressed as a policy for the MDP in Algorithm 2. (There does exist an MDP in which all three of the decision rules can be expressed as policies, but the state space for this model is exponentially larger.) Hence, for two of the rules, the processes that we use to generate data and the processes we use to learn are distinct.

As an initial set of experiments, we generated an equal number of agents from each rule. All the heuristic rules use the same parameter value. We have compared the method using statistical feature representations obtained from the raw decision trajectories and our IRL model-based method. We employ 10 fold cross-validation to obtain the average accuracy, and it is always 100% .

Table 2. NMI score for Secretary Problem

H	CR		SNCCR		CCR	
	Action	BayesIRL	Action	BayesIRL	Action	BayesIRL
1	0.0557	**0.5497**	0.0551	**0.1325**	0.0229	**0.2081**
11	0.3852	**0.6893**	0.2916	**0.7190**	0.1844	**0.4974**
21	0.6017	**0.7898**	0.4305	**0.8179**	0.2806	**0.5181**
31	0.7654	**0.8483**	0.5504	**0.8641**	0.4053	**0.6171**
41	0.8356	**0.9676**	0.5682	**0.9218**	0.4524	**0.6533**
51	0.8781	**0.9739**	0.5894	**0.9423**	0.5464	**0.6507**
61	0.9102	**0.9913**	0.5984	**0.9518**	0.5492	**0.6513**
71	0.9115	**0.9915**	0.6460	**0.9639**	0.6024	**0.6512**
81	0.9532	**1.0000**	0.6541	**0.9721**	**0.6708**	0.6563
91	0.9707	**1.0000**	0.6494	**0.9864**	**0.6884**	0.6544

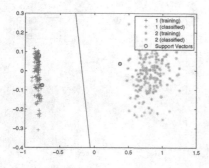 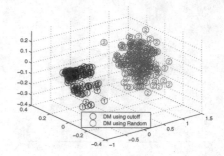

Fig. 6. Visualization of a binary classification problem for subjects using cutoff rule and random rules. The reward vectors are projected into 2(left)/3(right) dimensional subspace, which are spanned by the first 2/3 principal components.

Given that perfect classification performance was achieved by all algorithms, the problem of recognizing across decision rules appears to be quite easy. A more challenging problem is to recognize variations in strategy within a single decision rule. For each rule, we conducted recognition experiments in which 300 agents were simulated, 100 each for three distinct values of the rule parameter. Individuals were differentiated by adding random noise to the parameter. Here, we show the comparison of the clustering performance between the simple method called *FE* and our MDP model-based method. In Fig.5, the left figure displays an area marked "uncertainty" for the method called *FE*, while the right figure shows that the reward vectors have lower variance in the same group and higher variance between different groups.Fig.5 intuitively demonstrates that when the agents' behavior is represented in the reward space, the recognition problem becomes easier to solve.

Table 2 summarizes the NMI scores for using K-means clustering algorithm to recognize variations in strategy within one heuristic decision rule. We conduct experiments on three rules separately. The column called H in Table 2 records the number of decision trajectories that have been sampled for training. Table 2 indicates that the feature representation in reward space is almost always better than the representation with statistical features computed from the raw observation data. Moreover, the reward space can particularly better characterize the behavior when the scale of the observation data is small. Note that although none of the MDP policies can match the SNCCR and CCR rules, the reward vectors, which are recovered by IRL for the MDP model, still make the clustering problem easier to solve.

Fig.6 shows a binary classification result of using *PROJ* algorithm to learn the reward functions for the agents in Secretary problem and then categorize the agents into two groups. In this classification experiment, the users' ground truth label is either cutoff decision rule or random strategy that makes random decisions.

7   Conclusions

We have proposed the use of IRL to solve the agent recognition problem. The observed agent is not required to be rational in the decision-making process. However, we model the agent's behavior in an MDP environment and estimate the reward function by making the MDP policy match the observed behavior. Numerical experiments on *GridWorld* and the secretary problem suggest that the advantage that IRL enjoys over simple methods is more pronounced when observations are limited and incomplete. We also note that there seems to be a positive correlation between the success of IRL algorithms in apprenticeship learning (cf. [15]) and their success in the agent recognition problem. To some degree, this relationship parallels results from [11] [7], where apprenticeship learning benefits from a learning structure that based on sophisticated methods for task decomposition or hierarchical identification of skill trees. Exploration of IRL algorithms that consider subgoals and which of the algorithmic choices can help agent recognition is an avenue of future work.

Validation of the ideas proposed here can come only through experimentation with more difficult problems. Of particular importance would be problems involving human decision makers or other real-world scenarios, such as periodic investment, gambling, or stock trading.

References

1. Abbeel, P., Ng, A.Y.: Apprenticeship learning via inverse reinforcement learning. In: Proceedings of the Twenty-First International Conference on Machine Learning (2004)
2. Argall, B.D., Chernova, S., Veloso, M., Browning, B.: A survey of robot learning from demonstration. Robot. Auton. Syst. 57(5), 469–483 (2009)
3. Babes-Vroman, M., Marivate, V., Subramanian, K., Litman, M.: Apprenticeship learning about multiple intentions. In: The 28th International Conference on Machine Learning, WA, USA (2011)
4. Baker, C.L., Saxe, R., Tenenbaum, J.B.: Action understanding as inverse planning. Cognition 113, 329–349 (2009)
5. Boularias, A., Chaib-draa, B.: Bootstrapping apprenticeship learning. In: Advances in Neural Information Processing Systems 24. MIT Press (2010)
6. Choi, J., Kim, K.-E.: Map inference for bayesian inverse reinforcement learning. In: Advances in Neural Information Processing System, pp. 1989–1997 (2011)
7. Cobo, L.C., Isbell Jr., C.L., Thomaz, A.L.: Automatic task decomposition and state abstraction from demonstration. In: AAMAS, pp. 483–490 (2012)
8. Deepak, R., Eyal, A.: Bayesian inverse reinforcement learning. In: Proc. 20th International Joint Conf. on Artificial Intelligence (2007)
9. Duda, R.O., Hart, P.E., Stork, D.G.: Pattern Classification. Wiley (2001)
10. Dvijotham, K., Todorov, E.: Inverse optimal control with linearly-solvable mdps. In: Proc. 27th International Conf. on Machine Learning. ACM (2010)
11. Konidaris, G.D., Kuindersma, S.R., Grupen, R.A., Barto, A.G.: Robot learning from demonstration by constructing skill trees. International Journal of Robotics Research 31(3), 360–375 (2012)

12. Neu, G., Szepesvari, C.: Apprenticeship learning using inverse reinforcement learning and gradient methods. In: Proc. Uncertainty in Artificial Intelligence (2007)
13. Ng, A.Y., Russell, S.: Algorithms for inverse reinforcement learning. In: Proc. 17th International Conf. on Machine Learning, pp. 663–670. Morgan Kaufmann (2000)
14. Paddrik, M., Hayes, R., Todd, A., Yang, S., Beling, P., Scherer, W.: An agent based model of the e-mini s&p 500: Applied to flash crash analysis. In: 2012 IEEE Symposium on Computational Intelligence for Financial Engineering and Economics, CIFEr 2012 (2012)
15. Qiao, Q., Beling, P.A.: Inverse reinforcement learning via convex programming. In: Americon Control Conference (2011)
16. Ramirez, M., Geffner, H.: Plan recognition as planing. In: 21st Int'l Joint Conf. on Artificial Intelligence, pp. 1778–1783 (2009)
17. Ratliff, N.D., Bagnell, J.A., Zinkevich, M.A.: Maximum margin planning. In: Proceedings of the 23rd International Conference on Machine Learning (2006)
18. Schunk, D., Winter, J.: The relationship between risk attitudes and heuristics in search tasks: A laboratory experiment. Journal of Economic Behavior and Organization 71, 347–360 (2009)
19. Seale, D.A.: Sequential decision making with relative ranks: An experimental investigation of the 'secretary problem'. Organizational Behavior and Human Decision Process 69, 221–236 (1997)
20. Strehl, A., Ghosh, J.: Cluster ensembles? a knowledge reuse framework for combining multiple partitions. Journal of Machine Learning Research 3, 583–617 (2002)
21. Syed, U., Schapire, R.E.: A game-theoretic approach to apprenticeship learning. In: Advances in Neural Information Processing Systems, pp. 1449–1456. MIT Press (2008)
22. Tapia, E.M., Intille, S.S., Larson, K.: Activity recognition in the home using simple and ubiquitous sensors. In: Ferscha, A., Mattern, F. (eds.) PERVASIVE 2004. LNCS, vol. 3001, pp. 158–175. Springer, Heidelberg (2004)
23. Xu, L., Neufeld, J., Larson, B., Schuurmans, D.: Maximum margin clustering. In: Advanced Neural Information Process Systems, pp. 1537–1544 (2005)
24. Yang, S., Paddrik, M., Hayes, R., Todd, A., Kirilenko, A., Beling, P., Scherer, W.: Behavior based learning in identifying high frequency trading strategies. In: 2012 IEEE Symposium on Computational Intelligence for Financial Engineering and Economics, CIFEr 2012 (2012)

Learning Throttle Valve Control
Using Policy Search

Bastian Bischoff[1], Duy Nguyen-Tuong[1], Torsten Koller[1],
Heiner Markert[1], and Alois Knoll[2]

[1] Robert Bosch GmbH, Corporate Research, Robert-Bosch-Str. 2,
71701 Schwieberdingen, Germany
[2] TU Munich, Robotics and Embedded Systems, Boltzmannstr. 3,
85748 Garching at Munich, Germany

Abstract. The throttle valve is a technical device used for regulating a fluid or a gas flow. Throttle valve control is a challenging task, due to its complex dynamics and demanding constraints for the controller. Using state-of-the-art throttle valve control, such as model-free PID controllers, time-consuming and manual adjusting of the controller is necessary. In this paper, we investigate how reinforcement learning (RL) can help to alleviate the effort of manual controller design by automatically learning a control policy from experiences. In order to obtain a valid control policy for the throttle valve, several constraints need to be addressed, such as no-overshoot. Furthermore, the learned controller must be able to follow given desired trajectories, while moving the valve from any start to any goal position and, thus, multi-targets policy learning needs to be considered for RL. In this study, we employ a policy search RL approach, Pilco [2], to learn a throttle valve control policy. We adapt the Pilco algorithm, while taking into account the practical requirements and constraints for the controller. For evaluation, we employ the resulting algorithm to solve several control tasks in simulation, as well as on a physical throttle valve system. The results show that policy search RL is able to learn a consistent control policy for complex, real-world systems.

1 Introduction

The throttle valve, as shown in Figure 1, is an important and widely-used technical device for many industrial and automotive applications, such as for pressure control in gasoline combustion engines and flow regulation in air conditioning and heat pumps. Usually, the throttle valve system consists of a valve and an actuator, e.g. a DC-motor. The throttle valve control task is to move the valve from arbitrary positions to given desired positions by regulating the actuator inputs.

Controlling the throttle valve is a challenging task. Due to the spring-damper design of the valve system, we have a highly dynamic behavior. As many unknown nonlinearities are involved, such as complex friction, accurate physical models of the valve dynamics are hard to obtain. In practice, the valve needs to be

H. Blockeel et al. (Eds.): ECML PKDD 2013, Part I, LNAI 8188, pp. 49–64, 2013.
© Springer-Verlag Berlin Heidelberg 2013

controlled at a very high rate, e.g. 200Hz, and desired valve positions need to be reached as fast as possible. While requiring a fast control performance, no overshoot is allowed here, i.e. the valve position must not exceed the desired position. This requirement is essential for pressure control in gasoline combustion engines for automotive application, as addressed in this paper. Here, an open valve corresponds to a car acceleration and, thus, an overshoot during the valve control would result in undesirable jerks of engine torque. These constraints, e.g. unknown nonlinearities and fast control without overshoot, make the throttle valve controller design difficult in practice.

Fig. 1. Example of a throttle valve used in combustion engines for automotive applications

In the literature, several approaches are discussed to tackle challenges of throttle valve control based on methods of classical control theory [12–14]. These approaches usually involve tedious, manual tunning of controller parameters. Furthermore, profound knowledge of the physical system is required in order to obtain a good parametrization of the controller in this case. These limitations motivate the approach used in this study. We investigate how machine learning techniques, especially, Reinforcement Learning (RL), can be employed to successfully *learn* a control policy from experience, while incorporating required practical constraints. Beside the mentioned challenges, several RL problems need to be tackled, such as learning multi-targets and handling large data during the learning process. In this paper, we employ a probabilistic, model-based RL approach, e.g. the probabilistic inference for control algorithm (Pilco) [2], for learning the control policy. We modify Pilco taking in account the discussed requirements. The method is implemented and evaluated in simulation, as well as on a real throttle valve system. The evaluation shows the feasibility of the presented RL approach and, thus, indicates the suitability of RL for real-world, industrial applications.

The remainder of the paper is organized as follows: in the next section, we introduce the throttle valve system and motivate the use of RL. In Section 3, we briefly review the basic idea behind RL and introduce Pilco. Section 4 shows how probabilistic RL, especially, Pilco, can be modified to match the required constraints. Evaluation of our method in simulation, as well as on a real throttle valve, is provided in Section 5. Finally, a conclusion is given in Section 6.

2   The Throttle Valve System

Throttle valve systems are widely used in many industrial applications, ranging from semi-conductor manufacturing to cooling systems for nuclear power plants. However, one of the most important applications can be found in automotive control, where throttle valves are employed to regulate the flow of air entering a

combustion engine. As shown in Figure 1, the valve system basically consists of a
DC-motor, a spring and a flap with position sensors. Depending on the position
of the flap, the gasoline-to-air ratio in the combustion chamber is adjusted and,
subsequently, influences the torque generated by the engine. The dynamics of
the throttle valve system can be simplified by the model [10] to be

$$
\begin{bmatrix} \dot{\alpha}(t) \\ \dot{\omega}(t) \end{bmatrix} = \begin{bmatrix} 0 & 1 \\ -K_s & -K_d \end{bmatrix} \begin{bmatrix} \alpha(t) \\ \omega(t) \end{bmatrix} + \begin{bmatrix} 0 \\ C_s - K_f \operatorname{sgn}(\omega(t)) \end{bmatrix} + \begin{bmatrix} 0 \\ T(t) \end{bmatrix}, \tag{1}
$$

where α and ω are the flap angle and corresponding angular velocity, T is the actuator input. The parameters K_s, K_d, K_f and C_s are dynamics parameters and
need to be identified for a given system [18]. Model-based valve controllers rely
on this model [5] and, thus, identification of dynamics parameters is necessary.
However, parameter identification using sampled data can be time-consuming
and difficult. It it hard to create sufficiently rich data in order to obtain plausible dynamics parameters. Furthermore, the parameters that optimally fit a
data set, are often not physically consistent and, hence, physical consistency
constraints have to be imposed on the identification problem [17]. Using data-
based nonparametric models for RL — as employed in this paper — for learning
optimal control policies can help to overcome these limitations.

2.1   Requirements for Throttle Valve Control

As the throttle valve is a real-time system, precise and fast control is crucial
to provide optimal performance. In order to obtain fast control, we employ a
fixed radial-basis function structure as parametrization of the controller, which
can be evaluated in real-time. As shown in Section 5, the learned controller
can be evaluated at a frequency of about 200Hz. Furthermore, for learning the
dynamics used for model-based RL we employ a NARX-structure [9] to represent
the system dynamics, as described in Section 4.1. A well-approximated dynamics
model is prerequisite for learning a good control policy with model-based RL.

Typical RL problems are goal oriented [1], i.e. RL is formulated for reaching
single, desired goal positions. However, when employing throttle valve control,
trajectory tracking is inevitable. The learned policy needs to be able to follow
desired trajectory and, thus, multi-target RL as described in Section 4.2 is required here. It is shown in the evaluation that the learned policy can generalize
well for unknown goals and trajectories.

In addition to the fast control requirement, no overshoot of the trajectory is
essential. As an overshoot corresponds to undesirable jerks in engine torque, the
valve trajectory must not go beyond the desired valve position. On the other
hand, it is required that the valve moves to the desired position as close and
fast as possible. Taking in account these requirements, we design an appropriate
cost function in Section 4.3. The cost is defined such that the requirements are
accomplished and the resulting RL formulation remains solvable in closed form.

3   A Brief Review on RL

In this section, we provide a short review on the basic concepts of RL [1, 3]. Subsequently, we proceed to discuss the probabilistic policy search approach Pilco [2].

3.1   General Setting

In RL, we consider an agent and its interactions with the environment. The state of the learning agent is defined by $s \in S$. The agent can apply actions $a \in A$ and, subsequently, moves to a new state s' with probability given as $p(s'|s, a)$. The controller $\pi : S \to A$ determines in every state the action which should be used. If the controller is applied for T timesteps, we get a state-action sequence $\{s_0, a_0\}, \{s_1, a_1\}, \ldots, \{s_{T-1}, a_{T-1}\}, s_T$, which we call *rollout* of the controller. In case of uncertainty and noise, multiple rollouts will not be identical and the rollout must be described by probability distributions. The environment rates states s_i with a cost function $c : S \to \mathbb{R}$ (and, thus, gives a rating for the controller). The goal of the learning algorithm is to find a controller, which minimizes the expected sum $J(\pi)$ of collected cost, i.e.

$$\min_{\pi} J(\pi), \ J(\pi) = \sum_{t=0}^{T} E_{s_t}(c(s_t)), \tag{2}$$

where $p(s_0), \ldots, p(s_T)$ are the resulting state distributions on application of the controller π. The cost function c must be set according to the learning goal. The tuples s_i, a_i, s_{i+1} are saved as experience and are used to optimize the controller. RL algorithms differ in the way they use this experience to learn a new, improved controller π. Two important properties that characterize RL techniques, are model-free and model-based, as well as Policy Search and Value-function. Next we will shortly describe the approaches and examine their suitability for throttle valve control.

3.2   Approaches in Reinforcement Learning

Model-based RL describes algorithms, where the experience samples s_i, a_i, s_{i+1} are used to learn a dynamics model $f : S \times A \to S$ and the controller π is optimized using the dynamics model as internal simulation. *Model-free* RL algorithms, on the other hand, directly optimizes the controller π without usage of a dynamics model. In the last decades, model-free RL got much attention mainly because for discrete state and action sets, convergence guarantees can be given. However, often many trials are necessary to obtain a good controller. Model-based RL methods potentially use the data more efficient, but it is well known that *model bias* can strongly degrade the learning performance [4]. Here, a controller might succeed in simulation but fails when applied to the real system, if the model does not describe the complete system dynamics. To address this problem, it is important to incorporate uncertainty — which can result from a lack of experience or due to stochasticity in the system — and to employ probabilistic dynamics models.

Besides model-free and model-based, RL methods can be divided into *policy search* algorithms and *value-function* approaches. Policy search methods directly operate on the parameters θ of a controller π_θ to optimize the sum of expected cost given in Equation (2). Therefore, a parametrized control structure has to be defined by the user. This allows to include prior knowledge about good control strategies, but the structure also limits the set of strategies that can be learned. In contrast to policy search, value-function approaches try to estimate a long-term cost for each state.

3.3 Pilco: A Model-Based Probabilistic Policy Search

For the throttle valve control task — as for many physical systems — it is not possible to perform several thousands of experiments on the real system until the learning process converges. Thus, it is important to reduce interactions with the system. This favours model-based RL approaches. Furthermore, the throttle valve has high requirements on the control frequency due to the fast system dynamics. For example, evaluation of the controller given the current system state must take at most 5ms. Therefore, policy search RL, with a control structure that can be evaluated fast, seems to be an appropriate approach.

Pilco[1] is a model-based RL algorithm, which uses Gaussian processes (GP) for modeling the system dynamics [2]. Based on this probabilistic dynamics model, a control policy can be inferred. The controller can, for example, be parametrized as a radial basis function network (RBF) with Gaussian shaped basis function. Thus, the controller is given by

$$\pi(s) = \sum_{i=0}^{N} w_i \phi_i(s),$$

with $\phi_i(s) = \sigma_f^2 \exp(-\frac{1}{2}(s - s_i)^T \Lambda (s - s_i))$, $S = [s_0, s_1, \ldots, s_N]^T$ the set of support points and w representing the weight vector. The hyperparameters of the controller are given by $\Lambda = \mathrm{diag}(l_1^2, \ldots l_D^2)^{-1}$ and σ_f. The number N of support points is a free parameter that needs to be adjusted. The support points, the weight vector and the hyperparameters build the set θ of control parameters that need to be optimized during learning.

For learning the controller, we start with a Gaussian state distribution $p(s_0)$. The RBF network controller returns an action for every state, therefore we get a distribution $p(a_t) = \int p(a_t|s_t)p(s_t)ds_t$. This distribution is analytically approximated to a Gaussian. Given the state distribution $p(s_t)$ and the approximated Gaussian action distribution $p(a_t)$, the joint distribution $p(s_t, a_t)$ is approximated. The dynamics model takes this distribution as input and returns the distribution $p(s_{t+1})$ for the next state. The expected cost $J(\theta) = \sum_{t=0}^{T} E_{s_t}(c(s_t))$

[1] We thank Marc Deisenroth for providing us the Pilco code.

Algorithm 1. Pilco: model-based policy search

1: $D := D_{init}, \theta :=$ random ▷ initialize dynamics data set D and control parameters θ
2: **for** $e := 1$ **to** $Episodes$ **do**
3:    Learn dynamics model $GP_{dynamics} : S \times A \rightarrow S$
4:    Improve policy:
5:       Estimate rollout $p(s_0), p(s_1), \ldots, p(s_T)$ of π_θ using $GP_{dynamics}$
6:       Rate policy-parameters $J(\theta) = \sum_{t=0}^{T} E_{s_t}(c(s_t))$
7:       Adapt policy-parameters $\theta = \theta + \nabla J(\theta)$
8:    Apply controller π_θ on system, $D := D \cup \{s_0, \pi_\theta(s_0) = a_0, s_1, \ldots\}$
9: **end for**

---

can subsequently be computed from these rollout results. Based on generated cost values, the controller can now be optimized. The optimization step can be performed using gradient descend procedure, where analytical gradients are computed on the hyperparameters. The resulting algorithm [2] is summarized in Algorithm 1.

4   Learning Throttle Valve Control with RL

In Section 2, we described the throttle valve system and the task specific requirements for throttle valve control. In this section, we adapt the Pilco algorithm described in the previous section taking into account the desired requirements. First, we show how the system dynamics can be modeled using Gaussian processes while employing a NARX-structure with state and action feedback. Additionally, to handle the large amount of dynamics data occurring during learning, an information gain criterion is used for selecting informative data points. As the learned controller must be able to follow arbitrary trajectories, we describe the setting for multiple start and goal states. Finally, we define a cost function addressing the no-overshoot restriction, while the integrals involved in the Pilco learning process can still be solved analytically. The analytical gradients, which are essential for policy optimizing, will be provided.

4.1   Modeling Throttle Valve Dynamics

A dynamics model describes the behavior of a system, i.e. the probability $p(s'|s, a)$ that the system state changes to s' when action a is applied in state s. Here, the actions a correspond to the input voltage u of the DC-motor. The opening angle α of the valve changes dynamically depending on the input voltage. As shown in Equation (1), the dynamics of the valve can be approximated by a second-order system. Due to this insight, the RL state s_t is defined as $s_t = [\alpha_t, \alpha_{t-1}, \alpha_{t-2}, u_{t-1}, u_{t-2}]$. Thus, the resulting state s_t has the well-known *Nonlinear Autoregressive Exogenous* (NARX) structure [9], as shown in Figure 2. For modeling the dynamics, nonparametric Gaussian process regression is employed to predict α_{t+1} given a state s_t and the current action u_t.

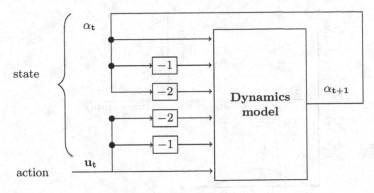

Fig. 2. NARX-structure with state and action feedback to model the throttle valve dynamics. The current valve opening α_t, as well as past valve openings $\alpha_{t-1}, \alpha_{t-2}$ and input voltage u_{t-1}, u_{t-2} are jointly used to predict α_{t+1}. For modeling the dynamics, GP regression is employed.

A further challenge in modeling the throttle valve dynamics is the amount of data generated by the system. Dynamics data is sampled at a high frequency, e.g. 200 samples per second, leading to massive data sets. Here, we employ an information gain criterion [8] to reduce the sampled data while retaining relevant dynamics information. This significantly reduces the overall learning time as well as the memory requirements. A GP is employed for modeling the dynamics, the information gain criterion can be computed analytically and efficiently. See [8] by Seeger et. al. for more details.

4.2   Multiple Start and Goal States

For throttle valve control, it is important that the learned controller can follow arbitrary trajectories. Thus, the controller must be able to move the valve from any given start position to any goal position. One approach would be to learn seperate controllers for a set of goal states and combine these controllers for arbitrary goal positions. However, this is complex for non-linear systems and may not be optimal globally. Instead, we include the goal position g as input to the controller, i.e. $u = \pi_\theta(s, g)$, as described in [11]. Now, one joint controller is learned for all goal positions. This formulation allows to set the goal state dynamically on controller application.

In the standard Pilco framework, the control parameters θ are optimized with respect to the sum of expected cost, $J(\theta) = \sum_{t=0}^{T} E(c(s_t))$, where $s_t \sim \mathcal{N}(\mu, \Sigma)$ is a Gaussian distribution. For multiple start states s_t^i and goal states g_i, the objective function can be modified (see [11]) to

$$J(\theta) = \sum_{i=0}^{|Z|} \sum_{t=0}^{T} E(c(s_t^i, g^i)), \qquad (3)$$

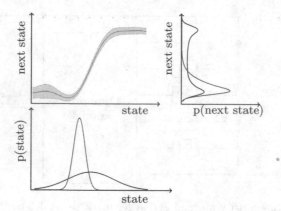

Fig. 3. The upper left figure shows the Gaussian process dynamics model for a fixed action. The lower picture shows two state distributions. The right upper plot shows the corresponding next state distributions, after mapping the state distributions through the dynamics GP. When the start distribution is too broad, the state distribution after mapping is too complicated to approximate by a Gaussian as performed in Pilco.

where Z represents pairs of start and goal states, e.g. $Z = \{(s_0^0, g^0), (s_0^1, g^1), \ldots\}$. The start distributions given by $s_0^i \in Z$ and σ_0 as well as the goal states g^i in Z are determined such that they cover the relevant state space parts. The variance of the start distributions $s_0^i \in Z$ needs to be chosen appropriately. Given a very broad start distribution s_0, the next state distribution — after mapping it through the dynamics GP model — can be difficult and, thus, more complicated when approximated by a Gaussian as done by Pilco. However, when the start variance is small, performance might me suboptimal for some start states not covered by the start distributions. Figure 3 illustrates the effects when mapping different state distributions through the GP dynamics model.

4.3   Cost Function for Learning Throttle Valve Control

In this section, we define an appropriate cost function for policy optimization. The saturated immediate cost [2] given by $c_d(s, g) = 1 - \exp(-\|s - g\|^2 / (2d^2))$ with goal state g is a general, task unspecific cost function. Here, the hyper-parameter d describes the width of the cost. However, this cost function is not appropriate for learning throttle valve control, as it does not avoid valve trajectory overshoot. Taking in account the no-overshoot restriction, we introduce an *asymmetric saturating cost function*

$$c(s, g) = \begin{cases} c_{d_1}(s, g), & \text{if } s \leq g \\ c_{d_2}(s, g), & \text{otherwise} \end{cases}, \tag{4}$$

where c_{d_i} is the saturating cost with width d_i and g is the goal state. This continuous, smooth function can be described as a saturating cost function with variable steepness on both sides of the goal depending on the parameters d_1, d_2.

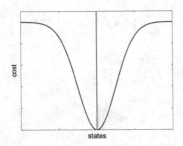 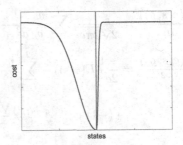

Fig. 4. The left figure shows the symmetric saturating cost over states given on the x-axis. The goal state is indicated by the red vertical line. The right figure shows the asymmetric saturating cost with variable steepness on both sides. Overshooting the goal state (here, states right of the goal state) implies high costs, while approaching the goal from a state left of the goal leads to decreasing cost.

In contrast to the usual symmetric saturating cost, this allows us to assign a decreasing cost when the state converges to the goal state while overshoot is punished, see Figure 4.

To estimate the cost for a given set of parameters in the probabilistics model-based RL framework, the expected cost $E(c(s,g))$ is required for a Gaussian state distribution $s \sim \mathcal{N}(\mu_s, \Sigma_s)$. For the asymmetric saturating cost in Equation (4), $E(c(s,g))$ is given as

$$E(c(s,g)) = \int_{-\infty}^{g} c_{d_1}(s,g)p(s)ds + \int_{g}^{\infty} c_{d_2}(s,g)p(s)ds$$

$$= \frac{1}{\sqrt{2\pi\sigma^2}}\left[\int_{-\infty}^{g} e^{\ell_1}ds + \int_{g}^{\infty} e^{\ell_2}ds\right] = \frac{1}{\sqrt{2\pi\sigma^2}}\left[\frac{w_1}{v_1}r_1 + \frac{w_2}{v_2}r_2\right]$$

with

$$\ell_i = -\left(1/d_i^2 + 1/\sigma^2\right)\left[s - \left(\frac{g\sigma^2 + \mu d_i^2}{\sigma^2 d_i^2}\right)\right]^2 - \frac{(g-\mu)^2}{d_i^2 + \sigma^2}, \quad w_i = e^{\left(\frac{-(g-\mu)^2}{d_i^2+\sigma^2}\right)}$$

$$v_i = \sqrt{\frac{1}{d_i^2} + \frac{1}{\sigma^2}}, \quad u_i = \frac{g\sigma^2 + \mu d_i^2}{\sigma^2 + d_i^2}, \quad r_1 = \int_{-\infty}^{v_1(g-u_1)} e^{-q^2}dq, \quad r_2 = \int_{v_2(g-u_2)}^{\infty} e^{-q^2}dq$$

Using the error function $\int_{-\infty}^{b} e^{-q^2}dq = \frac{\sqrt{\pi}}{2}(\mathrm{erf}(b) + 1)$, we have

$$r_1 = \frac{\sqrt{\pi}}{2}\left(\mathrm{erf}(v_1(g-u_1)) + 1\right), \quad r_2 = \frac{\sqrt{\pi}}{2}\left(1 - \mathrm{erf}(v_2(g-u_2))\right)$$

Given $E(c(s,g))$, the gradients for the policy optimization can be given as

$$\frac{\delta E(c(s,g))}{\delta \mu} = \frac{1}{2\sqrt{2\pi\sigma^2}} \sum_{i=1}^{2} \frac{w_i}{v_i} \frac{1}{d_i^2 + \sigma^2} \left[2r_i(g-\mu) + (-1)^i v_i e^{-(v_i(g-u_i))^2} \right],$$

$$\frac{\delta E(c(s,g))}{\delta \sigma^2} = \frac{-1}{2\sigma^2} E(c(s)) + \sum_{i=1}^{2} \frac{w_i}{v_i} \left[\frac{r_i}{2\sigma^4 v_i^2} + \frac{r_i(g-\mu)^2}{(\sigma^2 + d_i^2)^2} \right.$$

$$\left. + (-1)^i e^{-(v_i(g-u_i)^2)} \frac{\sqrt{2}}{\pi} (\delta_{v_i}(g-u_i) - \delta_{u_i} v_i) \right]$$

where

$$\delta_{u_i} = \frac{d_i^2(g-\mu)}{(\sigma^2 + d_i^2)^2}, \quad \delta_{v_i} = \frac{1}{2} v_i^{-1} \sigma^{-4}, \quad \delta_{w_i} = e^{\left(\frac{-(g-\mu)^2}{\sigma^2 + d_i^2} \right)} \frac{(g-\mu)^2}{(\sigma + d_i^2)^2}.$$

5   Evaluations

In this section, we evaluate the presented RL approach on a throttle valve simulation, as well as on a physical throttle valve system. The experiment setting and learning process are described in detail.

5.1   Simulation Results

First, learning is performed on a throttle valve simulator. We employed the Pilco algorithm as given in Algorithm 1 with an RBF control structure using 40 base functions. The controller can be evaluated in approximately 1ms, which allows for a control frequency of at most 1000Hz. In all experiments, a control frequency of 200Hz was used.

As a first step, we learn a controller for a single start and target state, i.e. a trajectory starting from $\alpha_0 = 70°$ with desired valve opening $g = 10°$. To obtain initial dynamics data D_{init}, random actions are applied twice for 0.16 seconds leading to two trajectories starting in α_0. A NARX-structure is established for modeling the dynamics (see Section 4.1). Optimization is performed with respect to the asymmetric saturating cost with width $d_1 = 0.5, d_2 = 3.5$. Figure 5 shows the learning process over several episodes. It can be seen that the learning converges to a near optimal solution after 4 Episodes. Next, we compare the learning result for the symmetric saturating cost with the asymmetric cost function introduced in Section 4.3. It can be seen in Figure 6 that the asymmetric cost function significantly reduces the overshoot (on the right) compared to the standard symmetric saturating cost (on the left).

So far, the learned controller was optimized for a single trajectory from a single start angle to a single goal angle. We now employ multiple start and goal states as described in Section 4.2. Here, we choose 10 different combinations of start positions and goal positions covering the state space equally, e.g.

$$\tilde{Z} = \{(\alpha_0^0, g^0), (\alpha_0^1, g^1), \ldots\}$$
$$= \{(30, 40), (30, 45), (30, 60), (40, 55), (45, 55),$$
$$(45, 60), (55, 40), (60, 30), (60, 45), (60, 50)\}.$$

Random trajectories   **Episode 2**   **Episode 4**

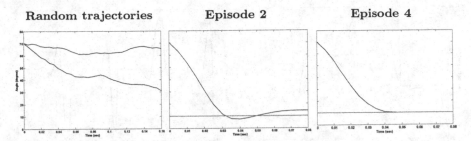

Fig. 5. The figure shows the learning process for the start angle $\alpha_0 = 70$ with goal angle $g = 10$. The left-most plot shows the two random trajectories used as initial dynamics data set. After 2 episodes, the learned controller gets close to the learning goal, but still overshoots and does not reach the goal angle accurately. After 4 episodes, the learning has converged and the resulting control policy shows a near optimal behavior.

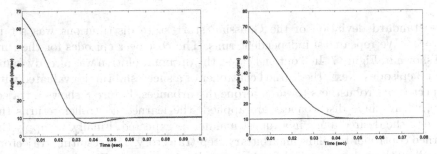

Fig. 6. On the left, the learning result for the symmetric saturating cost function is shown, the result for the asymmetric saturating cost is shown on the right. While the symmetric cost leads to significant overshoot, the behavior for the asymmetric cost is near optimal.

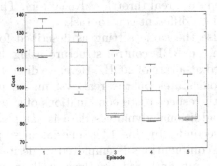

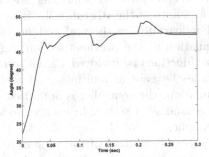

Fig. 7. The figure shows the cost over episodes of 4 independent learning attempts in simulation. In 3 of 4 cases, the learning converges to a near optimal solution after 3 episodes.

Fig. 8. The controller is applied to move the valve from 22 degree valve opening to 50 degree on the throttle valve simulator. At times 0.03, 0.13 and 0.23 torque disturbances are introduced. The controller handles the disturbance well and returns to the goal angle quickly.

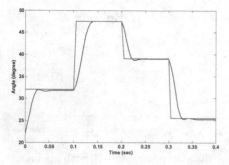 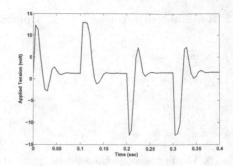

Fig. 9. The learned controller (blue) is applied on the simulated throttle valve with a step trajectory (red) as desired goal trajectory. The left plot shows the valve angle over time, while the right plot shows the applied voltage. The results show, that the learned controller successfully matches the desired trajectory.

The standard deviation of the Gaussian start state distributions was set to $\sigma_0 = 3°$. We repeated 4 independent runs. The cost over episodes for the runs are shown in Figure 7. In 3 out of 4 cases, the optimal solution was already found after 3 episodes. Next, the learned controller of a successful run is evaluated with respect to its robustness towards torque disturbances. Figure 8 shows a trajectory, where three disturbances are applied, the learned controller returns the valve to the desired angle in a robust manner as expected. Finally, we apply the learned controller to follow an arbitrary step trajectory. The resulting trajectory as well as the voltage applied by the controller is shown in Figure 9.

5.2 Real Throttle Valve Results

In this section, the RL controller is learned on a real throttle valve system. The performance of the RL controller is tested on different control tasks.

For learning on the real system, we use the same setting as described for simulation in the previous section. Again, an RBF control structure using 40 basis functions is employed, data is sampled at a rate of 200Hz. Here, we directly handle the case of multiple start and goal states on the real system. As in simulation, the controller is optimized with respect to 10 combinations of start states and goal states. In each episode, additional dynamics data is sampled by application of the controller for 1.2 seconds. In this 1.2s timeslot, a step trajectory consisting of the 10 start/goal combinations is employed as desired trajectory. The information gain criterion significantly reduces the overall size of the dynamics data set, e.g. the 1200 dynamics samples gathered after 4 episodes are reduced to a training set of only 300 elements.

Figure 10 shows the cost of controller application after each episode. The learning was stopped after 4 episodes, since the controller already reached near optimal performance. The controller learned after 4 episodes is evaluated on various desired trajectories. In Figure 11, the performance of two of the 10 learned trajectories is illustrated. Figure 12 shows the application of the learned controller on a more complex trajectory with arbitrary steps. Further, we used a

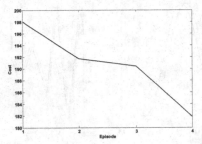

Fig. 10. In the learning process a rollout is performed after each learning episode. The figure shows the accumulated cost (see Equation (3)) for each episode rollout for the asymmetric saturating cost function and $d_1 = 2, d_2 = 0.5$.

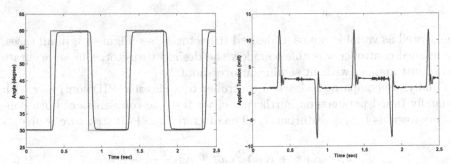

Fig. 11. The learned controller (blue) is applied on the physical throttle valve system with a step trajectory (red) as desired trajectory. While the left figure shows the valve angle over time, the voltage over time is shown on the right. As can be seen, the learned controller performs well and is able to follow the desired trajecotry in a robust manner.

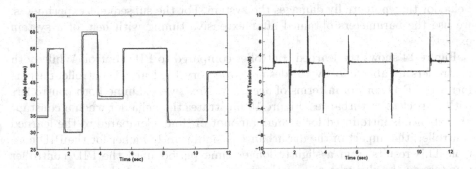

Fig. 12. As a next test case, we used a more complex step trajectory (red). The learned controller (blue) is able to follow the step trajectory, while the accurarcy varies with the goal angle in a range of approximately 1 degree.

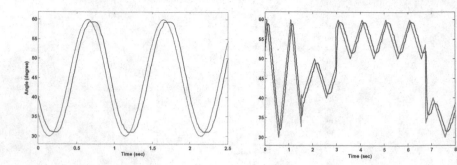

Fig. 13. The figure on the left shows the learned controller (blue) following a desired sine trajectory (red) on the physical valve. The resulting trajectory is shifted by a small amount of time, because of the time required to reach a new desired goal angle. On the right, the controller follows a ramp trajectory of varying steepness and amplitude.

sine as well as variable ramps as desired trajectories, see Figure 13. In all cases, the learned controller was able to follow the desired trajectories in an accurate and robust manner without significant overshoot.

Finally, we compare the learned controller to a classical PID controller with manually tuned parameters. Furthermore, we test the robustness of both controllers towards torque disturbances. The discrete time PID structure is give as

$$u_t = K_P e_t + K_I \sum_{i=0}^{t} e_i dt + K_D \frac{e_t - e_{t-1}}{dt}, \tag{5}$$

where T is the current time index, error $e_t = \alpha_t - g$, $1/dt$ equals the control frequency. The gains K_P, K_I, K_D are free parameters and need to be adjusted. This tuning involves several trade-offs, such as accuracy versus no overshoot. It must be kept in mind that inappropriate parameters lead to unstable control behavior that potentially damages the system. For the subsequent experiments, we use the parameters obtained after extensive tuning with help of a system expert.

Figure 14 shows the learned controller compared to PID control. While both controllers are able to follow the desired trajectory, the learned controller outperforms the PID control in terms of accuracy. Next, we examine both controllers with respect to disturbances. Figure 15 illustrates the behavior when a constant disturbance is introduced for a small amount of time. Compared to the learned controller, the impact of the disturbance is significantly higher for the PID control. This results from a slightly longer time period until the PID controller counteracts the disturbance. Furthermore, the accuracy of the PID control is significantly reduced after the disturbance due to the integration element I (see Equation (5)). More advanced methods of control theory help to improve the results compared to the standard PID control. However, this often increases the number of free parameters that need to be tuned.

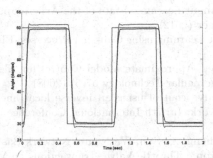

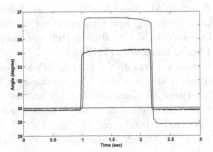

Fig. 14. The figure shows application of the learned controller (blue) and the PID controller (green) on the physical throttle valve. Both controller are able to follow the desired step trajectory (red), the accuracy of the learned controller exceeds the PID performance.

Fig. 15. At timestep 1s, a torque disturbance is introduced to the system until timestep 2.2s. The learned controller (blue) handles the disturbance well, while the PID control (green) shows a stronger impact of the disturbance and deteriorated performance afterwards.

A video was created to illustrate the learning process described in chapter 3, algorithm 1. A controller is learned over the course of 4 episodes and the learning progress is shown through rollouts on the throttle valve system on each episode: www.youtube.com/watch?v=-HpzKsxios4.

6 Conclusion

In this study, we investigate how throttle valve control can be learned from experience, while showing a practical application of probabilistics RL on a real-world problem. A throttle valve is an important industrial device to regulate flows of gas or fluids and has various application, e.g. pressure regulation in combustion engines. As analytical models are hard to obtain due to complex dynamics and unknown nonlinearities, model-based position control of the throttle valve is a challenging problem. In this paper, we modify the probabilistic inference for control algorithm (Pilco) to match the requirements of throttle valve control, such as no-overshoot restriction. We evaluated the approach in simulation, as well as on a real throttle valve system. The results show that policy search RL is able to learn a consistent control policy for complex, real-world systems.

References

1. Sutton, R.S., Barto, A.G.: Reinforcement Learning: An Introduction (Adaptive Computation and Machine Learning). The MIT Press (1998)
2. Deisenroth, P.M., Rasmussen, C.E.: PILCO: A Model-Based and Data-Efficient Approach to Policy Search. In: ICML, pp. 465–472 (2011)

3. Wiering, M., van Otterlo, M.: Reinforcement Learning: State-of-the-Art. Adaptation, Learning, and Optimization. Springer (2012)
4. Deisenroth, M.P.: Efficient Reinforcement Learning using Gaussian Processes. PhD Thesis, Karlsruhe (2010)
5. Yuan, X., Wang, Y., Wu, L.: SVM-Based Approximate Model Control for Electronic Throttle Valve. Transactions on Vehicular Technology 57(5) (2008)
6. Nentwig, M., Mercorelli, P.: Throttle valve control using an inverse local linear model tree based on a fuzzy neural network. In: 7th International Conference on Cybernetic Intelligent Systems (2008)
7. Yuan, X., Wang, Y., Lianghong, W., Xizheng, X., Sun, W.: Neural Network Based Self-Learning Control Strategy for Electronic Throttle Valve. Transactions on Vehicular Technology 59(8) (2010)
8. Seeger, M., Williams, C.K.I., Lawrence, N.D.: Fast Forward Selection to Speed Up Sparse Gaussian Process Regression. In: 9th International Workshop on Artificial Intelligence and Statistics (2003)
9. Leontaritis, I.J., Billings, S.A.: Input-output Parametric Models for Nonlinear Systems. International Journal of Control 41, 303–344 (1985)
10. Griffiths, P.G.: Embedded Software Control Design for an Electronic Throttle Body. Master's Thesis, Berkeley, California (2000)
11. Deisenroth, M.P., Fox, D.: Multiple-Target Reinforcement Learning with a Single Policy. In: ICML Workshop on Planning and Acting with Uncertain Models (2011)
12. Nakamura, H., Masashi, M.: Thottle valve positioning control apparatus. United States Patent 5,852,996 (1998)
13. Al-samarraie, S.A., Abbas, Y.K.: Design of Electronic Throttle Valve Position Control System using Nonlinear PID Controller. International Journal of Computer Applications 59, 27–34 (2012)
14. Wang, H., Yuan, X., Wang, Y., Yang, Y.: Harmony search algorithm-based fuzzy-PID controller for electronic throttle valve. Neural Computing and Applications 22, 329–336 (2013)
15. Deisenroth, M.P., Rasmussen, C.E., Fox, D.: Learning to Control a Low-Cost Manipulator using Data-Efficient Reinforcement Learning. RSS (2011)
16. Fisher Controls International LLC: Control Valve Handbook, 4th edn. (2005)
17. Ting, J., D'Souza, A., Schaal, S.: A Bayesian Approach to Nonlinear Parameter Identification for Rigid-Body Dynamics. Neural Networks (2009)
18. Garcia, C.: Comparison of Friction Models Applied to a Control Valve. Control Eng. Pract. 16(10), 1231–1243 (2008)

Model-Selection for Non-parametric Function Approximation in Continuous Control Problems: A Case Study in a Smart Energy System

Daniel Urieli and Peter Stone

Dept. of Computer Science,
The University of Texas at Austin,
Austin, TX, 78712 USA
{urieli,pstone}@cs.utexas.edu

Abstract. This paper investigates the application of value-function-based rein-forcement learning to a smart energy control system, specifically the task of con-trolling an HVAC system to minimize energy while satisfying residents' comfort requirements. In theory, value-function-based reinforcement learning methods can solve control problems such as this one optimally. However, since choos-ing an appropriate parametric representation of the value function turns out to be difficult, we develop an alternative method, which results in a practical algorithm for value function approximation in continuous state-spaces. To avoid the need to carefully design a parametric representation for the value function, we use a smooth non-parametric function approximator, specifically Locally Weighted Linear Regression (LWR). LWR is used within Fitted Value Iteration (FVI), which has met with several practical successes. However, for efficiency reasons, LWR is used with a limited sample-size, which leads to poor performance without careful tuning of LWR's parameters. We therefore develop an efficient meta-learning pro-cedure that performs online model-selection and tunes LWR's parameters based on the Bellman error. Our algorithm is fully implemented and tested in a realistic simulation of the HVAC control domain, and results in significant energy savings.

1 Introduction

This paper is motivated by a real-world *discrete-time continuous control problem* in which the state space is *continuous* and the action space is *discrete*. Specifically, we focus on the task of controlling an HVAC system's thermostat[1] in a house with 'heat', 'cool', 'auxiliary-heat' or 'off' actions, with the goal of reducing yearly energy con-sumption while satisfying temperature comfort requirements for the occupants. Such discrete-time continuous control problems commonly arise when a digital controller controls a physical system, and when the possible control actions constitute either a fi-nite set, or a low dimensional space that can be discretized without losing much control capability. Other examples of this class of problems are robot control, autonomous he-licopter control, and autonomous car control. When the controlled system's dynamics is unknown in advance, *model-based Reinforcement Learning* can be used to efficiently

---

[1] HVAC: Heating, Ventilation, and Air-conditioning.

H. Blockeel et al. (Eds.): ECML PKDD 2013, Part I, LNAI 8188, pp. 65–80, 2013.
© Springer-Verlag Berlin Heidelberg 2013

learn the dynamics first, and then solve the control problem, possibly by computing or approximating a value function[13].

When using such a method to learn fine-grained control actions, one of the most crucial choices is how to represent the value function. One reason that this choice is so crucial is that the cycle time between consecutive control actions is typically short, compared to the time-range, or the *horizon*, over which the overall system behavior is optimized. Therefore, a single action often has a relatively minor effect on both the state of the system and on the immediate cost/reward, so that the overall performance is a sum of large number of minor contributions. Consequently, the value of a state that results from a *suboptimal* action is typically close to the value of the state that results from an *optimal* action. To induce the optimal policy, a value function approximator must be able to capture these fine differences between state values. Note that taking a sub-optimal action may not seem like a problem when action effects are minor. However, since the problem's horizon can be orders of magnitude longer than the length of an action, the number of actions taken within the horizon is typically large, and repeatedly choosing suboptimal actions can accumulate to large losses.

Value function approximation is an active area of research: it is often unclear how to approximate the value function well enough so as to distinguish an optimal action from a suboptimal action. The three most common methods to approximate a value-function are lookup-tables, parametric methods, and non-parametric methods [16]. Lookup tables often suffer from Bellman's curse of dimensionality at the resolution levels that are required for continuous control problems. Parametric methods are typically computationally efficient, but assume that the value function takes some global, parametric form. Non-parametric methods make much weaker assumptions about the value-function's form, and therefore can, in principle, approximate any function. However, they typically require more data and computation than parametric methods.

One way to avoid the difficulties in approximating a value function in continuous spaces, is to use *direct policy search* methods, which directly optimize the parameters of some parametrized policy. Policy search methods have recently achieved several notable successes, e.g. [13,9,3], and have been gaining increased popularity for real-world control problems, perhaps due to the difficulties in approximating the value function in continuous spaces. However, if we could address the challenge of approximating the optimal value function well enough, we could gain some of the advantages of value-function-based methods over direct policy search methods, for instance aiming for global rather than local optimum, and requiring less interactions with the real-world due to bootstraping.

To address our HVAC control problem, we develop a general, practical, algorithm for approximating the value function in continuous state spaces. To avoid the need to carefully design a parametric representation for the value function, we use a smooth non-parametric function approximator, specifically Locally Weighted Linear Regression (LWR) (e.g. [2]). To compute the value function we use LWR within Fitted Value Iteration (FVI), an algorithm that has proven convergence properties and often performs well in practice [6,12]. However, being limited by a small sample size due to a run-time efficiency requirement on the system, we must tune LWR's parameters carefully, otherwise the system performs poorly. We therefore develop an efficient meta-learning procedure that performs online model-selection and tunes LWR's parameters. The model selection procedure is based on two main ideas, substantiated empirically through a large number of simulations. The first idea is that minimizing the empirical L_1 or L_∞ Bellman error of the approximate value function is correlated with optimizing performance on our task. It is shown that the same statement is not true for the L_2 Bellman error. The second idea is that minimizing the Bellman error by tuning LWR's parameters can be done efficiently. We note that while the Bellman error was used as a criterion for optimization by algorithms implementing generalized policy iteration using a *fixed* representation for the value function, and for tuning and generating basis functions in linear architectures, to the best of our knowledge it has not been used as an optimization criterion for tuning a non-parametric representation (see Sec. 6).

We apply our algorithm to the realistic control task of controlling a thermostat to optimize energy consumption while satisfying comfort requirements in a realistically simulated home. We build a complete Reinforcement Learning agent that uses our algorithm and show that (1) our agent outperforms the thermostat strategy that is deployed in practice, (2) our function approximation scheme leads to better performance than when using popular methods of value function discretization, linear function approximation with reasonable features, and non-parametric function approximation using equivalent computation with a much denser sample and without model-selection; and (3) our online model selection leads to performance that is close to that of an empirical upperbound achieved using a state-of-the-art optimization method (CMA-ES [7]) combined with a clairvoyant model evaluator that returns the actual future performance of a model. The result is an adaptive value-function approximation algorithm for continuous statespaces, which uses a non-parametric representation to minimize the assumptions about representation, and tunes it online to the specific environment in which it is deployed.

2 Preliminaries

2.1 Reinforcement Learning

In this paper we focus on solving control problems through Reinforcement Learning (RL) [19]. Reinforcement learning problems are often modeled as Markov Decision Processes (MDPs). An (episodic) Markov Decision Process (MDP) [18] is a tuple (S, A, P, R, T), where S is the set of states; A is a set of actions; $P : S \times A \times S \to [0, 1]$ is a state transition probability function where $P(s, a, s')$ denotes the probability of transitioning to state s' when taking action a from state s; $R : S \times A \to \mathbb{R}$ is a reward function; and $T \in S$ is a set of terminal states, where entering one of which terminates an *episode*. In the context of MDPs, the goal of RL is to learn an optimal *policy*, when the *model* (namely P and/or R) is initially unknown. A *policy* is a mapping $\pi : S \to A$ from states to actions. A policy π induces a *value* for each state $s \in S$, denoted as $V^\pi(s)$, defined as the expected sum of rewards obtained by the agent when starting in state s and following policy π: $V^\pi(s) = E\left[\sum_{t=0}^{N} R(s_t, a_t)|s_0 = s, s_N \in T, \pi\right]$. $V^\pi(s) : S \to \mathbb{R}$ is called a *value function*. For a given MDP, there exists an *optimal* policy π^* such that $V^{\pi^*}(s) \geq V^\pi(s)$ for every s.

While $V^{\pi^*}(s)$ is induced by the policy π^*, it also *induces* π^*. It can be shown that $\pi^*(s) = argmax_{a \in A} \sum_{s' \in S} R(s, a) + P(s, a, s') \cdot V^{\pi^*}(s')$. Therefore, given $V^{\pi^*}(s)$

(and R, P), an agent can act optimally using a one-step look-ahead from any given state. This is the premise of *value function based RL*. For a finite S, there are algorithms that provably find $V^{\pi^*}(s)$ and therefore the optimal policy. When S is infinite, in general we can only compute an approximation $\hat{V}^{\pi^*}(s)$ of the optimal value function, and the best methodology to do so is still an open research problem. As RL does not need to know the system dynamics (model) in advance, it is an appropriate approach to control problems when the system dynamics are either unknown or partially known, and when a system is controlled in an uncertain environment to which it needs to adapt.

2.2 The Challenge of Function Approximation

The choice of function approximator can be crucial to determining an RL algorithm's performance in control problems of the type we consider. We start by defining a sufficient condition for a function approximator to induce the optimal policy. Denote $E_{[s'|sa]}[V^{\pi^*}(s')] := \sum_{s' \in S} P(s, a, s') \cdot V^{\pi^*}(s')$. In a given state s, let ϵ_s be the smallest absolute difference between the expected values of states resulting from two different actions taken from s, where one action is optimal and the other is sub-optimal:

$$\epsilon_s := \min_{\substack{a^*, a \in A \\ a^* = \pi^*(s) \text{ is optimal} \\ a \text{ is sub-optimal}}} \{|E_{[s'|sa^*]}[V^{\pi^*}(s')] - E_{[s'|sa]}[V^{\pi^*}(s')]|\} \tag{1}$$

Suppose that the system is in state s_0 and an action needs to be chosen.[2] In general, if the function approximator is able to approximate $V^{\pi^*}(s)$ to within $\frac{\epsilon_{s_0}}{2}$, meaning

$$\max_{s \in S} |\hat{V}^{\pi^*}(s) - V^{\pi^*}(s)| < \frac{\epsilon_{s_0}}{2} \tag{2}$$

then greedy action selection based on $\hat{V}^{\pi^*}(s)$ is guaranteed to induce the optimal action from s_0, since:[3] $E_{[s'|sa^*]}[\hat{V}^{\pi^*}(s')] - E_{[s'|sa]}[\hat{V}^{\pi^*}(s')] > E_{[s'|sa^*]}[V^{\pi^*}(s') - \frac{\epsilon_{s_0}}{2}] - E_{[s'|sa]}[V^{\pi^*}(s') + \frac{\epsilon_{s_0}}{2}] \geq \epsilon_{s_0} - \frac{\epsilon_{s_0}}{2} - \frac{\epsilon_{s_0}}{2} = 0$. When condition 2 holds for every state $s_0 \in S$, the function approximator induces the optimal action in every state, and therefore the optimal policy. When it does not hold in every state, the function approximator may not induce the optimal policy.

Clearly, the smaller ϵ_s is, the harder it is to achieve the desired $\frac{\epsilon_s}{2}$ function approximation accuracy. Unfortunately, for the type of problems this paper is concerned with, namely real-world, discrete-time continuous control problems, ϵ_s can in fact be small for many states. This happens when the following two properties hold (for $S := R^n$):

– *Actions have "small" effect*: there exists some (relatively small) δ such that for every $s', s'' \in S$ that can result from taking actions at some state $s \in S$, it holds that $||s' - s''|| < \delta$. This can happen, for instance, when the cycle time between subsequent control actions is short compared to the problem's horizon.

---

[2] Note that we assume the existence of a sub-optimal action, since otherwise there is no decision of any consequence to be made, as all actions are optimal. Therefore $\epsilon_s > 0$.

[3] For simplicity of presentation, we neglect the 1-step reward, which could be incorporated into Equation (1) in a straightforward way.

– *The optimal value function is Lipschitz continuous*: there exists some $K > 0$ such that $\forall s_1, s_2 \in S : ||V(s_1) - V(s_2)|| < K||s_1 - s_2||$. This holds, for instance, when the reward and the transition functions are Lipschitz continuous.

Combining the two conditions, we get that $\forall s \in S : \epsilon_s < ||E[V(s')] - E[V(s'')]|| \leq E[|||V(s') - V(s'')|||] < E[K||s' - s''|||] < K\delta$, so that the smaller the action effect δ, the smaller ϵ_s is for any state s, and so the harder it is to achieve the desired approximation accuracy. In the results section we demonstrate how this results in repeated suboptimal actions, degrading performance on our thermostat control task.

3   Approximating the Value Function

RL algorithms that are based on value function approximation can roughly be divided into *model-free* algorithms, which are usually more computationally efficient, and *model-based* algorithms, which are usually more data efficient. As we are motivated by real-world problems, where gathering experience is often an expensive operation, we focus here on model-based RL. In model-based RL, an RL agent first explores the environment and learns an approximate model of it (namely P and R). Using this model, the agent simulates experiences and computes $\hat{V}^{\pi^*}(s)$. Since in this paper we focus on value-function approximation, we assume that an approximate model is either given, or was already learned by the agent. For instance, in the results section, our RL agent first learns an approximate model, and then uses it to compute $\hat{V}^{\pi^*}(s)$.

3.1   Approximate Dynamic Programming

For computing $\hat{V}^{\pi^*}(s)$, we start by using sampling-based Fitted Value Iteration (FVI) [6] (a detailed overview of its roots can be found at [12]). FVI is an approximate dynamic programming algorithm that computes $\hat{V}^{\pi^*}(s)$ by repeatedly scanning a finite sample of states $S_{FVI} := s^{(1)}, s^{(2)}, \ldots, s^{(m)}$, applying the following two steps:

$$\forall i \in 1, \ldots, m \tag{3}$$

$$y^{(i)} := max_a \left(R(s^{(i)}, a) + \gamma E_{[s'|s^{(i)}a]}[\hat{V}^{\pi^*}(s')] \right)$$

$$\hat{V}^{\pi^*}(s) := SL \left(\{ \langle s^{(i)}, y^{(i)} \rangle | i \in 1, \ldots, m \} \right) \tag{4}$$

where $\hat{V}^{\pi^*}(s^{(i)})$ is initialized arbitrarily; the expectation over the resulting state is approximated by Monte-Carlo sampling; and after each update scan, a supervised learning algorithm SL is used as a function approximator that approximates the value function over the complete state-space, based on the "labeled" examples $\langle s^{(i)}, y^{(i)} \rangle$. While FVI is not guaranteed to converge to $V^{\pi^*}(s)$, it often performs well in practice, and is theoretically well-behaved [12]. In addition, FVI is an *off-policy* algorithm, which means it can approximate an optimal policy before ever executing it.

3.2 Function Approximator

Inside FVI, the function approximator we use as SL is Locally Weighted Linear Regression (LWR). LWR is a non-parametric, smooth function approximator that uses only minimal representation assumptions, and that has been used successfully to model complex real-world dynamics [13]. Given a set of m labeled examples $(x^{(i)}, y^{(i)})$ and a query point x for which we want to predict a value, our version of LWR does the following:

1. $w^{(i)} := exp\left(-\frac{(x^{(i)}-x)^2}{2\tau}\right)$ for $i = 1, \ldots, m$ [compute a weight for each training example]
2. Fit θ that minimizes $\sum_{i=1}^{m} w^{(i)}(y^{(i)} - \theta^T x^{(i)})^2$ [use weights for weighted regression]
3. Output $\theta^T x$

Here τ is a "bandwidth" parameter that determines how quickly the weights decay. Small weights are typically truncated for computational efficiency reasons. Since the weights $w^{(i)}$ depend on the specific query point, LWR builds a local model around the query for every prediction it makes. Note that when used with FVI, $x^{(i)} := s^{(i)}$, where $s^{(i)} \in S_{FVI}$. In general, LWR results in smoother function approximation than simpler non-parametric methods such as nearest-neighbors, which is a desirable property for continuous control tasks. However, in general, LWR can extrapolate, and this can prevent FVI from converging [6], so to ensure convergence we trim LWR's predicted value to be within the range defined by its neighbors values.

4 Efficient Model Selection

Like most learning algorithms, LWR usually needs to be tuned to work well for a particular problem. LWR is typically tuned by adjusting the values of the bandwidth parameter τ and of distance-metric-related parameters $c_1, \ldots, c_n$, where c_i is a scalar that scales s_i, the i'th state attribute in a state s. Adjusting the values of $c_1, \ldots, c_n$ effectively changes the distance metric based on the relative importance of state attributes. While it is possible to make c_i a general function $c_i : S \to \mathbb{R}$ rather than a scalar, we take an approach of *global tuning* [2], in which c_i is a scalar. This keeps the number of LWR parameters at a total of $n + 1$, so that tuning is more computationally efficient and less susceptible to overfitting. A given set of parameter values is said to define a *model* to be used by the LWR function approximator, and the process of tuning these parameters is a form of *model selection*.[4] The goal of our model-selection process is to find a set of parameters, that when used by LWR inside FVI, results in a function approximation that is close to the optimal value function.

When model-selection is done in a supervised learning setup, each candidate model is typically evaluated using *cross-validation*. In our setup, using cross-validation by holding out subsets of S_{FVI} is problematic since (1) we don't have the *actual* values of states $s \in S_{FVI}$ as we have in supervised learning, but only the values that FVI *converges* to, and (2) as S_{FVI} is typically sparse (to keep the run-time of FVI acceptable), having a good cross-validation accuracy on S_{FVI} does not necessarily imply

---

[4] Note that LWR's model is different than, and should not be confused with, the MDP model.

good prediction accuracy over the rest of the state space $S \setminus S_{FVI}$. Therefore, we seek an alternative model evaluation measure. The ideal way of evaluating a model is by measuring the agent's performance when acting based on a value function that uses this model. However, evaluating the agent in the real-world with different models is often prohibitively time-consuming and expensive. Therefore, we only use it as an empirical upper-bound in our simulated domain.

Instead, we use a theoretically-founded model evaluation measure that is efficiently computed in practice: the value-function's *empirical max Bellman error*. In a given state s, the absolute *Bellman (optimality) error* of a function $\hat{V} : S \to \mathbb{R}$ is defined as:

$$BE_{\hat{V}}(s) := |\hat{V}(s) - max_a(R(s,a) + \gamma E_{[s'|sa]}[\hat{V}(s')])|$$

Next, for a function $\hat{V} : S \to \mathbb{R}$ the following holds:

$$\hat{V} \equiv V^{\pi^*} \iff \forall s \in S : BE_{\hat{V}(s)} = 0 \tag{5}$$

Furthermore, [21] (resp. [12]) establishes that for a full (resp. sample) Bellman backup:

$$|V^{\pi^*}(s) - \hat{V}(s)| \propto BE_{\hat{V}}(s) \tag{6}$$

Equations (5), (6) imply that ideally the Bellman error would be 0, or as close as possible to 0, for every state. *Note that while the convergence of FVI means that $\forall s \in S_{FVI}$: $BE_{\hat{V}(s)} \approx 0$, the Bellman error might still be large for states $s \in S \setminus S_{FVI}$.* In order to address that, we create a random sample of test states $\mathcal{T} := \{t^{(1)}, ..., t^{(m')}\}$, and define a vector of Bellman errors (overloading notation):

$$BE_{\hat{V}}(\mathcal{T}) := (BE_{\hat{V}}(t^{(1)}), \dots, BE_{\hat{V}} t^{(m')}) \tag{7}$$

Motivated by Equations (5) (6), we use the max Bellman error $\|BE_{\hat{V}}(\mathcal{T})\|_\infty$ as a model-evaluation measure when tuning LWR's parameters. This model evaluation measure is computed solely based on a value function computed by FVI, without needing more data or interactions with the environment. Our model selection process then becomes a continuous optimization problem of finding a set of LWR parameters $\psi \in \mathbb{R}^{n+1}$ that minimizes $\|BE_{\hat{V}}(\mathcal{T})\|_\infty$ where $\hat{V}$ is the resulting value function after running FVI with LWR using ψ. Ideally, the max Bellman error should be computed over all states in the state space, however since the state-space in non-enumerable, we take a practical approach and set $\mathcal{T}$ to be a (as dense as computationally possible) random sample of states. In the results section we show that (a) minimizing $\|BE_{\hat{V}}(\mathcal{T})\|_\infty$ is correlated with good actual performance, and that (b) minimizing it can be done efficiently.

Putting all of these components together, the main general contribution of the paper beyond the domain-specific results is the MSNP algorithm (Model Selection for Non-Parametric function approximation). As summarized in Algorithm 1, it executes the following steps. The algorithm's input is a learned MDP model, and an iterative continuous optimization algorithm, that finds a minimum of a function $F : \mathbb{R}^{n+1} \to \mathbb{R}$. Since we generally do not have the gradient of the Bellman error as a function of the LWR parameter set, we use gradient free optimization algorithms in our experiments. An interesting future extension would be comparing them with subgradient methods. MSNP starts by generating a sample $\mathcal{S}$ of states for FVI (step 1) and a sample of test states $\mathcal{T}$ over which it would compute the max Bellman Error (step 2). It then initializes a vector v of Bellman Errors (step 3) and a vector of LWR parameters ψ (step 5). In the main loop, it repeatedly runs the following steps until the max Bellman Error converges: run FVI (step 8), compute the resulting max Bellman Error (step 9-11), send it back to the optimization algorithm as the evaluation of the current parameter set (step 15), and get from the optimization algorithm a new set of LWR parameters to evaluate (steps 16-17).

Algorithm 1. MSNP(MDP-Model, OptimizationAlgorithm)

1: $\mathcal{S} \leftarrow \{s^{(1)}, s^{(2)}, \ldots, s^{(m)}\}$ the set of points used by *FittedValueIteration*
2: $\mathcal{T} \leftarrow \{t^{(1)}, s^{(2)}, \ldots, t^{(m')}\}$ test points sampled within the boundaries of $\mathcal{S}$
3: $v \leftarrow \{\infty, \ldots, \infty\} \in R^{m'}$
4: *PreviousError* $\leftarrow \infty$
5: $\psi \leftarrow$ *OptimizationAlgorithm.initializeParameters()*
6: *done* $\leftarrow$ *False*
7: **while** *not done* **do**
8:     $\hat{V} \leftarrow$ *FittedValueIteration*(MDP-Model, $\mathcal{S}, \psi$) // approximate value function
9:     **for** $i = 1 \rightarrow m'$ **do**
10:       $v_i \leftarrow BE_{\hat{V}}(t^{(i)})$ // the Bellman error in $t^{(i)}$
11:     **end for**
12:     *MaxBellmanError* $\leftarrow ||v||_\infty$
13:     **if** $|MaxBellmanError - PreviousError| < \epsilon$ **then**
14:       *done* $\leftarrow$ *True*
15:     **else**
16:       *OptimizationAlgorithm.observe(MaxBellmanError)*
17:       $\Delta\psi \leftarrow$ *OptimizationAlgorithm.step()*
18:       $\psi \leftarrow \psi + \Delta\psi$
19:       *PreviousError* $\leftarrow$ *MaxBellmanError*
20:     **end if**
21: **end while**

MSNP is an efficient algorithm for approximating the value function in continuous state spaces, that uses our model-selection procedure for tuning a Fitted Value Iteration with Locally Weighted Linear Regression, and which has the following properties:

1. *General*: It uses only minimal assumptions about a value function's representation, by using an non-parametric function approximator (LWR).
2. *Practical*: It tunes the representation of the LWR function approximator without the need to evaluate the agent in the environment but rather based on an internal property of the value function, namely the Max Bellman Error. In that sense, it is data efficient.
3. *Adaptive*: It tunes online the function approximator's representation to the environment it is deployed in.
4. *Effectively use computation*: Its model selection procedure achieves better performance than when using the same amount of computation on a larger S_{FVI} sample without any model-selection.

5 Results

In this section we investigate the application of MSNP to the task of controlling an HVAC's thermostat in a realistically simulated home.

5.1 Experimental Setup

Our experiments are run using GridLAB-D[5], an open-source smart-grid simulator that was developed for the U.S. Dept. of Energy. It models a residential home, including heat gains and losses and the effects of thermal mass, as a function of weather (temperature and solar radiation), occupant behavior (thermostat settings and internal heat gains from appliances), and heating/cooling system efficiencies. It uses meteorological data collected by the National Renewable Energy Laboratory[6] in cities across the USA. In our experiments, GridLAB-D simulates a residential home with a heat-pump based HVAC system, which is widely used due to its high efficiency.

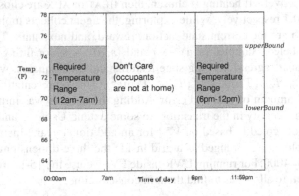

Fig. 1. Temperature Requirements Specification

We assume that occupants are at home between 6pm and 7am of the next day, and that the house is empty between 7am and 6pm (referred to as the *don't-care* period). Our goal is to (1) minimize the energy consumed by the HVAC system, while (2) keeping a desired temperature range of $69\text{-}75°(F)$ when the occupants are at home, and being indifferent to temperature otherwise (Figure 1). Due to uncertainty in future weather and in the house's environment, simple strategies fail to satisfy at least one of the requirements. For instance, turning the system off at 7am and turning it back on at a fixed time, such as 6pm, or even earlier, can fail to satisfy *both* requirements in cold winter days, since the temperature gets significantly out of range at 6pm and restoring it might take several hours, during which comfort is violated. Moreover, while doing so, an energy expensive auxiliary heater is used (since the heat-pump becomes inefficient), and the resulting energy is higher than when just keeping the temperature between $69\text{-}75°(F)$ throughout the day. We model the problem as an episodic MDP[7], as follows:

---

[5] http://www.gridlabd.org
[6] http://www.nrel.gov
[7] An action is taken every 2 minutes, as the simulator models a realistic lockout of the system.

- **S**: $\{\langle T_{in}, T_{out}, Time \rangle |$ T_{in} and T_{out} are the indoor and outdoor air temperatures (in Fahrenheit), and $Time$ is a 24-hour clock time (in minutes).$\}$
- **A**: $\{$COOL, OFF, HEAT, AUX$\}$. Namely, there are four possible actions for cooling, off, (heat-pump-)heating and auxiliary heating, respectively.
- **P**: computed by the GridLAB-D simulator and is initially unknown to the agent.
- **R**: $-$(the energy consumed by the last action) $-C_{6pm}$. Here, C_{6pm} is a quadratic cost applied when missing the temperature spec at 6pm.
- **T**: $\{s \in S | s.time == 23{:}59\text{pm}\}$

For testing the MSNP algorithm, we build a full RL agent that controls the thermostat. Since our focus is on value function approximation, we leave the problem of sample-efficient exploration and model-learning outside the scope of this work. Instead, to cover diverse weather conditions, the agent explores during don't-care periods of one simulated year, where the OFF action is chosen with probability of $(1 - \frac{currentTime - 7am}{6pm - 7am})$. Otherwise, cooling or heating is chosen, depending on whether the indoor temperature high or low, respectively. If heating is chosen, then HEAT or AUX are chosen with probabilities 0.9 and 0.1 respectively. While exploring, the agent collects tuples of the form $\langle s, a, r, s' \rangle$, which are the current state, action, reward, and next state. The agent uses these tuples as labeled examples $\langle s, a \rangle \rightarrow r$ and $\langle s, a \rangle \rightarrow s'$ for fitting the functions R and P with linear regression using state features and their squares, where s is represented as $\langle 1, T_{in}, T_{out}, Time, T_{in}{}^2, T_{out}{}^2, Time^2 \rangle$. This representation was chosen based on a small amount of trial and error. Adding the squares was intuitively aimed at addressing non-linearity in the transition, to some extent. Using P and R, the agent runs MSNP and acts greedily based on $\hat{V}^{\pi^*}$ for an additional year.[8] Inside MSNP, FVI uses a state sample S_{FVI} arranged as a grid inside the three-dimensional state space, of size 20x10x20=4000. For running LWR inside FVI we use the 15-nearest neighbors, and the grid structure allows us to find them in constant time.

5.2   Sensitivity to Errors in Function Approximation

Figure 2 demonstrates the difficulty of value function approximation in continuous domains with short actions, using the thermostat control task. The x-axis is the 24-hour time of day and the y-axis is the indoor temperature controlled by the actions of the agent, who acts greedily based on an approximate value function. The agent turns the system off during the don't-care period, letting the temperature rise, and eventually cools in advance to return the temperature back to range by 6pm. However before starting to cool, there are several heating actions that are physically wrong, chosen due to small approximation errors in the value function, in this case due to using LWR without tuning its parameters. Each suboptimal 2-minute action increases the daily consumption by only about 0.1%, but repeatedly taking them can increase consumption by 10%

---

[8] In practice, Gridlab-D only has one year of "average" weather data. We therefore used 9 of every 10 days during the training year, and the remaining days during the testing year so as to have separate training and testing data. Our reported results reflect the average of repeating this experiment 10 times with each different possible subset of "held-out" days.

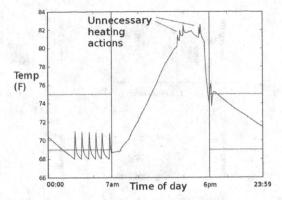

Fig. 2. Suboptimal policy due to func. approximation errors

and more. These suboptimalities and more severe ones happened when using discretized representations as well as linear value-function representations with reasonable features.

5.3 Using the Max Bellman Error for Model-Selection

Next, we investigate using the Bellman error as a criterion for model-selection of our LWR function approximator. The plots in Figure 3 show the correlation between empirical Bellman error in the approximate value function $\hat{V}^{\pi^*}$ and the agent's performance when acting based on $\hat{V}^{\pi^*}$. The plots summarize 10,000 experiments, each represented as a point. Each experiment tests a set of $n+1$ LWR parameters, which defines a *model* used by LWR, as was discussed in Section 4. In the thermostat domain there are three state attributes, and therefore four model parameters. We sweep the parameter space by setting each parameter to one of 10 possible values, and this gives $10 \times 10 \times 10 \times 10 = 10,000$ possible parameter sets. The empirical Bellman errors were measured as the L_1, L_2 or L_∞ norms of the vector of Bellman errors in $\hat{V}^{\pi^*}$ over a uniformly random sample of $|\mathcal{T}| = 256,000$ states. It can be seen that when the L_1 and L_∞ errors are smallest, performance is expected to be close to the best possible (lowest energy consumption). The same does not hold for the L_2 error, as minimizing the L_2 error results in consuming 4% more energy than the best result. Note that in general these plots clearly highlight the need for model-parameters tuning, as untuned parameters can consume about 25% more energy than the best possible parameters.

5.4 Efficiently Optimizing the Bellman Error

The previous section tested the first of two steps for creating an efficient model-selection algorithm. We saw that model-selection, or representation tuning, of the LWR function approximator can be done without the need to evaluate an agent in the environment, but rather based on an internal property of the value function, namely the L_∞ or L_1 Bellman errors, so in that sense it is data efficient. Next, we show that tuning the model parameters based on the max Bellman error can be done computationally efficiently. Note

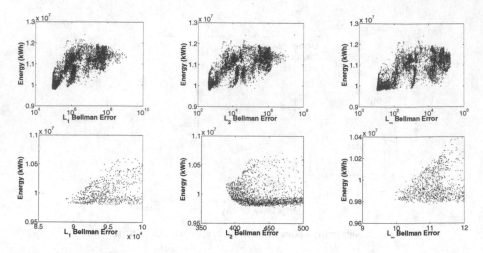

Fig. 3. Bellman errors (x-axes) vs. actual performance (y-axes, lower is better). Top row: full plots. Bottom row: zoom into the bottom-left corner (best performance) of each top row plot.

that once using the Bellman error for model evaluation, our model-selection problem becomes minimizing an objective function that maps an LWR parameter set to the max Bellman error in $\hat{V}^{\pi^*}$ computed using this parameter set by FVI with LWR. We compare several efficient local-search derivative-free algorithms for finding the minimum of a continuous function.[9] The algorithms we compare are Powell's method, Nelder–Mead method, also known as Amoeba, and a coordinate-descent algorithm in which we hold all parameters fixed and optimize one parameter at a time using Brent's method, using implementations from [17]. Results are shown in Figure 4. The horizontal gray line in the figure is the best max Bellman error that was achieved when using an offline, state-of-the-art, parallel optimization algorithm CMA-ES [7], when running it with 10,000 function evaluations (100 generations with a population-size of 100). It can be seen that after about 30 function evaluations all three methods get close to CMA-ES's value, and that the Brent's method-based coordinate-descent reaches there after about 15 function evaluations. Our FVI implementation converges in less than 2-minutes on a standard desktop machine, so that 15-30 function evaluations takes 30-60 minutes.

How robust is running local optimization for finding a global minimum of the max Bellman Error? To try to answer this question, we fixed the parameter values at $c_1 = c_2 = c_3 = 0.5$, $\tau = 0.0005$, and then changed one parameter at a time across its range ($c_i \in [0.05, 1]$, $\tau \in [0.00005, 0.0010]$) measuring the max Bellman error as a function of this parameter, where as usual, the max Bellman error was computed over the value function computed using FVI with LWR. Results are in Figure 5, and show that while the max Bellman error is not a convex function of representation parameters, it still has a relatively large basin of convergence.

---

[9] In general we do not have derivative information for the optimized function.

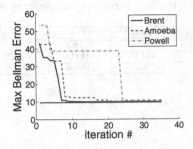

Fig. 4. Comparing different optimization algorithms on the task of finding a parameter set that minimizes the max Bellman error

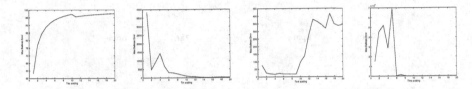

Fig. 5. Bellman Error basin of convergence: Bellman Error (y-axis) as a function of each LWR parameter, holding all the other parameters fixed. The x-axes from left to right: τ, $c_{T_{in}}$, $c_{T_{out}}$, c_{Time}

5.5 Performance of MSNP

Finally, Table 1 demonstrates the advantage of using MSNP, by testing it on our thermostat control problem. In these experiments, we use our RL agent, changing only the way the value function is approximated. The table shows the energy consumed by the HVAC system over one simulated year, in which the agent acts greedily based on each of the different approximate value functions. Simulations were run using real weather files from three different cities in the US. As a reference, the "Default" column shows the results of using a default heat-pump thermostat strategy that is used in real-world deployments, which just keeps the temperature between 69-75° (F) throughout the day. This strategy does not shut down the system during the don't-care period, since doing so without knowing how long in advance to turn the system back on can result in violating either or both requirements (1) and (2), as discussed above. The "Large-Sample" column was generated when using FVI with LWR to approximate the value function, but instead of using the model-selection like MSNP does, the agent spends the same amount of computation on just running FVI with LWR on a larger state sample S_{FVI} of 160x80x160=2048000 states without any model selection, and using default values for the LWR parameters, similar to the values used in Section 5.4: $c_i = 0.5$ for $i = 1, ..., n$ and $\tau = 0.0005$.[10] MSNP was run using a state sample S_{FVI} of 20x10x20 states and used $|\mathcal{T}| = 256,000$ states for computing the Bellman error, so that its

---

[10] The "LargeSample" results are actually slightly better than they should be because they did not use the 9 days of 10 methodology referenced in footnote[8]. It was trained on the full year of data.

Table 1. Performance of an agent using MSNP

City	Default (kWh)	LargeSample (kWh)	MSNP (kWh)	CMA-ES (kWh)	% Energy-Savings
New York City	11084.8	10923.5	**9859.3**	9816.3	**11.0%**
Boston	12277.1	12480.7	**11433.6**	11052.8	**6.9%**
Chicago	15172.5	14778.2	**14186**	13778.4	**6.5%**

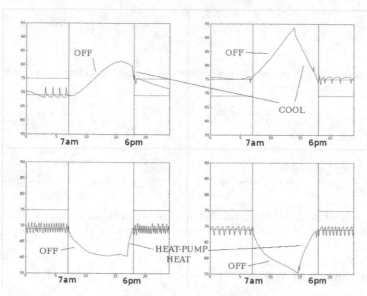

Fig. 6. Our agent controlling the temperature in a house in New York City area, in mild and hot summer days (top-left and top right, respectively), and mild and extreme winter days (bottom-left and bottom-right, respectively)

computation time, dominated by the number of LWR predictions was no larger than that of LargeSample's. The "CMA-ES" column serves as an empirical upper-bound on performance in our simulated domain. It was generated by running the state-of-the-art CMA-ES optimization method to perform model selection on top of FVI with LWR, using (1) 10,000 model evaluations (100 generations, each with a population size of 100), and (2) a clairvoyant model-evaluator that returns the *agent's actual future performance using a given model*, by running a one-year simulation using this model. It can be seen that MSNP performs better than "LargeSample", which demonstrates that online model-selection has an advantage over just increasing the density of the sample size. MSNP's performance is close to that of CMA-ES's, despite the fact that it uses only 40 function evaluations (instead of 10,000) and doesn't have access to the "real" model evaluation measure of the unknown future performance, that CMA-ES has. Note that while the MSNP and CMA-ES agents satisfied the temperature comfort requirements, the LargeSample agent frequently did not satisfy them. In Figure 6 we demonstrate how the RL agent controls the temperature in mild and extreme winter/summer days.

6 Related Work

RL has been applied to realistic control tasks, however recent successes frequently used policy search methods, rather than value-function-based methods(e.g. [13,9,3]). Value function based RL has had success on some robotic tasks [15], but there the assumption was that the value function can be represented as a predetermined set of basis functions, an assumption that does not necessarily hold in the general case. Non-parametric value function approximation methods have been suggested, e.g. in [4]. The idea of using the Bellman error as a criterion for optimization has been used by algorithms implementing generalized policy iteration, e.g. in [10,1]. The Bellman error has also been used for tuning and generating basis functions adaptation in linear function approximation architectures [11,8,14], while here we use it to tune a non-parametric representation. The model selection proposed here is different then the model selection done by [13]. There, the setup was offline, supervised learning for learning the transition function, while ours is an online reinforcement learning setup, for approximating the value function, where there are no labels over the data, but only the values to which FVI converge to, which could be different then the real state values. A paper that is closely related to ours is [5], which designs an abstract model-selection algorithm and proves theoretical guarantees about it. Similarly to here, they consider batch RL, in which a data set D of sampled transitions from the MDP is given, and is used for selecting a candidate value function by minimizing a Bellman error. In their case they abstract the way value function candidates are generated and assume they are independent of D, while here we actually use D to approximate the model and generate candidates using MSNP. Their theoretical guarantees are proved under a slightly different setup, and it would be interesting to explore whether they can be extended to our setup. The problem of thermostat control was addressed in [20], but there the focus was on solving the complete RL problem, including exploration, model learning and planning, and no value function approximation was used, while here the focus is on investigating the application of value-function based RL to the continuous, realistic domain of HVAC thermostat control.

7 Conclusion

This paper presents the application of value-function-based RL to the real-world smart-energy application of controlling an HVAC thermostat to minimize energy consumption while satisfying temperature comfort requirements, along with detailed empirical results and analysis. In addition, the paper introduces MSNP, which is a general, practical algorithm for approximating the value function for continuous control problems, using an efficient model-selection procedure based on the Bellman error.

This paper opens up several interesting directions for future work. For example, it is worth investigating the Bellman error's basin of convergence as a function of the model-parameters. Another interesting direction is exploring the use of subgradient methods for minimizing the Bellman error, and comparing them with the gradient-free methods we used. Finally, an important future direction is to expand MSNP's empirical analysis by including more domains and competing methods, and to evaluate it in higher-dimensional state spaces.

References

1. Antos, A., Szepesvári, C., Munos, R.: Learning near-optimal policies with bellman-residual minimization based fitted policy iteration and a single sample path. Mach. Learn. 71(1), 89–129 (2008)
2. Atkeson, C.G., Moore, A.W., Schaal, S.: Locally weighted learning (1997)
3. Deisenroth, M.P., Rasmussen, C.E.: PILCO: A Model-Based and Data-Efficient Approach to Policy Search. In: Getoor, L., Scheffer, T. (eds.) Proceedings of the 28th International Conference on Machine Learning, Bellevue, WA, USA (June 2011)
4. Engel, Y., Mannor, S., Meir, R.: Reinforcement learning with gaussian processes. In: Proc. of the 22nd International Conference on Machine Learning, pp. 201–208. ACM Press (2005)
5. Farahmand, A.M., Szepesvári, C.: Model selection in reinforcement learning. Mach. Learn. 85(3), 299–332 (2011)
6. Gordon, G.J.: Stable function approximation in dynamic programming. In: Machine Learning: Proceedings of the Twelfth International Conference. Morgan Kaufmann (1995)
7. Hansen, N.: The CMA Evolution Strategy: A Tutorial (January 2009)
8. Keller, P.W., Mannor, S., Precup, D.: Automatic basis function construction for approximate dynamic programming and reinforcement learning. In: Proceedings of the 23rd International Conference on Machine Learning, ICML 2006, pp. 449–456. ACM, New York (2006)
9. Kohl, N., Stone, P.: Machine learning for fast quadrupedal locomotion. In: The Nineteenth National Conference on Artificial Intelligence, pp. 611–616 (July 2004)
10. Lagoudakis, M.G., Parr, R.: Least-squares policy iteration. J. Mach. Learn. Res. 4, 1107–1149 (2003)
11. Menache, I., Mannor, S., Shimkin, N.: Basis function adaptation in temporal difference reinforcement learning. Annals of Operations Research 134, 215–238 (2005)
12. Munos, R., Szepesvári, C.: Finite time bounds for sampling based fitted value iteration. In: ICML, pp. 881–886 (2005)
13. Ng, A.Y., Kim, H.J., Jordan, M.I., Sastry, S.: Autonomous helicopter flight via reinforcement learning. In: Thrun, S., Saul, L., Schölkopf, B. (eds.) Advances in Neural Information Processing Systems 16. MIT Press, Cambridge (2004)
14. Parr, R., Painter-Wakefield, C., Li, L., Littman, M.: Analyzing feature generation for value-function approximation. In: Proceedings of the 24th International Conference on Machine Learning, ICML 2007, pp. 737–744. ACM, New York (2007)
15. Peters, J., Schaal, S.: Natural actor-critic. Neurocomputing 71(79), 1180–1190 (2008)
16. Powell, W.B.: Approximate Dynamic Programming: Solving the Curses of Dimensionality, 2nd edn. Wiley (2011)
17. Press, W.H., Teukolsky, S.A., Vetterling, W.T., Flannery, B.P.: Numerical Recipes, 3rd edn. The Art of Scientific Computing. Cambridge University Press, New York (2007)
18. Puterman, M.L.: Markov Decision Processes: Discrete Stochastic Dynamic Programming, 1st edn. John Wiley & Sons, Inc., New York (1994)
19. Sutton, R.S., Barto, A.G.: Reinforcement Learning: An Introduction. MIT Press, Cambridge (1998)
20. Urieli, D., Stone, P.: A learning agent for heat-pump thermostat control. In: Proceedings of the 12th International Conference on Autonomous Agents and Multiagent Systems (AAMAS) (May 2013)
21. Williams, R.J., Baird III, L.C.: Tight performance bounds on greedy policies based on imperfect value functions. In: Proceedings of the Tenth Yale Workshop on Adaptive and Learning Systems (1994), http://leemon.com/papers/1994wb.pdf

Learning Graph-Based Representations for Continuous Reinforcement Learning Domains

Jan Hendrik Metzen

Robotics Group, University of Bremen,
Robert-Hooke-Straße 5, 28359 Bremen, Germany
jhm@informatik.uni-bremen.de

Abstract. Graph-based domain representations have been used in discrete reinforcement learning domains as basis for, e.g., autonomous skill discovery and
representation learning. These abilities are also highly relevant for learning in
domains which have structured, continuous state spaces as they allow to decompose complex problems into simpler ones and reduce the burden of hand-
engineering features. However, since graphs are inherently discrete structures,
the extension of these approaches to continuous domains is not straight-forward.
We argue that graphs should be seen as discrete, generative models of continuous domains. Based on this intuition, we define the likelihood of a graph for a
given set of observed state transitions and derive a heuristic method entitled FIGE
that allows to learn graph-based representations of continuous domains with large
likelihood. Based on FIGE, we present a new skill discovery approach for continuous domains. Furthermore, we show that the learning of representations can be
considerably improved by using FIGE.

1 Introduction

Reinforcement Learning (RL) allows autonomous agents to learn to improve their performance with experience in an unknown environment. However, typically representations for policies and value functions need to be carefully hand-engineered for the
specific domain and learned knowledge is not efficiently reused in situations when an
agent has to solve several different but related tasks. Representation learning for RL [7]
and hierarchical RL [1] are approaches to alleviate these drawbacks.

Graph-based representations of the domain have been used as basis for both representation learning and hierarchical RL. For instance, Mahadevan and Maggioni [7] have
learned useful internal representations called proto-value functions based on a graph
representation of the domain. Such graphs have also been used to identify bottleneck
states of the environment [8, 10, 13] which are a common basis for skill discovery in
hierarchical RL. While these graph-based approaches have shown promising results in
domains with discrete state and action spaces, extending them to continuous domains is
not straight-forward. This is mainly due to the fact that graphs are intrinsically discrete
structures and thus cannot directly model a continuous environment.

Previous work on graph-based approaches in continuous domains has thus either
discretized the domain, i.e., placed graph nodes at a regular grid over the state space

H. Blockeel et al. (Eds.): ECML PKDD 2013, Part I, LNAI 8188, pp. 81–96, 2013.

[8], or placed graph nodes at a subset of the observed states [7]. While the former suffers from the curse-of-dimensionality, the later allows to exploit situations where the effective dimensionality of the state space is smaller. However, both approaches focus purely on covering the state space as uniformly as possible and neglect the dynamics of the environment. We argue that graph representations should take the dynamics into account since they can be seen as a model of the environment. That is, typical transitions encountered in the domain should be representable by the graph. The hypothesis evaluated in this paper is that a graph, which models the dynamics of its (continuous) environment well, will yield superior results with regard to representation learning and bottleneck identification. We propose a new heuristic called FIGE which allows to learn graph representations that explicitly aim at modeling the environment's dynamics.

The outline of the paper is as follows: In Section 2, we review graph-based RL methods and discuss how graph representations have been generated in these methods. In Section 3, we define the likelihood of a graph for a given 4 of transitions sampled according to the domain's dynamics. Thereupon, we propose the FIGE heuristic for learning graph representations of continuous environments, which is derived from the maximum graph likelihood formulation under simplifying assumptions. Furthermore, we compare FIGE with other graph learning heuristics empirically with regard to the graph likelihood. In Section 4, we propose and evaluate a new graph-based skill discovery method for continuous domains, which is based on the FIGE heuristic. Similarly, in Section 5, we present empirical evidence that Representation Policy Iteration [7] can benefit from using FIGE for graph generation in continuous domains. We summarize the results of this paper and provide an outlook in Section 6.

2 Graph-Based Reinforcement Learning

A Markov decision process (MDP) M can be formalized as a 4-tuple $M = (S, A, P_{ss'}^a, R_{ss'}^a)$ where S is a set of states, A is a set of actions, $P_{ss'}^a = P(s_{t+1} = s' | s_t = s, a_t = a)$ is the 1-step state transition probability also referred to as the "dynamics", and $R_{ss'}^a = E\{r_{t+1} | s_t = s, a_t = a, s_{t+1} = s'\}$ is the expected reward. In RL, these quantities are usually unknown to the agent but can be estimated based on samples collected during exploration. If both S and A are finite, we call M a discrete MDP, otherwise we call it a continuous MDP. If for all $s \in S, a \in A$ there exists one $s' \in S$ with $P_{ss'}^a = 1$, the MDP is called deterministic otherwise it is a stochastic MDP. The goal of RL is to learn without explicit knowledge of M a policy π^* such that some measure of the long-term reward is maximized. Learning is often based on approximating the optimal action-value function $Q^*(s, a) = \sum_{s'} P_{ss'}^a [R_{ss'}^a + \gamma \max_{a'} Q^*(s', a')]$, where $\gamma \in [0, 1]$ is a discount factor.

One way of representing a (finite) MDP is using a weighted labeled multigraph $G = (V, E, w)$. Assuming knowledge of M, we would set $V = S$ and $E = \{(s, s')_a | \forall s, s' \in S, a \in A : P_{ss'}^a > 0\}$. In a state transition graph [7, 10], edge weights encode transition probabilities between nodes. If we set the weight of edge $(s, s')_a$ to $w_{ss'}^a = P_{ss'}^a$, a state transition graph is just an other way of representing the domain's dynamics. However, the graph-based view of the MDP is particularly suited for representation learning [7] and for identifying bottlenecks of the MDP [8–10, 12, 13], which is a common prerequisite for skill discovery in hierarchical RL. In small and discrete domains, learning state transition graphs from experience for unknown MDPs is straightforward: one graph node is created for each observed domain state and an edge is created for any pair of states between which a direct transition has been observed. For domains with continuous state spaces $S \subset \mathbb{R}^d$, the situation is more complicated because graphs are inherently discrete structures and thus, there cannot be a 1-to-1 correspondence between states and graph nodes since there exists an infinite number of states. Thus, several states need to be aggregated into one node, i.e., $V \subsetneq S$. Accordingly, one has to choose how many nodes there should be and where in the state space these nodes should be placed.

Prior work on choosing the positions of the graph nodes has mainly focused on covering the state space uniformly with nodes and neglected the domain's dynamics $P_{ss'}^a$. Among these approaches are: (a) a heuristic which forms a uniform *grid* of v_{num} nodes over the state space with a grid resolution of $\lfloor \sqrt[d]{v_{num}} \rfloor$ per dimensions. This approach has been used in the context of graph-based skill discovery, e.g., by Mannor et al. [8]. An obvious disadvantage is that the approach will not scale to domains with many dimensions. (b) The *on-policy sampling* heuristic (also denoted as "random subsampling" by Mahadevan and Maggioni [7]), which samples v_{num} graph nodes uniform randomly from the set of states S' encountered during exploration. The heuristic is on-policy, i.e., regions of the state space that are often visited by the sampling policy are represented by more graph nodes. (c) The ε-*net* heuristic, also denoted as "trajectory-based subsampling" [7], which aims at covering the "effective state space" as uniformly as possible. It follows a greedy strategy for finding a locally maximal set of graph nodes $V \subset S'$ with pairwise distance at least ε. The advantage of this approach compared to the on-policy sampling method is that the effective state space is covered more uniformly. In order to parametrize the heuristic by v_{num} instead of ε, one can perform binary search for a value of ε that yields a set of graph nodes with cardinality v_{num}.

3   FIGE: Force-Based Iterative Graph Estimation

While the heuristics discussed in Section 2 focus on covering the state space uniformly, they do not take the domain's dynamics into account. Thus, for many valid state transitions $s \rightarrow s'$ of the domain, there may not be any pair of graph nodes $v_1, v_2 \in V$ such that $v_1 \rightarrow v_2$ is a good representation of $s \rightarrow s'$. Accordingly, the graph may not be able to capture the domain's dynamics $P_{ss'}^a$ accurately. In this section, we propose a generative process which defines how probable a set of observed transition has been generated from a transition graph. We then propose the heuristic FIGE which is derived from this generative process as maximum likelihood solution under simplifying assumptions.

3.1   Likelihood of Transition Graph

We propose to consider a graph as a generative model for transitions and to choose graph node positions such that the likelihood $p(T|G)$ of the resulting graph G for a set of observed transitions $T = \{(s_i, a_i, s_i')\}_{i=1}^n$ becomes maximal. We consider transitions to have been sampled from the graph using the following generative process: (1) Sample a graph node $v \in V$ uniform randomly, i.e., $p(v) = 1/|V|$. (2) Sample a state s for a given node v according to $p(s|v) = N_b \exp(-\frac{1}{b^2}\|s - v\|_2^2)$ where b controls how closely centered $p(s|v)$ is on v and N_b is a normalization constant, which only depends on b. (3) Sample an action a uniformly from the action space A, i.e, $p(a|s) = p(a) = 1/|A|$. (4) Sample the "successor node" v' according to the graph's edge weights, i.e., $p(v'|v,a) = w_{vv'}^a$. (5) Finally, sample the successor state s' according to the distribution $p(s'|v',v,s) = N_b \exp(-\frac{1}{b^2}\|s' - (v'-v+s)\|_2^2)$ with the same b and N_b as before. This distribution encourages that the state transition $s \to s'$ is close to parallel to the given node transition $v \to v'$. This 5-step generative process can be derived as follows:

$$p(T|G) = \prod_{i=1}^n p((s_i,a_i,s_i')|G) = \prod_{i=1}^n p_G(s_i)p(a_i|s_i)p_G(s_i'|s_i,a_i)$$

Under the independence assumptions $I = \{v \perp a|s; v' \perp s|v,a; s' \perp a|v',v',s\}$, we have

$$p_G(s'|s,a) = \sum_v p_G(v,s'|s,a) = \sum_v p_G(v|s,a)p_G(s'|v,s,a)$$

$$= \sum_v p_G(v|s,a)\sum_{v'} p_G(v',s'|v,s,a) = \sum_v p_G(v|s,a)\sum_{v'} p_G(v'|v,s,a)p_G(s'|v',v,s,a)$$

$$\stackrel{I}{=} \sum_v p_G(v|s)\sum_{v'} p(v'|v,a)p_G(s'|v',v,s)$$

Inserting this in $p(T|G)$ and using Bayes rule $p_G(v|s) = p(s|v)p(v)/p_G(s)$ yields

$$p(T|G) = \prod_{i=1}^n p_G(s_i)p(a_i|s_i)\left[\sum_{v\in V}\frac{p(s_i|v)p(v)}{p_G(s_i)}\sum_{v'\in V}p(v'|v,a)p(s_i'|v',v,s_i)\right]$$

$$= \prod_{i=1}^n p(a_i|s_i)\left[\sum_{v\in V}\underset{(1)}{p(v)}\underset{(2)}{p(s_i|v)}\sum_{v'\in V}\underset{(4)}{p(v'|v,a)}\underset{(5)}{p(s_i'|v',v,s_i)}\right].$$
$$\phantom{= \prod_{i=1}^n}\underset{(3)}{}$$

3.2 Method

Given the generative process discussed in the previous section, the optimal state transition graph for a given set of transitions T would be $G^* = \arg\max_G p(T|G)$. Unfortunately, solving this problem directly is hard; we propose the FIGE heuristic, which aims at finding close-to-optimal transition graphs iteratively and is computationally tractable. FIGE's update equations are derived from the maximum likelihood objective $G^* = \arg\max_G p(T|G)$ using two simplifying assumptions (see Appendix A): (A1) For each transition $(s,a,s') \in T$, assume $p(v'|v,a) = 1$ if $v = \text{NN}_V(s) \wedge v' = \text{NN}_V(s')$ else 0, where $\text{NN}_V(s) = \arg\min_{v\in V}\|s - v\|^2$. This assumption implies that whenever action a is executed in any state of the Voronoi cell $Vo(v) = \{s \in S|\text{NN}_V(s) = v\}$ the successor state will be with probability 1 in $Vo(v')$. (A2) Assume $p(T|V) = \prod_v p(T|v)$. This assumption implies that the choice of the positions of the graph nodes $v \in V$ can be made independently. Both assumptions are typically oversimplifying; A1 is more oversimplifying for domains whose dynamics are less locally smooth. A2 on the other hand is more simplifying in strongly connected domains where many transitions from the Voronoi cell of one node to the Voronoi cell of another node occur. To account for some of the errors made because of the oversimplifications of A1 and A2, FIGE iteratively refines the graph node positions by applying the derived update equations several times. Note that FIGE is a heuristic and no guarantee for converging to G^* is given.

Algorithm 1. Force-based Iterative Graph Estimation (FIGE)

1: **Input:** Transitions $T = \{(s_i, a_i, s_i')\}_{i=1}^n$, parameters v_{num}, K
2: # Choose initial node positions V from states in T s.t. distance of closest pair is maximized
3: $V = \text{INITIALIZE}(T, v_{num})$
4: **for** $i = 0$ to $K - 1$ **do**
5:   **for all** $v \in V$ **do**
6:     $S^V[v] = \{s \mid \exists (s, a, s') \in T : \text{NN}_V(s) = v\}$ # Observed states in Voronoi cell $Vo(v)$
7:     $F_S[v] = \text{MEAN}(S^V[v]) - v$ # Sample representation force
8:     $T^{\rightarrow}(v) = \{\text{NN}_V(s') - s' + s \mid \exists (s, a, s') \in T : \text{NN}_V(s) = v\}$ # Transitions starting in $Vo(v)$
9:     $T^{\leftarrow}(v) = \{\text{NN}_V(s) - s + s' \mid \exists (s, a, s') \in T : \text{NN}_V(s') = v\}$ # Transitions ending in $Vo(v)$
10:    $F_G[v] = 0.5 \cdot [\text{MEAN}(T^{\rightarrow}(v)) + \text{MEAN}(T^{\leftarrow}(v))] - v$ # Graph consistency force
11:   $V = V + \alpha_i \cdot 0.5(F_S[V] + F_G[V])$ # Update node positions (vector notation)
12: # Count transitions from Voronoi cell $Vo(v)$ to Voronoi cell $Vo(v')$ under action a
13: $N_{vv'}^a = |\{(s, s') \mid \exists (s, a, s') \in T : \text{NN}_V(s) = v \wedge \text{NN}_V(s') = v'\}|$
14: $E = \{(v, v')_a \mid v, v' \in V \; a \in A : N_{vv'}^a > 0\}$ # Edge between v and v' labeled with action a
15: $w_{vv'}^a = N_{vv'}^a / \sum_{\tilde{v}} N_{v\tilde{v}}^a$ # Edge weights are empirical transition probabilities on graph
16: **return** (V, E, w)

FIGE is summarized in Algorithm 1: The set of graph nodes V is initialized such that it covers the set of states contained in T uniformly by, e.g., maximizing the distance of the closest pair of graph nodes (line 3). Afterwards, for K iterations, the graph nodes are moved according to two kind of "forces" that act on them: The "sample representation" force (line 6-7) pulls each graph node v to the mean of all states S^V for which it is responsible, i.e., the states s for which it is the nearest neighbor $\text{NN}_V(s)$ in V. Thus, this force encourages node positions that capture the on-policy state distribution well and corresponds to an intrinsic k-means clustering. The "graph consistency" force (line 8-10) pulls each graph node v to a position where for all $(s, a, s') \in T$ with $\text{NN}_V(s) = v$ there is a vertex v' such that $v' - v$ is similar to $s' - s$, i.e., both vectors are close to parallel. Thus, this force encourages node positions which can represent the domain's dynamics well. The nodes are then moved according to the two forces (line 11), where the parameter $\alpha_i \in (0, 1]$ controls how greedily the node is moved to the position where the forces would become minimal. In order to ensure convergence of the graph nodes, α_i should go to 0 for i approaching K. If not explicitly stated, we use $\alpha_i = \lceil i/5 \rceil^{-1}$ and $K = 15$. An edge labeled with action a is added between two nodes v and v' if there exists at least one transition $(s, a, s') \in T$ with s being in the Voronoi cell of $Vo(v)$ and s in $Vo(v')$ (line 14). Furthermore, the edge weights are chosen as the empirical transition probabilities $\hat{P}_{vv'}^a$ from node v to v' under action a (line 15). It has recently been shown that this choice of edge weights is most robust under varying degrees of domain stochasticity and different exploration strategies of the agent [11].

3.3 Evaluation: Graph Likelihood

In this section, we present an empirical comparison of the different heuristics with regard to the obtained likelihood of the generated graphs in the mountain car domain (compare Chapter 8.4 of Sutton and Barto [15]). In the left graph of Figure 1, the graph

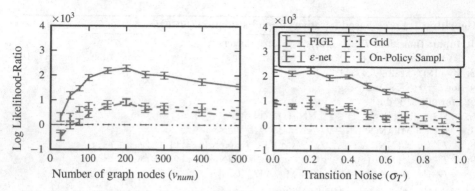

Fig. 1. Left graph: Graph Log Likelihood-Ratio relative to grid heuristic in deterministic mountain car for different values for the number of graph nodes. Right graph: Graph Log Likelihood-Ratio relative to grid heuristic in the stochastic mountain car domain for varying stochasticity σ_t and $v_{num} = 200$. Shown are mean and standard error of mean over 20 repetitions.

likelihood with $b = 0.02$ is evaluated for different heuristics and different number of graph nodes v_{num}. Since the likelihood depends on v_{num}, we plot the log of the ratio of the method's likelihood relative to the likelihood obtained by the data-independent grid heuristic. Note that the graph likelihood has been evaluated on test transitions T_{test} which were different from the training transitions T_{train} that were used for the optimization of graph node positions ($|T_{train}| = 5000$, and $|T_{test}| = 10^5$).

Regardless of v_{num}, the largest graph likelihood is achieved by FIGE and the smallest by the grid heuristic (for $v_{num} > 75$). The on-policy sampling heuristic performs slightly better than ε-net for $v_{num} < 150$ while both perform similarly for larger v_{num}. A possible explanation for the stronger deterioration of ε-net is that for small v_{num}, the minimal node distance ε gets larger than the typical distance of states and their successors and thus, $P_{ss'}^a$ cannot be represented in any part of the state space. In contrast, on-policy sampling allocates more graph nodes in densely sampled parts of the state space and thus allows to model at least these parts of the state space. FIGE can achieve a considerably larger graph likelihood than both by taking the domain's dynamics into account as well.

In a second experiment, we evaluate how robust the different heuristics are with regard to increasing stochasticity in the domain's state transition probability $P_{ss'}^a$. For this, we modify any transition from state s to s' governed by the domain's deterministic dynamics such that the i-the dimension of the actual successor state becomes $s_i' \leftarrow s_i' + \sigma_i(s_i' - s_i)$ with σ_i sampled uniformly from $[-\sigma_t, \sigma_t]$. Note that this is not purely observation noise since the actual internal state of the environment is altered. σ_t controls how "strong" the stochasticity of the domain is, with $\sigma_t = 0$ corresponding to the deterministic domain. The same amount of transition noise was also used for generating the test transitions T_{test}. The results are shown in the right graph in Figure 1. As expected, the grid-based heuristic deteriorates less with increasing stochasticity as it does not take the observed transitions into account. Nevertheless, the other data-dependent heuristics achieve better graph likelihood for $\sigma_t < 0.8$ with FIGE remaining the best heuristic for the whole investigated range of $\sigma_t \in [0, 1]$. This shows that FIGE is also suited for stochastic domains.

4 Graph-Based Skill Discovery

Next to continuous action spaces, scaling RL to real-world problems with large or continuous state spaces remains a challenge since the amount of experience the agent can collect is limited. One approach to alleviate this problem is hierarchical RL [1], which aims at dividing a problem into simpler subproblems, learning solutions for these subproblems, and encapsulate the acquired knowledge into so-called *skills* that can potentially be reused later on in the learning process. It has been shown that skills can help an agent to adapt to non-stationarity of the environment and to transfer knowledge between different but related tasks [3] and can increase the representability of the value function in continuous domains [5]. A major challenge in hierarchical RL is to identify what might constitute a useful skill, i.e., how the problem should be decomposed. Skills should be reusable, distinct, and easy to learn. The task of identifying such skills is called *skill discovery* [2, 4].

Most prior work on autonomous skill discovery is based on the concept of *bottleneck areas* in the state space. Informally, bottleneck areas have been described as the border states of densely connected areas in the state space [10] or as states that allow transitions to a different part of the environment [12]. Several heuristics have been proposed to identify bottlenecks. One class of heuristics are *frequency-based approaches* that compute local statistics of states like diverse density [9] and relative novelty [12]. An other class of heuristics that is typically more sample-efficient are *graph-based approaches* which are based on estimates of the domain's state transition graph (see Section 2). Graph-based approaches to skill discovery aim at partitioning this graph into subgraphs which are densely connected internally but only weakly connected with each other. Menache et al. [10] propose a top-down approach for partitioning the global transition graph based on the max-flow/min-cut heuristic. Şimşek et al. [13] follow a similar approach but partition local estimates of the global transition graph using a spectral clustering algorithm and use repeated sampling for identifying globally consistent bottlenecks. Mannor et al. [8] propose a bottom-up approach that partitions the global transition graph using agglomerative hierarchical clustering.

Relatively few works on autonomous skill discovery in domains with continuous state spaces exist. Mannor et al. [8] have evaluated their approach in the mountain car domain by uniformly discretizing the state space. However, such a uniform discretization is suboptimal since it does not scale well to higher dimensional state spaces. One skill discovery method that has been designed for continuous domains is "skill chaining" [5]. Skill chaining produces chains (or more general: trees) of skills such that each skill allows to reach a specific region of the state space, such as a terminal region or a region where an other skill can be invoked. Skill chaining requires to specify an area of interest (typically the terminal region of the state space) which is used as target for the skill at the root of the tree. For multi-task domains with several goal regions, it is unclear how the root of the skill tree should be chosen. In the next section, we present a generic algorithm for graph-based skill discovery in MDPs with continuous state spaces.

4.1 Approach: Skill Discovery by Clustering of Transition Graph

We adopt the *options framework* [14] for Hierarchical RL, in which skills are formalized as *options*: An option o consists of three components: the option's initiation set $I_o \subset S$

that defines the states in which the option may be invoked, the option's termination condition $\beta_o : S \to [0,1]$ which specifies the probability of option execution terminating in a given state, and the option's policy π_o which defines the probability of executing an action in a state under option o. In the options framework, the agent's policy π may in any state s decide not solely to execute a primitive action but also to call any of the options for which $s \in I_o$. If an option is invoked, the option's policy π_o is followed for several time steps until the option terminates according to β_o. The option's policy π_o is defined relative to an option-specific reward function R_o that may differ from the global external reward function. Skill discovery thus requires to choose I_o, β_o, and R_o.

For a given set of observed transitions T that have been sampled from the MDP, we can generate the state transition graph using any of the approaches discussed in this paper. Based on the generated transition graph, any of the graph-based skill discovery approaches for discrete MDPs could be used to identify skills. We adopt the concept of identifying densely connected subgraphs of the transition graph, which correspond to densely connected regions in the state space. In order to quantify to what extent the edges connecting two disjoint subgraphs form a bottleneck, a so-called *linkage criterion* is used. A linkage criterion is a function mapping two disjoint subgraphs $A, B \subset G$ onto a scalar, which is the larger the "stronger" the bottleneck between A and B in G is. In this paper, we adopt the off-policy $\hat{N}_{cut}$ linkage criterion that was proposed by Metzen [11]. The $\hat{N}_{cut}$ criterion is an approximation of the sum of probabilities that the agent transitions in one time step from a state in subgraph A to a state in subgraph B or vice versa. For identifying densely connected subgraphs of a graph G, we aim to determine a partition C^* of minimal cardinality of the graph nodes into disjoint sets c_i such that each induced subgraph does not contain a bottleneck. Formally:
$C^* = \arg\min_{C \in \mathscr{C}(V)} |C|$ s.t. $\max_{c_i \in C, d_i \subset c_i} \hat{N}_{cut}(c_i \setminus d_i, d_i) \leq \psi$, with $\mathscr{C}(V)$ being the set of all possible partitions of V and ψ a parameter controlling the granularity of the clustering. Note that the maximization goes over all possible ways of splitting c_i into two parts d_i and $c_i \setminus d_i$ and the constraint guarantees that there is no bottleneck with $\hat{N}_{cut} > \psi$ in any of the c_i. Since finding the optimal solution for this problem is NP-hard, we use an approximate approach that is based on agglomerative hierarchical clustering and similar to the one proposed by Mannor et al. [8]. This algorithm starts by assigning each node into a separate cluster and afterwards merges greedily the pair of clusters with minimal linkage until no pair remains with a linkage below ψ. As proposed by Mannor et al., only clusters which are connected in G can be merged.

For learning an option o based on a newly discovered skill, we need to choose appropriate I_o, β_o, and R_o based on the identified partitioning C_G of the transition graph. For this, the partition C_G of the transition graph is generalized to a partition of the entire state space C_S by a nearest-neighbor based approach, i.e., for all clusters $c_i \in C_G$: $C_S(c_i) = \{s \in S \mid \text{NN}_V(s) \in c_i\}$. For each cluster A, one skill is generated for each adjacent cluster B, where A and B are adjacent if there exists $v_a \in A, v_b \in B$ and action a such that $(v_a, v_b)_a \in E$. The corresponding skill prototype $(I_{A \to B}, \beta_{A \to B}, R_{A \to B})$ is defined as:

$$I_{A \to B} = C_S(A) \qquad \beta_{A \to B}(s) = 0 \text{ if } s \in I_{A \to B} \text{ else } 1$$
$$R_{A \to B}((s, a, r, s')) = -1 \text{ if } s' \in (C_S(A) \cup C_S(B)) \text{ else } r_p,$$

where r_p is a parameter of the algorithm that determines the penalty for failing to fulfill a skill's objective. Additionally, for each cluster that contains nodes in which an episode has terminated, a special skill is created that can be invoked in any state of the cluster, terminates successfully when the episode terminates, and terminates unsuccessfully (i.e., obtains the penalty reward) if the clusters is left. Note that in contrast to Mannor et al. [8], the generalization of the graph partition to the entire state space allows to perform the learning of skills and higher-level policies in the original MDP and not in a discretized version of it.

4.2 Evaluation

In this section, we present an empirical evaluation of the proposed skill discovery approach in the 2D Multi-Valley environment, which is an extension of the basic mountain car domain. The car the agent controls is not restrained to a one-dimensional surface, however, but to a two-dimensional surface. This two-dimensional surface consists of $2 \times 2 = 4$ valleys, whose borders are at $(\pi/6 \pm \pi/3, \pi/6 \pm \pi/3)$. The agent observes four continuous state variables: the positions in the two dimensions (x and y) and the two corresponding velocities (v_x and v_y). The agent can choose among the four discrete actions `northwest`, `northeast`, `southwest`, `southeast` which add $(\pm 0.001, \pm 0.001)$ to (v_x, v_y). In each time step, due to gravity $0.004\cos(3x)$ is added to v_x and $0.004\cos(3y)$ to v_y. The maximal absolute velocity in each dimension is restrained to 0.07. The agent is faced with a multi-task scenario: in each episode, the agent has to solve one out of twelve tasks. Each task is associated with a combination of two distinct valleys; e.g., in task $(0, 1)$ the agent starts in the floor[1] of valley 0 and has to navigate to the floor of valley 1 and reduce its velocity such that $\|(v_x, v_y)\|_2 \leq 0.03$. In each time step, the agent receives a reward of $r = -1$. Once a task is solved, the next episode starts with the car remaining at its current position and one of the tasks that starts in this valley is drawn at random. The current task is communicated as an additional state space dimension to the agent; the agent uses it for learning the top-level policy but ignores it during graph generation, graph clustering, and skill learning such that skills reusable in different tasks.

We present an empirical comparison of the learning performance of the entire hierarchical RL architecture with different graph node selection heuristics as base for skill discovery. We compare the performance to two baselines: (i) a monolithic approach which learns a flat policy for every task without using skills, and (ii) the same hierarchical RL framework but with predefined skills prototype (I_o, β_o, R_o). These prototypes have been generated in the same way as those discovered using graph clustering but are based on the ground-truth partition of the domain into the 4 valleys. Thus, baseline (ii) presents probably an upper boundary for the performance that any skill discovery method can achieve within the given hierarchical RL architecture.

Skill discovery has been performed after $n = 10^5$ state transitions have been observed in the environment and graphs with $v_{num} \in \{50, 100, 150, 500\}$ nodes have been generated. These have been clustered with the approach presented in Section 4.1 for $\psi = -0.03$. Each option's value function has been represented by an CMAC function

---

[1] The floor of valley 0 corresponds to the region $((-1/6 \pm 2/15)\pi, (-1/6 \pm 2/15)\pi)$.

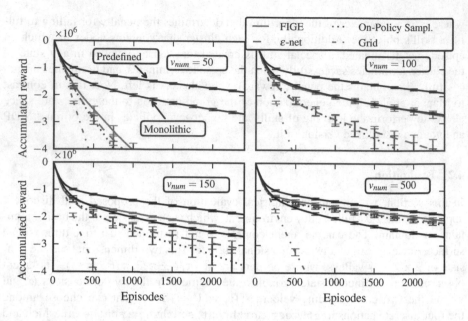

Fig. 2. Accumulated reward in the 2D Multi-Valley domain during 2400 episodes of learning for different graph generation heuristics and graph sizes. Baselines show performance of a monolithic learner and for the optimal predefined skill prototypes. Shown are mean and standard error of mean over 10 repetitions.

approximator consisting of 10 independent tilings with $7^2 \cdot 5^2$ tiles, where the higher resolutions have been used for the x and y dimensions. The penalty reward of an unsuccessful option has been set to $r_p = -1000$ and the value functions have been initialized to -100. For learning the compositional option π, a lower resolution of $5^2 \cdot 3^2$ tiles has been used and the value functions have been initialized to -1000. The discounting factor has been set to $\gamma = 1$ and all policies have been ε-greedy with $\varepsilon = 0.01$. The value functions were learned using Q-Learning and updated only for currently active options with a learning rate of 1. Episodes have been interrupted after 10^4 steps without solving the task and a new task was chosen at random. All parameters have been chosen based on preliminary investigations.

Figure 2 shows the accumulated reward obtained during the first 2400 learning episodes (the phase of learning during which the explorative bias provided by skills has the strongest impact) for different graph node selection heuristics and different number of graph nodes v_{num}. For too small v_{num}, e.g., $v_{num} = 50$, all graph node selection heuristic perform worse than learning a flat policy. Furthermore, one can see that the grid-based heuristic obtains poor results for any choice of v_{num}. When using many graph nodes ($v_{num} = 500$), no considerable differences between the other heuristics exist and the performance is considerably better than when learning a flat policy and only slightly worse than for predefined skills. However, for intermediate values of v_{num}, e.g., $v_{num} \in \{100, 150\}$, FIGE obtains significantly better results than the ε-net and the on-policy sampling heuristic ($p < 0.001$, Mann-Whitney U-test). Furthermore, the accumulated reward obtained by FIGE for $v_{num} = 150$ is considerably larger than when learning a flat policy, which is not the case for the other heuristics. In summary, FIGE allows to create skills that can provide a useful explorative bias during learning based on smaller graphs than other heuristics which allows to reduce computation time during graph clustering and learning.

5   Representation Learning

Representation Policy Iteration (RPI) is an approach proposed by Mahadevan and Maggioni [7] that aims at solving MDPs by jointly learning representations and optimal policies. In contrast to most other RL algorithms, RPI does not require an a-priori specification of basis functions. The main idea for learning basis functions is to first learn a state transition graph of the domain and to construct a symmetric diffusion operator on this graph. The normalized graph Laplacian $\mathcal{L} = D^{-1/2}(D - W)D^{-1/2}$ and the combinatorial Laplacian $L = D - W$ are examples for such diffusion operators—with W being the graph's symmetrized weight matrix and D being a diagonal matrix whose entries are the row sums of W. The smoothest eigenvectors of these operators, the so-called proto-value functions, are used as basis functions for representing value functions. Least-squares policy iteration (LSPI) as proposed by Lagoudakis and Parr [6] is used for control learning, i.e., for learning the parameters w^π of the action value function $Q^\pi = w^\pi \Phi$ of an ε-optimal policy π within the linear span of the basis functions Φ. In the original paper, at the end of each episode an additional set of samples is collected either on- or off-policy. We skip this additional sampling and use the samples collected during control learning also for representation learning to show that some graph node selection heuristics can deal with this better than others. In order to initialize representation and control learning, the agent explored the environment uniform randomly during the first 10 episodes. Thereupon, RPI was performed at the end of each episode and the policy obtained was followed ε-greedily.

RPI can also be used in MDPs with continuous state space. In such continuous domains, one challenge is to select the graph node position ("to subsample a set of states" in the terminology of Mahadevan and Maggioni). The authors discuss the usage of the on-policy subsampling and the ε-net heuristics; however we will show that considerable better results can be achieved by using FIGE. In RPI, each graph node is connected to its k nearest neighbor nodes in the euclidean space and the edge weight between nodes x_i and x_j is $W(i, j) = \tau(i) \exp(-\|x_i - x_j\|_2^2/\kappa)$ where $\tau(i)$ and κ are parameters to be specified. Note that this way of connecting graph nodes has the potential disadvantage that proximity of nodes in the euclidean space does not necessarily imply that a transition between these nodes is possible, e.g., if an obstacle lies between those states. Choosing graph edges based on observed transitions between states (compare Section 2) lessens this issue and seems thus preferable. However, for consistency with the original approach of Mahadevan and Maggioni [7], we adhere the "euclidean" connectivity.

5.1   Evaluation: Mountain Car

In a first experiment, we evaluate the performance of RPI for different graph node selection heuristics and different degrees of transition noise σ_t in mountain car (compare

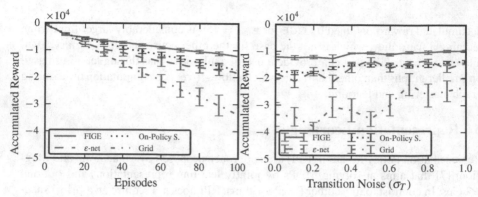

Fig. 3. Accumulated reward obtained by RPI in mountain car for different graph node selection heuristics. Left plot: learning curves in the deterministic domain. Right plot: reward accumulated after 100 episodes for different degrees of transition noise σ_T. Shown are mean and standard error of mean over 10 repetitions.

Section 3.3) for $v_{num} = 50$. For all node selection heuristics, we obtained the best results when setting the number of proto value-functions equal to the number of graph nodes, i.e., $p_{num} = v_{num}$. Furthermore, in accordance with Mahadevan and Maggioni [7] we set the discount factor γ to 0.99 and the exploration rate to $\varepsilon = 0.01$. We used the normalized Laplacian $\mathscr{L}$ as graph operator. The results are shown in Figure 3. The left plot shows learning curves of RPI in the deterministic mountain car domain: RPI performs best when combined with FIGE and worst when combined with the grid heuristic while on-policy sampling and ε-net achieve approximately the same results (no significant differences). This can be attributed to the fact that due to the randomly chosen start states of episodes, the on-policy state distribution (over several episodes) does not vary too strongly over the effective state space. Thus, sampling from the on-policy distribution yields in this domain graph nodes that cover the effective state space close to uniform. The worse results of the grid heuristic show that even for low-dimensional domains, a uniform discretization can be detrimental. The right plot shows how the reward accumulated after 100 episodes changes for different degrees of stochasticity of the domain. In general, increasing transition noise seems to make the task easier as the performance increase for all heuristics; probably because the value function becomes smoother and thus better representable. However, the relative order of different methods remains the same. This reinforces that FIGE can be used in stochastic domains as well.

5.2 Evaluation: Octopus Arm

In a second experiment, we investigate the performance of RPI using FIGE for graph node selection in the octopus arm domain [16]. The specific task is depicted in the left plot of Figure 4: the agent has to control the octopus arm such that it moves two food items (small yellow circles) into its mouth (large red circle). The base of the arm is restricted and cannot be actuated directly. The agent may control the arm in the following way: elongating or contracting the entire arm, bending the first half of the arm in either of the two directions, and bending the second half of the arm in either of the two

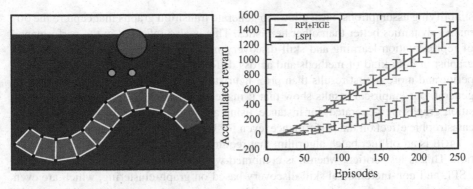

Fig. 4. Left plot: Visualization of the octopus arm task. Right plot: Accumulated reward obtained by LSPI and RPI using FIGE for graph node selection in the octopus arm domain. Shown are mean and standard error of mean over 20 repetitions.

directions. In each time step, the agent can control the elongation, the first half, and the second half of the arm independently, resulting in 8 discrete actions. The agent observes the positions x_i, y_i and velocities $\dot{x}_i$, $\dot{y}_i$ of the food items and of 24 selected parts of its arm (denoted by small black dots in Figure 4) and the angle and angular velocity of the arm's base. Thus, the state space is continuous and consists of 106 dimensions. Each dimension is normalized such that its values fall into $[0, 1]$. The agent obtains a reward of -0.01 per time step, a reward of 5 for moving the left food item into its mouth, and a reward of 7 for the right food item. The episode ends after 100 time steps or once both food items have been eaten. Because of the high-dimensional state space and the complex dynamics of the domain, the octopus arm problem is a challenging task.

We compare RPI combined with FIGE for $v_{num} = 75$ and $p_{num} = 5$ to LSPI using 75 radial basis functions (RBFs) as features ($\gamma = 0.99$, $\varepsilon = 0.01$). In the first 10 episodes, pure exploration without learning was conducted. The RBF centers c_i have been set to observed states such that the pairwise distance of the centers becomes maximal; the feature activation of center c_i for state s is computed as $\phi_i(s) = \exp(10\|c_i - s\|_2^2)$. The right plot of Figure 4 summarizes the results: using FIGE-based proto-value functions performs considerably better than standard RBF features; in particular, the agent learns in each run to move at least one food item into its mouth, which is not the case for LSPI. The main difference between the two approaches is that RBFs are local while proto-value functions can also capture more global properties. We suppose that since FIGE allows to capture the dynamics of a domain well, it allows to learn non-local proto-value functions that provide a useful bias to LSPI. In summary, the results suggest that FIGE can also support learning in high-dimensional problems.

6 Summary and Outlook

We have presented a new view on graph-based RL in continuous domains. Based on interpreting state transition graphs as generative models of the domain's dynamics, we have proposed a formulation for the likelihood of a graph for a given set of transitions. We have derived the new heuristic FIGE from the maximum likelihood objective under simplifying assumptions. FIGE allows to generate transition graphs that capture the domain's dynamics better than other heuristics. This is also reflected in the performance of representation learning and skill discovery methods that are built upon transition graphs: in both kind of methods and across different domains, FIGE has achieved superior and more robust results than prior heuristics for transition graph generation. In general, our empirical results show that it makes a considerable difference how transition graphs are generated; for instance, using a grid-based discretization often had a catastrophic effect on the performance, even for low-dimensional domains.

FIGE is an offline, batch algorithm that requires considerable amounts of computation. This is less critical when it is combined with other offline approaches like RPI, LSPI, and non-incremental skill discovery based on graph-clustering, which are even more expensive in terms of computation. However, for making use of transition graphs in online methods like, e.g., OGAHC [11] for skill discovery, it would be highly desirable to develop an online method for graph generation that aims at similar objectives as FIGE. A further direction for future research would be to extend FIGE to domains with continuous action spaces and to use it for learning a policy representation that can be used within direct policy search approaches.

Acknowledgment. This work was supported through a grant of the German Federal Ministry of Economics and Technology (BMWi, FKZ 50 RA 1217). The author would like to thank Yohannes Kassahun and the anonymous reviewers for many helpful comments.

References

1. Barto, A.G., Mahadevan, S.: Recent advances in hierarchical reinforcement learning. Discrete Event Dynamic Systems 13(4), 341–379 (2003)
2. Digney, B.L.: Emergent hierarchical control structures: Learning reactive/hierarchical relationships in reinforcement environments. In: From Animals to Animats: The 4th Conference on Simulation of Adaptive Behavior, Cambridge, MA, pp. 363–372 (1996)
3. Digney, B.L.: Learning hierarchical control structures for multiple tasks and changing environments. In: 5th Conference on the Simulation of Adaptive Behavior, pp. 321–330 (1998)
4. Kirchner, F.: Automatic decomposition of reinforcement learning tasks. In: AAAI 1995 Fall Symposium Series on Active Learning, Cambridge, MA, USA, pp. 56–59 (1995)
5. Konidaris, G., Barto, A.G.: Skill discovery in continuous reinforcement learning domains using skill chaining. In: NIPS, vol. 22, pp. 1015–1023 (2009)
6. Lagoudakis, M.G., Parr, R.: Least-squares policy iteration. Journal of Machine Learning Research 4, 1107–1149 (2003)
7. Mahadevan, S., Maggioni, M.: Proto-value functions: A laplacian framework for learning representation and control in markov decision processes. Journal of Machine Learning Research 8, 2169–2231 (2007)
8. Mannor, S., Menache, I., Hoze, A., Klein, U.: Dynamic abstraction in reinforcement learning via clustering. In: 21st International Conference on Machine Learning, pp. 560–567 (2004)
9. McGovern, A., Barto, A.G.: Automatic discovery of subgoals in reinforcement learning using diverse density. In: 18th International Conference on Machine Learning, pp. 361–368 (2001)

10. Menache, I., Mannor, S., Shimkin, N.: Q-Cut - dynamic discovery of sub-goals in reinforcement learning. In: Elomaa, T., Mannila, H., Toivonen, H. (eds.) ECML 2002. LNCS (LNAI), vol. 2430, pp. 295–306. Springer, Heidelberg (2002)
11. Metzen, J.H.: Online skill discovery using graph-based clustering. Journal of Machine Learning Research W&CP 24, 77–88 (2012)
12. Şimşek, Ö., Barto, A.G.: Using relative novelty to identify useful temporal abstractions in reinforcement learning. In: 21st International Conference on Machine Learning, pp. 751–758 (2004)
13. Şimşek, Ö., Wolfe, A.P., Barto, A.G.: Identifying useful subgoals in reinforcement learning by local graph partitioning. In: 22nd International Conference on Machine Learning, pp. 816–823 (2005)
14. Sutton, R.S., Precup, D., Singh, S.: Between MDPs and semi-MDPs: A framework for temporal abstraction in reinforcement learning. Artificial Intelligence 112, 181–211 (1999)
15. Sutton, R., Barto, A.: Reinforcement Learning: An Introduction. MIT Press (1998)
16. Yekutieli, Y., Sagiv-Zohar, R., Aharonov, R., Engel, Y., Hochner, B., Flash, T.: A dynamic model of the octopus arm. I. Biomechanics of the octopus reaching movement. Journal of Neurophysiology 5, 291–323 (2005)

A   Appendix: Derivation of FIGE's Update Equations

Let $T = \{(s_i, a_i s_i')\}_{i=1}^n$ be a set of n transitions. We aim at finding a state transition graph G^* with v_{num} nodes such that $G^* = \arg\max_G p(T|G)$. We derive the FIGE update equations as maximum likelihood solutions for $p(T|G)$ under two simplifying assumptions.

(A1) For $(s, a, s') \in T$, assume $p(v'|v, a) = 1$ if $v = \mathrm{NN}_V(s) \wedge v' = \mathrm{NN}_V(s')$ else 0.

Assumption A1 allows to effectively decouple the likelihood $p(T|G)$ from the graph's edges and their weights $w_{vv'}^a$, such that it depends solely on the graph node positions and can thus be written as $p(T|V)$. By using assumption A1, we obtain:

$$\log p(T|G) = \log \frac{1}{|A|^n |V|^n} \prod_{i=1}^n \left[\sum_{v \in V} p(s_i|v) \sum_{v' \in V} p(v'|v, a) p(s'|v', v, s_i) \right]$$

$$\overset{A1}{=} \log \frac{1}{|A|^n |V|^n} \prod_{i=1}^n p(s_i|\mathrm{NN}_V(s_i)) p(s_i'|\mathrm{NN}_V(s_i'), \mathrm{NN}_V(s_i), s_i) \hat{=} \log p(T|V)$$

$$\log p(T|V) = -n \log |A||V| + \sum_{i=1}^n \left[\log p(s_i|\mathrm{NN}_V(s_i)) + \log p(s_i'|\mathrm{NN}_V(s_i'), \mathrm{NN}_V(s_i), s_i) \right]$$

$$= -n \log |A||V| + n \log N_b - \frac{1}{b^2} D$$

with $D = \sum_{i=1}^n \left[\|s_i - \mathrm{NN}_V(s_i)\|_2^2 + \|(\mathrm{NN}_V(s_i') - \mathrm{NN}_V(s_i)) - (s_i' - s_i)\|_2^2 \right]$. For given V we create 2 partitions of T into v_{num} sets: $T^{\rightarrow}(v) = \{(s, s')|\exists(s, a, s') \in T : \mathrm{NN}_V(s) = v\}$ and $T^{\leftarrow}(v) = \{(s, s')|\exists(s, a, s') \in T : \mathrm{NN}_V(s') = v\}$. Furthermore, we create v_{num} sets

$S^V(v) = \{s \mid \exists (s,a,s') \in T : NN_V(s) = v\}$. For $|V| = v_{num} = const$, we can now maximize the log-likelihood $\log p(T|V)$ by minimizing D:

$$D = \sum_{i=1}^{n} \left[\|s_i - NN_V(s_i)\|_2^2 + 2\frac{1}{2}\|(NN_V(s_i') - NN_V(s_i)) - (s_i' - s_i)\|_2^2 \right]$$

$$= \sum_v \left[\sum_{s \in S^V(v)} \|s - v\|_2^2 + \frac{1}{2} \sum_{s,s' \in T^{\rightarrow}(v)} \|NN_V(s') - v - s' + s\|_2^2 \right.$$

$$\left. + \frac{1}{2} \sum_{s,s' \in T^{\leftarrow}(v)} \|v - NN_V(s) - s' + s\|_2^2 \right]$$

Each term of the outer sum corresponds to the contribution of node v's position to D; however, the terms cannot be minimized separately since they are coupled via $NN_V(s)$ and $NN_V(s')$. Minimizing them jointly is difficult because of the discontinuities of the nearest-neighbor terms. Thus, FIGE makes the following simplifying assumption:

(A2) Assume $p(T|V) = \prod_v p(T|v)$

This assumption implies that the couplings between the terms of D can be ignored and each v can be set greedily to the position where the respective term in the outer sum would become minimal as if all other $\tilde{v} \in V$ would remain unchanged. Finally, the greedy FIGE update equation which moves a node from position v_{old} to position v_{new} is

$$v_{new} = \arg\min_v \left[\sum_{s \in S^V(v_{old})} \|s - v\|_2^2 + \frac{1}{2} \sum_{s,s' \in T^{\rightarrow}(v_{old})} \|(NN_V(s') - s' + s) - v\|_2^2 \right.$$

$$\left. + \frac{1}{2} \sum_{s,s' \in T^{\leftarrow}(v_{old})} \|v - (NN_V(s) - s + s')\|_2^2 \right]$$

In this, the first sum is minimized by choosing $v_{new} = v_a = MEAN_{s \in S^V(v_{old})}(s)$, the second sum by choosing $v_{new} = v_b = MEAN_{s,s' \in T^{\rightarrow}(v_{old})}(NN_V(s') - s' + s)$, and the third by $v_{new} = v_c = MEAN_{s,s' \in T^{\leftarrow}(v_{old})}(NN_V(s) - s + s')$. By using forces that pull v_{new} to v_a, v_b, and v_c with the respective weights, we obtain the FIGE update rule

$$v_{new} = v_{old} + \alpha \left[\frac{1}{2}(v_a - v_{old}) + \frac{1}{4}(v_b - v_{old}) + \frac{1}{4}(v_c - v_{old}) \right].$$

Since $A2$ is oversimplifying, one sweep of the FIGE update equations will typically not find the maximum likelihood solution. Thus, FIGE performs several update iterations to account for couplings between nodes.

Regret Bounds for Reinforcement Learning with Policy Advice

Mohammad Gheshlaghi Azar[1], Alessandro Lazaric[2], and Emma Brunskill[1]

[1] Carnegie Mellon University, Pittsburgh, PA, USA
{ebrun,mazar}@cs.cmu.edu
[2] INRIA Lille - Nord Europe, Team SequeL, Villeneuve d'Ascq, France
alessandro.lazaric@inria.fr

Abstract. In some reinforcement learning problems an agent may be provided with a set of input policies, perhaps learned from prior experience or provided by advisors. We present a reinforcement learning with policy advice (RLPA) algorithm which leverages this input set and learns to use the best policy in the set for the reinforcement learning task at hand. We prove that RLPA has a sub-linear regret of $\widetilde{O}(\sqrt{T})$ relative to the best input policy, and that both this regret and its computational complexity are independent of the size of the state and action space. Our empirical simulations support our theoretical analysis. This suggests RLPA may offer significant advantages in large domains where some prior good policies are provided.

1   Introduction

In reinforcement learning an agent seeks to learn a high-reward policy for selecting actions in a stochastic world without prior knowledge of the world dynamics model and/or reward function. In this paper we consider when the agent is provided with an input set of potential policies, and the agent's objective is to perform as close as possible to the (unknown) best policy in the set. This scenario could arise when the general domain involves a finite set of types of RL tasks (such as different user models), each with known best policies, and the agent is now in one of the task types but doesn't know which one. Note that this situation could occur both in discrete state and action spaces, and in continuous state and/or action spaces: a robot may be traversing one of a finite set of different terrain types, but its sensors don't allow it to identify the terrain type prior to acting. Another example is when the agent is provided with a set of domain expert defined policies, such as stock market trading strategies. Since the agent has no prior information about which policy might perform best in its current environment, this remains a challenging RL problem.

Prior research has considered the related case when an agent is provided with a fixed set of input (transition and reward) models, and the current domain is an (initially unknown) member of this set [5,4,2]. This actually provides the agent with more information than the scenario we consider (given a model we can extract a policy, but the reverse is not generally true), but more significantly, we

H. Blockeel et al. (Eds.): ECML PKDD 2013, Part I, LNAI 8188, pp. 97–112, 2013.
© Springer-Verlag Berlin Heidelberg 2013

find substantial theoretical and computational advantages from taking a model-free approach. Our work is also closely related to the idea of policy reuse [6], where an agent tries to leverage prior policies it found for past tasks to improve performance on a new task; however, despite encouraging empirical performance, this work does not provide any formal guarantees. Most similar to our work is Talvitie and Singh's [14] AtEase algorithm which also learns to select among an input set of policies; however, in addition to algorithmic differences, we provide a much more rigorous theoretical analysis that holds for a more general setting.

We contribute a reinforcement learning with policy advice (RLPA) algorithm. RLPA is a model-free algorithm that, given an input set of policies, takes an optimism-under-uncertainty approach of adaptively selecting the policy that may have the highest reward for the current task. We prove the regret of our algorithm relative to the (unknown) best in the set policy scales with the square root of the time horizon, linearly with the size of the provided policy set, and is independent of the size of the state and action space. The computational complexity of our algorithm is also independent of the number of states and actions. This suggests our approach may have significant benefits in large domains over alternative approaches that typically scale with the size of the state and action space, and our preliminary simulation experiments provide empirical support of this impact.

2   Preliminaries

A Markov decision process (MDP) M is defined as a tuple $\langle \mathcal{S}, \mathcal{A}, P, r \rangle$ where $\mathcal{S}$ is the set of states, $\mathcal{A}$ is the set of actions, $P : \mathcal{S} \times \mathcal{A} \to \mathcal{P}(\mathcal{S})$ is the transition kernel mapping each state-action pair to a distribution over states, and $r : \mathcal{S} \times \mathcal{A} \to \mathcal{P}([0,1])$ is the stochastic reward function mapping state-action pairs to a distribution over rewards bounded in the $[0,1]$ interval.[1] A policy π is a mapping from states to actions. Two states s_i and s_j communicate with each other under policy π if the probability of transitioning between s_i and s_j under π is greater than zero. A state s is recurrent under policy π if the probability of reentering state s under π is 1. A recurrent class is a set of recurrent states that all communicate with each other and no other states. Finally, a Markov process is unichain if its transition matrix consists of a single recurrent class with (possibly) some transient states [12, Chap. 8].

We define the performance of π in a state s as its expected average reward

$$\mu^{\pi}(s) = \lim_{T \to \infty} \frac{1}{T} \mathbb{E}\left[\sum_{t=1}^{T} r(s_t, \pi(s_t)) \Big| s_0 = s \right], \tag{1}$$

where T is the number of time steps and the expectation is taken over the stochastic transitions and rewards. If π induces a unichain Markov process on M, then $\mu^{\pi}(s)$ is constant over all the states $s \in \mathcal{S}$, and we can define the bias function λ^{π} such that

$$\lambda^{\pi}(s) + \mu^{\pi} = \mathbb{E}\left[r(s, \pi(s)) + \lambda^{\pi}(s') \right]. \tag{2}$$

---

[1] The extension to larger bounded regions $[0, d]$ is trivial and just introduces an additional d multiplier to the resulting regret bounds.

Its corresponding span is $sp(\lambda^\pi) = \max_s \lambda^\pi(s) - \min_s \lambda^\pi(s)$. The bias $\lambda^\pi(s)$ can be seen as the total difference between the reward of state s and average reward.

In reinforcement learning [13] an agent does not know the transition P and/or reward r model in advance. Its goal is typically to find a policy π that maximizes its obtained reward. In this paper, we consider reinforcement learning in an MDP M where the learning algorithm is provided with an input set of m deterministic policies $\Pi = \{\pi_1, \ldots, \pi_m\}$. Such an input set of policies could arise in multiple situations, including: the policies may represent near-optimal policies for a set of m MDPs $\{M_1, \ldots, M_m\}$ which may be related to the current MDP M; the policies may be the result of different approximation schemes (i.e., approximate policy iteration with different approximation spaces); or they may be provided by m advisors. Our objective is to perform almost as well as the best policy in the input set Π on the new MDP M (with unknown P and/or r).

Our results require the following mild assumption:

Assumption 1. *There exists a policy $\pi^+ \in \Pi$, which induces a unichain Markov process on the MDP M, such that the average reward $\mu^+ = \mu^{\pi^+} \geq \mu^\pi(s)$ for any state $s \in S$ and any policy $\pi \in \Pi$. We also assume that $sp(\lambda^{\pi^+}) \leq H$, where H is a finite constant.*[2]

This assumption trivially holds when the optimal policy π^* is in the set Π. Also, in those cases that all the policies in Π induce some unichain Markov processes the existence of π^+ is guaranteed.[3]

A popular measure of the performance of a reinforcement learning algorithm over T steps is its regret relative to executing the optimal policy π^* in M. We evaluate the regret relative to the best policy π^+ in the input set Π,

$$\Delta(s) = T\mu^+ - \sum_{t=1}^{T} r_t, \qquad (3)$$

where $r_t \sim r(\cdot|s_t, a_t)$ and $s_0 = s$. We notice that this definition of regret differs from the standard definition of regret by an (approximation) error $T(\mu^* - \mu^+)$ due to the possible sub-optimality of the policies in Π relative to the optimal policy for MDP M. Further discussion on this definition is provided in Sec. 8.

3   Algorithm

In this section we introduce the Reinforcement Learning with Policy Advice (RLPA) algorithm (Alg. 1). Intuitively, the algorithm seeks to identify and use the policy in the input set Π that yields the highest average reward on the current MDP M. As the average reward of each $\pi \in \Pi$ on M, μ^π, is initially

---

[2] One can easily prove that the upper bound H always exists for any unichain Markov reward process (see [12, Chap. 8]).

[3] Note that Assumption 1 in general is a weaker assumption than assuming MDP M is ergodic or unichain, which would require that the induced Markov chains under *all* policies be recurrent or unichain, respectively: we only require that the best policy in the input set must induce a unichain Markov process.

Algorithm 1. Reinforcement Learning with Policy Advice (RLPA)

Require: Set of policies Π, confidence δ, span function f
1: Initialize $t = 0$, $i = 0$
2: Initialize $n(\pi) = 1$, $\widehat{\mu}(\pi) = 0$, $R(\pi) = 0$ and $K(\pi) = 1$ for all $\pi \in \Pi$
3: **while** $t \leq T$ **do**
4:     Initialize $t_i = 0$, $T_i = 2^i$, $\Pi_i = \Pi$, $\widehat{H} = f(T_i)$
5:     $i = i + 1$
6:     **while** $t_i \leq T_i$ & $\Pi_i \neq \emptyset$ **do** *(run trial)*
7:         $c(\pi) = (\widehat{H} + 1)\sqrt{48 \frac{\log(2t/\delta)}{n(\pi)}} + \widehat{H}\frac{K(\pi)}{n(\pi)}$
8:         $B(\pi) = \widehat{\mu}(\pi) + c(\pi)$
9:         $\widetilde{\pi} = \arg\max_\pi B(\pi)$
10:         $v(\widetilde{\pi}) = 1$
11:         **while** $t_i \leq T_i$ & $v(\widetilde{\pi}) < n(\widetilde{\pi})$ &
12:             $\widehat{\mu}(\widetilde{\pi}) - \frac{R(\widetilde{\pi})}{n(\widetilde{\pi})+v(\widetilde{\pi})} \leq c(\widetilde{\pi}) + (\widehat{H}+1)\sqrt{48 \frac{\log(2t/\delta)}{n(\widetilde{\pi})+v(\widetilde{\pi})}} + \widehat{H}\frac{K(\widetilde{\pi})}{n(\widetilde{\pi})+v(\widetilde{\pi})}$ **do**
13:             *(run episode)*
14:             $t = t + 1$, $t_i = t_i + 1$
15:             Take action $\widetilde{\pi}(s_t)$, observe s_{t+1} and r_{t+1}
16:             $v(\widetilde{\pi}) = v(\widetilde{\pi}) + 1$, $R(\widetilde{\pi}) = R(\widetilde{\pi}) + r_{t+1}$
17:         **end while**
18:         $K(\widetilde{\pi}) = K(\widetilde{\pi}) + 1$
19:         **if** $\widehat{\mu}(\widetilde{\pi}) - \frac{R(\widetilde{\pi})}{n(\widetilde{\pi})+v(\widetilde{\pi})} > c(\widetilde{\pi}) + (\widehat{H}+1)\sqrt{48 \frac{\log(2t/\delta)}{n(\widetilde{\pi})+v(\widetilde{\pi})}} + \widehat{H}\frac{K(\widetilde{\pi})}{n(\widetilde{\pi})+v(\widetilde{\pi})}$ **then**
20:             $\Pi_i = \Pi_i - \{\widetilde{\pi}\}$
21:         **end if**
22:         $n(\widetilde{\pi}) = n(\widetilde{\pi}) + v(\widetilde{\pi})$, $\widehat{\mu}(\widetilde{\pi}) = \frac{R(\widetilde{\pi})}{n(\widetilde{\pi})}$
23:     **end while**
24: **end while**

---

unknown, the algorithm proceeds by estimating these quantities by executing the different π on the current MDP. More concretely, RLPA executes a series of trials, and within each trial is a series of episodes. Within each trial the algorithm selects the policies in Π with the objective of effectively balancing between the exploration of all the policies in Π and the exploitation of the most promising ones. Our procedure for doing this falls within the popular class of "optimism in face uncertainty" methods. To do this, at the start of each episode, we define an upper bound on the possible average reward of each policy (Line 8): this average reward is computed as a combination of the average reward observed so far for this policy $\widehat{\mu}(\pi)$, the number of time steps this policy has been executed $n(\pi)$ and $\widehat{H}$, which represents a guess of the span of the best policy, H^+. We then select the policy with the maximum upper bound $\widetilde{\pi}$ (Line 9) to run for this episode. Unlike in multi-armed bandit settings where a selected arm is pulled for only one step, here the MDP policy is run for up to $n(\pi)$ steps, i.e., until its total number of execution steps is at most doubled. If $\widehat{H} \geq H^+$ then the confidence bounds computed (Line 8) are valid confidence intervals for the true best policy π^+; however, they may fail to hold for any other policy π whose span $sp(\lambda^\pi) \geq \widehat{H}$. Therefore, we can cut off execution of an episode when these confidence bounds fail to hold (the condition specified on Line 12), since the policy may not be an optimal one for the current MDP, if $\widehat{H} \geq H^+$.[4] In this case, we can eliminate the current policy $\widetilde{\pi}$ from the set of policies considered in this trial (see Line 20). After an episode terminates, the parameters of the current policy $\widetilde{\pi}$ (the number of steps $n(\pi)$ and average reward $\widehat{\mu}(\pi)$) are updated, new upper bounds on the policies are computed, and the next episode proceeds. As the average reward estimates converge, the better policies will be chosen more.

Note that since we do not know H^+ in advance, we must estimate it online: otherwise, if $\widehat{H}$ is not a valid upper bound for the span H^+ (see Assumption 1), a trial might eliminate the best policy π^+, thus incurring a significant regret. We address this by successively doubling the amount of time T_i each trial is run, and defining a $\widehat{H}$ that is a function f of the current trial length. See Sec. 4.1 for a more detailed discussion on the choice of f. This procedure guarantees the algorithm will eventually find an upper bound on the span H^+ and perform trials with very small regret in high probability. Finally, RLPA is an anytime algorithm since it does not need to know the time horizon T in advance.

4   Regret Analysis

In this section we derive a regret analysis of RLPA and we compare its performance to existing RL regret minimization algorithms. We first derive preliminary results used in the proofs of the two main theorems.

We begin by proving a general high-probability bound on the difference between average reward μ^π and its empirical estimate $\widehat{\mu}(\pi)$ of a policy π (throughout this discussion we mean the average reward of a policy π on a new MDP M). Let $K(\pi)$ be the number of episodes π has been run, each of them of length $v_k(\pi)$ $(k = 1, \ldots, K(\pi))$. The empirical average $\widehat{\mu}(\pi)$ is defined as

$$\widehat{\mu}(\pi) = \frac{1}{n(\pi)} \sum_{k=1}^{K(\pi)} \sum_{t=1}^{v_k(\pi)} r_t^k, \tag{4}$$

where $r_t^k \sim r(\cdot | s_t^k, \pi(s_t^k))$ is a random sample of the reward observed by taking the action suggested by π and $n(\pi) = \sum_k v_k(\pi)$ is the total count of samples. Notice that in each episode k, the first state s_1^k does not necessarily correspond to the next state of the last step $v_{k-1}(\pi)$ of the previous episode.

Lemma 1. *Assume that a policy π induces on the MDP M a single recurrent class with some additional transient states, i.e., $\mu^\pi(s) = \mu^\pi$ for all $s \in \mathcal{S}$. Then the difference between the average reward and its empirical estimate (Eq. 4) is*

$$|\widehat{\mu}(\pi) - \mu^\pi| \leq 2(H^\pi + 1)\sqrt{\frac{2\log(2/\delta)}{n(\pi)}} + H^\pi \frac{K(\pi)}{n(\pi)},$$

with probability $\geq 1 - \delta$, where $H^\pi = sp(\lambda^\pi)$ (see Eq. 2).

---

[4] See Sec. 4.1 for further discussion on the necessity of the condition on Line 12.

Proof. Let $r_\pi(s_t^k) = \mathbb{E}(r_t^k | s_t^k, \pi(s_t^k))$, $\epsilon_r(t, k) = r_t^k - r_\pi(s_t^k)$, and P^π be the state-transition kernel under policy π (i.e. for finite state and action spaces, P^π is the $|S| \times |S|$ matrix where the ij-th entry is $p(s_j | s_i, \pi(s_i))$). Then we have

$$\widehat{\mu}(\pi) - \mu^\pi = \frac{1}{n(\pi)} \left(\sum_{k=1}^{K(\pi)} \sum_{t=1}^{v_k(\pi)} (r_t^k - \mu^\pi) \right) = \frac{1}{n(\pi)} \left(\sum_{k=1}^{K(\pi)} \sum_{t=1}^{v_k(\pi)} (\epsilon_r(t,k) + r_\pi(s_t^k) - \mu^\pi) \right)$$

$$= \frac{1}{n(\pi)} \left(\sum_{k=1}^{K(\pi)} \sum_{t=1}^{v_k(\pi)} (\epsilon_r(t,k) + \lambda^\pi(s_t^k) - P^\pi \lambda^\pi(s_t^k)) \right),$$

where the second line follows from Eq. 2. Let $\epsilon_\lambda(t, k) = \lambda^\pi(s_{t+1}^k) - P^\pi \lambda^\pi(s_t^k)$. Then we have

$$\widehat{\mu}(\pi) - \mu^\pi = \frac{1}{n(\pi)} \left(\sum_{k=1}^{K(\pi)} \sum_{t=1}^{v_k(\pi)} (\epsilon_r(t,k) + \lambda^\pi(s_{t+1}^k) - \lambda^\pi(s_{t+1}^k) + \lambda^\pi(s_t^k) - P^\pi \lambda^\pi(s_t^k)) \right)$$

$$\leq \frac{1}{n(\pi)} \left(\sum_{k=1}^{K(\pi)} (H^\pi + \sum_{t=1}^{v_k(\pi)} \epsilon_r(t,k) + \sum_{t=1}^{v_k(\pi)-1} \epsilon_\lambda(t,k)) \right),$$

where we bounded the telescoping sequence $\sum_t (\lambda_{s_t^k}^\pi - \lambda^\pi(s_{t+1}^k)) \leq sp(\lambda^\pi) = H^\pi$. The sequences of random variables $\{\epsilon_r\}$ and $\{\epsilon_\lambda\}$, as well as their sums, are martingale difference sequences. Therefore we can apply Azuma's inequality and obtain the bound

$$\widehat{\mu}(\pi) - \mu^\pi \leq \frac{K(\pi)H^\pi + 2\sqrt{2n(\pi)\log(1/\delta)} + 2H^\pi \sqrt{2(n(\pi) - K(\pi))\log(1/\delta)}}{n(\pi)}$$

$$\leq H^\pi \frac{K(\pi)}{n(\pi)} + 2(H^\pi + 1)\sqrt{\frac{2\log(1/\delta)}{n(\pi)}},$$

with probability $\geq 1 - \delta$, where in the first inequality we bounded the error terms ϵ_r, each of which is bounded in $[-1, 1]$, and ϵ_λ, bounded in $[-H^\pi, H^\pi]$. The other side of the inequality follows exactly the same steps.   □

In the algorithm H^π is not known and at each trial i the confidence bounds are built using the guess on the span $\widehat{H} = f(T_i)$, where f is an increasing function. For the algorithm to perform well, it needs to not discard the best policy π^+ (Line 20). The following lemma guarantees that after a certain number of steps, with high probability the policy π^+ is not discarded in any trial.

Lemma 2. *For any trial started after $T \geq T^+ = f^{-1}(H^+)$, the probability of policy π^+ to be excluded from Π_A at anytime is less than $(\delta/T)^6$.*

Proof. Let i be the first trial such that $T_i \geq f^{-1}(H^+)$, which implies that $\widehat{H} = f(T_i) \geq H^+$. The corresponding step T is at most the sum of the length of all the trials before i, i.e., $T \leq \sum_{j=1}^{i-1} 2^j \leq 2^i$, thus leading to the condition $T \geq T^+ = f^{-1}(H^+)$. After $T \geq T^+$ the conditions in Lem. 1 (with Assumption 1) are satisfied for π^+. Therefore the confidence intervals hold with probability at least $1 - \delta$ and we have for $\widehat{\mu}(\pi^+)$

$$\widehat{\mu}(\pi^+) - \mu^+ \leq 2(H^+ + 1)\sqrt{\frac{2\log(1/\delta)}{n(\pi^+)}} + H^+ \frac{K(\pi^+)}{n(\pi^+)}$$

$$\leq 2(\widehat{H} + 1)\sqrt{\frac{2\log(1/\delta)}{n(\pi^+)}} + \widehat{H} \frac{K(\pi^+)}{n(\pi^+)},$$

where $n(\pi^+)$ is number of steps when policy π^+ has been selected until T. Using a similar argument as in the proof of Lem. 1, we can derive

$$\mu^+ - \frac{R(\pi^+)}{n(\pi^+) + v(\pi^+)} \leq 2(\widehat{H}+1)\sqrt{\frac{2\log(1/\delta)}{n(\pi^+) + v(\pi^+)}} + \widehat{H}\frac{K(\pi^+)}{n(\pi^+) + v(\pi^+)},$$

with probability at least $1-\delta$. Bringing together these two conditions, and applying the union bound, we have that the condition on Line 12 holds with at least probability $1-2\delta$ · and thus π^+ is never discarded. More precisely Algo. 1 uses slightly larger confidence intervals (notably $\sqrt{48\log(2t/\delta)}$ instead of $2\sqrt{2\log(1/\delta)}$), which guarantees that π^+ is discarded with at most a probability of $(\delta/T)^6$.   □

We also need the B-values (line 9) to be valid upper confidence bounds on the average reward of the best policy μ^+.

Lemma 3. *For any trial started after $T \geq T^+ = f^{-1}(H^+)$, the B-value of $\widetilde{\pi}$ is an upper bound on μ^+ with probability $\geq 1 - (\delta/T)^6$.*

Proof. Lem. 2 guarantees that the policy π^+ is in Π_A w.p. $(\delta/T)^6$. This combined with Lem. 1 and the fact that $f(T) > H^+$ implies that the B-value $B(\pi^+) = \widehat{\mu}(\pi^+) + c(\pi^+)$ is a high-probability upper bound on μ^+ and, since $\widetilde{\pi}$ is the policy with the maximum B-value, the result follows.   □

Finally, we bound the total number of episodes a policy could be selected.

Lemma 4. *After $T \geq T^+ = f^{-1}(H^+)$ steps of Algo. 1, let $K(\pi)$ be the total number of episodes π has been selected and $n(\pi)$ the corresponding total number of samples, then*

$$K(\pi) \leq \log_2(f^{-1}(H^+)) + \log_2(T) + \log_2(n(\pi)),$$

with probability $\geq 1 - (\delta/T)^6$.

Proof. Let $n_k(\pi)$ be the total number of samples at the beginning of episode k (i.e., $n_k(\pi) = \sum_{k'=1}^{k-1} v_{k'}(\pi)$). In each trial of Algo. 1, an episode is terminated when the number of samples is doubled (i.e., $n_{k+1}(\pi) = 2n_k(\pi)$), or when the consistency condition (last condition on Line 12) is violated and the policy is discarded or the trial is terminated (i.e., $n_{k+1} \geq n_k(\pi)$). We denote by $\overline{K}(\pi)$ the total number of episodes truncated before the number of samples is doubled, then $n(\pi) \geq 2^{K(\pi) - \overline{K}(\pi)}$. Since the episode is terminated before the number of samples is doubled only when either the trial terminates or the policy is discarded, in each trial this can only happen once per policy. Thus we can bound $\overline{K}(\pi)$ by the number of trials. A trial can either terminate because its maximum length T_i is reached or when all the polices are discarded (line 6). From Lem. 2, we have that after $T \geq f^{-1}(H^+)$, π^+ is never discarded w.h.p. and a trial only terminates when $t_i > T_i$. Since $T_i = 2^i$, it follows that the number of trials is bounded by $\overline{K}(\pi) \leq \log_2(f^{-1}(H^+)) + \log_2(T)$. So, we have $n(\pi) \geq 2^{K(\pi) - \log_2(f^{-1}(H^+)) - \log_2(T)}$, which implies the statement of the lemma.   □

Notice that if we plug this result in the statement of Lem. 1, we have that the second term converges to zero faster than the first term which decreases as $O(1/\sqrt{n(\pi)})$, thus in principle it could be possible to use alternative episode stopping criteria, such as $v(\pi) \leq \sqrt{n(\pi)}$. But while this would not significantly affect the convergence rate of $\widehat{\mu}(\pi)$, it may worsen the global regret performance in Thm. 1.

4.1   Gap-Independent Bound

We are now ready to derive the first regret bound for RLPA.

Theorem 1. *Under Assumption 1 for any $T \geq T^+ = f^{-1}(H^+)$ the regret of Algo. 1 is bounded as*

$$\Delta(s) \leq 24(f(T)+1)\sqrt{3Tm(\log(T/\delta))} + \sqrt{T} + 6f(T)m(\log_2(T^+) + 2\log_2(T)),$$

with probability at least $1 - \delta$ for any initial state $s \in \mathcal{S}$.

Proof. We begin by bounding the regret from executing each policy π. We consider the $k(\pi)$-th episode when policy π has been selected (i.e., π is the optimistic policy $\widetilde{\pi}$) and we study its corresponding total regret Δ_π. We denote by $n_k(\pi)$ the number of steps of policy π at the beginning of episode k and $v_k(\pi)$ the number of steps in episode k. Also at time step T, let the total number of episodes, $v_k(\pi)$ and n_k, for each policy π be denoted as $K(\pi)$, $v(\pi)$ and $n(\pi)$ respectively. We also let $\pi \in \Pi$, $B(\pi)$, $c(\pi)$, $R(\pi)$ and $\widehat{\mu}(\pi)$ be the latest values of these variables at time step T for each policy π. Let $\mathcal{E} = \{\forall t = f^{-1}(H^+), \ldots, T, \pi^+ \in \Pi_A \ \& \ \widetilde{\pi} \geq \mu^+\}$ be the event under which π^+ is never removed from the set of policies Π_A, and where the upper bound of the optimistic policy $\widetilde{\pi}$, $B(\widetilde{\pi})$, is always as large as the true average reward of the best policy μ^+. On the event $\mathcal{E}$, Δ_π can be bounded as

$$\Delta_\pi = \sum_{k=1}^{K(\pi)} \sum_{t=1}^{v_k(\pi)} (\mu^+ - r_t) \overset{(1)}{\leq} \sum_{k=1}^{K(\pi)} \sum_{t=1}^{v_k(\pi)} (B(\pi) - r_t) \leq (n(\pi) + v(\pi))(\widehat{\mu}(\pi) + c(\pi)) - R(\pi)$$

$$\overset{(2)}{\leq} (n(\pi) + v(\pi)) \left(3(f(T)+1)\sqrt{48\frac{\log(T/\delta)}{n(\pi)}} + 3f(T)\frac{K(\pi)}{n(\pi)} \right)$$

$$\overset{(3)}{\leq} 24(f(T)+1)\sqrt{3n(\pi)\log(T/\delta)} + 6f(T)K(\pi),$$

where in (1) we rely on the fact that π is only executed when it is the optimistic policy, and $B(\pi)$ is optimistic with respect to μ^+ according to Lem. 3. (2) immediately follows from the stopping condition at Line 12 and the definition of $c(\pi)$. (3) follows from the condition on doubling the samples (Line 12) which guarantees $v(\pi) \leq n(\pi)$.

We now bound the total regret Δ by summing over all the policies.

$$\Delta = \sum_{\pi \in \Pi} 24(f(T)+1)\sqrt{3n(\pi)\log(T/\delta)} + 6f(T)\sum_{\pi \in \Pi} K(\pi)$$

$$\overset{(1)}{\leq} 24(f(T)+1)\sqrt{3m\sum_{\pi \in \Pi} n(\pi)\log(T/\delta)} + 6f(T)\sum_{\pi \in \Pi} K(\pi)$$

$$\overset{(2)}{\leq} 24(f(T)+1)\sqrt{3mT\log(T/\delta)} + 6f(T)m(\log_2(f^{-1}(H^+)) + 2\log_2(T)),$$

where in (1) we use Cauchy-Schwarz inequality and (2) follows from $\sum_\pi n(\pi) \leq T$, Lem. 4, and $\log_2(n(\pi)) \leq \log_2(T)$.

Since T is an unknown time horizon, we need to provide a bound which holds with high probability uniformly over all the possible values of T. Thus we need to deal with the case when $\mathcal{E}$ does not hold. Based on Lem. 1 and by following similar lines to [7], we can prove that the total regret of the episodes in which the true model is discarded is bounded by $\sqrt{T}$ with probability at least $1 - \delta/(12T^{5/4})$. Due to space limitations, we omit the details, but we can then prove the final result by combining the regret in both cases (when $\mathcal{E}$ holds or does not hold) and taking union bound on all possible values of T.   □

A significant advantage of RLPA over generic RL algorithms (such as UCRL2) is that the regret of RLPA is independent of the size of the state and action spaces: in contrast, the regret of UCRL2 scales as $O(S\sqrt{AT})$. This advantage is obtained by exploiting the prior information that Π contains good policies, which allows the algorithm to focus on testing their performance to identify the best, instead of building an estimate of the current MDP over the whole state-action space as in UCRL2. It is also informative to compare this result to other methods using some form of prior knowledge. In [8] the objective is to learn the optimal policy along with a state representation which satisfies the Markov property. The algorithm receives as input a set of possible state representation models and under the assumption that one of them is Markovian, the algorithm is shown to have a sub-linear regret. Nonetheless, the algorithm inherits the regret of UCRL itself and still displays a $O(S\sqrt{A})$ dependency on states and actions. In [5] the Parameter Elimination (PEL) algorithm is provided with a set of MDPs. The algorithm is analyzed in the PAC-MDP framework and under the assumption that the true model actually belongs to the set of MDPs, it is shown to have a performance which does not depend on the size of the state-action space and it only has a $O(\sqrt{m})$ a dependency on the number of MDPs m.[5] In our setting, although no model is provided and no assumption on the optimality of π^* is made, RLPA achieves the same dependency on m.

The span $sp(\lambda^\pi)$ of a policy is known to be a critical parameter determining how well and fast the average reward of a policy can be estimated using samples (see e.g., [1]). In Thm. 1 we show that only the span H^+ of the best policy π^+ affects the performance of RLPA even when other policies have much larger spans. Although this result may seem surprising (the algorithm estimates the average reward for all the policies), it follows from the use of the third condition on Line12 where an episode is terminated, and a policy is discarded, whenever the empirical estimates are not consistent with the guessed confidence interval. Let us consider the case when $\widehat{H} > H^+$ but $\widehat{H} < sp(\lambda^\pi)$ for a policy which is selected as the optimistic policy $\widetilde{\pi}$. Since the confidence intervals built for π are not correct (see Lem. 1), $\widetilde{\pi}$ could be selected for a long while before selecting a different policy. On the other hand, the condition on the consistency of the observed rewards would discard π (with high probability), thus increasing the chances of the best policy (whose confidence intervals are correct) to be selected. We also note that H^+ appears as a constant in the regret through $\log_2(f^{-1}(H^+))$ and this suggests that the optimal choice of f is $f(T) = \log(T)$, which would lead to a bound of order (up to constants and logarithmic terms) $\widetilde{O}(\sqrt{Tm} + m)$.

---

[5] Notice that PAC bounds are always squared w.r.t. regret bounds, thus the original m dependency in [5] becomes $O(\sqrt{m})$ when compared to a regret bound.

4.2   Gap-Dependent Bound

Similar to [7], we can derive an alternative bound for RLPA where the dependency on T becomes logarithmic and the gap between the average of the best and second best policies appears. We first need to introduce two assumptions.

Assumption 2 (Average Reward). *Each policy $\pi \in \Pi$ induces on the MDP M a single recurrent class with some additional transient states, i.e., $\mu^\pi(s) = \mu^\pi$ for all $s \in S$. This implies that $H^\pi = sp(\lambda^\pi) < +\infty$.*

Assumption 3 (Minimum Gap). *Define the gap between the average reward of the best policy π^+ and the average reward of any other policy as $\Gamma(\pi, s) = \mu^+ - \mu^\pi(s)$ for all $s \in S$. We then assume that for all $\pi \in \Pi - \{\pi^+\}$ and $s \in S$, $\Gamma(\pi, s)$ is uniformly bounded from below by a positive constant $\Gamma_{\min} > 0$.*

Theorem 2 (Gap Dependent Bounds). *Let Assumptions 2 and 3 hold. Run Algo. 1 with the choice of $\delta = \sqrt[3]{1/T}$ (the stopping time T is assumed to be known here). Assume that for all $\pi \in \Pi$ we have that $H_\pi \leq H_{\max}$. Then the expected regret of Algo. 1, after $T \geq T^+ = f^{-1}(H^+)$ steps, is bounded as*

$$\mathbb{E}(\Delta(s)) = O\left(m \frac{(f(T) + H_{\max})(\log_2(mT) + \log_2(T^+))}{\Gamma_{\min}} \right), \tag{5}$$

for any initial state $s \in S$.

Proof. **(sketch)** Unlike for the proof of Thm. 1, here we need a more refined control on the number of steps of each policy as a function of the gaps $\Gamma(\pi, s)$. We first notice that Assumption 2 allows us to define $\Gamma(\pi) = \Gamma(\pi, s) = \mu^+ - \mu^\pi$ for any state $s \in S$ and any policy $\pi \in \Pi$. We consider the high-probability event $\mathcal{E} = \{\forall t = f^{-1}(H^+), \ldots, T, \pi^+ \in \Pi_A\}$ (see Lem. 2) where for all the trials run after $f^{-1}(H^+)$ steps never discard policy π^+. We focus on the episode at time t, when an optimistic policy $\widetilde{\pi} \neq \pi^+$ is selected for the $k(\pi)$-th time, and we denote by $n_k(\widetilde{\pi})$ the number of steps of $\widetilde{\pi}$ before episode k and $v_k(\pi)$ the number of steps during episode $k(\pi)$. The cumulative reward during episode k is $R_k(\widetilde{\pi})$ obtained as the sum of $\widehat{\mu}_k(\widetilde{\pi})n_k(\widetilde{\pi})$ (the previous cumulative reward) and the sum of $v_k(\widetilde{\pi})$ rewards received since the beginning of the episode. Let $\mathcal{E} = \{\forall t = f^{-1}(H^+), \ldots, T, \pi^+ \in \Pi_A \ \& \ \widetilde{\pi} \geq \mu^+\}$ be the event under which π^+ is never removed from the set of policies Π_A, and where the upper bound of the optimistic policy $\widetilde{\mu}$, $B(\widetilde{\pi})$, is always as large as the true average reward of the best policy μ^+. On event $\mathcal{E}$ we have

$$3(\widehat{H} + 1)\sqrt{48\frac{\log(t/\delta)}{n_k(\widetilde{\pi})}} + 3\frac{k(\pi)}{n_k(\widetilde{\pi})} \overset{(1)}{\geq} B(\widetilde{\pi}) - \frac{R_k(\widetilde{\pi})}{n_k(\widetilde{\pi}) + v_k(\widetilde{\pi})}$$

$$\overset{(2)}{\geq} \mu^+ - \frac{R_k(\widetilde{\pi})}{n_k(\widetilde{\pi}) + v_k(\widetilde{\pi})} \geq \mu^+ - \mu^{\widetilde{\pi}} + \frac{1}{n_k(\widetilde{\pi}) + v_k(\widetilde{\pi})} \sum_{t=1}^{n_k(\widetilde{\pi})+v_k(\widetilde{\pi})} (\mu^{\widetilde{\pi}} - r_t)$$

$$\overset{(3)}{\geq} \Gamma_{\min} + \frac{1}{n_k(\widetilde{\pi}) + v_k(\widetilde{\pi})} \sum_{t=1}^{n_k(\widetilde{\pi})+v_k(\widetilde{\pi})} (\mu^{\widetilde{\pi}} - r_t) \overset{(4)}{\geq} \Gamma_{\min} - H^{\widetilde{\pi}}\sqrt{48\frac{\log(t/\delta)}{n_k(\widetilde{\pi})}} - H^{\widetilde{\pi}}\frac{K(\widetilde{\pi})}{n_k(\widetilde{\pi})},$$

with probability $1 - (\delta/t)^6$. Inequality (1) is enforced by the episode stopping condition on Line 12 and the definition of $B(\pi)$, (2) is guaranteed by Lem. 3, (3) relies on the definition of gap and Assumption 3, while (4) is a direct application of Lem. 1. Rearranging the terms, and applying Lem. 4, we obtain

$$n_k(\widetilde{\pi})\Gamma_{\min} \leq (3\widehat{H} + 3 + H^{\widetilde{\pi}})\sqrt{n(\widetilde{\pi})}\sqrt{48\log(t/\delta)} + 4H^{\widetilde{\pi}}(2\log_2(t) + \log_2(f^{-1}(H^+))).$$

By solving the inequality w.r.t. $n_k(\widetilde{\pi})$ we obtain

$$\sqrt{n(\widetilde{\pi})} \leq \frac{(3\widehat{H} + 3 + H^{\widetilde{\pi}})\sqrt{48\log(t/\delta)} + 2\sqrt{H^{\widetilde{\pi}}\Gamma_{\min}(2\log_2(t) + \log_2(f^{-1}(H^+))}}{\Gamma_{\min}}, \quad (6)$$

w.p. $1 - (\delta/t)^6$. This implies that on the event $\mathcal{E}$, after t steps, RLPA acted according to a suboptimal policy π for no more than $O(\log(t)/\Gamma_{\min}^2)$ steps. The rest of the proof follows similar steps as in Thm. 1 to bound the regret of all the suboptimal policies in high probability. The expected regret of π^+ is bounded by H^+ and standard arguments similar to [7] are used to move from high-probability to expectation bounds.   □

Note that although the bound in Thm. 1 is stated in high-probability, it is easy to turn it into a bound in expectation with almost identical dependencies on the main characteristics of the problem and compare it to the bound of Thm. 2. The major difference is that the bound in Eq. 5 shows a $O(\log(T)/\Gamma_{\min})$ dependency on T instead of $O(\sqrt{T})$. This suggests that whenever there is a big margin between the best policy and the other policies in Π, the algorithm is able to accordingly reduce the number of times suboptimal policies are selected, thus achieving a better dependency on T. On the other hand, the bound also shows that whenever the policies in Π are very similar, it might take a long time to the algorithm before finding the best policy, although the regret cannot be larger than $O(\sqrt{T})$ as shown in Thm. 1.

We also note that while Assumption 3 is needed to allow the algorithm to "discard" suboptimal policies with only a logarithmic number of steps, Assumption 2 is more technical and can be relaxed. It is possible to instead only require that each policy $\pi \in \Pi$ has a bounded span, $H^\pi < \infty$, which is a milder condition than requiring a constant average reward over states (i.e., $\mu^\pi(s) = \mu^\pi$).

5   Computational Complexity

As shown in Algo. 1, RLPA runs over multiple trials and episodes where policies are selected and run. The largest computational cost in RLPA is at the start of each episode computing the B-values for all the policies currently active in Π_A and then selecting the most optimistic one. This is an $O(m)$ operation. The total number of episodes can be upper bounded by $2\log_2(T) + \log_2(f^{-1}(H^+))$ (see Lem. 4). This means the overall computational of RLPA is of $O(m(\log_2(T) + \log_2(f^{-1}(H^+))))$. Note there is no explicit dependence on the size of the state and action space. In contrast, UCRL2 has a similar number of trials, but requires solving extended value iteration to compute the optimistic MDP policy.

Extended value iteration requires $O(|S|^2|A|\log(|S|))$ computation per iteration: if D are the number of iterations required to complete extended value iteration, then the resulting cost would be $O(D|S|^2|A|\log(|S|))$. Therefore UCRL2, like many generic RL approaches, will suffer a computational complexity that scales quadratically with the number of states, in contrast to RLPA, which depends linearly on the number of input policies and is independent of the size of the state and action space.

6 Experiments

In this section we provide some preliminary empirical evidence of the benefit of our proposed approach. We compare our approach with two other baselines. As mentioned previously, UCRL2 [7] is a well known algorithm for generic RL problems that enjoys strong theoretical guarantees in terms of high probability regret bounds with the optimal rate of $O(\sqrt{T})$. Unlike our approach, UCRL2 does not make use of any policy advice, and its regret scales with the number of states and actions as $O(|\mathcal{S}|\sqrt{|\mathcal{A}|})$. To provide a more fair comparison, we also introduce a natural variant of UCRL2, Upper Confidence with Models (UCWM), which takes as input a set of MDP models $\mathcal{M}$ which is assumed to contain the actual model M. Like UCRL2, UCWM computes confidence intervals over the task's model parameters, but then selects the optimistic policy among the optimal policies for the subset of models in $\mathcal{M}$ consistent with the confidence interval. This may result in significantly tighter upper-bound on the optimal value function compared to UCRL2, and may also accelerate the learning process. If the size of possible models shrinks to one, then UCWM will seamlessly transition to following the optimal policy for the identified model. UCWM requires as input a set of MDP models, whereas our RLPA approach requires only input policies.

We consider a square grid world with 4 actions: up (a_1), down (a_2), right (a_3) and left (a_4) for every state. A *good* action succeeds with the probability 0.85, and goes in one of the other directions with probability 0.05 (unless that would cause it to go into a wall) and a *bad* action stays in the same place with probability 0.85 and goes in one of the 4 other directions with probability 0.0375. We construct four variants of this grid world $\mathcal{M} = \{M_1, M_2, M_3, M_4\}$. In model 1 (M_1) good actions are 1 and 4, in model 2 (M_2) good actions are 1 and 2, in model 3 good actions are 2 and 3, and in model 4 good actions are 3 and 4. All other actions in each MDP are bad actions. The reward in all MDPs is the same and is -1 for all states except for the four corners which are: 0.7 (upper left), 0.8 (upper right), 0.9 (lower left) and 0.99 (lower right). UCWM receives as input the MDP models and RLPA receives as input the optimal policies of $\mathcal{M}$.

We evaluate the performances of each algorithm in terms of the per-step regret, $\hat{\Delta} = \Delta/T$ (see Eq. 3). Each run is $T = 100000$ steps and we average the performance on 100 runs. The agent is randomly placed at one of the states of the grid at the beginning of each round. We assume that the true MDP model is M_4. Notice that in this case $\pi^* \in \Pi$, thus $\mu^+ = \mu^*$ and the regret compares to the optimal average reward. The identity of the true MDP is not known by

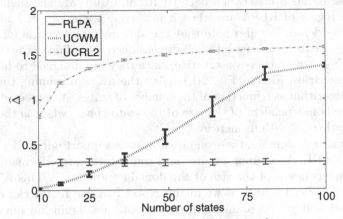

Fig. 1. Per-step regret versus number of states

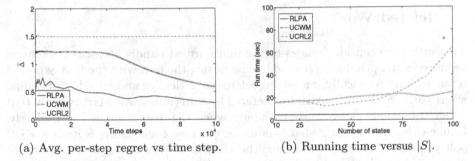

(a) Avg. per-step regret vs time step.   (b) Running time versus $|S|$.

Fig. 2. Transient behavior and run time

the agent. For RLPA we set $f(t) = \log(t)$.[6] We construct grid worlds of various sizes and compare the resulting performance of the three algorithms.

Fig. 1 shows per-step regret of the algorithms as the function of the number of states. As predicted by the theoretical bounds, the per-step regret $\widehat{\Delta}$ of UCRL2 significantly increases as the number of states increases, whereas the average regret of our RLPA is essentially independent of the state space size[7]. Although UCWM has a lower regret than RLPA for a small number of states, it quickly loses its advantage as the number of states grows. UCRL2's per-step regret plateaus after a small number of states since it is effectively reaching the maximum possible regret given the available time horizon.

To demonstrate the performance of each approach for a single task, Fig. 2(a) shows how the per-step regret changes with different time horizons for a grid-world with 64 states. RLPA demonstrates a superior regret throughout the run

---

[6] See Sec. 4.1 for the rational behind this choice.

[7] The RLPA regret bounds depend on the bias of the optimal policy which may be indirectly a function of the structure and size of the domain.

with a decrease that is faster than both UCRL and UCWM. The slight periodic increases in regret of RLPA are when a new trial is started, and all policies are again considered. We also note that the slow rate of decrease for all three algorithms is due to confidence intervals dimensioned according to the theoretical results which are often over-conservative, since they are designed to hold in the worst-case scenarios. Finally, Fig. 2(b) shows the average running time of one trial of the algorithm as a function of the number of states. As expected, RLPA's running time is independent of the size of the state space, whereas the running time of the other algorithms increases.

Though a simple domain, these empirical results support our earlier analysis, demonstrating RLPA exhibits a regret and computational performance that is essentially independent of the size of the domain state space. This is a significant advantage over UCRL2, as we might expect because RLPA can efficiently leverage input policy advice. Interestingly, we obtain a significant improvement also over the more competitive baseline UCWM.

7   Related Work

The setting we consider relates to the multi-armed bandit literature, where an agent seeks to optimize its reward by uncovering the arm with the best expected reward. More specifically, our setting relates to restless [9] and rested [15] bandits, where each arm's distribution is generated by a an (unknown) Markov chain that either transitions at every step, or only when the arm is pulled, respectively. Unlike either restless or rested bandits, in our case each "arm" is itself a MDP policy, where different actions may be chosen. However, the most significant distinction may be that in our setting there is a independent state that couples the rewards obtained across the policies (the selected action depends on both the policy/arm selected, and the state), in contrast to the rested and restless bandits where the Markov chains of each arm evolve independently.

Prior research has demonstrated a significant improvement in learning in a discrete state and action RL task whose Markov decision process model parameters are constrained to lie in a finite set. In this case, an objective of maximizing the expected sum of rewards can be framed as planning in a finite-state partially observable Markov decision process [10]: if the parameter set is not too large, off-the-shelf POMDP planners can be used to yield significant performance improvements over state-of-the-art RL approaches [2]. Other work [5] on this setting has proved that the sample complexity of learning to act well scales independently of the size of the state and action space, and linearly with the size of the parameter set. These approaches focus on leveraging information about the model space in the context of Bayesian RL or PAC-style RL, in contrast to our model-free approach that focuses on regret.

There also exists a wealth of literature on learning with expert advice (e.g. [3]). The majority of this work lies in supervised learning. Prior work by Diuk et al. [4] leverages a set of experts where each expert predicts a probabilistic concept (such as a state transition) to provide particularly efficient KWIK RL. In contrast, our approach leverages input policies, rather than models. Probabilistic policy reuse [6] also adaptively selects among a prior set of provided policies, but may also choose to create and follow a new policy. The authors present promising empirical results but no theoretical guarantees are provided. However, we will further discuss this interesting issue more in the future work section.

The most closely related work is by Talvitie and Singh [14], who also consider identifying the best policy from a set of input provided policies. Talvitie and Singh's approach is a special case of a more general framework for leveraging experts in sequential decision making environments where the outcomes can depend on the full history of states and actions [11]: however, this more general setting provides bounds in terms of an abstract quantity, whereas Talvitie and Singh provide bounds in terms of the bounds on mixing times of a MDP. There are several similarities between our algorithm and the work of Talvitie and Singh, though in contrast to their approach we take an optimism under uncertainty approach, leveraging confidence bounds over the potential average reward of each policy in the current task. However, the provided bound in their paper is not a regret bound and no precise expression on the bound is stated, rendering it infeasible to do a careful comparison of the theoretical bounds. In contrast, we provide a much more rigorous theoretical analysis, and do so for a more general setting (for example, our results do not require the MDP to be ergodic). Their algorithm also involves several parameters whose values must be correctly set for the bounds to hold, but precise expressions for these parameters were not provided, making it hard to perform an empirical comparison.

8   Future Work and Conclusion

In defining RLPA we preferred to provide a simple algorithm which allowed us to provide a rigorous theoretical analysis. Nonetheless, we expect the current version of the algorithm can be easily improved over multiple dimensions. The immediate possibility is to perform off-policy learning across the policies: whenever a reward information is received for a particular state and action, this could be used to update the average reward estimate $\widehat{\mu}(\pi)$ for all policies that would have suggested the same action for the given state. As it has been shown in other scenarios, we expect this could improve the empirical performance of RLPA. However, the implications for the theoretical results are less clear. Indeed, updating the estimate $\widehat{\mu}(\pi)$ of a policy π whenever a "compatible" reward is observed would correspond to a significant increase in the number of episodes $K(\pi)$ (see Eq. 4). As a result, the convergence rate of $\widehat{\mu}(\pi)$ might get worse and could potentially degrade up to the point when $\widehat{\mu}(\pi)$ does not even converge to the actual average reward μ^{π}. (see Lem. 1 when $K(\pi) \simeq n(\pi)$). We intend to further investigate this in the future.

Another very interesting direction of future work is to extend RLPA to leverage policy advice when useful, but still maintain generic RL guarantees if the input policy space is a poor fit to the current problem. More concretely, currently if π^+ is not the actual optimal policy of the MDP, RLPA suffers an additional linear regret to the optimal policy of order $T(\mu^* - \mu^+)$. If T is very large and π^+ is highly suboptimal, the total regret of RLPA may be worse than UCRL, which always eventually learns the optimal policy. This opens the question whether it is possible to design an algorithm able to take advantage of the small regret-to-best of RLPA when T is small and π^+ is nearly optimal and the guarantees of UCRL for the regret-to-optimal.

To conclude, we have presented RLPA, a new RL algorithm that leverages an input set of policies. We prove the regret of RLPA relative to the best policy scales sub-linearly with the time horizon, and that both this regret and the computational complexity of RLPA are independent of the size of the state and action space. This suggests that RLPA may offer significant advantages in large domains where some prior *good* policies are available.

References

1. Bartlett, P.L., Tewari, A.: Regal: A regularization based algorithm for reinforcement learning in weakly communicating mdps. In: UAI, pp. 35–42 (2009)
2. Brunskill, E.: Bayes-optimal reinforcement learning for discrete uncertainty domains. In: Abstract. Proceedings of the International Conference on Autonomous Agents and Multiagent System (2012)
3. Cesa-Bianchi, N., Freund, Y., Haussler, D., Helmbold, D.P., Schapire, R.E., Warmuth, M.K.: How to use expert advice. Journal of the ACM 44(3), 427–485 (1997)
4. Diuk, C., Li, L., Leffler, B.R.: The adaptive k-meteorologists problem and its application to structure learning and feature selection in reinforcement learning. In: ICML (2009)
5. Dyagilev, K., Mannor, S., Shimkin, N.: Efficient reinforcement learning in parameterized models: Discrete parameter case. In: European Workshop on Reinforcement Learning (2008)
6. Fernández, F., Veloso, M.M.: Probabilistic policy reuse in a reinforcement learning agent. In: AAMAS, pp. 720–727 (2006)
7. Jaksch, T., Ortner, R., Auer, P.: Near-optimal regret bounds for reinforcement learning. Journal of Machine Learning Research 11, 1563–1600 (2010)
8. Maillard, O., Nguyen, P., Ortner, R., Ryabko, D.: Optimal regret bounds for selecting the state representation in re inforcement learning. In: ICML, JMLR W&CP, Atlanta, USA, vol. 28(1), pp. 543–551 (2013)
9. Ortner, R., Ryabko, D., Auer, P., Munos, R.: Regret bounds for restless markov bandits. In: Bshouty, N.H., Stoltz, G., Vayatis, N., Zeugmann, T. (eds.) ALT 2012. LNCS (LNAI), vol. 7568, pp. 214–228. Springer, Heidelberg (2012)
10. Poupart, P., Vlassis, N., Hoey, J., Regan, K.: An analytic solution to discrete bayesian reinforcement learning. In: ICML (2006)
11. Pucci de Farias, D., Megiddo, N.: Exploration-exploitation tradeoffs for experts algorithms in reactive environments. In: Advances in Neural Information Processing Systems 17, pp. 409–416 (2004)
12. Puterman, M.L.: Markov Decision Processes: Discrete Stochastic Dynamic Programming, 1st edn. John Wiley & Sons, Inc., New York (1994)
13. Sutton, R.S., Barto, A.G.: Reinforcement Learning: An Introduction. MIT Press, Cambridge (1998)
14. Talvitie, E., Singh, S.: An experts algorithm for transfer learning. In: IJCAI (2007)
15. Tekin, C., Liu, M.: Online learning of rested and restless bandits. IEEE Transactions on Information Theory 58(8), 5588–5611 (2012)

Exploiting Multi-step Sample Trajectories for Approximate Value Iteration

Robert Wright[1,2], Steven Loscalzo[1], Philip Dexter[2], and Lei Yu[2]

[1] AFRL Information Directorate, Rome, NY, USA
{robert.wright,steven.loscalzo}@rl.af.mil
[2] Binghamton University, Binghamton, NY, USA
{pdexter1,lyu}@binghamton.edu

Abstract. Approximate value iteration methods for reinforcement learning (RL) generalize experience from limited samples across large state-action spaces. The function approximators used in such methods typically introduce errors in value estimation which can harm the quality of the learned value functions. We present a new batch-mode, off-policy, approximate value iteration algorithm called Trajectory Fitted Q-Iteration (TFQI). This approach uses the sequential relationship between samples within a trajectory, a set of samples gathered sequentially from the problem domain, to lessen the adverse influence of approximation errors while deriving long-term value. We provide a detailed description of the TFQI approach and an empirical study that analyzes the impact of our method on two well-known RL benchmarks. Our experiments demonstrate this approach has significant benefits including: better learned policy performance, improved convergence, and some decreased sensitivity to the choice of function approximation.

1 Introduction

Temporal Difference (TD) based value iteration methods solve reinforcement learning problems by estimating the optimal value function over the problem space [14]. This function describes the maximal expected long-term value of taking actions in any given state of the domain and can be used to extract an optimal policy. Representing value functions exactly is infeasible in all but the most trivial domains giving rise to Approximate Value Iteration (AVI) algorithms [8]. Function approximation provides a mechanism for efficiently representing value functions, however, by their very nature they introduce generalization errors. Typical AVI methods are dependent on the approximation model to derive and propagate the long-term value of a state through the function space. The generalization error introduced by function approximation adversely impacts the derivation of long-term values and as a consequence the policy described by the learned function.

Reinforcement Learning (RL) problems are multi-step and as such the data used by value iteration methods typically comes in the form of sequential sets of experience samples known as trajectories. Trajectories are actual observed

H. Blockeel et al. (Eds.): ECML PKDD 2013, Part I, LNAI 8188, pp. 113–128, 2013.
© Springer-Verlag Berlin Heidelberg 2013

paths through the problem space that describe how value is propagated through the state space without the error induced by function approximation. Despite the availability of this information, AVI based methods commonly ignore the sequential relationship between samples while deriving the value of a state and rely instead on the approximation model.

Figure 1 provides an illustration of the AVI generalization problem alluded to above. There are four sampled states, represented by the circles, that are part of two separate trajectories that were generated by arbitrary policies. In this example, assume that the immediate rewards for the transitions shown are zero and we are given the true optimal long-term values, V, for the transitions that go beyond states B and D, 4 and 2 respectively. The function approximation model here is a tiling abstraction represented by the rectangle, $\bar{S}$, that combines the value of states B and D to predict a generalized value, x where $2 < x < 4$, for any other state $\bar{S}$ contains. The value x may be a reasonable approximation of the value of unobserved states within $\bar{S}$, however it produces generalization errors for the values of B and D. With function approximation, standard AVI approaches will only use x to determine the long-term value for states A and C instead of the long-term values for the states that follow in their sampled trajectories, again 4 and 2 respectively. From the trajectories we can see that the value for state A in this example should be greater than that of state C's. But, due to the generalization error, value iteration will derive a value for those states based upon x that is an underestimate of the value of state A and, potentially, an overestimate of the value of state C.

Unfortunately this error will not just impact the derivation of the long-term value for the sampled states A and C. AVI methods calculate long-term value by back-propagating rewards along approximated transition paths. This error will therefore adversely affect the derivation of long-term value for all other sampled states that can reach states B and D. Errors such as this will propagate and reoccur through the function space and can reduce the quality of the learned policy. This form of generalization error is common among statistical function approximation, and is not limited to tiling abstractions.

The trajectory AB can be used to correct for the underestimation of the sampled value of A. Trajectory AB demonstrates that following some unknown policy from the sampled state B a value of at least 4 is attainable. Given this information, while updating the long-term value for the sampled state A we should use the long-term value implied by the trajectory, 4, rather than x because we know it is an underestimate. Unfortunately we cannot apply the same reasoning to correct for the potential overestimation in the sampled value of C. The trajectory CD does not provide definitive evidence that from state D a value greater than x is not achievable. Instead, the value of x should still be used while updating the long-term value estimate of C.

In this paper we present a new algorithm we call Trajectory Fitted Q-iteration (TFQI) that utilizes trajectory information in this way to reduce the impact of generalization error. It is a batch mode, model-free, off-policy RL algorithm based upon the Fitted Q-Iteration (FQI) framework [4]. FQI is a well-known

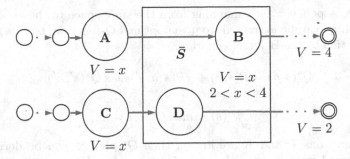

Fig. 1. A simple 4 state example scenario where AVI can produce misleading long-term value estimates. A, B, C, and D are states that have been sampled belonging to two sample trajectories AB and CD. $\bar{S}$ is a state abstraction that generalizes the value of the states it contains, B and D. The V's indicate the value derived for each state by standard AVI.

value iteration algorithm that has demonstrated great empirical performance on RL benchmarks [11]. In addition we provide an empirical analysis of this new approach that demonstrates TFQI learns policies that are superior to that of FQI using the same data and function approximation architecture with no significant additional computational costs beyond the original FQI formulation. TFQI not only demonstrates the ability to learn better policies, but also shows dramatically improved convergence and some decrease in the sensitivity to approximation model settings. The only additional assumption TFQI makes beyond FQI, and other off-policy TD methods, is the availability of trajectory based data which, as stated above, is the norm rather than the exception.

The remainder of the paper is structured as follows: Section 2 provides background and notation and a detailed description of FQI. Section 3 introduces TFQI and elaborates upon the ideas behind the new approach. In Section 4 we provide details for the setup of our empirical analysis. Section 5 presents the results as well as a discussion of their significance. Section 6 provides a brief overview of related work. Finally, in Section 7 we conclude and identify future directions for this research.

2   Background

Reinforcement learning is commonly discussed within the framework of Markov Decision Processes (MDP) [10]. An MDP, M, is a 5-tuple, $M = \langle S, A, P, R, \gamma \rangle$, where S is a set of states of the world, A is the set of actions, P is the state transition model such that $P(s, a, s') \in [0, 1]$ describes the probability of transitioning to state s' after taking action a in state s, R is the reward function such that $R(s, a) \in \mathbb{R}$ describes the immediate reward received for taking action a in state s, and γ is the discount factor on future rewards bound by $[0, 1)$.

The goal of RL is to derive an optimal policy, π^*, over M that maximizes the discounted long-term aggregate value that can be obtained starting from any given state. A policy, π, is a mapping from the state space to the action space, $\pi = S \mapsto A$. π^* can be extracted from the optimal Q-function, $Q^*(s,a)$, defined by the solution to the Bellman equation:

$$Q^*(s,a) = R(s,a) + \gamma P(s,a,s') \max_{a' \in A} Q^*(s',a') \tag{1}$$

$$\pi^*(s) = \max_{a \in A} Q^*(s,a) \tag{2}$$

If the functions P and R are known then Q^* and π^* can be derived in a straightforward fashion using dynamic programming. However, in the model-free RL context P and R are unknown and Q^* must instead be estimated from samples. Samples describe single-step observations of transitions taken in the domain. They are represented by tuples, (s,a,s',r), consisting of a state s, and action a, the state s' transitioned to by taking a in s, and r, the immediate reward for that transition. Samples are collected through interactions with the domain by the agent itself, a third party, or by a generative model when available.

Value iteration combined with Temporal Difference (TD) [14] provides a model-free way to estimate $Q^*(s,a)$ directly from samples. Given a set of samples to learn from, value iteration can learn estimates of the Q-values for each sample by iteratively applying the Bellman update rule to each sample:

$$Q_n(s,a) \leftarrow r + \gamma \max_{a' \in A} Q_{n-1}(s',a') \tag{3}$$

Update rule (3) states that the Q-value for a state-action pair at iteration n is equal to the expected value of immediate reward plus the discounted long-term value. Long-term value here is estimated by the current maximum Q-value for the successor state. Each iteration of the approach refines the long-term value estimate for the state-action pair by back propagating value from subsequent transitions that receive rewards. Through successive iterations this approach is guaranteed to converge toward Q^* in small domains, where states can be represented exactly, and with infinite sampling guarantees.

Non-trivial problems, however, have state and action spaces that are too large to define the Q-function explicitly, so we must use function approximation to represent an approximate Q-function, $\hat{Q}$. Linear function approximation is commonly used in practice and is the type of approximation we utilize in this paper. In a linear function approximation scheme the value-function or Q-function is represented by a weighted, w, linear combination of k features, ϕ, defined over the state-action space:

$$\hat{Q}(s,a) = \sum_{i \leftarrow 1}^{k} w_i \phi_i(s,a) \tag{4}$$

The types of features and their number is domain dependent and crucial to the success of the approach. Typically those parameters are chosen a priori by a domain expert. The Q-function is approximated by deriving an appropriate weight vector through linear regression.

Algorithm 1. FQI(D, γ, N)

Require: D: set of samples, γ: discount factor, N: number of iterations to complete
1: $\hat{Q}_0 \leftarrow 0$ //Initialize Q-function to zero everywhere
2: **for** $n = 1$ to N **do**
3:    **for all** $sample \in D$ **do**
4:       //$sample = (s_t, a_t, r_t, s_{t+1})$
5:       $Input_{sample} \leftarrow (s_t, a_t)$
6:       $Target_{sample} \leftarrow r_t + \gamma \max_{a \in A} \hat{Q}_{n-1}(s_{t+1}, a)$
7:    **end for**
8:    $\hat{Q}_n \leftarrow Regression(Input, Target)$
9: **end for**
10: **Return** $\hat{Q}_N$

Representing the Q-function this way is much more efficient and the choice of representation (features) provides valuable generalization. Generalization allows for effective learning in complex domains even under limited sampling conditions. However, while generalization is a useful benefit of approximation, it is limited by errors that can produce misleading value estimates as was shown in the example given in Section 1.

2.1 Fitted Q-Iteration

Fitted Q-Iteration (FQI) is a general AVI framework for solving RL problems. It derives $\hat{Q}$ through a sequence of standard supervised learning regression problems that iteratively converge upon the fixed point solution of the Bellman equation, equation 1. **Algorithm 1** provides a detailed outline of the FQI framework. First, the algorithm begins with a Q-function initialized to zero, $\hat{Q}_0 = 0$, and a set of provided samples, D. At each iteration FQI solves a regression problem that produces a more accurate Q-function model to be used in subsequent iterations. The regression problem is defined by using the state and action pairs, (s, a), from the samples in D as the input patterns and the current approximated Q-value of the pair, $\hat{Q}_n(s, a)$, as the regression target. After performing regression, the updated Q-function model, $\hat{Q}_{n-1}$, combined with the Bellman update rule (3) is used to generate the next iterations regression targets (see line 6 of **Algorithm 1**). Given an appropriate choice in approximation model, through successive iterations this process converges and produces a model that approximates Q^*.

FQI has several desirable traits that have made it a widely used algorithm among the RL community. First, it is an off-policy algorithm which enables FQI to effectively utilize samples collected by any means. This is an important feature when samples are difficult to obtain or cannot be simulated. FQI is a batch-mode algorithm which gives it favorable sample efficiency when compared to single sample update approaches. Because of the generality of the approach, it can be paired with a variety of regression models, and allows for the approach to be adapted to any given problem domain. Additionally, it has demonstrated competitive learning performance compared to that of other state-of-the-art RL algorithms [11].

However, FQI like other AVI methods suffers from the generalization problem discussed in Section 1. FQI's use of the Bellman update, shown on line 6 of **Algorithm** 1, explains why it is susceptible to this issue. The long-term value component of the update is based upon the maximal value that can be obtained from the successor state of the sample according to the function approximation model, $\hat{Q}_{n-1}$. It is not a strong assumption that the approximation model will contain generalization error, and as a result the derived long-term value of state-action pairs will be adversely impacted.

3   Trajectory Fitted Q-Iteration

AVI methods like FQI are sensitive to the generalization errors of an approximation because they assume that there is no other mechanism to propagate long-term reward through the function space than via the model. In most RL scenarios, however, this assumption is not true. Samples for RL are generally collected as part of a larger sequence known as a trajectory. These trajectories describe observed paths through the domain's state-action space that can also be used to propagate long-term reward without the generalization error of an approximation. In this section we present our approach, Trajectory Fitted Q-Iteration (TFQI) a new batch-mode, off-policy, AVI learning algorithm based on FQI that exploits trajectory data to improve the derivation of long-term value.

TFQI makes one additional assumption that other AVI approaches do not; the sample data to be learned over has been collected as sets of trajectories. A trajectory is a finite ordered sequence of samples that describe a series of successive transitions through the problem space. More formally, an n-step trajectory is comprised of the following ordered set of samples:

$$\{(s_{t_0}, a_{t_0}, r_{t_0}, s_{t_1})_1, (s_{t_1}, a_{t_1}, r_{t_1}, s_{t_2})_2, \ldots, (s_{t_{n-1}}, a_{t_{n-1}}, r_{t_{n-1}}, s_{t_n})_n\} \qquad (5)$$

Trajectories are collected through episodic multi-step interactions with the problem domain performed either by the agent itself or by a third-party. TFQI is an off-policy approach in that it makes no assumptions on the quality of the policy used to generate the trajectories or that they were all produced by the same policy. Additionally, the provided set of trajectories can consist of trajectories of varied length and my not end with transitions to terminal states.

The key insight behind TFQI is that the long-term value for any sample, as part of a trajectory, can be estimated as either the discounted value predicted by the approximation model given the successor state or by the value of the successor sample from the trajectory. The later option is more resilient to generalization errors of the approximation scheme because it is derived from the real value relationship between successive samples and not the approximation. However, the value derived for the successor sample is based on the action taken by the successor, which is not assumed to be the optimal action for the successor state, and therefore can be an underestimate of the optimal value for the sample.

Algorithm 2. TFQI($Traj, \gamma, N$)

Require: $Traj$: a set of trajectories, γ: discount factor, N: number of iterations
1: $\hat{Q}_0 \leftarrow 0$
2: **for** $n = 1$ to N **do**
3:   **for all** $trajectory \in Traj$ **do**
4:     **for** $i = |trajectory|$ downto 1 **do**
5:       $//sample_i = (s_t, a_t, r_t, s_{t+1})_i \in trajectory$
6:       $Input_{sample_i} \leftarrow (s_t, a_t)$
7:       $Target_{sample_i} \leftarrow r_t + \gamma \max(\max_{a \in A} \hat{Q}_{n-1}(s_{t+1}, a), Target_{sample_{i+1}})$
8:     **end for**
9:   **end for**
10:   $\hat{Q}_n \leftarrow Regression(Input, Target)$
11: **end for**
12: **Return** $\hat{Q}_N$

As such, TFQI uses the maximal value of the two values as the estimate for the long-term reward while performing updates.

Algorithm 2 provides a detailed outline of the TFQI algorithm. Structurally TFQI is very similar to the FQI algorithm. The differences are: TFQI accepts a set of trajectories, $Traj$, as opposed to a general set of samples. TFQI iterates through the samples of each trajectory in reverse sequential order while computing the regression targets. This is necessary to obtain an updated estimate of the successor sample's value, $Target_{sample_{i+1}}$. Finally, the update rule (line 7) has been expanded to utilize the maximum of either the approximation model or successor sample value to estimate the long-term value of a sample. [1]

These modifications to the original FQI formulation enables TFQI to harness trajectory data more effectively. The enhanced update rule is the most meaningful of the changes. It provides a mechanism for long-term value to propagate backward through trajectories. This one function has two beneficial consequences. First, it reduces the impact of generalization errors by correcting underestimates of long-term value for sampled states, as shown in the example given in Section 1. This ability can lead to more accurate $\hat{Q}$-functions and hopefully improved policies. And second, it propagates long-term value through entire trajectories at each iteration. This effect can speed how long-term value is learned throughout the function space and improve convergence especially in reward sparse domains.

TFQI retains the desirable traits (batch-mode and off-policy) of FQI with negligible additional computational costs and no additional overhead. Per iteration, TFQI only adds $|D|$ more comparisons, where D is the set of samples,

---

[1] One or both of the long-term value estimates will be undefined for the last sample of each trajectory. If the sample transitions to an absorbing state, r_t is taken as the value for that sample. If the sample does not transition to an absorbing state, the value predicted by the approximation model is used.

than that of FQI. In the subsequent section we provide an empirical analysis comparing FQI and TFQI to explore the potential benefits of this approach.

4 Experimental Setup

We performed an empirical comparison of TFQI and FQI over several experiments to assess the impact of the enhanced use of trajectory information for AVI. Our analysis, provided in the subsequent Section 5, evaluates both approaches based on final learned policy performance, convergence speed, and sensitivity to the function approximation model. This section provides the details for the setup of this comparison.

4.1 Domains

Mountain Car [4] (MC) and the Acrobot (ACRO) Swing-Up [6] problems are the two well-known RL domains that we chose to use in our analysis. For our experiments we used implementations of both domains provided by RL-Glue [15]. In MC the learner is tasked with learning a policy for driving an underpowered car from the bottom of a deep valley to the top of the forward hill. The ACRO problem challenges the learner to derive a policy for an underpowered robot, simulating a gymnast on a high-bar, that starting from a still straight position below the bar rotates the leg of the robot, at the hip, in such a way that the robot gathers enough momentum to swing and raise the end of its leg above the high bar position.

The objective for both problems is to reach a goal state in as few steps as possible. As such, the rewards for all transitions in both problems is -1, except for transitions to the goal state which have 0 rewards. While evaluating the learned policies we limit the number of steps a policy has to achieve the objective to 300 steps for MC and 1000 steps for ACRO. Policies that take longer than those limits are considered unsuccessful policies for the purpose of our evaluation.

Both problems require function approximation because of their continuous state-spaces. MC has a 2-dimensional state-space and ACRO has a 4 dimensional state-space. The two domains each have 2 discrete actions: apply forward or reverse throttle for MC and apply forward or reverse torque at the robot's hip for ACRO.

4.2 Function Approximation Models

We use two forms of linear function approximation in our experiments for representing $\hat{Q}$ that differ in the types of features. The first is Radial Basis Functions (RBF) which overlays several Gaussian curves uniformly over the state space and uses their activation as features. It is a common choice for RL and has good localized generalization properties [6]. The other is based on Fourier basis functions. Fourier basis functions have greater global generalization properties and have only recently been explored in an RL context [6].

Ridge regression is used for both variants to train the weight vectors of the models. It applies l_2 regularization on the objective function to prevent over-fitting. The amount of regularization is controlled by the shrinkage parameter, λ, that we manually set but can be tuned automatically using cross-validation. This form of function approximation paired with FQI is known to exhibit divergence behavior because errors in the approximation of value are potentially unbounded [4]. We are able to circumvent this issue for these specific problem domains by limiting the predicted value response of the model to be non-positive. Both domains have strictly non-positive rewards making it impossible for state-action pairs to have positive value.

4.3   Trajectory Generation

The NEAT [13] algorithm was used to generate large diverse sets of trajectories necessary for our analysis. NEAT solves RL problems via genetic algorithm that performs policy search. Each member of a NEAT population represents a potentially different policy. Trajectories were recorded by observing an evaluation of each policy on the problem domain. If more than one policy achieved the same aggregate value only one of the trajectories was recorded per NEAT run to maintain diversity. Several runs with varying random seeds for both problem domains generated the trajectory sets for our analysis. The recorded trajectories are all "complete" trajectories generated from successful policies. The trajectories are "complete" in that they all start from the same initial state and end with a transition to the goal state after some number of intermediate transitions.

There are roughly 1500 trajectories in the MC and 6000 trajectories in the ACRO data sets. The un-discounted aggregate values achieved by the trajectories in the sets ranges uniformly from -105 to -299 for MC and -70 to -999 for ACRO. In our experiments, for each run an identical set of trajectories is provided to both approaches. The trajectory sets vary from run to run and consist of randomly selected trajectories from the full generated sets.

4.4   Experiment Parameters

Unless stated otherwise, the learning parameters for all experiments are consistent for each domain. In our MC experiments RBF is used, and it is comprised of 25 features (k=25), $\lambda = 1.0$, and $\gamma = 0.9999$. For ACRO the Fourier basis functions are used, $k = 81$, $\lambda = 1.0$, and $\gamma = 0.9999$. These parameter settings were manually selected and were chosen because they produced the best observed overall learning performance for both methods across multiple experiments.

The reported results for all experiments are the average of 200 runs after 300 iterations (N=300). Additionally, we performed a paired t-test on the results from each approach to determine if differences in the results are statistically significant. We report the difference as being significant if the p-value < 0.05.

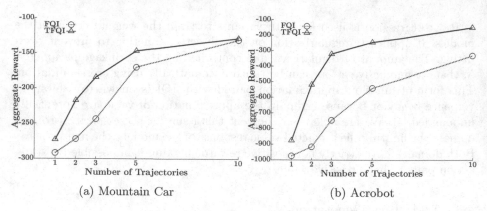

Fig. 2. Average converged policy performance for FQI and TFQI using increasing numbers of complete trajectories

5   Results and Discussion

5.1   Learned Policy Performance

The primary goal of RL is to learn effective policies, therefore the most direct way of evaluating both FQI and TFQI is to compare them based upon their learned policy performance over the same sets of samples. Here we report the results from two sets of experiments that measure average un-discounted aggregate reward achieved by the learned policies after N iterations for both approaches. The experiment sets differ in the type of sample trajectories that are provided.

Our first set of experiments provides both approaches with increasing numbers of complete trajectories. This scenario is arguably the most realistic for an off-policy batch-mode algorithm. The learner is presented with some number of trajectories that demonstrate how to achieve a goal to varying degrees of success and the learner must derive the most effective policy from these samples. We show experiments increasing the number of provided trajectories from 1 to 10.

From the results given in Figure 2, it can be seen the enhanced use of trajectory data gives TFQI an advantage in learned policy performance. TFQI learns better policies on average than FQI in all of these experiments using the same data. The performance differences are significant in all cases except for the MC 10 trajectories experiment where both methods performed similarly. TFQI's advantage is most distinct in the ACRO problem where its average policy performance can be more than twice as good as that of FQI.

The second set of experiments evaluates policy performance when the methods are provided with fragments of complete trajectories. Although full trajectories may be more common, neither approach is dependent on the availability of such data. This set of experiments is designed to demonstrate whether or not TFQI can effectively exploit this more general form of trajectory data. The trajectory sets in these experiments were generated by randomly selecting trajectory fragments of random length from the repository. Fragments are added until the

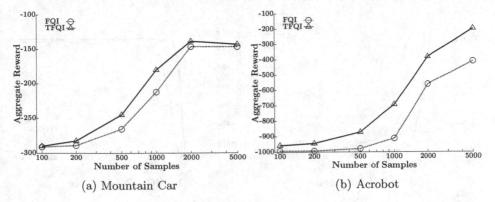

Fig. 3. Average converged policy performance for FQI and TFQI using increasing numbers of samples from trajectory fragments

aggregate number of samples reaches a desired threshold. Additionally, we guarantee that there are at least 3 fragments containing transitions to the goal state in each run. For this series of experiments we increased the aggregate size of the trajectory data incrementally from 100 to 5000 samples.

As shown in Figure 3 TFQI again demonstrates better learning performance than FQI on this more general form of trajectory data. All performance gaps shown in Figure 3b for the ACRO problem are significant as well as the 500, 1000, and 2000 sample runs for the MC problem. While the differences between the two approaches in these experiments are not quite as great as in the previous experiment there is still an improvement to be gained from utilizing this data and it can be significant.

Based on the results from these two experiments, when trajectory based data is available TFQI is clearly the preferable approach. The enhanced update allows TFQI to learn policies that are significantly better than FQI's on average. Improvements in performance are most pronounced when there are fewer samples, 2-5 trajectories or 500-1000 samples. Though we provide no theoretical bounds on sample efficiency of TFQI, these results suggest that our method can be more efficient than FQI in practice; a desirable trait when samples are costly to acquire.

5.2   Convergence

In addition to policy performance we compare FQI and TFQI on their convergence speed. Both methods iteratively refine $\hat{Q}$ by learning new sets of regression targets that estimate Q-values for the available samples. Through successive iterations, the difference in the regression targets from one iteration to the next, $\hat{Q}_n - \hat{Q}_{n-1}$, should approach zero and lead to a stable policy. Convergence speed is measured by the rate at which the difference approaches zero.

Figure 4 provides two sets of graphs for this analysis that show convergence of the $\hat{Q}$-function, Figures 4a and 4c, and average policy performance, Figure

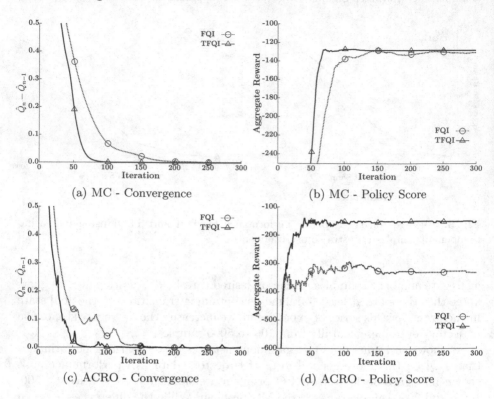

Fig. 4. Convergence, (a) and (c), and policy, (b) and (d), performance graphs for two experiments comparing FQI and TFQI in both domains. For the convergence graphs, lower values approaching zero indicate improved performance. Alternatively, higher values in the policy score graphs are better.

4b and 4d. The graphs report the results of two experiments where TFQI and FQI were given 10 complete trajectories for each domain. For Figures 4a and 4c, $\hat{Q}_n - \hat{Q}_{n-1}$ is measured by the mean squared difference of the regression targets from one iteration to the next. In both experiments the convergence graphs show that TFQI appears to converge nearly 100 iterations earlier than FQI, demonstrating a significant speedup. This result can be attributed to the way TFQI is able to propagate long-term value through an entire trajectory within a single iteration. FQI, alternatively, must wait for value to propagate one transition step at a time per iteration. Figure 4b shows the learning performance of the methods from the same 10 trajectory experiment reported in Figure 2a. Interestingly, despite not deriving a significantly better policy on average, TFQI is able to derive a similar policy in far fewer iterations. For the ACRO domain this result is even more impressive. TFQI is able to learn a far better policy in fewer iterations using the same exact set of sample trajectories. In Figure 4c TFQI's convergence does exhibit some oscillations, however they do not have a significant impact on the quality of the policy as seen in Figure 4d.

5.3   Model Parameter Sensitivity

The success of any machine learning approach depends on the selection of an appropriate model for a given domain. RL is certainly no exception to this statement. Selecting an appropriate representation is critical and is the subject of intense research to find automatic methods [7]. Determining the most appropriate settings for any given model is difficult because there are often many different interrelated parameters that must be tuned. Our approach, while still very much dependent on model selection, utilizes trajectory data to calculate value in a way that is partially independent of the approximation model. It is plausible then that this approach may exhibit some degree of insensitivity to model choice.

In this section we provide a preliminary non-exhaustive set of experiments that compare the sensitivity of FQI and TFQI to the settings of various model parameters. The parameters that we change are the type of feature, ϕ, the number of features, k, and the shrinkage parameter for ridge regression, λ. For both domains we use the best performing settings on 10 complete trajectories as a baseline. We then deviate one of the parameters from the baseline settings in each experiment and measure the difference in the average policy performance, Δ, to determine sensitivity.

Table 1. Approximation model parameter sensitivity experiment results for Mountain Car. The first entry shows the baseline settings and performance for both methods. Subsequent entries show the policy performance and deviation from the baseline performance for varying model parameters for both methods. Significant results are **bold**.

ϕ	k	λ	FQI	FQI Δ	TFQI	TFQI Δ
RBF	25	1.0	-130.700	NA	-128.320	NA
RBF	25	0.1	-156.250	-25.550	**-146.460**	-18.140
RBF	25	10.0	-297.264	-166.564	**-238.044**	-109.724
RBF	16	1.0	-286.519	-155.819	**-200.724**	-72.404
RBF	36	1.0	-143.370	-12.670	**-134.870**	-6.550
Fourier	25	1.0	-194.260	-63.560	**-178.455**	-50.135

Table 1 provides the complete results for these experiments on the MC domain. In general TFQI exhibits less of a degradation in policy performance than FQI, supporting our hypothesis. The difference in degradation is most significant when the model parameters are changed to increase the generality of the model. Generalization is increased here when the number of features is decreased to 16 and when the λ parameter is increased to 10. This observation is somewhat expected given that the use of trajectory data by the TFQI approach reduces the impact of generalization error.

The results for the ACRO problem can be seen in Table 2. They are not as conclusive as the ones reported for MC. FQI actually shows a performance improvement when λ is reduced to 0.1 while learning over 10 complete trajectories. $\lambda = 1.0$ was found to be a good for learning over other amounts of trajectories and this result re-emphasises the difficulty in finding the best parameter settings. Still, even at this setting FQI does not match the performance of TFQI. Again, we can see that increasing the generalization of the model has significantly less of an impact on TFQI. When the number of features is increased to 256 both perform comparably. Given the relative number of features and samples (about 5000) this experiment might be suffering from model over-fitting. Both approaches struggle to find a successful policy when RBFs are used. Recall we limit the number of steps a policy has to achieve the goal to 1000. FQI is completely incapable of deriving a successful policy at these settings over 200 runs, whereas TFQI occasionally learned a competent policy. This last result provides further evidence of our hypothesis and shows that TFQI can learn in some scenarios where FQI cannot.

6   Related Work

Our approach is similar to learning from demonstration [2,12] methods in that they also exploit trajectories. Learning from demonstration is a form of RL that takes examples, trajectories, demonstrated by known optimal or "good" behavior as input and attempts to use this data to derive policies that duplicate that behavior. While our approach could certainly take advantage of such data, it differentiates itself from most learning from demonstration work because no assumptions are made upon the quality of the policies used to generate the trajectories or that only a single policy was used to generate the trajectories.

The central idea behind our approach of using trajectory data to improve long-term value estimation is most similar to the idea of *augmented Bellman backups* described in [9]. In that work the authors define an update rule that uses either model induced value or value derived by a demonstration while performing a Bellman update of the value function. Our approach is a generalization of this idea and seamlessly intermixes the use of trajectory or value function model induced value in its update function. The intention of the *augmented Bellman backup* is to use demonstrations to accelerate the process of learning, whereas our intention is to improve the overall quality of learning. Like other learning from

Table 2. Approximation model parameter sensitivity experiment results for Acrobot. The first entry shows the baseline settings and performance for both methods. Subsequent entries show the policy performance and deviation from the baseline performance for varying model parameters for both methods. Significant results are **bold**.

ϕ	k	λ	FQI	FQI Δ	TFQI	TFQI Δ
Fourier	81	1.0	-331.455	NA	**-149.690**	NA
Fourier	81	0.1	-274.495	56.960	-254.660	-104.970
Fourier	81	10.0	-951.565	-620.110	**-390.295**	-240.605
Fourier	256	1.0	-389.345	-57.890	-396.943	-247.253
RBF	81	1.0	-999.000	-667.545	-967.535	-817.845

demonstration methods this approach assumes the availability of high quality demonstrations, unlike our approach.

Eligibility traces [14] are also similar in that they accelerate learning by applying TD updates to multiple samples along a trajectory. The idea was developed within an on-line on-policy context but it has been successfully ported to off-line off-policy algorithms. Nevertheless, the long-term values calculated by eligibility traces are derived exclusively from the approximated values of successor transitions, which makes them more susceptible to generalization error.

TFQI is not the first enhancement that has been made to the Fitted Q-Iteration framework since it was introduced in [4]. In [11] a multi-layer feed forward neural network architecture was substituted as the regression function, instead of the regression tree methods described in the original work, with improved empirical performance in standard RL benchmarks. More recently FQI was extended to work in continuous action space domains [1]. The extension proposed in this work is distinctly different and complimentary to these methods.

Finally, we must note that there are well known and explored theoretical issues when combing off-policy TD methods, such as FQI and TFQI, with function approximation. Off-policy TD methods are known to exhibit divergent behavior when paired with function approximators that do not provide bounds on approximation error [3]. Additionally, even when convergence can be ensured there are no known guaranteed bounds on the quality of the learned value function[5]. While the approach described in this paper does demonstrate empirical improvements, we do not claim to make a theoretical advancement regarding those issues.

7   Conclusion and Future Work

We have introduced TFQI and shown how its novel utilization of trajectory based data can reduce the impact of generalization error on the derivation of the value function. Our empirical analysis demonstrates how this generally available data can be further exploited to provide significant performance enhancements in the form of improved policies and quicker convergence. TFQI approach accomplishes all this while incurring no significant additional computational or memory costs over the standard FQI approach. One direction for future work is to explore the theoretical impact of this use of trajectory data.

Further, our empirical study found some evidence that trajectory based value updates provide some robustness to approximation model parameter settings. The value propagation information inherent in trajectories might then enable better modeling choices to be made a priori. We are additionally interested in investigating if trajectory data can similarly be harnessed to enhance automatic feature and model selection approaches.

Acknowledgements. This work is supported in part by grants from NSF (No. 0855204) and the AFRL Information Directorate's Visiting Faculty Research Program.

References

1. Antos, A., Munos, R., Szepesvári, C.: Fitted Q-iteration in continuous action-space mdps. In: NIPS (2007)
2. Argall, B.D., Chernova, S., Veloso, M., Browning, B.: A survey of robot learning from demonstration. Robot. Auton. Syst. 57(5), 469–483 (2009)
3. Boyan, J.A., Moore, A.W.: Generalization in reinforcement learning: Safely approximating the value function. In: Advances in Neural Information Processing Systems 7, pp. 369–376. MIT Press (1995)
4. Ernst, D., Geurts, P., Wehenkel, L., Littman, L.: Tree-based batch mode reinforcement learning. Journal of Machine Learning Research 6, 503–556 (2005)
5. Kolter, J.Z.Z.: The fixed points of off-policy td. In: Shawe-Taylor, J., Zemel, R., Bartlett, P., Pereira, F., Weinberger, K. (eds.) Advances in Neural Information Processing Systems 24, pp. 2169–2177 (2011)
6. Konidaris, G., Osentoski, S., Thomas, P.S.: Value function approximation in reinforcement learning using the Fourier basis. In: Proceedings of the Twenty-Fifth Conference on Artificial Intelligence, pp. 380–385 (August 2011)
7. Mahadevan, S.: Representation discovery in sequential decision making. In: Proceedings of the Twenty-Fourth AAAI Conference on Artificial Intelligence. AAAI Press (2010)
8. Munos, R.: Error bounds for approximate value iteration. In: Proceedings of the 20th National Conference on Artificial Intelligence, AAAI 2005, vol. 2, pp. 1006–1011. AAAI Press (2005)
9. Price, B., Boutilier, C.: Accelerating reinforcement learning through implicit imitation. Journal of Artificial Intelligence Research 19, 569–629 (2003)
10. Puterman, M.L.: Markov decision processes: discrete stochastic dynamic programming, vol. 414. Wiley-Interscience (2009)
11. Riedmiller, M.: Neural fitted Q iteration - first experiences with a data efficient neural reinforcement learning method. In: Gama, J., Camacho, R., Brazdil, P.B., Jorge, A.M., Torgo, L. (eds.) ECML 2005. LNCS (LNAI), vol. 3720, pp. 317–328. Springer, Heidelberg (2005)
12. Schaal, S.: Learning from demonstration. In: Advances in Neural Information Processing Systems 9. MIT Press (1997)
13. Stanley, K.O., Miikkulainen, R.: Evolving neural networks through augmenting topologies. Evolutionary Computation 10(2), 99–127 (2002)
14. Sutton, R.S., Barto, A.G.: Reinforcement Learning: An Introduction (Adaptive Computation and Machine Learning). The MIT Press (March 1998)
15. Tanner, B., White, A.: RL-Glue: Language-independent software for reinforcement-learning experiments. Journal of Machine Learning Research 10, 2133–2136 (2009)

Expectation Maximization for Average Reward Decentralized POMDPs

Joni Pajarinen[1] and Jaakko Peltonen[2]

[1] Department of Automation and Systems Technology, Aalto University, Finland
Joni.Pajarinen@aalto.fi
[2] Department of Information and Computer Science, Aalto University, Finland
Jaakko.Peltonen@aalto.fi

Abstract. Planning for multiple agents under uncertainty is often based on decentralized partially observable Markov decision processes (Dec-POMDPs), but current methods must de-emphasize long-term effects of actions by a discount factor. In tasks like wireless networking, agents are evaluated by average performance over time, both short and long-term effects of actions are crucial, and discounting based solutions can perform poorly. We show that under a common set of conditions expectation maximization (EM) for average reward Dec-POMDPs is stuck in a local optimum. We introduce a new average reward EM method; it outperforms a state of the art discounted-reward Dec-POMDP method in experiments.

Keywords: Dec-POMDP, average reward, expectation maximization, planning under uncertainty.

1 Introduction

Optimizing the behavior of several agents like robots [25,22] or wireless devices [7,18] is a crucial and hard problem, especially hard in an uncertain world where agents act using only noisy observations about the world and other agents. A decentralized partially observable Markov decision process (Dec-POMDP) can describe the optimal solution. Each agent gets observations on its own and decides its next action to optimize a shared goal. To plan actions, an agent must consider possible action-observation sequences of all agents, thus Dec-POMDP planning is computationally hard: finite-horizon Dec-POMDPs are NEXP-complete (doubly exponential), infinite-horizon Dec-POMDPs are undecidable [6].

In a Dec-POMDP, agents get a joint reward at each time step based on their actions and the world state. Finite-horizon Dec-POMDPs [22,14,23] maximize the sum of rewards over a fixed number of time steps and discounted infinite-horizon Dec-POMDPs [2,10,17] maximize the sum of discounted rewards over an infinite horizon; these objectives emphasize rewards closer to the first time steps, i.e., short-term effects of actions. However, in many Dec-POMDP problems it is natural to maximize *average reward* over an infinite horizon. In wireless networks [7] usual objectives are *average throughput* (average amount of transmitted data,

H. Blockeel et al. (Eds.): ECML PKDD 2013, Part I, LNAI 8188, pp. 129–144, 2013.

infinitely far into the future) or *average delay* (average time a data packet must wait). Such objectives emphasize short and long-term effects of actions equally. Usefulness of average rewards has been shown in robotics [25] and reinforcement learning [13]. Moreover, in finite-horizon and discounted reward methods the solution may depend heavily on the distribution for the first time step (initial belief), which may need to be designed by a domain expert. In many infinite-horizon problems a good initial belief depends on the optimal policy and vice versa (in wireless networks the amount of data in transmit buffers of devices depends on policy efficiency). In contrast, in an average-reward Dec-POMDP the solution does not depend on the initial belief, under certain conditions (see Section 3.1).

Optimizing average reward has been used in partially observable Markov decision processes (POMDPs) for one agent, and for special-case multiple agent problems, but solutions for generic multiple agent problems have not been given. Interaction of agents is essential e.g. in wireless network channel access [18]. We introduce a solution for multiple agents with partial observability: a Dec-POMDP method that optimizes average reward by a modified expectation-maximization (EM) algorithm. To our knowledge this is the *first general Dec-POMDP method for optimizing average reward*.

2   Related Work

We discuss related work on average reward Markov decision processes (MDPs), partially observable MDPs (POMDPs), and decentralized MDPs (Dec-MDPs). A fully observable POMDP or a single agent Dec-MDP is an MDP, a single agent Dec-POMDP is a POMDP, and a jointly fully observable Dec-POMDP is a Dec-MDP. The Dec-POMDP is the most general of these models. We know of previous work on average reward MDPs [13,21], average reward POMDPs [1,29,12], discounted reward POMDPs [20,8,2,17], transition and observation independent average reward Dec-MDPs [19], finite-horizon Dec-POMDPs [22,14,23], and discounted-reward Dec-POMDPs [24,5,4,3,2,10,17], but not on general average reward Dec-POMDPs. For *average reward MDPs*, policy iteration, value iteration, linear programming [21] and model-free methods [13] exist. Mahadevan et al. [13] showed average reward outperformed discounted reward in MDPs where an agent chose small short-term or large long-term rewards. Methods exist for *average reward POMDPs*: Li et al. [12] find memoryless policies, Yu et al. [29] use lower bound approximations, and Aberdeen [1] improves a finite state controller by gradient methods. For decentralized problems, Petrik et al. [19] show transition&observation independent *average reward Dec-MDPs* are NP-complete, and use bi-linear programming. Yagan et al. [28] minimize average cost in a transition&observation independent special-case Dec-POMDP where agents don't affect or sense the world state seen by other agents. In general Dec-POMDPs agents affect each other in complex ways. To our knowledge there is no research on general average reward Dec-POMDPs, but research on finite-horizon [22,14,23] and discounted reward Dec-POMDPs [24,5,4,3,2,10,17] exists.

Kakade et al. [9] showed MDP average reward could be approximated by discounting with large discount factor, but in our experiments real average reward optimization outperformed discounting.

3   Dec-POMDP

A Dec-POMDP is a solution to multi-agent planning under uncertainty about the world and other agents. It is defined by a set of N agents, the set of actions A, the set of states S, the set of observations O, the observation probability $P(o|s', a)$, the state transition probability $P(s'|s, a)$, and real valued immediate reward function $R(s, a)$. Here o denotes the observations $o_1, \ldots, o_N$ and a the actions $a_1, \ldots, a_N$ of all agents. In each time step, the world starts from state s, each agent i takes action a_i, and the world transitions to the next state s' with probability $P(s'|s, a)$. Agents then make their observations o with probability $P(o|s', a)$ and the action-observation cycle begins again. An agent does not sense actions, states or observations of other agents, so computational complexity of planning is high. In each time step the agents get immediate reward $R(s, a)$ depending on their actions a and the world state s. The finite-horizon objective is to maximize reward $E[\sum_{t=0}^{T} R_t(s, a)|\pi]$ where T is the horizon, π is the policy (consisting of the individual policies of all agents), and $R_t(s, a)|\pi$ is the reward at time step t following π. In the discounted reward case, expected discounted reward over an infinite-horizon $E[\sum_{t=0}^{\infty} \gamma^t R_t(s, a)|\pi]$ is maximized, with discount factor $0 < \gamma < 1$. With discounting, reward decreases geometrically with the horizon. Both finite-horizon and discounted reward objectives need an initial state probability distribution $b_0(s)$ called the initial belief.

Finite state controllers (FSCs) have been used as policy in POMDP [20,8,2,17] and infinite-horizon discounted reward Dec-POMDP [24,5,4,3,2;10,17] methods. The FSC of agent i consists of a set $\{q_i\}$ of FSC states q_i, an action probability distribution $P(a_i|q_i)$, and FSC state transition probability $P(q_i'|q_i, o_i)$. For simplicity, similar to the approach in [2], an agent starts in state $q_i = 1$. In each time step, agent i in state q_i takes action a_i with probability $P_{aq}^{(i)} = P(a_i|q_i)$. The world transitions to a new world state, the agent gets observation o_i about the world, and moves to a new FSC state q_i' with probability $P_{q'qo}^{(i)} = P(q_i'|q_i, o_i)$.

3.1   Average Reward Dec-POMDP

Intuitively average reward Dec-POMDPs optimize the policy to maximize average reward over an infinite horizon. Formally, they must maximize $R_{average} = E\left[\lim_{T \to \infty} \frac{1}{T} \sum_{t=0}^{T-1} R_t(s, a)|\pi\right]$. Unlike finite-horizon and discounted reward objectives, $R_{average}$ does not need a parameter controlling effective planning horizon and depending on the underlying Markov chain does not need an initial belief. In Dec-POMDPs, an agent needs the full observation history to make optimal decisions [6]. As average reward Dec-POMDPs run the policy for arbitrarily long times, we use FSCs as policies taking a fixed amount of memory. For a set of FSCs (one per agent), the world state s and the FSC states q_i

together form a state of a Markov chain as follows: given the current state (s, q), where $q = q_1, \ldots, q_N$, the **probability for the next time step state** (s', q') is $P_{s'q'sq} = P(s', q'|s, q) = \sum_{a,o} P(o|s', a)P(s'|s, a)\prod_i \left(P_{aq}^{(i)}P_{q'qo}^{(i)}\right)$. With initial belief $b_0(s)$ the **initial probability distribution** over (s, q) is $P_0(s, q) = b_0(s)\prod_i P(q_i)$ and $P_t(s, q)$ is the initial distribution projected t time steps into the future. The **expected immediate reward** for $P_t(s, q)$ is $\sum_{s,q,a} P_t(s, q)R(s, a)\prod_i P_{aq}^{(i)}$. The **optimization objective** is then $R_{\text{FSCs}} = \lim_{T \to \infty} \frac{1}{T}\sum_{t=0}^{T-1}\sum_{s,q,a}\left(P_t(s, q)R(s, a)\prod_i P_{aq}^{(i)}\right)$. Average reward Dec-POMDP problems can be grouped by properties of the above-described Markov chain. We consider fully stochastic policies, like FSCs with nonzero action and transition probabilities; the properties below don't depend on the policy as long as it is fully stochastic. Useful Markov chain classes (similar to [21]) are *Recurrent* - all states reachable from all states; *Periodic* - the greatest common divisor of the return time of one or more states is greater than one; *Aperiodic* - no state is periodic; *Unichain* - one set of recurrent states and a set of zero or more transient states; and *Multichain* - two or more closed irreducible sets of recurrent states and zero or more transient states. We focus on aperiodic problems. When the Markov chain is *aperiodic*, $P_t(s, q)$ converges to a stationary limiting distribution $P_*(s, q) = \lim_{t \to \infty} P_t(s, q)$. Since rewards are bounded, R_{FSCs} becomes

$$R_{\text{FSCs,aperiodic}} = \sum_{s,q,a} P_*(s, q)R(s, a)\prod_i P_{aq}^{(i)} . \tag{1}$$

For *multichain* Markov chains, the limiting distribution depends on initial belief: if e.g. a robot can enter one of two hallways but cannot switch later, its start position affects the limiting distribution. For *unichain* Markov chains the limiting distribution does not depend on initial belief. Average reward unichain models are of practical interest: in a wireless network case, agents' transmission buffer sizes are the world state and transmission policies must be optimized to keep buffers as empty as possible; the reward is the negative sum of buffer sizes and the initial belief is the distribution over buffer sizes. Generally initial belief influences the best achieved policy so the belief should be optimized with the policy, but for unichain Markov chains we need not optimize initial belief since the optimal policy always yields the optimal limiting distribution.

4   Expectation-Maximization Planning

Expectation-maximization (EM) has been used to optimize finite state controllers (FSCs) for discounted reward in MDPs and POMDPs [27], Dec-POMDPs [10,17], and factored Dec-POMDPs [16]. In EM the idea is to scale rewards into probabilities and, by inference, find FSC parameters maximizing the reward likelihood. EM has been extended to problems with huge [16] and continuous [27] state spaces. **We now introduce an average reward EM method for aperiodic Dec-POMDPs**. (For experiments, we introduce a nonlinear programming based alternative in the Appendix.) We first use a traditional EM approach and show the result is stuck in a local optimum under certain conditions (see Section 4.1 for more details); we then use it as a foundation for a modified approach and introduce a practical EM method that yields good results.

An EM approach scales the real valued reward function into a binary reward variable r. Denote with $\hat{R}_{sa} = \hat{R}(r = 1|s, a)$ the conditional probability for r to be one given actions a and state s. $\hat{R}_{sa}$ is computed by scaling the real valued reward function using the minimum R_{min} and maximum R_{max} rewards: $\hat{R}_{sa} = (R(s, a) - R_{min})/(R_{max} - R_{min})$. In average reward Dec-POMDPs, FSC parameters θ are optimized to maximize average reward over time, scaled as above into a likelihood of a binary reward:

$$P(r = 1|\theta) = \lim_{T_M \to \infty} \sum_{T=0}^{T_M-1} \frac{1}{T_M} \sum_{s,q,a} \hat{R}_{sa} P_T(s, q) \prod_i P_{aq}^{(i)} , \tag{2}$$

where the horizon T_M is taken to the limit. It can be shown (2) corresponds to the original average reward objective (1); moreover the continuous expected average reward $R(\theta)$ can be extracted from the likelihood of the binary reward as $R(\theta) = P(r = 1|\theta)(R_{max} - R_{min}) + R_{min}$. Each EM iteration consists of an E- and M-step: the E-step computes alpha and beta messages with old FSC parameters to compute the log likelihood function, and the M-step finds new FSC parameters that maximize the log likelihood.

E-step. Based on the current policy parameters θ, the E-step computes alpha α_t^{sq} and beta β_t^{sq} messages:

$$\alpha_0^{sq} = P_0(s, q) , \quad \alpha_t^{s'q'} = \sum_{s,q} P_{s'q'sq} \alpha_{t-1}^{sq} , \tag{3}$$

$$\beta_0^{s,q} = \sum_a \hat{R}_{sa} \prod_i P_{aq}^{(i)} , \quad \beta_{t+1}^{sq} = \sum_{s',q'} P_{s'q'sq} \beta_t^{s'q'} . \tag{4}$$

M-step. Let L_t denote a sequence of world states and FSC state, observation, and action variables of all agents from time $t = 0$ to T, so that $L_T = \{(s_t, q_{1,t}, \ldots, q_{N,t}, o_{1,t}, \ldots, o_{N,t}, a_{1,t}, \ldots, a_{N,t})\}_{t=0}^T$. Moreover, use $P_{os'sa}^{(t)}$ to denote $P(o_{t+1}, s_{t+1}|s_t, a_t)$ and $\hat{R}_{sa}^{(t)}$ to denote $\hat{R}(r_t = 1|s_t, a_t)$, and for agent i denote $P(a_{i,t}|q_{i,t})$ with $P_{aq}^{(i,t)}$ and $P(q_{i,t+1}|q_{i,t}, o_{i,t+1})$ with $P_{q'qo}^{(i,t)}$. Denote the set of current FSC parameters (action and transition probabilities) by θ and the set of new parameters by $\acute{\theta}$. In the M-step, EM maximizes the expected log likelihood $Q(\theta, \acute{\theta})$ denoted here with Q with respect to the new FSC parameters $\acute{\theta}$:

$$Q = \lim_{T_M \to \infty} \sum_{T=0}^{T_M-1} \sum_{L_T} P_{\theta,T_M}^{r,L_T,T} \log P_{\acute{\theta},T_M}^{r,L_T,T} , \tag{5}$$

where

$$\log P_{\theta,T_M}^{r,L_T,T} = \log P(r=1, L_T, T|\hat\theta, T_M) = \log \hat{R}_{sa}^{(T)} + \log P(s_0, \boldsymbol{q}_0)$$

$$+ \sum_{t=1}^{T} \log P_{os'sa}^{(t-1)} + \sum_i (\sum_{t=0}^{T} \log \acute{P}_{aq}^{(i,t)} + \sum_{t=1}^{T} \log \acute{P}_{q'qo}^{(i,t-1)}) - \log T_M \quad (6)$$

is the log-probability to receive binary reward $r = 1$ after the latent sequence of actions and states L_T. (6) shows that the new FSC probabilities of agent i (action probability $\acute{P}_{aq}^{(i,t)}$ and FSC transition probability $\acute{P}_{q'qo}^{(i,t-1)}$) do not depend on the new distributions of other agents. For brevity, denote sets of sum indices as $V_{\neq i} = \{s, q_{j\neq i}, a_{j\neq i}\}$ and $W_{\neq i} = \{s, q_{j\neq i}, a_{j\neq i}, s', \boldsymbol{o}, \boldsymbol{q}'\}$.

We now construct $\tilde{Q}_i$, the part of Q affecting the new action probability $\acute{P}_{aq}^{(i)}$ for agent i. We use $\acute{P}_{aq}^{(i,t)}$ to denote $\acute{P}_{aq}^{(i)}$ at time step t and use $P(r=1, a_{i,t} = a_i, q_{i,t} = q_i|T, \theta) = \sum_{W_{\neq i}} \alpha_t^{sq} P_{os'sa} P_{q'qo}^{(i)} \prod_{j\neq i} P_{aq}^{(j)} P_{q'qo}^{(j)} \beta_{T-t-1}^{s'q'}$ to denote the probability of a binary reward $r = 1$ at time T, when at $t \leq T$ agent i takes action a_i and the FSC state is q_i. Since $\sum_{L_T} P_{\theta,T_M}^{r,L_T,T} \sum_{t=0}^{T} \log \acute{P}_{aq}^{(i,t)} = \sum_{t=0}^{T} \sum_{a_i,q_i} P(r=1, a_{i,t} = a_i, q_{i,t} = q_i|T, \theta) \log \acute{P}_{aq}^{(i)}$, inserting (6) into (5) yields

$$\tilde{Q}_i = \lim_{T_M \to \infty} \sum_{T=0}^{T_M-1} \sum_{t=0}^{T} \sum_{W_{\neq i}} \frac{\alpha_t^{sq}}{T_M} P_{os'sa} P_{q'qo}^{(i)} (\prod_{j\neq i} P_{aq}^{(j)} P_{q'qo}^{(j)}) \beta_{T-t-1}^{s'q'} \log \acute{P}_{aq}^{(i)} . \quad (7)$$

In (7), breaking the sum over t into $t = T$ and $t = 0, \ldots, T-1$ yields $\tilde{Q}_i = \sum_{a_i,q_i} P_{aq}^{(i)} \log \acute{P}_{aq}^{(i)} \lim_{T_M \to \infty} \sum_{T=0}^{T_M-1} \left[\sum_{V_{\neq i}} \frac{\hat{R}_{sa}}{T_M} (\prod_{j\neq i} P_{aq}^{(j)}) \alpha_T^{sq} + \right.$

$$\left. \sum_{t=0}^{T-1} \sum_{W_{\neq i}} \frac{\alpha_t^{sq}}{T_M} P_{os'sa} P_{q'qo}^{(i)} (\prod_{j\neq i} P_{aq}^{(j)} P_{q'qo}^{(j)}) \beta_{T-t-1}^{s'q'} \right].$$

Because $\acute{P}_{aq}^{(i)}$ is normalized over a_i, maximizing $\tilde{Q}_i$ with respect to $\acute{P}_{aq}^{(i)}$ yields $\acute{P}_{aq}^{(i)} = P_{aq}^{(i)} \lim_{T_M \to \infty} \frac{1}{C_{q_i}} \sum_{T=0}^{T_M-1} \left[\sum_{V_{\neq i}} \hat{R}_{sa} \prod_{j\neq i} P_{aq}^{(j)} \alpha_T^{sq} \right.$

$$\left. + \sum_{t=0}^{T-1} \sum_{W_{\neq i}} \alpha_t^{sq} P_{os'sa} P_{q'qo}^{(i)} \prod_{j\neq i} P_{aq}^{(j)} P_{q'qo}^{(j)} \beta_{T-t-1}^{s'q'} \right],$$

where C_{q_i} is a normalizing constant.

Similarly to [26,10], we separate sums over alpha and beta messages using $\lim_{T_M \to \infty} \sum_{T=0}^{T_M-1} \sum_{t=0}^{T-1} \frac{\alpha_t^{sq} \beta_{T-t-1}^{s'q'}}{T_M} = \lim_{T_M \to \infty} \sum_{t=0}^{T_M-1} \alpha_t^{sq} \sum_{\tau=0}^{T_M-1} \frac{\beta_\tau^{s'q'}}{T_M}$, where $\tau = T - t - 1$. The action probability update becomes

$$\acute{P}_{aq}^{(i)} = P_{aq}^{(i)} \lim_{T_\alpha,T_\beta \to \infty} \frac{1}{C_{q_i}} \left[\sum_{V_{\neq i}} \hat{R}_{sa} (\prod_{j\neq i} P_{aq}^{(j)}) \sum_{T=0}^{T_\alpha-1} \alpha_T^{sq} + \right.$$

$$\left. \sum_{W_{\neq i}} \sum_{t=0}^{T_\alpha-1} \alpha_t^{sq} P_{os'sa} P_{q'qo}^{(i)} (\prod_{j\neq i} P_{aq}^{(j)} P_{q'qo}^{(j)}) \sum_{\tau=0}^{T_\beta-1} \beta_\tau^{s'q'} \right], \quad (8)$$

where we have used alpha and beta horizons, T_α and T_β, in place of T_M, to be used in later discussions.

4.1   Analysis: Stuck in a Local Optimum

We now prove that the action probability update of the traditional EM approach is stuck in a local optimum under certain conditions. The proof that the FSC transition probability updates are stuck is similar and is omitted.

The proof requires stochastic FSCs and that each closed irreducible state set has at least one state with a non-zero reward probability. These conditions are common. Firstly, the policy is usually stochastic, because a deterministic policy is always stuck, even in discounted POMDPs/Dec-POMDPs, because of the multiplicative nature of EM parameter updates. Secondly, the reward probability condition is common. If all sets of irreducible states have zero reward probability, then only transient states have non-zero reward probability. Therefore, the reward probability approaches zero at distant time steps and the need for taking long-term effects of actions into account, the motivation behind average reward, disappears. There may be multichain problems where some of the irreducible closed state sets have, and others do not have, non-zero reward probabilities. We are not aware of such problems, but this may need further investigation.

Note that the proof applies also to average reward POMDPs. The proof for POMDPs is obtained by just setting the number of agents to one. We do not claim the proof to hold in problems without stochastic controllers (e.g. it is possible to use EM in MDPs so that the action probability depends directly on the world state). In particular, we assume in the proof that $\sum_s \alpha_*^{sq} > 0$ for all q, which is true for stochastic controllers.

Preliminary. Recall that $\alpha_t = \{\alpha_t^{sq}\}$ is a projection of the initial belief for t steps following the current policy. To measure difference between a probability distribution and the limiting distribution, we use the *total variation distance* D_{TV} [11], defined as the largest absolute difference of the probability of the same state in two distributions. The distance between distribution α_t at time step t and the limiting distribution α_* is $D_{TV}(\alpha_t, \alpha_*) = \max_{s,q} |\alpha_t^{sq} - \alpha_*^{sq}|$. In aperiodic Markov chains, total variation distance decreases exponentially[1] with time t:

$$D_{TV}(\alpha_t, \alpha_*) \leq C_\epsilon \epsilon^t; 0 < \epsilon < 1 \,, \tag{9}$$

where $C_\epsilon > 0$ and ϵ are constants. In unichains the limiting distribution is unique, but in multichains it depends on the starting distribution. We will not denote the dependence on the starting distribution explicitly but we refer to it when necessary.

Theorem 1. *In unichain and multichain aperiodic Dec-POMDPs, the EM action probability update never changes finite state controller (FSC) parameter values, when each closed irreducible state set has at least one state for which a non-zero reward probability exists, and when the FSC policy is fully stochastic.*

---

[1] Theorem 4.9 in [11] shows this for aperiodic irreducible Markov chains. It is straightforward to modify the proof of the theorem to also apply to aperiodic unichains and multichains, which may have transient states in addition to irreducible communicating classes of states: the equilibrium distribution π in [11] is just replaced with a limiting distribution, which has a zero probability for each transient state.

Proof. We write (8) as $\acute{P}_{aq}^{(i)} = P_{aq}^{(i)}\left[H_{aq}^{(i)} + J_{aq}^{(i)}\right]$, where $H_{aq}^{(i)}$ is the expected sum of reward probabilities gained over all time in situations where agent i was in state q_i and took action a_i, scaled by a normalization term, and $J_{aq}^{(i)}$ is the expected sum of reward probabilities over the future from such situations, again scaled by the normalization term. We have $H_{aq}^{(i)} = \lim_{T_\alpha, T_\beta \to \infty} \frac{1}{C_{q_i}} \tilde{H}_{aq}^{(i)}$ and $J_{aq}^{(i)} = \lim_{T_\alpha, T_\beta \to \infty} \frac{1}{C_{q_i}} \tilde{J}_{aq}^{(i)}$ where $C_{q_i} = \sum_{a_i} P_{aq}^{(i)}(\tilde{H}_{aq}^{(i)} + \tilde{J}_{aq}^{(i)})$ is the normalization term, and we denoted $\tilde{H}_{aq}^{(i)} = \sum_{V_{\neq i}} \hat{R}_{sa} \prod_{j \neq i} P_{aq}^{(j)} \sum_{T=0}^{T_\alpha - 1} \alpha_T^{sq}$ and also denoted $\tilde{J}_{aq}^{(i)} = \sum_{q_{j \neq i}, a_{j \neq i}, s, s', q'} \sum_{t=0}^{T_\alpha - 1} \alpha_t^{sq} P_{s'q'a_i sq}^{(i)} \sum_{\tau=0}^{T_\beta - 1} \beta_\tau^{s'q'}$. The term $P_{s'q'sqa_i}^{(i)} = \sum_{o, a_{j \neq i}} P_{os'sa} P_{q'qo}^{(i)} \prod_{j \neq i} P_{aq}^{(j)} P_{q'qo}^{(j)}$ is the probability that the world and agents will transition to states (s', q') given their current states $(s, q_{j \neq i})$ and a specific action a_i and controller state q_i of the ith agent. For convenience, define $\hat{J}_{aq}^{(i)}$ as

$$\hat{J}_{aq}^{(i)} = \lim_{T_\alpha, T_\beta \to \infty} \frac{T_\alpha T_\beta}{\hat{C}_{q_i}} \sum_{s_\tau, a_\tau, q_\tau} \hat{R}_{s_\tau a_\tau} \alpha_*^{s_\tau, q_\tau} \prod_j P_{a_\tau q_\tau}^{(j)} = \lim_{T_\alpha, T_\beta \to \infty} \frac{T_\alpha T_\beta}{\hat{C}_{q_i}} \cdot \text{const} ,$$

here $\hat{C}_{q_i} = \sum_{a_i} T_\alpha T_\beta \sum_{s_\tau, a_\tau, q_\tau} \hat{R}_{s_\tau a_\tau} \alpha_*^{s_\tau, q_\tau} \prod_j P_{a_\tau q_\tau}^{(j)}$ is another normalizing term. We now prove that $J_{aq}^{(i)} = \hat{J}_{aq}^{(i)}$ and $H_{aq}^{(i)} = 0$. We will then show $\hat{J}_{aq}^{(i)}$ converges to a constant and that the action update is thus stuck.

To prove $J_{aq}^{(i)} = \hat{J}_{aq}^{(i)}$ we show that $|J_{aq}^{(i)} - \hat{J}_{aq}^{(i)}| = 0$. Expand the recursive form of beta messages as

$$\beta_\tau^{s,q} = \sum_{s_\tau, q_\tau, a_\tau} P(s_\tau, q_\tau | s_0 = s, q_0 = q) \hat{R}_{s_\tau a_\tau} \prod_j P_{a_\tau q_\tau}^{(j)} ,$$

where $P_{s,q}^{s_\tau, q_\tau} = P(s_\tau, q_\tau | s_0 = s, q_0 = q)$ is the probability to arrive at world and controller states s_τ, q_τ in τ steps when starting from s, q. Use the expanded form to compute an upper bound on $|J_{aq}^{(i)} - \hat{J}_{aq}^{(i)}|$:

$$\left| \lim_{T_\alpha, T_\beta \to \infty} \left[\frac{1}{C_{q_i}} \sum_{\substack{q_{j \neq i}, a_{j \neq i} \\ s, s', q'}} \sum_{t=0}^{T_\alpha - 1} \alpha_t^{sq} P_{s'q'a_i sq}^{(i)} \sum_{\tau=0}^{T_\beta - 1} \beta_\tau^{s'q'} - \frac{T_\alpha T_\beta}{\hat{C}_{q_i}} \sum_{s_\tau, a_\tau, q_\tau} \left(\right. \right. \right.$$

$$\left. \left. \left. \hat{R}_{s_\tau a_\tau} \alpha_*^{s_\tau, q_\tau} \prod_j P_{a_\tau q_\tau}^{(j)} \right) \right] \right| \leq \left| \lim_{T_\alpha, T_\beta \to \infty} T_\alpha \left[\frac{1}{C_{q_i}} \sum_{\substack{q_{j \neq i}, a_{j \neq i} \\ s, s', q'}} \alpha_*^{sq} P_{s'q'a_i sq}^{(i)} \right. \right.$$

$$\left. \left. \sum_{\tau=0}^{T_\beta - 1} \sum_{s_\tau, q_\tau, a_\tau} P_{s', q'}^{s_\tau, q_\tau} \hat{R}_{s_\tau a_\tau} \prod_j P_{a_\tau q_\tau}^{(j)} - \frac{1}{\hat{C}_{q_i}} \sum_{\tau=0}^{T_\beta - 1} \sum_{s_\tau, a_\tau, q_\tau} \hat{R}_{s_\tau a_\tau} \alpha_*^{s_\tau, q_\tau} \prod_j P_{a_\tau q_\tau}^{(j)} \right] \right|$$

$$\leq \lim_{T_\alpha, T_\beta \to \infty} \left[\sum_{\tau=0}^{T_\beta - 1} \sum_{s_\tau, a_\tau} \frac{\hat{R}_{s_\tau a_\tau} P_{a_\tau q_\tau}^{(i)} T_\alpha}{\min(C_{q_i}, \hat{C}_{q_i})} \left| \sum_{\substack{q_{j \neq i}, a_{j \neq i} \\ s, s', q'}} \alpha_*^{sq} P_{s'q'a_i sq}^{(i)} P_{s', q'}^{s_\tau, q_\tau} - \alpha_*^{s_\tau q_\tau} \right| \right]$$

$$\leq \lim_{T_\alpha, T_\beta \to \infty} \frac{T_\alpha}{\min(C_{q_i}, \hat{C}_{q_i})} \sum_{\tau=0}^{T_\beta - 1} C_\epsilon \epsilon^\tau = \lim_{T_\alpha, T_\beta \to \infty} \frac{T_\alpha}{\min(C_{q_i}, \hat{C}_{q_i})} \frac{C_\epsilon}{1 - \epsilon} = 0 . \quad (10)$$

The last equality follows because $\min(C_{q_i}, \hat{C}_{q_i})$ approaches infinity quadratically: omitting all nonessential notation, C_{q_i} contains a double sum over terms $\alpha_t^{sq}\beta_\tau^{s'q'}$, from $t = 0$ to $T_\alpha - 1$ and from $\tau = 0$ to $T_\beta - 1$. Since the FSCs are fully stochastic, for each q the marginal limit probability is nonzero and thus one state (s, q) must have nonzero limit probability α_*^{sq} (and probability close to the limit for an infinite number of terms), i.e. it is a recurrent state. By assumption (see theorem) one recurrent state must have nonzero reward; such states are visited an infinite number of times, thus the double sum grows faster than $T_\alpha T_\beta \cdot \text{const}$ for some constant. $\hat{C}_{q_i}$ has similar terms and also grows quadratically.

The first inequality in (10) comes from exponential decrease of $D_{TV}(\alpha_t, \alpha_*)^2$. In the second inequality we bounded terms $P_{a_\tau q_\tau}^{(j)}$ by 1 for $j \neq i$. The third inequality   follows   from   using   (9)   to   upper   bound   the   term $\left| \sum_{q_{j\neq i}, a_{j\neq i}, s, s', q'} \alpha_*^{sq} P_{s'q'a_i sq}^{(i)} P(s_\tau, q_\tau | s_0 = s', q_0 = q') - \alpha_*^{s_\tau q_\tau} \right|$. To apply (9), $\alpha_*^{sq} P_{s'q'a_i sq}^{(i)} P(s_\tau, q_\tau | s_0 = s', q_0 = q')$ must converge in the limit $\tau \to \infty$ to $\alpha_*^{s_\tau q_\tau}$, we show this. Define $P_0(s', q' | a_i, q_i) = \sum_{q_{j\neq i}, a_{j\neq i}, s} \alpha_*^{sq} P_{s'q'a_i sq}^{(i)}$ and $P_\tau(s_\tau, q_\tau | a_i, q_i) = \sum_{s', q'} P(s_\tau, q_\tau | s_0 = s', q_0 = q') P_0(s', q' | a_i, q_i)$.

In a *unichain*, the starting distribution does not affect the limiting distribution. Hence, $\lim_{\tau \to \infty} P_\tau(s_\tau, q_\tau | a_i, q_i) = \alpha_*^{s_\tau q_\tau}$. In a *multichain* the limiting distribution depends on the starting distribution, however, in α_*^{sq} and thus in $P_0(s', q' | a_i, q_i)$, all transient Markov chain states have zero probability (easy to verify from the definition of a transient state) and the probability mass is distributed among closed irreducible classes in the exactly same proportion as in $\alpha_*^{s_\tau q_\tau}$. Further forward projection of the Markov chain does not change this probability mass distribution (as the irreducible classes are closed), thus, similarly to the unichain case, the Markov chain starting from $P_0(s', q' | a_i, q_i)$ converges to $\alpha_*^{s_\tau q_\tau}$. Next, we show that $H_{aq}^{(i)}$ is zero.

We have $H_{aq}^{(i)} = \lim_{T_\alpha, T_\beta \to \infty} \frac{T_\alpha}{C_{q_i}} \sum_{V_{\neq i}} \hat{R}_{sa} \prod_{j \neq i} P_{aq}^{(j)} \frac{1}{T_\alpha} \sum_{T=0}^{T_\alpha - 1} \alpha_T^{sq} = \lim_{T_\alpha, T_\beta \to \infty} \frac{1}{C_{q_i}} T_\alpha \sum_{V_{\neq i}} \hat{R}_{sa} \prod_{j \neq i} P_{aq}^{(j)} \alpha_*^{s,q}$, because $\lim_{T_\alpha \to \infty} \frac{1}{T_\alpha} \sum_{T=0}^{T_\alpha - 1} \alpha_T^{sq} = \lim_{T_\alpha \to \infty} \frac{1}{T_\alpha} T_\alpha \alpha_*^{sq} = \alpha_*^{sq}$. Because $\frac{T_\alpha}{C_{q_i}}$ becomes zero in the limit, by the same argument as $\frac{T_\alpha}{\min(C_{q_i}, \hat{C}_{q_i})}$ becomes zero in (10), and because other terms are finite in the limit, $H_{aq}^{(i)}$ is zero. Since $H_{aq}^{(i)}$ is zero and $J_{aq}^{(i)}$ converges to a constant, the probability update multiplies all action probabilities by the same constant; this concludes the proof and $\acute{P}_{aq}^{(i)} = P_{aq}^{(i)} \cdot \text{const}$.

Theorem 1 may be surprising as the discounted reward EM methods [26,10,16] improve the policy in each EM iteration so that the discounted reward never decreases. Getting stuck is a consequence of the average reward setting, where the entire future must be fully taken into account. We next give a practical EM approach for average reward Dec-POMDPs that allows policy improvement.

---

[2] See http://users.ics.aalto.fi/jpajarin/avgrew/supplement.pdf for details.

4.2   A Practical EM Method

The average reward EM described above is always stuck in a local optimum. To force a change to FSC parameters in the M-step, one could try to use fixed instead of infinite horizons. Fixing a horizon induces an approximation error to parameter updates that decreases with a larger horizon. Discounted reward EM methods effectively fix both T_α and T_β to the same horizon by using discounted rewards. This has at least three drawbacks in average reward problems: 1) an initial belief is needed in optimization, 2) discounting rewards increases approximation error compared to uniform rewards, 3) limiting both T_α and T_β increases approximation error more than limiting only one of them.

We now give update rules with an infinite T_α, and propose to set only T_β to a fixed value which is doubled during optimization whenever the current policy value would decrease. This has several advantages. By not limiting T_α we do not need an initial belief in unichain problems and can compute the sum of alpha messages efficiently as detailed later in this Section. Furthermore, the approach allows to reduce the approximation error in parameter updates until the policy value increases. The adaptation of T_β is necessary not only because we know a priori that a too low T_β may not always yield increased value, but also because the approximation error that a specific T_β causes is problem dependent: the mixing rate of the Dec-POMDP determines how fast a distribution converges to the stationary distribution and this in turn determines how high the approximation error for a certain T_β is. In short, this kind of approach is necessary to adapt T_β to the specific Dec-POMDP problem.

Since $\lim_{T_\alpha \to \infty} \frac{1}{T_\alpha} \sum_{t=0}^{T_\alpha - 1} \alpha_t^{sq} = \lim_{T_\alpha \to \infty} \alpha_{T_\alpha}^{sq} = \alpha_*^{sq}$, the **action probability update** is derived from (8) and becomes $\dot{P}_{aq}^{(i)} =$

$$\frac{P_{aq}^{(i)}}{C_{q_i}} \sum_{s,q_{j\neq i}} \alpha_*^{sq} \sum_{a_{j\neq i}} \left[\hat{R}_{sa} \prod_{j\neq i} P_{aq}^{(j)} + \sum_{s',o,q'} P_{os'sa} P_{q'qo}^{(i)} \Big(\prod_{j\neq i} P_{aq}^{(j)} P_{q'qo}^{(j)} \Big) \sum_{\tau=0}^{T_\beta} \beta_\tau^{s'q'} \right].$$

(11)

The **transition probability update** is derived similarly to the action probability update resulting in

$$\dot{P}_{q'qo}^{(i)} = \frac{P_{q'qo}^{(i)}}{C_{oq}^{(i)}} \sum_{s,q_{j\neq i},\mathbf{a},s',o_{j\neq i},q'_{j\neq i}} \left[\alpha_*^{sq} P_{os'sa} P_{aq}^{(i)} \prod_{j\neq i} \Big(P_{aq}^{(j)} P_{q'qo}^{(j)} \Big) \sum_{\tau=0}^{T_\beta} \beta_\tau^{s'q'} \right]. \quad (12)$$

We propose the practical EM algorithm as follows: set T_β to an initial value (we use 32), then apply E- and M-steps in turn until the policy value does not increase or until any other stopping criterion is satisfied.

In the E-step the algorithm computes beta messages up to the horizon T_β using (4) and the limiting distribution α_*^{sq} for alpha messages either projecting until convergence using (3) or by solving a system of linear equations. Because the EM algorithm gradually improves the policy, the limiting distribution from the previous EM iteration is likely close to the new limiting distribution. An efficient unichain implementation thus starts projecting from the limiting distribution of the previous EM iteration (in multichain problems projecting must start from the initial belief). This saves much computation: for example, in the "long fire fighting" experiment, iteration 1 needed 5000 projections, next iterations only 3-100 projections.

In the M-step the algorithm computes new FSC action and transition probabilities by (11) and (12). *After the M-step* the algorithm checks whether the value of a policy decreased: if it did, the algorithm multiplies T_β by 2 and recomputes the beta messages and performs the M-step again, until the value does not decrease (for $T_\beta \to \infty$ this would yield the original EM we derived; we limit T_β to a maximum of 32768). In the experiments, T_β needed duplication only rarely.

Our practical EM is better than naive bounding/discounting both alpha and beta. We efficiently compute exact infinite-horizon alphas, using the limiting distribution from the previous iteration as the start of propagation, whereas discounted EM would need to choose a discount factor and propagate alpha to large horizons. Our EM is intuitive and easy to implement.

5   Experiments

We evaluate the average reward on two different sets of benchmark problems. The first set consists of benchmark problems, used previously for evaluating discounted reward Dec-POMDP methods [10,3,2,16]. The second set consists of two new average reward benchmark problems, which emphasize long-term effects of actions.

For all problems, we compare the new expectation maximization (EM) average reward DEC-POMDP method (denoted "AvgEM") of Section 4.2, against a baseline and loose upper bounds of performance. We use a uniformly random policy as baseline. For (loose) upper bounds we compute the optimal solution to the average reward MDP underlying the DEC-POMDP with linear programming [21]; this upper bound corresponds to agents that have full knowledge of the environment and each other. We also show AvgEM outperforms an alternative new non-linear programming approach (denoted "AvgNLP") which we introduce in the Appendix. We compare AvgEM with a state of the art discounted reward EM (denoted "DiscEM") method [10] on different discount factors 0.9, 0.99, and 0.999; we show that AvgEM outperforms DiscEM in average reward problems and has equal or better performance in benchmark problems from the discounted reward literature. Optimization of a controller using the EM methods, optimization of the random baseline, and optimization of the MDP upper bound were run in Matlab on a single processor core. Methods were stopped if the change in the policy value between iterations was under a small threshold. EM methods had a time limit of one hour. Non-linear programs were solved with the SNOPT solver on the publicly available NEOS server.

Benchmark problems from the discounted reward literature. The first six problems in Table 1 (denoted "Disc. Prob.") have been used to evaluate discounted reward methods [17], but as we evaluate methods by average reward, the earlier evaluations based on discounted reward are not directly comparable. The problems are: DecTiger (2,3,2), Recycling robots (4,3,2), 2x2 Grid meeting (16,5,2),

Wireless network (64,2,6), Box pushing (100,4,5), and Mars rovers (256,6,8), where for each problem we list (number of states, number of actions, number of observations). For each problem AvgEM, AvgNLP, and DiscEM optimized different size FSCs in parallel over 10 random FSC initializations. Table 1 shows also the average reward for the random policy and for the MDP upper bound. AvgEM performs well, in "recycling robots" it is even close to the full-knowledge upper bound. AvgEM outperforms AvgNLP and performs as well as "DiscEM 0.9". "DiscEM 0.9" outperforms "DiscEM 0.99" and "DiscEM 0.999" demonstrating that, in these problems, good results are already obtained with a small discount factor. Next, we will discuss two new average reward problems with long-term effects of actions.

Wireless network with overhead ($|S| = 64$, $|A_i| = 2$, $|O_i| = 6$). In the wireless networking problem, [16] two wireless agents try to keep their transmit buffers, modeled with four states, as empty as possible. Each buffer gets data from a two-state source model. Buffer fullness is modeled as few states at rough intervals; insertions/transmissions have a probability to change the buffer state. If both agents transmit simultaneously both transmissions fail and data is not removed from the buffers. The world state is the cross product of the transmit buffers and source models, in total 64 states. In [16] the objective corresponded to minimizing delay. In the new problem, successful transmissions are rewarded, corresponding to maximizing throughput. In real wireless networks, decisions are made at 10 microsecond intervals; to reflect this, we multiplied the probability to transition from one buffer state to another and the probability to insert data into a buffer with 0.01. As overhead from packet headers etc. is proportionally smaller for larger packets, the new wireless problem allows transmission of more data, when the buffer is fuller: for buffer size x, $y = 2x/(x + 1)$ data units are transmitted (probability to change buffer state is proportional to y).

Long fire fight ($|S| = 27$, $|A_i| = 2$, $|O_i| = 2$). In the fire fighting problem [15] two robots try to extinguish three houses and receive negative reward for higher house fire levels (see [15] for details). In the new *long fire fighting* version a house can also start burning on its own with probability 0.1. To make a single Dec-POMDP time step correspond to a shorter time in the real application, we multiplied all transition probabilities between fire levels with 0.01. In this version a fire takes longer to put out, and it takes longer for fire levels to increase.

Table 1 shows results for the wireless network with overhead (denoted "Long Wirel.") and the long fire fight (denoted "Long FF") problems (FSC size was fixed to 3). In both problems AvgEM converged rapidly and got highest average reward. Figure 1 shows convergence of the EM methods. Results for the discounted method DiscEM agree with the observations in Section 4.2 about the negative effect of discounting alpha and beta messages. DiscEM converges with a low discount factor 0.9 to suboptimal solutions and with a large 0.999 factor too slowly. Interestingly, in fire fighting "DiscEM 0.9" convergences to a bad local optimum where both agents only try to extinguish the middle house, showing the necessity of adapting optimization parameters to the specific Dec-POMDP

Table 1. Expected average reward of a uniformly random policy ("Random"), a MDP based upper bound ("MDP"), the average reward nonlinear programming method ("AvgNLP"), the discounted expectation maximization method ("DiscEM") for discount factors 0.9, 0.99, and 0.999, and the average reward expectation maximization method ("AvgEM") in benchmark problems used in discounted method research [10,3,2,16] ("Disc. Prob.") and in new average reward benchmarks ("Avg. Prob."). A result is bolded, when the 95% confidence interval of the best result contains the result or vice versa. AvgEM outperforms AvgNLP, performs as well or better as DiscEM in discounted reward problems, and outperforms DiscEM in the average reward problems.

Disc. Prob.	Random	MDP	AvgNLP	DiscEM 0.9	DiscEM 0.99	DiscEM 0.999	AvgEM
DecTiger	−46.22	20.00	**−2.00**	**−1.375**	**−1.80**	−2.19	**−1.79**
Rec. robots	0.45	3.27	1.24	**3.08**	**3.08**	2.59	**3.08**
2x2 Grid	0.25	1.00	0.28	0.80	**0.83**	0.56	0.75
Wireless	−3.04	−1.46	−3.00	**−1.96**	−2.07	−2.86	−2.05
Box pushing	−1.37	20.35	−0.19	3.69	3.45	0.28	**3.75**
Mars rovers	−1.21	2.88	1.05	**1.77**	0.80	−0.315	**1.55**
Avg. Prob.	Random	MDP	AvgNLP	DiscEM 0.9	DiscEM 0.99	DiscEM 0.999	AvgEM
Long Wirel.	0.0063	0.0099	**0.0089**	0.0081	0.0085	0.0066	**0.0093**
Long FF	−1.85	−0.20	−3.00	−4.00	−1.095	−1.44	**−0.91**

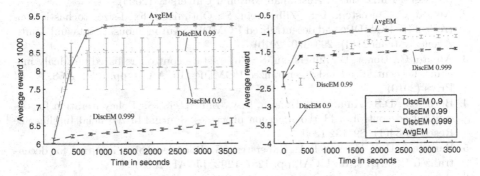

Fig. 1. Expected average reward of discounted reward EM (DiscEM) with 3 discount factors and our average reward EM method (AvgEM), for "wireless network with overhead" (left) and "long fire fighting" (right). Error bars are 95% confidence intervals from bootstrapping.

problem. In fact, for most EM iterations AvgEM held parameter T_β (see Section 4.2) between 32 and 512 in "wireless network with overhead" and at 32 in "long fire fight".

6   Conclusions

Average reward is a useful criterion for planning under uncertainty with multiple agents; it has real-life importance in wireless networks and other domains.

We showed that traditional expectation maximization is stuck in average reward Dec-POMDPs (and POMDPs) under certain conditions and provided a new EM based method for average reward Dec-POMDPs. Our new EM method yields good performance, outperforming a state of the art discounted reward EM method in average reward problems. We also introduced two average reward benchmark problems, long fire fighting and wireless network with overhead. **To our knowledge this is the first general Dec-POMDP method for optimizing average reward.**

Acknowledgements. The work was supported by the Academy of Finland, decisions 252845, 271394, and 251170 (Finnish CoE in Computational Inference Research COIN). The work was also supported by Nokia. J. Peltonen belongs to COIN and Helsinki Institute for Information Technology HIIT. J. Pajarinen carried out most of the work at the Department of Information and Computer Science, Aalto University.

References

1. Aberdeen, D.: Policy-gradient algorithms for partially observable Markov decision processes. Ph.D. thesis, Australian National University (2003)
2. Amato, C., Bernstein, D.S., Zilberstein, S.: Optimizing fixed-size stochastic controllers for POMDPs and decentralized POMDPs. Autonomous Agents and Multi-Agent Systems 21(3), 293–320 (2010)
3. Amato, C., Bonet, B., Zilberstein, S.: Finite-state controllers based on Mealy machines for centralized and decentralized POMDPs. In: AAAI, pp. 1052–1058. AAAI Press (2010)
4. Bernstein, D.S., Amato, C., Hansen, E.A., Zilberstein, S.: Policy iteration for decentralized control of Markov decision processes. Journal of Artificial Intelligence Research 34(1), 89–132 (2009)
5. Bernstein, D.S., Hansen, E.A., Zilberstein, S.: Bounded policy iteration for decentralized POMDPs. In: IJCAI, pp. 1287–1292. IJCAI (2005)
6. Bernstein, D., Givan, R., Immerman, N., Zilberstein, S.: The complexity of decentralized control of Markov decision processes. Mathematics of Operations Research, 819–840 (2002)
7. Bianchi, G., Fratta, L., Oliveri, M.: Performance evaluation and enhancement of the CSMA/CA MAC protocol for 802.11 wireless LANs. In: PIMRC, vol. 2, pp. 392–396. IEEE (1996)
8. Ji, S., Parr, R., Li, H., Liao, X., Carin, L.: Point-based policy iteration. In: AAAI, vol. 22, pp. 1243–1249. AAAI Press (2007)
9. Kakade, S.: Optimizing average reward using discounted rewards. In: Helmbold, D.P., Williamson, B. (eds.) COLT 2001 and EuroCOLT 2001. LNCS (LNAI), vol. 2111, pp. 605–615. Springer, Heidelberg (2001)
10. Kumar, A., Zilberstein, S.: Anytime planning for decentralized POMDPs using Expectation Maximization. In: UAI, pp. 294–301. AUAI Press (2010)
11. Levin, D., Peres, Y., Wilmer, E.: Markov chains and mixing times. American Mathematical Society (2009)

12. Li, Y., Yin, B., Xi, H.: Finding optimal memoryless policies of POMDPs under the expected average reward criterion. European Journal of Operational Research 211(3), 556–567 (2011)
13. Mahadevan, S.: Average reward reinforcement learning: Foundations, algorithms, and empirical results. Machine Learning 22(1), 159–195 (1996)
14. Oliehoek, F.: Value-Based Planning for Teams of Agents in Stochastic Partially Observable Environments. Ph.D. thesis, Informatics Institute, University of Amsterdam (February 2010)
15. Oliehoek, F., Spaan, M., Whiteson, S., Vlassis, N.: Exploiting locality of interaction in factored DEC-POMDPs. In: AAMAS, vol. 1, pp. 517–524. IFAAMAS (2008)
16. Pajarinen, J., Peltonen, J.: Efficient planning for factored infinite-horizon DEC-POMDPs. In: IJCAI, pp. 325–331. AAAI Press (2011)
17. Pajarinen, J., Peltonen, J.: Periodic finite state controllers for efficient POMDP and DEC-POMDP planning. In: NIPS, pp. 2636–2644 (2011)
18. Pajarinen, J., Hottinen, A., Peltonen, J.: Optimizing spatial and temporal reuse in wireless networks by decentralized partially observable Markov decision processes. IEEE Transactions on Mobile Computing (2013) (preprint)
19. Petrik, M., Zilberstein, S.: Average reward decentralized Markov decision processes. In: IJCAI, pp. 1997–2002 (2007)
20. Poupart, P., Boutilier, C.: Bounded finite state controllers. In: NIPS, pp. 823–830. MIT Press (2004)
21. Puterman, M.L.: Markov decision processes: discrete stochastic dynamic programming. Wiley (2005)
22. Seuken, S., Zilberstein, S.: Formal models and algorithms for decentralized decision making under uncertainty. Autonomous Agents and Multi-Agent Systems 17(2), 190–250 (2008)
23. Spaan, M., Oliehoek, F., Amato, C.: Scaling up optimal heuristic search in DEC-POMDPs via incremental expansion. In: IJCAI. AAAI Press (2011)
24. Szer, D., Charpillet, F.: An optimal best-first search algorithm for solving infinite horizon DEC-POMDPs. In: Gama, J., Camacho, R., Brazdil, P.B., Jorge, A.M., Torgo, L. (eds.) ECML 2005. LNCS (LNAI), vol. 3720, pp. 389–399. Springer, Heidelberg (2005)
25. Tangamchit, P., Dolan, J., Khosla, P.: The necessity of average rewards in cooperative multirobot learning. In: ICRA, vol. 2, pp. 1296–1301. IEEE (2002)
26. Toussaint, M., Harmeling, S., Storkey, A.: Probabilistic inference for solving (PO)MDPs. Tech. rep., University of Edinburgh (2006)
27. Toussaint, M., Storkey, A.: Probabilistic inference for solving discrete and continuous state Markov decision processes. In: ICML, pp. 945–952. ACM (2006)
28. Yagan, D., Tham, C.: Coordinated reinforcement learning for decentralized optimal control. In: ADPRL, pp. 296–302. IEEE (2007)
29. Yu, H., Bertsekas, D.P.: Discretized approximations for POMDP with average cost. In: UAI, pp. 619–627. AUAI Press (2004)

Appendix: Non-linear Programming for Average Reward Dec-POMDPs

A non-linear programming (NLP) approach has been used in recent discounted reward POMDP and Dec-POMDP research [2]. To study whether a NLP approach is suitable for average reward cases, we introduce a new NLP based method as an alternative to the expectation-maximization approach that we recommend. We do not claim that the method below is the only possible NLP approach to average reward Dec-POMDPs, but to our knowledge no other NLP methods for average reward Dec-POMDPs have been presented so far, therefore we use our method below as a first proxy.

Motivated by the linear programming solution for average reward MDPs [21] we use the same basic idea that the limiting distribution remains the same over successive time steps. Note that the discounted reward NLP approach in [2] uses the Bellman equation to recursively define the optimal value function over world and FSC states, but the approach requires a discount factor and is not directly applicable to average reward problems. Instead we use the limiting distribution as the basis for optimization.

Table 2. Non-linear program for an aperiodic unichain average reward Dec-POMDP. The program maximizes the immediate reward of the limiting distribution $P_*(s, \boldsymbol{q})$, which corresponds to maximizing the average reward. The program solves for the FSC parameters $P^{(i)}_{q'qo}$ and $P^{(i)}_{aq}$ of each agent i.

Variables: $P_*(s, \boldsymbol{q})$ and for each agent i: $P^{(i)}_{q'qo}$, $P^{(i)}_{aq}$

Optimization goal: Maximize $\sum_{s,a} R_{sa} \sum_{\boldsymbol{q}} P_*(s, \boldsymbol{q}) \prod_i P^{(i)}_{aq}$

Subject to the following **constraints**:

$P_*(s', \boldsymbol{q}') - \sum_{s,q} P_{s'q'sq} P_*(s, \boldsymbol{q}) = 0$, $\sum_{s,q} P_*(s, \boldsymbol{q}) = 1$, $\quad P_*(s, \boldsymbol{q}) \geq 0 \ \forall s \ \forall \boldsymbol{q}$

$\sum_{a_i} P^{(i)}_{aq} = 1 \ \forall q_i$, $P^{(i)}_{aq} \geq 0 \ \forall q_i \ \forall a_i$, $\quad \sum_{q_{i'}} P^{(i)}_{q'qo} = 1 \ \forall q_i \ \forall o_i$, $P^{(i)}_{q'qo} \geq 0 \ \forall q_i \ \forall o_i \ \forall q'_i$

Table 2 shows the non-linear program for solving aperiodic unichain average reward Dec-POMDPs. In Table 2 we have kept the notation for probability distributions used throughout the paper, one may use functions instead of distributions for notational purposes. We now discuss the program from top to bottom. **Variables:** The limiting distribution $P_*(s, \boldsymbol{q})$ and the FSC parameters of each agent i, $P^{(i)}_{q'qo}$ and $P^{(i)}_{aq}$, are the variables to solve for. **Optimization goal:** The optimization goal of the non-linear program is to maximize the average reward $\sum_{s,a} R_{sa} \sum_{\boldsymbol{q}} P_*(s, \boldsymbol{q}) \prod_i P^{(i)}_{aq}$. **First constraint:** The first constraint $P_*(s', \boldsymbol{q}') - \sum_{s,q} P_{s'q'sq} P_*(s, \boldsymbol{q}) = 0$ forces $P_*(s, \boldsymbol{q})$ to be a limiting distribution. **Other constraints:** The remaining constraints force the probability distributions to be positive and to sum to one. In the experiments non-linear programs were solved with the SNOPT solver on the publicly available NEOS server.

Properly Acting under Partial Observability with Action Feasibility Constraints

Caroline P. Carvalho Chanel[1,2] and Florent Teichteil-Königsbuch[1]

[1] Onera – The French Aerospace Lab, Toulouse, France
{caroline.carvalho,florent.teichteil}@onera.fr
[2] ISAE – Institut Supérieur de l'Aéronautique et de l'Espace, Toulouse, France

Abstract. We introduce Action-Constrained Partially Observable Markov Decision Process (AC-POMDP), which arose from studying critical robotic applications with damaging actions. AC-POMDPs restrict the optimized policy to only apply feasible actions: each action is feasible in a subset of the state space, and the agent can observe the set of applicable actions in the current hidden state, in addition to standard observations. We present optimality equations for AC-POMDPs, which imply to operate on α-vectors defined over many different belief subspaces. We propose an algorithm named PreCondition Value Iteration (PCVI), which fully exploits this specific property of AC-POMDPs about α-vectors. We also designed a relaxed version of PCVI whose complexity is exponentially smaller than PCVI. Experimental results on POMDP robotic benchmarks with action feasibility constraints exhibit the benefits of explicitly exploiting the semantic richness of action-feasibility observations in AC-POMDPs over equivalent but unstructured POMDPs.

Keywords: sequential decision-making, partially observable Markov decision processes, safe robotics, action feasibility constraints, action preconditions.

1 Introduction

In automated planning, dealing with action preconditions – those feasibility constraints modeling the set of states where a given action is applicable – is an usual standard [1–4]. They allow planning problems' designers to explicitly express properties about feasible actions as logic formulas, which is of first importance in real-life systems or robots. Feasible actions are defined as [5]: neither physically impossible (e.g. flying to Prague from a city without airport), nor forbidden for safety reasons (taking off without sufficient fuel), nor suboptimal and thus useless (flying from Toulouse to Prague via São Borja).

When constructing a solution plan, deciding whether an action is feasible in the current state of the system is obvious if states are fully observable, by testing if the current state is in the set of states where the action is feasible. However, if the agent cannot know its current state with perfect precision, it must rather reason about its *belief state*, that encodes all the different possible states in which the agent can be [6–8]. Thus, solution strategies are defined over belief states but not states, whereas action feasibility constraints are still defined over states. Therefore, additional information

H. Blockeel et al. (Eds.): ECML PKDD 2013, Part I, LNAI 8188, pp. 145–161, 2013.

from the environment is required to disambiguate the belief state insomuch the set of feasible actions to insert in the plan can be deduced [9].

For example, consider an autonomous coast guard robot navigating along a cliff with abysses, as shown in Figure 1(a). This example is a slight variation of the Hallway problem [10], where surrounding walls are replaced by cliffs from which the robot may fall down. The goal is to reach the star while being certain (i.e. with probability 1) not to fall down a cliff: in the states near abysses, actions that might make the robot fall down *have to* be prohibited because they are unsafe. Imagine that the belief state of the robot includes states 1, 2, 3 depicted in Figure 1(a). Without sensing the configuration of surrounding abysses, i.e. the feasible actions in the current hidden state, there is no way for the robot to go south in order to sense the presence of the goal. The modeling solution that guarantees to reach the goal while applying only safe actions is well-known by researchers on planning under partial observability: it consists in adding information about applicable actions in the agent's observations, and in assigning near-infinite costs to infeasible state-action pairs. Doing this, we are guaranteed that the maximum-reward policy will: (i) sufficiently disambiguate the belief state in order to reach high interesting rewards ; (ii) only apply feasible actions.

Despite the existence of well-known modeling principles to deal with action feasibility constraints in partially observable planning, there is place for improvements by noting that the set of observations has a specific structure in many real-life or robotic applications: namely, the set of observations is factored in the form of $\Omega = \mathcal{O} \times \Theta$, where, in the current hidden state, the agent can receive "standard" observations randomly from $\mathcal{O}$, and "feasibility" observations deterministically from Θ. For instance in the coast guard problem, a laser or camera sensor will imperfectly locate the agent in the grid, but provide a unique deterministic configuration of surrounding abysses, i.e. set of feasible actions. Thus, this paper aims at benefiting from the specific structure of the observation set to significantly reduce the complexity of finding an optimal policy. We conduct this study with probabilistic settings, in the context of Partially Observable Markov Decision Processes (POMDPs) [6, 7]. Our proposal is oriented towards the exploitation of the specific structure of the problem, whereas standard algorithms can still solve the problem without using this information but far less efficiently.

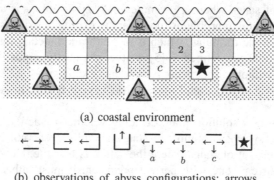

(a) coastal environment

(b) observations of abyss configurations; arrows represent feasible actions in each configuration

Fig. 1. Coast guard robotic problem

The remainder of the paper is organized as follows. First, we make explicit and formalize action feasibility constraints in a new model named Action-Constrained POMDPs (AC-POMDPs), which is a subset of POMDPs. This richer model provides structured observation sets and functions, as well as a sound optimization criterion, which properly selects only policies whose actions are feasible in the current hidden state of the system. Most importantly, this criterion reveals that *optimizing AC-POMDPs can be reduced to handle α-vectors that are defined over many different small belief subspaces*, thus significantly reducing computations. Then, we present a point-based algorithm named PreCondition Value Iteration (PCVI), which takes advantage of the specific structure of AC-POMDPs by implementing α-vector procedures that operate over different small belief subspaces. In comparison, standard algorithms like PBVI - Point Based Value Iteration - [11] operate on the full belief space. Finally, we propose a relaxed version of PCVI, which computes a lower bound on the value function that totally removes the set of action feasibility observations from computations, yielding additional exponential-time speedups. Our experimental results on many benchmarks and on a real aerial robotic problem, where action feasibility constraints are essential for safety reasons, highlight the computational benefits of explicitly dealing with action feasibility semantics in POMDPs.

1.1   Related Work

Recently, researchers proposed a structured POMDP model, named Mixed-Observable Markov Decision Processes (MOMDPs, see [12, 13]), which divides the observation space Ω in visible and hidden parts: $\Omega = \Omega_v \times \Omega_h$. MOMDPs exploit the specific structure of the observation set to reduce the dimension of the belief space, resulting in significant computation gains. However, in our approach, the semantics of observation variables are totally different: we assume $\Omega = \mathcal{O} \times \Theta$, with $\Theta \subseteq 2^A$ being a set of applicable actions from the set A of actions. Among algorithmic differences, MOMDPs' α-vectors are all defined on the same subspace, whereas AC-POMDPs' α-vectors are each defined on different subspaces (see later). This noticeable difference suggests that AC-POMDPs can not be simply viewed as MOMDPs for which visible observations would be sets of feasible actions. Further work is needed to explicitly exploit the specific semantics of action-feasibility observations, as actually proposed in this paper.

Our work is also related to POMDP models that incorporate constraints on states or on execution paths. Goal POMDPs [14] require the optimized policy to reach a set of goal states from a given initial state. Constrained POMDPs [15] search for a policy maximizing the value function for a given reward function subject to inequality constraints on value functions for different reward functions, which can be often interpreted as constraining the optimized policy to some areas of the belief space. Thus, these models put state-based constraints on the optimized policy, which are not directly related to properties of feasible actions. On the contrary, our AC-POMDP model forces the optimized policy to apply only actions that are feasible in a given belief state, knowing constraints on feasible state-action pairs. Our action-feasibility constraints are weaker than Constrained POMDPs' ones, which allows us to use a modified dynamic programming schema that does not include the cumulative cost in the state space, contrary to

Constrained POMDPs. As a result, the complexity of solving Constrained POMDPs is much higher than in AC-POMDPs.

2 Theoretical Backgrounds

Our work is built upon Partially Observable Markov Decision Processes (POMDPs), which offer a sound mathematical model for sequential decision-making under probabilistic partial observability. A POMDP [6, 7] is a tuple $\langle S, A, \Omega, T, O, R, b_0 \rangle$, where: S is the set of states; A is the set of actions; Ω is the set of observations; $T : S \times A \times S \rightarrow [0, 1]$ is the transition function, such that: $T(s, a, s') = p(s_{t+1} = s' | s_t = s, a_t = a)$; $O : \Omega \times A \times S \rightarrow [0, 1]$ is the observation function such that: $O(o, a, s') = p(o_{t+1} = o | s_{t+1} = s', a_t = a)$; $R : S \times A \times S \rightarrow \mathbb{R}$ is the reward function associated with transitions; b_0 is the initial probability distribution over states. We denote $\Delta \subset [0; 1]^{|S|}$ the (continuous) set of probability distributions over states, named *belief space*. Figure 2(a) depicts the dynamic influence diagram of a POMDP.

At each time step, the agent updates its current *belief* b according to the performed action and the received observation, using Bayes' rule :

$$b_a^o(s') = \frac{O(o, a, s') \sum_{s \in S} T(s, a, s') b(s)}{\sum_{s \in S} \sum_{s'' \in S} O(o, a, s'') T(s, a, s'') b(s)} \tag{1}$$

Solving a POMDP consists in finding a policy function $\pi : \Delta \rightarrow A$ that maximizes a performance criterion. The expected discounted reward from any initial belief $V^\pi(b) = E_\pi \left[\sum_{t=0}^{\infty} \gamma^t r(b_t, \pi(b_t)) \mid b_0 = b \right]$ is usually optimized. The value of an optimal policy π^* is defined by the optimal value function V^* that satisfies the Bellman optimality equation:

$$V^*(b) = \max_{a \in A} \left[r(b, a) + \gamma \sum_{o \in \Omega} p(o|a, b) V^*(b_a^o) \right] \tag{2}$$

where $r(b, a) = \sum_{s \in S} b(s) \sum_{s' \in S} T(s, a, s') R(s, a, s')$. This value function is proven to be piecewise linear and convex over the belief space [6], so that at n^{th} optimization stage, the value function V_n can be parametrized as a set of hyperplanes over Δ named α-vectors. An α-vector and the associated action $a(\alpha_n^i)$ define a region of the belief space for which this vector maximizes V_n. Thus, the value of a belief b can be defined as: $V_n(b) = \max_{\alpha_n^i \in V_n} b \cdot \alpha_n^i$. The optimal policy at this step is then: $\pi_n(b) = a(\alpha_n^b)$.

3 Action-Constrained POMDPs

In this section, we propose a more expressive POMDP model, named Action-Constrained POMDP (AC-POMDP), which makes explicit the semantics of feasible actions in the model. Namely, as common in robotic applications, it assumes that observation symbols are factored in 2 parts: probabilistic observations informing about the hidden state, and deterministic observations informing about the set of actions that are feasible in the current hidden state. We will see that the second part implies to maximize the value function over different belief subspaces, yielding computational savings over traditional POMDP solvers.

3.1 AC-POMDPs: Model and Optimization Criterion

An Action-Constrained POMDP is defined as a tuple $\langle \mathcal{S}, (\mathcal{A}_s)_{s \in \mathcal{S}}, \Omega, T, O, \mathbb{I}, R, b_0, \Theta_0 \rangle$, where, in contrast to POMDPs: $(\mathcal{A}_s)_{s \in \mathcal{S}}$ is the set of applicable action sets, such that $\mathcal{A}_s$ is the set of actions that are feasible in a given state s; $\Omega = \mathcal{O} \times \Theta$ is the set of observations, such that $\Theta \subseteq 2^{\mathcal{A}}$; observations in $\mathcal{O}$ and in Θ are independent given any next state and applied action ; $O : \mathcal{O} \times \mathcal{A} \times \mathcal{S} \to [0, 1]$ is the observation function such that: $O(o, a, s') = p(o_{t+1} = o | s_{t+1} = s', a_t = a)$; $\mathbb{I} : \Theta \times \mathcal{S} \to \{0, 1\}$ is the feasibility function: $\mathbb{I}(\theta, s') = p(\theta_{t+1} = \theta \mid s_{t+1} = s') = 1$ if $\theta = \mathcal{A}_{s'}$, otherwise 0 ; Θ_0 is the initial set of applicable actions, observed before applying the first action. Like similar approaches in non-deterministic settings [9], Θ_0 is required to safely apply the first action. For convenience, we also define the action feasibility function $\mathbb{F}$ such that $\mathbb{F}(a, s) = 1$ if and only if $a \in \mathcal{A}_s$.

Figure 2(b) represents an AC-POMDP as a controlled stochastic process. The action a_t executed at time t is constrained to belong to the observed set of feasible actions θ_t. The next observed set θ_{t+1} is equal to the set of actions $\mathcal{A}_{s_{t+1}}$ that are feasible in the hidden state s_{t+1}, which stochastically results from applying action a_t in state s_t. The other part of observations (o_t and o_{t+1}) are stochastically received in the same way as in POMDPs.

Contrary to POMDPs, policies of AC-POMDPs are constrained to only execute actions that are feasible in the current hidden state. Given the history $h_t = (\omega_0 = (o_0, \theta_0), \cdots, \omega_t = (o_t, \theta_t))$ of observations up to time t, we define the set of feasible policies as:

$$\Pi^{h_t} = \{\pi \in \mathcal{A}^\Delta : \forall 0 \leqslant i \leqslant t, \pi(b_i(o_0, \cdots, o_i)) \in \theta_i\}$$

where $b_i(o_0, \cdots, o_i)$ is the belief state resulting from observing $(o_0, \cdots, o_i)$. Thus, solving an AC-POMDP consists in finding a policy π^* such that, for all $b \in \Delta$ and $\theta \in \Theta_0$:

$$\pi^*(b, \theta) \in \underset{\pi \in \Pi^{h_\infty}}{\mathrm{argmax}}\, E\left[\sum_{t=0}^{+\infty} \gamma^t r(b_t, \pi(b_t)) \mid b_0 = b, \theta_0 = \theta\right] \tag{3}$$

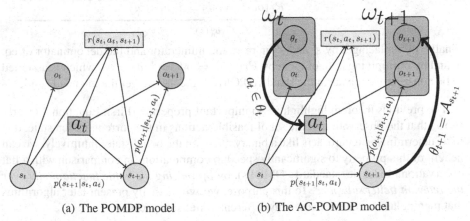

(a) The POMDP model   (b) The AC-POMDP model

Fig. 2. Dynamic influence diagrams of POMDPs and AC-POMDPs

3.2 Belief State Update

As in POMDPs, we should compute the new belief state of the agent, noted b_a^ω, after applying an action a in belief state b and receiving an observation $\omega = (o, \theta)$.

Theorem 1. *Let b be the belief state at a given time step, a the action applied by the agent at this time step, and $\omega = (o, \theta)$ the pair of observations immediately received. The next belief state, for all possible next state s', is:*

$$b_a^{(o,\theta)}(s') = \frac{\mathbb{I}(\theta, s')b_a^o(s')}{\sum_{s'' \in \mathcal{S}} \mathbb{I}(\theta, s'')b_a^o(s'')} \tag{4}$$

with b_a^o equal to the expression given in eq. 1.

Proof.

$$
\begin{aligned}
b_a^{(o,\theta)}(s') &= Pr(s_{t+1} = s'|o_{t+1} = o, \theta_{t+1} = \theta, b_t = b, a_t = a) \\
&= \frac{Pr(s_{t+1} = s', o_{t+1} = o, \theta_{t+1} = \theta|b_t = b, a_t = a)}{Pr(o_{t+1} = o, \theta_{t+1} = \theta|b_t = b, a_t = a)} \\
&= \frac{U(s', o, \theta|b, a)}{\sum_{s'' \in \mathcal{S}} U(s'', o, \theta|b, a)}
\end{aligned}
\tag{5}
$$

with $U(s, o, \theta|b, a) = Pr(s_{t+1} = s, o_{t+1} = o, \theta_{t+1} = \theta|b_t = b, a_t = a)$. As observation symbols o and θ are assumed to be independent given any next state and applied action a, we can factorize U in the form of:

$$
U(s, o, \theta|b, a) \times \frac{1}{Pr(o_{t+1} = o|b_t = b, a_t = a)} =
$$
$$
\underbrace{Pr(\theta_{t+1} = \theta|s_{t+1} = s, b_t = b, a_t = a)}_{\mathbb{I}(\theta, s)} \times
$$
$$
\underbrace{\frac{Pr(o_{t+1} = o|s_{t+1} = s, b_t = b, a_t = a) \times Pr(s_{t+1} = s|b_t = b, a_t = a)}{Pr(o_{t+1} = o|b_t = b, a_t = a)}}_{b_a^o(s)}
$$

Finally, by replacing s by s' and s'' in resp. the numerator and the denominator of eq. 5, and by multiplying both of them by $Pr(o_{t+1} = o|b_t = b, a_t = a)$, which is assumed to be non-zero exactly as in the standard POMDP theory, we get the intended result.

The previous theorem highlights two important properties. First, Equation 4 clearly shows that the observation of the set of feasible actions in the current hidden state, due to its deterministic nature, acts like a binary mask on the belief state. Intuitively, we can benefit of this property to significantly speedup computations in comparison with a flat observation model (ie. standard POMDPs), by *optimizing the value function only over the relevant belief subspace*. To this purpose, we will actually present later algorithms that manipulate α-vectors over many different belief subspaces.

Second, following from eq. 4, we will prove that at any given time step, all states s' whose belief is non-zero have the same set of feasible actions. This property shows that executing policies in AC-POMDPs is coherent with the proposed framework. Most importantly, at optimization time, we can deduce without ambiguity the set of actions over which we maximize the value function for the current belief state. Note that *this primordial property would not be true if the agent would not observe the set of feasible actions*, which gives a theoretical justification of observing them in real robotic applications. More formally, let us define the support of the belief state as: $\sigma(b_a^{(o,\theta)}) = \{s' \in \mathcal{S} : b_a^{(o,\theta)}(s') > 0\}$, used in the following theorem.

Theorem 2. *Let $b_a^{(o,\theta)}$ be the belief at a given time step. For any two states s_1' and s_2' in $\sigma(b_a^{(o,\theta)})$, we have: $\mathcal{A}_{s_1'} = \mathcal{A}_{s_2'}$.*

Proof. Suppose that $\mathcal{A}_{s_1'} \neq \mathcal{A}_{s_2'}$. Thus, by definition, $\mathbb{I}(\theta, s_1') \neq \mathbb{I}(\theta, s_2')$. If $\mathbb{I}(\theta, s_1') = 0$, then $b_a^{(o,\theta)}(s_1') = 0$ according to eq. 4, which contradicts $s_1' \in \sigma(b_a^{(o,\theta)})$. Thus, $\mathbb{I}(\theta, s_1') = 1$, but then $\mathbb{I}(\theta, s_2') = 0$, so that $b_a^{(o,\theta)}(s_2') = 0$: again, this is a contradiction with $s_2' \in \sigma(b_a^{(o,\theta)})$.

3.3 Optimality Equation

Theorem 2 allows us to adapt dynamic programming equations of POMDPs to AC-POMDPs, by "just" maximizing the value function over the set of feasible actions in the current belief state, instead of considering all actions. This adaptation may seem very simplistic in appearance, but it is absolutely not if we consider that it would definitely not be possible without Theorem 2, ie., by deduction, without observing the set of feasible actions at any decision epoch. Specifically, since all states in the support of the belief state have the same set of feasible actions, we can deduce *the* set of feasible actions from a given belief state b, noted $\mathcal{A}_b$ without ambiguity: $\forall s \in \sigma(b), \mathcal{A}_b = \mathcal{A}_s$. Therefore, AC-POMDP policies defined in eq. 3 can be functions of only b by abusing the notation: $\pi(b) = \pi(b, \mathcal{A}_b)$.

Theorem 3.

$$V^*(b) = \max_{a \in \mathcal{A}_b} \left[r(b,a) + \gamma \sum_{\substack{o \in \mathcal{O} \\ \theta \in \Theta}} p(o, \theta \mid a, b) V^* \left(b_a^{(o,\theta)} \right) \right] \tag{6}$$

with $b_a^{(o,\theta)}$ given in eq. 4 and:

$$p(o, \theta \mid a, b) = \sum_{s' \in \mathcal{S}} \mathbb{I}(\theta, s') O(o, a, s') \sum_{s \in \mathcal{S}} T(s, a, s') b(s) \tag{7}$$

Proof. According to Theorem 2, the set of applicable actions observed just before computing b, i.e. θ_{t-1}, can be deduced from b without ambiguity from the support of b: $\theta_{t-1} = \mathcal{A}_b = \mathcal{A}_s$ for any $s \in \sigma(b)$. Since optimal policies are constrained to apply only applicable actions (see eq. 3), the candidate greedy actions that maximize the value function must be chosen in $\mathcal{A}_b$. Then, eq. 6 can be obtained in a similar way to POMDPs, considering (o, θ) as a joint observation. Eq. 7 is proven using a similar reasoning to the proof of Theorem 1:

$$p(o, \theta | a, b) = Pr(\theta | a, b) Pr(o | a, b) = \sum_{s' \in \mathcal{S}} Pr(\theta | s') Pr(s' | o, a, b) Pr(o | a, b)$$
$$= \sum_{s' \in \mathcal{S}} Pr(\theta | s') Pr(s', o | a, b) = \sum_{s' \in \mathcal{S}} Pr(\theta | s') Pr(o | s', a) Pr(s' | a, b)$$
$$= \sum_{s' \in \mathcal{S}} \mathbb{I}(\theta, s') O(o, a, s') \sum_{s \in \mathcal{S}} T(s, a, s') b(s)$$

4   PreCondition Value Iteration

We implemented a point-based algorithm to solve AC-POMDPs, which can be viewed as an adaptation of PBVI [11] to the update equations of Theorem 3. The ideas behind this adaptation are yet independent from PBVI, and could have been applied to generalize any modern α-vector-based POMDP planner, like Perseus [16], HSVI2 [17], or SARSOP [18]. The pseudo-code is given in Algorithm 1.

The expansion of $\mathcal{B}$ is performed in a way similar to PBVI (see Line 9 of the Algorithm 1). First, for each point $b \in \mathcal{B}$, a state s is drawn from the belief distribution b. Second, for each action a in $\mathcal{A}_b := \mathcal{A}_s, \forall s \in \sigma(b)$, a successor state s' is drawn from the transition model $T(s, a, s)$, and a pair of observations (θ, o) is drawn using $\mathbb{I}(\theta, s')$, $p(o | s', a)$ and $p(s' | s, a)$. Knowing (b, a, θ, o), we can calculate the new belief set $\{b_{a_0}, ..., b_{a_j}\}$. Finally, the farthest point from all points already in $\mathcal{B}$ is chosen and integrated into $\mathcal{B}$.

Apart from the fact that observations are structured in the form of pairs of "standard" observations o and "feasible action set" observations θ, and that the expansion of $\mathcal{B}$ is

---

Algorithm 1. PreCondition Value Iteration (PCVI)

1 $k \leftarrow 0$;   Initialize $V_{k=0} \leftarrow \emptyset$;   Initialize $\mathcal{B}$ with b_0;
2 **repeat**
3      $k \leftarrow k+1$; $V_k \leftarrow \emptyset$;
4   **for** $a \in (\mathcal{A}_b)_{b \in \mathcal{B}}$ **and** $(o, \theta) \in \mathcal{O} \times \Theta$ **do**
5          $\Gamma^{a, (o, \theta)} \leftarrow \alpha_i^{a, (o, \theta)}(s) =$
             $\gamma \mathbb{F}(a, s) \sum_{s' : \mathbb{I}(\theta, s') \neq 0} T(s, a, s') O(o, a, s') \alpha_i'(s'), \forall \alpha_i' \in V_{k-1}, a_{\alpha_i} \in \theta$;
6   **for** $b \in \mathcal{B}, a \in \mathcal{A}_b$ **do**
7          $\Gamma_b^a \leftarrow \Gamma^{a, *} + \sum_{\substack{o \in \mathcal{O} \\ \theta \in \Theta}} \operatorname*{argmax}_{\alpha \in \Gamma^{a, (o, \theta)}} (\alpha \cdot b), \forall a \in \mathcal{A}_b$;
8          $V_k \leftarrow \operatorname*{argmax}_{\alpha_b^a \in \Gamma_b^a, \forall a \in \mathcal{A}_b} (\alpha_b^a \cdot b), \forall b \in \mathcal{B}$;
9   Expand $\mathcal{B}$ as in PBVI [11];
10 **until** $\left\| \max_{\alpha_k \in V_k} \alpha_k \cdot b - \max_{\alpha_{k-1} \in V_{k-1}} \alpha_{k-1} \cdot b \right\|_{b \in \mathcal{B}} < \epsilon$;

performed mostly as in PBVI, there are mainly two differences between our PCVI algorithm and standard point-based algorithms. First, the projections $\Gamma^{a,(o,\theta)}$ are computed *only for actions in* $\mathcal{A}_b, \forall b \in \mathcal{B}$, which can save many projections compared with PBVI that generates them for all $a \in \cup_{s \in \mathcal{S}} \mathcal{A}_s$. In the same vein, the backup value function V_k for a given belief $b \in \mathcal{B}$, is computed only for actions in $\mathcal{A}_b$ (see Line 8), contrary to PBVI that uses all actions of the problem. Remember that, according to Theorem 2, $\mathcal{A}_b$ is computed at optimization time by choosing any state $s \in \sigma(b)$ and assigning $\mathcal{A}_b = \mathcal{A}_s$.

The second difference to standard approaches is much more significant in terms of complexity improvements, and is the biggest benefit of reasoning with explicit action feasibility constraints. By explicitly exploiting the semantics of the AC-POMDP model, PCVI is able to operate α-vectors defined on many different *reduced* belief subspaces. Namely, since a given action a is defined only over a subset of states $\mathcal{S}_a = \{s \in \mathcal{S} : \mathbb{F}(a, s) = 1\}$, its corresponding α-vectors $\alpha^{a,(o,\theta)}$ are **defined only over a reduced belief subspace** $\Delta_a \subset [0; 1]^{\mathcal{S}_a} \subset [0; 1]^{\mathcal{S}}$ (see Figure 3(b)). In comparison, PBVI (or any other algorithm for solving POMDPs) works with α-vectors that are defined over the *full* belief space $\Delta \subset [0; 1]^{\mathcal{S}}$ (see Figure 3(a)). To this respect, the recent MOMDP model by [12] can be considered as a simpler algorithmic subclass of AC-POMDPs, because MOMDPs deal with α-vectors that are all defined over the *same* belief subspace, whereas AC-POMDPs' α-vectors operate over *different* belief subspaces.

More precisely, the set $\Gamma^{a,(o,\theta)}$ of α-vectors of a given action a and observation (o, θ), knowing the previously computed value function V_{k-1}, is defined as:

$$\Gamma^{a,(o,\theta)} \leftarrow \alpha_i^{a,(o,\theta)}(s) = \gamma \mathbb{F}(a, s) \sum_{s':\mathbb{I}(\theta,s')\neq 0} T(s, a, s') \times$$

$$O(o, a, s')\alpha_i'(s'), \forall \alpha_i' \in V_{k-1}, a_{\alpha_i} \in \theta \quad (8)$$

This equation fundamentally differs from standard POMDP algorithms. Action feasibility constraints allow us to apply binary masks on the value function, in order to ensure that values of α-vectors are computed only for the states where the corresponding

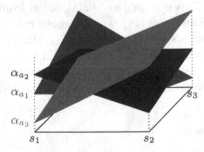
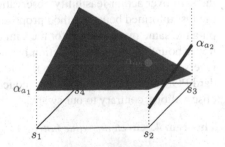

(a) POMDPs: α-vectors are all defined over the entire belief space.

(b) AC-POMDPs: α-vectors are defined over *different* belief **sub**spaces: a_1 feasible in all states; a_2 feasible only in s_2 and s_3; a_3 feasible only in s_2.

Fig. 3. Domain of definition of α-vectors in POMDPs versus AC-POMDPs

actions are defined. Equation 8 shows that two masks are applied: (1) the mask $\mathbb{I}(\theta, s')$ restricts the sum over next states s' to the states where the action $a_{\alpha'_i}$ is feasible (the agent observes the set θ of feasible actions, thus necessarily $a_{\alpha'_i} \in \theta$) ; (2) the mask $\mathbb{F}(a, s)$ guarantees to compute the values of $\alpha_i^{a,(o,\theta)}$ only for the states s where it is defined, i.e. where a is feasible.

Note that this masking mechanism is very *different from the masks used in HSVI2* [17]. In HSVI2, so-called masked α-vectors are just sparse vectors that compute and record only the entries of α-vectors corresponding to non-zeros of b. In our case, we explicitly mask (not by just using sparse representations of vectors) the *irrelevant* entries of α-vectors that correspond to infeasible state-action pairs. In AC-POMDPs, the entries of an α-vector, where the corresponding action is not feasible, are not simply equal to zero but can have arbitrary irrelevant values that must be explicitly pruned by the $\mathbb{F}(a, s)$ masking function. In fact, HSVI2's masking mechanism and ours can be independently applied together.

More precisely, HSVI2's masks automatically prune zero rewards using sparse vectors. Yet, note that modeling infeasible actions in standard POMDP semantics requires to assign near-infinite costs to infeasible actions. Thus, HSVI2 would still have to evaluate these near-infinite costs, which are non-zero and not pruned by its masking mechanism. On the contrary, PCVI's masks automatically prune infeasible actions' costs, which are irrelevant in AC-POMDP semantics, via the function $\mathbb{F}(a, s)$. As a result, masks of HSVI2 and PCVI are totally different, in such a way that infeasible actions' costs are automatically discarded by PCVI but not by HSVI2.

4.1   Relaxed Lower Bound Computation

By studying eq. 6 more in depth, we are able to further benefit from the specific structure of AC-POMDPs' observation model in order to provide a computationally efficient lower bound on the value function. This lower bound, proposed in the following theorem, depends only on the observation set $\mathcal{O}$, instead of the full observation set $\mathcal{O} \times \Theta$. As a result, the computational gains are potentially exponential in the number of actions, since $\Theta \subseteq 2^{\mathcal{A}}$. The idea consists in swapping the max operator over α-vectors and the sum over action-feasibility observations θ. This is related to – but different from – the fast informed bound method proposed by Hauskrecht [19], which consists in swapping the same max operator for the sum over states s in standard POMDPs, yielding an upper bound on the value function but not a lower bound (since Hauskrecht's swap is reversed in comparison with ours). Note that Hauskrecht's swap was designed for standard POMDPs, so that it does not reduce the complexity induced by action-feasibility observations, contrary to our swap.

Theorem 4. *Given the value function* V_n, *we have:*

$$V_{n+1}(b) \geq \max_{a \in \mathcal{A}_b} \left[r(b, a) + \gamma \sum_{o \in \mathcal{O}} \max_{\alpha_n \in V_n} \sum_{\substack{s \in \mathcal{S} \\ s' \in \mathcal{S}}} b(s)O(o, a, s')T(s, a, s')\alpha_n(s') \right] \quad (9)$$

Proof.

$$V_{n+1}(b) = \max_{a \in \mathcal{A}_b} \left[r(b,a) + \sum_{\substack{o \in \mathcal{O} \\ \theta \in \Theta}} p(o,\theta|a,b) V_n \left(b_a^{(o,\theta)} \right) \right]$$

$$= \max_{a \in \mathcal{A}_b} \left[r(b,a) + \sum_{\substack{o \in \mathcal{O} \\ \theta \in \Theta}} p(o,\theta|a,b) \max_{\alpha_n \in V_n} \sum_{s' \in \mathcal{S}} b_a^{(o,\theta)}(s') \alpha_n(s') \right]$$

$$\geqslant \max_{a \in \mathcal{A}_b} \left[r(b,a) + \sum_{o \in \mathcal{O}} \max_{\alpha_n \in V_n} \sum_{\substack{s' \in \mathcal{S} \\ s \in \mathcal{S}}} O(o,a,s') T(s,a,s') b(s) \alpha_n(s') \underbrace{\sum_{\theta \in \Theta} \mathbb{I}(\theta,s')}_{=1} \right]$$

The computation of this lower bound is equivalent to ignoring projections on observations θ of feasible actions in Lines 5 and 7 of Algorithm 1, *as if* feasible actions were not observed by the agent. In this way, projections are only computed for the observation set $\mathcal{O}$, instead of the full observation set $\mathcal{O} \times \Theta$, which potentially yields an exponential gain. Note that α-vectors are still defined only for the states where the corresponding actions are defined, so that the relaxed PCVI algorithm is yet not equivalent to the standard PBVI algorithm. To emphasize this point, we give below the update equation of the α-vectors that make up the lower bound value function. The set of α-vectors $\Gamma^{a,o}$ only depends on o, but each α-vector $\alpha^{a,o}$ of the set is computed by using the feasibility function $\mathbb{F}(a,s)$:

$$\Gamma^{a,o} \leftarrow \alpha^{a,o}(s) = \gamma \sum_{s' \in \mathcal{S}} T(s,a,s') \mathbb{F}(a,s) O(o,a,s') \alpha_i'(s'), \forall \alpha_n \in V_n \qquad (10)$$

The lower bound relaxation of PCVI uses Eq. 10 in place of Eq. 8 in Line 5 of Alg. 1. Line 7 is replaced with the following update: $\Gamma_b^a \leftarrow \Gamma^{a,*} + \sum_{o \in \mathcal{O}} \text{argmax}_{\alpha \in \Gamma^{a,o}}(\alpha \cdot b), \forall a \in \mathcal{A}_b$. The resulting algorithm has the same complexity as standard POMDPs, while guaranteeing that α-vectors and the optimized policy use only feasible actions.

5   Experimental Evaluations

We tested various robotic-like planning problems with action feasibility constraints, which we modeled as AC-POMDPs and solved using our PCVI algorithm. In the subsequent figures, the unrelaxed and relaxed versions of PCVI are respectively noted PCVI1 and PCVI2. We compared our approach with equivalent standard POMDP models, as defined by practitioners to deal with action feasibility constraints: observations of feasible actions are incorporated in the set of observations, but the resulting observation set is treated as an unstructured flat observation set; near-infinite costs are assigned to infeasible state-action pairs in order to prevent the optimized policy from containing illegal actions. Otherwise, there are no guarantees that the optimized policies of standard POMDP models do not apply infeasible actions. Standard POMDP models are solved by PBVI [11] or HSVI2 [17]. We first prove that this translation is sound, before presenting the actual experimental comparisons. We studied four performance criteria depending on benchmarks: 1) the size of value function in terms of the number of α-vectors it contains; 2) the evolution of Bellman error during computation of an optimal policy; 3) the planning time up to convergence at $\epsilon = 0.5$; 4) the statistical expected accumulated rewards from the initial belief state by running 1000 simulations of the optimized policy.

5.1   Translating AC-POMDPs into Equivalent POMDPs

Let $\mathcal{M} = \langle \mathcal{S}, (\mathcal{A}_s)_{s \in \mathcal{S}}, \Omega = \mathcal{O} \times \Theta, T, O, \mathbb{I}, R, b_0, \Theta_0 \rangle$ be a given AC-POMDP. Consider POMDP $\widetilde{\mathcal{M}} = \langle \mathcal{S}, \widetilde{\mathcal{A}}, \Omega = \mathcal{O} \times \Theta, T, \widetilde{O}, \widetilde{R}, b_0, \Theta_0 \rangle = \Psi(\mathcal{M})$ where:

- $\widetilde{\mathcal{A}} = \bigcup_{s \in \mathcal{S}} \mathcal{A}_s$;
- $\widetilde{O} : (\mathcal{O} \times \Theta) \times \widetilde{\mathcal{A}} \times S \to [0; 1]$ is the aggregated observation function, such that $\widetilde{O}((o, \theta), a, s') = O(o, a, s') \mathbb{I}(\theta, s')$;
- $\widetilde{R} : \mathcal{S} \times \widetilde{\mathcal{A}} \times \mathcal{S} \to \mathbb{R}$ is the modified reward function, such that $\widetilde{R}(s, a, s') = R(s, a, s')$ if $a \in \mathcal{A}_s$, otherwise $\widetilde{R}(s, a, s') = -\infty$.

Then, based on POMDPs' and AC-POMDPs' optimality equations, we can easily prove that any optimal policy for POMDP $\widetilde{\mathcal{M}}$ is optimal for the original AC-POMDP $\mathcal{M}$.

Theorem 5. *Let $\mathcal{M}$ be an AC-POMDP and $\widetilde{\mathcal{M}} = \Psi(\mathcal{M})$ its POMDP translation. Then any optimal policy for $\widetilde{\mathcal{M}}$ is optimal for $\mathcal{M}$.*

Proof. Let π^* be an optimal policy for $\mathcal{M}$. According to Bellman eq. 2, we have:

$$\pi^*(b) \in \operatorname*{argmax}_{a \in \widetilde{\mathcal{A}}} \left\{ \tilde{r}(b, a) + \gamma \sum_{\substack{o \in \mathcal{O} \\ \theta \in \Theta}} \tilde{p}((o, \theta) \mid a, b) V^{\pi^*}(b_a^{(o, \theta)}) \right\}$$

with $\tilde{r}(b, a) = \sum_{s \in \mathcal{S}} b(s) \sum_{s' \in \mathcal{S}} T(s, a, s') \widetilde{R}(s, a, s')$ and:

$$\tilde{p}((o, \theta) \mid a, b) = \sum_{s' \in \mathcal{S}} \widetilde{O}((o, \theta), a, s') \sum_{s \in \mathcal{S}} T(s, a, s') b(s)$$

$$= \sum_{s' \in \mathcal{S}} O(o, a, s') \mathbb{I}(\theta, s') \sum_{s \in \mathcal{S}} T(s, a, s') b(s)$$

$$= p(o, \theta \mid a, b)$$

as defined in eq. 7.

Moreover, for $a \notin \mathcal{A}_b, \tilde{r}(b, a) = -\infty$: indeed, by definition of $\mathcal{A}_b$, it means that there is a state $s \in \sigma(b)$, i.e. $b(s) > 0$, such that $a \notin \mathcal{A}_s$ and thus $\widetilde{R}(s, a, s') = -\infty$ for all next state s'. Consequently, the maximum value of the above max operator is necessarily obtained for an action $a^* \in \mathcal{A}_b$. Finally, for all $a \in \mathcal{A}_b$ and states $s \in \sigma(b)$ and $s' \in \mathcal{S}$, $\widetilde{R}(s, a, s') = R(s, a, s')$, so that $\tilde{r}(b, a) = r(b, a)$. Putting it all together, we have:

$$\pi^*(b) \in \operatorname*{argmax}_{a \in \mathcal{A}_b} \left\{ r(b, a) + \gamma \sum_{\substack{o \in \mathcal{O} \\ \theta \in \Theta}} p((o, \theta) \mid a, b) V^{\pi^*}(b_a^{(o, \theta)}) \right\}$$

which means that V^{π^*} is solution of the optimality equation of AC-POMDPs (eq. 6).

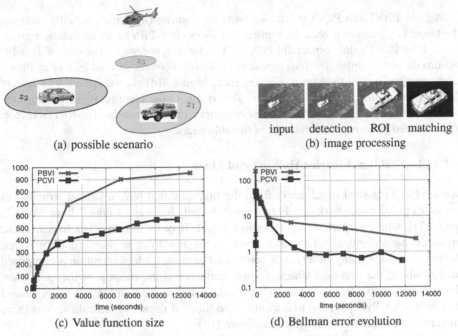

(a) possible scenario

input   detection   ROI   matching

(b) image processing

(c) Value function size

(d) Bellman error evolution

Fig. 4. Multi-target detection, identification & inspection

5.2 Multi-target Detection, Identification and Inspection

We first present a real robotic mission, which we solved and actually achieved with real aerial robots. This mission, sketched in Figure 4, is especially interesting because robot's actions are feasible only on a subset of states for safety reasons (accident risk, regulations specific to the test terrain). An autonomous helicopter has to detect, identify then inspect a specific car in an environment composed of different zones, which can possibly contain cars of different models (see Figure 4(a)). The helicopter can receive "standard" observations from an image processing algorithm [20] (see Figure 4(b)): no car detected, car detected but not identified, car identified as another model. Four different actions can be performed: go to a given zone, feasible *only at* altitude 40 meters; land, feasible *only at* altitude 30 meters and *requiring* that the helicopter can land in the zone over which it flies; increase the view angle of the observed car by 45 degrees, feasible *only at* altitude 30 meters; change altitude, without constraints. The action feasibility constraints only depend on the helicopter's altitude and on the fact that the zone below the helicopter is safe for landing (no obstacles). Thus, the helicopter is equipped of a laser that gives the current altitude, which is directly interpreted as a set of applicable actions for this altitude. Similarly, a simple image processing algorithm based on texture analysis allows the helicopter to known whether the landing action is feasible or not. Note that these "feasibility" observations are totally independent from the ones that give information about cars' detection and identification.

We ran PBVI and PCVI on the scenario of Figure 4(a), which actually contains the target car, parked in zone z_2. Figure 4(c) shows that PBVI's value function grows faster than PCVI's one, especially PCVI2. The latter ignores observations of feasible actions during optimization, thus producing much less α-vectors than PCVI1 or PBVI. Concerning planning time up to convergence, Figure 4(d) shows that PBVI's rate of convergence is lower than PCVI's ones. This is due to the fact that PCVI backups the value function using a smaller number of α-vectors, and uses masks to restrict α-vectors to be defined only over their relevant belief subspaces.

5.3   Classical Benchmarks: Hallway and Maze

We modified classical benchmarks from the literature that have a similar structure to our coast guard benchmark (see Figure 1(a)), namely hallway and the 2-floor "4x5x2" maze [21], where we forbade the robot to hit walls in order to prevent damages. These problems have identical actions and observations, but differ in the number of states. Actions consist in going either north, south, east or west. Each observation is composed of 2 symbols: the first one, which is noisy, indicates if the robot is at the goal state; the second is perfectly sensed and informs the robot of the topology of walls around it, which is totally equivalent to informing the robot of the set of feasible actions in its current hidden state, as represented in Figure 1(b).

Results are presented in Figures 5 and 6. On the maze domain, the value function's sizes of PCVI1 and PCVI2 are nearly the same, yet much less than PBVI (see Figure 5(a)). Since PCVI2 ignores observations of feasible actions when operating on α-vectors, this result suggests that there are a few number of such possible observations in this domain. However, Figure 5(b) shows that PCVI2 converges significantly faster than PCVI1, which is itself more efficient than PBVI. Remember that PCVI1 solves the exact same problem as PBVI, yet by explicitly exploiting the semantics of feasible actions, whereas PCVI2 solves a relaxed simpler problem. PBVI can not reason about action feasibility constraints, which are lost in the equivalent but flat unstructured POMDP model.

Concerning the hallway domain, PCVI2 outperforms PCVI1, which is itself better than PBVI, both in terms of value function size and convergence rates. In comparison with the maze domain, PCVI2 is now able to generate significantly less α-vectors than PCVI1, because the number of different possible sets of feasible observations that can be observed (and ignored by PCVI2) is quite large.

5.4   Larger Navigation Problems

Finally, we tested random navigation problems whose domain is identical to hallway and maze, except that many cells are obstacles that can damage the robot. The problems have also many more states. The robot can observe obstacles around it, using for instance a circular laser sensor. This time, we compared PCVI with HSVI2 [17], which is a heuristic point-based planner that proved to be very efficient in many domains of the literature. As PCVI is adapted from PBVI, which is generally outperformed by HSVI2, we could expect that PVCI would perform poorly in comparison with HSVI2. But Figure 7(a) shows that PCVI2's planning time is actually comparable to HSVI2 for the

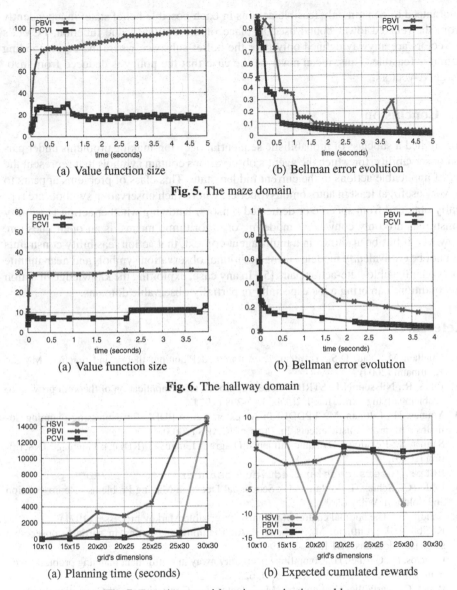

(a) Value function size

(b) Bellman error evolution

Fig. 5. The maze domain

(a) Value function size

(b) Bellman error evolution

Fig. 6. The hallway domain

(a) Planning time (seconds)

(b) Expected cumulated rewards

Fig. 7. Experiments with various navigation problems

largest problems, and is even 10 times faster on 2 problems (logarithmic scale). Moreover, HSVI2's policies are very poor on many problems (see Figure 7(b)), especially for the "25x30" problem for which PCVI gets a similar planning time.

Most interestingly, PCVI1 and PCVI2 get similar expected cumulated rewards, as shown in Figure 7(b). Thus, in grid-like problems, on which many robotic applications are based, ignoring the observation of feasible actions at optimization time has no impact on the quality of the optimized policy. Note that PCVI2's solution policy is still guaranteed to contain only feasible actions in each possible belief state: independently from the relaxed lower bound used in place of the optimal value function, PCVI2's α-vectors are anyway defined only over the belief subspace where the corresponding action is feasible. Thus, it can never be the case that the policy is deduced from incoherent α-vectors.

6   Conclusion

We studied a subclass of probabilistic sequential decision-making problems under partial observability, for which the agent's observations contain symbols that represent the set of applicable actions in the current hidden state. This class of problems appears to be very useful at least in autonomous robotics, where such observation symbols are typically obtained from specific or dedicated sensors. Knowing whether action feasibility constraints are only convenient modeling or algorithmic means, is an open question. However, it has been shown in a multi-agent context, that action feasibility constraints can not be equivalently modeled using additional observation symbols and near-infinite costs on infeasible state-action pairs [5]. In any case, exploiting the knowledge of action preconditions can bring a lot, especially in partially observable domains.

References

1. Ghallab, M., Nau, D.S., Traverso, P.: Automated Planning: theory and practice. Morgan Kaufmann (2004)
2. Fikes, R., Nilsson, N.J.: STRIPS: A new approach to the application of theorem proving to problem solving. Artif. Intell. 2(3/4), 189–208 (1971)
3. Younes, H., Littman, M.: PPDDL1.0: An extension to PDDL for expressing planning domains with probabilistic effects. In: Proc. of ICAPS (2003)
4. Sanner, S.: Relational Dynamic Influence Diagram Language (RDDL): Language Description (2010),
   http://users.cecs.anu.edu.au/~ssanner/IPPC_2011/RDDL.pdf
5. Pralet, C., Schiex, T., Verfaillie, G.: Sequential Decision-Making Problems - Representation and Solution. Wiley (2009)
6. Sondik, E.: The optimal control of partially observable Markov processes (1971)
7. Kaelbling, L., Littman, M., Cassandra, A.: Planning and acting in partially observable stochastic domains. AIJ 101(1-2) (1998)
8. Palacios, H., Geffner, H.: Compiling uncertainty away in conformant planning problems with bounded width. J. Artif. Int. Res. 35(1), 623–675 (2009)
9. Pralet, C., Verfaillie, G., Lemaître, M., Infantes, G.: Constraint-based Controller Synthesis in Non-Deterministic and Partially Observable Domains. In: Proc. of ECAI (2010)
10. Littman, M.L., Cassandra, A.R., Kaelbling, L.P.: Learning policies for partially observable environments: Scaling up. In: ICML, pp. 362–370 (1995)
11. Pineau, J., Gordon, G., Thrun, S.: Point-based value iteration: An anytime algorithm for POMDPs. In: Proc. of IJCAI (2003)
12. Ong, S.C.W., Png, S.W., Hsu, D., Lee, W.S.: Planning under uncertainty for robotic tasks with mixed observability. Int. J. Rob. Res. 29(8), 1053–1068 (2010)
13. Araya-Lopez, M., Thomas, V., Buffet, O., Charpillet, F.: A Closer Look at MOMDPs. In: Proc. of the 22nd IEEE International Conference on Tools with Artificial Intelligence, ICTAI 2010, vol. 02, pp. 197–204. IEEE Computer Society, Washington, DC (2010)

14. Bonet, B., Geffner, H.: Solving pomdps: Rtdp-bel vs. point-based algorithms. In: Proceedings of the 21st International Jont Conference on Artifical Intelligence, IJCAI 2009, pp. 1641–1646. Morgan Kaufmann Publishers Inc., San Francisco (2009)
15. Kim, D., Lee, J., Kim, K.E., Poupart, P.: Point-based value iteration for constrained pomdps. In: IJCAI, pp. 1968–1974 (2011)
16. Spaan, M., Vlassis, N.: Perseus: Randomized point-based value iteration for POMDPs. JAIR 24, 195–220 (2005)
17. Smith, T., Simmons, R.G.: Point-based POMDP algorithms: Improved analysis and implementation. In: Proc. UAI (2005)
18. Kurniawati, H., Hsu, D., Lee, W.: SARSOP: Efficient point-based POMDP planning by approximating optimally reachable belief spaces. In: Proc. RSS (2008)
19. Hauskrecht, M.: Value-function approximations for partially observable markov decision processes. J. Artif. Intell. Res. (JAIR) 13, 33–94 (2000)
20. Saux, B., Sanfourche, M.: Robust vehicle categorization from aerial images by 3d-template matching and multiple classifier system. In: 7th International Symposium on Image and Signal Processing and Analysis (ISPA), pp. 466–470 (2011)
21. Cassandra, A.R.: POMDP's Homepage (1999), http://www.pomdp.org/pomdp/index.shtml

Iterative Model Refinement of Recommender MDPs Based on Expert Feedback

Omar Zia Khan[1], Pascal Poupart[1], and John Mark Agosta[2,*]

[1] David R. Cheriton School of Computer Science,
University of Waterloo, Ontario, Canada
{ozkhan,ppoupart}@uwaterloo.ca
[2] Toyota InfoTechnology Center, Mountain View, CA, USA
jmagosta@us.toyota-itc.com

Abstract. In this paper, we present a method to iteratively refine the parameters of a Markov Decision Process by leveraging constraints implied from an expert's review of the policy. We impose a constraint on the parameters of the model for every case where the expert's recommendation differs from the recommendation of the policy. We demonstrate that consistency with an expert's feedback leads to non-convex constraints on the model parameters. We refine the parameters of the model, under these constraints, by partitioning the parameter space and iteratively applying alternating optimization. We demonstrate how the approach can be applied to both flat and factored MDPs and present results based on diagnostic sessions from a manufacturing scenario.

1 Introduction

Markov decision processes (MDPs) provide a natural and principled framework for sequential decision making under uncertainty. They are used in a multitude of domains from robotic control to recommender systems. A frequent bottleneck for the deployment of systems based on MDPs is the acquisition of the model i.e., the transition and reward functions. To that effect, reinforcement learning provides numerous approaches to optimize a policy from data (sequences of state-action-reward triples). However, depending on the application, data may be difficult to obtain. For instance, consider the class of recommender systems where the actions recommended by a system are to be executed by a user. Whenever humans are involved in the execution of actions, it is challenging to obtain a significant amount of data because users may be difficult to recruit and each trial can take a while (users may need anywhere from a few seconds to months to execute an action). Furthermore, some application domains such as fault detection/diagnostics offer few cases to collect data since faults are rare events to start with. In other domains, it is also desirable to obtain a good policy before deployment to ensure good performance, but this restricts the amount of data available for training.

In this paper we consider the problem of refining the transition function of a Markov decision process based on user feedback. Such feedback may be implicit by noting the actions followed by an expert during a trial or explicit when an expert directly

---

* This work was done when the author was associated with Intel Labs.

H. Blockeel et al. (Eds.): ECML PKDD 2013, Part I, LNAI 8188, pp. 162–177, 2013.

confirms or corrects the actions to be executed in some states by inspecting a policy. Such feedback provides valuable information to adjust the transition model of an MDP that may be imprecise due to a lack of data. We formulate the refinement of a transition function as an optimization problem and incorporate expert feedback as constraints. We also show how to exploit certain properties of recommender systems to partition the variables and optimize them in alternation. We demonstrate the approach with a diagnostic scenario in manufacturing.

The paper is structured as follows. Section 2 reviews Markov decision processes and some important properties of recommender systems. Section 3 explains how this work relates to other work. Section 4 describes our approach to refine a transition function based on expert feedback. We first explain how to do this with flat MDPs and then factored MDPs. Section 5 demonstrates the approach for recommender applications with a real-world diagnostic scenario in manufacturing. Finally, Section 6 concludes and suggests some future work.

2 Background

2.1 Markov Decision Processes

A Markov Decision Process (MDP) is defined by the tuple $M = \langle S, A, T, R, \gamma \rangle$ where S is the set of states s, A is the set of actions a, $T : S \times S \times A \rightarrow \mathbb{R}$ is the transition function which indicates the probability $\Pr(s'|s, a)$ of reaching s' by executing a in s, $R : S \times S \times A \rightarrow \mathbb{R}$ is the reward function which indicates the reward $R(s', s, a)$ of executing a in s and reaching s', γ is the discount factor (value between 0 and 1, with a lower value indicating a greater preference for an immediate reward). Note that we can rewrite the reward function as $R : S \times A \rightarrow \mathbb{R}$, where $R(s, a) = \sum_{s' \in S} R(s', s, a) \Pr(s'|s, a)$. We shall use these equivalent notations for the reward function inter-changeably. A policy $\pi : S \rightarrow A$ for an MDP provides a mapping from states to actions. Techniques such as value iteration can then be used to compute optimal policies for MDPs in which the Bellman's optimality equation (Eq 1) is used as an update rule and is applied iteratively.

$$V^{\pi^*}(s) = \max_a \left[R(s, a) + \gamma \sum_{s'} \Pr(s'|s, a) V^{\pi^*}(s') \right] \tag{1}$$

$V^{\pi^*}(s)$ denotes the value of executing the optimal policy π^* when starting in state s and is equal to the expected discounted sum of all rewards accumulated by executing it when starting in state s. For a policy to be considered optimal, it means that $V^{\pi^*}(s) \geq V^{\pi}(s) \forall s, \pi$. The notation $Q^{\pi}(s, a)$ is used to represent the value of executing a, starting in s and following the policy π from thereon. This can be considered a function that assigns a value to every state-action pair and can be computed using Eq. 2.

$$Q^{\pi}(s, a) = R(s, a) + \gamma \sum_{s' \in S} \Pr(s'|s, a) V^{\pi}(s') \tag{2}$$

In practice, the state space of many MDPs is defined by the cross product of the domain of several variables (or features). Such MDPs are often referred to as factored MDPs since the transition function is the product of several factors, each corresponding to the conditional distribution of a variable given its parents. Optimizing the policy of a factored MDP is notoriously difficult due to the exponential number of states corresponding to all possible joint assignments of the state variables. In this work, we will adapt the Monte-Carlo Value Iteration algorithm [5] (originally developed for POMDPs) to factored MDPs. The key idea in this work is the observation that value iteration implicitly builds a policy graph. Hence, instead of representing the value function over exponentially many states, a policy graph is incrementally constructed. The value function of a policy graph can be approximately evaluated at a given state by Monte Carlo sampling. Hence, approximate value iteration is performed by incrementally constructing a policy graph that provides a sufficient and compact representation from which the value function can be reconstructed.[1]

2.2 MDPs for Recommender Systems

In this work, we focus on recommender systems where an MDP recommends an action to a user at each step. Examples of recommender MDPs include diagnostics, course advising, and so on. Recommender systems lend themselves naturally to a factored representation. The state contains one variable for each of the possible actions to record the value, i.e., tests and grades in courses. The actions are recommendations for the next diagnostic test or the next course to register. Furthermore, we assume that repeating an action does not change the result.

Figure 1 presents the flat representation of a toy diagnostic MDP with three state variables and four actions. Each node represents a state, each arc represents a transition from one state to another via an action corresponding to the label of the arc. In this example, there are two test variables with domain $\{T, F, \_\}$ and one cause variable with domain $\{C_1, C_2, \_\}$ where the value $\_$ indicates that the variable has not been observed yet. The cause variable records the cause identified by the decision maker (if any) instead of the true underlying cause. We do not use any variable to encode the underlying cause since the test variables already encode all the information that would normally be used to express a distribution over the underlying cause. The actions consist of performing one of the tests or identifying a cause. More generally, recommender MDPs can be structured in a similar way with variables that can take n values corresponding to $n - 1$ observations or the null value $\_$.

The states can be organized in levels, where each level groups all the states with the same number of variables instantiated. For instance, at level 0, no variable has been observed and only state is part of this level. All actions are available at this level. At level 1, each state has one variable observed, so the number of actions available at level 1 is two since the action corresponding to the observed variable is no longer available. Similarly, at level 3, three variables have been observed and no further actions are available with all variables already observed. We shall use the concept of levels to enforce

---

[1] Although the algorithm builds a policy graph, it is not a policy iteration algorithm, but definitely an approximate form of value iteration since the policy graph only serves as a compact representation from which the value function can be evaluated.

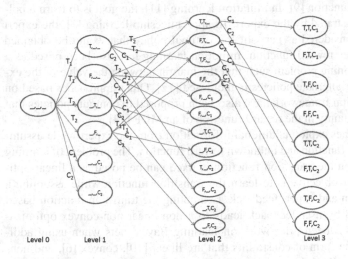

Fig. 1. Sample Flat Recommender MDP

a partial ordering on the states such that all states in level 0 are ordered lower than all states in level 1, and all states in level 1 are ordered lower than all states in level 2, and so on. This ordering also makes it clear that states are never visited more than once.

2.3 Problem Statement

The expert designs the MDP by defining the state variables, the actions, and then esti-mating/specifying transition and reward functions, as well as a discount factor. In this paper, we assume the reward function R and the discount factor γ are specified accu-rately, while the transition function is imprecise. Let us denote the imprecise transition function as $\tilde{T}$ and the resulting imprecise MDP as $\tilde{M} = \langle S, A, \tilde{T}, R, \gamma \rangle$. Let the true un-derlying MDP be denoted as $M = \langle S, A, T, R, \gamma \rangle$, where T is the actual transition func-tion. Since $\tilde{T}$ is imprecise, the optimal policy for $\tilde{M}$, $\tilde{\pi}^*$, may also not be truly optimal, *i.e.*, we are not guaranteed that $\tilde{\pi}^* = \pi^*$. As the expert reviews the policy for $\tilde{M}$, she can point out non-optimal actions and specify true optimal actions for those states, which would reflect π^*. These observations from experts can be treated as constraints, where each constraint is represented as a state-action pair, $\langle s, a \rangle$, which indicates the true op-timal action for that state. Our objective in refining the transition function is to modify $\tilde{T}$ to $\hat{T}$ such that the optimal policy $\hat{\pi}^*$ for this new MDP $\hat{M} = \langle S, A, \hat{T}, R, \gamma \rangle$ obeys all constraints and matches the true optimal policy for these states, *i.e.*, $\hat{\pi}^*(s) = \pi^*(s)$.

3 Related Work

The idea of learning and refining an MDP model or a policy based on expert feedback or demonstration has been widely used, but the focus has mostly been to learn reward function or otherwise learn the optimal policy without learning the reward function. Inverse reinforcement learning deals with recovering a reward function using a known policy and transition function [9]. In imitation learning [11], the goal is to learn a policy as good as demonstrated by the expert. In apprenticeship learning [1], the expert demonstrations are considered as parts of the optimal policy that would be obtained using the unknown true reward function. Imitation learning has also been posed as a maximum margin planning problem such that the margin between the value of the expert's policy and other alternate policies is increased [12]. Other approaches based on the preference elicitation framework have also been proposed to compute policies that are robust to the uncertainty in the reward function of an MDP [13].

The above approaches exploit additional information from the expert while assuming a known transition function and unknown reward function. The problem of learning a reward function when the transition function is fixed can be posed as a linear optimization problem. Our objective is to learn the transition function while assuming a known reward function and expert feedback. Estimating the transition function based on constraints implied by user feedback leads to a non-linear non-convex optimization problem. There has been prior work on learning Bayes' nets when using additional knowledge in the form of constraints that are linear [10], convex [6], and non-convex [8]. However, in Bayes' nets the constraints only provide information about the immediate action whereas in MDPs, the policies are sequential in nature and need to account for possible future plans. Constrained reinforcement learning [7] and constrained MDPs [4] have been proposed to handle multi-objective scenarios, but the constraints in these cases are often of the form which limit the value of a policy. In our case, the constraints that arise from expert feedback are imposed on the Q function instead of the policy which makes the problem non-convex and harder to solve. Abbeel and Ng [2] present a technique to learn the dynamics of a system after observing multiple expert trajectories. Their technique involves running several trials using the expert's policy and then using a maximum likelihood technique on these state-action trajectories to estimate the transition function. Such approaches assume the availability of significant feedback from experts which may be fine for control problems in robotics but not for cases where feedback from expert is very limited (such as diagnostics).

4 Model Refinement

Let Γ be the set of constraints obtained from expert feedback of the form $\langle s, a^* \rangle$, which means that executing a^* should have a value at least as high as any other action in s.

$$Q^{\hat{\pi}^*}(s, a^*) \geq Q^{\hat{\pi}^*}(s, a) \ \forall a$$

We explain how to refine the transition model based on such constraints for "flat" MDPs (Section 4.1) and then for "factored" MDPs (Section 4.2).

4.1 Flat Model Refinement

We can setup an optimization problem to find a refined transition model $\hat{T}$ that maximizes the gap δ between optimal and non-optimal Q-values as specified by the expert's constraints.

$$\max_{\hat{T}, \delta} \ \delta \quad \text{s.t.} \quad Q^{\hat{\pi}}(s, a^*) \geq Q^{\hat{\pi}}(s, a) + \delta \quad \forall \langle s, a^* \rangle \in \Gamma, \forall a \tag{3}$$

When δ is non-negative, the refined model satisfies the expert's constraints. If the user's constraints are inconsistent, we will simply find a model that minimizes the degree of violation for all constraints. For problems with a finite horizon h, we can rewrite the Q function as a sum of expected rewards

$$Q^\pi(s_0, a_0) = R(s_0, a_0) + \sum_{t=1}^{h} \gamma^t \sum_{s_t} \Pr(s_t|s_0, a_0, \pi) R(s_t, \pi(s_t)) \qquad (4)$$

where the probability $\Pr(s_t|s_0, a_0)$ is obtained by a product of transition probabilities.

$$\Pr(s_t|s_0, a_0, \pi) = \sum_{s_{1..t-1}} \Pr(s_1|s_0, a_0) \prod_{i=2}^{t} \Pr(s_i|s_{i-1}, \pi(s_{i-1})) \qquad (5)$$

In a flat MDP, the transition probabilities are the transition parameters. Hence, we will denote by θ the vector of transition parameters $\theta_{s'|s,a} = \Pr(s'|s,a)$. We can then rewrite the optimization problem (3) in terms of θ by substituting Equations 4 and 5:

$$\max_{\hat{\theta},\delta} \delta \quad \text{s.t.} \quad \sum_{s'} \hat{\theta}_{s'|s,a} = 1 \ \forall s, a \qquad \hat{\theta}_{s'|s,a} \geq 0 \ \forall s, a, s' \qquad (6)$$

$$R(s, a^*) + \sum_{t=1}^{h} \gamma^t \sum_{s_{1..t}} \Pr(s_1|s, a^*) \prod_{i=2}^{t} \theta_{s_i|s_{i-1}, \pi(s_{i-1})} R(s_t, a_t) \geq$$

$$R(s, a) + \sum_{t=1}^{h} \gamma^t \sum_{s_{1..t}} \Pr(s_1|s, a) \prod_{i=2}^{t} \theta_{s_i|s_{i-1}, \pi(s_{i-1})} R(s, a) + \delta \quad \forall \langle s, a^* \rangle \in \Gamma, \forall a$$

The optimization problem is non-linear (and in fact non-convex) due to the product of θ's in the last constraint.

We propose to tackle the problem by alternating optimization where we iteratively optimize a subset of the parameters while keeping the remaining parameters fixed. We take advantage of the fact that states are organized in levels to do this. As explained earlier, states are never visited twice since at each step one more test variable is observed. Since each transition parameter $\theta_{s'|s,a}$ is associated with a state s, the transition parameters can also be partitioned into levels and the same transition parameter won't occur more than once in any state trajectory. Hence, if we vary only the parameters in level l while keeping the other parameters fixed, we can write the Q function of the state-action pair of any constraint before level l as a linear function of the θ's in level l.

$$Q(s, a^*) = c(nil) + \sum_{s_l, a_l, s_{l+1}} c(s_l, a_l, s_{l+1}) \theta_{s_{l+1}|s_l, a_l}$$

Here, $c(s_l, a_l, s_{l+1})$ is the coefficient of $\theta_{s_{l+1}|s_l, a_l}$ and $c(nil)$ is a constant. Algorithm 1 describes how to compute the coefficients of the parameters at level l for the Q function at level $j \leq l$. First, the value function at level $l+1$ is computed by value iteration, then the coefficients for the Q function at level l are initialized and finally the coefficients of the Q functions at previous levels are computed by dynamic programming.

Algorithm 1. Linear dependence of the Q function at level j on the θ's at level l

---

LEVELLINEARDEPENDENCE(j, l, π)

    Compute $V^\pi(s_{l+1}) \; \forall s_{l+1}$

1   $V^\pi(s_h) = R(s_h, \pi(s_h)) \; \forall s_h$

2   **for** $t = h - 1$ down to $l + 1$

3      $V^\pi(s_t) \leftarrow R(s_t, \pi(s_t)) + \gamma \sum_{s_{t+1}} \theta_{s_{t+1}|s_t, \pi(s_t)} V^\pi(s_{t+1}) \; \forall s_t$

    Initialize the coefficients for the Q function at level l

4   **for** each s_l, a_l

5      $c_{s_l, a_l}(nil) \leftarrow R(s_l, a_l)$

6      $c_{s_l, a_l}(s_l, a_l, s_{l+1}) \leftarrow \gamma V(s_{l+1}) \; \forall s_{l+1}$

7      $c_{s_l, a_l}(s, a, s') \leftarrow 0 \; \forall \langle s, a \rangle \neq \langle s_l, a_l \rangle, \forall s'$

    Compute the coefficients for the Q function at levels before l

8   **for** $t = l - 1$ down to j

9   **for** each s_t, a_t

10        $c_{s_t, a_t}(nil) \leftarrow R(s_t, \pi(s_t)) + \gamma \sum_{s_{t+1}} \theta_{s_{t+1}|s_t, \pi(s_t)} c_{s_{t+1}, \pi(s_{t+1})}(nil)$

11        $c_{s_t, a_t}(s_l, a_l, s_{l+1}) \leftarrow \gamma \sum_{s_{t+1}} \theta_{s_{t+1}|s_t, \pi(s_t)} c_{s_{t+1}, \pi(s_{t+1})}(s_l, a_l, s_{l+1}) \; \forall s_l, a_l, s_{l+1}$

12 **return** c

If we restrict the optimization problem (6) to the parameters at level l, we obtain a linear program (7) since the last constraint expresses an inequality between pairs of Q functions that are linear combinations of the coefficients at level l.

$$\max_{\hat\theta, \delta} \delta \quad \text{s.t.} \quad \sum_{s'} \hat\theta_{s_{l+1}|s_l, a_l} = 1 \; \forall s, a \qquad \hat\theta_{s_{l+1}|s_l, a_l} \geq 0 \; \forall s_l, a_l, s_{l+1} \tag{7}$$

$$c_{s, a^*}(nil) + \sum_{s_l, a_l, s_{l+1}} c_{s, a^*}(s_l, a_l, s_{l+1}) \hat\theta_{s_{l+1}|s_l, a_l} \geq$$

$$c_{s, a}(nil) + \sum_{s_l, a_l, s_{l+1}} c_{s, a}(s_l, a_l, s_{l+1}) \hat\theta_{s_{l+1}|s_l, a_l} + \delta \quad \forall \langle s, a^* \rangle \in \Gamma$$

To summarize, instead of directly solving the non-linear optimization problem (6), we propose an alternating optimization technique (Algorithm 2) that solves a sequence of linear programs (7) that varies only the parameters at one level. The algorithm continues until the gap δ is non-negative or until convergence. There is no guarantee that a feasible solution will be found, but each iteration ensures that δ will increase or remain constant, meaning that the degree of inconsistency is monotonically reduced. Given the non-convex nature of the optimization, random restarts are employed to increase the chances of finding a model that is as consistent as possible with the expert's constraints.

4.2 Factored Model Refinement

The approach described in the previous section assumes that we flatten the Markov decision process. This will only scale for small problems with a few test variables since the number of states grows exponentially with the number of tests. We now consider

---

Algorithm 2. Alternating optimization to reduce the degree of inconsistency of the transition model with the expert's constraints in flat MDPs

---

ALTERNATINGOPT

1   **repeat**
2   Initialize θ randomly
3   **repeat**
4   **for** $l = 1$ to h
5   Compute coefficients for level l according to Algorithm 1
6               $\delta, \{\theta_{s_{l+1}|a_l,s_l}\} \leftarrow$ solve LP (7) for level l
7   **until** convergence
8   **until** $\delta \geq 0$
9   **return** θ

---

a variant for problems with a large number of tests that avoids flattening by working directly with a factored model. We assume that the transition function is factored into a product of conditional distributions for each variable X_i' given its parents $par(X_i')$.

$$Pr(s'|s,a) = \prod_i Pr(X_i'|par(X_i'))$$

Furthermore, we assume that the parents of each variable are a small subset of all the variables. For instance, in a course advising domain, the grade of a course may depend only on the grades of the pre-requisites. As a result, the total number of parameters for the transition function shall be polynomial in the number of variables even though the number of states is exponential. We denote by $\theta_{X_i'|par(X_i)}$ the family of parameters defining the conditional distribution $Pr(X_i'|par(X_i))$.

We need to deal with two issues in factored domains. First, we cannot perform dynamic programming to compute the Q-values at each state in polynomial time. We will use Monte Carlo Value Iteration [5] to approximate Q-values at a sample of reachable states. Second, even though the same state is not revisited in any trajectory, the same transition parameters will be used at each stage of the process. So instead of partitioning the parameters by levels, we will partition them by families corresponding to different conditional distributions. This will allow us to alternate between a sequence of linear programs as before.

We first explain how to do approximate dynamic programming by adapting the Monte Carlo Value Iteration technique [5] (originally designed for continuous POMDPs) to factored discrete MDPs. Instead of storing an exponentially large Q-function at each stage, we store a policy graph $G = \langle N, E \rangle$. The nodes $n \in N$ of policy graphs are labeled with actions, and the edges $e \in E$ are labeled with observations (i.e., values for the test corresponding to the previous action). A policy graph $G = \langle \phi, \psi \rangle$ is parameterized by a mapping $\phi : N \to A$ from nodes to actions and a mapping $\psi : E \to N$ from edges to next nodes. Since each edge is rooted at a node and labeled with an observation, we will also refer to ψ as a mapping from node-observation pairs to next nodes (i.e. $\psi : N \times O \to N$). Here an observation is the result of a test. A useful operation on

Algorithm 3. Evaluate G at s

EVALGRAPH(G, s)
1   Let N be the set of nodes for $G = \langle \phi, \psi \rangle$
2   **for** each $n \in N$
3       $V(n) \leftarrow 0$
4   **repeat** k times
5           $V(n) \leftarrow V(n) + $ EVALTRAJECTORY$(G, s, n)/k$
6   $n^* \leftarrow argmax_{n \in N} V(n)$
7   **return** $V(n^*)$ and n^*

EVALTRAJECTORY(G, s, n)
8   Let $G = \langle \phi, \psi \rangle$
9   **if** n does not have any edge
10   **return** $R(s, \phi(n))$
11   **else**
12   Sample $o \sim \Pr(o|s, \phi(n))$
13   Let s' be the state reached when observing o after executing $\phi(n)$ in s
14   **return** $R(s, \phi(n)) + \gamma$ EVALTRAJECTORY$(G, s', \psi(n, o))$

policy graphs will be to determine the best value that can be achieved at a given state by starting in any node. Algorithm 3 describes how to compute this by Monte Carlo sampling. k trajectories are sampled starting in each node. The node with the highest value is returned along with its value.

The main purpose of the policy graph is to provide a succinct and implicit representation of a value function. More precisely, we can estimate the value of a state by calling EVALGRAPH(G, s). While we could also use the policy graph as a controller, we will do a one step look ahead to infer the best action to execute at each step in the same way that it would be done if we had an explicit value function and we wanted to extract a policy. In other words, if we have a value function V, we can extract the best action a^* for any state s by computing

$$a^* = \arg\max_a R(s, a) + \gamma \sum_{s'} \Pr(s'|s, a)V(s')$$

Similarly, we will extract the best action to execute at each time step when in state s based on policy graph G by computing

$$a^* = \arg\max_a R(s, a) + \gamma \sum_{s'} \Pr(s'|s, a)\text{EVALGRAPH}(G, s')$$

Algorithm 4 describes how to construct a policy graph G by approximate value iteration. Here, value iteration is performed by approximate backups that compute and store a policy graph instead of a value function at each step. Figures 2, 3, and 4 present a sample trace of how the policy graph may appear after each iteration of the for loop in Algorithm 4 on line 2. Initially, all actions are present as disconnected nodes. As

---

Algorithm 4. Monte Carlo Value Iteration

MCVI(*setOfStates*, *horizon*)

1   Initialize G with no edge and $|A|$ nodes such that ϕ maps each node to a different action
2   **for** $t = 1$ to *horizon*
3   **for** each $s \in$ *setOfStates*
4   **for** each $a \in A$
5            $Q(s,a) \leftarrow R(s,a)$
6   **for** each o observable from s after executing a
7   Let s' be the state reached when observing o after executing a in s
8               $[V(s'), n_{a,o}] \leftarrow$ EVALGRAPH(G, s')
9               $Q(s,a) \leftarrow Q(s,a) + \gamma Pr(o|s,a)V(s')$
10         $a^* \leftarrow argmax_a Q(s,a)$
11   Add new node n to G such that $\phi(n) = a^*$ and $\psi(n,o) = n_{a^*,o}$
12 **return** G

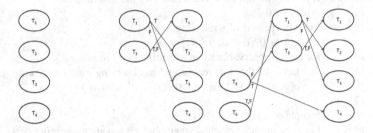

Fig. 2. Sample Policy Graph after 1 iteration of Algorithm 4

Fig. 3. Sample Policy Graph after 2 iterations of Algorithm 4

Fig. 4. Sample Policy Graph after 3 iterations of Algorithm 4

more iterations are completed, more nodes are added to the graph. Each node represents an action and each arrow represents the observation obtained after executing that action. The arrow links to another node that indicates the next action to execute after an observation for a given action.

Point-based backups are performed only at a set of states *setOfStates*. This set of states can be obtained in several ways. It should be representative of the reachable states and allow for the construction of a good set of conditional plans. As we will see later, it is desirable to include in *setOfStates* all the states s' that are reachable from the states s for which we have constraints $\langle s, a^* \rangle$. At each iteration, a new node is added to the policy graph for each state in *setOfStates*. Although not shown in Algorithm 4, redundant nodes could be pruned from the policy graph to improve efficiency.

Similar to flat MDPs, we would like to optimize the parameters of the conditional distributions to satisfy the expert's constraints. We can approximate the Q-values on which we have constraints by the EVALGRAPH procedure.

$$Q^G(s,a) = R(s,a) + \gamma \sum_{s'} Pr(s'|s,a) \text{EVALGRAPH}(G, s')$$

Algorithm 5. Linear dependency of $Q^G(s, a)$ on parameters of $\Pr(X_i'|par(X_i))$ when executing a in s and following G thereon. This function returns the coefficients c of $\Pr(X_i'|par(X_i))$ based on k sampled trajectories of G.

LINEARDEPENDENCE(G, s, a, i)

1     $c(nil) \leftarrow R(s, a)$ and $c(o, x) \leftarrow 0\ \forall o \in dom(X_i'), v \in dom(par(X_i))$

2   **repeat** k times

3   Sample s' from $\Pr(s'|s, a)$

4   Let n' be the node created in G for $s' \in setOfStates$

5         $c \leftarrow c + \gamma$LINEARDEPENDENCERECURSIVE(G, s', n', i)$/k$

6   **return** c

LINEARDEPENDENCERECURSIVE(G, s, n, i)

7   **if** n does not have any edge

8         $c(nil) \leftarrow R(s, \phi(n))$

9         $c(o, x) \leftarrow 0\ \forall o \in dom(V_i'), v \in dom(par(V_i'))$

10   **else if** $\phi(n) = a_i$ and $\phi(n)$ is executed for the first time

11        $c(nil) \leftarrow 0$

12   Let x be the part of s referring to $par(X_i)$

13        $c(o, x') \leftarrow 0\ \forall o \in dom(X_i), x' \neq x$

14   **for** each o observable when executing $\phi(n)$ in s

15   Let s' be the state reached when observing o after executing $\phi(n)$ in s

16            $c(o, x) =$ EVALTRAJECTORY($G, s', \psi(n, o)$)

17   **else**

18   Sample $o \sim \Pr(o|s, \phi(n))$

19   Let s' be the state reached when observing o after executing $\phi(n)$ in s

20        $c \leftarrow \gamma$ LINEARDEPENDENCERECURSIVE($G, s', \psi(n, o), i$)

21        $c(nil) \leftarrow R(s, \phi(n)) + c(nil)$

22   **return** c

Since the Q-function has a non-linear dependence on the transition parameters, we partition the parameters in families $\theta_{X_i'|par(X_i)}$ corresponding to conditional distributions $\Pr(X_i'|par(X_i))$ for each test variable X_i with the corresponding action a_i that selects to observe X_i. Alternating between the optimization of different families of parameters ensures that the optimization is linear. In any trajectory, a variable X_i is observed at most once and therefore at most one transition parameter for the observation of X_i participates in the product of probabilities of the entire state trajectory. Hence, we can write the Q function as a linear combination of the parameters of a given family

$$Q(s, a) = c(nil) + \sum_{o,x} c(o, x) \Pr(X_i' = o|par(X_i) = x) \qquad (8)$$

where $c(nil)$ denotes a constant and $c(o, x)$ is the coefficient of the probability of observing outcome o for X_i' given that the joint value of the parent variables of X_i' is x. Algorithm 5 shows how to compute the linear dependency on the parameters of $\Pr(X_i'|par(X_i))$. More precisely, it computes a vector c of coefficients by sampling k

trajectories in G and averaging the linear coefficients of those trajectories. In each trajectory, a recursive procedure computes the coefficients based on three cases: i) when n is a leaf node (i.e., no edges), it returns the reward as a constant in $c(nil)$; ii) when a is executed for the first time, it returns the value of each o in $c(o, x)$; iii) otherwise, it recursively calls itself and adds the reward in $c(nil)$.

Similar to the linear program (7) for flat MDPs, we can define a linear program to optimize the transition parameters of a single family subject to linear constraints on Q-values as defined in Equation 8. We can also alternate between the optimization of different families similar to Algorithm 2, but for factored MDPs.

5 Evaluation and Experiments

5.1 Evaluation Criteria

Formally, for M, the true MDP that we aim to learn, the optimal policy π^* determines the choice of best next test as the one with the highest value function. If the correct choice for the next test is known (such as demonstrated by an expert), we can use this information to include a constraint on the model. We denote by Γ^+ the set of observed constraints and by Γ^* the set of all possible constraints that hold for M. Having only observed Γ^+, our technique will consider any $M^+ \in \mathbf{M}^+$ as a possible true model, where $\mathbf{M}^+$ is the set of all models that obey Γ^+. We denote by $\mathbf{M}$ the set of all models that are *constraint equivalent* to M (i.e., obey Γ^*), by $\tilde{M}$ the initial model that we start with, and by $\hat{M}_{\Gamma^+}$ the particular model obtained by iterative model refinement based on the constraints Γ^+.

Ideally we would like to find the true underlying model M, hence we will report the KL-divergence$(M, \hat{M}_{\Gamma^+})$. However, other constraint equivalent models may recommend the same actions as M and thus have similar constraints, so we also report *test consistency* with $\mathbf{M}$ (i.e., # of states in which optimal actions are the same) and the simulated value of the policy of $\hat{M}_{\Gamma^+}$ with respect to the true transition function T.

Given a consistent set of constraints Γ and sufficient time (for random restarts), our technique for model refinement will choose a model $\hat{M}_\Gamma \in \mathbf{M}$ by construction. If the constraints specified by the expert are inconsistent (i.e., do not correspond to any possible model), our approach minimizes the violation of the constraints as much as possible through alternating optimization combined with random restarts. We report the best solution found after exhausting the time quota to perform refinement.

5.2 Experimental Results on Synthetic Problems

We start by presenting our results on a 4-test recommender system. We want to discover the transition model of some model $M \in \mathbf{M}$. We select M by randomly sampling its transition and reward functions. Given this model M, we sample a set of constraints Γ^+ and use our technique to find $\hat{M}_{\Gamma^+}$. To evaluate $\hat{M}_{\Gamma^+}$, we first compute the constraints Γ^* for M and estimate the set of constraint-equivalent models $\mathbf{M}$ by sampling 100 models from $\mathbf{M}$. We then compare these constraint equivalent models with $\hat{M}_{\Gamma^+}$.

We compute the KL-divergence between each constraint-equivalent model and the refined model KL-DIV$(M_i, \hat{M}_{\Gamma^+})$, and take its ratio with the KL-divergence between

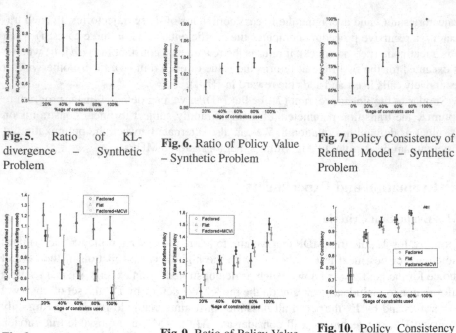

Fig. 5. Ratio of KL-divergence – Synthetic Problem

Fig. 6. Ratio of Policy Value – Synthetic Problem

Fig. 7. Policy Consistency of Refined Model – Synthetic Problem

Fig. 8. Ratio of KL-divergence – Diagnostic Problem

Fig. 9. Ratio of Policy Value – Diagnostic Problem

Fig. 10. Policy Consistency of Refined Model – Diagnostic Problem

the constraint equivalent model and the initial model KL-DIV$(M_i, \tilde{M})$ as shown in Figure 5. A lower value of this ratio indicates that the refined model $\hat{M}_{\Gamma^+}$ is closer to the true model M than the initial model $\tilde{M}$. We can also see that the mean KL-divergence decreases as the number of constraints in Γ^+ increases since the feasible region becomes smaller. Figures 6 and 7 show similar trends for test consistency and simulated value of the policy. We observed similar trends for KL-divergence, test consistency and simulated value of policy when increasing the number of variables.

5.3 Experimental Results on Diagnostic Problems

We also evaluate our technique on diagnostic MDPs. To construct such MDPs, we choose the number of tests and causes. The total number of actions in the MDP is the sum of the tests and causes with an action either being the option to execute a test and observe its value or make a diagnostic prediction regarding the cause. Executing a test has a small negative reward. The diagnostic prediction has a high positive reward if the correct cause is diagnosed and a high negative reward for an incorrect diagnosis. No discount factor is used as it is a finite horizon problem.

Diagnostic MDPs are better represented as factored MDPs as executing a test only affects a part of the state space. While diagnostic MDPs can be encoded with a flat representation, a factored representation allows a more succinct representation with fewer parameters to be learned for the transition function.

We present the results of model refinement on the same diagnostic MDP represented as a flat MDP, a factored MDP with exact value iteration and a factored MDP with Monte Carlo Value Iteration (MCVI) in Figures 8, 9, and 10. These results are shown for a 4-cause and 4-test network. We see that the factored representation yields better results than the flat representation. This is because the factored representation exploits the inherent structure of the diagnostic MDP, whereas the flat representation is unable to preserve this structure after refinement. This is clearly evident in the case of KL-divergence where the resulting model does obey the constraints, but is in fact farther away from the true model than the starting model. We also see that considering a sub-set of states for $setOfStates$ in MCVI (states reachable from constraints with 50% of remaining states), the results for KL-divergence, test consistency and value of policy deteriorate in comparison to the exact factored case. In separate experiments, we ob-served that increasing the size of $setOfStates$ results in improved refined models and decreasing them results in refined models that are not as good. For the purpose of this work, we are using MCVI as a method to solve factored MDPs and demonstrate our technique for refinement on a large problem. We leave the question of determining an optimal $setOfStates$ for MCVI as future work, though we note that this question has been extensively studied in point-based value iteration algorithms for POMDPs [14].

5.4   Experimental Results on Large Scale Diagnostic Problems

We evaluate our technique on a real-world diagnostic network collected and reported by Agosta et al. [3], where the authors collected detailed session logs over a period of seven weeks in which the entire diagnostic sequence was recorded. The sequences intermingle model building and querying phases. The model network structure was inferred from an expert's sequence of positing causes and tests. Test-ranking constraints were deduced from the expert's test query sequences once the network structure is established.

The logs captured 157 sessions over seven weeks that resulted in a model with 115 tests and 82 root causes. The network consists of several disconnected sub-networks, each identified with a symptom represented by the first test in the sequence, and all subsequent tests applied within the same subnet. There were 20 sessions in which more than two tests were executed, resulting in a total of 32 test constraints. We pruned our diagnostic network to remove the sub-networks with no constraints to get 54 tests and 30 causes, divided in 7 sub-networks.We apply our model refinement technique to learn the parameters for each sub-network separately. The largest sub-network has 15 tests and 10 causes resulting in 25 actions and more than 14 million states. We use MCVI for these larger networks as it would not be possible to solve them exactly otherwise.

We use the 32 constraints extracted from the session logs to represent a feasible region from which we sample 100 true models. We sample 1000 states in addition to the states reachable by the constraints to form the $setOfStates$ used by MCVI. The approximation in MCVI often results in situations where no feasible model is available during refinement. In such a case, we stop the experiments after an allocated amount of time and report the model that violates the constraints the least among those computed so far. For the experiments in this section, the refinement process was terminated after 10 random restarts of the alternating optimization problem, i.e., randomly perturbing

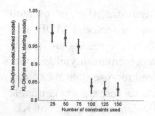

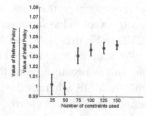

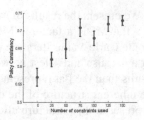

Fig. 11. Ratio of KL-divergence – Large Scale Diagnostic Problem

Fig. 12. Ratio of Policy Value – Large Scale Diagnostic Problem

Fig. 13. Policy Consistency of Refined Model – Large Scale Diagnostic Problem

the parameters 10 times after the solution had locally converged before choosing the best solution available till that time.

Figures 11, 12, and 13 show the results for KL-divergence, simulated value of policy and policy consistency respectively for the real world diagnostic network provided by our industrial partner. Since the total number of constraints is exponential, we randomly sampled a subset of constraints and show the results using these subsets instead of a percentage of all possible constraints. Similarly, the policy consistency is also computed by randomly sampling 100 states and then comparing optimal actions in those states. We can see that using a small subset of constraints and a small number of states as input to MCVI yields benefits in moving closer to the original model.

6   Conclusion and Future Work

In summary, we presented an approach to refine the transition function of an MDP based on feedback from an expert. While several approaches address the problem of learning the reward function based on expert knowledge, this paper makes a novel contribution by tackling the problem of refining transition functions. This is particularly useful in scenarios where the amount of data (state-action-state triples) is limited. Our work makes three important contributions. First, we demonstrate how to use feedback from an expert to define constraints on the parameters of the transition function. This feedback may be implicit when obtained from logs of diagnostic sessions performed by a domain expert. Second, we design an approach to handle non-convex constraints that arise when expert feedback on optimal actions for different states is available. Third, our approach is easily applicable for flat and factored MDPs, and we demonstrate that it can be used in conjunction with approximate Monte Carlo techniques that are necessary to solve large real-world MDPs. We present results of refined models for synthetic recommender systems and a real-world diagnostic scenario from the manufacturing domain. We show that our technique not only helps in getting closer to the true transition function, but also improves policy consistency and the value of the policy.

In the future, it would be interesting to generalize this work to Partially Observable MDPs and see if the transition and observation functions can be refined simultaneously. Another possibility is to estimate transition functions from both Q-value constraints implied by user feedback and observed state transitions (i.e., state-action-state triples) by combining this work with model-based reinforcement learning approaches.

Acknowledgements. This work was supported by a grant from Intel Corporation.

References

1. Abbeel, P., Ng, A.Y.: Apprenticeship learning via inverse reinforcement learning. In: Twenty-First International Conference on Machine Learning, ICML (2004)
2. Abbeel, P., Ng, A.Y.: Exploration and apprenticeship learning in reinforcement learning. In: Twenty Second International Conference on Machine Learning, ICML (2005)
3. Agosta, J.M., Khan, O.Z., Poupart, P.: Evaluation results for a query-based diagnostics application. In: Fifth Workshop on Probabilistic Graphical Models, PGM (2010)
4. Altman, E.: Constrained Markov Decision Processes. CRC Press (1999)
5. Bai, H., Hsu, D., Lee, W.S., Ngo, V.A.: Monte carlo value iteration for continuous-state POMDPs. In: Hsu, D., Isler, V., Latombe, J.-C., Lin, M.C. (eds.) Algorithmic Foundations of Robotics IX. STAR, vol. 68, pp. 175–191. Springer, Heidelberg (2010)
6. de Campos, C.P., Ji, Q.: Improving Bayesian network parameter learning using constraints. In: International Conference in Pattern Recognition, ICPR (2008)
7. Geibel, P.: Reinforcement learning for MDPs with constraints. In: Fürnkranz, J., Scheffer, T., Spiliopoulou, M. (eds.) ECML 2006. LNCS (LNAI), vol. 4212, pp. 646–653. Springer, Heidelberg (2006)
8. Khan, O.Z., Poupart, P., Agosta, J.M.: Automated refinement of Bayes networks' parameters based on test ordering constraints. In: Neural Information Processing Systems, NIPS (2011)
9. Ng, A.Y., Russell, S.J.: Algorithms for inverse reinforcement learning. In: Seventeenth International Conference on Machine Learning, ICML (2000)
10. Niculescu, R.S., Mitchell, T.M., Rao, R.B.: Bayesian network learning with parameter constraints. Journal of Machine Learning Research (JMLR) 7, 1357–1383 (2006)
11. Price, B., Boutilier, C.: Accelerating reinforcement learning through implicit imitation. Journal of Artificial Intelligence Research (JAIR) 19, 569–629 (2003)
12. Ratliff, N., Andrew (Drew) Bagnell, J., Zinkevich, M.: Maximum margin planning. In: Twenty Third International Conference on Machine Learning, ICML (2006)
13. Regan, K., Boutilier, C.: Robust policy computation in reward-uncertain MDPs using non-dominated policies. In: Twenty-Fourth Conference on Artificial Intelligence, AAAI (2010)
14. Shani, G., Pineau, J., Kaplow, R.: A survey of point-based POMDP solvers. Autonomous Agents and Multi-Agent Systems 27, 1–51 (2013)

Solving Relational MDPs with Exogenous Events and Additive Rewards

Saket Joshi[1], Roni Khardon[2], Prasad Tadepalli[3], Aswin Raghavan[3], and Alan Fern[3]

[1] Cycorp Inc., Austin, TX, USA
[2] Tufts University, Medford, MA, USA
[3] Oregon State University, Corvallis, OR, USA

Abstract. We formalize a simple but natural subclass of *service domains* for relational planning problems with object-centered, independent exogenous events
and additive rewards capturing, for example, problems in inventory control. Focusing on this subclass, we present a new symbolic planning algorithm which is
the first algorithm that has explicit performance guarantees for relational MDPs
with exogenous events. In particular, under some technical conditions, our planning algorithm provides a monotonic lower bound on the optimal value function.
To support this algorithm we present novel evaluation and reduction techniques
for generalized first order decision diagrams, a knowledge representation for real-
valued functions over relational world states. Our planning algorithm uses a set
of focus states, which serves as a training set, to simplify and approximate the
symbolic solution, and can thus be seen to perform learning for planning. A preliminary experimental evaluation demonstrates the validity of our approach.

1 Introduction

Relational Markov Decision Processes (RMDPs) offer an attractive formalism to study
both reinforcement learning and probabilistic planning in relational domains. However,
most work on RMDPs has focused on planning and learning when the only transitions
in the world are a result of the agent's actions. We are interested in a class of problems
modeled as *service domains*, where the world is affected by exogenous service requests
in addition to the agent's actions. In this paper we use the inventory control (IC) domain as a motivating running example and for experimental validation. The domain
models a retail company faced with the task of maintaining the inventory in its shops to
meet consumer demand. Exogenous events (service requests) correspond to arrival of
customers at shops and, at any point in time, any number of service requests can occur
independently of each other and independently of the agent's action. Although we focus
on IC, independent exogenous service requests are common in many other problems,
for example, in fire and emergency response, air traffic control, and service centers such
as taxicab companies, hospitals, and restaurants. Exogenous events present a challenge
for planning and reinforcement learning algorithms because the number of possible next
states, the "stochastic branching factor", grows exponentially in the number of possible
simultaneous service requests.

In this paper we consider symbolic dynamic programming (SDP) to solve RMDPs,
as it allows to reason more abstractly than what is typical in forward planning and reinforcement learning. The SDP solutions for propositional MDPs can be adapted to

H. Blockeel et al. (Eds.): ECML PKDD 2013, Part I, LNAI 8188, pp. 178–193, 2013.

RMDPs by grounding the RMDP for each size to get a propositional encoding, and then using a "factored approach" to solve the resulting planning problem, e.g., using algebraic decision diagrams (ADDs) [5] or linear function approximation [4]. This approach can easily model exogenous events [2] but it plans for a fixed domain size and requires increased time and space due to the grounding. The relational (first order logic) SDP approach [3] provides a solution which is independent of the domain size, i.e., it holds for any problem instance. On the other hand, exogenous events make the first order formulation much more complex. To our knowledge, the only work to have approached this is [17,15]. While Sanner's work is very ambitious in that it attempted to solve a very general class of problems, the solution used linear function approximation, approximate policy iteration, and some heuristic logical simplification steps to demonstrate that some problems can be solved and it is not clear when the combination of ideas in that work is applicable, both in terms of the algorithmic approximations and in terms of the symbolic simplification algorithms.

In this paper we make a different compromise by constraining the class of problems and aiming for a complete symbolic solution. In particular, we introduce the class of service domains, that have a simple form of independent object-focused exogenous events, so that the transition in each step can be modeled as first taking the agent's action, and then following a sequence of "exogenous actions" in any order. We then investigate a relational SDP approach to solve such problems. The main contribution of this paper is a new symbolic algorithm that is proved to provide a lower bound approximation on the true value function for service domains under certain technical assumptions. While the assumptions are somewhat strong, they allow us to provide the first complete analysis of relational SDP with exogenous events which is important for understanding such problems. In addition, while the assumptions are needed for the analysis, they are not needed for the algorithm that can be applied in more general settings. Our second main contribution provides algorithmic support to implement this algorithm using the GFODD representation of [8]. GFODDs provide a scheme for capturing and manipulating functions over relational structures. Previous work has analyzed some theoretical properties of this representation but did not provide practical algorithms. In this paper we develop a model evaluation algorithm for GFODDs inspired by variable elimination (VE), and a model checking reduction for GFODDs. These are crucial for efficient realization of the new approximate SDP algorithm. We illustrate the new algorithm in two variants of the IC domain, where one satisfies our assumptions and the other does not. Our results demonstrate that the new algorithm can be implemented efficiently, that its size-independent solution scales much better than propositional approaches [5,19], and that it produces high quality policies.

2   Preliminaries: Relational Symbolic Dynamic Programming

We assume familiarity with basic notions of Markov Decision Processes (MDPs) and First Order Logic [14,13]. Briefly, a MDP is given by a set of states S, actions A, transition function $Pr(s'|s, a)$, immediate reward function $R(s)$ and discount factor $\gamma < 1$. The solution of a MDP is a policy that maximizes the expected discounted total reward obtained by following that policy starting from any state. The Value Iteration algorithm

(VI), calculates the optimal value function V^* by iteratively performing Bellman back-ups $V_{i+1} = T[V_i]$ defined for each state s as,

$$V_{i+1}(s) \leftarrow \max_a \{R(s) + \gamma \sum_{s'} Pr(s'|s,a)V_i(s')\}. \tag{1}$$

Relational MDPs: Relational MDPs are simply MDPs where the states and actions are described in a function-free first order logical language. In particular, the language allows a set of logical constants, a set of logical variables, a set of predicates (each with its associated arity), but no functions of arity greater than 0. A state corresponds to an *interpretation* in first order logic (we focus on finite interpretations) which specifies (1) a finite set of n domain elements also known as objects, (2) a mapping of constants to domain elements, and (3) the truth values of all the predicates over tuples of domain elements of appropriate size (to match the arity of the predicate). Atoms are predicates applied to appropriate tuples of arguments. An atom is said to be ground when all its arguments are constants or domain elements. For example, using this notation $empty(x_1)$ is an atom and $empty(shop23)$ is a ground atom involving the predicate $empty$ and object $shop23$ (expressing that the shop $shop23$ is empty in the IC domain). Our notation does not distinguish constants and variables as this will be clear from the context. One of the advantages of relational SDP algorithms, including the one in this paper, is that the number of objects n is not known or used at planning time and the resulting policies generalize across domain sizes.

The state transitions induced by agent actions are modeled exactly as in previous SDP work [3]. The agent has a set of action types $\{A\}$ each parametrized with a tuple of objects to yield an action template $A(x)$ and a concrete ground action $A(o)$ (e.g. template $unload(t,s)$ and concrete action $unload(truck1, shop2)$). To simplify notation, we use x to refer to a single variable or a tuple of variables of the appropriate arity. Each agent action has a finite number of action variants $A_j(x)$ (e.g., action success vs. action failure), and when the user performs $A(x)$ in state s one of the variants is chosen randomly using the state-dependent action choice distribution $Pr(A_j(x)|A(x))$.

Similar to previous work we model the reward as some additive function over the domain. To avoid some technical complications, we use average instead of sum in the reward function; this yields the same result up to a multiplicative factor.

Relational Expressions and GFODDs: To implement planning algorithms for relational MDPs we require a symbolic representation of functions to compactly describe the rewards, transitions, and eventually value functions. In this paper we use the GFODD representation of [8] but the same ideas work for any representation that can express open-expressions and closed expressions over interpretations (states). An expression represents a function mapping interpretations to real values. An open expression $f(x)$, similar to an open formula in first order logic, can be evaluated in interpretation I once we substitute the variables x with concrete objects in I. A closed expression $(aggregate_x f(x))$, much like a closed first order logic formula, aggregates the value of $f(x)$ over all possible substitutions of x to objects in I. First order logic limits $f(x)$ to have values in $\{0,1\}$ (i.e., evaluate to *false* or *true*) and provides the aggregation max (corresponding to existential quantification) and min (corresponding to universal quantification) that can be used individually on each variable in x.

Expressions are more general allowing for additional aggregation functions (for example, average) so that aggregation generalizes quantification in logic, and allowing $f(x)$ to take numerical values. On the other hand, our expressions require aggregation operators to be at the front of the formulas and thus correspond to logical expressions in prenex normal form. This enables us to treat the aggregation portion and formula portion separately in our algorithms. In this paper we focus on average and max aggregation. For example, in the IC domain we might use the expression: "$\max_t, \text{avg}_s$, (if $\neg empty(s)$ then 1, else if $tin(t, s)$ then 0.1, else 0)". Intuitively, this awards a 1 for any non-empty shop and at most one shop is awarded a 0.1 if there is a truck at that shop. The value of this expression is given by picking one t which maximizes the average over s.

GFODDs provide a graphical representation and associated algorithms to represent open and closed expressions. A GFODD is given by an aggregation function, exactly as in the expressions, and a labeled directed acyclic graph that represents the open formula portion of the expression. Each leaf in the GFODD is labeled with a non-negative numerical value, and each internal node is labeled with a first-order atom (allowing for equality atoms) where we allow atoms to use constants or variables as arguments. As in propositional diagrams [1], for efficiency reasons, the order over nodes in the diagram must conform to a fixed ordering over node labels, which are first order atoms in our case. Figure 1(a) shows an example GFODD capturing the expression given in the previous paragraph.

Given a diagram $B = (\text{aggregate}_x f(x))$, an interpretation I, and a substitution of variables in x to objects in I, one can traverse a path to a leaf which gives the value for that substitution. The values of all substitutions are aggregated exactly as in expressions. In particular, let the variables as ordered in the aggregation function be $x_1, \ldots, x_n$. To calculate the final value, $\text{map}_B(I)$, the semantics prescribes that we enumerate all substitutions of variables $\{x_i\}$ to objects in I and then perform the aggregation over the variables, going from x_n to x_1. We can therefore think of the aggregation as if it organizes the substitutions into blocks (with fixed value to the first $k - 1$ variables and all values for the k'th variable), and then aggregates the value of each block separately, repeating this from x_n to x_1. We call the algorithm that follows this definition directly *brute force evaluation*. A detailed example is shown in Figure 3(a). To evaluate the diagram in Figure 3(a) on the interpretation shown there we enumerate all $3^3 = 27$ substitutions of 3 objects to 3 variables, obtain a value for each, and then aggregate the values. In the block where $x_1 = a$, $x_2 = b$, and x_3 varies over a, b, c we get the values $3, 2, 2$ and an aggregated value of $7/3$. This can be done for every block, and then we can aggregate over substitutions of x_2 and x_1. The final value in this case is $7/3$.

Any binary operation *op* over real values can be generalized to open and closed expressions in a natural way. If f_1 and f_2 are two closed expressions, f_1 *op* f_2 represents the function which maps each interpretation w to $f_1(w)$ *op* $f_2(w)$. We follow the general convention of using $\oplus$ and $\otimes$ to denote $+$ and $\times$ respectively when they are applied to expressions. This provides a definition but not an implementation of binary operations over expressions. The work in [8] showed that if the binary operation is *safe*, i.e., it distributes with respect to all aggregation operators, then there is a simple algorithm (the Apply procedure) implementing the binary operation over expressions. For example $\oplus$

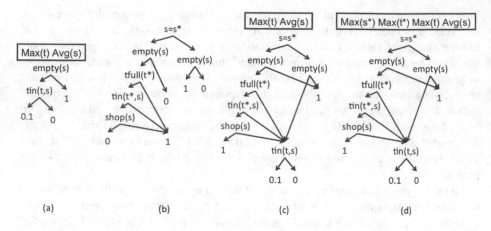

Fig. 1. IC Dynamics and Regression (a) An example GFODD. (b) TVD for $empty(s)$ under the deterministic action $unload(t^*, s^*)$. (c) Regressing the GFODD of (a) over $unload(t^*, s^*)$. (d) Object Maximization. In these diagrams and throughout the paper, left-going edges represent the true branch out of the node and right-going edges represent the false branch.

is safe w.r.t. max aggregation, and it is easy to see that $(\max_x f(x)) \oplus (\max_x g(x)) = \max_x \max_y f(x) + g(y)$, and the open formula portion (diagram portion) of the result can be calculated directly from the open expressions $f(x)$ and $g(y)$. The Apply procedure [20,8] calculates a diagram representing $f(x) + g(y)$ using operations over the graphs representing $f(x)$ and $g(y)$. Note that we need to standardize apart, as in the renaming of $g(x)$ to $g(y)$ for such operations.

SDP for Relational MDPs: SDP provides a symbolic implementation of the value iteration update of Eq (1) that avoids state enumeration implicit in that equation. The SDP algorithm of [8] generalizing [3] calculates one iteration of value iteration as follows. As input we get (as GFODDs) closed expressions V_n, R (we use Figure 1(a) as the reward in the example below), and open expressions for the probabilistic choice of actions $Pr(A_j(x)|A(x))$ and for the dynamics of deterministic action variants.

The action dynamics are specified by providing a diagram (called truth value diagram or TVD) for each variant $A_j(x)$ and predicate template $p(y)$. The corresponding TVD, $T(A_j(x), p(y))$, is an open expression that specifies the truth value of $p(y)$ *in the next state* when $A_j(x)$ has been executed *in the current state*. Figure 1(b) shows the TVD of $unload(t^*, s^*)$ for predicates $empty(s)$. Note that in contrast to other representations of planning operators (but similar to the successor state axioms of [3]) TVDs specify the truth value after the action and not the change in truth value. Since unload is deterministic we have only one variant and $Pr(A_j(x)|A(x)) = 1$. We illustrate probabilistic actions in the next section. Following [20,8] we require that $Pr(A_j(x)|A(x))$ and $T(A_j(x), p(y))$ have no aggregations and cannot introduce new variables, that is, the first refers to x only and the second to x and y but no other variables. This implies that the regression and product terms in the algorithm below do not change the aggregation function and therefore enables the analysis of the algorithm.

The SDP algorithm of [8] implements Eq (1) using the following 4 steps. We denote this as $V_{i+1} = SDP^1(V_i)$.

1. **Regression:** The n step-to-go value function V_n is regressed over every deterministic variant $A_j(x)$ of every action $A(x)$ to produce $Regr(V_n, A_j(x))$. Regression is conceptually similar to goal regression in deterministic planning but it needs to be done for all (potentially exponential number of) paths in the diagram, each of which can be thought of as a goal in the planning context. This can be done efficiently by replacing every atom in the open formula portion of V_n (a node in the GFODD representation) by its corresponding TVD without changing the aggregation function.

   Figure 1(c) illustrates the process of block replacement for the diagram of part (a). Note that $tin()$ is not affected by the action. Therefore its TVDs simply repeats the predicate value, and the corresponding node is unchanged by block replacement. Therefore, in this example, we are effectively replacing only one node with its TVD. The TVD leaf valued 1 is connected to the left child (true branch) of the node and the 0 leaf is connected to the right child (false branch). To maintain the diagrams sorted we must in fact use a different implementation than block replacement; the implementation does not affect the constructions or proofs in the paper and we therefore refer the reader to [20] for the details.

2. **Add Action Variants:** The Q-function $Q_{V_n}^{A(x)} = R \oplus [\gamma \otimes \oplus_j(Pr(A_j(x)) \otimes Regr(V_n, A_j(x)))]$ for each action $A(x)$ is generated by combining regressed diagrams using the binary operations $\oplus$ and $\otimes$ over expressions.

   Recall that probability diagrams do not refer to additional variables. The multiplication can therefore be done directly on the open formulas without changing the aggregation function. As argued by [20], to guarantee correctness, both summation steps ($\oplus_j$ and $R\oplus$ steps) must standardize apart the functions before adding them.

3. **Object Maximization:** Maximize over the action parameters $Q_{V_n}^{A(x)}$ to produce $Q_{V_n}^A$ for each action $A(x)$, thus obtaining the value achievable by the best ground instantiation of $A(x)$ in each state. This step is implemented by converting action parameters x in $Q_{V_n}^{A(x)}$ to variables, each associated with the max aggregation operator, and appending these operators to the head of the aggregation function.

   For example, if object maximization were applied to the diagram of Figure 1(c) (we skipped some intermediate steps) then $t*, s*$ would be replaced with variables and given max aggregation so that the aggregation is as shown in part (d) of the figure. Therefore, in step 2, $t*, s*$ are constants (temporarily added to the logical language) referring to concrete objects in the world, and in step 3 we turn them into variables and specify the aggregation function for them.

4. **Maximize over Actions:** The $n + 1$ step-to-go value function $V_{n+1} = \max_A Q_{V_n}^A$, is generated by combining the diagrams using the binary operation max over expressions.

The main advantage of this approach is that the regression operation, and the binary operations over expressions $\oplus$, $\otimes$, max can be performed symbolically and therefore the final value function output by the algorithm is a closed expression in the same language. We therefore get a completely symbolic form of value iteration. Several instantiations of this idea have been implemented [11,6,18,20]. Except for the work of [8,18] previous work has handled only max aggregation. Previous work [8] relies on the fact that the binary operations $\oplus$, $\otimes$, and max are safe with respect to max, min aggregation to provide a GFODD based SDP algorithm for problems where the reward function has max and min aggregations . In this paper we use reward functions with max and avg aggregation. The binary operations $\oplus$ and $\otimes$ are safe with respect to avg but the binary operation max is not. For example $2 + \text{avg}\{1,2,3\} = \text{avg}\{2+1,2+2,2+3\}$ but $\max\{2,\text{avg}\{1,2,3\}\} \neq \text{avg}\{\max\{2,1\},\max\{2,2\},\max\{2,3\}\}$. To address this issue we introduce a new implementation for this case in the next section.

3   Model and Algorithms for Service Domains

We now proceed to describe our extensions to SDP to handle exogenous events. Exogenous events refer to spontaneous changes to the state without agent action. Our main modeling assumption, denoted **A1**, is that we have *object-centered exogenous actions* that are automatically taken in every time step. In particular, for every object i in the domain we have action $E(i)$ that acts on object i and the conditions and effects of $\{E(i)\}$ are such that they are mutually non-interfering: given any state s, all the actions $\{E(i)\}$ are applied simultaneously, and this is equivalent to their sequential application in any order. We use the same GFODD action representation described in the previous section to capture the dynamics of $E(i)$.

Example: IC Domain. We use a simple version of the inventory control domain (IC) as a running example, and for some of the experimental results. In IC the objects are a depot, a truck and a number of shops. A shop can be empty or full, i.e., the inventory has only two levels and the truck can either be at the depot or at a shop. The reward is the fraction (average) of non-empty shops. Agent actions are deterministic and they capture stock replacement. In particular, a shop can be filled by *unload*ing inventory from the truck in one step. The truck can be *load*ed in a depot and *drive*n from any location (shop or depot) to any location in one step. The exogenous action $E(i)$ has two variants; the success variant $E_{succ}(i)$ (customer arrives at shop i, and if non-empty the inventory becomes empty) occurs with probability 0.4 and the fail variant $E_{fail}(i)$ (no customer, no changes to state) occurs with probability 0.6. Figure 2 parts (a)-(d) illustrate the model for IC and its GFODD representation. In order to facilitate the presentation of algorithmic steps, Figure 2(e) shows a slightly different reward function (continuing previous examples) that is used as the reward in our running example.

For our analysis we make two further modeling assumptions. **A2**: we assume that exogenous action $E(i)$ can only affect unary properties of the object i. To simplify the presentation we consider a single such predicate $sp(i)$ that may be affected, but any number of such predicates can be handled. In IC, the special predicate $sp(i)$ is $empty(i)$ specifying whether the shop is empty. **A3**: we assume that $sp()$ does not appear in the precondition of any agent action. It follows that $E(i)$ only affects $sp(i)$ and that $sp(i)$ can appear in the precondition of $E(i)$ but cannot appear in the precondition of any other action.

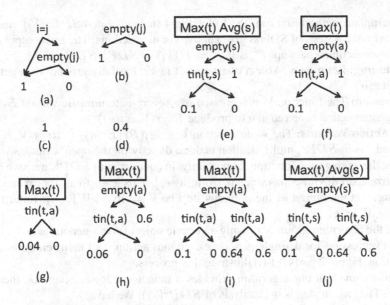

Fig. 2. Representation and template method for IC. (a) TVD for $empty(j)$ under action variant $E_{succ}(i)$. (b) TVD for $empty(j)$ under action variant $E_{fail}(i)$. (c) A specialized form of (a) under $i = j$. This is simply the value 1 and is therefore a GFODD given by a single leaf node. (d) $Pr(E_{succ}(i)|E(i))$ which is simply the value 0.4. (e) A simple reward function. (f) Grounding (e) using Skolem constant a. (g) Regressing (f) over $E_{succ}(a)$ and multiplying with the probability diagram in (d). (h) Regressing (f) over $E_{fail}(a)$ and multiplying by its probability diagram. (i) Adding (g) and (h) without standardizing apart. (j) Reintroducing the Avg aggregation.

3.1 The Template Method

Extending SDP to handle exogenous events is complicated because the events depend on the objects in the domain and on their number and exact solutions can result in complex expressions that require counting formulas over the domain [17,15]. A possible simple approach would explicitly calculate the composition of the agent's actions with all the exogenous events. But this assumes that we know the number of objects n (and thus does not generalize) and results in an exponential number of action variants, which makes it infeasible. A second simple approach would be to directly modify the SDP algorithm so that it sequentially regresses the value function over each of the ground exogenous actions before performing the regression over the agent actions, which is correct by our assumptions. However, this approach, too, requires us to know n and because it effectively grounds the solution it suffers in terms of generality.

We next describe the *template method*, one of our main contributions, which provides a completely abstract approximate SDP solution for the exogenous event model. We make our final assumption, **A4**, that the reward function (and inductively V_i) is a closed expression of the form $\max_x \text{avg}_y V(x, y)$ where x is a (potentially empty) set of variables and y is a single variable, and in $V(x, y)$ the predicate $sp()$ appears instantiated only as $sp(y)$. The IC domain as described above satisfies all our assumptions.

The template method first runs the following 4 steps, denoted $SDP^2(V_i)$, and then follows with the 4 steps of SDP as given above for user actions. The final output of our approximate Bellman backup, T', is $V_{i+1} = T'(V_i) = SDP^1(SDP^2(V_i))$.

1. **Grounding:** Let a be a Skolem constant not in V_i. Partially ground V to get $V = \max_x V(x, a)$

2. **Regression:** The function V is regressed over every deterministic variant $E_j(a)$ of the exogenous action centered at a to produce $Regr(V, E_j(a))$.

3. **Add Action Variants:** The value function $V = \oplus_j(Pr(E_j(a)) \otimes Regr(V, E_j(a)))$ is updated. As in SDP^1, multiplication is done directly on the open formulas without changing the aggregation function. Importantly, in contrast with SDP^1, here we do not standardize apart the functions when performing $\oplus_j$. This leads to an approximation.

4. **Lifting:** Let the output of the previous step be $V = \max_x W(x, a)$. Return $V = \max_x \mathrm{avg}_y W(x, y)$.

Thus, the algorithm grounds V using a generic object for exogenous actions, it then performs regression for a single generic exogenous action, and then reintroduces the aggregation. Figure 2 parts (e)-(j) illustrate this process.

We now show that our algorithm provides a monotonic lower bound on the value function. The crucial step is the analysis of $SDP^2(V_i)$. We have:

Lemma 1. *Under assumptions* **A1, A2, A4** *the value function calculated by* $SDP^2(V_i)$ *is a lower bound on the value of regression of* V_i *through all exogenous actions.*

Due to space constraints the complete proof is omitted and we only provide a sketch. This proof and other omitted details can be found in the full version of this paper [10].

Proof. (sketch) The main idea in the proof is to show that, under our assumptions, the result of our algorithm is equivalent to sequential regression of all exogenous actions, where in each step the action variants are not standardized apart.

Recall that the input value function V_i has the form $V = \max_x \mathrm{avg}_y V(x, y) = \max_x \frac{1}{n}[V(x, 1) + V(x, 2) + \ldots + V(x, n)]$. To establish this relationship we show that after the sequential algorithm regresses $E(1), \ldots, E(k)$ the intermediate value function has the form $\max_x \frac{1}{n}[W(x, 1) + W(x, 2) + \ldots + W(x, k) + V(x, k+1) + \ldots + V(x, n)]$. That is, the first k portions change in the same structural manner into a diagram W and the remaining portions retain their original form V. In addition, $W(x, \ell)$ is the result of regressing $V(x, \ell)$ through $E(\ell)$ which is the same form as calculated by step 3 of the template method. Therefore, when all $E(\ell)$ have been regressed, the result is $V = \max_x \mathrm{avg}_y W(x, y)$ which is the same as the result of the template method.

The sequential algorithm is correct by definition when standardizing apart but yields a lower bound when not standardizing apart. This is true because for any functions f^1 and f^2 we have $[\max_{x_1} \mathrm{avg}_{y_1} f^1(x_1, y_1)] + [\max_{x_2} \mathrm{avg}_{y_2} f^2(x_2, y_2)] \geq \max_x [\mathrm{avg}_{y_1} f^1(x, y_1) + \mathrm{avg}_{y_2} f^2(x, y_2)] = \max_x \mathrm{avg}_y[(f^1(x, y) + f^2(x, y))]$ where the last equality holds because y_1 and y_2 range over the same set of objects. Therefore, if f^1 and f^2 are the results of regression for different variants from step 2, adding them without standardizing apart as in the last equation yields a lower bound. □

The lemma requires that V_i used as input satisfies **A4**. If this holds for the reward function, and if SDP^1 maintains this property then **A4** holds inductively for all V_i.

Put together this implies that the template method provides a lower bound on the true Bellman backup. It therefore remains to show how SDP^1 can be implemented for $\max_x$ avg_y aggregation and that it maintains the form **A4**.

First consider regression. If assumption **A3** holds, then our algorithm using regression through TVDs does not introduce new occurrences of $sp()$ into V. Regression also does not change the aggregation function. Similarly, the probability diagrams do not introduce $sp()$ and do not change the aggregation function. Therefore **A4** is maintained by these steps. For the other steps we need to discuss the binary operations $\oplus$ and max.

For $\oplus$, using the same argument as above, we see that $[\max_{x_1} \text{avg}_{y_1} f^1(x_1, y_1)] + [\max_{x_2} \text{avg}_{y_2} f^2(x_2, y_2)] = \max_{x_1} \max_{x_2} [\text{avg}_y f^1(x_1, y) + f^2(x_2, y)]$ and therefore it suffices to standardize apart the x portion but y can be left intact and **A4** is maintained.

Finally, recall that we need a new implementation for the binary operation max with avg aggregation. This can be done as follows: to perform $\max\{[\max_{x_1} \text{avg}_{y_1} f^1(x_1, y_1)], [\max_{x_2} \text{avg}_{y_2} f^2(x_2, y_2)]\}$ we can introduce two new variables z_1, z_2 and write the expression: "$\max_{z_1, z_2} \max_{x_1} \max_{x_2} \text{avg}_{y_1} \text{avg}_{y_2}$ (if $z_1 = z_2$ then $f^1(x_1, y_1)$ else $f^2(x_2, y_2)$)". This is clearly correct whenever the interpretation has at least two objects because z_1, z_2 are unconstrained. Now, because the branches of the if statement are mutually exclusive, this expression can be further simplified to "$\max_{z_1, z_2} \max_x \text{avg}_y$ (if $z_1 = z_2$ then $f^1(x, y)$ else $f^2(x, y)$)". The implementation uses an equality node at the root with label $z_1 = z_2$, and hangs f^1 and f^2 at the true and false branches. Crucially it does not need to standardize apart the representation of f^1 and f^2 and thus **A4** is maintained. This establishes that the approximation returned by our algorithm, $T'[V_i]$, is a lower bound of the true Bellman backup $T[V_i]$.

An additional argument (details available in [10]) shows that this is a monotonic lower bound, that is, for all i we have $T[V_i] \geq V_i$ where $T[V]$ is the true Bellman backup. It is well known (e.g., [12]) that if this holds then the value of the greedy policy w.r.t. V_i is at least V_i (this follows from the monotonicity of the policy update operator T_π). The significance is, therefore, that V_i provides an immediate certificate on the quality of the resulting greedy policy. Recall that $T'[V]$ is our approximate backup, $V_0 = R$ and $V_{i+1} = T'[V_i]$. We have:

Theorem 1. *When assumptions* **A1**, **A2**, **A3**, **A4** *hold and the reward function is non-negative we have for all* i: $V_i \leq V_{i+1} = T'[V_i] \leq T[V_i] \leq V^*$.

As mentioned above, although the assumptions are required for our analysis, the algorithm can be applied more widely. Assumptions **A1** and **A4** provide our basic modeling assumption per object centered exogenous events and additive rewards. It is easy to generalize the algorithm to have events and rewards based on object tuples instead of single objects. Similarly, while the proof fails when **A2** (exogenous events only affect special unary predicates) is violated the algorithm can be applied directly without modification. When **A3** does not hold, $sp()$ can appear with multiple arguments and the algorithm needs to be modified. Our implementation introduces an additional approximation and at iteration boundary we unify all the arguments of $sp()$ with the average variable y. In this way the algorithm can be applied inductively for all i. These extensions of the algorithm are demonstrated in our experiments.

Relation to Straight Line Plans: The template method provides symbolic way to calculate a lower bound on the value function. It is interesting to consider what kind of lower bound this provides. Recall that the *straight line plan approximation* (see e.g., discussion in [2]) does not calculate a policy and instead at any state it seeks the best linear plan with highest expected reward. As the next observation argues (proof available in [10]) the template method provides a related approximation. We note, however, that unlike previous work on straight line plans our computation is done symbolically and calculates the approximation for all start states simultaneously.

Observation 1. *The template method provides an approximation that is related to the value of the best straight line plan. When there is only one deterministic agent action template we get exactly the value of the straight line plan. Otherwise, the approximation is bounded between the value of the straight line plan and the optimal value.*

4   Evaluation and Reduction of GFODDs

The symbolic operations in the SDP algorithm yield diagrams that are redundant in the sense that portions of them can be removed without changing the values they compute. Recently, [8,7] introduced the idea of model checking reductions to compress such diagrams. The basic idea is simple. Given a set of "focus states" S, we evaluate the diagram on every interpretation in S. Any portion of the diagram that does not "contribute" to the final value in of the interpretations is removed. The result is a diagram which is exact on the focus states, but may be approximate on other states. We refer the reader to [8,7] for further motivation and justification. In that work, several variants of this idea have been analyzed formally (for max and min aggregation), have been shown to perform well empirically (for max aggregation), and methods for generating S via random walks have been developed. In this section we develop the second contribution of the paper, providing an efficient realization of this idea for $\max_x \text{avg}_y$ aggregation.

The basic reduction algorithm, which we refer to below as brute force model checking for GFODDs, is: (1) Evaluate the diagram on each example in our focus set S marking all edges that actively participate in generating the final value returned for that example. Because we have $\max_x \text{avg}_y$ this value is given by the "winner" of max aggregation. This is a block of substitutions that includes one assignment to x and all possible assignments to y. For each such block collect the set of edges traversed by any of the substitutions in the block. When picking the max block, also collect the edges traversed by that block, breaking ties by lexicographic ordering over edge sets. (2) Take the union of marked edges over all examples, connecting any edge not in this set to 0.

Consider again the example of evaluation in Figure 3(a), where we assigned node identifiers 1,2,3. We identify edges by their parent node and its branch so that the left-going edge from the root is edge $1t$. In this case the final value $7/3$ is achieved by multiple blocks of substitutions, and two distinct sets of edges $1t2f3t3f$ and $1f3t3f$. Assuming $1<2<3$ and $f<t$, $1f3t3f$ is lexicographically smaller and is chosen as the marked set. This process is illustrated in the tables of Figure 3(a). Referring to the reduction procedure, if our focus set S includes only this interpretation, then the edges $1t, 2t, 2f$ will be redirected to the value 0.

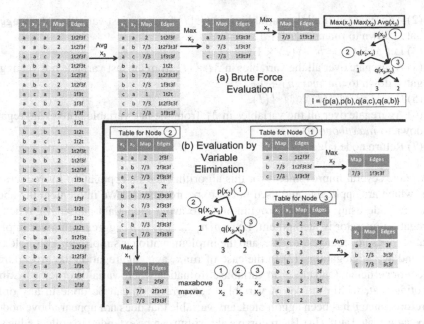

Fig. 3. GFODD Evaluation (a) Brute Force method. (b) Variable Elimination Method.

Efficient Model Evaluation and Reduction: We now show that the same process of evaluation and reduction can be implemented more efficiently. The idea, taking inspiration from variable elimination, is that we can aggregate some values early while calculating the tables. However, our problem is more complex than standard variable elimination and we require a recursive computation over the diagram.

For every node n let $n.lit = p(x)$ be the literal at the node and let $n_{\downarrow f}$ and $n_{\downarrow t}$ be its false and true branches respectively. Define $above(n)$ to be the set of variables appearing above n and $self(n)$ to be the variables in x. Let $maxabove(n)$ and $maxself(n)$ be the variables of largest index in $above(n)$ and $self(n)$ respectively. Finally let $maxvar(n)$ be the maximum between $maxabove(n)$ and $maxself(n)$. Figure 3(b) shows $maxvar(n)$ and $maxabove(n)$ for our example diagram. Given interpretation I, let $bl^{n_{\downarrow t}}(I)$ be the set of bindings a of objects from I to variables in x such that $p(a) \in I$. Similarly $bl^{n_{\downarrow f}}(I)$ is the set of bindings a such that $\neg p(a) \in I$. The two sets are obviously disjoint and together cover all bindings for x. For example, for the root node in the diagram of Figure 3(b), $bl^{n_{\downarrow t}}(I)$ is a table mapping x_2 to a, b and $bl^{n_{\downarrow f}}(I)$ is a table mapping x_2 to c. The evaluation procedure, Eval(n), is as follows:

1. If n is a leaf:
   (1) Build a "table" with all variables implicit, and with the value of n.
   (2) Aggregate over all variables from the last variable down to $maxabove(n) + 1$.
   (3) Return the resulting table.

2. Otherwise n is an internal node:
   (1) Let $M^{\downarrow t}(I) = bl^{n_{\downarrow t}}(I) \times$ Eval($n_{\downarrow t}$), where $\times$ is the join of the tables.

(2) Aggregate over all the variables in $M^{\downarrow t}(I)$ from the last variable not yet aggregated down to $maxvar(n) + 1$.

(3) Let $M^{\downarrow f}(I) = bl^{n_{\downarrow f}}(I) \times \text{Eval}(n_{\downarrow f})$

(4) Aggregate over all the variables in $M^{\downarrow f}(I)$ from the last variable not yet aggregated down to $maxvar(n) + 1$.

(5) Let $M = M^{\downarrow t}(I) \cup M^{\downarrow f}(I)$.

(6) Aggregate over all the variables in M from the last variable not yet aggregated down to $maxabove(n) + 1$.

(7) Return node table M.

We note several improvements for this algorithm and its application for reductions, all of which are applicable and used in our experiments. (I1) We implement the above recursive code using dynamic programming to avoid redundant calls. (I2) When an aggregation operator is idempotent, i.e., $op\{a, \ldots, a\} = a$, aggregation over implicit variables does not change the table, and the implementation is simplified. This holds for max and avg aggregation. (I3) In the case of $\max_x \text{avg}_y$ aggregation the procedure is made more efficient (and closer to variable elimination where variable order is flexible) by noting that, within the set of variables x, aggregation can be done in any order. Therefore, once y has been aggregated, any variable that does not appear above node n can be aggregated at n. (I4) The recursive algorithm can be extended to collect edge sets for winning blocks by associating them with table entries. Leaf nodes have empty edge sets. The join step at each node adds the corresponding edge (for true or false child) for each entry. Finally, when aggregating an average variable we take the union of edges, and when aggregating a max variable we take the edges corresponding to the winning value, breaking ties in favor of the lexicographically smaller set of edges.

A detailed example of the algorithm is given in Figure 3(b) where the evaluation is on the same interpretation as in part (a). We see that node 3 first collects a table over x_2, x_3 and that, because x_3 is not used above, it already aggregates x_3. The join step for node 2 uses entries (b, a) and (c, a) for (x_1, x_2) from the left child and other entries from the right child. Node 2 collects the entries and (using I3) aggregates x_1 even though x_2 appears above. Node 1 then similarly collects and combines the tables and aggregates x_2. The next theorem is proved by induction over the structure of the GFODD (details available in [10]).

Theorem 2. *The value and max block returned by the modified Eval procedure are identical to the ones returned by the brute force method.*

5   Experimental Validation

In this section we present an empirical demonstration of our algorithms. To that end we implemented our algorithms in Prolog as an extension of the FODD-PLANNER [9], and compared it to SPUDD [5] and MADCAP [19] that take advantage of propositionally factored state spaces, and implement VI using propositional algebraic decision diagrams (ADD) and affine ADDs respectively. For SPUDD and MADCAP, the domains were specified in the Relational Domain Description Language (RDDL) and translated into

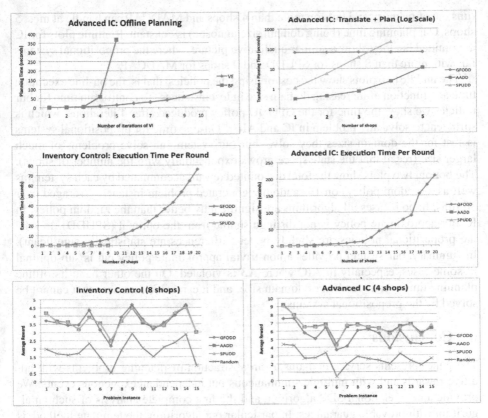

Fig. 4. Experimental Results

propositional descriptions using software provided for the IPPC 2011 planning competition [16]. All experiments were run on an Intel Core 2 Quad CPU @ 2.83GHz. Our system was given 3.5Gb of memory and SPUDD and MADCAP were given 4Gb.

We tested all three systems on the IC domain as described above where shops and trucks have binary inventory levels (empty or full). We present results for the IC domain, because it satisfies all our assumptions and because the propositional systems fare better in this case. We also present results for a more complex IC domain (advanced IC or AIC below) where the inventory can be in one of 3 levels 0,1 and 2 and a shop can have one of 2 consumption rates 0.3 and 0.4. AIC does not satisfy assumption **A3**. As the experiments show, even with this small extension, the combinatorics render the propositional approach infeasible. In both cases, we constructed the set of focus states to include all possible states over 2 shops. This provides exact reduction for states with 2 shops but the reduction is approximate for larger states as in our experiments.

Figure 4 summarizes our results, which we discuss from left to right and top to bottom. The top left plot shows runtime as a function of iterations for AIC and illustrates that the variable elimination method is significantly faster than brute force evaluation and that it enables us to run many more iterations. The top right plot shows the total time (translation from RDDL to a propositional description and off-line planning for 10 iterations of VI) for the 3 systems for one problem instance per size for AIC. SPUDD runs out of memory and fails on more than 4 shops and MADCAP can handle at most 5 shops. Our planning time (being domain size agnostic) is constant. Runtime plots for IC are omitted but they show a similar qualitative picture, where the propositional systems fail with more than 8 shops for SPUDD and 9 shops for MADCAP.

The middle two plots show the cost of using the policies, that is, the on-line execution time as a function of increasing domain size in test instances. To control run time for our policies we show the time for the GFODD policy produced after 4 iterations, which is sufficient to solve any problem in IC and AIC.[1] On-line time for propositional systems is fast for the domain sizes they solve, but our system can solve problems of much larger size (recall that the state space grows exponentially with the number of shops). The bottom two plots show the total discounted reward accumulated by each system (as well as a random policy) on 15 randomly generated problem instances averaged over 30 runs. In both cases all algorithms are significantly better than the random policy. In IC our approximate policy is not distinguishable from the optimal (SPUDD). In AIC the propositional policies are slightly better (differences are statistically significant). In summary, our system provides a non-trivial approximate policy but is sub-optimal in some cases, especially in AIC where **A3** is violated. On the other hand its offline planning time is independent of domain size, and it can solve instances that cannot be solved by the propositional systems.

6 Conclusions

The paper presents service domains as an abstraction of planning problems with additive rewards and with multiple simultaneous but independent exogenous events. We provide a new relational SDP algorithm and the first complete analysis of such an algorithm with provable guarantees. In particular our algorithm, the template method, is guaranteed to provide a monotonic lower bound on the true value function under some technical conditions. We have also shown that this lower bound lies between the value of straight line plans and the true value function. As a second contribution we introduce new evaluation and reduction algorithms for the GFODD representation, that in turn facilitate efficient implementation of the SDP algorithm. Preliminary experiments demonstrate the viability of our approach and that our algorithm can be applied even in situations that violate some of the assumptions used in the analysis. The paper provides a first step toward analysis and solutions of general problems with exogenous events by focusing on a well defined subset of such models. Identifying more general conditions for existence of compact solutions, representations for such solutions, and associated algorithms is an important challenge for future work. In addition, the problems involved in evaluation and application of diagrams are computationally demanding. Techniques to speed up these computations are an important challenge for future work.

Acknowledgements. This work was partly supported by NSF under grants IIS-0964457 and IIS-0964705 and the CI fellows award for Saket Joshi. Most of this work was done when Saket Joshi was at Oregon State University.

---

[1] Our system does not achieve structural convergence because the reductions are not comprehensive. We give results at 4 iterations as this is sufficient for solving all problems in this domain. With more iterations, our policies are larger and their execution is slower.

References

1. Bahar, R., Frohm, E., Gaona, C., Hachtel, G., Macii, E., Pardo, A., Somenzi, F.: Algebraic decision diagrams and their applications. In: Proceedings of the IEEE/ACM International Conference on Computer-Aided Design, pp. 188–191 (1993)
2. Boutilier, C., Dean, T., Hanks, S.: Decision-theoretic planning: Structural assumptions and computational leverage. Journal of Artificial Intelligence Research 11, 1–94 (1999)
3. Boutilier, C., Reiter, R., Price, B.: Symbolic dynamic programming for first-order MDPs. In: Proceedings of the International Joint Conference of Artificial Intelligence, pp. 690–700 (2001)
4. Guestrin, C., Koller, D., Parr, R., Venkataraman, S.: Efficient solution algorithms for factored MDPs. Journal of Artificial Intelligence Research 19, 399–468 (2003)
5. Hoey, J., St-Aubin, R., Hu, A., Boutilier, C.: SPUDD: Stochastic planning using decision diagrams. In: Proceedings of Uncertainty in Artificial Intelligence, pp. 279–288 (1999)
6. Hölldobler, S., Karabaev, E., Skvortsova, O.: FluCaP: a heuristic search planner for first-order MDPs. Journal of Artificial Intelligence Research 27, 419–439 (2006)
7. Joshi, S., Kersting, K., Khardon, R.: Self-Taught decision theoretic planning with first-order decision diagrams. In: Proceedings of the International Conference on Automated Planning and Scheduling, pp. 89–96 (2010)
8. Joshi, S., Kersting, K., Khardon, R.: Decision theoretic planning with generalized first order decision diagrams. Artificial Intelligence 175, 2198–2222 (2011)
9. Joshi, S., Khardon, R.: Probabilistic relational planning with first-order decision diagrams. Journal of Artificial Intelligence Research 41, 231–266 (2011)
10. Joshi, S., Khardon, R., Tadepalli, P., Raghavan, A., Fern, A.: Solving relational MDPs with exogenous events and additive rewards. CoRR abs/1306.6302 (2013), http://arxiv.org/abs/1306.6302
11. Kersting, K., van Otterlo, M., De Raedt, L.: Bellman goes relational. In: Proceedings of the International Conference on Machine Learning, pp. 465–472 (2004)
12. McMahan, H.B., Likhachev, M., Gordon, G.J.: Bounded real-time dynamic programming: RTDP with monotone upper bounds and performance guarantees. In: Proceedings of the International Conference on Machine Learning, pp. 569–576 (2005)
13. Puterman, M.L.: Markov Decision Processes: Discrete Stochastic Dynamic Programming. Wiley (1994)
14. Russell, S., Norvig, P.: Artificial Intelligence: A Modern Approach. Prentice Hall Series in Artificial Intelligence (2002)
15. Sanner, S.: First-order decision-theoretic planning in structured relational environments. Ph.D. thesis, University of Toronto (2008)
16. Sanner, S.: Relational dynamic influence diagram language (RDDL): Language description (2010), http://users.cecs.anu.edu.au/~sanner/IPPC2011/RDDL.pdf
17. Sanner, S., Boutilier, C.: Approximate solution techniques for factored first-order MDPs. In: Proceedings of the International Conference on Automated Planning and Scheduling, pp. 288–295 (2007)
18. Sanner, S., Boutilier, C.: Practical solution techniques for first-order MDPs. Artificial Intelligence 173, 748–788 (2009)
19. Sanner, S., Uther, W., Delgado, K.: Approximate dynamic programming with affine ADDs. In: Proceeding of the International Conference on Autonomous Agents and Multiagent Systems, pp. 1349–1356 (2010)
20. Wang, C., Joshi, S., Khardon, R.: First-Order decision diagrams for relational MDPs. Journal of Artificial Intelligence Research 31, 431–472 (2008)

Continuous Upper Confidence Trees
with Polynomial Exploration – Consistency

David Auger[1], Adrien Couëtoux[2], and Olivier Teytaud[2]

[1] AlCAAP, Laboratoire PRiSM, Bât. Descartes,
Université de Versailles Saint-Quentin-en-Yvelines,
45 avenue des États-Unis, F-78035 Versailles Cedex, France
david.auger@prism.uvsq.fr
[2] TAO, Lri, UMR CNRS 8623, Bat. 490,
Université Paris-Sud, F-91405 Orsay Cedex
{adrien.couetoux,olivier.teytaud}@gmail.com

Abstract. Upper Confidence Trees (UCT) are now a well known algorithm for sequential decision making; it is a provably consistent variant of Monte-Carlo Tree Search. However, the consistency is only proved in a the case where the action space is finite. We here propose a proof in the case of fully observable Markov Decision Processes with bounded horizon, possibly including infinitely many states, infinite action space and arbitrary stochastic transition kernels. We illustrate the consistency on two benchmark problems, one being a legacy toy problem, the other a more challenging one, the famous energy unit commitment problem.

Keywords: Upper Confidence Trees, Consistency Proof, Infinite Markov Decision Process, Unit commitment.

1 State of the Art and Outline of the Paper

It is known that partially observable Markov Decision Processes are undecidable, even with finite state space (see [15]). With full observation, they become decidable; Monte-Carlo Tree Search (MCTS, [9]) is a recent well known solver for this case, with impressive results in many cases, in particular the game of Go [14]. Its most famous variant, Upper Confidence Trees [13] provides provably consistent solvers in the finite case. We here show that Upper Confidence Tree can be slightly adapted to become consistent in the more general finite horizon case, even with infinite state space and infinite action space.

Recent impressive results in the field of planning with MCTS variants in continuous MDP have been published; most of them, as far as we know, rely on a discretization of the action space. This the case of HOOT [16] and HOLOP [17] that both rely on the HOO algorithm, introduced in [5]. HOO is a bandit algorithm that deals with continuous arms by using a tree of coverings of the action space. Other notable contributions using a discretization of the action space are [12] and [1]. What these methods have in common is the assumption that the action space is continuous, but that we have enough knowledge about

H. Blockeel et al. (Eds.): ECML PKDD 2013, Part I, LNAI 8188, pp. 194–209, 2013.

it to divide it in a certain number of equally spaced actions. Or, in the case of HOO, it is required to have a compact action space with known bounds. In toy benchmark problems like inverted pendulum, this is straightforward. However, in more realistic applications, this can be difficult. This is the case of the unit commitment problem, as described in [3] , where the agent needs to decide at each time step how to use a wide array of energy production facilities: water stocks, thermal plants, nuclear plants, etc. This problem has an action space that cannot be easily discretized. First, it has both discrete and continuous components (some power plants having a minimal energy output). Second, there are many operational constraints, making the action space non convex, and the bounds hard to find. In practice, finding feasible actions can come down to adding noise to the objective function of a simplified version of the problem, applying a Linear Programming method on said simplified problem, and using the result as a feasible action. There are many other options to sample a feasible action, but raw discretization is not one of them.

In this work, we investigate the consistency of a method that does not require any knowledge about the action space itself. The only assumption made is that we have access to a black box action sampler. Further details on the assumptions made are found below.

Section 2 introduces notations and specifies the setting of the Markov Decision Processes that we consider. In Section 3, we define our PUCT (Polynomial Upper Confidence Trees) algorithm. Section 4 gives the main consistency result, with convergence rate. The proof of this result is divided in three parts, which are Sections 5, 6 and 7. Section 8 presents experimental results. Section 9 concludes.

2   Specification of the Makov Decision Tree Setting

We use the classical terminology of Markov Decision Processes. In this frame-work, a *player* has to make sequential decisions until the process stops: he is then given a *reward*. As usual, the goal of the player is to maximize the ex-pected reward. This paper considers the general case where the process, also called transition, is a fully observable MDP, with finite horizon, and no cycles. *In this setting, the only things available to the agent are a simulator, or transition function, and an action sampler.*

As per usual in this setting, there is a state space and an action space. To build a tree in the stochastic setting, we choose to build it with two distinct and alternated types of nodes:

- decision nodes, where a decision needs to be made, are generally noted z. The intuition is that they correspond to a certain state where the agent might be.
- random nodes, where the transition can be called, are noted $w = (z, a)$. They correspond to the case where the agent was in state z and decided to take action a (sometimes called post-decision state).

The tree will have a unique root decision node r, the initial state where the agent starts. We define the depth of a node as half the distance from this node to the root in the tree. Hence decision nodes have integer depth while random nodes have semi-integer depth, e.g. to access a node of depth 2 we have the sequence of nodes root=decision(depth 0) - random (0.5) - decision (1) - random (1.5) - decision (2). Leaves are assumed to all have the same integer depth, denoted $d_{\max}$, and bear some deterministic *reward* $r(z)$.

It is well known [2] that for each node z, there exists a value $V^*(z)$, termed optimal Bellman value, frequently used as a criterion to select the best action in sequential decision making problems. In this paper, we will use this value as a measure of optimality for actions. Given our distinction between decision nodes and random nodes, we use a natural notation for optimal Bellman values for both categories of nodes.

Let $w = (z, a)$ be a random node, and $P(z'|z, a)$ be the probability of being in node z' after taking action a in node z. Then, its optimal value is:

$$V^*(z, a) = \int_{z'} dP(z'|z, a)V^*(z') \tag{1}$$

Let z be a decision node. Then, its optimal value is defined as follows:

$$V^*(z) = \begin{cases} sup_a V^*(z, a) & \text{if } z \text{ is not a leaf,} \\ r(z) & \text{if } z \text{ is a leaf} \end{cases} \tag{2}$$

In particular, we formally define optimality of actions as follows:

Definition 1. *Let z be a non-leaf decision node, $w = (z, a)$ be a child of z, and $\epsilon > 0$. We say that the action a, i.e. the selection of node w, is optimal with precision ϵ if and only if $V(w) \geq V^*(z) - \epsilon$.*

There may be no optimal action since the number of children is infinite.

Regularity Hypothesis for Decision Nodes. This is the assumption that for any $\Delta > 0$, there is a non zero probability to sample an action that is optimal with precision Δ. More precisely, there is a $\theta > 0$ and a $p > 1$ (which remain the same during the whole simulation) such that for all $\Delta > 0$,

$$V(w = (z, a)) \geq V^*(z) - \Delta \text{ with probability at least } \min(1, \theta\Delta^p). \tag{3}$$

3   Specification of the Polynomial Upper Confidence Tree Algorithm

We refer to [13] for the detailed specification of Upper Confidence Tree; we here define our variant PUCT (Polynomial Upper Confidence Trees).

In PUCT, we sequentially repeat episodes of the MDP and use information from previous episodes in order to explore and find optimal actions in the subsequent episodes. We denote by $n(z)$, for any decision node z, the total number of times that node z has been visited after the nth episode. Hence a node z has been encountered at episode n if $n(z) \geq 1$, and we always have $n = n(r)$. The notation is identical for random nodes.

We denote by $\hat{V}(z)$ the empirical average of a decision node z and $\hat{V}(z, a)$ the empirical average of a random node $w = (z, a)$. Note that if PUCT works properly, $\hat{V}(z)$ should converge to $V^*(z)$ when $n(z)$ goes to infinity.

How we select and construct children of a given node depends on two sequences of coefficients: α_d, the *progressive widening coefficient*, defined for all integer and semi-integer depths d, and e^d, the *exploration coefficient*, defined only for integer depths (i.e. decision nodes). These coefficients are defined according to Table 1. We sometimes indicate, as on Table 1, by a small "R" or "D" if a coefficient corresponds to a random or decision node, but otherwise it should be clear from the context.

PUCT algorithm
Input: a root node r, a transition function, an action sampler, a time budget, a depth d_{max}, parameters α and e for each layer
Output: an action a
while time budget not exhausted **do**
    while current node is not final **do**
        if current node is a decision node z **then**
            if $\lfloor n(z)^\alpha \rfloor > \lfloor (n(z) - 1)^\alpha \rfloor$ **then**
                we call the action sampler and add a child $w = (z, a)$ to z
            else
                we choose as an action among the already visited children (z, a) of z, the one that maximizes its score, defined by:

$$\hat{V}(z, a) + \sqrt{\frac{n(z)^{e(d)}}{n(z, a)}}. \tag{4}$$

            end if
        else
            if $\lfloor n(w)^\alpha \rfloor = \lfloor (n(w) - 1)^\alpha \rfloor$ **then**
                we select the child of z that was least visited during the simulation
            else
                we construct a new child (i.e. we call the transition function with argument w)
            end if
        end if
    end while
    we reached a final node z with reward $r(z)$; we back propagate all the information in the constructed nodes, and we go back to the root node r.
end while
Return the most simulated child of r.

With this algorithm, we see that if a decision node z at depth d has been visited n times, then we have visited during the simulation exactly $\lfloor n^{\alpha_d^D} \rfloor$ of its children, a number which depends on the *progressive widening constant* α_d^D. This is the so-called progressive widening trick [10].

For a random node z, we actually have the same property, depending on the *double progressive widening constant* α_d^R: this is the so-called double progressive widening trick ([7]; see also [11]).

4 Main Result

Definition 2 (Exponentially Sure in n). *We say that some property (P) depending on an integer n is* exponentially sure in n *(denoted e.s.) if there exists positive constants C, h, η such that the probability that (P) holds is at least*

$$1 - C\exp(-hn^\eta).$$

Theorem 1. *Define all exploration coefficients e_d and all progressive widening coefficients α_d as in Table 1. There is a constant $C > 0$, only depending on $d_{\max}$, such that after n episodes of PUCT, for every node z at depth d we have*

$$|\hat{V}(z) - V^*(z)| \leq \frac{C}{n(z)^{\gamma_d}} \ \textit{e.s. in } n(z) \tag{5}$$

Additionally, for every node $w = (z, a)$ at depth $d + \frac{1}{2}$ we have

$$|\hat{V}(w) - V^*(w)| \leq \frac{C}{n(w)^{\gamma_{d+\frac{1}{2}}}} \ \textit{e.s. in } n(w) \tag{6}$$

Corollary 1. *After n episodes, let $w_n(r)$ be the most simulated child node of r. Then,*

$$w_n(r) \textit{ is optimal with precision } O\left(n^{-\frac{1}{10d_{\max}}}\right) \ \textit{e.s. in } n \tag{7}$$

Table 1. Definition of coefficients and convergence rates

Decision Node (d integer)	Random Node (d semi-integer)
$\alpha_d^D := \dfrac{1}{10(d_{\max} - d) - 3}$ for $d \leq d_{\max} - 1$	$\alpha_d^R := \begin{cases} \frac{3}{10(d_{\max}-d)-3} & \text{for } d \leq d_{\max} - \frac{3}{2} \\ 1 & \text{for } d = d_{\max} - \frac{1}{2} \end{cases}$
$e^d := \dfrac{1}{2p}\left(1 - \dfrac{3}{10(d_{\max} - d)}\right)$ for $d \leq d_{\max} - 1$	
$\gamma_d^D := \dfrac{1}{10(d_{\max} - d)}$ for $d \leq d_{\max} - 1$	$\gamma_d^R := \dfrac{1}{10(d_{\max} - d) - 2}$ for $d \leq d_{\max} - \frac{1}{2}$

The proof is based on an induction on the following property and is detailed in the following three sections. Let us define this property.

Definition 3 (Induction Property $Cons(\gamma_d, d)$)
   There is a $C_d > 0$ such that for all nodes at integer depth d,

$$|\hat{V}(z) - V^*(z)| \leq C_d n(z)^{-\gamma_d} \ \textit{e.s. in } n(z)$$

and for all nodes w at semi integer depths $d + \frac{1}{2}$,

$$|\hat{V}(w) - V^*(w)| \leq C_{d+\frac{1}{2}} n(w)^{-\gamma_{d+\frac{1}{2}}} \ \textit{e.s. in } n(w)$$

In Section 5, we show that if $Cons(\gamma_d, d)$ holds for $d \geq 1$, i.e. for decision nodes in one given layer, then $Cons(\gamma_{d-\frac{1}{2}}, d - \frac{1}{2})$ holds, i.e. holds for the random nodes in the above layer. In Section 6, we show that if $Cons(\gamma_{d+\frac{1}{2}}, d + \frac{1}{2})$ holds for $d \geq 0$, i.e. for random nodes in one given layer, then $Cons(\gamma_d, d)$ holds, i.e. holds for the decision nodes in the above layer. Finally, we establish in Section 7 that $Cons(\gamma, d)$ holds for maximal depth d_{max}, which will settle the proof of Theorem 5.

5   From Decision Nodes to Random Nodes

In this section we consider a random node w with semi-integer depth $d - \frac{1}{2} \geq 0$. We suppose that there exist a $\gamma_d^D > 0$ such that $Cons(\gamma_d^D, d)$ holds for any child node z of w. Recall that all nodes at this depth have $\lfloor n^{\alpha_{d-\frac{1}{2}}^R} \rfloor$ constructed children when they have been visited n times. We will show that we can define $\alpha_{d-\frac{1}{2}}^R$ so that $Cons(\gamma_{d-\frac{1}{2}}^R, d - \frac{1}{2})$ holds. For convenience, if w is a random node, we will refer to the ith child z_i of w by its index i directly. Then, the number of visits in z_i after the nth iteration of PUCT will be simply called $n(i)$ instead of $n(z_i)$. Similarly, the empirical value of this node will be noted $\hat{V}(i)$ instead of $\hat{V}(z_i)$.

5.1   Children of Random Nodes Are Selected almost the Same Number of Times

With our politics for dealing with random nodes, described in section 3, the kth child of a random node w is constructed at episode $\lceil k^{\frac{1}{\alpha}} \rceil$. We now show that all constructed children of w but the last one are visited almost the same number of times.

Lemma 1. *Let w be a random node with progressive widening coefficient $\alpha \in$ $]0; 1[$. Then after the nth visit of w in the simulation, all children z_i, z_j of w with $1 \leq i, j < \lfloor n^\alpha \rfloor$ satisfy*

$$|n(i) - n(j)| \leq 1. \tag{8}$$

In fact, in the next section, we will only use the following consequence of Lemma 1.

Corollary 2. *When a random node z is visited for the nth time, all children of z have been selected at most $\frac{n}{\lceil n^\alpha \rceil - 1}$ times, and all children of z but the last one have been selected at least $\frac{n}{\lceil n^\alpha \rceil} - 1$ times.*

Proof. For length reasons, we only provide the following sketch of the proof:

Let us write k the kth child of w for all $k \geq 1$, and $n_k = \lceil k^{\frac{1}{\alpha}} \rceil$ the number of visits in w when child k was introduced. Remark the statement of Lemma 1 is equivalent to:

(8) is satisfied *for all children* of z at every time step $n_k - 1$ for $k \geq 2$.
Then, prove the above statement by induction, by proving the following equivalent formulation: $n_{k+1} - n_k \geq \lfloor \frac{n_k - 1}{k-1} - 1 \rfloor$. $\square$

5.2   Consistency of Random Nodes

Lemma 2 (Random Nodes are Consistent). *If there is a* $1 \geq \gamma_d > 0$ *such that for any child* z *of the random node* w *we have* $Cons(\gamma_d, d)$, *then we have* $Cons(\gamma_{d-\frac{1}{2}}, d - \frac{1}{2})$, *with* $\gamma_{d-\frac{1}{2}} = \frac{\gamma_d}{1+3\gamma_d}$ *if we define the progressive widening coefficient* $\alpha^R_{d-\frac{1}{2}}$ *by* $\alpha^R_{d-\frac{1}{2}} = \frac{3\gamma_d}{1+3\gamma_d}$.

Proof. From now on, w is fixed in order to simplify notation; therefore, we simply denote $\alpha^R_{d-\frac{1}{2}}$ by α, and $n(w)$ by n.

Fix n such that $n^\alpha \geq 3$. Define $i_0 = \lfloor n^\alpha \rfloor$ as the last constructed child of node w, and $r = \lfloor n^\alpha \rfloor - 1 = i_0 - 1$. To prove the result, we need to prove an upper bound on the following quantity, that holds exponentially surely in n:

$$|\hat{V}(w) - V^*(w)| = |\left(\sum_{1 \leq i < i_0} \frac{n(i)}{n}\hat{V}(i) + \frac{n(i_0)}{n}\hat{V}(i_0) \right) - V^*(w)|$$

Decompose this as

$$|\hat{V}(w) - V^*(w)| \leq |\sum_{1 \leq i < i_0} \left(\frac{n(i)}{n} - \frac{1}{r} \right)\hat{V}(i)| \tag{9}$$

$$+ |\sum_{1 \leq i < i_0} \frac{1}{r}\left(\hat{V}(i) - V^*(i) \right)| \tag{10}$$

$$+ |\sum_{1 \leq i < i_0} \frac{1}{r}(V^*(i) - V^*(w))| \tag{11}$$

$$+ |\frac{n(i_0)}{n}\hat{V}(i_0)| \tag{12}$$

First consider (9). By Lemma 1, there is a integer p such that all children $i = 1, \cdots, i_0 - 1$ have been selected p or $p+1$ times, with $p = O\left(n^{1-\alpha}\right)$. So, we have for all $i = 1, 2, \cdots, i_0 - 1$,

$$|\frac{n(i)}{n} - \frac{1}{\lfloor n^\alpha \rfloor - 1}| \leq |\frac{p}{n} - \frac{1}{\lfloor n^\alpha \rfloor - 1}| + \frac{1}{n}$$

The definition of p gives $(i_0 - 1)p \leq n \leq i_0(p+1)$, so that

$$|\frac{p}{n} - \frac{1}{i_0 - 1}| \leq \frac{i_0 + p}{(i_0 - 1)n} = O\left(\frac{1}{n} + \frac{1}{n^{2\alpha}}\right)$$

so that in the end for (9) we have

$$|\sum_{1 \leq i < i_0} \left(\frac{n(i)}{n} - \frac{1}{r} \right)\hat{V}(i)| = O\left(n^\alpha\left(\frac{1}{n} + \frac{1}{n^{2\alpha}} + \frac{1}{n}\right)\right) = O\left(\frac{1}{n^{1-\alpha}} + \frac{1}{n^\alpha}\right)$$

Consider now (10). $Cons(\gamma_d, d)$ holds, so for each child $i = 1, 2, \cdots, \lfloor n^\alpha \rfloor - 1$ of w, Lemma 1 leads to:

$$|\hat{V}(i) - V(i)| \le C_d p^{-\gamma_d} \le C_d \frac{1}{\lfloor n^{1-\alpha} \rfloor^{\gamma_d}} \text{ e.s. in } n^{1-\alpha}$$

Finally for (10) it is exponentially sure in n that

$$|\sum_{0 \le i < i_0} \frac{1}{\lfloor n^\alpha \rfloor - 1} \left(\hat{V}(i) - V^*(i) \right)| = O\left(\frac{1}{n^{(1-\alpha)\gamma_d}} \right). \tag{13}$$

Now we turn to (11). Since w is a random node, the value $V^*(i)$ of each new child i of w constructed by the algorithm is given by a random law whose mean is $V^*(w)$. Thus we can apply Hoeffding's inequality to the sum in (11) and we obtain that for $t > 0$,

$$|\sum_{0 \le i < i_0} \frac{1}{\lfloor n^\alpha \rfloor - 1} (V^*(i) - V^*(w))| \le t \tag{14}$$

with probability at least $1 - 2\exp\left(-2t^2 \left(\lfloor n^\alpha \rfloor - 1\right)\right) = 1 - 2\exp(-Cn^{\frac{\gamma_d}{1+3\gamma_d}})$

with $t := n^{-\frac{\gamma'}{1+3\gamma_d}}, \alpha = \frac{3\gamma_d}{1+3\gamma_d}$, and $C > 0$. This proves that (14) is e.s. in n.

Finally consider (12): since the last child of w has been selected at most p times, we have

$$\left| \frac{n(i_0)}{n} \hat{V}(i_0) \right| = \frac{1}{n} \times O\left(\frac{n}{n^\alpha} \right) = O\left(\frac{1}{n^\alpha} \right).$$

All in all, we have have shown that it is exponentially sure in $n = n(w)$ that

$$|\hat{V}(w) - V^*(w)| = O\left(\underbrace{\frac{1}{n^{1-\alpha}}}_{(9)} + \underbrace{\frac{1}{n^\alpha}}_{} + \underbrace{\frac{1}{n^{(1-\alpha)\gamma_d}}}_{(10)} + \underbrace{\frac{1}{n^{\frac{\gamma_d}{1+3\gamma_d}}}}_{(11)} + \underbrace{\frac{1}{n^\alpha}}_{(12)} \right). \tag{15}$$

With $\alpha = \frac{3\gamma_d}{1+3\gamma_d}$ and $\gamma_d \le 1$, it is straightforward to check that the smallest exponent is $\frac{\gamma_d}{1+3\gamma_d}$, so that $Cons(\gamma_{d-\frac{1}{2}}, d - \frac{1}{2})$ is true with $\gamma_{d-\frac{1}{2}} = \frac{\gamma_d}{1+3\gamma_d}$ $\qquad\square$

6 From Random Nodes to Decision Nodes

Let z be a non leaf decision node at depth d. In this section, we will show that if the induction property holds for all random nodes at depth $d + \frac{1}{2}$, it will hold for z.

Lemma 3 (Children of Decision Nodes are Selected Infinitely Often).
*Let f be a non-decreasing map from $\mathbb{N}$ to $\mathbb{N}$. Consider a stochastic bandit setting
with a countable set of children, progressive widening coefficient α and exploration function f, i.e. the score at time n of a child i is computed by*

$$sc_n(i) = \hat{V}_n(i) + \sqrt{\frac{f(n)}{n(i)}}.$$

Then if i denotes the ith constructed child, for all $n \geq i^{\frac{1}{\alpha(1-\alpha)}}$ we have

$$n(i) \geq \frac{1}{4}\min(f(n^{1-\alpha}), n^{1-\alpha}).$$

*In particular, all constructed children are selected infinitely often provided that
$\lim_{+\infty} f = +\infty$.*

Proof. Fix n and consider the child i_0 maximizing $n(i_0)$, i.e. the most selected
child at time n. Let n' be the last time i_0 has been selected. Since there are at
most n^α children at time n we have

$$n'(i_0) = n(i_0) \geq \frac{n}{n^\alpha} = n^{1-\alpha} \tag{16}$$

where (i) $n'(i_0)$ is the number of times i_0 has been drawn before time n'; (ii)
$n(i_0)$ is the number of times i_0 has been drawn before time n. Thus we also have

$$n' \geq n'(i_0) \geq n^{1-\alpha}. \tag{17}$$

Consider now any child i already constructed at time n'. Since i_0 was selected
at time n' we must have

$$\sqrt{\frac{f(n')}{n'(i)}} \leq sc_{n'}(i) \leq sc_{n'}(i_0) \leq 1 + \sqrt{\frac{f(n')}{n'(i_0)}}. \tag{18}$$

Rewriting 18 and using 16 leads to

$$\frac{1}{\sqrt{n'(i)}} \leq \frac{1}{\sqrt{n^{1-\alpha}}} + \frac{1}{\sqrt{f(n')}} \leq \frac{2}{\sqrt{\min(f(n'), n^{1-\alpha})}} \tag{19}$$

so that for all children i at time n existing at time n' we have

$$n(i) \geq n'(i) \geq \frac{1}{4}\min\left(f(n^{1-\alpha}), n^{1-\alpha}\right)$$

as announced. Finally, note that a child i existed at time n' if $i \leq (n^{1-\alpha})^\alpha \leq n'^\alpha$,
which leads to the prescribed condition. $\qquad\square$

Corollary 3. *For the exploration function $f(n) = n^e$ with $0 < e < 1$ we obtain*

$$n(i) \geq \frac{1}{4}n^{e(1-\alpha)} \text{ if } i \leq n^{\alpha(1-\alpha)}.$$

Lemma 4 (Decision Nodes are Consistent). *If there is a $\frac{1}{2} > \gamma_{d+\frac{1}{2}} > 0$ such that for any child w of the decision node z we have $Cons(\gamma_{d+\frac{1}{2}}, d+\frac{1}{2})$, then we have $Cons(\gamma_d, d)$ with $\gamma_d = \frac{\gamma_{d+\frac{1}{2}}}{1+7\gamma_{d+\frac{1}{2}}}$ if we define the progressive widening coefficient α_d^D by $\alpha_d^D = \frac{\gamma_{d+\frac{1}{2}}}{1+4\gamma_{d+\frac{1}{2}}}$.*

Proof. Let z be a decision node at depth $d \geq 0$. For simplicity, we note $\alpha_d = \alpha$ and $e_d = e$. Suppose that there is a $\frac{1}{2} > \gamma_{d+\frac{1}{2}} > 0$ such that for all random nodes w at depth $d + \frac{1}{2}$, $Cons(\gamma_{d+\frac{1}{2}}, d + \frac{1}{2})$ is true. To show $Cons(\gamma_d, d)$, we will proceed in two steps: first we establish an upper bound on $\hat{V}(z) - V^*(z)$, and then a lower bound.

Upper Bound. First we obtain an upper bound on $\hat{V}(z)-V^*(z)$. Let $\epsilon < 1-\alpha$ to be fixed later. We partition the children of z in two classes:

- class I : children i such that $n(i) \leq n(z)^{1-\alpha-\epsilon}$;
- class II : other children;

$$\hat{V}(z) - V^*(z) = \sum_{i \text{ in class } I} \frac{n(i)}{n(z)}(V(\hat{i}) - V^*(z)) + \sum_{i \text{ in class } II} \frac{n(i)}{n(z)}(V(\hat{i})-V^*(z))$$

$$\leq \sum_{i \text{ in class } I} \frac{n(i)}{n(z)} + \sum_{i \text{ in class } II} \frac{n(i)}{n(z)}(\hat{V}(i) - V^*(i))$$

$$\leq \frac{n^\alpha \times n^{1-\alpha-\epsilon}}{n} + C_{d+\frac{1}{2}}(n)^{-\gamma_{d+\frac{1}{2}}(1-\alpha-\epsilon)} \text{ e.s. in } n^{-\gamma_{d+\frac{1}{2}}(1-\alpha-\epsilon)}$$

by induction

$$\leq n^{-\epsilon} + C_{d+\frac{1}{2}}n^{-\gamma_{d+\frac{1}{2}}(1-\alpha-\epsilon)}.$$

We now choose $\epsilon = \frac{\gamma_{d+\frac{1}{2}}(1-\alpha)}{1+\gamma_{d+\frac{1}{2}}}$ and obtain

$$\hat{V}(z) - V(z) \leq (1 + C_{d+\frac{1}{2}})n^{-\gamma_{d+\frac{1}{2}}\frac{1-\alpha}{1+\gamma_{d+\frac{1}{2}}}} \text{ e.s. in } n \tag{20}$$

Lower Bound
We assumed that there exists a constant θ such that when we pick a new child for z, it has a value satisfying $V(i) \geq V^*(z) - \Delta$ with probability at least $\min(1, \theta\Delta^p)$.

The induction hypothesis on the next level gives us a fixed coefficient $\gamma_{d+\frac{1}{2}} \in]0; 0.5[$ such that all children w of z verify e.s. in $n(w)$:

$$\left|V^*(w) - \hat{V}(w)\right| \leq C_{d+\frac{1}{2}}n(w)^{-\gamma_{d+\frac{1}{2}}}.$$

The parameters to be fixed on this level are

- the progressive widening coefficient $\alpha := \frac{\gamma_{d+\frac{1}{2}}}{1+4\gamma_{d+\frac{1}{2}}}$;

– the exploration coefficient $e := \frac{1}{1+4\gamma_{d+\frac{1}{2}}} - \frac{1}{\gamma_{d+\frac{1}{2}}}(1-\frac{1}{2p})\alpha = \frac{1}{2p(1+4\gamma_{d+\frac{1}{2}})}$.

To these coefficients we add a parameter ξ which we define by

$$\xi := \frac{1}{1 + e\gamma_{d+\frac{1}{2}}(1-\alpha)} \tag{21}$$

and let $\quad \Delta := \left(\frac{1}{4}n^{\xi e(1-\alpha)}\right)^{-\gamma_{d+\frac{1}{2}}}. \tag{22}$

First step : exponentially surely in n there exists at time $\lceil n^{\xi(1-\alpha)}\rceil$ a child i_0 of z such that

$$V(i_0) \geq V(z) - \Delta \text{ and } i_0 \leq n^{\xi(1-\alpha)\alpha}. \tag{23}$$

At time step $\lceil n^{\xi(1-\alpha)}\rceil$, the number of children of z is a at least $\lfloor n^{\xi(1-\alpha)\alpha}\rfloor$. The (true hidden optimal) values of these children being given randomly and independently, the probability there is not a single child i_0 with $V(i_0) \geq V^*(z) - \Delta$ at time $\lceil n^{\xi(1-\alpha)}\rceil$ is at most

$$p_n := (1 - \theta\Delta^p)^{\lfloor n^{\xi(1-\alpha)\alpha}\rfloor}$$

$$\log p_n \sim_n n^{\xi(1-\alpha)\alpha} \log(1 - \theta\Delta^p)$$

$$\sim_n -n^{\xi(1-\alpha)\alpha}\theta\left(\frac{1}{4}n^{\xi e(1-\alpha)}\right)^{-\gamma_{d+\frac{1}{2}}p}$$

$$\sim_n -4^{\gamma_{d+\frac{1}{2}}p}\theta n^{\xi(1-\alpha)(\alpha - e\gamma_{d+\frac{1}{2}}p)}$$

$$\sim_n -4^{\gamma_{d+\frac{1}{2}}p}\theta n^{\xi(1-\alpha)0.5\alpha}.$$

The exponent of n in this quantity being positive, we deduce that the existence of i_0 is exponentially sure in n.

Second step: e.s. in n, all children selected at a time n' between n^ξ and n have a high score.

Let n' be such that $n^\xi \leq n' \leq n$. Then $n'^{\alpha(1-\alpha)} \geq n^{\xi(1-\alpha)\alpha} \geq i_0$. And, by Corollary 3,

$$n'(i_0) \geq \frac{1}{4}n'^{e(1-\alpha)} \geq \frac{1}{4}n^{\xi e(1-\alpha)}.$$

Hence there exists a $C' > 0$ by the induction hypothesis such that we have, as long as $n^\xi \leq n' \leq n$,

$$\hat{V}(i_0) \geq V^*(i_0) - C'\left(\frac{1}{4}n^{\xi e(1-\alpha)}\right)^{-\gamma_{d+\frac{1}{2}}} \text{ e.s. in } n'$$

$$\geq V^*(z) - (1+C')\Delta \text{ e.s. in } n'.$$

Consider any child i_1 chosen by the algorithm at a time $n' \geq n^\xi$, i.e. the one which has the greatest score at time n'. All values being considered at time n', we have

$$\hat{V}(i_1) + \sqrt{\frac{n'^e}{n'(i_1)}} \geq \hat{V}(i_0) + \sqrt{\frac{n'^e}{n'(i_0)}},$$

hence $\hat{V}(i_1) + \sqrt{\dfrac{n^e}{n'(i_1)}} \geq V^*(z) - (1 + C')\Delta$ e.s. in n'. $\qquad$ (24)

To conclude this part, all we have to do is to show that some property exponentially sure in n' is also exponentially sure in n. This easily follows from the fact that $n' \geq n^\xi$ and that ξ, is bounded below by some constant. One can easily check from the definition of ξ that $\xi \geq \frac{2}{3}$, since $e \leq \frac{1}{2}$.

Third step : lower bound on $\hat{V}(z)$.
Consider a child i_1 selected after n^ξ. By the previous step, exponentially surely in n, this child must either satisfy

$$\sqrt{\frac{n^e}{n(i_1)}} \geq \Delta \qquad (25)$$

$$\text{or } \hat{V}(i_1) \geq V(z) - (2 + C')\Delta. \qquad (26)$$

Under this hypothesis we can split the children of z in three categories:

1. children i_1 visited only before time n^ξ ;
2. children i_1 visited after n^ξ satisfying (25) ;
3. children i_1 visited after n^ξ satisfying (26) .

Let us use this decomposition to lower bound the sum

$$\hat{V}(z) - V^*(z) = \sum_{i=1\cdots\lfloor n^\alpha\rfloor} \frac{n(i)}{n}(\hat{V}(i) - V^*(z)).$$

For the children in the first category, we have

$$\left| \sum_{i_1 \text{in cat.1}} \frac{n(i_1)}{n}(\hat{V}(i_1) - V^*(z)) \right| \leq \frac{\sum_{i_1 \text{in cat.1}} n(i_1)}{n} \leq \frac{n^\xi}{n}.$$

For children in the second category, since there are at most n^α of these children, we have

$$\left| \sum_{i_1 \text{in cat.2}} \frac{n(i_1)}{n}(\hat{V}(i_1) - V^*(z)) \right| \leq \frac{\sum_{i_1 \text{in cat.2}} n(i_1)}{n} \leq \frac{n^\alpha}{n}\frac{n^e}{\Delta^2} = \frac{n^{\alpha+e-1}}{\Delta^2}.$$

Finally, using (26) for the third category of children, we see that

$$\hat{V}(z) - V^*(z) \geq -(2 + C')\Delta(1 - n^{\xi-1}) - n^{\xi-1} - \frac{n^{\alpha+e-1}}{\Delta^2}.$$

Now we compare the three terms

$$\Delta, n^{\xi-1} \text{ and } \frac{n^{\alpha+e-1}}{\Delta^2}. \tag{27}$$

By (21) we have $\xi - 1 = -\xi e \gamma_{d+\frac{1}{2}}(1-\alpha)$, thus by (22), $n^{\xi-1} = 4^{-\gamma_{d+\frac{1}{2}}}\Delta \leq \Delta$.

This implies that the term $n^{\xi-1}$ in the three terms (Eq. 27) is $O(\Delta)$. We now compare the two other terms; from the definition of Δ, we see that we must compare $n^{\alpha+e-1}$ and $\Delta^3 = 4^{3\gamma_{d+\frac{1}{2}}}n^{-3\xi e(1-\alpha)\gamma_{d+\frac{1}{2}}}$. Using the definitions of ξ, e and α, one can check that:

$$1 - e - \alpha \geq \frac{3\gamma_{d+\frac{1}{2}} + \frac{1}{2}}{1 + 4\gamma_{d+\frac{1}{2}}} \geq \frac{1}{2}$$

and, using $\xi \leq 1$, $(1 - \alpha) \leq 1$, $e\gamma_{d+\frac{1}{2}} = \frac{\gamma_{d+\frac{1}{2}}}{2(1+4\gamma_{d+\frac{1}{2}})} \leq \frac{1}{8}$,

$$3\xi e(1-\alpha)\gamma_{d+\frac{1}{2}} \leq \frac{3}{8}$$

$$n^{e+\alpha-1} \leq n^{-3\xi e(1-\alpha)\gamma_{d+\frac{1}{2}}} = 4^{-3\gamma_{d+\frac{1}{2}}}\Delta^3 \leq \Delta^3$$

so that $\hat{V}(z) - V(z) \geq -(5+C')\Delta$. Finally, one can check that $\xi e(1-a)\gamma_{d+\frac{1}{2}} \geq \frac{\gamma_{d+\frac{1}{2}}}{1+7\gamma_{d+\frac{1}{2}}}$ so that $\hat{V}(z) - V^*(z) \geq -(5+C')4^{\gamma_{d+\frac{1}{2}}}n^{\frac{\gamma_{d+\frac{1}{2}}}{1+7\gamma_{d+\frac{1}{2}}}}$ which can now be written $\hat{V}(z)-V^*(z) \geq -Cn^{-\gamma}$ with $C := (5+C')4^{\gamma_{d+\frac{1}{2}}}$ and $\gamma = \frac{\gamma_{d+\frac{1}{2}}}{1+7\gamma_{d+\frac{1}{2}}}$. $\square$

7   Base Step, Initialization and Conclusion of the Proof

Let w be a random node of depth $d_{\max} - \frac{1}{2}$. Its children are leaf nodes, and all have a fixed reward in $[0; 1]$. These children form a ensemble of independent and identically distributed variables, all following the random distribution associated with w, of mean $V^*(w)$. Hoeffding's inequality gives, for $t > 0$,

$$\mathbb{P}\left(|\frac{1}{n}\sum_{z_i child of w} V^*(z_i) - V^*(w)| \geq t\right) \leq 2\exp(-2t^2 n).$$

Setting the exploration coefficient $\alpha_{d_{\max}-\frac{1}{2}}$ to 1 (since there is no point in selecting again children with a constant reward) and t to $n^{-\frac{1}{3}}$, we obtain

$$\mathbb{P}\left(|V\hat{(}w) - V^*(w)| \geq n^{-\frac{1}{3}}\right) \leq 2\exp(-2n^{\frac{1}{3}})$$

so that $|V\hat{(}w) - V^*(w)| \leq n^{-\frac{1}{3}}$ is exponentially sure in n, i.e. $Cons(\frac{1}{3}, d_{\max} - \frac{1}{2})$ holds. Of course one can consider a coefficient different from $\frac{1}{3}$ for t, as long as it is less than $\frac{1}{2}$ – we just aim so as to simplify the definition of coefficients. This gives a singular value of $\alpha^R_{d_{\max}-\frac{1}{2}} = 1$ and an initialization of the convergence rate as $\gamma^R_{d_{\max}-\frac{1}{2}} = \frac{1}{3}$. It is now elementary to check this value of $\frac{1}{3}$ for γ at depth $d_{\max} - \frac{1}{2}$, together with recursive definitions of coefficients derived in Lemmas 2 and 4, yield the values given on Table 1. This concludes the proof of Theorem 5.

8   Experimental Validation

In this section, we show some experimental results, by implementing PUCT on two tests problems. We used fixed parameters α and e, quickly tuned by hand. We added a custom default policy, as seen in [6] and [8], that is computed offline using Direct Policy Search (DPS), once per problem instance. We also gave heavier weights to the decisions with high average value when computing the empirical value of a state, as it showed increased performances in practice. There are many ways to finely tune PUCT that we did not explore. Our goal was simply to check that our PUCT has a satisfying behaviour, to verify our theoretical results. We acknowledge that depending on implementation subtleties, results can vary. Our source code is available upon request.

Cart Pole. We used the well known benchmark of cart pole, and more precisely the version presented in [17]. As our code uses time budget, and not a limit in the number of iterations, we only approximated their limit of 200 roll outs (on our machine, 0.001 second per action. We took HOLOP as a baseline, that yields an average reward of -47.45[17]. Our results are shown in Table 2. Though cart pole is not as challenging as real world applications, these results are encouraging and supporting our theoretical results of consistency.

Unit Commitment. We used a unit commitment problem, inspired by ongoing work with an industrial partner. The agent owns 2 water reservoirs and 5 power plants. Each reservoir is a free but limited source of energy. Each power plant has a fixed capacity, has a fixed cost to be turned on, as well as quadratic running costs that change over time. The time horizon was fixed to 6 time steps. At each time step, the agent decides how to produce energy in order to satisfy a varying demand, and the water reservoirs receive a random inflow. Failure to satisfy the demand incurs a prohibitive cost. This problem is challenging for many reasons, including: the action space is non convex, the objective function is non linear and discontinuous, there are binary and continuous variables, and finally, the action space can be subject to operational constraints that make a discretization by hand very tedious. The purpose of PUCT in this application is not to solve all of it, but rather to improve existing solvers. This is an especially promising method, with the many powerful heuristics available for this problem. The results are shown in Table 2. PUCT manages to reliably improve the actions suggested by DPS, and its performances increase with the time budget it is given.

Table 2. Left: Cart Pole results; episodes are 200 time steps long. Right: Unit Commitment results, with 2 stocks, 5 plants, and 6 time steps.

Budget (s)	0.001	0.004
HOLOP	-47.45 ± .	.
DPS	-838.7 ± 78.0	-511.0 ± 100.0
PUCT+DPS	-13.84 ± 0.80	-11.11 ± 0.95

Budget (s)	0.04	0.16	0.64
DPS	-8.02 ± 0.98	-7.06 ± 0.024	-6.98 ± 0.03
PUCT+DPS	-7.23 ± 0.45	-6.69± 0.03	6.57 ± 0.02

9   Conclusion

[13] have shown the consistency of the UCT approach for finite Markov Decision Processes. We have shown the consistency of our modified version, with polynomial exploration and double progressive widening, for a more general case. [7] have shown that the classical UCT is not consistent in this case and already proposed double progressive widening; we here give a proof of the consistency of this approach, when we use polynomial exploration; [7] was using logarithmic exploration.

Some extensions of our work are straightforward. We considered trees, but the extension to MDP with possibly two distinct paths leading to the same node is straightforward. Also, we assumed, only for simplifying notation, that the probability that a random node leads twice to the same decision node (when drawn independently with the probability distribution of the random node) is zero, but the extension is possible. On the other hand, we point out two deeper limitations of our work: (i) We do not know if similar results can be derived without switching to polynomial exploration. (ii) The general case of a possibly cyclic MDP with unbounded horizon is not covered by our result.

We have shown consistency in the sense that Bellman values are properly estimated. This does not explain which decision should be actually made when PUCT has been performed for generating episodes and estimating V values. Our result implies that choosing the action by empirical distribution of play (i.e. randomly draw a decision with probability equal to the frequency at which it was simulated during episodes; see discussion in [4]) is asymptotically consistent. Also, choosing the most simulated child is consistent (this is a classical method in UCT), as well as selecting the child with best $\hat{V}$ among child nodes of the root of class II; our results do not show the superiority of one or another of these recommendation methodologies.

Our experimental results on the classical Cart pole problem show that PUCT outperforms HOLOP; PUCT also outperformed a specialized DPS on a unit commitment problem. This last empirical result is especially interesting because unit commitment problems are, in practice, highly non Markovian. And, even though we worked in the framework of MDP to relate to its abundant literature, our algorithm does not actually need the random process to be Markovian, as the history is naturally embedded in the tree structure. Hence, PUCT could be a way to approach difficult and more general non Markovian continuous sequential decision making problems.

References

1. Auer, P., Ortner, R., Szepesvári, C.: Improved rates for the stochastic continuum-armed bandit problem. In: Bshouty, N.H., Gentile, C. (eds.) COLT 2007. LNCS (LNAI), vol. 4539, pp. 454–468. Springer, Heidelberg (2007)
2. Bellman, R.: Dynamic Programming. Princeton Univ. Press (1957)
3. Bertsimas, D., Litvinov, E., Sun, X.A., Zhao, J., Zheng, T.: Adaptive robust optimization for the security constrained unit commitment problem 28(1), 52–63 (2013)
4. Bourki, A., Coulm, M., Rolet, P., Teytaud, O., Vayssière, P.: Parameter Tuning by Simple Regret Algorithms and Multiple Simultaneous Hypothesis Testing. In: ICINCO 2010, Funchal, Madeira, Portugal, p. 10 (2010)
5. Bubeck, S., Munos, R., Stoltz, G., Szepesvári, C.: Online optimization in x-armed bandits. In: Koller, D., Schuurmans, D., Bengio, Y., Bottou, L. (eds.) NIPS, pp. 201–208. Curran Associates, Inc. (2008)
6. Buffet, O., Lee, C., Lin, W., Teytaud, O.: Optimistic heuristics for minesweeper. In: International Computer Symposium, p. 9 (2012)
7. Couëtoux, A., Hoock, J.-B., Sokolovska, N., Teytaud, O., Bonnard, N.: Continuous Upper Confidence Trees. In: Coello Coello, C.A. (ed.) LION 2011. LNCS, vol. 6683, pp. 433–445. Springer, Heidelberg (2011)
8. Couetoux, A., Teytaud, O., Doghmen, H.: Learning a move-generator for upper confidence trees. In: Chang, R.-S., Jain, L.C., Peng, S.-L. (eds.) Advances in Intelligent Systems and Applications. SIST, vol. 20, pp. 209–218. Springer, Heidelberg (2013)
9. Coulom, R.: Efficient Selectivity and Backup Operators in Monte-Carlo Tree Search. In: van den Herik, H.J., Ciancarini, P., Donkers, H.H.L.M(J.) (eds.) CG 2006. LNCS, vol. 4630, pp. 72–83. Springer, Heidelberg (2007)
10. Coulom, R.: Computing elo ratings of move patterns in the game of go. In: Computer Games Workshop, Amsterdam, The Netherlands (2007)
11. Gerevini, A., Howe, A.E., Cesta, A., Refanidis, I. (eds.): Proceedings of the 19th International Conference on Automated Planning and Scheduling, ICAPS 2009, Thessaloniki, Greece, September 19-23. AAAI (2009)
12. Kleinberg, R.D.: Nearly tight bounds for the continuum-armed bandit problem. In: NIPS (2004)
13. Kocsis, L., Szepesvári, C.: Bandit based Monte-Carlo planning. In: Fürnkranz, J., Scheffer, T., Spiliopoulou, M. (eds.) ECML 2006. LNCS (LNAI), vol. 4212, pp. 282–293. Springer, Heidelberg (2006)
14. Lee, C.-S., Wang, M.-H., Chaslot, G., Hoock, J.-B., Rimmel, A., Teytaud, O., Tsai, S.-R., Hsu, S.-C., Hong, T.-P.: The Computational Intelligence of MoGo Revealed in Taiwan's Computer Go Tournaments. IEEE Transactions on Computational Intelligence and AI in Games (2009)
15. Madani, O., Hanks, S., Condon, A.: On the undecidability of probabilistic planning and related stochastic optimization problems. Artif. Intell. 147(1-2), 5–34 (2003)
16. Mansley, C.R., Weinstein, A., Littman, M.L.: Sample-based planning for continuous action markov decision processes. In: Bacchus, F., Domshlak, C., Edelkamp, S., Helmert, M. (eds.) ICAPS. AAAI (2011)
17. Weinstein, A., Littman, M.L.: Bandit-based planning and learning in continuous-action markov decision processes. In: McCluskey, L., Williams, B., Silva, J.R., Bonet, B. (eds.) ICAPS. AAAI (2012)

A Lipschitz Exploration-Exploitation Scheme for Bayesian Optimization

Ali Jalali, Javad Azimi, Xiaoli Fern, and Ruofei Zhang

Turn Inc., Microsoft Inc., Oregon State University and Microsoft Inc.
ajalali@turn.com, xfern@eecs.oregonstate.edu,
{jaazimi,bzhang}@microsoft.com

Abstract. The problem of optimizing unknown costly-to-evaluate functions has been studied extensively in the context of Bayesian optimization. Algorithms in this field aim to find the optimizer of the function by requesting only a few function evaluations at carefully selected locations. An ideal algorithm should maintain a perfect balance between exploration (probing unexplored areas) and exploitation (focusing on promising areas) within the given evaluation budget. In this paper, we assume the unknown function is Lipschitz continuous. Leveraging the Lipschitz property, we propose an algorithm with a distinct exploration phase followed by an exploitation phase. The exploration phase aims to select samples that shrink the search space as much as possible, while the exploitation phase focuses on the reduced search space and selects samples closest to the optimizer. We empirically show that the proposed algorithm significantly outperforms the baseline algorithms.

Keywords: Bayesian Optimization, Exploration, Exploitation, Lipschitz Continuity.

1 Introduction

In many applications, we would like to optimize an unknown function $f(\cdot)$ that is costly to evaluate over a compact input space. Classic optimization methods, such as gradient descent, cannot be applied to this type of problems since they need to evaluate the function frequently. In contrast, Bayesian Optimization (BO) [1,2] algorithms try to solve this problem with a small number of function evaluations. Bayesian optimization algorithms, generally, have two key components: 1) A posterior model to predict the output value of the function at any arbitrary input point, and 2) A selection criterion to determine which point to be evaluated next.

The first step of a BO algorithm is to learn a posterior probabilistic model over unobserved points of the function. Gaussian processes (GP) [3] have been used in the literature of Bayesian optimization as the probabilistic posterior model. GP models the function output for any unobserved point in the input space as a normal random variable, whose mean and variance depend on the location of the point in relation to a set of given observed samples. Based on

H. Blockeel et al. (Eds.): ECML PKDD 2013, Part I, LNAI 8188, pp. 210–224, 2013.
© Springer-Verlag Berlin Heidelberg 2013

the learned posterior model, a selection criterion is then used to choose the next sample to be evaluated. A number of selection criteria have been proposed in the literature of Bayesian optimization. They typically work by selecting an example that optimizes some objective function designed to balance between exploring unobserved area and exploiting areas that are promising based on existing observations. Maximum probability of improvement [4,5] and maximum expected improvement (EI) [6] are two successful examples.

In this paper, we focus on the design of the selection criterion for Bayesian optimization. In particular, we study BO in a sequential setting [1,7], where the samples are chosen sequentially and a selection is made only after the function evaluations of the previous samples are revealed. We make a mild assumption that the unknown function is Lipschitz-continuous. Leveraging the Lipschitz property, we design a selection algorithm that operates in two distinct phases: the exploration phase and the exploitation phase. In general, in the context of Bayesian optimization [1] and bandit problems [8], the exploration phase selects sample from unexplored area while the exploitation focuses on promising area. In this paper, we introduce a new interpretation of exploration and exploitation.

The exploration phase of the proposed algorithm, at each step, selects a sample that eliminates the largest possible portion of the input space while guaranteeing, with high probability, that the eliminated part does not include the maximizer of the function. Hence, the exploration stage of the algorithm tries to shrink the search space of the function as much as possible. In contrast, the exploitation phase of our algorithm selects the point which is believed to be the closest sample to the optimal point with high probability.

Experimental results over 8 real and synthetic benchmarks indicate that the proposed approach is able to outperform the Expected Improvement (EI) criterion, one of the current state-of-the-art BO selection methods. In particular, we show that our algorithm is better than EI both in terms of the mean and variance of the performance. We also investigate whether combining our exploration stage with EI can boost the performance of EI. However, the results were negative. Sometimes it helps and sometimes it hurts and on average we observe little to no improvement to EI. This is possibly because our exploration method actively aims to eliminate regions from the input space and the EI criterion does not take that into consideration when selecting samples.

The remainder of the paper is organized as follows. In Section 2, we motivate the use of exploration-exploitation Bayesian optimization by analyzing the behavior of EI. Section 3 introduces our algorithm and provides insights into both theoretical and practical aspects of the algorithm. Experimental evaluation of our algorithm is shown in Section 4. Finally, the paper is concluded in Section 5.

2   Motivating Observation

In this section, we motivate our approach by revealing a key observation about the well known Expected Improvement (EI) algorithm [6]. The original EI is defined as

$$EI(x) = \mathbb{E}\left[(f(x) - y_{\max})\,\mathbb{I}_{\{f(x) - y_{\max} > 0\}}\right], \tag{1}$$

where $\mathbb{I}_{\{\cdot\}}$ is the indicator function. Hence, it measures the expected improvement of the choice of x over the current maximum function evaluations $y_{\max}$ over observed samples. Using Gaussian Process (GP) [3] as the posterior model of the unknown function, the EI objective can be represented by

$$EI(x|\mathcal{O}) = (\mu_{x|\mathcal{O}} - y_{\max})\Phi\left(\frac{\mu_{x|\mathcal{O}} - y_{\max}}{\sigma_{x|\mathcal{O}}}\right) + \sigma_{x|\mathcal{O}}\,\phi\left(\frac{\mu_{x|\mathcal{O}} - y_{\max}}{\sigma_{x|\mathcal{O}}}\right), \qquad (2)$$

where, $\mu_{x|\mathcal{O}}$ and $\sigma_{x|\mathcal{O}}$ are the mean and standard deviation associated with the point x by GP, and, $\Phi(\cdot)$ and $\phi(\cdot)$ are standard Gaussian CDF and PDF, respectively. Here, $\mathcal{O} = \{(x_i, f(x_i))\}_{i=1}^n$ is the set of n observed samples $x_{\mathcal{O}}$ with their function evaluations $f(x_{\mathcal{O}})$ and define $y_{\max} = \max_{x_i \in x_{\mathcal{O}}} f(x_i)$. Further, the means and variances are defined as follows:

$$\mu_{x|\mathcal{O}} = k(x, x_{\mathcal{O}})\,k(x_{\mathcal{O}}, x_{\mathcal{O}})^{-1}\,f(x_{\mathcal{O}})$$
$$\sigma_{x|\mathcal{O}}^2 = k(x, x) - k(x, x_{\mathcal{O}})\,k(x_{\mathcal{O}}, x_{\mathcal{O}})^{-1}\,k(x_{\mathcal{O}}, x),$$

where $k(\cdot, \cdot)$ is some kernel function. In this paper, we consider Gaussian kernel $k(x_1, x_2) = \exp(-\frac{1}{\ell}\|x_1 - x_2\|_2^2)$.

EI has been widely used and studied; however, there has been always a concern about balancing the exploration and exploitation of EI. The main reason for this concern is that even though the asymptotic convergence of EI is guaranteed under certain conditions [9], EI tries to exploit the information and potentially can request a lot of samples if it hits a local optimum region, while we have a limited number of experiments. There has been some attempts in the literature to address this concern with varying degrees of success, which we briefly discuss here.

(a) Considering the original definition of EI, researchers have proposed to replace $y_{\max}$ with a smaller value to make EI more exploitative and with a larger value to make it more explorative. In particular, [10] suggested $y_{\max} + \xi$ and [11] suggested $(1 + \xi)y_{\max}$ to replace $y_{\max}$. However, this approach has not seen much empirical success. [10] showed that starting with large values of ξ (to be explorative in the beginning) and cooling it down (to make it more and more exploitative) makes little or no difference in the performance of EI.

(b) On a separate line of work, [12] proposed to consider a surrogate function

$$EI_\xi(x) = \mathbb{E}\left[(f(x) - y_{\max})^\xi \mathbb{I}_{\{f(x) - y_{\max} > 0\}}\right].$$

For $\xi = 1$, this objective tries to improve over $y_{\max}$ (exploiting mode) and if we decrease ξ it starts to explore uncertain areas (exploration mode). This method is very sensitive to small changes in ξ and except for very specific setup like the one used in [13], there is no systematic way to choose ξ. This makes it nearly impossible to use this method.

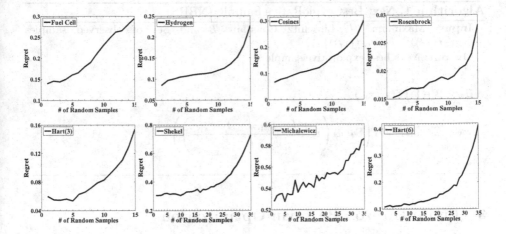

Fig. 1. Plot of regret versus the number of random exploration for EI algorithm. For a fixed budget n_b, we run a number of experiments as follows: first we consider the case where there are 1 random samples followed by $n_b - 1$ EI samples, next we consider the case where there are 2 random samples followed by $n_b - 2$ EI samples and so on. For 2D and 3D functions, we let $n_b = 15$ and for high-dimensional functions, we let $n_b = 35$. This result shows that the best EI performance is when we do not do random exploration.

(c) The third proposal is to have a "random" exploration phase proceeding EI. In this approach, we take a number of random samples before switching to EI. We analyzed this method in Fig. 1. For a fixed budget n_b, we run n_b experiments as follows: first we consider the case where there is 1 random sample followed by $n_b - 1$ samples selected by the EI criterion, next we consider the case where there are 2 random samples followed by $n_b - 2$ EI samples and so on. The purpose of this investigation is to understand whether exploring with random samples prior to selecting with EI can improve the performance of EI, and if so how much exploring is necessary. We run this experiments on a number of different functions introduced in Section 4. These experiments reveal that "random" exploration never helps EI, since the regret monotonically increases as we increase the number of random samples from 1 to n_b. One possible explanation for this behavior is that the values of the function are highly correlated and hence, uniform sampling does not efficiently represent the skewness of the data points.

Based on the existing literature as well as our empirical investigation of EI discussed above, we would like to know whether or not it is possible to design an algorithm that operates in two naturally defined phases of exploration and exploitation and achieves consistently better performance than EI. We devote the next section to answer this question and introduce our proposed algorithm.

Algorithm 1. Next Best exploRative Sample (NBRS)

Input: Maximum M, Lipschitz Constant L and Set of observed samples $\{(x_1, f(x_1)), \ldots, (x_t, f(x_t))\}$
Output: Next best explorative sample x

$$\mathbb{D}_t = \mathbb{D} - \bigcup_{i=1}^{t} \mathbb{S}(x_i, r_{x_i})$$

$$x \longleftarrow \underset{x \in \mathbb{D}_t}{\operatorname{argmax}} \ \mathbf{Vol}\left(\mathbb{D}_t \cap \mathbb{S}\left(x, \frac{|M - \mu_{x|\mathcal{O}}| - 1.5\sigma_{x|\mathcal{O}}}{L}\right)\right)$$

---

3 Finite Horizon Bayesian Optimization

Not being able to balance the exploration-exploitation, EI might have poor performance especially when the query budget is small. In this section, we propose a two-phase exploration/exploitation algorithm that outperforms EI with its smart exploration and exploitation.

3.1 Exploration

Generally, a good exploration algorithm should be able to shrink the search space, so that we are left with a small region to focus on during the exploit stage. Let $\mathbb{D} = \bigotimes[a_i, b_i] \in \mathbb{R}^d$ be the Cartesian product of intervals $[a_i, b_i]$ for some $a_i < b_i$ and $i \in \{1, 2, \ldots, d\}$. Suppose the unknown function $f : \mathbb{D} \mapsto [m, M]$ (with $f(x^*) = M$) is a Lipschitz function over $\mathbb{D}$ with constant L, that is for all $x_1, x_2 \in \mathbb{D}$, we have

$$|f(x_1) - f(x_2)| \leq L\|x_1 - x_2\|_2.$$

Notice that if the function is not Lipschitz, then there is no hope that we can find the global optimum of $f(\cdot)$ even with infinitely countable evaluations. Thus, the Lipschitz continuity assumption is not a strong assumption. Moreover, functions with larger L are harder to optimize since they change more abruptly over the space.

For any point $x \in \mathbb{D}$, let $r_x = \frac{M - f(x)}{L}$ be the associated radius to the point x. By Lipschitz continuity assumption, we know that $x^* \notin \mathbb{S}(x, r_x)$, where, $\mathbb{S}(x, r_x)$ is the set of all points inside the sphere (or circle) with radius r_x centered at x (and single point x if $r_x \leq 0$); otherwise, the Lipschitz assumption is violated. This means if we have a sample at point x, then we do not need any more samples inside $\mathbb{S}(x, r_x)$.

The expected value of r_x satisfies $\mathbb{E}[r_x] = \frac{|M - \mu_x|}{L}$. Since $f(x)$ is a normal random variable $\mathcal{N}(\mu_x, \sigma_x^2)$, using Hoeffding inequality for all $\epsilon > 0$, we have

$$\mathbb{P}\left[r_x < \frac{|M - \mu_x|}{L} - \epsilon\right] \leq \exp\left(-\frac{2\epsilon^2 L^2}{\sigma_x^2}\right).$$

Replacing ϵ with $1.5\frac{\sigma_x}{L}$, the above inequality entails that with high probability (99%), $r_x \geq \frac{|M-\mu_x|-1.5\sigma_x}{L}$. Hence, a "good" algorithm for exploration should try to find x that maximizes the lower bound on r_x. This choice of x will remove a large volume of points from the search space. Note, however, if x is close to the boundaries of $\mathbb{D}$, then it might be the case that most of the volume of the sphere lies outside $\mathbb{D}$. Also, the sphere associated with x might have significant overlap with spheres of other points that are already selected. To fix this issue, we pick the point whose sphere has the largest intersection with unexplored search space in terms of its volume. The pseudo code of this method is described in Algorithm 1, which we refer to as the Next Best exploRative Sample (NBRS) algorithm. NBRS achieves the optimal exploration in the sense that it maximizes the *expected explored* volume.

The value of $|M - \mu_x| - 1.5\sigma_x$ might be negative, especially for large values of σ_x. This artifact happens at points x that are "far" from previously observed samples. To prevent/minimize this, we need to make sure that the observed samples affect the mean and variance of all points in the space. For example, if we use the Gaussian kernel $k(x_1, x_2) = \exp(-\frac{1}{\ell_r}\|x_1 - x_2\|_2^2)$ for exploration, then we need to choose ℓ_r large enough to make sure each observed sample affects all the points in the space, e.g., $\ell_r \geq \sum_{i=1}^d (b_i - a_i)^2$. If we pick small ℓ_r, then the exploration algorithm starts exploring around the previous samples and extend the explored area gradually to reach to the other side of the search space. This strategy is not optimal if we have limited samples for exploration.

To implement NBRS, we need to maximize the volume

$$g(x) = \mathbf{Vol}\left(\mathbb{D}_t \cap \mathbb{S}\left(x, \frac{|M - \mu_{x|\mathcal{O}}| - 1.5\sigma_{x|\mathcal{O}}}{L}\right)\right)$$

where $\mathbb{D}_t$ represents the current unexplored input space. To evaluate $g(x)$, we take a large number of points N inside the sphere $\mathbb{S}(x, \frac{|M-\mu_{x|\mathcal{O}}|-1.5\sigma_{x|\mathcal{O}}}{L})$ uniformly at random. Then, for each point, we check if it crosses the borders $[a_i, b_i]$ or falls into the spheres of previously observed samples. If not, we count that point as a newly explored point. Finally, if there are n newly explored points, then we set $g(x) \approx \frac{n}{N}\left(\frac{|M-\mu_{x|\mathcal{O}}|-1.5\sigma_{x|\mathcal{O}}}{L}\right)^d$.

To optimize $g(x)$, one can use deterministic and derivative free optimizers like DIRECT [14]. The problem is that DIRECT only optimizes Lipschitz continuous functions; however, $g(x)$ is not necessarily Lipschitz continuous. In our implementation, we take a large number of points inside $\mathbb{D}_t$ and evaluate $g(\cdot)$ at those points and pick the maximum. This method might be slower than DIRECT, but avoids inaccurate results of DIRECT especially when $\mathbb{D}_t$ describes a small region.

3.2   Exploitation

In the exploitation phase of the algorithm, we would like to use the information gained in the exploration phase to find the optimal point of $f(\cdot)$. Suppose we

---

Algorithm 2. Next Best exploItive Sample (NBIS)

Input: Maximum M, Lipschitz Constant L and Set of observed samples $\{(x_1, f(x_1)), \ldots, (x_q, f(x_q))\}$
Output: Next best exploitive sample x

$$\mathbb{D}_q = \mathbb{D} - \bigcup_{i=1}^{q} \mathbb{S}(x_i, r_{x_i})$$

$$x \longleftarrow \underset{x \in \mathbb{D}_q}{\operatorname{argmin}} \; \mathbf{Vol}\left(\mathbb{S}\left(x, \frac{\left|M - \mu_{x|\mathcal{O}}\right| + 1.5\sigma_{x|\mathcal{O}}}{L}\right)\right)$$

---

have explored the search space with t samples and we want to find $x^* \in \mathbb{D}_t$. In order to exploit, we would like to find points x whose sphere is small. The reason is that if $r_x = \frac{M - f(x)}{L} \leq \gamma$ is small enough, then by *local* strong convexity of $f(\cdot)$ around x^*, for some constant κ we have

$$\frac{\kappa}{2}\|x - x^*\|_2^2 \leq M - f(x) \leq L\gamma.$$

Following the argument in Section 3.1, we estimate r_x by its mean $\mathbb{E}[r_x] = \frac{|M - \mu_x|}{L}$. By Hoeffding inequality, for all $\epsilon > 0$, we have

$$\mathbb{P}\left[r_x > \frac{|M - \mu_x|}{L} + \epsilon\right] \leq \exp\left(-\frac{2\epsilon^2 L^2}{\sigma_x^2}\right).$$

Similarly, replacing ϵ with $1.5\frac{\sigma_x}{L}$, the above inequality entails that with high probability (99%), $r_x \leq \frac{|M - \mu_x| + 1.5\sigma_x}{L}$. Hence, a "good" algorithm for exploitation should try to find the point x that minimizes the upper bound on r_x. This choice of x introduces the expected closest point to x^*. We present the pseudo code of this method in Algorithm 2.

The optimization in Algorithm 2 is nothing but minimizing

$$h(x) = \frac{|M - \mu_{x|\mathcal{O}}| + 1.5\sigma_{x|\mathcal{O}}}{L}.$$

To optimize $h(x)$, again we take a large number of points in $\mathbb{D}_q$ (the current unexplored space) uniformly at random and evaluate $h(\cdot)$ on those and pick the minimum.

3.3 Exploration-Exploitation Trade-Off

The main algorithm consists of an initial exploration phase followed by exploitation. Notice that we are using GP as an estimate of the unknown function and our method, like EI, highly relies on the quality of this estimation. On a high level, if the function is very complex, i.e., has large Lipschitz constant L, then

Table 1. Benchmark Functions

Cosines(2)	$1-(u^2+v^2-0.3\cos(3\pi u)-0.3\cos(3\pi v))$ $u=1.6x-0.5, v=1.6y-0.5$	Rosenbrock(2)	$10-100(y-x^2)^2-(1-x)^2$
Hartman(3,6)	$\sum_{i=1}^{4}\Omega_i\exp\left(-\sum_{j=1}^{d}A_{ij}(x_j-P_{ij})^2\right)$ $\Omega_{1\times4}, A_{4\times d}, P_{4\times d}$ are constants	Michalewicz(5)	$-\sum_{i=1}^{5}\sin(x_i)\sin\left(\frac{i\,x_i^2}{\pi}\right)^{20}$
Shekel(4)	$\sum_{i=1}^{10}\frac{1}{\omega_i+\Sigma_{j=1}^{4}(x_j-B_{ji})^2}$	$\omega_{1\times10}, B_{4a\times10}$ are constants	

we need more exploration to fit better with GP. Small values of L correspond to flatter functions that are easier to optimize. Thus, in general, we expect the number of exploration steps to scale up with L. As a rule of thumb, functions we normally deal with satisfy $2 < L < 20$, for which we spend 20% of our budget in exploration and the rest in exploitation.

We use different kernel widths for the exploration and exploitation phases. In the case of exploration for complex functions, if we have enough budget (and hence, enough explorative samples), the kernel width can be set to a small value to fit a better local GP model. However, if we do not have enough budget, we need to take the kernel width to be large. In the case of exploitation, we pick the kernel width under which EI achieves its best performance.

Note that the choice of M and L plays a crucial role in this algorithm. If we pick L larger than the true Lipschitz function, then the radius of our spheres shrink and hence we might need more budget to achieve a certain performance. Choosing L smaller than the true Lipschitz is dangerous since it makes the spheres large and increases the chance of including the optimal point in a sphere and hence removing it. Thus, it is better to choose L slightly larger than our estimate of the true Lipschitz to be on the safe side.

The method is less sensitive to the choice of M, since the derivative of the radius with respect to M is proportional to $\frac{1}{L}$. Thus, as long as we do not over estimate M significantly, the $\frac{1}{L}$ factor prevents the spheres to become very large (and include/remove the optimal point). Small values of M, make the spheres smaller and hence, if we underestimate M, we would need more budget to achieve certain performance. However, if M is significantly (proportional to L) smaller than the true maximum of the function, then the algorithm will look for the point that achieves M and hence will perform poorly.

4 Experimental Results

In this section, we compare our algorithm with EI under different scenarios for different functions. We consider six well-known synthetic benchmark functions:

(1,2) Cosines [15] and Rosenbrock [16] over $[0,1]^2$
(3,4) Hartman(3,6) [17] over $[0,1]^{3,6}$
 (5) Shekel [17] over $[3,6]^4$
 (6) Michalewicz [18] over $[0,\pi]^5$

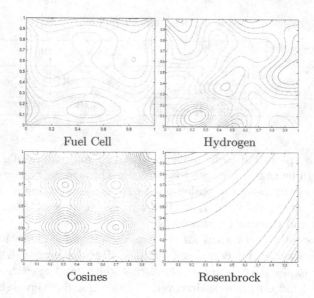

Fuel Cell   Hydrogen

Cosines   Rosenbrock

Fig. 2. The contour plots for the four 2−dimension proposed benchmarks

The mathematical expression of these functions are shown in Table 1. Moreover, we use two benchmarks derived from real-world applications:

(1) Hydrogen [19] over $[0,1]^2$
(2) Fuel Cell [20] over $[0,1]^2$

The contour plots of these two benchmarks along with the Cosines and Rosenbrock benchmarks are shown in Fig 2. The Fuel Cell benchmark is based on optimizing electricity output of microbial fuel cell by modifying some nano structure properties of the anodes. In particular, the inputs that we try to adjust are the average area and average circularity of the nano tube and the output that we try to maximize is the power output of the fuel cell. We fit a regression model on a set of observed samples to simulate the underlying function $f(\cdot)$ for evaluation. The Hydrogen benchmark is based on maximizing the Hydrogen production of a particular bacteria by varying the PH and Nitrogen levels of its growth medium. A GP is fitted to a set of observed samples to simulate the underlying function $f(\cdot)$. We consider a Lipschitz constant $L \approx 3$ for all of the benchmarks, except for Cosines and Michalewicz with $L \approx 6$ and Rosenbrock with $L \approx 45$. For the sake of comparison, we consider the normalized versions of all these functions and hence $M = 1$ in all cases. As mentioned previously, we spend 20% of the budget on exploration and 80% on exploitation.

4.1   Comparison to EI

In the first set of experiments, we would like to compare our algorithm with the best possible performance of EI. For each benchmark, we search over different

Table 2. Comparison of the best results of EI, NBRS+EI and NBRS+NBIS. This result shows that our algorithm outperforms the other two counterparts significantly in most cases both in terms of the mean and variance of the performance.

	EI	EI$_M$	NBRS+EI	NBRS+NBIS
Cosines	.0736 ± .016	.2938 ± .020	.1057 ± .029	**.0270 ± .009**
Fuel Cell	.1366 ± .006	.2232 ± .007	.1357 ± .004	**.0965 ± .004**
Hydrogen	.0902 ± .004	.1689 ± .012	.1149 ± .004	**.0475 ± .006**
Rosen	.0134 ± .001	.0153 ± .003	.0163 ± .001	**.0034 ± .000**
Hart(3)	.0618 ± .006	.0837 ± .001	.0450 ± .003	**.0384 ± .003**
Shekel	.3102 ± .017	.4104 ± .021	**.3011 ± .018**	.3240 ± .030
Michal	.5173 ± .010	.5210 ± .008	.5011 ± .010	**.4554 ± .019**
Hart(6)	.1212 ± .002	.2207 ± .006	.1235 ± .002	**.1020 ± .003**

values of the kernel width and find the one that optimizes EI's performance. Fig. 1 is plotted using these optimal kernel widths and shows that the best performance of EI happens when we take only one random sample from a given budget. This performance is then used as the baseline for comparison in Table 2. In addition to EI, we introduced a new version of EI, called EI$_M$. Instead of taking the expectation of improvement I from 0 to infinity, (equation 2), we calculate the expectation of improvement from 0 to $M - y_{max}$ assuming M is given. This simple change decreases the level of exploration of EI and changes its behavior to be more exploitative than explorative. Using GP as our posterior model, the following lemma represents the EI$_M$. The proof is in supplementary document.

Lemma 1. *Let* $u_1 = \frac{y_{max}-\mu_x}{\sigma_x}$ *and* $u_2 = \frac{M-\mu_x}{\sigma_x}$, *then*

$$
\begin{aligned}
EI_M(x) &= \mathbb{E}\left[(f(x) - y_{\max})\mathbb{I}_{\{0 \le f(x)-y_{\max} \le M-y_{max}\}}\right] \\
&= \sigma(x)\left(-u_1\Phi(u_2) + u_1\Phi(u_1) + \phi(u_1)\right).
\end{aligned}
\tag{3}
$$

In light of the results of Fig. 1, we are also interested in whether our exploration algorithm can be used to improve the performance of EI. To this end, we replace the proposed exploitation algorithm with EI to examine if our exploration strategy helps EI. We refer to this setting as NBRS+EI.

Table 2 summarizes the mean and variance of the performance, measured as the "Regret"$= M - \max f(x_\mathcal{O})$, for different benchmarks estimated over 1000 random runs. Interestingly, EI can consistently outperform the EI$_M$ in all benchmarks. This shows that decreasing the exploration rate of EI could degrade the performance.

It is easy to see that in all benchmarks, our algorithm (NBRS+NBIS) outperforms EI consistently except for the Shekel benchmark where EI and NBRS+EI have slightly better performances. We suspect this is due to the fact that we have not optimized our kernel widths, where as the EI kernel width is optimized.

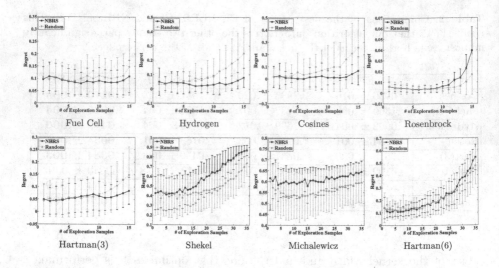

Fig. 3. Plot of regret versus the number of explorations for NBIS algorithm. For a fixed budget n_b, we run a number of experiments as follows: first we consider the case where there are 1 explorative sample (either random or NBRS) followed by $n_b - 1$ EI samples, next we consider the case where where there are 2 explorative samples followed by $n_b - 2$ EI samples and so on. For 2D and 3D functions, we let $n_b = 15$ and for high-dimensional functions, we let $n_b = 35$. This result shows that in most cases, our exploration is a) better than random, and b) necessary, since the regret achieves its minimum somewhere apart from zero. On average, we need to explore 20% of our budget, however, this portion can be optimized if we consider any specific function. The error bar here is the variance of the regret over different runs. This shows that our regret variance is smaller.

We also note that NBRS+EI does not lead to any consistent improvement over EI. This is possibly due to the fact that EI does not take advantage of the reduced search space produced by NBRS during selection.

4.2   Exploration Analysis

In the second set of experiments, we would like to compare our exploration algorithm NBRS with random exploration when using NBIS for exploitation. As discussed previously, both random exploration and NBRS fail to produce better performance when used with EI. Thus, it is interesting to see whether they can help NBIS in terms of the overall regret, and if so which one is more effective. Figure 3 summarizes this result for all benchmarks. For a fixed budget n_b, we start with 1 explorative sample (either using NBRS or random) followed by $n_b - 1$ NBIS samples; next, we start with 2 explorative samples followed by $n_b - 2$ NBIS samples and so on. In each case, we average the regret over 1000 runs. The black line corresponds to the NBRS exploration and the green line corresponds to the random exploration. We will discuss each function in more details later, but in general, this result shows that our exploration algorithm is a) better than random exploration and b) necessary. To see why it is necessary, notice that the minimum regret on all curves is achieved for a non-zero number of NBRS samples. This means unlike EI, our exploitation algorithm benefits from NBRS.

Looking closer into the results, we see that NBRS always lead to a smaller regret comparing to the random exploration. On the Shekel benchmark, we see that random exploration has better performance if we spend majority of the budget to *explore*. However, for a *reasonable* amount of exploration that leads to the minimum regret (5 to 10 experiments), random exploration and NBRS achieve similar performance.

On our 6-dimensional benchmark Hartman(6), we notice that random exploration and NBRS behave very similarly. This shows that the input space is so large that no matter how clever you explore, you will not likely to improve the performance for the limited budget of 35.

NBRS starts from an initial point and explores the input space step by step. Imagine you are in a dark room with a torch in your hand and you want to explore the room. You start from an initial point and little by little walk through the space until you explore the whole space. This is exactly how NBRS does the exploration. Roughly speaking, NBRS minimizes $\mu_{x|\mathcal{O}} + 1.5\sigma_{x|\mathcal{O}}$ and hence, if a point is far from previous observations, i.e., $\sigma_{x|\mathcal{O}}$ is large, it is unlikely to be chosen. We see this effect in all functions, but most clearly in the Michalewicz benchmark. When the number of explorative samples is smaller than 10, the step-by-step explore procedure cannot explore the whole space and the exploitation can be trapped in local minima. For 10   15 explorative samples, NBRS can walk through the entire space fairly well and hence we get a minimum regret. For more than 15 explorative samples, since the space is well explored, we are wasting the samples that could be potentially used to improve our exploitation and hence, the performance becomes worse.

Finally, this investigation suggests that the result in Table 2 can be further improved by taking different number of explorative samples for different functions. To minimize parameter tuning, we chose to explore 20% of our budget. In general, this ratio can be adjusted according to the property of the function (e.g., the Lipschitz constant).

5   Conclusion

In this paper, we consider the problem of maximizing an unknown costly-to-evaluate function when we have a small evaluation budget. Using the Bayesian optimization framework, we proposed a two-phase exploration-exploitation algorithm that finds the maximizer of the function with few function evaluations by leveraging the Lipschitz property of the unknown function. In the exploration phase, our algorithm tries to remove as many points as possible from the search space and hence shrinks the search space. In the exploitation phase, the algorithm tries to find the point that is closest to the optimal. Our empirical results show that our algorithm outperforms EI (even in its best condition).

References

1. Jones, D.R.: A taxonomy of global optimization methods based on response surfaces. Journal of Global Optimization 21, 345–383 (2001)
2. Brochu, E., Cora, M., de Freitas, N.: A tutorial on Bayesian optimization of expensive cost functions, with application to active user modeling and hierarchical reinforcement learning. Technical Report TR-2009-23, Department of Computer Science, University of British Columbia (2009)
3. Rasmussen, C.E., Williams, C.K.I.: Gaussian Processes for Machine Learning. MIT (2006)
4. Elder IV, J.F.: Global rd optimization when probes are expensive: the grope algorithm. In: IEEE International Conference on Systems, Man and Cybernetics, pp. 577–582 (1992)
5. Stuckman, B.E.: A global search method for optimizing nonlinear systems. IEEE Transactions on Systems, Man, and Cybernetic 18, 965–977 (1988)
6. Locatelli, M.: Bayesian algorithms for one-dimensional global optimization. Journal of Global Optimization 10(1), 57–76 (1997)
7. Moore, A., Schneider, J., Boyan, J., Lee, M.S.: Q2: Memory-based active learning for optimizing noisy continuous functions. In: ICML, pp. 386–394 (1998)
8. Li, W., Wang, X., Zhang, R., Cui, Y., Mao, J., Jin, R.: Exploitation and exploration in a performance based contextual advertising system. In: Proceedings of the 16th ACM SIGKDD International Conference on Knowledge Discovery and Data Mining, KDD 2010, pp. 27–36. ACM (2010)
9. Vazquez, E., Bect, J.: Convergence properties of the expected improvement algorithm with fixed mean and covariance functions. Journal of Statistical Planning and Inference 140(11), 3088–3095 (2010)
10. Lizotte, D.: Practical Bayesian Optimization. PhD thesis, University of Alberta, Edmonton, Alberta, Canada (2008)
11. Azimi, J., Fern, A., Fern, X.: Batch bayesian optimization via simulation matching. In: NIPS (2010)
12. Schonlau, M.: Computer Experiments and Global Optimization. PhD thesis, University of Waterloo, Waterloo, Ontario, Canada (1997)
13. Sasena, M.J.: Flexibility and Efficiency Enhancement for Constrained Global Design Optimization with Kriging Approximations. PhD thesis, University of Michigan, Michigan, MI (2002)
14. Jones, D.R., Perttunen, C.D., Stuckman, B.E.: Lipschitzian optimization without the lipschitz constant. Journal of Optimization Theory and Applications 79(1), 157–181 (1993)
15. Anderson, B.S., Moore, A., Cohn, D.: A nonparametric approach to noisy and costly optimization. In: ICML (2000)
16. Brunato, M., Battiti, R., Pasupuleti, S.: A memory-based rash optimizer. In: AAAI 2006 Workshop on Heuristic Search, Memory Based Heuristics and Their Applications (2006)
17. Dixon, L., Szeg, G.: The Global Optimization Problem: An Introduction Toward Global Optimization. North-Holland, Amsterdam (1978)
18. Michalewicz, Z.: Genetic algorithms + data structures = evolution programs, 2nd edn. Springer-Verlag New York, Inc., New York (1994)

19. Burrows, E.H., Wong, W.K., Fern, X., Chaplen, F.W., Ely, R.L.: Optimization of ph and nitrogen for enhanced hydrogen production by synechocystis sp. pcc 6803 via statistical and machine learning methods. Biotechnology Progress 25, 1009–1017 (2009)
20. Azimi, J., Fern, X., Fern, A., Burrows, E., Chaplen, F., Fan, Y., Liu, H., Jaio, J., Schaller, R.: Myopic policies for budgeted optimization with constrained experiments. In: AAAI (2010)

Appendix: Proof of Lemma 1

Let $f(x)$ be our function prediction at any point x distributed as a normal random variable with mean μ_x and variance σ_x^2; i.e $f(x) \sim \mathcal{N}(\mu(x, \sigma_x^2))$ where μ_x and σ_x^2 obtained from Gaussian process. Suppose y_{max} is the best current observation, the probability of improvement of $I \in [0, M - y_{max}]$ can be calculated as $p(f(x) = y_{max} + I)$:

$$p\left(f(x) = y_{max} + I\right) = \frac{1}{\sqrt{2\pi}\sigma_x} \exp\left(-\frac{(y_{max} + I - \mu_x)^2}{2\sigma_x^2}\right). \tag{4}$$

Therefore we define $EI_M(x)$ as is simply the expectation of likelihood over $I \in [0, M]$ at any given point x:

$$EI_M(x) = \int_{I=0}^{I=M-y_{max}} I\left\{\frac{1}{\sqrt{2\pi}\sigma_x} \exp\left(-\frac{(y_{max} + I - \mu_x)^2}{2\sigma_x^2}\right)\right\} dI$$

$$= \frac{1}{\sqrt{2\pi}\sigma_x} \exp\left(-\frac{(y_{max} - \mu_x)^2}{2\sigma_x^2}\right) \int_0^{M-y_{max}} I \exp\left(-\frac{2I(y_{max} - \mu_x) + I^2}{2\sigma_x^2}\right) dI. \tag{5}$$

Let define

$$T = \exp\left(-\frac{2I(y_{max} - \mu_x) + I^2}{2\sigma_x^2}\right) \tag{6}$$

$$\frac{\partial T}{\partial I} = -\frac{1}{\sigma_x^2}\left(IT + (y_{max} - \mu_x T)\right),$$

therefore we can get

$$IT = -(y_{max} - \mu_x)T - \frac{\partial T}{\partial I}\sigma_x^2. \tag{7}$$

Using equations 7,6,5 we can get

$$EI_M(x) = \frac{1}{\sqrt{2\pi}\sigma_x} \exp\left(-\frac{(y_{max} - \mu_x)^2}{2\sigma_x^2}\right) \int_0^{M-y_{max}} IT \, dI$$

$$= \sigma_x \phi\left(\frac{y_{max} - \mu_x}{\sigma_x}\right)$$

$$- (y_{max} - \mu_x)\int_0^{M-y_{max}} \frac{1}{\sqrt{2\pi}\sigma_x} \exp\left(-\frac{1}{2}\left(\frac{y_{max} + I - \mu_x}{\sigma_x}\right)^2\right) dI. \tag{8}$$

Let

$$I^* = \frac{y_{max} + I - \mu_x}{\sigma_x}, \qquad then \qquad dI^* = \frac{dI}{\sigma_x}, \qquad (9)$$

then the equation 8 can be written as

$$
EI_M(x) = \sigma_x \phi \left(\frac{y_{max} - \mu_x}{\sigma_x} \right)
$$
$$
- (y_{max} - \mu_x) \int_{\frac{y_{max} - \mu_x}{\sigma_x}}^{\frac{M - \mu_x}{\sigma_x}} \frac{1}{\sqrt{2\pi}} \exp \left(-\frac{1}{2} I^{*2} \right) dI^*
$$
$$
= \sigma_x \phi \left(\frac{y_{max} - \mu_x}{\sigma_x} \right) - \left[(y_{max} - \mu_x) \left(\Phi \left(\frac{M - \mu_x}{\sigma_x} \right) - \Phi \left(\frac{y_{max} - \mu_x}{\sigma_x} \right) \right) \right].
$$
$$(10)$$

Let

$$u_1 = \frac{y_{max} - \mu_x}{\sigma_x}, u_2 = \frac{M - \mu_x}{\sigma_x},$$

then we can finally drive the maximum expected improvement at any given point x as

$$MEI(x) = \sigma_x \left(-u_1 \Phi(u_2) + u_1 \Phi(u_1) + \phi(u_1) \right), \qquad (11)$$

where $\Phi(\cdot)$ is the normal cumulative distribution function and $\phi(\cdot)$ is the standard nomal distribution.

Parallel Gaussian Process Optimization with Upper Confidence Bound and Pure Exploration

Emile Contal, David Buffoni, Alexandre Robicquet, and Nicolas Vayatis

CMLA, ENS Cachan, CNRS, 61 Avenue du Président Wilson, F-94230 Cachan
{contal,buffoni,vayatis}@cmla.ens-cachan.fr,
alexandre.robicquet@ens-cachan.fr

Abstract. In this paper, we consider the challenge of maximizing an unknown function f for which evaluations are noisy and are acquired with high cost. An iterative procedure uses the previous measures to actively select the next estimation of f which is predicted to be the most useful. We focus on the case where the function can be evaluated in parallel with batches of fixed size and analyze the benefit compared to the purely sequential procedure in terms of cumulative regret. We introduce the Gaussian Process Upper Confidence Bound and Pure Exploration algorithm (GP-UCB-PE) which combines the UCB strategy and Pure Exploration in the same batch of evaluations along the parallel iterations. We prove theoretical upper bounds on the regret with batches of size K for this procedure which show the improvement of the order of $\sqrt{K}$ for fixed iteration cost over purely sequential versions. Moreover, the multiplicative constants involved have the property of being dimension-free. We also confirm empirically the efficiency of GP-UCB-PE on real and synthetic problems compared to state-of-the-art competitors.

1 Introduction

Finding the maximum of a non-convex function by means of sequential noisy observations is a common task in numerous real world applications. The context of a high dimensional input space with expensive evaluation cost offers new challenges in order to come up with efficient and valid procedures. This problem of sequential global optimization arises for example in industrial system design and monitoring to choose the location of a sensor to find out the maximum response, or when determining the parameters of a heavy numerical code designed to maximize the output. The standard objective in this setting is to minimize the cumulative regret R_T, defined as the sum $\sum_{t=1}^{T} \left(f(x^\star) - f(x_t) \right)$ of the differences between the values of f at the points queried x_t and the true optimum of f noted $x^\star$. For a fixed horizon T, we refer to [1]. In the context where the horizon T is unknown, the query selection has to deal with the exploration/exploitation tradeoff. Successful algorithms have been developed in different settings to address this problem such as experimental design [2], Bayesian optimization [3–8], active learning [9,10], multiarmed bandit [11–17] and in particular Hierarchical Optimistic Optimization algorithm, HOO [18] for bandits in a generic space,

H. Blockeel et al. (Eds.): ECML PKDD 2013, Part I, LNAI 8188, pp. 225–240, 2013.
© Springer-Verlag Berlin Heidelberg 2013

namely $\mathcal{X}$-Armed bandits. In some cases, it is possible to evaluate the function in parallel with batches of K queries with no increase in cost. This is typically the case in the sensors location problem if K sensors are available at each iteration, or in the numerical optimization problem on a cluster of K cores. Parallel strategies have been developed recently in [19, 20]. In the present paper, we propose to explore further the potential of parallel strategies for noisy function optimization with unknown horizon aiming simultaneously at practical efficiency and plausible theoretical results. We introduce a novel algorithm called GP-UCB-PE based on the Gaussian process approach which combines the benefits of the UCB policy with Pure Exploration queries in the same batch of K evaluations of f. The Pure Exploration component helps to reduce the uncertainty about f in order to support the UCB policy in finding the location of the maximum, and therefore in increasing the decay of the regret R_t at every timestep t. In comparison to other algorithms based on Gaussian processes and UCB such as GP-BUCB [19], the new algorithm discards the need for the initialization phase and offers a tighter control on the uncertainty parameter which monitors over-confidence. As a result, the derived regret bounds do not suffer from the curse of dimensionality since the multiplicative constants obtained are dimension free in contrast with the doubly exponential dependence observed in previous work. We also mention that Monte-Carlo simulations can be proposed as an alternative and this idea has been implemented in the *Simulation Matching* with UCB policy (SM-UCB) algorithm [20] which we also consider for comparison in the present paper. Unlike GP-BUCB, no theoretical guarantees for the SM-UCB algorithm are known for the bounds on the number of iterations needed to get close enough to the maximum, therefore the discussion will be reduced to empirical comparisons over several benchmark problems. The remainder of the paper is organized as follows. We state the background and our notations in Section 2. We formalize the Gaussian Process assumptions on f, and give the definition of regret in the parallel setting. We then describe the GP-UCB-PE algorithm and the main concepts in Section 3. We provide theoretical guarantees through upper bounds for the cumulative regret of GP-UCB-PE in Section 4. We finally show comparisons of our method and the related algorithms through a series of numerical experiments on real and synthetic functions in Section 5.[1]

2   Problem Statement and Background

2.1   Sequential Batch Optimization

We address the problem of finding in the lowest possible number of iterations the maximum of an unknown function $f : \mathcal{X} \to \mathbb{R}$ where $\mathcal{X} \subset \mathbb{R}^d$, denoted by :

$$f(x^*) = \max_{x \in \mathcal{X}} f(x) .$$

---

[1] The documented source codes and the assessment data sets are available online at http://econtal.perso.math.cnrs.fr/software/

The arbitrary choice of formulating the optimization problem as a maximization is without loss of generality, as we can obviously take the opposite of f if the problem is a minimization one. At each iteration t, we choose a batch of K points in $\mathcal{X}$ called the queries $\{x_t^k\}_{0 \leqslant k < K}$, and then observe simultaneously the noisy values taken by f at these points,

$$y_t^k = f(x_t^k) + \epsilon_t^k ,$$

where the ϵ_t^k are independent Gaussian noise $\mathcal{N}(0, \sigma^2)$.

2.2 Objective

Assuming that the horizon T is unknown, a strategy has to be good at any iteration. We denote by $r_t^{(k)}$ the difference between the optimum of f and the point queried x_t^k,

$$r_t^{(k)} = f(x^\star) - f(x_t^k) .$$

We aim to minimize the batch cumulative regret,

$$R_T^K = \sum_{t < T} r_t^K ,$$

which is the standard objective with these formulations of the problem [21]. We focus on the case where the cost for a batch of evaluations of f is fixed. The loss r_t^K incurred at iteration t is then the simple regret for the batch [22], defined as

$$r_t^K = \min_{k < K} r_t^{(k)} .$$

An upper bound on R_T^K gives an upper bound of $\frac{R_T^K}{T}$ on the minimum gap between the best point found so far and the true maximum. We also provide bounds on the full cumulative regret,

$$R_{TK} = \sum_{t < T} \sum_{k < K} r_t^{(k)} ,$$

which model the case where all the queries in a batch should have a low regret.

2.3 Gaussian Processes

In order to analyze the efficiency of a strategy, we have to make some assumptions on f. We want extreme variations of the function to have low probability.

Modeling f as a sample of a Gaussian Process (GP) is a natural way to formalize the intuition that nearby location are highly correlated. It can be seen as a continuous extension of multidimensional Gaussian distributions. We say that a random process f is Gaussian with mean function m and non-negative definite covariance function (kernel) k written :

$$f \sim GP(m, k) ,$$

$$\text{where } m : \mathcal{X} \to \mathbb{R}$$

$$\text{and } k : \mathcal{X} \times \mathcal{X} \to \mathbb{R}^+ ,$$

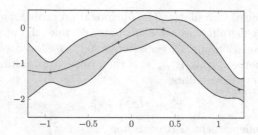

Fig. 1. Gaussian Process inference of the posterior mean $\hat{\mu}$ (blue line) and deviation $\hat{\sigma}$ based on four realizations (blue crosses). The high confidence region (area in grey) is delimited by $\hat{f}^+$ and $\hat{f}^-$.

when for any finite subset of locations the values of the random function form a multivariate Gaussian random variable of mean vector μ and covariance matrix $\mathbf{C}$ given by the mean m and the kernel k of the GP. That is, for all finite n and $x_1, \ldots, x_n \in \mathcal{X}$,

$$(f(x_1), \ldots, f(x_n)) \sim \mathcal{N}(\mu, \mathbf{C}) ,$$
$$\text{with } \mu[x_i] = m(x_i)$$
$$\text{and } \mathbf{C}[x_i, x_j] = k(x_i, x_j) .$$

If we have the prior knowledge that f is drawn from a GP with zero mean[2] and known kernel, we can use Bayesian inference conditioned on the observations after T iterations to get the closed formulae for computing the posterior [23], which is a GP of mean and variance given at each location $x \in \mathcal{X}$ by :

$$\hat{\mu}_{T+1}(x) = \mathbf{k}_T(x)^\top \mathbf{C}_T^{-1} \mathbf{Y}_T \tag{1}$$
$$\text{and } \hat{\sigma}_{T+1}^2(x) = k(x, x) - \mathbf{k}_T(x)^\top \mathbf{C}_T^{-1} \mathbf{k}_T(x) , \tag{2}$$

$\mathbf{X}_T = \{x_t^k\}_{t<T, k<K}$ is the set of queried locations, $\mathbf{Y}_T = [y_t^k]_{x_t^k \in \mathbf{X}_T}$ is the vector of noisy observations, $\mathbf{k}_T(x) = [k(x_t^k, x)]_{x_t^k \in \mathbf{X}_T}$ is the vector of covariances between x and the queried points, and $\mathbf{C}_T = \mathbf{K}_T + \sigma^2 \mathbf{I}$ with $\mathbf{K}_T = [k(x, x')]_{x, x' \in \mathbf{X}_T}$ the kernel matrix and $\mathbf{I}$ stands for the identity matrix.

The three most common kernel functions are:

- the polynomial kernels of degree $\alpha \in \mathbb{N}$, $k(x_1, x_2) = (x_1^\top x_2 + c)^\alpha$, $c \in \mathbb{R}$,
- the (Gaussian) Radial Basis Function kernel (RBF or Squared Exponential) with length-scale $l > 0$, $k(x_1, x_2) = \exp\left(-\frac{\|x_1, x_2\|^2}{2l^2}\right)$,
- the Matérn kernel, of length-scale l and parameter ν,

$$k(x_1, x_2) = \frac{2^{1-\nu}}{\Gamma(\nu)} \left(\frac{\sqrt{2\nu}\, \|x_1, x_2\|}{l}\right)^\nu K_\nu\left(\frac{\sqrt{2\nu}\, \|x_1, x_2\|}{l}\right) , \tag{3}$$

where K_ν is the modified Bessel function of the second kind and order ν.

---

[2] This is without loss of generality as the kernel k can completely define the GP [23].

The Bayesian inference is represented on Figure 1 in a sample problem in dimension 1. The posteriors are based on four observations of a Gaussian Process. The vertical height of the grey area is proportional to the posterior deviation at each point.

3   Parallel Optimization Procedure

3.1   Confidence Region

A key property from the GP framework is that the posterior distribution at a location x has a normal distribution $\mathcal{N}(\hat{\mu}_T(x), \hat{\sigma}_T^2(x))$. We can then define a upper confidence bound $\hat{f}^+$ and a lower confidence bound $\hat{f}^-$, such that f is included in the interval with high probability,

$$\hat{f}_T^+(x) = \hat{\mu}_T(x) + \sqrt{\beta_T}\hat{\sigma}_T(x) \tag{4}$$

$$\text{and } \hat{f}_T^-(x) = \hat{\mu}_T(x) - \sqrt{\beta_T}\hat{\sigma}_T(x) , \tag{5}$$

with $\beta_T \in \mathcal{O}(\log T)$ defined in Section 4.

$\hat{f}^+$ and $\hat{f}^-$ are illustrated on Figure 1 respectively by the upper and lower envelope of the grey area. The region delimited in that way, the high confidence region, contains the unknown f with high probability. This statement will be a main element in the theoretical analysis of the algorithm in Section 4.

3.2   Relevant Region

We define the relevant region $\mathfrak{R}_t$ being the region which contains x^* with high probability. Let $y_t^\bullet$ be our lower confidence bound on the maximum,

$$y_t^\bullet = \hat{f}_t^-(x_t^\bullet), \text{ where } x_t^\bullet = \underset{x \in \mathcal{X}}{\text{argmax}} \, \hat{f}_t^-(x) .$$

$y_t^\bullet$ is represented by the horizontal dotted green line on Figure 2. $\mathfrak{R}_t$ is defined as :

$$\mathfrak{R}_t = \left\{ x \in \mathcal{X} \mid \hat{f}_t^+(x) \geqslant y_t^\bullet \right\} .$$

$\mathfrak{R}_t$ discard the locations where x^* does not belong with high probability. It is represented in green on Figure 2. We refer to [24] for related work in the special case of deterministic Gaussian Process Bandits.

In the sequel, we will use a modified version of the relevant region which also contains $\text{argmax}_{x \in \mathcal{X}} \hat{f}_{t+1}^+(x)$ with high probability. The novel relevant region is formally defined by :

$$\mathfrak{R}_t^+ = \left\{ x \in \mathcal{X} \mid \hat{\mu}_t(x) + 2\sqrt{\beta_{t+1}}\hat{\sigma}_t(x) \geqslant y_t^\bullet \right\} . \tag{6}$$

Using $\mathfrak{R}_t^+$ instead of $\mathfrak{R}_t$ guarantees that the queries at iteration t will leave an impact on the future choices at iteration $t + 1$.

Algorithm 1: GP-UCB-PE

for $t = 0, \ldots, T$ **do**

    Compute $\hat{\mu}_t$ and $\hat{\sigma}_t$ with Eq.1 and Eq.2

    $x_t^0 \leftarrow \text{argmax}_{x \in \mathcal{X}} \, \hat{f}_t^+(x)$

    Compute $\mathfrak{R}_t^+$ with Eq.6

    for $k = 1, \ldots, K - 1$ **do**

        Compute $\hat{\sigma}_t^{(k)}$ with Eq.2

        $x_t^k \leftarrow \text{argmax}_{x \in \mathfrak{R}_t^+} \, \hat{\sigma}_t^{(k)}(x)$

    Query $\{x_t^k\}_{k < K}$

3.3 GP-UCB-PE

We present here the Gaussian Process Upper Confidence Bound with Pure Exploration algorithm, GP-UCB-PE, a novel algorithm combining two strategies to determine the queries $\{x_t^k\}_{k<K}$ for batches of size K. The first location is chosen according to the GP-UCB rule,

$$x_t^0 = \underset{x \in \mathcal{X}}{\text{argmax}} \, \hat{f}_t^+(x) \, . \tag{7}$$

This single rule is enough to tackle the exploration/exploitation tradeoff. The value of β_t balances between exploring uncertain regions (high posterior variance $\hat{\sigma}_t^2(x)$) and focusing on the supposed location of the maximum (high posterior mean $\hat{\mu}_t(x)$). This policy is illustrated with the point x^0 on Figure 2.

The $K - 1$ remaining locations are selected via Pure Exploration restricted to the region $\mathfrak{R}_t^+$. We aim to maximize $I_t(\mathbf{X}_t^{K-1})$, the information gain about f by the locations $\mathbf{X}_t^{K-1} = \{x_t^k\}_{1 \leqslant k < K}$ [25]. Formally, $I_t(\mathbf{X})$ is the reduction of entropy when knowing the values of the observations $\mathbf{Y}$ at $\mathbf{X}$, conditioned on $\mathbf{X}_t$ the observations we have seen so far,

$$I_t(\mathbf{X}) = H(\mathbf{Y}) - H(\mathbf{Y} \mid \mathbf{X}_t) \, . \tag{8}$$

Finding the $K - 1$ points that maximize I_t for any integer K is known to be NP-complete [26]. However, due to the submodularity of I_t [4], it can be efficiently approximated by the greedy procedure which selects the points one by one and never backtracks. For a Gaussian distribution, $H(\mathcal{N}(\mu, \mathbf{C})) = \frac{1}{2} \log \det(2\pi e \mathbf{C})$. We thus have $I_t(\mathbf{X}) \in \mathcal{O}(\log \det \mathbf{\Sigma})$, where $\mathbf{\Sigma}$ is the covariance matrix of $\mathbf{X}$. For GP, the location of the single point that maximizes the information gain is easily computed by maximizing the posterior variance. For all $1 \leqslant k < K$ our greedy strategy selects the following points one by one,

$$x_t^k = \underset{x \in \mathfrak{R}_t^+}{\text{argmax}} \, \hat{\sigma}_t^{(k)}(x) \, , \tag{9}$$

where $\hat{\sigma}_t^{(k)}$ is the updated variance after choosing $\{x_t^{k'}\}_{k' < k}$. We use here the fact that the posterior variance does not depend on the values y_t^k of the observations,

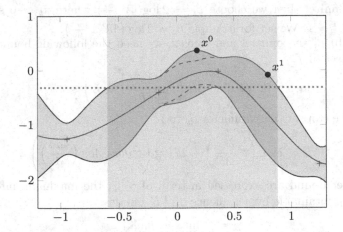

Fig. 2. Two queries of GP-UCB-PE on the previous example. The lower confidence bound on the maximum is represented by the horizontal dotted green line at $y_t^\bullet$. The relevant region $\mathfrak{R}$ is shown in light green (without edges). The first query x^0 is the maximizer of $\hat{f}^+$. We show in dashed line the upper and lower bounds with the update of $\hat{\sigma}$ after having selected x^0. The second query x^1 is the one maximizing the uncertainty inside $\mathfrak{R}^+$, an extension of $\mathfrak{R}$ which is not illustrated here.

but only on their position x_t^k. One such point is illustrated with x^1 on Figure 2. These $K-1$ locations reduce the uncertainty about f, improving the guesses of the UCB procedure by x_t^0. The overall procedure is shown in Algorithm 1.

3.4 Numerical Complexity

Even if the numerical cost of GP-UCB-PE is insignificant in practice compared to the cost of the evaluation of f, the complexity of the exact computations of the variances (Eq.2) is in $\mathcal{O}(n^3)$ and might by prohibitive for large $n = tK$. One can reduce drastically the computation time by means of Lazy Variance Calculation [19], built on the fact that $\hat{\sigma}_t(x)$ always decreases when t increases for all $x \in \mathcal{X}$. We further mention that efficient approximated inference algorithms such as the EP approximation and MCMC sampling [27] can be used in order to face the challenge of large n.

4 Regret Bounds

4.1 Main Result

The main theoretical result of this article is the upper bound on the regret formulated in Theorem 1. We need to adjust the parameter β_t such that $f(x)$ is contained by the high confidence region for all iterations t with probability at least $1 - \delta$ for a fixed $0 < \delta < 1$.

- If $\mathcal{X}$ is finite, then we choose $\beta_t = 2\log(|\mathcal{X}|\frac{\pi_t}{\delta})$ where $\pi_t > 0$ such that $\sum_{t=0}^{\infty} \pi_t^{-1} = 1$. We set for example $\beta_t = 2\log\left(|\mathcal{X}|\,t^2\frac{\pi^2}{6\delta}\right)$.
- If $\mathcal{X} \subset [0,r]^d$ is compact and convex, we need the following bounds on the derivatives of f,

$$\exists a, b > 0, \ \forall j \leqslant d, \ \Pr\left(\sup_{x\in\mathcal{X}}\left|\frac{\partial f}{\partial x_j}\right| > L\right) \leqslant ae^{-\frac{L^2}{b^2}} .$$

Then, we can set the parameter β_t to :

$$\beta_t = 2\log\left(t^2\frac{2\pi^2}{3\delta}\right) + 2d\log\left(t^2 dbr\sqrt{\log\left(\frac{4da}{\delta}\right)}\right) .$$

The regret bound are expressed in term of γ_{TK}, the maximum information gain (Eq. 8) obtainable by a sequence of TK queries,

$$\gamma_t = \max_{\mathbf{X}\subset\mathcal{X},|\mathbf{X}|=t} I_0(\mathbf{X}) .$$

Under these assumptions, we obtain the following result.

Theorem 1. *Fix* $0 < \delta < 1$ *and consider the calibration of* β_t *defined as above, assuming* $f \sim GP(0,k)$ *with bounded variance,* $\forall x \in \mathcal{X}$, $k(x,x) \leqslant 1$, *then the batch cumulative regret incurred by* **GP-UCB-PE** *on* f *is bounded by* $\mathcal{O}\left(\sqrt{\frac{T}{K}\beta_T\gamma_{TK}}\right)$ *whp, More precisely, with* $C_1 = \frac{4}{\log(1+\sigma^{-2})}$, *and* $C_2 = \frac{\pi}{\sqrt{6}}$, $\forall T$,

$$\Pr\left(R_T^K \leqslant \sqrt{C_1\frac{T}{K}\beta_T\gamma_{TK}} + C_2\right) \geqslant 1 - \delta .$$

For the full cumulative regret R_{TK} *we obtain similar bounds with* $C_1 = \frac{36}{\log(1+\sigma^{-2})}$

$$\Pr\left(R_{TK} \leqslant \sqrt{C_1 TK\beta_T\gamma_{TK}} + C_2\right) \geqslant 1 - \delta .$$

4.2 Discussion

When $K \ll T$, the upper bound for R_T^K is better than the one of sequential GP-UCB by an order of $\sqrt{K}$, and equivalent for R_{TK}, when the regrets for all the points in the batch matter. Compared to [19], we remove the need of the initialization phase. GP-UCB-PE does not need either to multiply the uncertainty parameter β_t by $\exp(\gamma_{TK}^{\text{init}})$ where $\gamma_{TK}^{\text{init}}$ is equal to the maximum information gain obtainable by a sequence of TK queries after the initialization phase. The improvement can be doubly exponential in the dimension d in the case of RBF Kernels. To the best of our knowledge, no regret bounds have been proven for the *Simulation Matching* algorithm.

The values of γ_{TK} for different common kernel are reported in Table 1, where d is the dimension of the space considered and $\alpha = \frac{d(d+1)}{2\nu+d(d+1)} \leqslant 1$, ν being the Matérn parameter. We also compare on Table 1 the general forms of the bounds for the regret obtained by GP-UCB-PE and GP-BUCB up to constant terms. The cumulative regret we obtained with RBF Kernel is of the form $\tilde{\mathcal{O}}\left(\sqrt{\frac{T}{K}(\log TK)^d}\right)$ against $\tilde{\mathcal{O}}\left(\exp((\frac{2d}{e})^d)\sqrt{\frac{T}{K}(\log TK)^d}\right)$ for GP-BUCB.

Table 1. General Forms of Regret Bounds for GP-UCB-PE and GP-BUCB

	GP-UCB-PE	GP-BUCB
R_T^K	$\sqrt{\frac{T \log T}{K}} \gamma_{TK}$	$C \sqrt{\frac{T \log TK}{K}} \gamma_{TK}$

Kernel	Linear	RBF	Matérn
γ_{TK}	$d \log TK$	$(\log TK)^{d+1}$	$(TK)^\alpha \log TK$
C	$\exp(\frac{2}{e})$	$\exp((\frac{2d}{e})^d)$	e

4.3   Proofs of the Main Result

In this section, we analyze theoretically the regret bounds for the GP-UCB-PE algorithm. We provide here the main steps for the proof of Theorem 1. On one side the UCB rule of the algorithm provides a regret bounded by the information we have on f conditioned on the values observed so far. On the other side, the Pure Exploration part gathers information and therefore accelerates the decrease in uncertainty. We refer to [19] for the proofs of the bounds for GP-BUCB.

For the sake of concision, we introduce the notations σ_t^k for $\hat{\sigma}_t^{(k)}(x_t^k)$ and σ_t^0 for $\hat{\sigma}_t(x_t^0)$. We simply bound r_t^K the regret for the batch at iteration t by the simple regret $r_t^{(0)}$ for the single query chosen via the UCB rule. We then give a bound for $r_t^{(0)}$ which is proportional to the posterior deviations σ_t^0. Knowing that the sum of all $(\sigma_t^k)^2$ is not greater than $C_1 \gamma_{TK}$, we want to prove that the sum of the $(\sigma_t^0)^2$ is less than this bound divided by K. The arguments are based on the fact that the posterior for $f(x)$ is Gaussian, allowing us to choose β_t such that :

$$\forall x \in \mathcal{X}, \forall t < T, \ f(x) \in [\hat{f}_t^-(x), \hat{f}_t^+(x)]$$

holds with high probability. Here and in the following, "with high probability" or *whp* means "with probability at least $1 - \delta$" for any $0 < \delta < 1$, the definition of β_t being dependent of δ.

Lemma 1. *For finite $\mathcal{X}$, we have $r_t^K \leqslant r_t^{(0)} \leqslant 2\sqrt{\beta_t}\sigma_t^0$, and for compact and convex $\mathcal{X}$ following the assumptions of Theorem 1, $r_t^K \leqslant r_t^{(0)} \leqslant 2\sqrt{\beta_t}\sigma_t^0 + \frac{1}{t^2}$, holds with probability at least $1 - \delta$.*

We refer to [6] (Lemmas 5.2, 5.8) for the detailed proof of the bound for $r_t^{(0)}$.

Now we show an intermediate result bounding the deviations at the points x_{t+1}^0 by the one at the points x_t^{K-1}.

Lemma 2. *The deviation of the point selected by the UCB policy is bounded by the one for the last point selected by the PE policy at the previous iteration, whp, $\forall t < T, \ \sigma_{t+1}^0 \leqslant \sigma_t^{K-1}$*

Proof. By the definitions of x_{t+1}^0 (Eq.7), we have $\hat{f}_{t+1}^+(x_{t+1}^0) \geqslant \hat{f}_{t+1}^+(x_t^\bullet)$. Then, we know with high probability that $\forall x \in \mathcal{X}, \forall t < T$, $\hat{f}_{t+1}^+(x) \geqslant \hat{f}_t^-(x)$. We can therefore claim *whp* $\hat{f}_{t+1}^+(x_{t+1}^0) \geqslant y_t^\bullet$, and thus that $x_{t+1}^0 \in \mathfrak{R}_t^+$ *whp*.

We have as a result by the definition of x_t^{k-1} (Eq.9) that $\hat{\sigma}_t^{(k-1)}(x_{t+1}^0) \leqslant \hat{\sigma}_t^{(k-1)}(x_t^{k-1})$ *whp*. Using the "Information never hurts" principle [28], we know that the entropy of $f(x)$ for all location x decreases while we observe points x_t. For GP, the entropy is also a non-decreasing function of the variance, so that :

$$\forall x \in \mathcal{X}, \ \hat{\sigma}_{t+1}^{(0)}(x) \leqslant \hat{\sigma}_t^{(k-1)}(x) \ .$$

We thus prove $\sigma_{t+1}^0 \leqslant \sigma_t^{k-1}$.

Lemma 3. *The sum of the deviations of the points selected by the UCB policy are bounded by the one for all the selected points divided by K, whp,*

$$\sum_{t=0}^{T-1} \sigma_t^0 \leqslant \frac{1}{K} \sum_{t=0}^{T-1} \sum_{k=0}^{K-1} \sigma_t^k \ .$$

Proof. Using Lemma 2 and the definitions of the x_t^k, we have that $\sigma_{t+1}^0 \leqslant \sigma_t^k$ for all $k \geqslant 1$. Summing over k, we get for all $t \geqslant 0$, $\sigma_t^0 + (K-1)\sigma_{t+1}^0 \leqslant \sum_{k=0}^{K-1} \sigma_t^k$. Now, summing over t and with $\sigma_0^0 \geqslant 0$ and $\sigma_T^0 \geqslant 0$, we obtain the desired result.

Next, we can bound the sum of all posterior variances $(\sigma_t^k)^2$ via the maximum information gain for a sequence of TK locations.

Lemma 4. *The sum of the variances of the selected points are bounded by a constant factor times γ_{TK}, $\exists C_1' \in \mathbb{R}$, $\sum_{t<T} \sum_{k<K}(\sigma_t^k)^2 \leqslant C_1' \gamma_{TK}$ where γ_{TK} is the maximum information gain obtainable by a sequential procedure of length TK.*

Proof. We know that the information gain for a sequence of T locations x_t can be expressed in terms of the posterior variances $(\hat{\sigma}_{t-1}(x_t))^2$. The deviations σ_t^k being independent of the observations y_t^k, the same equality holds for the posterior variances $(\hat{\sigma}_t^{(k)}(x_t^k))^2$. See Lemmas 5.3 and 5.4 in [6] for the detailed proof, giving $C_1' = \frac{2}{\log(1+\sigma^{-2})}$.

Lemma 5. *The cumulative regret can be bound in terms of the maximum information gain, whp, $\exists C_1, C_2 \in \mathbb{R}$,*

$$\sum_{t<T} r_t^K \leqslant \sqrt{\frac{T}{K}C_1\beta_T\gamma_{TK} + C_2} \ .$$

Proof. Using the previous lemmas and the fact that $\beta_t \leqslant \beta_T$ for all $t \leqslant T$, we have in the case of finite $\mathcal{X}$, *whp*,

$$\sum_{t<T} r_t^K \leqslant \sum_{t<T} 2\sqrt{\beta_t}\sigma_t^0 \text{ , by Lemma 1}$$

$$\leqslant 2\sqrt{\beta_T}\frac{1}{K}\sum_{t<T}\sum_{k<K}\sigma_t^k \text{ , by Lemma 3}$$

$$\leqslant 2\sqrt{\beta_T}\frac{1}{K}\sqrt{TK\sum_{t<T}\sum_{k<K}(\sigma_t^k)^2} \text{ , by Cauchy-Schwarz}$$

$$\leqslant 2\sqrt{\beta_T}\frac{1}{K}\sqrt{TKC_1'\gamma_{TK}} \text{ , by Lemma 4}$$

$$\leqslant \sqrt{\frac{T}{K}C_1\beta_T\gamma_{TK}} \text{ with } C_1 = \frac{4}{\log(1+\sigma^{-2})} \text{ .}$$

For compact and convex $\mathcal{X}$, a similar reasoning gives :

$$R_T^K \leqslant \sqrt{\frac{T}{K}C_1\beta_T\gamma_{TK}} + C_2 \text{ with } C_2 = \frac{\pi}{\sqrt{6}} < 2 \text{ .}$$

Lemma 5 conclude the proof of Theorem 1 for the regret R_T^K. The analysis for R_{TK} is simpler, using the Lemma 6 which bounds the regret for the Pure Exploration queries, leading to $C_1 = \frac{36}{\log(1+\sigma^{-2})}$.

Lemma 6. *The regret for the queries x_t^k selected by Pure Exploration in $\mathfrak{R}_t^+$ are bounded whp by, $6\sqrt{\beta_t}\sigma_t^k$.*

Proof. As in Lemma 1, we have *whp*, for all $t \leqslant T$ and $k \geqslant 1$,

$$r_t^{(k)} \leqslant \hat{\mu}_t(x^\star) + \sqrt{\beta_t}\hat{\sigma}_t(x^\star) - \hat{\mu}_t(x_t^k) + \sqrt{\beta_t}\sigma_t^k$$

$$\leqslant \hat{f}_t^-(x_t^\bullet) + 2\sqrt{\beta_t}\hat{\sigma}_t(x^\star) - \hat{\mu}_t(x_t^k) + \sqrt{\beta_t}\sigma_t^k \text{ by definition of } x_t^\bullet$$

$$\leqslant \hat{\mu}_t(x_t^k) + 2\sqrt{\beta_{t+1}}\sigma_t^k + 2\sqrt{\beta_t}\sigma_t^k - \hat{\mu}_t(x_t^k) + \sqrt{\beta_t}\sigma_t^k \text{ by definition of } \mathfrak{R}_t^+$$

$$\leqslant 3\sqrt{\beta_t}\sigma_t^k + 2\sqrt{\beta_t}\sigma_t^k + \sqrt{\beta_t}\sigma_t^k \text{ by definition of } \beta_{t+1}$$

$$\leqslant 6\sqrt{\beta_t}\sigma_t^k \text{ .}$$

To conclude the analysis of R_{TK} and prove Theorem 1, it suffices to use then the last three steps of Lemma 5.

5   Experiments

5.1   Protocol

We compare the empirical performances of our algorithm against the state of the art of global optimization by batches, GP-BUCB [19] and SM-UCB [20]. The tasks used for assessment come from three real applications and two synthetic

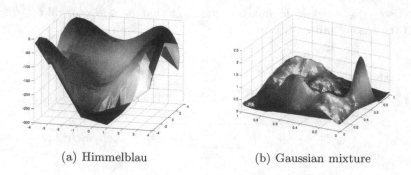

(a) Himmelblau   (b) Gaussian mixture

Fig. 3. Visualization of the synthetic functions used for assessment

problems described here. The results are shown in Figure 4. For all datasets and algorithms, the size of the batches K was set to 10 and the learners were initialized with a random subset of 20 observations (x_i, y_i). The curves on Figure 4 show the evolution of the regret R_t^K in term of iteration t. We report the average value with the confidence interval over 64 experiments. The parameters for the prior distribution, like the bandwidth of the RBF Kernel, were chosen by maximization of the marginal likelihood.

5.2   Description of Data Sets

Generated GP. The `Generated GP` functions are random GPs drawn from a Matérn kernel (Eq. 3) in dimension 2, with the kernel bandwidth set to $\frac{1}{4}$, the Matérn parameter $\nu = 3$ and noise variance σ^2 set to 1.

Gaussian Mixture. This synthetic function comes from the addition of three 2-D Gaussian functions. at $(0.2, 0.5)$, $(0.9, 0.9)$, and the maximum at $(0.6, 0.1)$. We then perturb these Gaussian functions with smooth variations generated from a Gaussian Process with Matérn Kernel and very few noise. It is shown on Figure 3(b). The highest peak being thin, the sequential search for the maximum of this function is quite challenging.

Himmelblau Function. The `Himmelblau` task is another synthetic function in dimension 2. We compute a slightly tilted version of the Himmelblau's function, and take the opposite to match the challenge of finding its maximum. This function presents four peaks but only one global maximum. It gives a practical way to test the ability of a strategy to manage exploration/exploitation tradeoffs. It is represented in Figure 3(a).

Mackey-Glass Function. The Mackey-Glass delay-differential equation[3] is a chaotic system in dimension 6, but without noise. It models real feedback systems

---

[3] http://www.scholarpedia.org/article/Mackey-Glass_equation

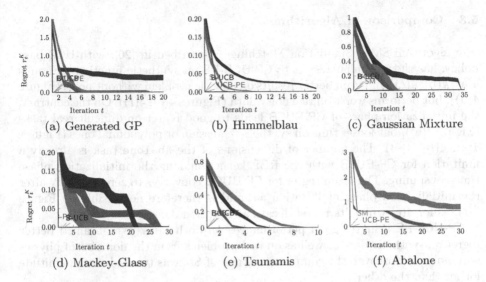

Fig. 4. Experiments on several real and synthetics tasks. The curves show the decay of the mean of the simple regret r_t^K with respect to the iteration t, over 64 experiments. We show with the translucent area the confidence intervals.

and is used in physiological domains such as hematology, cardiology, neurology, and psychiatry. The highly chaotic behavior of this function makes it an exceptionally difficult optimization problem. It has been used as a benchmark for example by [29].

Tsunamis. Recent post-tsunami survey data as well as the numerical simulations of [30] have shown that in some cases the run-up, which is the maximum vertical extent of wave climbing on a beach, in areas which were supposed to be protected by small islands in the vicinity of coast, was significantly higher than in neighboring locations. Motivated by these observations [31] investigated this phenomenon by employing numerical simulations using the VOLNA code [32] with the simplified geometry of a conical island sitting on a flat surface in front of a sloping beach. Their setup was controlled by five physical parameters and their aim was to find with confidence and with the least number of simulations the maximum run-up amplification on the beach directly behind the island, compared with the run-up on a lateral location, not influenced by the presence of the island. Since this problem is too complex to treat analytically, the authors had to solve numerically the Nonlinear Shallow Water Equations.

Abalone. The challenge of the `Abalone` dataset is to predict the age of a specie of sea snails from physical measurements. It comes from the study by [33] and it is provided by the UCI Machine Learning Repository.[4] We use it as a maximization problem in dimension 8.

---

[4] http://archive.ics.uci.edu/ml/datasets/Abalone

5.3 Comparison of Algorithms

The algorithm SM —Simulation Matching— described in [20], with UCB base policy, has shown similar results to GP-UCB-PE on synthetic functions (Figures 4(a), 4(b), 4(c)) and even better results on chaotic problem without noise (Figure 4(d)), but performs worse on real noisy data (Figures 4(e), 4(f)). On the contrary, the initialization phase of GP-BUCB leads to good regret on difficult real tasks (Figure 4(e)), but looses time on synthetic Gaussian or polynomial ones (Figures 4(a), 4(b), 4(c)). The number of dimensions of the `Abalone` task is already a limitation for GP-BUCB with the RBF kernel, making the initialization phase time-consuming. The mean regret for GP-BUCB converges to zero abruptly after the initialization phase at iteration 55, and is therefore not visible on Figure 4(f), as for 4(c) where its regret decays at iteration 34.

GP-UCB-PE achieves good performances on both sides. We obtained better regret on synthetic data as well as on real problems from the domains of physics and biology. Moreover, the computation time of SM was two order of magnitude longer than the others.

6 Conclusion

We have presented the GP-UCB-PE algorithm which addresses the problem of finding in few iterations the maximum of an unknown arbitrary function observed via batches of K noisy evaluations. We have provided theoretical bounds for the cumulative regret obtained by GP-UCB-PE in the Gaussian Process settings. Through parallelization, these bounds improve the ones for the state-of-the-art of sequential GP optimization by a ratio of $\sqrt{K}$, and are strictly better than the ones for GP-BUCB, a concurrent algorithm for parallel GP optimization. We have compared experimentally our method to GP-BUCB and SM-UCB, another approach for parallel GP optimization lacking of theoretical guarantees. These empirical results have confirmed the effectiveness of GP-UCB-PE on several applications.

The strategy of combining in the same batch some queries selected via Pure Exploration is an intuitive idea that can be applied in many other methods. We expect for example to obtain similar results with the Maximum Expected Improvement policy (MEI). Any proof of regret bound that relies on the fact that the uncertainty decreases with the exploration should be easily adapted to a paralleled extension with Pure Exploration.

On the other hand, we have observed in practice that the strategies which focus more on exploitation often lead to faster decrease of the regret, for example the strategy that uses K times the GP-UCB criterion with updated variance. We conjecture that the regret for this strategy is unbounded for general GPs, justifying the need for the initialization phase of GP-BUCB. However, it would be relevant to specify formally the assumptions needed by this greedy strategy to guarantee good performances.

References

1. Hennig, P., Schuler, C.J.: Entropy search for information-efficient global optimization. Journal of Machine Learning Research 13, 1809–1837 (2012)
2. Fedorov, V.V.: Theory of Optimal Experiments. Academic Press (1972)
3. Chen, B., Castro, R., Krause, A.: Joint optimization and variable selection of high-dimensional gaussian processes. In: Proceedings of ICML. ACM (2012)
4. Guestrin, C., Krause, A., Singh, A.: Near-optimal sensor placements in Gaussian processes. In: Proceedings of ICML, pp. 265–272. ACM (2005)
5. Grünewälder, S., Audibert, J.Y., Opper, M., Shawe-Taylor, J.: Regret Bounds for Gaussian Process Bandit Problems. In: Proceedings of AISTATS, pp. 273–280. MIT Press (2010)
6. Srinivas, N., Krause, A., Kakade, S.M., Seeger, M.W.: Information-theoretic regret bounds for gaussian process optimization in the bandit setting. IEEE Transactions on Information Theory 58(5), 3250–3265 (2012)
7. Mockus, J.: Bayesian approach to global optimization: theory and applications. Mathematics and its applications (Kluwer Academic Publishers). Soviet series, Kluwer Academic (1989)
8. Mes, M.R., Powell, W.B., Frazier, P.I.: Hierarchical knowledge gradient for sequential sampling. Journal of Machine Learning Research 12, 2931–2974 (2011)
9. Carpentier, A., Lazaric, A., Ghavamzadeh, M., Munos, R., Auer, P.: Upper-confidence-bound algorithms for active learning in multi-armed bandits. In: Kivinen, J., Szepesvári, C., Ukkonen, E., Zeugmann, T. (eds.) ALT 2011. LNCS, vol. 6925, pp. 189–203. Springer, Heidelberg (2011)
10. Chen, Y., Krause, A.: Near-optimal batch mode active learning and adaptive submodular optimization. In: Proceedings of ICML. ACM (2013)
11. Auer, P., Cesa-Bianchi, N., Fischer, P.: Finite-time analysis of the multiarmed bandit problem. Machine Learning 47(2-3), 235–256 (2002)
12. Auer, P., Ortner, R., Szepesvári, C.: Improved rates for the stochastic continuum-armed bandit problem. In: Proceedings of COLT, pp. 454–468. Omnipress (2007)
13. Coquelin, P.A., Munos, R.: Bandit algorithms for tree search. In: Proceedings of UAI, pp. 67–74. AUAI Press (2007)
14. Kleinberg, R.: Nearly tight bounds for the continuum-armed bandit problem. In: Advances in NIPS, pp. 697–704. MIT Press (2004)
15. Kocsis, L., Szepesvári, C.: Bandit based monte-carlo planning. In: Fürnkranz, J., Scheffer, T., Spiliopoulou, M. (eds.) ECML 2006. LNCS (LNAI), vol. 4212, pp. 282–293. Springer, Heidelberg (2006)
16. Audibert, J.Y., Bubeck, S., Munos, R.: Bandit view on noisy optimization. In: Optimization for Machine Learning, pp. 431–454. MIT Press (2011)
17. Sutton, R.S., Barto, A.G.: Reinforcement Learning: An Introduction. A Bradford Book (1998)
18. Bubeck, S., Munos, Stoltz, G., Szepesvári, C.: Online optimization in x-armed bandits. In: Advances in NIPS, pp. 201–208. Curran Associates, Inc. (2008)
19. Desautels, T., Krause, A., Burdick, J.: Parallelizing exploration-exploitation tradeoffs with gaussian process bandit optimization. In: Proceedings of ICML. icml.cc/Omnipress (2012)
20. Azimi, J., Fern, A., Fern, X.: Batch bayesian optimization via simulation matching. In: Advances in NIPS, pp. 109–117. Curran Associates, Inc. (2010)
21. Bubeck, S., Cesa-Bianchi, N.: Regret analysis of stochastic and nonstochastic multi-armed bandit problems. Foundations and Trends in Machine Learning 5(1), 1–122 (2012)

22. Bubeck, S., Munos, R., Stoltz, G.: Pure exploration in multi-armed bandits problems. In: Gavaldà, R., Lugosi, G., Zeugmann, T., Zilles, S. (eds.) ALT 2009. LNCS, vol. 5809, pp. 23–37. Springer, Heidelberg (2009)
23. Rasmussen, C.E., Williams, C.: Gaussian Processes for Machine Learning. MIT Press (2005)
24. de Freitas, N., Smola, A.J., Zoghi, M.: Exponential regret bounds for gaussian process bandits with deterministic observations. In: Proceedings of ICML. icml.cc/Omnipress (2012)
25. Cover, T.M., Thomas, J.A.: Elements of Information Theory. Wiley-Interscience (1991)
26. Ko, C., Lee, J., Queyranne, M.: An exact algorithm for maximum entropy sampling. Operations Research, 684–691 (1995)
27. Kuss, M., Pfingsten, T., Csató, L., Rasmussen, C.E.: Approximate inference for robust gaussian process regression (2005)
28. Krause, A., Guestrin, C.: Near-optimal nonmyopic value of information in graphical models. In: Proceedings of UAI, pp. 324–331. AUAI Press (2005)
29. Flake, G.W., Lawrence, S.: Efficient svm regression training with smo. Machine Learning 46(1-3), 271–290 (2002)
30. Hill, E.M., Borrero, J.C., Huang, Z., Qiu, Q., Banerjee, P., Natawidjaja, D.H., Elosegui, P., Fritz, H.M., Suwargadi, B.W., Pranantyo, I.R., Li, L., Macpherson, K.A., Skanavis, V., Synolakis, C.E., Sieh, K.: The 2010 mw 7.8 mentawai earthquake: Very shallow source of a rare tsunami earthquake determined from tsunami field survey and near-field gps data. J. Geophys. Res. 117, B06402 (2010)
31. Stefanakis, T.S., Dias, F., Vayatis, N., Guillas, S.: Long-wave runup on a plane beach behind a conical island. In: Proceedings of WCEE (2012)
32. Dutykh, D., Poncet, R., Dias, F.: The VOLNA code for the numerical modelling of tsunami waves: generation, propagation and inundation. European Journal of Mechanics B/Fluids 30, 598–615 (2011)
33. Nash, W., Tasmania. Marine Research Laboratories: The population biology of abalone (haliotis species) in tasmania: Blacklip abalone (h. rubra) from the north coast and the islands of bass strait. Technical report, Tasmania. Sea Fisheries Division (1994)

Greedy Confidence Pursuit: A Pragmatic Approach to Multi-bandit Optimization

Philip Bachman and Doina Precup

McGill University, School of Computer Science
phil.bachman@gmail.com, dprecup@cs.mcgill.ca

Abstract. We address the practical problem of maximizing the number of high-confidence results produced among multiple experiments sharing an exhaustible pool of resources. We formalize this problem in the framework of bandit optimization as follows: given a set of multiple multi-armed bandits and a budget on the total number of trials allocated among them, select the top-m arms (with high confidence) for as many of the bandits as possible. To solve this problem, which we call *greedy confidence pursuit*, we develop a method based on *posterior sampling*. We show empirically that our method outperforms existing methods for top-m selection in single bandits, which has been studied previously, and improves on baseline methods for the full greedy confidence pursuit problem, which has not been studied previously.

1 Introduction

Clinical and scientific teams often pursue multiple research objectives on a fixed budget. To obtain as many significant results as possible, they must intelligently allocate their limited resources among one or more concurrent experiments. The machine learning community has developed ways to formulate and address variations on this problem. For example, budgeted learning [12] and subsequent work considers the problem of active learning when a fixed budget is given for probing which model among a collection of models is best for a given task.

In this paper, we adopt the framework provided by bandit problems [3] to address resource allocation among multiple concurrent tasks. Bandits offer a simple way of formalizing many decision problems, e.g. deciding which among a set of drugs most effectively treats a particular disease. In the standard formulation a bandit has multiple arms with unknown expected payoffs and one must probingly pull the arms in order to find the best one. Most bandit optimization problems focus on regret minimization, i.e. minimizing some measure of loss incurred over the course of an experiment. The goal in practical experimental settings, e.g. clinical trials, is often different: one typically has a fixed budget for acquiring patients to be treated, and the goal is to identify the best treatment option *at the end* of the experiment. Hence, payoffs during the experiment are not counted, in contrast to regret minimization, and the objective is solely to maximize the (statistical) confidence with which the best action can be selected

H. Blockeel et al. (Eds.): ECML PKDD 2013, Part I, LNAI 8188, pp. 241–256, 2013.

after the experiment is over. In a recent series of papers, this idea has been developed under the label "pure exploration" in multi-armed bandits [4, 2, 8].

The problem of selecting the best arm with high confidence using a minimum number of trials has also been tackled by [13] and [7]. In [9], the authors extended the approach of [7] to the case in which one wants to select not just the best arm, but the m arms with highest payoffs. In recent work [10], the same authors provided an alternative algorithm with stronger PAC guarantees. Note that best-arm selection is the special case of top-m selection where $m = 1$.

In the clinical trial setting, significant interest is currently directed towards personalized medicine based on treatments which only work for specific sub-populations. For example, it is understood that diseases like cancer may evolve differently based on certain genetic mutations, and thus any treatment for such a disease may only benefit certain types of patients. In such cases, no budgeted clinical trial can hope to show that a treatment is universally effective; instead, one should try to identify sub-populations within which the treatment works with high confidence. One can naturally describe this problem using multiple bandits (i.e. sub-populations) each comprising multiple arms (i.e. available treatments). Typically, a fixed total number of patients can be enrolled (corresponding to a fixed total number of trials). Hence, patients should be recruited and allocated among the sub-populations and treatments to maximize the number of sub-populations for which an effective treatment is confidently identified.

We formalize this problem as multi-bandit top-m selection: given a set of n multi-armed bandits, a trial budget T, and a target confidence τ, maximize the number of top-m groups identified with confidence $\rho > \tau$ after performing T trials. We refer to this general problem as *greedy confidence pursuit*, as it preferentially directs resources (i.e. trials) towards experiments (i.e. bandits) in which confident results are easiest to achieve. Work in [8] addresses a related problem which focuses, roughly speaking, on minimizing the probability of incorrectly identifying any top-m group. We will discuss the relation between [8] and our own work in detail. Similarly, work in [6] considers a multi-bandit objective which focuses on minimizing the maximum uncertainty among the estimated per-arm returns. In contrast, we propose the more pragmatic objective of maximizing the number of confident results achieved on a fixed budget[1].

In Section 2 of this paper, we define greedy confidence pursuit and contrast it with objectives previously considered in the multi-bandit setting. In Section 3 we develop an algorithm for intra-bandit top-m selection in bandits with Bernoulli-distributed returns. In Section 5 we develop an algorithm for inter-bandit trial allocation which completes our approach to greedy confidence pursuit. In Sections 4 and 6, we compare the performance of our algorithms with existing algorithms across a range of problems, illustrating the power of our approach and highlighting the differences between greedy confidence pursuit and other

---

[1] Our objective is pragmatic as many practical scenarios (e.g. scientific publication) require surpassing some confidence threshold for capturing *any* value, with extra confidence beyond the threshold providing rapidly diminishing additional value.

objectives previously considered in the multi-bandit setting. We conclude the paper and discuss future work in Section 7.

2 Motivating and Formulating Our Objective

Consider a pharmaceutical developer evaluating a new drug for potential use in multiple sub-populations of patients. Given a fixed budget for processing trial patients, the developer may seek to maximize the number of sub-populations for which their proposed drug is identified as significantly better than existing treatments[2]. We can formalize this problem as follows:

- Each sub-population is represented by a bandit b_i.
- Each bandit has a set of n arms $A_i = \{a_{i1}, ..., a_{in}\}$, with arm a_{i1} representing the new drug and the rest representing existing treatments.
- The random variables $\mathbf{R}_i = \{\mathbf{r}_{i1}, ..., \mathbf{r}_{in}\}$ give the per-trial outcomes for b_i.
- The objective is to maximize the number of bandits b_i for which we find $\mathbb{E}[\mathbf{r}_{i1}] > \max_{j \neq 1} \mathbb{E}[\mathbf{r}_{ij}]$ with confidence $\rho_i > \tau$.

In the above scenario, only confident results involving a particular target arm (i.e. the pharmaceutical developer's proposed drug) are considered worth pursuing. This represents a variant of the general greedy confidence pursuit problem, in which confident results involving any best arm are pursued equally.

We approach greedy confidence pursuit by decomposing trial allocation into three stages: bandit selection, arm selection, and belief updates based on the trial outcome. Methods for these stages can be combined "a la carte", which facilitates algorithm development and eases comparison with existing work.

In previous work [8], given a set of N bandits $B = \{b_1, ...b_N\}$, Gabillon et. al proposed the following objective for multi-bandit subset selection:

$$\text{maximize } \mathbb{E}_H[\min_i \rho_i], \tag{1}$$

where ρ_i measures the confidence that the top-m group selected for bandit b_i is correct. In contrast, our objective can be written as follows:

$$\text{maximize } \mathbb{E}_H[\sum_i \mathbb{I}\{\rho_i > \tau\}], \tag{2}$$

where ρ_i is as above, τ is a confidence threshold and $\mathbb{I}$ is the indicator function. The expectations are over *histories* (i.e. sequences of observed trial outcomes). Intuitively, (1) maximizes a lower bound on the per-bandit confidences and (2) maximizes the number of bandits for which the top-m group can be selected with high confidence. The precise confidence measure we use is given in (3).

For both objectives (1) and (2), the trials allocated to bandit b_i should be distributed among its arms to maximize ρ_i. Hence, good arm selection for (1) will

---

[2] While human drug trials are slow to adopt novel experimental designs, one could analogously consider trials of a new consumer product across multiple potential target demographics, or exploratory drug trials in non-human model systems.

also be good for (2). However, methods for optimizing these objectives will select bandits quite differently. Intuitively, methods optimizing (1) will tend to allocate trials to bandits with relatively low confidence, while methods optimizing (2) will tend to allocate trials to bandits with relatively low expected *completion cost* (out of the bandits b_i for which $\rho_i \leq \tau$). The practical differences between (1) and (2) are most striking when some bandit b_i is effectively intractable with respect to the operative confidence measure and trial budget; an algorithm optimizing (1) will still sink its budget into (hopelessly) pursuing improvements in ρ_i, while algorithms optimizing (2) will ignore b_i in favor of lower-hanging fruit.

The completion cost is a critical concept when working with (2), which we define as follows: a bandit b_i has completion cost c_i if efficiently allocating c_i trials among the arms of b_i is expected to push ρ_i above τ. Note that if each c_i were deterministic and known a priori, an optimal trial allocation policy for (2) would be to sort the bandits such that $c_1 \leq c_2 \leq ... \leq c_N$, then allocate c_1 trials to b_1, c_2 trials to b_2 etc., until budget exhaustion. This greedy policy maximizes the number of tasks completed on a fixed budget when each task has a known cost. The difficulty in our case is that each c_i is neither known a priori nor deterministic. Thus, a balance between exploring (to better estimate each c_i) and exploiting (to push each ρ_i past τ) must be struck.

In Section 5 we describe an estimator for the completion costs c_i and discuss how to use these estimates for inter-bandit trial allocation during greedy confidence pursuit. Next, we present our method for intra-bandit top-m selection.

3   Bayesian Top-m Selection

Our intra-bandit subset selection algorithm uses Bayesian estimates of the per-arm returns and follows a general approach called posterior sampling, of which Thompson sampling [16] is perhaps the best-known example. The notation introduced in this section will be reused throughout the remainder of this paper.

3.1   Definitions and Notation

For a set B of N bandits, where each $b_i \in B$ has a set A_i of n arms with Bernoulli-distributed returns, our algorithm maintains its beliefs about the return of each arm $a_{ij} \in A_i$ using a beta distribution $\mathcal{B}_{ij} = \mathcal{B}(\alpha_{ij}, \beta_{ij})$, where α_{ij} and β_{ij} count the observed successes and failures for arm a_{ij}, respectively. We set priors over the returns by initializing all parameters α_{ij} and β_{ij} to a common value (e.g. we set them to 1 in all of our tests). The belief for arm a_{ij} is updated by incrementing α_{ij} or β_{ij} following each trial allocated to a_{ij}. The MAP estimate of the return of a_{ij} is given by $\alpha_{ij}(\alpha_{ij} + \beta_{ij})^{-1}$.

For a bandit b_i with current MAP return estimates $\bar{R}_i = \{\bar{r}_{i1}, ..., \bar{r}_{in}\}$, we define its current MAP gap location $\bar{\gamma}_i$ as $\frac{1}{2}(\bar{r}_{im} + \bar{r}_{i(m+1)})$, in which $\bar{r}_{im}$ and $\bar{r}_{i(m+1)}$ refer to the mth and $(m+1)^{th}$ largest MAP return estimates respectively. Given $\bar{\gamma}_i$, we define the current MAP per-arm gaps $\bar{\Gamma}_i = \{\bar{\gamma}_{i1}, ...\bar{\gamma}_{in}\}$ such that $\bar{\gamma}_{ij} = |\bar{r}_{ij} - \bar{\gamma}_i|$. We also refer to a bandit's true returns and gaps (R_i, Γ_i) as its parameters $\theta \in \Theta$, where Θ spans all bandits permitted by the prior.

We associate each bandit b_i with a confidence ρ_i, which should give the probability that its current top-m group (based on the MAP return estimates) is correct. Since this is impractical to compute exactly, we use a lower bound[3]. For a bandit b_i with current MAP return estimates $\bar{R}_i = \{\bar{r}_{i1}, ..., \bar{r}_{in}\}$, we compute this bound as follows:

$$\bar{\rho}_i = 1 - \sum_{j=1}^{n} 1 - \Phi\left(\frac{\sqrt{t_{ij}}\, |\bar{r}_{ij} - \bar{\gamma}_i|}{\bar{\sigma}_{ij}}\right), \tag{3}$$

where $\bar{\sigma}_{ij} = \sqrt{\bar{r}_{ij}(1 - \bar{r}_{ij})}$ is the current MAP estimate of the standard deviation of the return for arm a_{ij}, t_{ij} is the number of trials previously allocated to arm a_{ij}, $\bar{\gamma}_i$ is the MAP gap location derived from $\bar{R}_i$, and Φ is the CDF for a standard normal distribution. This bound uses a normal approximation to the posterior distribution of the return estimate for each arm and computes a union bound on the probability that all MAP return estimates are on the same side of $\bar{\gamma}_i$ as their true values. When (3) is negative, we define $\bar{\rho}_i = 0$.

The algorithms presented in this paper all sample from the current posterior over a bandit's returns and gaps (i.e. its parameters $\theta \in \Theta$) as follows: sample a return for each arm from its current Beta distribution, compute the gap location implied by the sampled returns, and compute the per-arm gaps using the sampled returns and the computed gap location.

3.2 Posterior Sampling and Its Merits

Posterior sampling, or randomized probability matching, is a flexible approach to sequential optimization problems drawing increasing interest from the theoretical and applied sides of machine learning [1, 11, 15, 5]. For bandit problems, posterior sampling policies π^p select arms as follows:

$$\pi^p(a_{ij}|H) \propto p\left(a_{ij} = \arg\max_{a_{kl}} f_\theta^H(a_{kl}) \,\middle|\, H\right), \tag{4}$$

in which $\pi^p(a_{ij}|H)$ is the probability of π^p selecting a_{ij} given H, the trial history H records the outcomes of all previous trials, $\theta \in \Theta$ is an unobserved parameter specifying the distribution of the bandit's returns, and f_θ^H is any deterministic function with bounded range. The remaining component of any posterior sampling policy π^p is the conditional distribution $p(\theta|H)$, which describes the posterior over $\theta \in \Theta$ after observing the trials recorded in H. Alg. (1) gives the general form followed by posterior sampling algorithms.

While the f_θ^H used in (4) must be deterministic given particular values for θ and H, its use in posterior sampling induces a stochastic policy by virtue of our imperfect knowledge of θ, which we observe only through the trials recorded in

---

[3] The true (Bayesian) confidence for a bandit can be computed to arbitrary precision by repeatedly sampling from the joint posterior over its per-arm returns and observing the frequency with which its MAP top-m group appears as the top-m group among the sampled sets of returns.

H. Thus, while f_θ^H must be deterministic, its value for a particular arm a_{ij} given a particular history H is stochastic, with stochasticity provided by entropy in the posterior $p(\theta|H)$.

Algorithm 1. PostSample(f_θ^H, $p(\theta|H)$, H, T)

1: **for** $1 \leq t \leq T$:
3:    Sample $\hat\theta \in \Theta$ from the posterior given by $p(\hat\theta|H)$
4:    Let $\hat a_{ij}^* = \arg\max_{a_{kl}} f_{\hat\theta}^H(a_{kl})$
5:    Pull arm $\hat a_{ij}^*$ and update H based on the outcome
6: **end for**

The performance of a posterior sampling policy π^p is most naturally measured by its *Bayes risk* with respect to f_θ^H, which can be written as follows:

$$\mathbb{E}_\theta \sum_{t=1}^T \left[f_\theta^H(a_{ij}^*) - f_\theta^H(\pi_t^p) \right] , \tag{5}$$

in which π_t^p indicates an arm selected according to the probabilities given by $\pi^p(a_{ij}|H)$ and a_{ij}^* is an arm which maximizes f_θ^H. The Bayes risk describes the sub-optimality of π^p with respect to an optimal policy π^* that always knows a_{ij}^*, with respect to a prior over Θ chosen a priori. Based on work in [14], we decompose the Bayes risk for posterior sampling policies as follows:

$$(5) = \mathbb{E}_H \mathbb{E}_\theta \sum_{t=1}^T \left[f_\theta^H(a_{ij}^*) - f_\theta^H(\pi_t^p) \right]$$

$$= \mathbb{E}_H \mathbb{E}_\theta \sum_{t=1}^T \left[f_\theta^H(a_{ij}^*) - U_t^H(\pi_t^p) + U_t^H(\pi_t^p) - f_\theta^H(\pi_t^p) \right]$$

$$= \mathbb{E}_H \mathbb{E}_\theta \sum_{t=1}^T \left[f_\theta^H(a_{ij}^*) - U_t^H(a_{ij}^*) + U_t^H(\pi_t^p) - f_\theta^H(\pi_t^p) \right]$$

$$= \mathbb{E}_\theta \sum_{t=1}^T \left[f_\theta^H(a_{ij}^*) - U_t^H(a_{ij}^*) \right] + \mathbb{E}_\theta \sum_{t=1}^T \left[U_t^H(\pi_t^p) - f_\theta^H(\pi_t^p) \right]$$

in which U_t^H is any function that is deterministic and bounded given H. The key step in this decomposition relies on the property that $\mathbb{E}_{\theta|H}[U_t^H(a_{ij}^*)] = \mathbb{E}_{\theta|H}[U_t^H(\pi_t^p)]$, which results from the posterior sampling construction of π^p according to (4), which makes the distributions $\pi^p(a_{ij}|H)$ and $p(a_{ij} = a_{ij}^*|H)$ identical. We emphasize that this decomposition is valid for any π^p based on posterior sampling for any f_θ^H and U_t^H meeting the stated contraints.

Analyses of the Bayes risk for UCB policies follow a decomposition parallel to that for posterior sampling, with a final step that results in:

$$\mathbb{E}_\theta \sum_{t=1}^T \left[f_\theta^H(a_{ij}^*) - U_t^H(a_{ij}^*) \right] + \mathbb{E}_\theta \sum_{t=1}^T \left[U_t^H(\pi_t^u) - f_\theta^H(\pi_t^u) \right] ,$$

in which U_t^H meets the same constraints as for posterior sampling and π_t^u is the arm selected by a UCB policy π^u based on U_t^H, i.e. one where $\pi_t^u = \arg\max_{a_{ij}} U_t^H(a_{ij})$. The key step in the Bayes risk decomposition for UCB policies relies on the fact that $U_t^H(\pi_t^u) \geq U_t^H(a_{ij}^*)$ for all t, due to the UCB construction of π^u.

The parallel decompositions of the Bayes risks for posterior sampling and UCB algorithms show that, if for some f_θ^H there exists an upper bound U_t^H which produces a UCB policy π^u with provably good Bayes risk, then substituting that U_t^H into the decomposed Bayes risk for the policy π^p which performs posterior sampling with respect to f_θ^H proves an equivalent Bayes risk for π^p. Thus, the Bayes risk of posterior sampling with respect to any f_θ^H is upper-bounded by the lowest Bayes risk upper bound for any π^u constructed from any upper bound U_t^H on f_θ^H. For detailed coverage of this result and its implications, see [14].

3.3   Top-m Selection via Posterior Sampling

Motivated by the preceding result, we derive a function f_θ^H for which good Bayes risk ensures good subset selection performance. We begin by restating an efficient static allocation policy π^s for subset selection described in detail by [8]:

$$\pi_\theta^s(a_{ij}) = \frac{Tb^2}{\gamma_{ij}^2 \sum_{kl} \frac{b^2}{\gamma_{kl}^2}} , \tag{6}$$

in which $\pi_\theta^s(a_{ij})$ gives the number of trials to allocate to a_{ij} assuming the gaps γ_{ij} are known a priori (the gaps are determined by the bandit parameters θ), T gives the total number of trials to allocate, and b is a bound on the range of the returns (e.g., $b = 1$ for Bernoulli bandits). The policy induced by (6) is optimal with respect to a lower bound on selection confidence analogous to that in (3). Next, for any trial history H, define $H(a_{ij})$ as the number of trials recorded for a_{ij} in H[4]. Finally, for history H and bandit parameters θ, define the log misallocation ratio as:

$$f_\theta^H(a_{ij}) = \log\left(\frac{\pi_\theta^s(a_{ij})}{H(a_{ij})}\right) . \tag{7}$$

Note that this f_θ^H implicitly depends on the desired subset size m through the definition of the per-arm gaps used in computing $\pi_\theta^s(a_{ij})$ for each arm and that it is bounded by $\pm\log(T)$. This f_θ^H provides a particularly interesting target for posterior sampling because we only ever observe it indirectly, through the information recorded in H over the course of an experiment.

Intuitively, posterior sampling with respect to (7) will select arms in proportion to their posterior probability of being most under-sampled relative to their sample density in the optimal static policy π_θ^s. Any policy whose Bayes risk with respect to (7) grows sublinearly in T has performance asymptotically equivalent

---

[4] Without loss of generality, we assume all arms have at least one trial in H.

to that of π_θ^s for the true θ as $T \to \infty$. And, from the earlier result, the existence of any UCB policy with good Bayes risk with respect to (7) suggests good Bayes risk for posterior sampling with respect to (7).

We perform intra-bandit top-m selection by posterior sampling with respect to the value in (7). Alg. (2) describes how our algorithm allocates trials at each round. While our full approach to greedy confidence pursuit calls Alg. (2) one round at a time, it can also be iterated following the form of Alg. (1) for application to single bandit subset selection problems.

Algorithm 2. SelectArm(bandit b_i, trial history H)

1: Sample $\hat{\theta}_i = (\hat{R}_i, \hat{\Gamma}_i)$ according to $p(\hat{\theta}_i | H)$.
2: Compute $\pi_{\hat{\theta}_i}^s(a_{ij})$ for each $a_{ij} \in A_i$ according to (6).
3: Compute $\hat{a}_{ij}^* = \arg\max_{a_{ij}} f_{\hat{\theta}_i}^H(a_{ij})$, with $f_{\hat{\theta}_i}^H$ as in (7).
4: Return $\hat{a}_{ij}^*$.

As further justification for our algorithm, consider the relation:

$$\arg\max_{a_{ij}} \log \left(\frac{\pi_\theta^s(a_{ij})}{H(a_{ij})} \right) = \arg\min_{a_{ij}} \frac{\sqrt{H(a_{ij})}\gamma_{ij}}{b}, \tag{8}$$

which follows from a straightforward derivation. If one were to model all arms using the same bound b on their standard deviation, then the values in the argmin above are equivalent to the values passed to Φ in (3) when computing the contribution of each arm to a bandit's confidence ρ_i. Thus, by posterior sampling with respect to (7), our algorithm selects arms according to their posterior probability of having the lowest confidence in (3). This can be interpreted as stochastic greedy maximization of the following lower bound on ρ_i:

$$\rho_i \geq 1 - n \left(1 - \min_j \left[\Phi \left(\frac{\sqrt{t_{ij}} |\bar{r}_{ij} - \bar{\gamma}_i|}{\bar{\sigma}_{ij}} \right) \right] \right). \tag{9}$$

4   Testing Top-m Selection

This section empirically compares our subset selection algorithm with two existing methods. The first one [10] offers a standard PAC guarantee on sample complexity and success probability that matches a theoretical lower bound on the optimal samples/accuracy tradeoff (up to constant factors). The second method is based on the optimally efficient (up to constant factors) method for best arm selection presented in [8], which we adapt for use in subset selection. We refer to our arm selection method as Bayes and the respective baseline methods as PAC and UCB. We now describe the PAC and UCB methods as used in our tests.

Using the notation from the previous section, both the PAC and UCB methods rely primarily on the current MAP estimates of the gaps (i.e. $\{\bar{\gamma}_{i1}, ..., \bar{\gamma}_{in}\}$) for

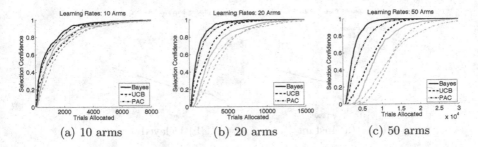

Fig. 1. These plots show average confidence lower bounds as a function of trials allocated for three different arm selection methods and two subset sizes at each of three arm counts. To generate each line, confidence lower bounds were averaged over 100 tests using bandits generated as described in the main text. Methods are indicated by line style. In each subfigure, the darker lines correspond to selecting the best arm and the lighter lines correspond to selecting the top half of the arms.

each of the arms in bandit b_i. Both methods allocate the next trial to an arm a_{ij} such that $-\bar{\gamma}_{ij} + \beta_{ij} = \max_k[-\bar{\gamma}_{ik} + \beta_{ik}]$, in which the negative gap $-\bar{\gamma}_{ij}$ encourages a focus on arms near the boundary and the term β_{ij} encourages exploration to improve the per-arm gap estimates. The PAC and UCB methods differ only in their computation of the β_{ij} term.

The PAC method, referred to in [10] as LUCB1, computes β_{ij} as follows:

$$\beta_{ij} = \sqrt{\frac{1}{2t_{ij}} \ln\left(\frac{5nt^4}{4\delta}\right)}, \qquad (10)$$

in which t_{ij} gives the number of trials previously allocated to a_{ij}, t gives the total number of trials previously allocated, n is the number of bandit arms, and $(1 - \delta)$ is the desired probability of correct subset selection (we set $\delta = 0.05$ in our tests). UCB computes β_{ij} as follows:

$$\beta_{ij} = \sqrt{\frac{2\kappa_i \bar{\sigma}_{ij}^2}{t_{ij}}} + \frac{7\kappa_i \nu_i}{3(t_{ij} - 1)}, \qquad (11)$$

in which $\bar{\sigma}_{ij}^2$ is the current empirical (i.e. MAP) estimate of the variance of the return of a_{ij}, t_{ij} is as in (10), and κ_i/ν_i are constants computed from continuously updated empirical estimates of the complexity of bandit b_i. A full description of the κ_i/ν_i computations is beyond the scope of this paper and appears in [8][5].

All tests underlying Figures 1 and 2 used bandits with return distributions generated by the same process. Four parameters determined the return distribution of each bandit used in these tests: the minimum allowed gap γ_{min}, the

---

[5] For those familiar with the source material, we have implemented AGapE-V with the per-arm gaps Δ_{mk} redefined to permit top-m selection. This redefinition of the gaps permits simpler notation, while effecting only a constant shift in all gap values, thus leaving the selection process unchanged when $m = 1$.

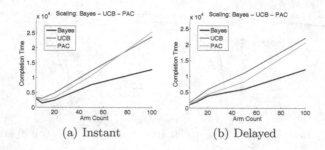

Fig. 2. This plot compares best arm selection performance of the Bayes, PAC, and UCB algorithms. The lines show the median completion times achieved by each method over 100 tests at each arm count in $\{5, 10, 20, 50, 100\}$, with bandits generated as described in the text. Tests were considered complete when a confidence ≥ 0.98 was maintained for at least 100 rounds. Feedback in (a) was instant, while feedback in (b) was delayed 100 trials.

maximum allowed gap γ_{max}, the number of arms n, and the number of top arms to select m. Without loss of generality, we assume that the arms are sorted in order of descending returns. We generated a random set of returns meeting the constraints imposed by these parameters by generating sets of n returns uniformly distributed over $[0.1...0.9]$ until the gap between the mth and $(m+1)^{th}$ largest returns was in the range $[\gamma_{min}...\gamma_{max}]$. For the tests in this section, and those in the remaining sections, for a given set of per-arm returns (i.e. a bandit), we presented each algorithm with matching sequences of trial outcomes. This allowed us to expose all methods tested to problems of equivalent difficulty. For tests in this section we set $\gamma_{min} = 0.05$ and $\gamma_{max} = 0.15$.

In Figure 1 we show the results of running the Bayes, PAC, and UCB methods on bandits with various arm counts when selecting either the best arm or the top half of the arms. We plot the average learning curves over 100 bandits for each arm count/subset size pair. The confidence values plotted in these curves were computed according to (3). Confidence curves for all other tests in this paper were computed similarly. The tests in Figure 1 show our method consistently outperforming existing methods over all arm counts and subset sizes.

Figure 2 compares Bayes, PAC, and UCB methods across a larger range of arm counts, in the context of best arm selection. For these tests, we compute *completion time* as the first round at which the confidence bound $\bar{\rho}_i$ was at least 0.98 for the previous 100 trials. Our method clearly has a large advantage as the number of arms increases. While the absolute advantage at arm counts ≤ 10 is smaller, it still represents a $10\% - 20\%$ reduction in completion time.

5   Bayesian Greedy Confidence Pursuit

Recall that, if the completion cost c_i for each bandit b_i were deterministic and known a priori, an optimal policy for greedy confidence pursuit would be to complete bandits in order of increasing completion costs, until budget exhaustion.

To compensate for the uncertain completion costs encountered in practical scenarios, we address greedy confidence pursuit by posterior sampling with respect to an approximate per-bandit completion cost.

For use in greedy confidence pursuit, an approximate completion cost need only predict the relative ranking of a set of bandits in terms of their true completion costs, as this permits mimicking the optimal greedy policy for known completion costs, which depends only on the cost-induced bandit order. For a bandit b_i with true returns R_i and gaps Γ_i, we use the following cost estimate:

$$\hat{c}_i = \sum_{j=1}^{n} \frac{\left(\sigma_{ij} + \sqrt{\sigma_{ij}^2 + (16/3)\gamma_{ij}}\right)^2}{\gamma_{ij}^2}, \tag{12}$$

in which σ_{ij} is the standard deviation associated with the return r_{ij}. The value in (12) comes from a bandit complexity measure described in [8]. Figure 3 supports the predictiveness of (12) with respect to relative empirical costs.

Note that, if one assumes the same target confidence τ for all bandits $b_i \in B$, then accounting for τ in $\hat{c}_i$ would not affect the ordering of bandits according to $\hat{c}_i$, as an "easier" bandit according to (12) would also have a smaller expected completion cost for any value of τ. By using (12), we also ignore the effort previously expended on a given bandit. While considering the number of trials already spent on a bandit could improve on the performance of (12), it would require steps to avoid the "sunk-cost" fallacy of economics, as manifested by premature commitment to bandits wrongly identified as "easy".

We perform bandit selection for greedy confidence pursuit by (minimum) posterior sampling with respect to $f_\theta^H(b_i) = \hat{c}_i$. The resulting algorithm is given in Alg. (3). Note that (12) captures dependence on the subset size m through its use of Γ_i and that $\hat{c}_i$ becomes stochastic when sampled with respect to the

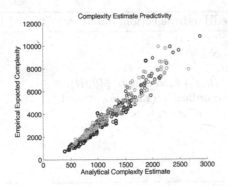

Fig. 3. This figure examines the predictive power of the completion cost in (12). From darkest to lightest the points represent selecting the top 1, 3, and 5 arms of a 10-armed bandit. Points correspond to particular bandits for which 20 runs of our Bayesian subset selection were performed, using independently generated trial outcomes for each run. The x coordinate of each point is the value of (12) for the true returns and gaps underlying its runs, while the y coordinate is the mean completion time for its runs.

per-bandit posteriors over returns and gaps. A theoretical analysis of our allocation process is beyond the scope of this paper, but the properties of posterior sampling described in Section 3 suggest it will efficiently direct trials towards the bandits with minimal $\hat{c}_i$. Section 6 empirically supports the design of this approach.

Algorithm 3. SelectBandit(bandit set B, trial history H)

1: **for each** $b_i \in B$:
2:     Sample $\hat{\theta}_i = (\hat{R}_i, \hat{\Gamma}_i)$ according to $p(\hat{\theta}_i|H)$.
3:     Compute $\hat{c}_i$ according to (12) using $\hat{R}_i$ and $\hat{\Gamma}_i$.
4: **end for**
5: Let $b_i^* = \arg\min_{b_i : \rho_i < \tau} \hat{c}_i$.
6: Return b_{i*}.

5.1   Greedy Confidence Pursuit for "Targeted" Tasks

Now, consider the following problem:

- Given a finite trial budget T and N bandits b_i with returns $R_i = \{r_{i1}, ... r_{in}\}$
- Maximize the number of bandits b_i for which we can say with confidence ρ_i greater than τ that (without loss of generality) $r_{i1} > \max_{j \neq 1} r_{ij}$,

which reformulates the example scenario from Section 2. The twist in this scenario is that we only care about bandits for which a specific arm is best.

We address this problem by extending our algorithm for bandit selection in general greedy confidence pursuit. Intuitively, we sample bandits in proportion to their probability of having the lowest completion cost among bandits in which the targeted arm is best. Alg. (4) describes our extension of Alg. (3).

Algorithm 4. TargetedBanditSelection(bandit set B, trial history H)

01: $\forall i$, set $\hat{c}_i = \infty$.
02: **while** $(\min_i \hat{c}_i == \infty)$
03:     **for each** $b_i \in B$:
04:         Sample $\hat{\theta}_i = (\hat{R}_i, \hat{\Gamma}_i)$ according to $p(\hat{\theta}_i|H)$.
05:         Compute $\hat{c}_i$ according to (12) using $\hat{R}_i$ and $\hat{\Gamma}_i$.
06:         If $\hat{r}_{i1} < \max_{j \neq 1} \hat{r}_{ij}$, set $\hat{c}_i = \infty$
08:     **end for**
09: **end while**
10: Let $b_i^* = \arg\min_{b_i : \rho_i < \tau} \hat{c}_i$.
11: Return b_{i*}.

In the next section, we empirically support the value of Alg. (4) in situations where one is focused on maximizing "positive" results involving specific arms. For practical reasons, we upper bound the number of runs through the "resampling" loop of lines $02 - 09$. If, prior to reaching the upper bound, no bandit has been found for which $\hat{r}_{i1} > \max_{j \neq 1} \hat{r}_{ij}$, we select a bandit according to Alg. (3).

6   Testing Greedy Confidence Pursuit

We begin our empirical examination of greedy confidence pursuit with tests supporting (12) as an approximate completion cost. These tests were based on selecting the top 1, 3, and 5 arms of 10-armed bandits, with returns distributed as in Section 4. For each test, we generated a bandit, computed its cost according to (12) using the true returns, and then ran our Bayes arm selection 20 times on the bandit, using independently simulated trials for each run. Each point in Figure 3 corresponds to the analytically computed cost and the empirical expected cost for a particular bandit, with empirical completion costs measured as for Figure 2. These tests show that the cost estimates given by (12) are highly predictive with respect to the behavior of our algorithm.

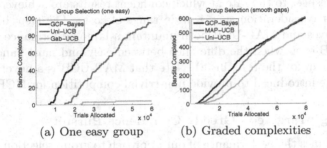

(a) One easy group   (b) Graded complexities

Fig. 4. This figure examines the performance of our method for greedy confidence pursuit, when selecting best arms. Curves in (a) were computed over 100 tests, each of which used 20 10-armed bandits, with the gap for one bandit set to 0.1 and the remaining gaps set to 0.01. Curves in (b) were computed over 100 tests, each of which used 15 10-armed bandits, with the gaps for the bandits evenly spaced on a log scale from 0.01 to 0.1. Bandit generation for the tests in (a) and (b) is described in the text. The curves in (a) and (b) show the number of bandits confidently completed prior to a given trial, aggregated across the relevant tests, with completion defined as for Fig. 2.

For the tests underlying Figures 4 and 5, the per-bandit objective was best arm selection. These tests compared our method for greedy confidence pursuit (tag: GCP-Bayes), comprising the bandit selection described in Section 5 and the arm selection described in Section 3, to three baseline methods. The first baseline was Uni-UCB, which uniformly selected bandits and then applied the UCB arm selection described in Section 4. The second baseline was Gab-UCB, which used UCB arm selection applied jointly over the bandits as described for GapE-V in [8][6]. The final baseline was provided by MAP-UCB, which selected bandits stochastically in inverse proportion to estimates of their completion costs computed by plugging MAP estimates of the relevant values into (12), and then used UCB for intra-bandit arm selection.

---

[6] Note that Gab-UCB is designed to optimize (1) rather than (2). By selecting jointly over all arms/bandits, our Bayesian approach to top-m selection can also be applied towards (1).

Figure 4 examines whether our approach to greedy confidence pursuit can improve the rate at which confident results are achieved. In each test underlying (a), 20 bandits were generated such that one had gap 0.1 and the rest had gap 0.01. For each test underlying (b), 15 bandits were generated to have gaps evenly spaced on a logarithmic scale over [0.01...0.10]. Given the desired gap size γ for each bandit, the best arm was set to return $0.5 + \gamma$, and the remaining returns were set uniformly at random in [0.0...0.5] and then uniformly shifted such that the second best arm had return 0.5. The curves in Figure 4 show the cumulative confident results achieved by each method prior to a given trial, computed based on 100 independently generated sets of test bandits for both (a) and (b).

Overall, Figure 4 shows that, in comparison to Uni-UCB and Gab-UCB, our method significantly accelerates the achievement of confident results. The tests in (a) show that Gab-UCB, which optimizes the objective described in (1), performs poorly with respect to the rate at which confident results are achieved when the bandits under consideration span a wide range of costs. The tests in (b) show that GCP-Bayes and MAP-UCB both maintain a large performance advantage over Uni-UCB even when the difference between easy and hard bandits is less pronounced than for the tests in (a). Note that MAP-UCB is a novel algorithm which we have introduced to provide non-trivial competition for GCP-Bayes.

6.1   Testing "Targeted" Greedy Confidence Pursuit

Figure 5 examines the performance of our approach to group selection for greedy confidence pursuit in the context of the targeted scenario from Section 2. In each test underlying the plots, 20 10-armed bandits were generated with gaps distributed uniformly at random over [0.01...0.10]. In each test, 5 bandits had their target arm best and the other 15 bandits had some other arm best. Given the gap size and best arm index for each bandit, the per-arm returns were set as for the tests underlying Figure 4.

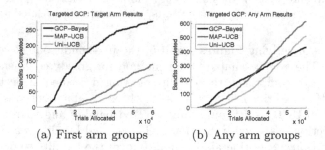

(a) First arm groups   (b) Any arm groups

Fig. 5. This figure gives two views of the cumulative number of best arms confidently selected prior to a given trial, similar to Figure 4. Each of the 100 tests on which these plots are based used 20 10-armed bandits, of which 5 had the target arm best while the remaining 15 had some other arm best. Gaps for all bandits were set uniformly at random in [0.01...0.10]. Bandit generation for these tests is described in the text. Plot (a) shows the cumulative number of bandits completed among those whose target arm was best, while (b) shows cumulative completions among all bandits.

The curves in Figure 5 show the rate at which each considered method achieved confident results, as measured by the number of bandits confidently completed prior to a given round, aggregated over 100 independently generated sets of bandits. For (a), only completed bandits among those with their target arm best were considered when computing the plotted curves. For (b), all completed bandits were considered when computing the plotted curves.

The curves in (a) show that, in comparison to both Uni-UCB and MAP-UCB, the targeted version of GCP-Bayes from Section 5.1 dramatically increases the rate at which confident results are achieved among bandits with their target arm best. The curves in (b) show that the increased focus of this version of GCP-Bayes on a particular subset of the bandits also modestly increases the initial rate at which confident results are achieved among all bandits, but that this early advantage fades as easy target-arm-best results are exhausted. After completing the easiest target-arm-best results, GCP-Bayes falls behind MAP-UCB, which greedily and impartially pursues all easy results.

7   Conclusion and Future Work

We presented a new multi-bandit optimization objective, called greedy confidence pursuit, which captures the general problem of maximizing the number of significant results achieved among a set of experiments sharing a finite pool of fungible resources. We derived algorithms for optimizing this objective in the context of top-m arm identification, both for single and multiple bandits. Our methods compare favorably to existing UCB-style algorithms in terms of empirical performance. In particular, for subset selection, our method scales much better with increasing arm counts than existing algorithms, which suggests its applicability in domains frequently involving numerous actions, such as online advertising and Monte-Carlo tree search for games with high branching factors.

While we used Bernoulli bandits in this paper, our methods directly extend to other return types e.g. normally-distributed continuous returns, through a simple change of priors. Structured priors, e.g. Gaussian processes, can also be used to capture both inter-bandit and intra-bandit relationships between returns. We used bandits with homogenous arm counts, but our methods handle heterogenous arm counts with no changes. With minor modifications, our methods can be used with bandits that share arms and for tasks other than subset selection, e.g. estimating quantiles or rank-ordering all returns. For practical applications, it may also be useful to account for variability in the value of completing each bandit. Such extensions are beyond the scope of the current paper, but provide rich material for future work. We gave one brief illustration of the flexibility granted by our use of posterior sampling by transforming Alg. (3) into Alg. (4), for application to problems in which only specific confident results are pursued.

References

[1] Agrawal, S., Goyal, N.: Analysis of thompson sampling for the multi-armed bandit problem. In: COLT (2012)

[2] Audibert, J.-Y., Bubeck, S., Munos, R.: Best arm identification in multi-armed bandits. In: COLT (2010)

[3] Berry, D.A., Fristedt, B.: Bandit Problems. Chapman and Hall Ltd. (1985)

[4] Bubeck, S., Munos, R., Stoltz, G.: Pure exploration in multi-armed bandits problems. In: Gavaldà, R., Lugosi, G., Zeugmann, T., Zilles, S. (eds.) ALT 2009. LNCS, vol. 5809, pp. 23–37. Springer, Heidelberg (2009)

[5] Chappelle, O., Li, L.: An empirical evaluation of thompson sampling. In: Advances in Neural Information Processing Systems (2011)

[6] Deng, K., Pineau, J., Murphy, S.: Active learning for personalizing treatment. In: IEEE Symposium on Adaptive Dynamic Programming and Reinforcement Learning (2011)

[7] Even-Dar, E., Mannor, S., Mansour, Y.: Action elimination and stopping conditions for the multi-armed bandit and reinforcement learning problems. Journal of Machine Learning Research 7, 1079–1105 (2006)

[8] Gabillon, V., Ghavamzadeh, M., Lazaric, A., Bubeck, S.: Multi-bandit best arm identification. In: Advances in Neural Information Processing Systems (2011)

[9] Kalyanakrishnan, S., Stone, P.: Efficient selection of multiple bandit arms: Theory and practice. In: International Conference on Machine Learning (2010)

[10] Kalyanakrishnan, S., Tewari, A., Auer, P., Stone, P.: Pac subset selection in stochastic multi-armed bandits. In: International Conference on Machine Learning (2012)

[11] Li, L., Chappelle, O.: Open problem: Regret bounds for thompson sampling. In: COLT (2012)

[12] Madani, O., Lizotte, D.J., Greiner, R.: The budgeted multi-armed bandit problem. In: COLT (2004)

[13] Mannor, S., Tsitsiklis, J.N.: The sample complexity of exploration in the multi-armed bandit problem. Journal of Machine Learning Research 5, 623–648 (2004)

[14] Russo, D., Van Roy, B.: Learning to optimize via posterior sampling. arXiv:1301.2609v1 [cs.LG] (2013)

[15] Scott, S.L.: A modern bayesian look at the multi-armed bandit. Applied Stochastic Models in Business and Industry 26, 639–658 (2010)

[16] Thompson, W.R.: On the likelihood that one unknown probability exceeds another in view of the evidence of two samples. Biometrika 25(3-4), 285–294 (1933)

A Time and Space Efficient Algorithm
for Contextual Linear Bandits

José Bento[1], Stratis Ioannidis[2], S. Muthukrishnan[3], and Jinyun Yan[3]

[1] Stanford University
jbento@stanford.edu
[2] Technicolor
stratis.ioannidis@technicolor.com
[3] Rutgers University
{muthu,jinyuny}@cs.rutgers.edu

Abstract. We consider a multi-armed bandit problem where payoffs are a linear function of an observed stochastic contextual variable. In the scenario where there exists a gap between optimal and suboptimal rewards, several algorithms have been proposed that achieve $O(\log T)$ regret after T time steps. However, proposed methods either have a computation complexity per iteration that scales linearly with T or achieve regrets that grow linearly with the number of contexts $|\mathcal{X}|$. We propose an ϵ-greedy type of algorithm that solves both limitations. In particular, when contexts are variables in $\mathbb{R}^d$, we prove that our algorithm has a constant computation complexity per iteration of $O(poly(d))$ and can achieve a regret of $O(poly(d)\log T)$ even when $|\mathcal{X}| = \Omega(2^d)$. In addition, unlike previous algorithms, its space complexity scales like $O(Kd^2)$ and does not grow with T.

Keywords: Contextual Linear Bandits, Space and Time Efficiency.

1 Introduction

The contextual multi-armed bandit problem is a sequential learning problem [17,13]. At each time step, a learner has to chose among a set of possible actions/arms $\mathcal{A}$. Prior to making its decision, the learner observes some additional side information $x \in \mathcal{X}$ over which he has no influence. This is commonly referred to as the *context*. In general, the reward of a particular arm $a \in \mathcal{A}$ under context $x \in \mathcal{X}$ follows some unknown distribution. The goal of the learner is to select arms so that it minimizes its expected *regret*, *i.e.*, the expected difference between its cumulative reward and the reward accrued by an optimal policy, that knows the reward distributions.

Langford and Zhang [17] propose an algorithm called *epoch-Greedy* for general contextual bandits. Their algorithm achieves an $O(\log T)$ regret in the number of timesteps T in the *stochastic* setting, in which contexts are sampled from an unknown distribution in an i.i.d. fashion. Unfortunately, the proposed algorithm and subsequent improvements [13] have high computational complexity. Selecting an arm at time step t requires making a number of calls to a so-called *optimization oracle* that grows polynomially in T. In addition, the cost of an implementation of this optimization oracle can grow linearly in $|\mathcal{X}|$ in the worst case; this is prohibitive in many interesting cases, including

H. Blockeel et al. (Eds.): ECML PKDD 2013, Part I, LNAI 8188, pp. 257–272, 2013.
© Springer-Verlag Berlin Heidelberg 2013

the case where $|\mathcal{X}|$ is exponential in the dimension of the context. In addition, both algorithms proposed in [17] and [13] require keeping a history of observed contexts and arms chosen at every time instant. Hence, their space complexity grows linearly in T.

In this paper, we show that the challenges above can be addressed when rewards are linear. In the above contextual bandit set up, this means that $\mathcal{X}$ is a subset of $\mathbb{R}^d$, and the expected reward of an arm $a \in \mathcal{A}$ is an unknown linear function of the context x, *i.e.*, it has the form $x^\dagger \theta_a$, for some unknown vector θ_a. This is a case of great interest, arising naturally when, conditioned on x, rewards from different arms are uncorrelated:

Example 1. *(Processor Scheduling)* A simple example is assigning incoming jobs to a set of processors $\mathcal{A}$, whose processing capabilities are not known *a priori*. This could be the case if, *e.g.*, the processors are machines in the cloud or, alternatively, humans offering their services through, *e.g.*, Mechanical Turk. Each arriving job is described by a set of attributes $x \in \mathbb{R}^d$, each capturing the work load of different types of sub-tasks this job entails, *e.g.*, computation, I/O, network communication, *etc*. Each processor's unknown feature vector θ_a describes its processing capacity, *i.e.*, the time to complete a sub-task unit, in expectation. The expected time to complete a task x is given by $x^\dagger \theta_a$; the goal of minimizing the delay (or, equivalently, maximizing its negation) brings us in the contextual bandit setting with linear rewards. □

Example 2. *(Display Ad Placement)* In the online ad placement problem, online users are visiting a website, which must decide which ad to show them selected from a set $\mathcal{A}$. Each online user visiting the website is described by a set of attributes $x \in \mathbb{R}^d$ capturing, *e.g.*, its geo-location, its previous viewing history, or any information available through a tracking service like BlueKai. Each ad $a \in \mathcal{A}$ has a probability of being clicked that is of the form $x^\dagger \theta_a$, where $\theta_a \in \mathbb{R}^d$ an unknown vector describing each ad. The system objective is to maximize the number of clicks, falling again under the above contextual bandit setting. □

Example 3. *(Group Activity Selection)* Another motivating example is maximizing group satisfaction, observed as the outcome of a secret ballot election. In this setup, a subset of d users congregate to perform a joint activity, such as, *e.g.*, dining, rock climbing, watching a movie, *etc*. The group is dynamic and, at each time step, the vector $x \in \{0,1\}^d$, is an indicator of present participants. An arm (*i.e.*, a joint activity) is selected; at the end of the activity, each user votes whether they liked the activity or not in a secret ballot, and the final tally is disclosed. In this scenario, the unknown vectors $\theta_a \in \mathbb{R}^d$ indicate the probability a given participant will enjoy activity a, and the goal is to select activities that maximize the aggregate satisfaction among participants present at the given time step. □

Our contributions are as follows.

- We isolate and focus on linear payoff case of stochastic multi-armed bandit problems, and design a simple arm selection policy which does not recourse to sophisticated oracles inherent in prior work.
- We prove that our policy achieves an $O(\log T)$ regret after T steps in the stochastic setting, when the expected rewards of each arm are well separated. This meets the regret bound of best known algorithms for contextual multi-armed bandit problems.

In addition, for many natural scenarios, it scales as $O(poly(d) \log T)$, which we believe we are the first to prove under arm separation and for an efficient algorithm.
- We show that our algorithm has $O(|\mathcal{A}|d^2)$ computational complexity per step and its expected space complexity scales like $O(|\mathcal{A}|d^2)$. For algorithms that achieve similar regrets, this is a significant improvement over known contextual multi-armed bandit problems, as well as for bandits specialized for linear payoffs.

Our algorithm is inspired by the work of [3] on the ϵ-greedy algorithm and the use of linear regression to estimate the parameters θ_a. The main technical innovation is the use of matrix concentration bounds to control the error of the estimates of θ_a in the stochastic setting. We believe that this is a powerful realization and may ultimately help us analyze richer classes of payoff functions.

The remainder of this paper is organized as follows: in Section 2 we compare our results with existing literature. In Section 3 we describe the set up of our problem in more detail. In Section 4 we state our main results and prove them in Section 5. Section 6 is devoted to exemplifying the performance and limitations of our algorithm by means of simple numerical simulations. We discuss challenges in dealing with an adversarial setting in Section 7 and draw our conclusions in Section 8.

2 Related Work

The original paper by Langford and Zhang [17] assumes that the context $x \in \mathcal{X}$ is sampled from a probability distribution $p(x)$ and that, given an arm $a \in \mathcal{A}$, and conditioned on the context x, rewards r are sampled from a probability distribution $p_a(r \mid x)$. As is common in bandit problems, there is a tradeoff between *exploration, i.e.*, selecting arms to sample rewards from the distributions $\{p_a(r \mid x)\}_{a \in \mathcal{A}}$ and learn about them, and *exploitation*, whereby knowledge of these distributions based on the samples is used to select an arm that yields a high payoff.

In this setup, a significant challenge is that, though contexts x are sampled independently, they are not independent conditioned on the arm played: an arm will tend to be selected more often in contexts in which it performs well. Hence, learning the distributions $\{p_a(r \mid x)\}_{a \in \mathcal{A}}$ from such samples is difficult. The epoch-Greedy algorithm [17] deals with this by separating the exploration and exploitation phase, effectively selecting an arm uniformly at random at certain time slots (the exploration "epochs"), and using samples collected only during these epochs to estimate the payoff of each arm in the remaining time slots (for exploitation). Our algorithm uses the same separation in "epochs". Langford and Zhang [17] establish an $O(T^{2/3}(\ln |\mathcal{X}|)^{1/3})$ bound on the regret for epoch-Greedy in their stochastic setting. They further improve this to $O(\log T)$ when a lower bound on the gap between optimal and suboptimal arms in each context exists, *i.e.*, under *arm separation*.

Unfortunately, the price of the generality of the framework in [17] is the high computational complexity when selecting an arm during an exploitation phase. In a recent improvement [13], this computation requires a $poly(T)$ number of calls to an optimization oracle. Most importantly, even in the linear case we study here, there is no clear way to implement this oracle in sub-exponential time in d, the dimension of the context. As Dudik *et al.* [13] point out, the optimization oracle solves a so-called cost-sensitive classification problem. In the particular case of linear bandits, the oracle thus reduces to finding the "least-costly" linear classifier. This is hard, even in the case of only two arms: finding the linear classifier with the minimal number of errors is NP-hard [15], and remains NP hard even if an approximate solution is required [7]. As such, a different approach is warranted under linear rewards.

Contextual bandits with linear rewards is a special case of the classic linear bandit setup [4,9,18,19]. In this setup, the arms themselves are represented as vectors, *i.e.*, $\mathcal{A} \subset \mathbb{R}^d$, and, in addition, the set $\mathcal{A}$ can change from one time slot to the next. The expected payoff of an arm a with vector x_a is given by $x_a^\dagger \theta$, for some unknown vector $\theta \in \mathbb{R}^d$, *common among all arms*.

There are several different variants of the above linear model. Auer [4], Li *et al.* [18], and Chu *et al.* [9], and Li a study this problem in the adversarial setting, assuming a finite number of arms $|A|$. In the adversarial setting, contexts are not sampled i.i.d. from a distribution but can be an arbitrary sequence, for example, chosen by an adversary that has knowledge of the algorithm and its state variables.Both algorithms studied, LinRel and LinUCB, are similar to ours in that they use an upper confidence bound and both estimate the unknown parameters for the linear model using a least-square-error type method. In addition, both methods apply some sort of regularization. LinRel does it by truncating the eigenvalues of a certain matrix and LinUCB by using ridge regression. In the adversarial setting, and with no arm separation, the regret bounds obtained of the form $O(\sqrt{T}polylog(T))$.

Dani *et al.* [12], Rusmevichientong and Tsitsiklis [19], and Abbasi-Yadkori *et al.* [1] study contextual linear bandits in the stochastic setting, in the case where $\mathcal{A}$ is a fixed but possibly uncountable bounded subset of $\mathbb{R}^d$. Dani *et al.* [12] obtain regret bounds of $O(\sqrt{T})$ for an infinite number of arms; under arm separation, by introducing a gap constant Δ, their bound is $O(d^2(\log T)^3)$. Rusmevichientong and Tsitsiklis [19] also study the regret under arm separation and obtain a $O(\log(T))$ bound that depends exponentially on d. Finally, Abbasi-Yadkori *et al.* [1] obtain a $O(poly(d)\log^2(T))$ bound under arm separation.

Our problem can be expressed as a special case of the linear bandits setup by taking $\theta = [\theta_1; \ldots; \theta_K] \in \mathbb{R}^{Kd}$, where $K = |\mathcal{A}|$, and, given context x, associating the i-th arm with an appropriate vector of the form $x_{a_i} = [0 \ldots x \ldots 0]$. As such, all of the bounds described above [4,18,9,12,19,1] can be applied to our setup. However, in our setting, arms are uncorrelated; the above algorithms do not exploit this fact. Our algorithm indeed exploits this to obtain a *logarithmic* regret, while also scaling well in terms of the dimension d.

Several papers study contextual linear bandits under different notions of regret. For example, Dani *et al.* [11] define regret based on the worst sequence of loss vectors. In our setup, this corresponds to the rewards coming from an arbitrary temporal sequence and not from adding noise to $x^\dagger \theta_a$, resembling the 'worst-case' regret definition of [5]. Abernethy *et al.* [2] assume a notion of regret with respect to a best choice fixed in time that the player can make from a fixed set of choices. However, in our case, the best choice changes with time t via the current context. This different setup yields worse bounds than the ones we seek: for both stochastic and adversarial setting the regret is $O(\sqrt{T}polylog(T))$.

Recent studies on multi-class prediction using bandits [16,14,10] have some connections to our work. In this setting, every context x has an associated label y that a learner tries to predict using a linear classifier of the type $\hat{y} = \arg\max_a \theta_a^\dagger x$. Among algorithms proposed, the closest to ours is by Crammer and Gentile [10], which uses an estimator for $\{\theta_a\}$ that is related to LinUCB, LinRel and our algorithm. However, the multi-class prediction problem differs in many ways from our setting. To learn the vectors θ_a, the learner receives a one-bit feedback indicating whether the label predicted is correct (*i.e.*, the arm was maximal) or not. In contrast, in our setting, the learner directly observes $\theta_a^\dagger x$, possibly perturbed by noise, without learning if it is maximal.

Finally, bandit algorithms relying on experts such as EXP4 [6] and EXP4.P [8] can also be applied to our setting. These algorithms require a set of policies (experts) against which the regret is measured. Regret bounds grow as $\log^C N$, where N is the number of experts and C a constant. The trivial reduction of our problem to EXP4(.P) assigns an expert to each possible context-to-arm mapping. The 2^d contexts in our case lead to K^{2^d} experts, an undesirable exponential growth of regret in d; a better choice of experts is a new problem in itself.

3 Model

In this section, we give a precise definition of our linear contextual bandit problem.

Contexts. At every time instant $t \in \{1, 2, ...\}$, a context $x_t \in \mathcal{X} \subset \mathbb{R}^d$, is observed by the learner. We assume that $\|x\|_2 \leq 1$; as the expected reward is linear in x, this assumption is without loss of generality (w.l.o.g.). We prove our main result (Theorem 2) in the stochastic setting where x_t are drawn i.i.d. from an unknown multivariate probability distribution $\mathcal{D}$. In addition, we require that the set of contexts is finite *i.e.*, $|\mathcal{X}| < \infty$. We define $\Sigma_{\min} > 0$ to be the smallest non-zero eigenvalue of the covariance matrix $\Sigma \equiv \mathbb{E}\{x_1 x_1^\dagger\}$.

Arms and Actions. At time t, after observing the context x_t, the learner decides to play an arm $a \in \mathcal{A}$, where $K \equiv |\mathcal{A}|$ is finite. We denote the arm played at this time by a_t. We study *adaptive* arm selection policies, whereby the selection of a_t depends only on the current context x_t, and on all past contexts, actions and rewards. In other words, $a_t = a_t\left(x_t, \{x_\tau, a_\tau, r_\tau\}_{\tau=1}^{t-1}\right)$.

Payoff. After observing a context x_t and selecting an arm a_t, the learner receives a payoff r_{a_t, x_t} which is drawn from a distribution p_{a_t, x_t} independently of all past contexts, actions or payoffs. We assume that the expected payoff is a linear function of the context. In other words,

$$r_{a_t, x_t} = x_t^\dagger \theta_a + \epsilon_{a,t} \tag{1}$$

where $\{\epsilon_{a,t}\}_{a\in\mathcal{A}, t\geq 1}$ are a set of independent random variables with zero mean and $\{\theta_a\}_{a\in\mathcal{A}}$ are unknown parameters in $\mathbb{R}^d$. Note that, w.l.o.g, we can assume that $Q = \max_{a\in\mathcal{A}} \|\theta_a\|_2 \leq 1$. This is because if $Q > 1$, as payoffs are linear, we can divide all payoffs by Q; the resulting payoff is still a linear model, and our results stated below apply. Recall that Z is a sub-gaussian random variable with constant L if $\mathbb{E}\{e^{\gamma Z}\} \leq$

$e^{\gamma^2 L^2}$. In particular, sub-gaussianity implies $\mathbb{E}\{Z\} = 0$. We make the following technical assumption.

Assumption 1. *The random variables $\{\epsilon_{a,t}\}_{a \in \mathcal{A}, t \geq 1}$ are sub-gaussian random variables with constant $L > 0$.*

Regret. Given a context x, the optimal arm is $a_x^* = \arg\max_{a \in \mathcal{A}} x^\dagger \theta_a$. The expected cumulative regret the learner experiences over T steps is defined by

$$R(T) = \mathbb{E}\Big\{ \sum_{t=1}^{T} x_t^\dagger (\theta_{a_{x_t}^*} - \theta_{a_t}) \Big\}. \qquad (2)$$

The expectation above is taken over the contexts x_t. The objective of the learner is to design a policy $a_t = a_t \left(x_t, \{x_\tau, a_\tau, r_\tau\}_{\tau=1}^{t-1} \right)$ that achieves as low expected cumulative regret as possible. In this paper we are also interested in arm selection policies having a low computational complexity. We define $\Delta_{\max} \equiv \max_{a,b \in \mathcal{A}} \|\theta_a - \theta_b\|_2$, and $\Delta_{\min} \equiv \inf_{x \in \mathcal{X}, a: x^\dagger \theta_a < x^\dagger \theta_{a_x^*}} x^\dagger (\theta_{a_x^*} - \theta_a) > 0$. Observe that, by the finiteness of $\mathcal{X}$ and $\mathcal{A}$, the defined infimum is attained (*i.e.*, it is a minimum) and is indeed positive.

4   Main Results

We now present a simple and *efficient* on-line algorithm that, under the above assumptions, has expected *logarithmic* regret. Specifically, its computational complexity, at each time instant, is $O(Kd^2)$ and the expected memory requirement scales like $O(Kd^2)$. As far as we know, our analysis is the first to show that a simple and *efficient* algorithm for the problem of linearly parametrized bandits can, under reward separation and i.i.d. contexts, achieve logarithmic expected cumulative regret that simultaneously can scale like $polylog(|\mathcal{X}|)$ for natural scenarios.

Before we present our algorithm in full detail, let us give some intuition about it. Part of the job of the learner is to estimate the unknown parameters θ_a based on past actions, contexts and rewards. We denote the estimate of θ_a at time t by $\hat{\theta}_a$. If $\theta_a \approx \hat{\theta}_a$ then, given an observed context, the learner will more accurately know which arm to play to incur in small regret. The estimates $\hat{\theta}_a$ can be constructed based on a history of past rewards, contexts and arms played. Since observing a reward r for arm a under context x does not give information about the magnitude of θ_a along directions orthogonal to x, it is important that, for each arm, rewards are observed and recorded for a rich class of contexts. This gives rise to the following challenge: If the learner tries to build this history while trying to minimize the regret, the distribution of contexts observed when playing a certain arm a will be biased and potentially not rich enough. In particular, when trying to achieve a small regret, conditioned on $a_t = a$, it is more likely that x_t is a context for which a is optimal.

We address this challenge using the following idea, also appearing in the epoch-Greedy algorithm of [17]. We partition time slots into *exploration* and *exploitation* *epochs*. In exploration epochs, the learner plays arms uniformly at random, independently of the context, and records the observed rewards. This guarantees that in the history of past events, each arm has been played along with a sufficiently rich set of

---

Algorithm 1. Contextual ϵ -greedy

For all $a \in A$, set $A_a \leftarrow 0_{d \times d}$; $n_a \leftarrow 0$; $b_a \leftarrow 0_d$
for $t = 1$ to p **do**
    $a \leftarrow 1 + (t \mod K)$; Play arm a
    $n_a \leftarrow n_a + 1; b_a \leftarrow b_a + r_t x_t; A_a \leftarrow A_a + x_t x_t^\dagger$
end for
for $t = p + 1$ to T **do**
    $e \leftarrow$ Bernoulli(p/t)
    if $e = 1$ **then**
        $a \leftarrow$ Uniform$(1/K)$; Play arm a
        $n_a \leftarrow n_a + 1; b_a \leftarrow b_a + r_t x_t; A_a \leftarrow A_a + x_t x_t^\dagger$
    else
        for $a \in \mathcal{A}$ **do**
            Get $\hat{\theta}_a$ as the solution to the linear system: $\left(\lambda_{n_a} I + \frac{1}{n_a} A_a \right) \hat{\theta}_a = \frac{1}{n_a} b_a$
        end for
        Play arm $a_t = \arg\max_{a \in \mathcal{A}} x_t^\dagger \hat{\theta}_a$
    end if
end for

---

contexts. In exploitation epochs, the learner makes use of the history of events stored during exploration to estimate the parameters θ_a and determine which arm to play given a current observed context. The rewards observed during exploitation are not recorded.

More specifically, when exploiting, the learner performs two operations. In the first operation, for each arm $a \in \mathcal{A}$, an estimate $\hat{\theta}_a$ of θ_a is constructed from a simple ℓ_2-regularized regression, as in in [4] and [9]. In the second operation, the learner plays the arm a that maximizes $x_t^\dagger \hat{\theta}_a$. Crucially, in the first operation, only information collected during exploration epochs is used. In particular, let $\mathcal{T}_{a,t-1}$ be the set of exploration epochs up to and including time $t - 1$ (*i.e.*, the times that the learner played an arm a uniformly at random (u.a.r.)). Moreover, for any $\mathcal{T} \subset \mathbb{N}$, denote by $r_\mathcal{T} \in \mathbb{R}^n$ the vector of observed rewards for all time instances $t \in \mathcal{T}$, and $X_\mathcal{T} \in \mathbb{R}^{n \times d}$ is a matrix of $\mathcal{T}$ rows, each containing one of the observed contexts at time $t \in \mathcal{T}$. Then, at time t the estimator $\hat{\theta}_a$ is the solution of the following convex optimization problem.

$$\min_{\theta \in \mathbb{R}^d} \frac{1}{2n} \| r_\mathcal{T} - X_\mathcal{T} \theta \|_2^2 + \frac{\lambda_n}{2} \| \theta \|_2^2. \tag{3}$$

where $\mathcal{T} = \mathcal{T}_{a,t-1}$, $n = |\mathcal{T}_{a,t-1}|$, $\lambda_n = 1/\sqrt{n}$. In other words, the estimator $\hat{\theta}_a$ is a (regularized) estimate of θ_a, based only on observations made during exploration epochs. Note that the solution to (3) is given by $\hat{\theta}_a = \left(\lambda_n I + \frac{1}{n} X_\mathcal{T}^\dagger X_\mathcal{T} \right)^{-1} \frac{1}{n} X_\mathcal{T}^\dagger r_\mathcal{T}$.

An important design choice is the above process selection of the time slots at which the algorithm explores, rather than exploits. Following the ideas of [20], we select the exploration epochs so that they occur approximately $\Theta(\log t)$ times after t slots. This guarantees that, at each time step, there is enough information in our history of past events to determine the parameters accurately while only incurring in a regret of $O(\log t)$. There are several ways of achieving this; our algorithm explores at each time step with probability $\Theta(t^{-1})$.

The above steps are summarized in pseudocode by Algorithm 1. Note that the algorithm contains a scaling parameter p, which is specified below, in Theorem 2. Because there are K arms and for each arm $(x_t, r_{a,t}) \in \mathbb{R}^{d+1}$, the expected memory required by the algorithm scales like $O(Kd^2)$. In addition, both the matrix $X_{\mathcal{T}}^\dagger X_{\mathcal{T}}$ and the vector $X_{\mathcal{T}}^\dagger r_{\mathcal{T}}$ can be computed in an online fashion in $O(d^2)$ time: $X_{\mathcal{T}}^\dagger X_{\mathcal{T}} \leftarrow X_{\mathcal{T}}^\dagger X_{\mathcal{T}} + x_t x_t^\dagger$ and $X_{\mathcal{T}}^\dagger r_{\mathcal{T}} \leftarrow X_{\mathcal{T}}^\dagger r_{\mathcal{T}} + r_t x_t$. Finally, the estimate of $\hat{\theta}_a$ does not require full matrix inversion but only solving a linear system (see Algorithm 1), which can be done in $O(d^2)$ time. The above is summarized in the following theorem.

Theorem 1. *Algorithm 1 has computational complexity of $O(Kd^2)$ per iteration and its expected space complexity scales like $O(Kd^2)$.*

We now state our main theorem that shows that Algorithm 1 achieves $R(T) = O(\log T)$.

Theorem 2. *Under Assumptions 1, the expected cumulative regret of algorithm 1 satisfies,*

$$R(T) \leq p\Delta_{\max}\sqrt{d} + 14\Delta_{\max}\sqrt{d}Ke^{Q/4} + p\Delta_{\max}\sqrt{d}\log T.$$

for any

$$p \geq \frac{CKL'^2}{(\Delta'_{\min})^2(\Sigma'_{\min})^2}. \tag{4}$$

Above, C is a universal constant, $\Delta'_{\min} = \min\{1, \Delta_{\min}\}$, $\Sigma'_{\min} = \min\{1, \Sigma_{\min}\}$ and $L' = \max\{1, L\}$.

Algorithm 1 requires the specification of the constant p. In Section 4.2, we give two examples of how to efficiently choose a p that satisfies (4). In Theorem 2, the bound on the regret depends on p - small p is preferred - and hence it is important to understand how the right hand side (r.h.s.) of (4) might scale when K and d grow. In Section 4.1, we show that, for a concrete distribution of contexts and choice of expected rewards θ_a, and assuming (4) holds, $p = O(K^3 d^5)$ [1]. There is nothing special about the concrete details of how contexts and θ_a's are chosen and, although not included in this paper, for many other distributions, one also obtains $p = O(poly(d))$. We can certainly construct pathological cases where, for example, p grows exponentially with d. However, we do not find these intuitive. Specially when interpreting these having in mind real applications as the ones introduced in Examples 1- 3.

4.1 Example of Scaling of p with d and K

Assume that contexts are obtained by normalizing a d-dimensional vector with i.i.d. entries as Bernoulli random variables with parameter w. Assume in addition that every θ_a is obtained i.i.d. from the following prior distribution: every entry of θ_a is drawn i.i.d. from a uniform distribution and then θ_a is normalized. Finally, assume that the payoffs are given by $r_{a,t} = x_t^\dagger \Theta_a$, where $\Theta_a \in \mathbb{R}^d$ are random variables that fluctuate around $\theta_a = \mathbb{E}\{\Theta_a\}$ with each entry fluctuating by at most F.

Under these assumptions the following is true:

---

[1] This bound holds with probability converging to 1 as K and d get large.

- $\Sigma_{\min} = \Omega(d^{-1})$. In fact, the same result holds asymptotically independently of $w = w(d)$ if, for example, we assume that on average groups are roughly of the same size, M, with $w = M/d$;
- $L = O(\sqrt{d})$. This holds because $\epsilon_{a,t} = r_{a,t} - \mathbb{E}\{r_{a,t}\} = x_t^\dagger(\Theta_a - \theta_a)$ are bounded random variables with zero mean and $\|x_t^\dagger(\Theta_a - \theta_a)\}\|_\infty = O(\sqrt{d})$.
- $\Delta_{\min} = \Omega(1/(Kd\sqrt{w})$ with high-probability (for large K and d). This can be see as follows, if $\Delta_{\min} = x^\dagger(\theta_a - \theta_b)$ for some x, a and b, then it must be true that θ_a and θ_b differ in a component for which x is non-zero. The minimum difference between components among all pairs of θ_a and θ_b is lower bounded by $\Omega(1/(K\sqrt{d}))$ with high probability (for large K and d). Taking into account that each entry of x is $O(1/\sqrt{dw})$ with high-probability, the bound on $\Delta_{\min}$ follows.

If we want to apply Theorem 2 then (4) must hold and hence putting all the above calculations together we conclude that $p = O(K^3 d^5)$ with high probability for large K and p.

4.2   Computing p in Practice

If we have knowledge of an a priori distribution for the contexts, for the expected pay-offs and for the variance of the rewards then we can quickly compute the value of $\Sigma_{\min}$, L and a typical value for $\Delta_{\min}$. An example of this was done above (Section 4.1). There, the values were presented only in order notation but exact values are not hard to obtain for that and other distributions. Since a suitable p only needs to be larger then the r.h.s. of (4), by introducing an appropriate multiplicative constant, we can produce a p that satisfied (4) with high probability.

If we have no knowledge of any model for the contexts or expected payoffs, it is still possible to find p by estimating $\Delta_{\min}$, $\Sigma_{\min}$ and L from data gathered while running Algorithm 1. Notice again that, since all that is required for our theorem to hold is that p is greater then a certain function of these quantities, an exact estimation is not necessary. This is important because, for example, accurately estimating $\Sigma_{\min}$ is hard when matrix $\mathbb{E}\{x_1 x_1^\dagger\}$ has a large condition number.

Not being too concerned about accuracy, $\Sigma_{\min}$ can be estimated from $\mathbb{E}\{x_1 x_1^\dagger\}$, which can be estimated from the sequence of observed x_t. $\Delta_{\min}$ can be estimated from Algorithm 1 by keeping track of the smallest difference observed until time t between $\max_b x^\dagger \hat{\theta}_b$ and the second largest value of the function being maximized. Finally, the constant L can be estimated from the variance of the observed rewards for the same (or similar) contexts. Together, these estimations do not incur in any significant loss in computational performance of our algorithm.

5   Proof of Theorem 2

The general structure of the proof of our main result follows that of [3]. The main technical innovation is the realization that, in the setting when the contexts are drawn i.i.d. from some distribution, a standard matrix concentration bound allows us to treat $\lambda_n I + n^{-1}(X_\mathcal{T}^\dagger X_\mathcal{T})$ in Algorithm 1 as a deterministic positive-definite symmetric matrix, even as $\lambda_n \to 0$.

Let $\mathcal{E}_T$ denote the time instances for $t > p$ and until time T in which the algorithm took an exploitation decision. Recall that, by Cauchy-Schwarz inequality, $x_t^\dagger(\theta_{a_{x_t}^*} - \theta_a) \leq \|x_t\|_1\|(\theta_{a_{x_t}^*} - \theta_a)\|_\infty \leq \sqrt{d}\|x_t\|_2\|(\theta_{a_{x_t}^*} - \theta_a)\|_\infty \leq \sqrt{d}\Delta_{\max}$. In addition, recall that $\sum_{t=2}^T 1/t \leq \log T$. For $R(T)$ the cumulative regret until time T, we can write

$$R(T) = \mathbb{E}\{\sum_{t=1}^T x_t^\dagger(\theta_{a_{x_t}^*} - \theta_a)\} \leq p\Delta_{\max}\sqrt{d} + \Delta_{\max}\sqrt{d}\mathbb{E}\{\sum_{t=p+1}^T \mathbb{1}\{x_t^\dagger\theta_a < x_t^\dagger\theta_{a_{x_t}^*}\}\}$$

$$\leq p\Delta_{\max}\sqrt{d} + \Delta_{\max}\sqrt{d}\mathbb{E}\{|\mathcal{E}_T|\} + \Delta_{\max}\sqrt{d}\mathbb{E}\{\sum_{t\in\mathcal{E}_T} \mathbb{1}\{x_t^\dagger\theta_a < x_t^\dagger\theta_{a_{x_t}^*}\}\}$$

$$\leq p\Delta_{\max}\sqrt{d} + p\Delta_{\max}\sqrt{d}\log T + \Delta_{\max}\sqrt{d}\mathbb{E}\{\sum_{t\in\mathcal{E}_T} \mathbb{1}\{x_t^\dagger\theta_a < x_t^\dagger\theta_{a_{x_t}^*}\}\}$$

$$\leq p\Delta_{\max}\sqrt{d} + p\Delta_{\max}\sqrt{d}\log T + \Delta_{\max}\sqrt{d}\mathbb{E}\{\sum_{t\in\mathcal{E}_T}\sum_{a\in\mathcal{A}} \mathbb{1}\{x_t^\dagger\hat{\theta}_a > x_t^\dagger\hat{\theta}_{a_{x_t}^*}\}\}.$$

In the last line we used the fact that when exploiting, if we do not exploit the optimal arm $a_{x_t}^*$, then it must be the case that the estimated reward for some arm a, $x_t^\dagger\hat{\theta}_a$, must exceed that of the optimal arm, $x_t^\dagger\hat{\theta}_{a_{x_t}^*}$, for the current context x_t.

We can continue the chain of inequalities and write,

$$R(T) \leq p\Delta_{\max}\sqrt{d} + p\Delta_{\max}\sqrt{d}\log T + \Delta_{\max}\sqrt{d}K\sum_{t=1}^T \mathbb{P}\{x_t^\dagger\hat{\theta}_a > x_t^\dagger\hat{\theta}_{a_{x_t}^*}\}.$$

The above expression depends on the value of the estimators for time instances that might or might not be exploitation times. For each arm, these are computed just like in Algorithm 1, using the most recent history available. The above probability depends on the randomness of x_t and on the randomness of recorded history for each arm.

Since $x_t^\dagger(\theta_{a_{x_t}^*} - \theta_a) \geq \Delta_{\min}$ we can write

$$\mathbb{P}\{x_t^\dagger\hat{\theta}_a > x_t^\dagger\hat{\theta}_{a_{x_t}^*}\} \leq \mathbb{P}\Big\{x_t^\dagger\hat{\theta}_a \geq x_t^\dagger\theta_a + \frac{\Delta_{\min}}{2}\Big\} + \mathbb{P}\Big\{x_t^\dagger\hat{\theta}_{a_{x_t}^*} \leq x_t^\dagger\theta_{a_{x_t}^*} - \frac{\Delta_{\min}}{2}\Big\}.$$

We now bound each of these probabilities separately. Since their bound is the same, we focus only on the first probability.

Substituting the definition of $r_a(t) = x_t^\dagger\theta_a + \epsilon_{a,t}$ into the expression for $\hat{\theta}_a$ one readily obtains,

$$(\hat{\theta}_a - \theta_a) = \Big(\lambda_n I + \frac{1}{n}X_\mathcal{T}^\dagger X_\mathcal{T}\Big)^{-1}\Big(\frac{1}{n}\sum_{\tau\in\mathcal{T}} x_\tau\epsilon_{a,\tau} - \lambda_n\theta_a\Big).$$

We are using again the notation $\mathcal{T} = \mathcal{T}_{a,t-1}$ and $n = |\mathcal{T}|$. From this expression, an application of Cauchy-Schwarz's inequality and the triangular inequality leads to,

$$|x_t^{\dagger}(\hat{\theta}_a - \theta_a)| = \left|x_t^{\dagger}\left(\lambda_n I + \frac{1}{n}X_{\mathcal{T}}^{\dagger}X_{\mathcal{T}}\right)^{-1}\left(\frac{1}{n}\sum_{\tau \in \mathcal{T}}x_{\tau}\epsilon_{a,\tau} - \lambda_n\theta_a\right)\right|$$

$$\leq \sqrt{x_t^{\dagger}\left(\lambda_n I + \frac{1}{n}X_{\mathcal{T}}^{\dagger}X_{\mathcal{T}}\right)^{-2}x_t}\left(\left|\frac{1}{n}\sum_{\tau \in \mathcal{T}}x_t^{\dagger}x_{\tau}\epsilon_{a,\tau}\right| + \lambda_n|x_t^{\dagger}\theta_a|\right).$$

We introduce the following notation

$$c_{a,t} \equiv \sqrt{x_t^{\dagger}\left(\lambda_n I + \frac{1}{n}X_{\mathcal{T}}^{\dagger}X_{\mathcal{T}}\right)^{-2}x_t}. \tag{5}$$

Note that, given a and t both n and $\mathcal{T}$ are well specified.

We can now write,

$$\mathbb{P}\left\{x_t^{\dagger}\hat{\theta}_a \geq x_t^{\dagger}\theta_a + \frac{\Delta_{\min}}{2}\right\} \leq \mathbb{P}\left\{\left|\frac{1}{n}\sum_{\tau \in \mathcal{T}}x_t^{\dagger}x_{\tau}\epsilon_{a,\tau}\right| \geq \frac{\Delta_{\min}}{2c_{a,t}} - \lambda_n|x_t^{\dagger}\theta_a|\right\}$$

$$\leq \mathbb{P}\left\{\left|\frac{1}{n}\sum_{\tau \in \mathcal{T}}x_t^{\dagger}x_{\tau}\epsilon_{a,\tau}\right| \geq \frac{\Delta_{\min}}{2c_{a,t}} - \lambda_n Q\right\}.$$

Since $\epsilon_{a,\tau}$ are sub-gaussian random variables with sub-gaussian constant upper bounded by L and since $|x_t^{\dagger}x_{\tau}| \leq 1$, conditioned on x_t, $\mathcal{T}$ and $\{x_{\tau}\}_{\tau \in \mathcal{T}}$, each $x_t^{\dagger}x_{\tau}\epsilon_{a,\tau}$ is a sub-gaussian random variable and together they form a set of i.i.d. sub-gaussian random variables. One can thus apply standard concentration inequality and obtain,

$$\mathbb{P}\left\{\left|\frac{1}{n}\sum_{\tau \in \mathcal{T}}x_t^{\dagger}x_{\tau}\epsilon_{a,\tau}\right| \geq \frac{\Delta_{\min}}{2c_{a,t}} - \lambda_n Q\right\} \leq \mathbb{E}\left\{2e^{-\frac{n}{2L^2}\left(\frac{\Delta_{\min}}{2c_{a,t}} - \lambda_n Q\right)^{+2}}\right\}. \tag{6}$$

where both n and $c_{a,t}$ are random quantities and $z^+ = z$ if $z \geq 0$ and zero otherwise.

We now upper bound $c_{a,t}$ using the following fact about the eigenvalues of any two real-symmetric matrices M_1 and M_2: $\lambda_{\max}(M_1^{-1}) = 1/\lambda_{\min}(M_1)$ and $\lambda_{\min}(M_1 + M_2) \geq \lambda_{\min}(M_1) - \lambda_{\max}(M_2) = \lambda_{\min}(M_1) - \|M_2\|$.

$$c_{a,t} \leq \left(\lambda_n + \lambda_{\min}^+(\mathbb{E}\{x_1^{\dagger}x_1\}) - \left\|\frac{1}{n}X_{\mathcal{T}}^{\dagger}X_{\mathcal{T}} - \mathbb{E}\{x_1^{\dagger}x_1\}\right\|^+\right)^{-1}.$$

Both the eigenvalue and the norm above only need to be computed over the subspace spanned by the vectors x_t that occur with non-zero probability. We use the symbol $^+$ to denote the restriction to this subspace. Now notice that $\|.\|^+ \leq \|.\|$ and, since we defined $\Sigma_{\min} \equiv \min_{i:\lambda_i > 0}\lambda_i(\mathbb{E}\{X_1 X_1^{\dagger}\})$, we have that $\lambda_{\min}^+(\mathbb{E}\{X_1 X_1^{\dagger}\}) \geq \Sigma_{\min}$. Using the following definition, $\Delta\Sigma_n \equiv n^{-1}X_{\mathcal{T}}^{\dagger}X_{\mathcal{T}} - \mathbb{E}\{X_1 X_1^{\dagger}\}$, this leads to, $c_{a,t} \leq (\lambda_n + \Sigma_{\min} - \|\Delta\Sigma_n\|)^{-1} \leq (\Sigma_{\min} - \|\Delta\Sigma_n\|)^{-1}$.

We now need the following Lemma.

Lemma 1. *Let $\{X_i\}_{i=1}^n$ be a sequence of i.i.d. random vectors of 2-norm bounded by 1. Define $\hat{\Sigma} = \frac{1}{n}\sum_{i=1}^n X_i X_i^{\dagger}$ and $\Sigma = \mathbb{E}\{X_1 X_1^{\dagger}\}$. If $\epsilon \in (0,1)$ then,*

$$\mathbb{P}(\|\hat{\Sigma} - \Sigma\| > \epsilon\|\Sigma\|) \leq 2e^{-C\epsilon^2 n},$$

where $C < 1$ is an absolute constant.

For a proof see [21] (Corollary 50).

We want to apply this lemma to produce a useful bound on the r.h.s. of (6). First notice that, conditioning on n, the expression inside the expectation in (6) depends through $c_{a,t}$ on n i.i.d. contexts that are distributed according to the original distribution. Because of this, we can write,

$$\mathbb{P}\left\{\left|\frac{1}{n}\sum_{\tau\in\mathcal{T}}x_t^\dagger x_\tau \epsilon_{a,\tau}\right| \geq \frac{\Delta_{\min}}{2c_{a,t}} - \lambda_n Q\right\} \leq \mathbb{E}\left\{2e^{-\frac{n}{2L^2}\left(\frac{\Delta_{\min}}{2c_{a,t}} - \lambda_n Q\right)^{+2}}\right\}$$

$$\leq \sum_{n=1}^{t}\left(\mathbb{P}\{|\mathcal{T}_{a,t-1}| = n\} \times \mathbb{E}\left\{2e^{-\frac{n}{2L^2}\left(\frac{\Delta_{\min}}{2c_{a,t}} - \lambda_n Q\right)^{+2}}\Big| |\mathcal{T}_{a,t-1}| = n\right\}\right).$$

Using the following algebraic relation: if $z, w > 0$ then $(z-w)^{+2} \geq z^2 - 2zw$, we can now write,

$$\mathbb{E}\left\{e^{-\frac{n}{2L^2}\left(\frac{\Delta_{\min}}{2c_{a,t}} - \lambda_n Q\right)^{+2}}\Big| |\mathcal{T}_{a,t-1}| = n\right\}$$

$$\leq \mathbb{P}\{|\Delta\Sigma_n| > \Sigma_{\min}/2|\ |\mathcal{T}_{a,t-1}| = n\} + e^{-\frac{n}{2L^2}\left(\frac{\Sigma_{\min}\Delta_{\min}}{4} - \lambda_n Q\right)^{+2}}$$

$$\leq \mathbb{P}\{|\Delta\Sigma_n| > \Sigma_{\min}/2|\ |\mathcal{T}_{a,t-1}| = n\} + e^{\frac{Q\Delta_{\min}\Sigma_{\min}}{4L^2}}e^{-\frac{n(\Delta_{\min})^2(\Sigma_{\min})^2}{32L^2}}$$

Using Lemma 1 we can continue the chain of inequalities,

$$\mathbb{E}\left\{e^{-\frac{n}{2L^2}\left(\frac{\Delta_{\min}}{2c_{a,t}} - \lambda_n Q\right)^{+2}}\Big| |\mathcal{T}_{a,t-1}| = n\right\} \leq 2e^{-C(\Sigma_{\min})^2 n/4} + e^{\frac{Q\Delta_{\min}\Sigma_{\min}}{4L^2}}e^{-\frac{n(\Delta_{\min})^2(\Sigma_{\min})^2}{32L^2}}.$$

Note that $||\Sigma|| \leq 1$ follows from our non-restrictive assumption that $||x_t||_2 \leq 1$ for all x_t. Before we proceed we need the following lemma:

Lemma 2. If $n_c = \frac{p}{2k}\log t$, then $\mathbb{P}\{|\mathcal{T}_{a,t-1}| < n_c\} \leq t^{-\frac{p}{16K}}$.

Proof. First notice that $|\mathcal{T}_{a,t-1}| = \sum_{i=1}^{t-1} z_i$ where $\{z_i\}_{i=1}^{t-1}$ are independent Bernoulli random variables with parameter $p/(Ki)$. Remember that we can assume that $i > p$ since in the beginning of Algorithm 1 we play each arm p/K times.

Note that $\mathbb{P}(X > c) \leq \mathbb{P}(X+q > c)$ is always true for any r.v. X, c and $q > 0$. Now write,

$$\mathbb{P}(|\mathcal{T}_{a,t-1}| < n_c) = \mathbb{P}\left(\sum_{i=1}^{t-1} z_i < n_c\right) = \mathbb{P}\left(\sum_{i=1}^{t-1}(z_i - p/(Ki)) < n_c - (p/K)\sum_{i=1}^{t-1} 1/i\right)$$

$$\leq \mathbb{P}\left(\sum_{i=1}^{t-1}(-z_i + p/i) > -n_c + (p/K)\sum_{i=1}^{t-1} 1/i\right)$$

$$\leq \mathbb{P}\left(\sum_{i=1}^{t-1}(-z_i + p/i) > (p/K)\log t - n_c\right). \tag{7}$$

Since $\sum_{i=1}^{t-1}\mathbb{E}\{(z_i - p/(Ki))^2\} = \sum_{i=p+1}^{t-1}(1 - p/(Ki))(p/(Ki)) \leq \frac{p}{K}\log t$, we have that $\{-z_i + p/i\}_{i=1}^{t-1}$ are i.i.d. random variables with zero mean and sum of variances upper bounded by $(p/K)\log t$. Replacing $n_c = (p/2K)\log t$ in (7) and applying Bernstein inequality we get, $\mathbb{P}(|\mathcal{T}_{a,t-1}| < n_c) \leq e^{-\frac{\frac{1}{2}(p/(2K))^2 \log^2 t}{\frac{p}{K}\log t + \frac{1}{3}(p/(2K))\log t}} \leq t^{-\frac{p}{16K}}$. $\qquad \square$

We can now write, by splitting the sum in $n < n_c$ and $n \geq n_c$

$$\mathbb{P}\left\{\left|\frac{1}{n}\sum_{\tau \in \mathcal{T}} x_t^\dagger x_\tau \epsilon_{a,\tau}\right| \geq \frac{\Delta_{\min}}{2c_{a,t}} - \lambda_n Q\right\}$$

$$\leq \sum_{n=1}^{t} \mathbb{P}\{|\mathcal{T}_{a,t-1}| = n\}\mathbb{E}\left\{2e^{-\frac{n}{2L^2}\left(\frac{\Delta_{\min}}{2c_{a,t}} - \lambda_n Q\right)^{+2}}\Big| |\mathcal{T}_{a,t-1}| = n\right\}$$

$$\leq \mathbb{P}\{|\mathcal{T}_{a,t-1}| < n_c\} + 4e^{-C(\Sigma_{\min})^2 n_c/4} + 2e^{\frac{Q\Delta_{\min}\Sigma_{\min}}{4L^2}}e^{-\frac{n_c(\Delta_{\min})^2(\Sigma_{\min})^2}{32L^2}}$$

$$\leq t^{-\frac{p}{16K}} + 4t^{-\frac{Cp(\Sigma_{\min})^2}{8K}} + 2e^{\frac{Q\Delta_{\min}\Sigma_{\min}}{4L^2}}t^{-\frac{p(\Delta_{\min})^2(\Sigma_{\min})^2}{64KL^2}}.$$

We want this quantity to be summable over t. Hence we require that,

$$p \geq \frac{128KL^2}{(\Delta_{\min})^2(\Sigma_{\min})^2}, p \geq \frac{16K}{C(\Sigma_{\min})^2}, p \geq 32K. \tag{8}$$

It is immediate to see that our proof also follows if $\Delta_{\min}$, $\Sigma_{\min}$ and L are replaced by $\Delta'_{\min} = \min\{1, \Delta_{\min}\}$, $\Sigma'_{\min} = \min\{1, \Sigma_{\min}\}$ and $L' = \max\{1, L\}$ respectively. If this is done, it is easy to see that conditions (8) are all satisfied by the p stated in Theorem 2. Since $\sum_{t=1}^{\infty} 1/t^2 \leq 2$, gathering all terms together we have,

$$R(T) \leq p\Delta_{\max}\sqrt{d} + p\Delta_{\max}\sqrt{d}\log T + \Delta_{\max}\sqrt{d}K\left(4e^{\frac{Q\Delta'_{\min}\Sigma'_{\min}}{4L'^2}} + 10\right)$$

$$\leq p\Delta_{\max}\sqrt{d} + 14\Delta_{\max}\sqrt{d}Ke^{Q/4} + p\Delta_{\max}\sqrt{d}\log T. \qquad \square$$

6 Numerical Results

In Theorem 2, we showed that, in the stochastic setting, Algorithm 1 has an expected regret of $O(\log T)$. We now illustrate this point by numerical simulations and, most importantly, exemplify how violating the stochastic assumption might degrade its performance. Figure 1 (a) shows the average cumulative regret (in semi-log scale) over 10 independent runs of Algorithm 1 for $T = 10^5$ and for the following setup. The context variables $x \in \mathbb{R}^3$ and at each time step $\{x_t\}_{t \geq 1}$ are drawn i.i.d. in the following way: (a) set each entry of x to 1 or 0 independently with probability 1/2; (b) normalize x. We consider $K = 6$ arms with corresponding parameters θ_a generated independently from a standard multivariate gaussian distribution. Given a context x and an arm a, rewards were random and independently generated from a uniform distribution $U([0, 2x^\dagger\theta_a])$. As expected, the regret is logarithmic. Figure 1 (a) shows a straight line at the end.

To understand the effect of the stochasticity of x on the regret, we consider the following scenario: with every other parameter unchanged, let $\mathcal{X} = \{x, x'\}$. At every time step $x = [1, 1, 1]$ appears with probability $1/I$, and $x' = [1, 0, 1]$ appears with probability $1 - (1/I)$. Figure 1 (b) shows the dependency of the expected regret on the context distribution for $I = 5$, 10 and 100. One can see that an increase of I causes a proportional increase in the regret.

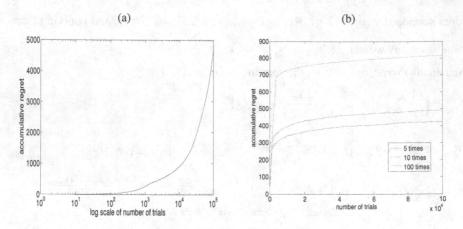

Fig. 1. (a) Regret over T when x_t is from i.i.d. (b) Regret over T when x_t is not from i.i.d.

7 Adversarial Setting

In the stochastic setting, the richness of the subset of $\mathbb{R}^d$ spanned by the observed contexts is related to the skewness of the distribution $\mathcal{D}$. The fact that the bound in Theorem 2 depends on $\Sigma_{\min}$ and that the regret increases as this value becomes smaller indicates that our approach does not yield a $O(\log T)$ regret for the adversarial setting, where an adversary choses the contexts and can, for example, generate $\{x_t\}$ from a sequence of stochastic processes with decreasing $\Sigma_{\min}(t)$.

In particular, the main difficulty in using a linear regression, and the reason why our result depends on $\Sigma_{\min}$, is related to the dependency of our estimation of $x_t^\dagger \theta_a$ on $\frac{1}{|\mathcal{T}_{a,t-1}|} X_{\mathcal{T}_{a,t-1}}^\dagger X_{\mathcal{T}_{a,t-1}}$. It is not hard to show that the error in approximating $x_t^\dagger \theta_a$ with $x_t^\dagger \hat{\theta}_a$ is proportional to

$$\sqrt{x_t^\dagger \left(\lambda_n I + \frac{1}{n} X_{\mathcal{T}}^\dagger X_{\mathcal{T}} \right)^{-2} x_t}. \tag{9}$$

This implies that, even if a given context has been observed relatively often in the past, the algorithm can "forget" it because of the *mean* over contexts that is being used to produce estimates of $x_t^\dagger \theta_a$ (the mean shows up in (9) as $\frac{1}{n} X_{\mathcal{T}}^\dagger X_{\mathcal{T}}$).

The effect of this phenomenon on the performance of Algorithm 1 can be readily seen in the following pathological example. Assume that $\mathcal{X} = \{(1,1),(1,0)\} \subset \mathbb{R}^2$. Assume that the contexts arrive in the following way: $(1,1)$ appears with probability $1/I$ and $(1,0)$ appears with probability $1-1/I$. The correlation matrix for this stochastic process is $\{(1,1/I),(1/I,1/I)\}$ and its minimum eigenvalue scales like $O(1/I)$. Hence, the regret scales as $O(I^2 \log T)$. If I is allowed to slowly grow with t, we expect that our algorithm will not be able to guarantee a logarithmic regret (assuming that our upper bound is tight). In other words, although $(1,1)$ might have appeared a sufficient number of times for us to be able to predict the expected reward for this context, Algorithm 1 performs poorly since the mean (9) will the 'saturated' with the context $(1,0)$ and forget about $(1,1)$.

Algorithm 2. Contextual UCB

for $t = 1$ to p **do**
    $a \leftarrow 1 + (t \mod K)$; Play arm a; $\mathcal{T}_{a,t} \leftarrow \mathcal{T}_{a,t-1} \cup \{t\}$
end for
for $t = p + 1$ to T **do**
    for $a \in \mathcal{A}$ **do**

$$c_{a,t} \leftarrow \min_{\mathcal{T} \subset \mathcal{T}_{a,t-1}} \frac{\log t}{|\mathcal{T}|} x_t^\dagger \left(\lambda_n I + \frac{1}{n} X_\mathcal{T}^\dagger X_\mathcal{T} \right)^{-2} x_t$$

        $\mathcal{T}^* \leftarrow$ subset of $\mathcal{T}_{a,t-1}$ that achieves the minimum; $n \leftarrow |\mathcal{T}^*|$

        Get $\hat{\theta}_a$ as the solution to the linear system: $\left(\lambda_n I + \frac{1}{n} X_\mathcal{T}^\dagger X_\mathcal{T} \right) \hat{\theta}_a = \left(\frac{1}{n} X_\mathcal{T}^\dagger r_\mathcal{T} \right)$

    end for
    Play arm $a_t = \arg\max_a x_t^\dagger \hat{\theta}_a + \sqrt{c_{a,t}}$; Set $\mathcal{T}_{a,t} \leftarrow \mathcal{T}_{a,t-1} \cup \{t\}$
end for

One solution for this problem is to ignore some past contexts when building an estimate for $x_t^\dagger \theta_a$, by including in the mean (9) past contexts that are closer in direction to the current context x_t. Having this in mind, and building on the ideas of [4], we propose the UCB-type Algorithm 2.

It is straightforward to notice that this algorithm cannot be implemented in an efficient way. In particular, the search for $\mathcal{T}^* \subset \mathcal{T}_{a,t-1}$ has a computational complexity exponential in t. The challenge is to find an efficient way of approximating $\mathcal{T}^*$ efficiently. This can be done by either reducing the size of $\mathcal{T}_{a,t-1}$ – the history from which one wants to extract $\mathcal{T}_{a,t-1}$ – by not storing all events in memory (for example, if we can guarantee that $|\mathcal{T}_{a,t}| = O(\log t)$ then the complexity of the above algorithm at time step t is $O(t)$), or by finding an efficient algorithm of approximating the minimization over the $\mathcal{T}_{a,t-1}$ (or both). It remains an open problem to find such an approximation scheme and to prove that it achieves $O(\log T)$ regret for a setting more general than the i.i.d. contexts considered in this paper.

8 Conclusions

We introduced an ϵ-greedy type of algorithm that provably achieves logarithmic regret for the contextual multi-armed bandits problem with linear payoffs in the stochastic setting. Our online algorithm is both fast and uses small space. In addition, our bound on the regret scales nicely with dimension of the contextual variables, $O(poly(d) \log T)$. By means of numerical simulations we illustrate how the stochasticity of the contexts is important for our bound to hold. In particular, we show how to construct a scenario for which our algorithm does not give logarithmic regret. The reason for this amounts to the fact that the mean $n^{-1} X_\mathcal{T}^\dagger X_\mathcal{T}$ that is used in estimating the parameters θ_a can "forget" previously observed contexts. Because of this, it remains an open problem to show that there are efficient algorithms that achieve $O(poly(d) \log T)$ under reward separation ($\Delta_{\min} > 0$) in the non-stochastic setting. We believe that a possible solution might be constructing a variant of our algorithm where in $n^{-1} X_\mathcal{T}^\dagger X_\mathcal{T}$ we use a more careful average of past observed contexts give the current observed context. In addition, we leave it open to produce simple and efficient online algorithms for multi-armed bandit problems under rich context models, like the one we have done here for linear payoff.

References

1. Abbasi-Yadkori, Y., Pál, D., Szepesvári, C.: Improved algorithms for linear stochastic bandits. In: Advances in Neural Information Processing Systems (2011)
2. Abernethy, J., Hazan, E., Rakhlin, A.: Competing in the dark: An efficient algorithm for bandit linear optimization. In: Proceedings of the 21st Annual Conference on Learning Theory (COLT), vol. 3, p. 3 (2008)
3. Auer, P., Cesa-Bianchi, N., Fischer, P.: Finite-time analysis of the multiarmed bandit problem. Machine Learning 47(2), 235–256 (2002)
4. Auer, P.: Using confidence bounds for exploitation-exploration trade-offs. The Journal of Machine Learning Research 3, 397–422 (2003)
5. Auer, P., Cesa-Bianchi, N., Freund, Y., Schapire, R.E.: Gambling in a rigged casino: The adversarial multi-armed bandit problem. In: Proceedings of the 36th Annual Symposium on Foundations of Computer Science, pp. 322–331. IEEE (1995)
6. Auer, P., Cesa-Bianchi, N., Freund, Y., Schapire, R.E.: The nonstochastic multiarmed bandit problem. SIAM Journal on Computing 32(1), 48–77 (2002)
7. Bartlett, P., Ben-David, S.: Hardness results for neural network approximation problems. In: Fischer, P., Simon, H.U. (eds.) EuroCOLT 1999. LNCS (LNAI), vol. 1572, pp. 50–62. Springer, Heidelberg (1999)
8. Beygelzimer, A., Langford, J., Li, L., Reyzin, L., Schapire, R.E.: Contextual bandit algorithms with supervised learning guarantees. In: Proceedings of the International Conference on Artificial Intelligence and Statistics, AISTATS (2011)
9. Chu, W., Li, L., Reyzin, L., Schapire, R.E.: Contextual bandits with linear payoff functions. In: Proceedings of the International Conference on Artificial Intelligence and Statistics, AISTATS (2011)
10. Crammer, K., Gentile, C.: Multiclass classification with bandit feedback using adaptive regularization. In: Proceedings of the 28th International Conference on Machine Learning (2011)
11. Dani, V., Hayes, T.P., Kakade, S.M.: The price of bandit information for online optimization. In: Advances in Neural Information Processing Systems 20, pp. 345–352 (2008)
12. Dani, V., Hayes, T.P., Kakade, S.M.: Stochastic linear optimization under bandit feedback. In: Proceedings of the 21st Annual Conference on Learning Theory (COLT), pp. 355–366 (2008)
13. Dudik, M., Hsu, D., Kale, S., Karampatziakis, N., Langford, J., Reyzin, L., Zhang, T.: Efficient optimal learning for contextual bandits. In: UAI (2011)
14. Hazan, E., Kale, S.: Newtron: an efficient bandit algorithm for online multiclass prediction. In: Advances in Neural Information Processing Systems, NIPS (2011)
15. Johnson, D.S., Preparata, F.P.: The densest hemisphere problem. Theoretical Computer Science 6(1), 93–107 (1978)
16. Kakade, S.M., Shalev-Shwartz, S., Tewari, A.: Efficient bandit algorithms for online multiclass prediction. In: Proceedings of the 25th International Conference on Machine Learning, pp. 440–447. ACM (2008)
17. Langford, J., Zhang, T.: The epoch-greedy algorithm for contextual multi-armed bandits. In: Advances in Neural Information Processing Systems 20, pp. 1096–1103 (2007)
18. Li, L., Chu, W., Langford, J., Schapire, R.: A contextual-bandit approach to personalized news article recommendation. In: Proceedings of the 19th International Conference on World Wide Web, pp. 661–670. ACM (2010)
19. Rusmevichientong, P., Tsitsiklis, J.: Linearly parameterized bandits. Mathematics of Operations Research 35(2) (2010)
20. Sutton, B.: Reinforcement learning, and introduction. MIT Press, Cambdrige (1998)
21. Vershynin, R.: Introduction to the non-asymptotic analysis of random matrices. In: Compressed Sensing, Theory and Applications, ch. 5 (2012)

Knowledge Transfer for Multi-labeler Active Learning

Meng Fang[1], Jie Yin[2], and Xingquan Zhu[1]

[1] QCIS, University of Technology, Sydney
[2] CSIRO ICT Centre, Australia
Meng.Fang@student.uts.edu.au, Jie.Yin@csiro.au,
Xingquan.Zhu@uts.edu.au

Abstract. In this paper, we address multi-labeler active learning, where data labels can be acquired from multiple labelers with various levels of expertise. Because obtaining labels for data instances can be very costly and time-consuming, it is highly desirable to model each labeler's expertise and only to query an instance's label from the labeler with the best expertise. However, in an active learning scenario, it is very difficult to accurately model labelers' expertise, because the quantity of instances labeled by all participating labelers is rather small. To solve this problem, we propose a new probabilistic model that transfers knowledge from a rich set of labeled instances in some auxiliary domains to help model labelers' expertise for active learning. Based on this model, we present an active learning algorithm to simultaneously select the most informative instance and its most reliable labeler to query. Experiments demonstrate that transferring knowledge across related domains can help select the labeler with the best expertise and thus significantly boost the active learning performance.

Keywords: Active Learning, Transfer Learning, Multi-Labeler.

1 Introduction

Active learning is an effective tool for reducing the labeling costs by choosing the most informative instance to label for supervised classification. Traditional active learning research has primarily relied on a single omniscient labeler to provide a correct label for each queried instance. This is particularly true for applications involving a handful of well-trained professional labelers. Recent advances in Web 2.0 technology have fostered a new active learning paradigm [16,21], which involves multiple (non-experts) labelers, aiming to label collections of large-scale and complex data. For example, crowdsourcing services (*i.e.*, Amazon Mechanical Turk[1]) allow a large number of labelers around the world to collaborate on annotation tasks at low cost. In such settings, data can be accessed by different labelers, who annotate the instances based on their own expertise and knowledge. Given multiple (possibly noisy) labels, majority vote is a simple but

---

[1] https://www.mturk.com/

H. Blockeel et al. (Eds.): ECML PKDD 2013, Part I, LNAI 8188, pp. 273–288, 2013.

popular approach widely used by crowdsourcing services to generate the most reliable label for each instance.

In multi-labeler scenarios, labelers tend to have different but hidden competence for a given task, depending on their background knowledge and expertise. Therefore, it is unlikely that all labelers are able to provide accurate labels for all instances, and labels provided by less competent labelers might be more error-prone. As a result, taking majority vote without considering the reliability of different labelers would deteriorate the classification models. More importantly, active learning starts with a small amount of labeled instances, with very few annotations from each labeler. The limited number of labeled data gives very little information to model labelers' expertise, which may incur incorrect labels and degrade classification accuracy. Therefore, accurately modeling labelers' expertise using a limited number of data poses a main challenge for active learning.

While labeled data is either costly to obtain, or easy to be outdated in a given domain, there often exists some labeled data from a different but related domain. This is often the case when the labeled data is out-of-date, but new data continuously arrives from fast evolving sources. For example, there may often be very few Blog documents annotated for certain Blog types, but there may be a lot of newsgroup documents labeled by numerous information sources. The newsgroup and Blog documents are in two different domains, but share common features (*i.e.* topics). Another example is text classification in online mainstream news. The model trained from old news articles may easily become outdated, and its classification accuracy would decrease dramatically over time. It would be very time-consuming to obtain annotations for new documents. Therefore, one important question is, how can we transfer useful knowledge from related domains to accurately model labelers' expertise in order to boost the active learning performance?

In this paper, we propose a novel probabilistic model to address the multi-labeler active learning problem. The proposed model can transfer knowledge from a related domain to help model labelers' expertise for active learning. We use a multi-dimensional topic distribution to represent a labeler's knowledge, which determines the labeler's reliability in labeling an instance. This approach provides a high-level abstraction of the labeled data in a low dimensional space, which reveals the labelers' hidden areas of expertise. More importantly, our model opens opportunity to find "good" latent topics shared by two related domains and further transfer such knowledge for improving the estimation of the labelers' expertise in a unified probabilistic framework. Based on this probabilistic model, we present a new active learning algorithm that simultaneously decides which instance should be labeled next and which labeler should be queried to maximally benefit the active learning performance. Compared with existing multi-labeler active learning methods, the advantage of our proposed method is that it can accurately model the labelers' expertise via transferring knowledge from related domains, and can thus select the labeler with the best expertise to label a queried instance. This advantage eventually leads to a higher classification accuracy for active learning.

2   Related Work

According to the query strategies, existing active learning techniques can be roughly categorized into three categories: 1) uncertainty sampling [8,18], which focuses on selecting the instances that the current classifier is most uncertain about; 2) query by committee [6,9], which considers the most informative instance to be the one that a committee of classifiers disagree most; 3) expected error reduction [13], which aims to query instances which can maximally reduce the model loss reduction of the current classifier once labeled. Most of existing works have mainly focused on a single domain and assumed that an omniscient oracle exists to provide an accurate label for each query.

Recently, learning from crowds has drawn a lot of research attention in the presence of multiple labelers [12]. Different from conventional supervised learning in which the annotations for data instances are provided by a single omniscient labeler, a given learning task seeks to collect labels from multiple labelers via crowdsourcing services at low cost, *e.g.*, Amazon Mechanical Turk. Since labelers have different knowledge or expertise, the resultant labels are inherently subjective (possibly noisy) with substantial variations among different annotators. Majority vote is one simple but popular way for integrating multiple noisy labels from crowdsourcing systems. Some research works have attempted to improve the overall quality of labeling from noisy labels. Sheng *et al.* [16] proposed to use repeated labeling strategies to improve the label quality inferred via majority vote. Donmez and Carbonell [4] introduced different costs to the labelers and solved a utility optimization problem to select an optimal labeler-instance pair subject to a budget constraint, in which expensive labelers are assumed to provide high-quality labels. Wallace *et al.* [19] furthered this work and proposed instance allocation strategies to better balance the workload between novice and stronger experts. These works have assumed that the labelers' levels of expertise are known through available domain information such as associated costs or expert salaries. However, the challenge of explicitly estimating each labeler's reliability has not been properly addressed.

In multi-labeler settings, active learning has focused on intelligently selecting the most reliable labelers to reduce the labeling costs. One line of research has tried to build a classifier for each labeler and approximate the labelers' expertise using confidence scores [2,11]. Other works have proposed to estimate the reliability of labelers based on a small sample of instances labeled by all participating experts. Yan *et al.* [22] directly used raw features of instances to represent the labelers' expertise. Fang *et al.* [5] modeled the reliability of the labelers via a Gaussian mixture model with respect to some concepts. However, these methods have relied on a small set of labeled data to estimate the dependency between labelers' reliability and original instances. Instead, in our work, we model the expertise of a labeler by using a multi-dimensional topic distribution, which, at an abstract level, better represents the labeler's expertise, thus enabling each queried instance to be labeled by a labeler with the best knowledge.

Transfer learning is another learning paradigm designed to save the labeling cost for supervised classification. Given an oracle and a lot of labeled data from a source domain, some researchers have proposed to combine transfer learning and active learning to train an accurate classifier for a target domain. Saha *et al.* [15] proposed to use the source domain classifier as one free oracle, which answers the target domain queries that appear similar to the source domain data. Similarly, Shi *et al.* [17] used the classifier from the source domain to answer the queries as often as possible, and the target-domain labelers are queried only when necessary. These methods assume that the target and source domains have the exactly same labeling problem, that is, the oracle/classifier in the source domain shares a same set of labels with the target domain. Different from these works, we do not require the labeling problems in the two domains to be the same, and there is also no need to involve the source domain oracles/labelers in the active learning process. More importantly, we consider multiple labelers in the target domain and focus on transferring knowledge from the labeled source data to help estimate the expertise of labelers. To the best of our knowledge, our work is the first to leverage transfer learning to help model labelers' expertise for multi-labeler active learning problem.

3   Problem Definition and Framework

We consider active learning in a multiple labeler setting with a target data set $\mathcal{X} = \{\mathbf{x}_1, \cdots, \mathbf{x}_N\}$ and a source data set $\mathcal{X}_s = \{\mathbf{x}_{s_1}, \cdots, \mathbf{x}_{s_{N_s}}\}$. In the target domain, there are a total of M labelers $(l_1, \cdots, l_M)$ to provide labeling information for instances $\mathcal{X}$. For any selected instance $\mathbf{x}_i$, we denote the label provided by labeler l_j as $y_{i,j}$, and its ground truth (unknown) label as z_i. In the source domain, each instance $\mathbf{x}_{s_i}$ is annotated with a label $c_i \in \{c_1, \ldots, c_D\}$ by one or multiple labelers. In this paper, we assume that the labeling problems in the source and target domain can be different. Once the data in the source domain are all labeled, there is no need to involve source domain labelers in the active learning process.

To characterize a labeler's labeling capability, we assume that each labeler's reliability of labeling an instance $\mathbf{x}_i$ is determined by whether the labeler has the expertise with respect to the latent topics, which the instance $\mathbf{x}_i$ belongs to. Formally, we give specific definitions as follows.

Definition 1 Topic: A topic t represents the semantic categorization of a set of instances. Each instance is then modeled as an infinite mixture over a set of latent topics. For example, *sports* is a common topic of a set of documents (*i.e.* instances) related to sports. A document contains words, such as "win", "games", "stars", which may belong to multiple topics, such as *sports* and *music*.

Definition 2 Expertise: The expertise of a labeler l_j, denoted by $\mathbf{e}_j$, is represented as a multinomial distribution over a set of topics $\mathcal{T}$. For example, a labeler may have expertise on two topics $\{t_1 = sports, t_2 = music\}$, with probabilities 0.8 and 0.6, respectively.

Given M labelers in the target domain, and a set of labeled data $\mathcal{X}_s$ from the source domain, the **aim** of active learning is to select the most informative instance from the target data pool $\mathcal{X}$, and to query the most reliable labeler to label the selected instance, such that the classifier trained from labeled instances has the highest classification accuracy in the target domain.

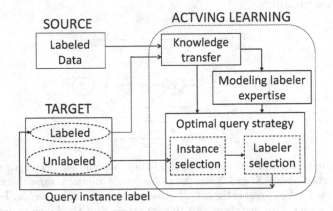

Fig. 1. An overview of the proposed framework. "Knowledge transfer module" uses source data to help model each labeler's expertise in the target domain. During active learning process, the most information instance is selected to be labeled by a labeler with the best expertise.

Proposed Framework. The overview of the proposed framework is shown in Figure 1. Our goal is to select the most informative instances and find the labelers with the best expertise to label the instances. Because labeled instances are rather limited and insufficient to characterize the labelers, we leverage the data from some source domains to strengthen the active learning process. In the following, we first describe the modeling of multiple labelers by using knowledge transfer in Section 4, and then detail the active learning algorithm in Section 5.

4   Modeling Expertise of Multiple Labelers

This section details our proposed model for modeling multiple labelers and describes transfer learning techniques used to estimate labelers' expertise.

4.1   Probabilistic Model

The main aim of modeling multiple labelers is to enable the selection of a labeler with the best expertise to label a queried instance. Given an instance $\mathbf{x}$ selected for labeling and a number of labelers, each having his/her own expertise, we assume that the label y provided by each labeler to instance $\mathbf{x}$ is subject to labeler's expertise with respect to some latent topics and the ground truth label z of $\mathbf{x}$. Therefore, we propose a probabilistic graphical model, as shown in Figure 2. The two variables X, where $\mathbf{x}_i \in \mathcal{X}$, represents an instance, and Y, where $y_{i,j}$ denotes the label provided by labeler l_j to instance $\mathbf{x}_i$, are directly observable. All other variables – the topic distribution $\mathbf{t}_i$ of an instance $\mathbf{x}_i$, the ground truth label z_i, and a labeler's expertise $\mathbf{e}_j$ – are hidden, so their values must be inferred from observed variables.

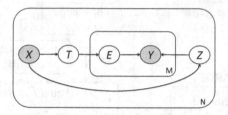

Fig. 2. Probabilistic graphical model for modeling multiple labelers with different expertise. The gray nodes X and Y are two observable random variables denoting instances and their labels, respectively. All other nodes are unobservable. For instances X, the latent topics T of instances and the expertise E of the labelers determine X's labels Y provided by the labelers, which are assumed to be an offset, subject to a Gaussian distribution, with respect to X's genuine labels Z.

This probabilistic graphical model can be represented using the joint probability distribution as follows

$$p = \prod_{i}^{N} p(z_i|\mathbf{x}_i)p(\mathbf{t}_i|\mathbf{x}_i) \prod_{j}^{M} p(e_{i,j}|\mathbf{t}_i)p(y_{i,j}|z_i, e_{i,j}). \tag{1}$$

In our model, we allow different labelers to have varying levels of expertise. That is, the expertise of a labeler depends on the topic distribution $\mathbf{t}_i$ of the instance $\mathbf{x}_i$. Because an instance can belong to one or multiple latent topics, we use $p(t_k|\mathbf{x}_i)$ to represent $\mathbf{x}_i$'s membership probability of belonging to topic t_k. Given an instance's topic distribution $\mathbf{t}_i = \{p(t_1|\mathbf{x}_i), \cdots, p(t_k|\mathbf{x}_i)\}$, we use logistic regression to define the expertise of labeler l_j with respect to $\mathbf{t}_i$ as a probability distribution given by

$$p(e_{i,j}|\mathbf{t}_i) = (1 + \exp(-\sum_{k=1}^{K} e_j^k p(t_k|\mathbf{x}_i) - \nu_j))^{-1}. \tag{2}$$

Our model assumes that the ground truth label z_i of instance $\mathbf{x}_i$ is solely dependent on the instance itself. To capture the relationships between $\mathbf{x}_i$ and z_i, any probabilistic model can be used. For simplicity, we use a logistic regression model to compute the conditional probability $p(z_i|\mathbf{x}_i)$ as

$$p(z_i|\mathbf{x}_i) = (1 + \exp(-\gamma^T \mathbf{x}_i - \lambda))^{-1}. \tag{3}$$

For an instance x_i, the actual label $y_{i,j}$ provided by the labeler l_j is assumed to depend on both the labeler's expertise $e_{i,j}$ and the ground truth label z_i of x_i. We model the offset between the actual label $y_{i,j}$ provided by the labeler and the instance's genuine label z as a Gaussian distribution

$$p(y_{i,j}|e_{i,j}, z_i) = N(z_i, e_{i,j}^{-1}). \tag{4}$$

Intuitively, if a labeler has a higher reliability $e_{i,j}$ of labeling instance x_i, the variance $e_{i,j}^{-1}$ of the Gaussian distribution would be smaller. That is, the actual label $y_{i,j}$ provided by the labeler would be closer to x's ground truth label z_i.

So far, we have discussed the calculation of probabilities $p(z_i|x_i)$, $p(e_{i,j}|t_i)$, and $p(y_{i,j}|z_i, e_{i,j})$ in Eq.(1). We now focus on estimating the distribution of latent topics of the instances, i.e., $p(t|x)$. Given a set of instances $\mathcal{X}$, the information about the latent topics is usually unavailable. A simple approach would be conducting latent semantic analysis on an initial set of labeled data. However, since the number of initially labeled data for active learning is very small, the accuracy of the induced model is largely limited. Therefore, we resort to leveraging labeled data from a related domain, which is detailed in the next subsection.

4.2   Transferring Knowledge

Given labeled data from a related domain, the basic idea is to exploit transfer learning to help discover latent topics of the instances in the target domain. That is, we aim to find common "good" latent topics to minimize the divergence of the two domains, through which we can estimate a more accurate topic distribution $p(t|x)$, thus improving the accuracy in estimating labelers' expertise, as defined in Eq. (2). Formally, given a source data $\mathcal{X}_s$, where each instance x_s is annotated with a corresponding label $c \in \{c_1, \ldots, c_D\}$, our objective is to estimate a topic distribution $p(t|x)$ for the target data $\mathcal{X}$.

For this task, we employ probabilistic latent semantic analysis (PLSA) to model the instances (*i.e.* documents) in the two domains [7]. PLSA aims to map the high-dimensional feature vectors of documents into a low dimensional representation in a latent semantic space. This abstraction offers an ideal way to represent labelers' expertise with respect to latent topics. Following the assumption that two related domains share similar topics from the terms in [20], we bridge the two domains through common latent topics, denoted by random variable T, as illustrated in Figure 3.

Specifically, we perform PLSA on the two domains. Thus, we have

$$p(x_s|w) = \sum_t p(x_s|t)p(t|w), \tag{5}$$

for the source data set, and

$$p(x|w) = \sum_t p(x|t)p(t|w), \tag{6}$$

for the target data set. In the above equations, both decompositions share the same term-specific mixing part $p(t|w)$ and relate the conditional probabilities for

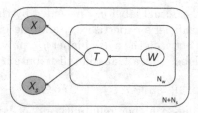

Fig. 3. PLSA model for bridging two related domains. The target data $\mathcal{X}$ and source data $\mathcal{X}_s$ are linked through latent variables T (topics) and W (terms). By transferring knowledge from the source data, a more accurate topic distribution can be obtained, which further improves the estimation of the labelers' expertise.

the two domains; each topic has a different probability of generating a document, $p(\mathbf{x}_s|t)$, in the source domain, and $p(\mathbf{x}|t)$, in the target domain, respectively.

To fully make use of the label information in the source domain, we also enforce must-link constrains and cannot-link constrains used in semi-supervised clustering [1]. For two instances having the same label, we define the must-link constraint as

$$\text{same}(\mathbf{x}_{s_i}, \mathbf{x}_{s_j}) = \log \sum p(\mathbf{x}_{s_i}|t)p(\mathbf{x}_{s_j}|t), \tag{7}$$

and for any two instances having different labels, we define the cannot-link constraint as

$$\text{diff}(\mathbf{x}_{s_i}, \mathbf{x}_{s_j}) = \log \sum_{t_i \neq t_j} p(\mathbf{x}_{s_i}|t_i)p(\mathbf{x}_{s_j}|t_j) \tag{8}$$

Therefore, we define our objective function to maximize the log-likelihood of the data with two penalty terms:

$$L = \sum_w \left\{ \sum_{\mathbf{x}} \log \sum_t p(\mathbf{x}|t)p(t|w) \right.$$
$$\left. + \sum_{\mathbf{x}_s} \log \sum_t p(\mathbf{x}_s|t)p(t|w) \right\}$$
$$+ \beta_1 \text{diff}(\mathbf{x}_{s_i}, \mathbf{x}_{s_j}) + \beta_2 \text{same}(\mathbf{x}_{s_i}, \mathbf{x}_{s_j}), \tag{9}$$

where β_1 and β_2 are two hyper-parameters that control the weights of the must-link and cannot-link constrains and the question of how they would affected the accuracy of active learning will be empirically investigated.

To solve this optimization problem, we adopt a standard EM algorithm detailed as follows.

– E-step:

$$p(t|\mathbf{x}_s, w) = \frac{p(\mathbf{x}_s|t)p(t|w)}{p(\mathbf{x}_s|w)} \tag{10}$$

$$p(t|\mathbf{x}, w) = \frac{p(\mathbf{x}|t)P(t|w)}{p(\mathbf{x}|w)} \tag{11}$$

– M-step:

$$p(\mathbf{x}|t) \propto \sum_w n(w, \mathbf{x})p(t|\mathbf{x}, w) \tag{12}$$

$$p(\mathbf{x}_s|t) \propto \sum_w n(w, \mathbf{x}_s)p(t|\mathbf{x}_s, w)$$

$$+\beta_1 \sum_{\mathbf{x}_s, c_i = c_j} \frac{p(\mathbf{x}_{s_i}|t)p(\mathbf{x}_{s_j}|t)}{\sum_t p(\mathbf{x}_{s_i}|t)p(\mathbf{x}_{s_j}|t)}$$

$$+\beta_2 \sum_{\mathbf{x}_s, c_i \neq c_j} \frac{p(\mathbf{x}_{s_i}|t)p(\mathbf{x}_{s_j}|t_j)}{\sum_{t_j \neq t} p(\mathbf{x}_{s_i}|t)p(\mathbf{x}_{s_j}|t_j)} \tag{13}$$

$$p(t|w) \propto \sum_{\mathbf{x}_s} n(w, \mathbf{x}_s)p(t|\mathbf{x}_s, w) + \sum_{\mathbf{x}} n(w, \mathbf{x})p(t|\mathbf{x}, w) \tag{14}$$

Finally, for the target domain, we can calculate the latent topic distribution $p(\mathbf{t}|\mathbf{x})$ using Eq. (11).

4.3 Parameter Estimation

Now we discuss the learning process to estimate the parameters of our proposed graphical model. Given observed variables – instances, their labels provided by labelers, and topic distribution of instances estimated via transfer learning (described in Section 4.2), we would like to infer two groups of hidden variables $\Omega = \{\Theta, \Phi\}$, where $\Theta = \{\gamma, \lambda\}$, $\Phi = \{\mathbf{e}_j, \nu_j\}_{j=1}^M$. This learning task can be solved by using a Bayesian style of EM algorithm [3].

E-step: We compute the expectation of the data log likelihood with respect to the distribution of the hidden variables derived from the current estimates of model parameters. Given current parameter estimates, we compute the posterior on the estimated ground truth:

$$\hat{p}(z_i) = p(z_i|\mathbf{x}_i, \mathbf{t}_i, \mathbf{e}_i, \mathbf{y}_i) \propto p(z_i, \mathbf{t}_i, \mathbf{e}_i, \mathbf{y}_i|\mathbf{x}_i), \tag{15}$$

where

$$p(z_i, \mathbf{t}_i, \mathbf{e}_i, \mathbf{y}_i|\mathbf{x}_i) = p(z_i|\mathbf{x}_i)p(\mathbf{t}_i|\mathbf{x}_i)\prod_j^M p(e_{i,j}|\mathbf{x}_i)p(y_{i,j}|z_i, e_{i,j}). \tag{16}$$

M-step: To estimate the model parameters, we maximize the expectation of the logarithm of the posterior on z with respect to $\hat{p}(z_i)$ from E-step:

$$\Omega^* = \underset{\Omega}{argmax}\, \mathcal{Q}(\Omega, \hat{\Omega}), \tag{17}$$

where $\hat{\Omega}$ is the estimate from the previous iteration, and

$$
\begin{aligned}
\mathcal{Q}(\Omega, \hat{\Omega}) &= \mathbb{E}_{\hat{p}(z_i)} \left[\sum_i \log p(\mathbf{x}_i, \mathbf{t}_i, \mathbf{y}_i \mid z_i) \right] \\
&= \sum_{i,j} \mathbb{E}_{\hat{p}(z_i)} [\log p(e_{i,j} | \mathbf{x}_i) + \log p(y_{i,j} | z_i, e_{i,j}) \\
&\quad + \log p(z_i | \mathbf{x}_i) + \log p(\mathbf{t}_i | \mathbf{x}_i)].
\end{aligned}
\tag{18}
$$

To solve the above optimization problem, we compute the updated parameters by using the L-BFGS quasi-Newton method [10].

5   Knowledge Transfer for Active Learning

Based on our probabilistic model, multi-labeler active learning seeks to select the most informative instance and the most appropriate labeler, with respect to the selected instance, to query for its label.

Instance Selection. The goal of active learning is to learn the most accurate classifier with the least number of labeled instances. We thus employ a commonly used uncertainty sampling strategy, by using the posteriori probability $p(z|\mathbf{x})$ from our graphical model, to select the most informative instance:

$$
\mathbf{x}^* = \underset{\mathbf{x}_i \in \mathcal{X}}{argmax} \, H(z_i | \mathbf{x}_i),
\tag{19}
$$

where

$$
H(z_i | \mathbf{x}_i) = - \sum_{z_i} p(z_i | \mathbf{x}_i) \log (z_i | \mathbf{x}_i).
\tag{20}
$$

Since the calculation of the posteriori probability $p(z|\mathbf{x})$ takes multiple labelers and their expertise into consideration, the instance selected using Eq.(19) represents the most informative instance from all labelers' perspectives.

Labeler Selection. Given an instance selected using Eq.(19), labeler selection aims to identify the labeler who can provide the most accurate label for the queried instance. For each selected instance $\mathbf{x}_i$, we first calculate the latent topic distribution $p(\mathbf{t}_i | \mathbf{x}_i)$ using Eq. (11), and then compute the confidence of each labeler as follows:

$$
e_{i,j}(\mathbf{x}_i) = \sum_{k=1}^{K} e_j^k p(t_k | \mathbf{x}_i) + \nu_j.
\tag{21}
$$

Accordingly, we rank the confidence values from Eq.(21) and select the labeler with the highest confidence score to label the selected instance

$$
j^* = \underset{j \in M}{argmax} \, e_{i,j}(\mathbf{x}_i).
\tag{22}
$$

After selecting the best instance and labeler, we make a query to the labeler for the instance. The active learning algorithm is summarized in Algorithm 1.

Algorithm 1. Knowledge Transfer for Active Learning

Input: (1) Target data set $\mathcal{X}$; (2) Multiple labelers $l_1, \cdots, l_M$; (3) Source data set $\mathcal{X}_s$; and (4) Labeling budget: *budget*
Output: Labeled instance set $\mathcal{L}$, Parameters Ω
1: Train an initial model with the labeled target data $\mathcal{L}$ and source data $\mathcal{X}_s$;
2: Perform transfer learning from $\mathcal{X}_s$ to calculate topic distribution $p(\mathbf{t}|\mathbf{x})$ for each instance $\mathbf{x}$ (Eq.(11));
3: $numQueries \leftarrow 0$;
4: **while** $numQueries \leq budget$ **do**
5:   $\mathbf{x}^* \leftarrow$ the most informative instance from pool $\mathcal{X}$ (Eq.(19));
6:   $j^* \leftarrow$ the most reliable labeler for instance $\mathbf{x}^*$ (Eq.(22));
7:   $(\mathbf{x}^*, y_{\mathbf{x}^*, j^*}) \leftarrow$ query instance $\mathbf{x}^*$'s label from labeler l_{j^*};
8:   $\mathcal{L} \leftarrow \mathcal{L} \cup (\mathbf{x}^*, y_{\mathbf{x}^*, j^*})$;
9:   $\Omega \leftarrow$ retrain the model using the updated labeled data;
10:   $numQueries \leftarrow numQueries + 1$.
11: **end while**

6 Experiments

To validate the effectiveness of our proposed algorithm, we conduct experiments on both synthetic data and real-world data. Our proposed algorithm is referred to as **AL+kTrM**. For comparison, we use five other algorithms as baselines:

- **RD+MV** is a baseline method that randomly selects an instance to query. It collects all labels provided by multiple labelers and then uses majority vote to generate the label for the queried instance.
- **AL+MV** uses the same strategy as our proposed AL+kTrM algorithm to select an instance but it relies on majority vote to generate the label for the queried instance.
- **AL+rM** is a state-of-the-art multi-labeler active learning method [21]. It uses raw features of the instances to calculate the reliability of labelers.
- **AL+gM** models a labeler's reliability using a Gaussian mixture model (GMM) with respect to some concepts, as proposed by [5].
- **AL+kM** uses the same probabilistic model as our proposed AL+kTrM algorithm, but does not utilize transfer learning to improve the estimation of labelers' expertise. By comparing with this baseline, we can validate whether the transfer learning module in our AL+kTrM algorithm can help improve active learning to achieve a higher accuracy.

In our experiments, all results are based on 10-fold cross-validation. In each round, we initially started with a small labeled data set (3% of train data), and then made queries by using different active learning strategies. The reported results are averaged over 10 rounds. We used logistic regression as the base classifier for classification, and evaluated algorithms by comparing their accuracies on the same test data. For our proposed AL+kTrM algorithm, the two parameters in Eq.(9) were set as $\beta_1 = 60$, $\beta_2 = 40$, and their impact on the classification accuracy will be empirically studied in Section 6.3.

6.1   Results on Real-World Data

We carried out experiments on a real-world data set, which is a publicly available corpus of scientific texts annotated by multiple annotators [14]. The inconsistency between multiple labelers makes this data set an ideal test-bed for evaluating the proposed algorithm. This corpus consists of two parts; we used the first part of 1,000 sentences annotated by five labelers as the target data, and the second part of 10,000 sentences annotated by eight labelers as source data for transfer learning. During the original labeling process, each expert broke a sentence into a number of fragments and provided a label to each fragment.

For the target data, we used the *focus, evidence, polarity* labels and considered a binary classification problem for each label. We set the fragments as the instances and their annotations were treated as labels. We removed the fragments whose number of characters is less than 10 and only kept the fragments segmented by all five labelers in the same way. The fragments were pre-processed by removing stopwords. As a result, we had 504 instances containing 3828 features (words) in the target data set. In the source data, the labels, including *generic, methodology,* and *science* were used to form the constraints in Eq.(9).

Figure 4 compares the classification accuracy of different algorithms with respect to the number of queries. Figure 4(a)-4(c) clearly show that AL+kTrM outperforms other baselines and achieves the highest accuracy. Particularly, at the beginning of querying, its accuracy is much higher than others. This indicates that, when there is a limited number of labeled data, transferring knowledge from a related domain boosts the accuracy of active learning. AL+kM and AL+gM perform slightly better than AL+rM, although their performance is close to each other in Figure 4(b). AL+MV and RD+MV perform worst, because they use majority vote to aggregate the labels but do not consider the expertise and reliability of different labelers. AL+MV performs better than RD+MV because it selects the most informative instance to query. Overall, AL+kTrM achieves the highest classification accuracy during the active learning process.

6.2   Results on Synthetic Data

Since real-world data does not have the ground truth information about labelers' expertise, we also evaluated the effectiveness of our algorithm using synthetic data in which we can construct different expertise domains for labelers and explicitly evaluate the accuracy of labeler selection. The synthetic data we used is based on the 20 Newsgroups[2]. This data set contains 16,242 postings that are tagged by four high-level domains: *comp, rec, sci* and *talk*. To simulate the labelers, we assume that each labeler knows the ground truth labels of two tagged sub data sets, and gives a random guess for the rest of the data. In this way, we constructed five labelers of different expertise and formulated a binary classification problem. For each domain, we selected 150 instances as the target data, and used the rest as the source data for transfer learning. We started

---

[2] http://kdd.ics.uci.edu/databases/20newsgroups/20newsgroups.html

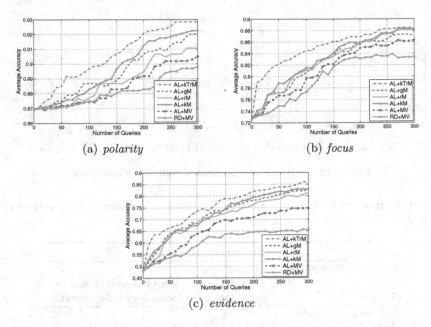

(a) *polarity*

(b) *focus*

(c) *evidence*

Fig. 4. Accuracy comparison of different algorithms on scientific text data for the *polarity, focus* and *evidence* labels

all active learning algorithms with an initially labeled set and made queries to improve the classification accuracy.

Figure 5(a) compares the accuracy of different algorithms with respect to the number of queries. We can see that, AL+kTrM is superior to other baselines, and RD+MV performs worst. RD+MV randomly selects the instances thus leading to the worst performance. AL+MV improves RD+MV because it selects the most informative instance to label. However, the two methods, RD+MV and AL+MV, rely on majority vote to aggregate the labels for instances without considering the reliability of labelers. AL+gM and AL+kM achieve higher accuracy than AL+rM, while AL+kM performs slightly better than AL+gM. This is because, both AL+gM and AL+kM model the expertise of labelers in terms of some topics at an abstract level, which better reveals the labelers' areas of knowledge. However, since AL+gM has a strong assumption that the expertise model follows a GMM distribution, its performance is limited in complex text data. Furthermore, by utilizing labeled data from a related domain, our AL+kTrM algorithm yields the highest classification accuracy in the active learning process.

In order to better understand how AL+kTrM models the expertise of labelers, Table 1 shows the correlation between the top two latent topics discovered by our algorithm and the domain of expertise of two labelers we constructed: Labeler 1 which is simulated to have expertise in *comp* and *rec* domain, and Labeler 2 to have expertise in *sci* and *talk* domain. The results clearly show that for Labeler 1, Topic 2 is related to *rec.sport* domain, and Topic 4 is related to *comp.sys*

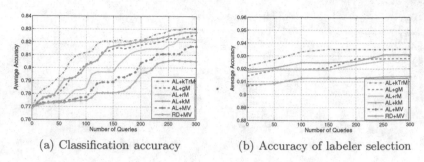

(a) Classification accuracy   (b) Accuracy of labeler selection

Fig. 5. Performance comparison of different algorithms

Table 1. Correlation between the labelers' expertise and the latent topics discovered by our AL+kTrM algorithm

Latent topic	Top correlated words
Labeler 1 (*comp, rec*)	
Topic 2	win,team,games,players,baseball,season
Topic 4	drive,data,card,technology,video,driver
Labeler 2 (*sci, talk*)	
Topic 1	world,law,children,jews,religion,fact,war
Topic 10	question,state,research,earth,space,orbit

domain; for Labeler 2, Topic 1 and Topic 10 are correlated to *talk.religion* and *sci.space* domain, respectively. This well explains our motivation that discovering latent topics can better reveal labelers' areas of expertise.

To further demonstrate the advantage of our AL+kTrM algorithm, we explicitly compared different algorithms with respect to their abilities to select the best labelers to label the queried instances. Figure 5(b) reports the accuracy of labeler selection in terms of different numbers of queries. We can observe that, AL+MV and RD+MV performs much worse than other methods. This is because they use majority vote to aggregate labels without considering the reliability of labelers. In contrast, by modeling the labelers' expertise, multi-labeler active learning methods significantly improve majority vote. Among them, our AL+kTrM algorithm can be observed to yield the highest accuracy for selecting the best labelers to label the queried instances.

6.3 Study on the Impact of β_1 and β_2

Now we study the impact of the two parameters β_1 and β_2 in our AL+kTrM algorithm on classification accuracy. Parameters β_1 and β_2 are two coefficients in the knowledge transfer module that controls the contribution of the must-link and cannot-link constrains, as defined in Eq.(9). Specifically, we fixed the value of one parameter at 50, and studied the impact of the other parameter by varying its value from 0 to 100. We analyzed the impact of β_1 and β_2 on both synthetic data and real data. Due to the space limit, we used the synthetic data as a case study, because similar observations are obtained for real data.

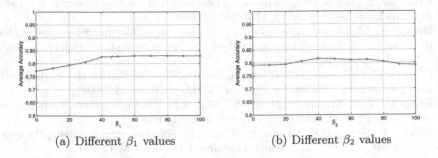

(a) Different β_1 values   (b) Different β_2 values

Fig. 6. Classification accuracy with different β_1 and β_2 values

Figure 6 shows the classification accuracy by varying the values of β_1 and β_2, respectively. From Figure 6(a), we can see that, as the value of β_1 increases, AL+kTrM gradually achieves higher accuracy. When β_1 reaches the value of 60, the accuracy becomes relatively saturated. From Figure 6(b), we can observe that, the accuracy is not very sensitive to different values of β_2. Overall, the impact of the must-link constraint (controlled by β_1) seems to be larger than that of the cannot-link constraint (controlled by β_2).

7   Conclusion

This paper proposed a new probabilistic model to address active learning involving multiple labelers. We argued that when labelers have different levels of expertise, it is important to properly characterize the knowledge of each labeler to ensure the label quality. In active learning scenarios, the quantity of instances labeled by all participating labelers is very small, which raises a challenge to model each labeler's strength and weakness. So we proposed to utilize data from a related domain to help estimate labelers' expertise. Using the proposed model, our active learning algorithm can always select the most informative instance and query its label using a single labeler with the best expertise with respect to the queried instance. Experiments demonstrated that our method significantly outperforms existing multi-labeler active learning methods, and transferring knowledge from a related domain can indeed help improve active learning.

Acknowledgments. This research is sponsored by an Australian Research Council (ARC) Future Fellowship under grant No. FT100100971 and Discovery Project under grant No. DP130102748, as well as a supplementary postgraduate scholarship from CSIRO.

References

1. Cohn, D., Caruana, R., McCallum, A.: Semi-supervised clustering with user feedback. Constrained Clustering: Advances in Algorithms, Theory, and Applications 4(1), 17 (2003)

2. Crammer, K., Kerns, M., Wortman, J.: Learning from multiple sources. Journal of Machine Learning Research 9, 1757–1774 (2008)
3. Dempster, A.P., Laird, N.M., Rubin, D.B.: Maximum likelihood from incomplete data via the EM algorithm. Journal of the Royal Statistical Society. Series B (Methodological), 1–38 (1977)
4. Donmez, P., Carbonell, J.G.: Proactive learning: Cost-sensitive active learning with multiple imperfect oracles. In: Proc. of CIKM, pp. 619–628 (2008)
5. Fang, M., Zhu, X., Li, B., Ding, W., Wu, X.: Self-taught active learning from crowds. In: Proc. of ICDM, pp. 858–863. IEEE (2012)
6. Freund, Y., Seung, H.S., Shamir, E., Tishby, N.: Selective sampling using the query by committee algorithm. Machine Learning 28(2), 133–168 (1997)
7. Hofmann, T.: Unsupervised learning by probabilistic latent semantic analysis. Machine Learning 42(1), 177–196 (2001)
8. MacKay, D.J.C.: Information-based objective functions for active data selection. Neural Computation 4(4), 590–604 (1992)
9. Melville, P., Mooney, R.: Diverse ensembles for active learning. In: Proc. of ICML, pp. 584–591 (2004)
10. Nocedal, J., Wright, S.J.: Numerical optimization. Springer (1999)
11. Raykar, V.C., Yu, S., Zhao, L.H., Jerebko, A., Florin, C., Valadez, G.H., Bogoni, L., Moy, L.: Supervised learning from multiple experts: Whom to trust when everyone lies a bit. In: Proc. of ICML, pp. 889–896 (2009)
12. Raykar, V.C., Yu, S., Zhao, L.H., Valadez, G.H., Florin, C., Bogoni, L., Moy, L.: Learning from crowds. Journal of Machine Learning Research 11, 1297–1322 (2010)
13. Roy, N., McCallum, A.: Toward optimal active learning through sampling estimation of error reduction. In: Proc. of ICML, pp. 441–448 (2001)
14. Rzhetsky, A., Shatkay, H., Wilbur, W.J.: How to get the most out of your curation effort. PLoS Computational Biology 5(5), e1000391 (2009)
15. Saha, A., Rai, P., Daumé III, H., Venkatasubramanian, S., DuVall, S.L.: Active supervised domain adaptation. In: Gunopulos, D., Hofmann, T., Malerba, D., Vazirgiannis, M. (eds.) ECML PKDD 2011, Part III. LNCS, vol. 6913, pp. 97–112. Springer, Heidelberg (2011)
16. Sheng, V.S., Provost, F., Ipeirotis, P.G.: Get another label? improving data quality and data mining using multiple, noisy labelers. In: Proc. of SIGKDD, pp. 614–622. ACM (2008)
17. Shi, X., Fan, W., Ren, J.: Actively transfer domain knowledge. In: Daelemans, W., Goethals, B., Morik, K. (eds.) ECML PKDD 2008, Part II. LNCS (LNAI), vol. 5212, pp. 342–357. Springer, Heidelberg (2008)
18. Tong, S., Koller, D.: Support vector machine active learning with applications to text classification. Journal of Machine Learning Research 2, 45–66 (2002)
19. Wallace, B.C., Small, K., Brodley, C.E., Trikalinos, T.A.: Who should label what? instance allocation in multiple expert active learning. In: Proc. of SDM (2011)
20. Xue, G.R., Dai, W., Yang, Q., Yu, Y.: Topic-bridged plsa for cross-domain text classification. In: Proc. of SIGIR, pp. 627–634. ACM (2008)
21. Yan, Y., Rosales, R., Fung, G., Dy, J.: Active learning from crowds. In: Proc. of ICML, pp. 1161–1168 (2011)
22. Yan, Y., Rosales, R., Fung, G., Schmidt, M., Hermosillo, G., Bogoni, L., Moy, L., Dy, J., Malvern, P.A.: Modeling annotator expertise: Learning when everybody knows a bit of something. In: Proc. of AISTATS, vol. 9, pp. 932–939 (2010)

Spectral Learning of Sequence Taggers over Continuous Sequences[*]

Adrià Recasens and Ariadna Quattoni

Universitat Politècnica de Catalunya,
Jordi Girona 1-3, 08034 Barcelona
arecasens@gmail.com, aquattoni@lsi.upc.edu

Abstract. In this paper we present a spectral algorithm for learning weighted finite-state sequence taggers (WFSTs) over paired input-output sequences, where the input is continuous and the output discrete. WFSTs are an important tool for modelling paired input-output sequences and have numerous applications in real-world problems. Our approach is based on generalizing the class of weighted finite-state sequence taggers over discrete input-output sequences to a class where transitions are linear combinations of elementary transitions and the weights of the linear combination are determined by dynamic features of the continuous input sequence. The resulting learning algorithm is efficient and accurate.

1 Introduction

Weighted Finite-state Sequence Taggers (WFSTs) are an important tool for modelling paired input-output sequences and have found numerous applications in areas such as natural language processing and computational biology (e.g. part of speech tagging, NP-chunking, entity recognition and protein folding prediction, to mention a few). The problem of modelling paired input-output sequences is usually referred in the literature as sequence tagging. In sequence tagging the goal is to predict a tag (i.e. a discrete output) for each symbol in the input sequence. And thus different from the general transduction problem where input and output sequences might be of different lengths, in the sequence tagging problem both input and output sequences are 'aligned' and have the same length. Still the problem of learning sequence taggers with latent states remains a challenging task.

The most popular methods for learning sequence taggers with hidden states are based on gradient-based or EM optimizations [12,13], but these can be computationally expensive and are susceptible to local optima issues. Recently, an emerging line of work on spectral methods has proposed algorithms for latent-variable structure modelling that overcome some of the limitations of EM [11,16,21,4,15,24,3,8,6,2]. Of these, [5] proposed a spectral method for learning

---

[*] This work has been partially funded by: the European Commission for project XLike (FP7-288342); the Spanish Government for project BASMATI (TIN2011-27479-C04-03); and the ERA-Net CHISTERA project VISEN.

H. Blockeel et al. (Eds.): ECML PKDD 2013, Part I, LNAI 8188, pp. 289–304, 2013.

WFSTs that learns distributions where both inputs and outputs are sequences from a discrete alphabet.

However, many real world problems require tagging sequences where the inputs are not discrete but continuous sequences. For example [22,28,20] consider the problem of human gesture or action recognition where given a video sequence the task is to predict the gesture that is been performed at each frame. Clearly, this can be framed as a sequence tagging problem where the continuous inputs correspond to real-valued features of the video sequence and the discrete outputs correspond to the gestures been performed at each point in time.

In this paper we extend the previous line of work on spectral learning for continuous sequences to handle the task of tagging sequences of continuous inputs. Our approach is based on generalizing the class of weighted finite-state sequence taggers over discrete input-output sequences to a class where transitions are linear combinations of elementary transitions and the weights of the linear combinations are determined by dynamic features of the continuous input sequence. One intuitive way to understand our approach is to think that we are learning a basis of the vectorial space of transition functions and that the weights of the linear combination depend only on the given features of the continuous input sequence.

Similar to [19,6], we develop a spectral method for our model from forward-backward recursions which are used to derive useful matrix decompositions of observable statistics. These matrix decompositions are then in turn exploited to induce the relationships between observations and latent state dynamics. As with previous spectral methods our algorithm for learning finite-state sequence taggers is simple and efficient. It reduces to estimating simple statistics from samples of paired input-output sequences, performing a singular value decomposition and inversion of some matrices.

In summary the main contributions of this paper are: (1) We present a latent state model for sequence tagging over continuous inputs; (2) We derive an efficient spectral learning algorithm for this model from forward-backward recursions; and (3) We present experiments on a real-task that validate the effectiveness of our approach.

2   Models for Sequences of Continuous Inputs and Discrete Outputs

2.1   Preliminary: Weighted Finite-State Sequence Taggers

We start by defining a class of functions over pairs of discrete sequences. More specifically, let $x = x_1, \ldots, x_T$ be an input sequence and $y = y_1, \ldots, y_T$ be an output sequence, where $x \in \Delta^*$ and $y \in \Sigma^*$. Here both Δ and Σ are assumed to be discrete alphabets. We follow [5] in that we assume that x and y have the same length (i.e we model aligned sequences). Defining spectral learning algorithms over pairs of sequences of different lengths would require handling unobserved alignments which is outside the scope of this paper.

A weighted finite-state sequence tagger (WFST) over $\Delta \times \Sigma$ with m states can be defined as a tuple $A = \langle \alpha_1, \alpha_\infty, A_\delta^\sigma \rangle$ where $\alpha_1, \alpha_\infty \in \mathbb{R}^m$ are the initial and final weight vectors and $A_\delta^\sigma \in \mathbb{R}^{m \times m}$ are the $|\Delta \times \Sigma|$ transition matrices associated to each pair of symbols $\langle \delta, \sigma \rangle \in \Delta \times \Sigma$. The function f_A realized by a WFST is defined as:

$$f_A(x, y) = \alpha_1^\top A_{x_1}^{y_1} \cdots A_{x_T}^{y_T} \alpha_\infty . \tag{1}$$

The above equation is an algebraic representation of the computation performed by a WFST on a pair of sequences $\langle x, y \rangle$. To see this consider a state vector $s_t \in \mathbb{R}^m$ where the ith entry represents the sum of the weights of all the state paths that generate the prefix $\langle x_{1:t}, y_{1:t} \rangle$ and end in state i. Initially, $s_1 = \alpha_1$, and then $s_{t+1}^\top = s_t^\top A_{x_t}^{y_t}$ updates the state distribution by simultaneously emiting the symbol $\langle x_t, y_t \rangle$ and transitioning to the next state vector.

Notice that since x and y are aligned sequences we could regard a WFST as a weighted finite-state automata (WFA) over a combined alphabet $\Gamma = \Delta \times \Sigma$. The reason why we maintain separate alphabets will become evident in the next sections when we will consider modelling pairs of *continuous* input sequences and discrete outputs.

We say that a WFST is stochastic if the function f_A is a probability distribution over $(\Delta \times \Sigma)^*$. That is, if $f_A(x, y) > 0$ for all $\langle x, y \rangle \in (\Delta \times \Sigma)^*$ and $\sum_{\langle x,y \rangle \in (\Delta \times \Sigma)^*} f_A(x, y) = 1$. When x is continuous we have the analogous condition: $\int_{(x,y) \in (\Delta \times \Sigma)^*} f_A(x, y) \, dx \, dy = 1$. To make it clear that $f_A(x, y)$ represents the probability of pairs of sequences $\langle x, y \rangle$ we will sometimes write it as $\mathbb{P}[x, y]$.

2.2 Sequence Taggers over Continuous Sequences

We will now consider the case in which the input sequences are not discrete but continuous. More specifically, let $\mathcal{X}$ be an arbitrary domain of input symbols (possibly infinite) and $\Phi = \{\phi_1, \ldots, \phi_k\}$ be a set of feature functions over $\mathcal{X}$, where $\phi_i : \mathcal{X} \to \mathbb{R}$. For any symbol $a \in \mathcal{X}$ we regard the vector $\Phi(a) = [\phi_1(a), \ldots, \phi_k(a)] \in \mathbb{R}^k$ as the real representation of a under the $\mathcal{X} \to \mathbb{R}^k$ mapping induced by Φ. When necessary we will use $\Phi(\mathcal{X})$ to refer to the range of this mapping.

We could attempt to define a WFST over $(\mathcal{X} \times \Sigma)^*$ as:

$$f_A(x, y) = \alpha_1^\top A_{\Phi(x_1)}^{y_1} \cdots A_{\Phi(x_T)}^{y_T} \alpha_\infty. \tag{2}$$

Clearly, there is a problem with the above formulation because there are an infinite number of transition matrices (i.e. one for each member of $\Phi(\mathcal{X}) \times \Sigma$), thus we need to impose some further restrictions on f_A. The first observation is that instead of regarding $A_{\Phi(a)}^\sigma$ as a matrix in $\mathbb{R}^{m \times m}$ we can define it as a function:

$$A(\Phi(a), \sigma) : \mathbb{R}^k \times \Sigma \to \mathbb{R}^{m \times m} \tag{3}$$

We can now restrict f_A by restricting A, in particular we will assume that:

$$A(\Phi(a), \sigma) = \sum_{l=1}^{k} \phi_l(a) O_l^\sigma \tag{4}$$

where $O_l^\sigma \in \mathbb{R}^{m \times m}$ is an operator associated with each of the k functions of Φ and each output symbol $\sigma \in \Sigma$. Thus for each output symbol we restrict our transition function to be a linear combination of a set of k elementary operators. The weights of the linear combination are those induced by Φ.

In summary, a *Continuous Weighted Finite-state Sequence Tagger (CWFST)* over $(\Phi(\mathcal{X}) \times \Sigma)^*$ with m states can be defined as a tuple $A = \langle \Phi, \alpha_1, \alpha_\infty, O_l^\sigma \rangle$ where Φ is a set of k functions, $\alpha_1, \alpha_\infty \in \mathbb{R}^m$ are the initial and final weight vectors, and $O_l^\sigma \in \mathbb{R}^{m \times m}$ are the $k \times |\Sigma|$ operator matrices associated with each each symbol in Σ and each function in Φ. The function f_A realized by a CWFST is defined as:

$$f_A(x, y) = \alpha_1^\top \, A(\Phi(x_1), y_1) \cdots A(\Phi(x_T), y_T) \, \alpha_\infty \tag{5}$$

$$= \alpha_1^\top \left(\sum_{l=1}^k \phi_l(x_1) O_l^{y_1} \right) \cdots \left(\sum_{l=1}^k \phi_l(x_T) O_l^{y_T} \right) \alpha_\infty \tag{6}$$

2.3   Some Examples

We give now some examples of classes of functions that can be computed by CWFSTs.

A WFST as a CWFST. We start by considering WFSTs as they were defined in the previous section. It is easy to see that if we have a WFST defined over $\Delta \times \Sigma$ we can construct a CWFST that will compute the same function. The construction is very simple, to map a WFST $A = \langle \alpha_1, \alpha_\infty, A_\delta^\sigma \rangle$ to a CWFST $A' = \langle \Phi, \alpha_1', \alpha_\infty', O_l^\sigma \rangle$ we perform the following construction: (1) Define one indicator function $\phi_\delta : \Delta \to \mathbb{R}$ for each $\delta \in \Delta$ as: $\phi_\delta(a) = 1$ if $a = \delta$ and 0 otherwise; (2) Set the $|\Delta| \times |\Sigma|$ operators to $O_l^\sigma = A_\delta^\sigma$; (3) Define $\alpha_1' = \alpha_1$ and $\alpha_\infty' = \alpha_\infty$. Clearly, the CWFST A' resulting from this construction will compute the same function as A since by construction $A(\Phi(\delta), \sigma) = A_\delta^\sigma$.

Transitions as Mixture Models. We will now describe a more interesting case of a distribution over $(\mathcal{X} \times \Sigma)^*$ that can be represented as a CWFST. To motivate this example consider a gesture recognition problem where given a sequence of video frames we wish to predict the gesture been performed at each point in time.

One of the challenges in the gesture recognition task is that each video frame lies in a high-dimensional space which makes generalization to unseen samples difficult. To alleviate this problem we could consider a two step process where in the first step we induce a mapping from the high-dimensional space to a lower dimensional semantic space.

For example in the first step, like in [10] we could learn a visual topic model [7] over frames and represent each frame as a posterior distribution over visual topics. In the second step we need to be able to learn a sequence model from the topic space to gesture labels. To define such a model we will make use of some intermediate latent variables.

More precisely, let $H = \{c_1, \ldots, c_m\}$ be a set of m hidden states and Z be a k dimensional multinomial random variable. In the gesture recognition example Z would correspond to the latent topic variable for each video frame. Consider now the following distribution over paired $\langle x, y \rangle$ sequences:

$$\mathbb{P}[x, y] = \sum_{h \in H^{T+1}} \mathbb{P}[x, y, h] \tag{7}$$

$$= \sum_{h \in H^{T+1}} \mathbb{P}[h_0] \prod_{t=1}^{T-1} \mathbb{P}[h_{t+1}, x_t, y_t \mid h_t] \tag{8}$$

$\mathbb{P}[h_{t+1}, x_t, y_t | h_t]$ is the probability of emiting a pair of symbols (x, y) at time t and transitioning to a new state. Since x might lie in a high-dimensional space, to ease modelling of this conditional distribution we will define it as a mixture of k elementary conditional distributions:

$$\{ \mathbb{P}_1[h_{t+1}, y_t \mid h_t], \ldots, \mathbb{P}_k[h_{t+1}, y_t \mid h_t] \} \tag{9}$$

More precisely, we define the transition function as:

$$\mathbb{P}[h_{t+1}, x_t, y_t \mid h_t] = \sum_{l=1}^{k} \mathbb{P}_l[h_{t+1}, y_t \mid h_t] \, \mathbb{P}[z = l \mid x_t] \, \mathbb{P}[x_t] \tag{10}$$

Thus, in this model the emission of an output symbol y is conditioned on z which is itself conditioned on the input variable x. Intuitively, we can think of $\mathbb{P}[z = l|x]$ as the probability of x taking discrete label l. In the gesture example, this would correspond to the posterior probability of a topic l given some input x. An alternative interpretation is that Z induces a soft partition of $\mathcal{X}$. The model exploits this partition to induce a better mapping between inputs and outputs.

Finally, we show how to construct a CWFST that realizes $\mathbb{P}[x, y]$. The idea is quite simple, we define a feature function for each of the k possible values that Z can take. More precisely, we define a CWFST A in the following manner: (1) Define one feature function $\phi_l(x)$ for each possible value of Z as $\mathbb{P}[z = l \mid x] \, \mathbb{P}[x]$; (2) Define the $k \times |\Sigma|$ operators as $O_l^\sigma(i, j) = \mathbb{P}_l[h_{t+1} = i, \sigma \mid h_t = j]$; (3) Define $\alpha_1(i) = \mathbb{P}[h_0 = i]$ and $\alpha_\infty = 1$. It is easy to see that A computes $\mathbb{P}[x, y]$ since by definition $A(\Phi(\delta), \sigma) = \mathbb{P}[h_{t+1}, \delta, \sigma \mid h_t]$.

We would like to end this section with a note on the limitations of the model. One of the key assumptions is that the feature functions depend only on the input. This means that the feature function needs to capture enough information so that at any point in time the output y_t can be predicted knowing the current state vector (which is a summary of the $[x_{1:t-1}; y_{1:t-1}]$ history) and the input features at time t. In other words, there must be enough information in the feature functions to explain the dynamics of the output symbols.

3   Spectral Learning of Stochastic CWFSTs

Recall that a stochastic CWFST computes a function f_A that is a probability distribution over $(\Delta \times \Sigma)^*$. In this section we will derive a learning algorithm for inducing the parameters of the CWFST from samples. We begin by defining some expectation matrices induced by f_A. We continue by presenting a duality result between a subclass of stochastic CWFSTs and factorizations of these matrices. Finally, we describe the spectral method for CWFSTs which is a statistically robust implementation of the arguments used in the duality proof.

3.1 Duality and Minimal CWFST

Let $\mathbb{P}$ be the function computed by f_A, we define functions: $\phi_{ij}(aa') = \phi_i(a)\phi_j(a')$ and $\phi^\sigma_{ilj}(aa'a'', \sigma'\sigma''\sigma''') = \phi_i(a)\phi_l(a')\phi_j(a'')\mathbb{P}(a')$ if $\sigma'' = \sigma$ and 0 otherwise. Using these functions we construct the observable statistics matrices $H_1 \in \mathbb{R}^k$, $H_2 \in \mathbb{R}^{k \times k}$, $H^\sigma_l \in \mathbb{R}^{k \times k}$ and $C \in \mathbb{R}^{k \times k}$ as:

$$H_1(i) = \mathbb{E}_\mathbb{P}[\phi_i(a)] \tag{11}$$

$$H_2(i,j) = \mathbb{E}_\mathbb{P}[\phi_{ij}(aa')] \tag{12}$$

$$H^\sigma_l(i,j) = \mathbb{E}_\mathbb{P}[\phi^\sigma_{ilj}(aa'a'', \sigma'\sigma''\sigma''')] \tag{13}$$

$$C(i,j) = \mathbb{E}_\mathbb{P}[\phi_i(a)\phi_j(a)] \tag{14}$$

We say that a CWFST $A = \langle \Phi, \alpha_1, \alpha_\infty, O^\sigma_l \rangle$ with $O^\sigma_l \in \mathbb{R}^{m \times m}$ for $l = 1 : k$ is minimal for f_A if $\mathrm{rank}(H_2) = m$ and $\mathrm{rank}(C) = k$.

The rank-m restriction on H_2 is analogous to the restriction that O and T be rank m in [15,6]. There are ways to relax this assumption by considering higher order expectations, but this is outside the scope of the paper. Now we will show a relationship between minimal A and some useful rank-m factorizations of H_2. Under appropriate stationary assumptions:

Lemma 1. *Let H_1, H_2, H^σ_l and C be the expectation matrices induced by an m-state minimal A. There exist matrices $F \in \mathbb{R}^{k \times m}$ and $B \in \mathbb{R}^{m \times k}$ such that the following holds:*

$$H_2 = FB \tag{15}$$

$$H^\sigma_l = F \sum_{i=1}^k O^\sigma_i C(l,i) B \tag{16}$$

$$H_1 = F\alpha_\infty = \alpha_1^\top B \tag{17}$$

Proof. Define a *forward* vector $f_l \in \mathbb{R}^m$ and a *backward* vector $b_l \in \mathbb{R}^m$ for each feature function:

$$f_l = \alpha_1^\top \int_{(x,y)\in(\mathcal{X}\times\Sigma)^*} A(x,y)\phi_l(x) \; dx \; dy \tag{18}$$

$$b_l = \int_{(x,y)\in(\mathcal{X}\times\Sigma)^*} A(x,y)\phi_l(x) \; dx \; dy \; \alpha_\infty \tag{19}$$

where we use the shorthand notation $A(x, y) = A(\Phi(x_1), y_1) \cdots A(\Phi(x_T), y_T))$. It is not hard to see that H_2 can be written as:

$$
\begin{aligned}
H_2(i, j) &= E[\phi_i(a)\phi_j(a')] \\
&= \int_{(xa, y\sigma)} \int_{(a'x', \sigma'y')} f_A(xaa'x', y\sigma\sigma'y') \; \phi_i(a)\phi_j(a') \\
&= \int_{(xa, y\sigma)} \int_{(a'x', \sigma'y')} \alpha_1^\top A(xa, y\sigma) A(a'x', \sigma'y') \alpha_\infty \; \phi_i(a)\phi_j(a') \\
&= \langle f_i, b_j \rangle
\end{aligned}
$$

Thus if we define a forward matrix $F \in \mathbb{R}^{k \times m}$ where each row corresponds to a forward vector and a backward matrix $B \in \mathbb{R}^{m \times k}$ where each column corresponds to a backward vector we get $H_2 = FB$ as desired. For the next claim we have:

$$
\begin{aligned}
H_l^\sigma(i, j) &= \mathbb{E}_{\mathbb{P}}[\phi_{ilj}^\sigma(a'aa'', \sigma''\sigma'\sigma''')] \\
&= \int_{(xa', y\sigma'')} \int_a \int_{(a''x', \sigma'''y')} f_A(xa'aa''x', y\sigma''\sigma\sigma'''y') \phi_i(a')\phi_l(a)\phi_j(a'')\mathbb{P}(a) \\
&= f_i^t \int_a A(a, \sigma)\phi_l(a)\mathbb{P}(a) \; b_j \\
&= f_i^t \sum_{r=1}^k O_r^\sigma C(l, r) \; b_j
\end{aligned}
$$

A few extra algebraic manipulations using F and B give us the remaining claims.   □

We now develop the opposite direction of the duality between factorizations and minimal CWFSTs, which is the key to understand the spectral learning algorithm. The following theorem shows that given any rank factorization of H_2 we can compute a CWFST for f.

Theorem 1. *Let H_1, H_2, H_l^σ and C be the expectation matrices of some function f computed by a minimal CWFST and let $H_2 = FB$ be a rank factorization, then $A = \langle \Phi, \alpha_1, \alpha_\infty, O_l^\sigma \rangle$ defined as:*

$$\alpha_\infty = F^+ H_1 \tag{20}$$

$$\alpha_1^\top = H_1 B^+ \tag{21}$$

$$[O_1^\sigma(i, j), \ldots, O_k^\sigma(i, j)]^\top = C^{-1}[Q_1^\sigma(i, j), \ldots, Q_k^\sigma(i, j)] \tag{22}$$

$$Q_l^\sigma = F^+ H_l^\sigma B^+ \tag{23}$$

computes f.

Proof. Let $\widetilde{A} = \langle \Phi, \widetilde{\alpha}_1, \widetilde{\alpha}_\infty, \widetilde{O}_l^\sigma \rangle$ be a minimal CWFST for f that induces rank factorization $H_2 = \widetilde{F}\widetilde{B}$. We first show that there exists an invertible matrix M such that for all $(x, y) \in (\mathcal{X} \times \Sigma)^*$ we have that: $M^{-1}\widetilde{A}(x, y)M = A(x, y)$. Define

$M = \widetilde{B}B^+$, we have that: $F^+\widetilde{F}\widetilde{B}B^+ \implies F^+H_2B^+ = I \implies M^{-1} = F^+\widetilde{F}$. Thus M is invertible. We also have that for every σ and every l the following holds:

$$\sum_{i=1}^{k} O_i^\sigma C(l,i) = F^+H_l^\sigma B^+ = F^+\widetilde{F}\sum_{i=1}^{k}\widetilde{O}_i^\sigma C(l,i)\widetilde{B}B^+$$

$$= M^{-1}\sum_{i=1}^{k}\widetilde{O}_i^\sigma C(l,i)M \tag{24}$$

For each σ we have k different equations, one per feature l:

$$\sum_{i=1}^{k} O_i^\sigma C(l,i) = M^{-1}\sum_{i=1}^{k}\widetilde{O}_i^\sigma C(l,i)M \tag{25}$$

Fixing the right part of the equation we observe that it is a system of k equations with k variables O_i^σ. Since f is minimal C is invertible and we can perform Gauss elimination and end up with a unique solution for the system. Since $O_i^\sigma = M^{-1}\widetilde{O}_i^\sigma M$ is a solution we must have that

$$\forall (a,\sigma) \in (\mathcal{X} \times \Sigma) \;:\; \sum_{i=1}^{k} O_i^\sigma \phi_i(a) = M^{-1}\sum_{i=1}^{k}\widetilde{O}_i^\sigma \phi_i(a)M. \tag{26}$$

Some algebraic manipulations give: $\alpha_1^\top = \widetilde{\alpha}_1^\top M$ and $\alpha_\infty = M^{-1}\widetilde{\alpha}_\infty$. Therefore, we can compute f as:

$$f_A(x,y) = \alpha_1^\top \left(\sum_{l=1}^{k} O_l^{y_1}\phi_l(x_1)\right) \cdots \left(\sum_{l=1}^{k} O_l^{y_t}\phi_l(x_t)\right) \alpha_\infty$$

$$= \widetilde{\alpha}_1^\top MM^{-1}\left(\sum_{l=1}^{k} \widetilde{O}_l^{y_1}\phi_l(x_1)\right) M \cdots M^{-1}\left(\sum_{l=1}^{k} \widetilde{O}_l^{y_t}\phi_l(x_t)\right) MM^{-1}\widetilde{\alpha}_\infty$$

$$= f_{\widetilde{A}}(x,y) = f \tag{27}$$

$\square$

This result shows that there exists a duality between rank factorizations of H_2 and minimal CWFST for f. A consequence of this is that minimal CWFSTs for a function f with covariance C are related to each other via some change of basis.

Corollary 1. *Let $A = \langle \Phi, \alpha_1, \alpha_\infty, O_l^\sigma \rangle$ and $\widetilde{A} = \langle \Phi, \widetilde{\alpha}_1, \widetilde{\alpha}_\infty, \widetilde{O}_l^\sigma \rangle$ be two minimal CWFSTs for some f of rank m with covariance C. Then there exists an invertible matrix $M \in \mathbb{R}^{m \times m}$ such that $\alpha_1^\top = \widetilde{\alpha}_1^\top M$, $\alpha_\infty = M^{-1}\widetilde{\alpha}_\infty$ and*

$$\forall (a,\sigma) \in (\mathcal{X} \times \Sigma) \;:\; \sum_{l=1}^{k} O_l^\sigma \phi_l(a) = M^{-1}\sum_{l=1}^{k}\widetilde{O}_l^\sigma \phi_l(a)M.$$

In practice, we do not observe H_2, H_l^σ and C but we can estimate them from n training samples (x, y). The errors in the estimation can be bounded using the Hoeffding inequality, for example for H_2 we would get:

$$P(|\widehat{\mathbb{E}}[\phi_i(a)\phi_j(a')] - \mathbb{E}[\phi_i(a)\phi_j(a')]| > \epsilon) \leq 2 \exp^{\frac{-2n\epsilon^2}{(\mu-\lambda)^2}}$$

where a and a' are any two input symbols, and λ and μ are bounds on the minimum and maximum values for $\phi_i(a) \cdot \phi_j(a')$.

The spectral learning algorithm that we present in the next section uses the singular value decomposition of H_2; this choice of low rank decomposition is appealing because its robustness to noise in the estimation of H_2. Using results from the stability of the singular value decomposition it is possible to show that the CWFST obtained from an approximate H_2 will be close to the one obtained using the exact statistics and thus the algorithm is statistically consistent. Also, it is not hard to see that one could use techniques similar to [15,4,6] to prove sample complexity bounds that depend linearly in the number of input features and $|\Sigma|$.

3.2   Spectral Algorithm

In this section we present a learning algorithm for stochastic CWFST based on spectral decompositions of observable statistics. Given samples from the joint distribution of paired input-output sequences $\mathbb{P}[x, y]$ and feature functions Φ, the task is to induce a CWFST: $A = \langle \Phi, \alpha_1, \alpha_\infty, O_l^\sigma \rangle$ that approximates $\mathbb{P}$.

More precisely, we are given:

- A set of n training samples $S = \{(x^1, y^1), \ldots, (x^n, y^n)\}$ of input-output sequences (where $x \in \mathcal{X}^T$ and $y \in \Sigma^T$ for some T), sampled from $\mathbb{P}[x, y]$
- A set of k feature functions $\Phi = \{\phi_1(a), \ldots, \phi_k(a)\}$
- The desired number of states m

We first use the training samples to compute estimates of H_1, H_2, H_l^σ and C. Recall that the compact SVD of a $k \times k$ matrix of rank m is given by: $H_2 = U\Lambda V^\top$ where $U \in k \times m$ and $V \in k \times m$ are orthogonal matrices. Clearly, $H_2 = (U\Lambda)V^\top$ is a rank m factorization of H_2, note that since $VV^\top = I$ this factorization is equivalent to $H_2 = (H_2V)V^\top$. Applying the ideas of the duality theorem for the factorization $F = H_2V$ and $B = V^\top$ we get that the model parameters are given by:

$$\alpha_\infty = (HV)^+ H_1 \tag{28}$$

$$\alpha_1^\top = H_1 V \tag{29}$$

$$[O_1^\sigma(i,j), \ldots O_k^\sigma(i,j)]^\top = C^{-1}[Q_1^\sigma(i,j), \ldots Q_k^\sigma(i,j)] \tag{30}$$

$$Q_l^\sigma = (HV)^+ H_l^\sigma V \tag{31}$$

Computing the expectation matrices is linear in the size of the training set and the number of features and output symbols. The cost of the algorithm is dominated by the singular value decomposition of the $k \times k$ matrix H_2 and therefore the overall cost is at most cubic on the number of features.

4   Related Work

Modelling continuous sequences with spectral methods has been studied in the context of HMMs [26], where a spectral algorithm for this case was derived. Their approach builds on previous work [27] on Hilbert Space Embeddings of conditional distributions. The main idea is first to map continuous distributions to points in a Hilbert Space and then derive a spectral method that works directly in the embedded space. [25] proposed an alternative spectral algorithm for continuous HMMs which is based on kernels. In spirit, our algorithm shares some similarities with all these methods since all of them work by embedding the transition function in some vectorial space.

Modelling continuous sequences has also been addressed in the original work by Jaeger [17,18] on observable operator models (OOMs). Similar to that approach, we also consider operators that can be written as linear combinations of some *basis operators*. The main difference is that while they consider modelling continuous sequences, we consider the special case of *tagging* continuous sequences, that is, modelling paired sequences of continuous inputs and discrete outputs. Furthermore, we study the case in which the weights of the linear combination are provided in the form of feature functions that depend only on the continuous input.

More closely related to our approach is the work by [9] on transformed predictive state representations (TPSR). Although they do not directly address the sequence tagging problem (they are interested in predicting the conditional output of a dynamical system), implicitly they do consider paired sequences which can be sampled from a continuous space. Furthermore, they also use feature representations and operators that can be seen as linear combinations of elementary operators. One of the main difference between our work is that we focus on the case in which the following holds: (1) one of the two sequences comes from a small discrete alphabet; and (2) the weights of the linear operators depend on features of the continuous sequence only (in their case the feature function depends on both sequences). We show that for this special case the observable statistics on past and future events that are used to compute the basis of the operators depend only on the continuous input sequence. In their case, all observable statistics depend on both sequences. In this sense the difference between our work and theirs is analogous to the difference of a vanilla approach for computing joint distributions of discrete paired input/output sequences versus the work by [5], where they show that the basis can be computed from one of the two sequences alone.

Thus, although the learning algorithms might seem similar at first hand, the observable statistics on which they rely are quite different and thus they both have different properties. For example, we can consider cases in which we can easily estimate the statistics of the input distribution needed to compute the basis but in which estimating the joint input/output statistic might be hard. Another property of our model is that since some observable statistics depend only on the input we could easily use unlabeled samples (i.e. samples for which the output sequences are unknown) to better estimate them.

Apart from the differences mentioned above, the techniques that we use to prove the correctness of our algorithm are different. We derive the algorithm directly from a duality between low-rank factorizations of certain observable statistics and the parameters of the model. Finally, at the experimental level the two works are quite different. We test the accuracy of our learning algorithm in sequence tagging tasks while they test their model in tasks that involve predicting the future state of a dynamic system conditioned on the observed history.

5   Experiments

We conducted experiments on the Wall-following Navigation dataset of the UCI repository [1]. Given a sequence of sensor readings, the task is to predict an appropriate movement action out of a set of discrete actions. There are four possible actions: move-right, move-left, right-turn, left-turn. The sensor readings are the outputs of 24 ultrasound sensors sampled at a rate of 9 samples per second. When we frame this task as a sequence prediction problem over continuous inputs we have that x consists of sequences of sensor readings and y consists of sequences of appropriate actions.

The dataset consists of one long sequence of sensor readings and corresponding robot actions. For our experiments we split this sequence into 150 contiguous sequences of approximately 4 seconds each (36 contiguous samples per sequence). We then randomly partition these sequences and use 100 sequences as training data 25 sequences as validation and the remaining 25 sequences as test. When we report optimal performance for a given model, the validation sequences were used to pick the optimal number of states and to choose the optimal parameters of the feature functions.

5.1   Feature Functions

In general the feature functions can be validated using a held-out validation data. The goal of the first set of experiments is to test the robustness of our method with respect to different feature functions.

In kernel learning one usually assumes that a kernel function is provided, analogously a natural way to define feature functions in our setting it so assume that we are provided with some distance function between elements in $\mathcal{X}$. Once we have the distance function we can obtain centroids on the input space by performing vector-quantization (e.g. k-means) using the given distance. If a kernel was provided instead we could also perform kernel k-means to obtain centroids. Finally, we compute features as similarities to each of the centroids. More specifically, to obtain features for these experiments we do the following: (1) Perform k-means (with the provided distance function) on the input training samples to obtain k cluster centroids; and (2) For each cluster centroid c define the corresponding feature function: $\phi_c(x) = \frac{\exp\frac{-d(c,x)}{\tau}}{z}$. Here $-d(x,x')$ is the provided distance function, we will compare three distance functions: (1) Square Euclidean; (2) Correlation, computed as 1-sample correlation between points; and (3) Cosine, computed as 1-the cosine of the included angle between points. The other parameter τ defines the width of the kernel function and z is a normalization constant. A small τ will result in sparse feature vectors for each point, where most of the mass will be concentrated around a few features. To compare against the discrete WFST we create a discrete alphabet by mapping each point to its closest cluster centroid according to the provided distance function.

In all experiments as a performance metric we report the accuracy on predicting actions for the test sequences. To predict the most probable sequence of actions y for a given test sequence x we must compute:

$$\operatorname{argmax}_y \mathbb{P}(y|x) = \operatorname{argmax}_y \mathbb{P}(x; y) \tag{32}$$

Due to the presence of the latent state variables the above computation is known to be untractable. Instead we use the standard approximation of maximizing the marginal probability at each time, that is we compute:

$$\operatorname{argmax}_{y_t} \sum_{y_{1:t-1}, y_{t+1:T}} \mathbb{P}(x_{1:T}, y_{1:t-1} y_t y_{t+1:T}) \tag{33}$$

In the next section we validate the accuracy of this approximation.

Figure 1 shows the accuracy of CWFST and WFST as a function of the number of latent states m for the Euclidean, Correlation and Cosine feature functions. The number of features for these graphs is 80. As we can see CWFST outperforms WFST for all feature functions. In the three figures we can see the performance of CWFST for different values of τ (i.e. different feature functions). Larger values of τ result in feature functions that induce a softer partition of the input space. Thus we expect that for small τ values CWFST and WFST give similar performance, and this seems to be the case.

CWFST seems to be quite robust to the particular choice of feature function and what seems to change in each case is the optimal kernel width τ. For the cosine and correlation functions sparser feature vectors seem to be preferred (i.e. smaller τ) than for the Euclidean distance function. WFST on the other hand seems to be less robust to the choice of distance function (used in this case to discretize the inputs) and Cosine and Correlation seem to perform significantly better than the Euclidean distance.

Figure 2 (Left) shows accuracy as a function of the number of features (i.e. for optimal number of states and τ). As we can see CWFST significantly outperforms WFST for any number of features. This seems to suggest that working with a soft partition of the input space always results in better performance, regardless of the number of partitions. This appears to be true independent of the particular choice of feature function.

We end this subsection with a note on how to pick the optimal number of states. In general, one should use a validation set to pick the optimal number of states. One advantage of spectral learning algorithms is that they are very fast, hence parameter validation is cost-less. Still, we can use information of the spectrum of H_2 to guide our search for optimal m. Figure 2 (Center) shows

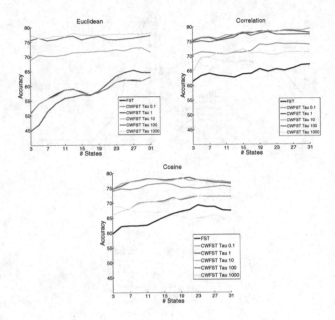

Fig. 1. Accuracy as a function of the number of states for different feature functions

sorted singular singular values of H_2 for the correlation model with 80 features. As we can see the singular values drop to almost 0 after the 34th singular vector. Most likely, the optimal number of states for the spectral method will be less than 34 and probably in between 30 and 35 states.

5.2   Comparison with Other Methods

In the second set of experiments we fix the feature function to be Correlation and compare CWFST against two other methods:

- (EM) We train a model as defined in section 2.3.2 using expectation maximization. The models were run for a maximum of 400 iterations but the actual stopping criteria was chosen using the held-out validation data. That is we picked the model resulting from the iteration that performed best in validation data, which was less than 400 iterations (see table 2).
- (Bayesian) As a second model to compare we choose the winner algorithm of a recent probabilistic automata competition [1]. The winner algorithm [23] was a Bayesian method that implements Collapsed Gibbs Sampling [14]. Since this method assumes discrete inputs, we discretize the continuous inputs following the same approach that was discussed in the previous section for WFST.

---

[1] http://ai.cs.umbc.edu/icgi2012/challenge/Pautomac

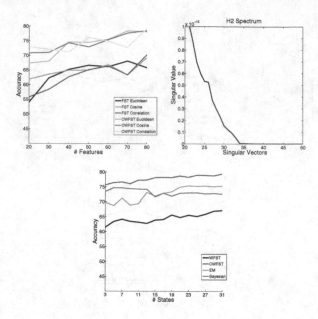

Fig. 2. (Left) Accuracy as a function of the number of features. (Center) Singular Values of H_2. (Right) Accuracy as a function of number of states for different methods.

Figure 2 (Right) compares the performance of CWFST, EM and Bayesian as a function of the number of states. As we can see CWFST outperforms both the EM and Bayesian algorithms. The Bayesian algorithm seems to be able to provide more compact models than EM (i.e. fewer number of states). Table 1 (Left) shows the performance of the best models for each learning algorithm. Recall that we resorted to approximate max marginal inference. Given that

Table 1. Comparison with other methods

	#states	Acc Marginals	Acc Exact
EM:	23	75.36%	76.32%
Bayesian:	5	74.67%	75.90%
CWFST:	31	79.36%	81.12%
FST:	31	67.09%	68.06%

Table 2. Training time (in seconds) and accuracy for Expectation Maximization for optimal model with 23 states

iters:	1	20	60	80	140	180	200	210	400
time:	14s	300s	1100s	2000s	4200s	5400s	6000s	6400s	10000s
acc.:	68%	74%	73.8%	74.8%	75%	75.36*%	75.36%	75.31%	75.17%

the average length of each sequence in the test sample is 10 it is still possible (though costly) to perform exact inference. That is to compute: $\text{argmax}_y \mathbb{P}(y|x) = \text{argmax}_y \mathbb{P}(x; y)$ by doing exhaustive search. The last row of Table 1 (Left) shows the accuracy of each model when the approximate inference is replaced by exact inference. In all cases we see an improvement in between 1 % and 2 %. This seems to suggest that the approximation is a good trade-off of accuracy vs inference time. Table 2 shows accuracy of EM as a function of training time (for optimal m and τ). For time comparison, the spectral training algorithm takes less than 30 seconds to train.

6 Conclusions

In this paper we presented a novel spectral learning algorithm that allows to exploit the representational power of latent variables to solve sequence tagging problems where the input is a continuous sequence and the output is discrete. Our approach is based on regarding the transition function of a weighted finite-state sequence tagger as a linear combination of atomic transition functions. We derive a spectral learning algorithm for this model from forward-backward mappings. The resulting algorithm is both simple and fast. Intuitively, the atomic transition functions operate on a soft partition of the input space. Experiments on a real task have shown the effectiveness of the method and its ability to take full advantage of these soft partitions.

References

1. Asuncion, A., Newman, D.J.: UCI machine learning repository (2007), http://www.ics.uci.edu/~mlearn/MLRepository.html
2. Anandkumar, A., Hsu, D., Kakade, S.M.: A method of moments for mixture models and hidden markov models. CoRR abs/1203 0683 (2012)
3. Bailly, R.: Quadratic weighted automata: Spectral algorithm and likelihood maximization. Journal of Machine Learning Research (2011)
4. Bailly, R., Denis, F., Ralaivola, L.: Grammatical inference as a principal component analysis problem. In: Proc. ICML (2009)
5. Balle, B., Quattoni, A., Carreras, X.: A spectral learning algorithm for finite state transducers. In: Gunopulos, D., Hofmann, T., Malerba, D., Vazirgiannis, M. (eds.) ECML PKDD 2011, Part I. LNCS, vol. 6911, pp. 156–171. Springer, Heidelberg (2011)
6. Balle, B., Quattoni, A., Carreras, X.: Local loss optimization in operator models: A new insight into spectral learning. In: Proceedings of ICML, pp. 1879–1886 (2012)
7. Blei, D.M., Ng, A.Y., Jordan, M.I., Lafferty, J.: Latent dirichlet allocation. Journal of Machine Learning Research 3, 2003 (2003)
8. Boots, B., Siddiqi, S., Gordon, G.: Closing the learning planning loop with predictive state representations. I. J. Robotic Research (2011)
9. Boots, B., Gordon, G.J.: An online spectral learning algorithm for partially observable nonlinear dynamical systems. In: Proceedings of the 25th National Conference on Artificial Intelligence (2001)

10. Bosch, A., Zisserman, A., Muñoz, X.: Scene classification via pLSA. In: Leonardis, A., Bischof, H., Pinz, A. (eds.) ECCV 2006, Part IV. LNCS, vol. 3954, pp. 517–530. Springer, Heidelberg (2006)
11. Chang, J.T.: Full reconstruction of markov models on evolutionary trees: Identifiability and consistency. Mathematical Biosciences 137, 51–73 (1996)
12. Clark, A.: Partially supervised learning of morphology with stochastic transducers. In: Proc. of NLPRS, pp. 341–348 (2001)
13. Eisner, J.: Parameter estimation for probabilistic finite-state transducers. In: Proc. of ACL, pp. 1–8 (2002)
14. Gao, J., Johnson, M.: A comparison of Bayesian estimators for unsupervised Hidden Markov Model POS taggers. In: Proceedings of EMNLP, pp. 344–352 (2008)
15. Hsu, D., Kakade, S.M., Zhang, T.: A spectral algorithm for learning hidden markov models. In: Proc. of COLT (2009)
16. Jaeger, H.: Observable operator models for discrete stochastic time series. Neural Computation 12, 1371–1398 (2000)
17. Jaeger, H.: Characterizing distributions of stochastic processes by linear operators. Tech. Rep. 62, German National Research Center for Information Technology (1999)
18. Jaeger, H.: Modeling and learning continuous-valued stochastic processes with ooms. Tech. Rep. 102, German National Research Center for Information Technology (2001)
19. Luque, F., Quattoni, A., Balle, B., Carreras, X.: Spectral learning in nondeterministic dependency parsing. In: EACL (2012)
20. Morency, L.P., Quattoni, A., Darrell, T.: Latent-dynamic discriminative models for continuous gesture recognition. In: CVPR (2007)
21. Mossel, E., Roch, S.: Learning nonsingular phylogenies and hidden markov models. In: Proc. of STOC (2005)
22. Quattoni, A., Wang, S., Morency, L., Collins, M., Darrell, T.: Hidden-state conditional random fields. Pattern Analysis and Machine Intelligence (2007)
23. Shibata, C., Yoshinaka, R.: Marginalizing out transition probabilities for several subclasses of pfas. In: JMLR Workshop and Conference Proceedings, ICGI 2012, vol. 21, pp. 259–263 (2012)
24. Siddiqi, S.M., Boots, B., Gordon, G.J.: Reduced-Rank Hidden Markov Models. In: Proc. AISTATS, pp. 741–748 (2010)
25. Siddiqi, S., Boots, B., Gordon, G.J.: Reduced-rank hidden Markov models. In: Proceedings of the Thirteenth International Conference on Artificial Intelligence and Statistics, AISTATS 2010 (2010)
26. Song, L., Boots, B., Siddiqi, S.M., Gordon, G.J., Smola, A.J.: Hilbert space embeddings of hidden Markov models. In: Proc. 27th Intl. Conf. on Machine Learning, ICML (2010)
27. Song, L., Huang, J., Smola, A., Fukumizu, K.: Hilbert space embeddings of conditional distributions with applications to dynamical systems (2009)
28. Wang, S.B., Quattoni, A., Morency, L.P., Demirdjian, D.: Hidden conditional random fields for gesture recognition. In: Proceedings of the 2006 IEEE Computer Society Conference on Computer Vision and Pattern Recognition, CVPR 2006, vol. 2, pp. 1521–1527 (2006)

Fast Variational Bayesian Linear State-Space Model

Jaakko Luttinen

Aalto University, Espoo, Finland
jaakko.luttinen@aalto.fi

Abstract. This paper presents a fast variational Bayesian method for linear state-space models. The standard variational Bayesian expectation-maximization (VB-EM) algorithm is improved by a parameter expansion which optimizes the rotation of the latent space. With this approach, the inference is orders of magnitude faster than the standard method. The speed of the proposed method is demonstrated on an artificial dataset and a large real-world dataset, which shows that the standard VB-EM algorithm is not suitable for large datasets because it converges extremely slowly. In addition, the paper estimates the temporal state variables using a smoothing algorithm based on the block LDL decomposition. This smoothing algorithm reduces the number of required matrix inversions and avoids a model augmentation compared to previous approaches.

Keywords: variational Bayesian methods, linear state-space models, parameter expansion.

1 Introduction

Linear state-space models (LSSM) are widely used in time-series analysis [1, 2]. They assume that the observations are generated linearly from a latent linear dynamical system. Although many real-world processes are non-linear, the linearity makes the model easy to analyze and efficient to estimate. In addition, many non-linear systems can be approximated using linear models, thus the LSSM is an important tool for time-series applications.

The Bayesian framework offers a principled way to estimate the model parameters from data. However, the estimation is analytically intractable making approximations necessary. This paper focuses on the variational Bayesian (VB) approximation, which can be computed using the variational Bayesian expectation-maximization (VB-EM) algorithm. The VB-EM algorithm assumes that the variables are independent and updates the approximate posterior distributions of the variables one at a time [3, 4].

The standard VB-EM algorithm may converge extremely slowly if the variables are strongly coupled. Because the variables are updated one at a time, the updates to each variable may be small and this results in zigzagging. This effect can be reduced by using parameter expansion to add auxiliary variables which

H. Blockeel et al. (Eds.): ECML PKDD 2013, Part I, LNAI 8188, pp. 305–320, 2013.

reduce the coupling between some variables [5, 6]. This can be seen as a parameterized joint optimization of multiple variables. Because the expansion and the effect on the speed of the algorithm depends.on the model, it is important to examine efficient parameter expansions for different models (see, e.g., [7–9]).

This paper derives a parameter expansion for the linear state-space model and shows experimentally that the VB-EM algorithm can be unusable for large datasets if the expansion is not used. The proposed parameter expansion is based on the rotation of the latent space, thus reducing the coupling between the states and the system parameters. Similar parameter expansion has been applied to canonical correlation and factor analysis models [7, 8]. However, those results cannot be applied directly to the LSSM because the rotation of the dynamics adds extra complexity.

In addition to convergence speed problems, the estimation of the state variables is not trivial, because the standard Rauch-Tung-Striebel [10] smoother cannot be applied directly as noted in [11]. This has been previously solved by using a parallel variant of the smoother in [11] and a model augmentation in [12]. This paper provides another perspective on the smoothing algorithm by deriving it from the Cholesky, or LDL, decomposition of a block-banded matrix. This results in a smoothing algorithm which requires less matrix inversions than the previous approaches, avoids the cost of the model augmentation and can be extended to other Markov random fields with different graph structure.

The paper is organized as follows: Section 2 defines the linear state-space model used in the paper. Section 3 briefly summarizes the standard VB-EM algorithm for the model. Section 4 derives the proposed smoothing algorithm. Section 5 presents the parameter expansion for the model. Section 6 presents experimental results that show the effect of the parameter expansion. Section 7 ends the paper with conclusions.

2   Model

In linear state-space models a sequence of M-dimensional observations $\mathbf{Y} = (\mathbf{y}_1, \ldots, \mathbf{y}_N)$ is assumed to be generated from latent D-dimensional states $\mathbf{X} = (\mathbf{x}_1, \ldots, \mathbf{x}_N)$ which follow a first-order Markov process:

$$\mathbf{x}_n = \mathbf{A}\mathbf{x}_{n-1} + \text{noise}, \tag{1}$$

$$\mathbf{y}_n = \mathbf{C}\mathbf{x}_n + \text{noise}, \tag{2}$$

where the noise is Gaussian, $\mathbf{A}$ is the $D \times D$ state dynamics matrix and $\mathbf{C}$ is the $M \times D$ loading matrix. Usually, the latent space dimensionality D is assumed to be much smaller than the observation space dimensionality M in order to model the dependencies of high-dimensional observations efficiently.

The equations defining the linear state-space model can be used to construct a Bayesian model [11]. The likelihood function is

$$p(\mathbf{Y}|\mathbf{C}, \mathbf{X}, \boldsymbol{\tau}) = \prod_{n=1}^{N} \mathcal{N}(\mathbf{y}_n|\mathbf{C}\mathbf{x}_n, \text{diag}(\boldsymbol{\tau})^{-1}), \tag{3}$$

where $\mathcal{N}(\mathbf{x}|\boldsymbol{\mu}, \boldsymbol{\Sigma})$ is the probability density function of the Gaussian distribution of variable $\mathbf{x}$ with mean $\boldsymbol{\mu}$ and covariance $\boldsymbol{\Sigma}$. The covariance matrix in (3) is diagonal, that is, the noise is independent for each observed element of $\mathbf{y}_n$. The probability of the states is given as

$$p(\mathbf{X}|\mathbf{A}) = \mathcal{N}(\mathbf{x}_0|\mathbf{m}_0, \boldsymbol{\Lambda}_0^{-1}) \prod_{n=1}^{N} \mathcal{N}(\mathbf{x}_n|\mathbf{A}\mathbf{x}_{n-1}, \mathbf{I}), \tag{4}$$

where $\mathbf{x}_0$ is an auxiliary initial state with mean $\mathbf{m}_0$ and precision $\boldsymbol{\Lambda}_0$. The noise of the process in (4) has unit covariance matrix without loss of generality, because the latent space can be rotated arbitrarily by compensating it in the parameters $\mathbf{A}$ and $\mathbf{C}$. The parameters of the process are given the following priors:

$$p(\mathbf{A}|\boldsymbol{\alpha}) = \prod_{i=1}^{D}\prod_{j=1}^{D} \mathcal{N}(a_{ij}|0, \alpha_j^{-1}), \qquad p(\boldsymbol{\alpha}) = \prod_{d=1}^{D} \mathcal{G}(\alpha_d|a_\gamma, b_\gamma), \tag{5}$$

$$p(\mathbf{C}|\boldsymbol{\gamma}) = \prod_{m=1}^{M}\prod_{d=1}^{D} \mathcal{N}(c_{md}|0, \gamma_d^{-1}), \qquad p(\boldsymbol{\gamma}) = \prod_{d=1}^{D} \mathcal{G}(\gamma_d|a_\gamma, b_\gamma), \tag{6}$$

$$p(\boldsymbol{\tau}) = \prod_{m=1}^{M} \mathcal{G}(\tau_m|a_\tau, b_\tau), \tag{7}$$

where a_{ij} is the element on the i-th row and j-th column of the matrix $\mathbf{A}$, α_d is the d-th element of the vector $\boldsymbol{\alpha}$, and $\mathcal{G}(\alpha|a, b)$ is the probability density function of the gamma distribution with shape a and rate b. The variables $\boldsymbol{\alpha}$ and $\boldsymbol{\gamma}$ are automatic relevance determination (ARD) parameters, which prune out components that are not significant enough. The hyperparameters a_α, b_α, a_γ, b_γ, a_τ and b_τ can be set to small values (e.g., 10^{-5}) to give broad priors. The above model definition is similar to [11, 12] and the details can be modified without affecting the main results of this paper.

3   Posterior Approximation

As the posterior distribution of the variables is analytically intractable, it is approximated using variational Bayesian (VB) framework [4]. The approximation is assumed to factorize with respect to the variables as

$$p(\mathbf{X}, \mathbf{A}, \boldsymbol{\alpha}, \mathbf{C}, \boldsymbol{\gamma}, \boldsymbol{\tau}|\mathbf{Y}) \approx q(\mathbf{X}, \mathbf{A}, \boldsymbol{\alpha}, \mathbf{C}, \boldsymbol{\gamma}, \boldsymbol{\tau}) = q(\mathbf{X})q(\mathbf{A})q(\boldsymbol{\alpha})q(\mathbf{C})q(\boldsymbol{\gamma})q(\boldsymbol{\tau}). \tag{8}$$

The approximation is optimized by minimizing the Kullback-Leibler divergence from the true posterior, which is equivalent to maximizing the lower bound of the marginal log likelihood

$$\mathcal{L}(\mathbf{Y}) = \langle \log p(\mathbf{Y}|\mathbf{C}, \mathbf{X}, \boldsymbol{\tau}) \rangle + \left\langle \log \frac{p(\mathbf{X}|\mathbf{A})}{q(\mathbf{X})} \right\rangle + \left\langle \log \frac{p(\mathbf{A}|\boldsymbol{\alpha})}{q(\mathbf{A})} \right\rangle$$
$$+ \left\langle \log \frac{p(\boldsymbol{\alpha})}{q(\boldsymbol{\alpha})} \right\rangle + \left\langle \log \frac{p(\mathbf{C}|\boldsymbol{\gamma})}{q(\mathbf{C})} \right\rangle + \left\langle \log \frac{p(\boldsymbol{\gamma})}{q(\boldsymbol{\gamma})} \right\rangle + \left\langle \log \frac{p(\boldsymbol{\tau})}{q(\boldsymbol{\tau})} \right\rangle, \tag{9}$$

where $\langle \cdot \rangle$ is the expectation with respect to the approximate posterior distribution q.

For conjugate-exponential models, the approximation can be optimized by using the variational Bayesian expectation-maximization (VB-EM) algorithm [3]. In VB-EM, the posterior approximation is updated for the variables one at a time and iterated until convergence. The approximate distributions have the following forms:

$$q(\mathbf{X}) = \mathcal{N}([\mathbf{X}]_: | \boldsymbol{\mu}_x, \boldsymbol{\Sigma}_x), \qquad q(\boldsymbol{\tau}) = \prod_{m=1}^{M} \mathcal{G}(\tau_m | \bar{a}_\tau^{(m)}, \bar{b}_\tau^{(m)}), \qquad (10)$$

$$q(\mathbf{A}) = \prod_{d=1}^{D} \mathcal{N}(\mathbf{a}_d | \boldsymbol{\mu}_a^{(d)}, \boldsymbol{\Sigma}_a^{(d)}), \qquad q(\boldsymbol{\alpha}) = \prod_{d=1}^{D} \mathcal{G}(\alpha_d | \bar{a}_\alpha^{(d)}, \bar{b}_\alpha^{(d)}), \qquad (11)$$

$$q(\mathbf{C}) = \prod_{m=1}^{M} \mathcal{N}(\mathbf{c}_m | \boldsymbol{\mu}_c^{(m)}, \boldsymbol{\Sigma}_c^{(m)}), \qquad q(\boldsymbol{\gamma}) = \prod_{d=1}^{D} \mathcal{G}(\gamma_d | \bar{a}_\gamma^{(d)}, \bar{b}_\gamma^{(d)}), \qquad (12)$$

where $\mathbf{a}_d$ and $\mathbf{c}_m$ are the row vectors of $\mathbf{A}$ and $\mathbf{C}$, respectively, and $[\mathbf{X}]_:$ is a vector obtained by stacking the vectors $\mathbf{x}_n$. It is straightforward to derive the following update equations of the variational parameters:

$$\boldsymbol{\Sigma}_a^{(d)} = \left(\langle \mathrm{diag}(\boldsymbol{\alpha}) \rangle + \sum_{n=1}^{N} \langle \mathbf{x}_{n-1} \mathbf{x}_{n-1}^{\mathrm{T}} \rangle \right)^{-1}, \qquad \boldsymbol{\mu}_a^{(d)} = \boldsymbol{\Sigma}_a^{(d)} \sum_{n=1}^{N} \langle x_{dn} \mathbf{x}_{n-1} \rangle, \quad (13)$$

$$\bar{a}_\alpha^{(d)} = a_\alpha + \frac{D}{2}, \qquad \bar{b}_\alpha^{(d)} = b_\alpha + \frac{1}{2} \sum_{i=1}^{D} \langle a_{id}^2 \rangle, \qquad (14)$$

$$\boldsymbol{\Sigma}_c^{(m)} = \left(\langle \mathrm{diag}(\boldsymbol{\gamma}) \rangle + \sum_{n \in \mathcal{O}_{m:}} \langle \tau_m \rangle \langle \mathbf{x}_n \mathbf{x}_n^{\mathrm{T}} \rangle \right)^{-1}, \qquad \boldsymbol{\mu}_c^{(m)} = \boldsymbol{\Sigma}_c^{(m)} \sum_{n \in \mathcal{O}_{m:}} y_{mn} \langle \tau_m \rangle \langle \mathbf{x}_n \rangle,$$

$$(15)$$

$$\bar{a}_\gamma^{(d)} = a_\gamma + \frac{M}{2}, \qquad \bar{b}_\gamma^{(d)} = b_\gamma + \frac{1}{2} \sum_{m=1}^{M} \langle c_{md}^2 \rangle, \qquad (16)$$

$$\bar{a}_\tau^{(m)} = a_\tau + \frac{N_m}{2}, \qquad \bar{b}_\tau^{(m)} = b_\tau + \frac{1}{2} \sum_{n \in \mathcal{O}_{m:}} \xi_{mn}, \qquad (17)$$

where $\mathcal{O}_{m:}$ is the set of time instances n for which the observation y_{mn} is not missing, N_m is the size of the set $\mathcal{O}_{m:}$, and $\xi_{mn} = \langle (y_{mn} - \mathbf{c}_m^{\mathrm{T}} \mathbf{x}_n)^2 \rangle$. Gaussian and gamma distributed variables have the following expectations:

$$\text{For } \mathbf{x} \sim \mathcal{N}(\boldsymbol{\mu}, \boldsymbol{\Sigma}), \qquad \langle \mathbf{x} \rangle = \boldsymbol{\mu} \quad \text{and} \quad \langle \mathbf{x}\mathbf{x}^{\mathrm{T}} \rangle = \boldsymbol{\mu}\boldsymbol{\mu}^{\mathrm{T}} + \boldsymbol{\Sigma}. \qquad (18)$$

$$\text{For } \alpha \sim \mathcal{G}(a, b), \qquad \langle \alpha \rangle = \frac{a}{b} \quad \text{and} \quad \langle \log \alpha \rangle = \psi(a) - \log(b). \qquad (19)$$

The formula for updating $q(\mathbf{X})$ is discussed in the following section.

4 Smoothing Algorithm

The approximate posterior distribution $q(\mathbf{X})$ can be updated using filtering and smoothing algorithms. However, the standard Rauch-Tung-Striebel (RTS) smoother cannot be applied straightforwardly because the required expectations under $q(\mathbf{A})$ are difficult to compute [11]. Previous approaches have solved this by using a parallel variant of the smoother [11] or augmenting the model to be able to apply standard RTS smoother [12]. Although these are working methods, this section presents another view on the smoothing problem and derives an algorithm which can be applied easily and efficiently in the VB framework.

Instead of trying to apply standard filters and smoothers directly, the smoothing can be computed equivalently from a Cholesky decomposition perspective [13]. The smoothing can be seen as a multiplication by the inverse of a large block-banded matrix. Utilizing the block-banded structure of the matrix, the computational complexity of the inversion is $\mathcal{O}(D^3 N)$ instead of $\mathcal{O}(D^3 N^3)$, where $D \ll N$. The inverse is computed using the block LDL decomposition and inverting this decomposition in two parts. The resulting smoothing algorithm is similar to the standard Kalman filter [14] and RTS smoother although not exactly identical.

The smoothing algorithm can be derived from the standard update equations of $q(\mathbf{X})$. Computational aspects aside, the update equation of the covariance matrix $\mathbf{\Sigma}_x$ is

$$\mathbf{\Sigma}_x = \begin{bmatrix} \mathbf{\Sigma}_{1,1} & \cdots & \mathbf{\Sigma}_{1,N} \\ \vdots & \ddots & \vdots \\ \mathbf{\Sigma}_{N,1} & \cdots & \mathbf{\Sigma}_{N,N} \end{bmatrix} = \begin{bmatrix} \mathbf{\Psi}_{0,0} & \mathbf{\Psi}_{0,1} & & \\ \mathbf{\Psi}_{0,1}^{\mathrm{T}} & \mathbf{\Psi}_{1,1} & \ddots & \\ & \ddots & \ddots & \mathbf{\Psi}_{N-1,N} \\ & & \mathbf{\Psi}_{N-1,N}^{\mathrm{T}} & \mathbf{\Psi}_{N,N} \end{bmatrix}^{-1} = \mathbf{\Psi}^{-1}, \quad (20)$$

where the block-banded matrix $\mathbf{\Psi}$ is defined as

$$\mathbf{\Psi} = \begin{bmatrix} \mathbf{\Lambda}_0 + \langle \mathbf{A}^{\mathrm{T}}\mathbf{A} \rangle & \langle \mathbf{A}_1 \rangle^{\mathrm{T}} & & \\ \langle \mathbf{A} \rangle & \mathbf{I} - \langle \mathbf{A}^{\mathrm{T}}\mathbf{A} \rangle & \ddots & \\ & \ddots & \ddots & \langle \mathbf{A} \rangle^{\mathrm{T}} \\ & & \langle \mathbf{A} \rangle & \mathbf{I} - \langle \mathbf{A}^{\mathrm{T}}\mathbf{A} \rangle \end{bmatrix}$$
$$+ \begin{bmatrix} \mathbf{0} & & & \\ & \sum_{m \in \mathcal{O}_{:1}} \langle \tau_m \rangle \langle \mathbf{c}_m \mathbf{c}_m^{\mathrm{T}} \rangle & & \\ & & \ddots & \\ & & & \sum_{m \in \mathcal{O}_{:N}} \langle \tau_m \rangle \langle \mathbf{c}_m \mathbf{c}_m^{\mathrm{T}} \rangle \end{bmatrix}, \quad (21)$$

and $\mathcal{O}_{:n}$ is the set of dimensions m for which the observation y_{mn} is not missing. The first matrix term in the sum (21) comes from the prior (4) and the second matrix term comes from the likelihood (3). Note that although $\mathbf{\Psi}$ is block-banded, $\mathbf{\Sigma}$ is dense in general. The posterior mean parameter is updated as

---

Algorithm 1. Forward pass.

Input: $\{\Psi_{n,n}\}_{n=0}^{N}$, $\{\Psi_{n,n+1}\}_{n=0}^{N-1}$, $\{\mathbf{v}_n\}_{n=0}^{N}$

$\quad \tilde{\Sigma}_{0,0} \leftarrow \Psi_{0,0}^{-1}$

$\quad \tilde{\mu}_0 \leftarrow \tilde{\Sigma}_{0,0}\mathbf{v}_0$

$\quad$ **for** $n = 0 \to N - 1$ **do**

$\qquad \tilde{\Sigma}_{n,n+1} \leftarrow \tilde{\Sigma}_{n,n}\Psi_{n,n+1}$

$\qquad \tilde{\Sigma}_{n+1,n+1} \leftarrow \left(\Psi_{n+1,n+1} - \tilde{\Sigma}_{n,n+1}^{T}\Psi_{n,n+1} \right)^{-1}$

$\qquad \tilde{\mu}_{n+1} \leftarrow \tilde{\Sigma}_{n+1,n+1} \left(\mathbf{v}_{n+1} - \tilde{\Sigma}_{n,n+1}^{T}\tilde{\mu}_n \right)$

$\quad$ **end for**

Output: $\{\tilde{\Sigma}_{n,n}\}_{n=0}^{N}$, $\{\tilde{\Sigma}_{n,n+1}\}_{n=0}^{N-1}$, $\{\tilde{\mu}_n\}_{n=0}^{N}$

$$
\mu_x = \begin{bmatrix} \mu_o \\ \mu_1 \\ \vdots \\ \mu_N \end{bmatrix} = \Sigma_x \begin{bmatrix} \mathbf{v}_0 \\ \mathbf{v}_1 \\ \vdots \\ \mathbf{v}_N \end{bmatrix} = \Sigma_x \mathbf{v}, \quad \text{where } \mathbf{v} = \begin{bmatrix} \Lambda_0 \mathbf{m}_0 \\ \sum_{m \in \mathcal{O}_{:1}} y_{mn}\langle \tau_m \rangle \langle \mathbf{c}_m \rangle \\ \vdots \\ \sum_{m \in \mathcal{O}_{:N}} y_{mn}\langle \tau_m \rangle \langle \mathbf{c}_m \rangle \end{bmatrix}. \quad (22)
$$

The vector $\mathbf{v}_0$ comes from the prior and the vectors $\{\mathbf{v}_n\}_{n=1}^{N}$ come from the likelihood.

Instead of computing the full matrix Σ_x, it is sufficient for the VB-EM algorithm to compute only the diagonal blocks $\Sigma_{n,n}$, the first super-diagonal blocks $\Sigma_{n,n+1}$, and the mean μ_x. These terms can be computed efficiently by writing the parameters as $\Sigma_x = \Psi^{-1}\mathbf{I}$ and $\mu_x = \Psi^{-1}\mathbf{v}$, and utilizing the block-banded structure of Ψ. Because Ψ is a symmetric positive-definite matrix, it can be decomposed using the block LDL decomposition $\Psi = \mathbf{LDL}^{T}$, where $\mathbf{D}$ is a block-diagonal matrix and $\mathbf{L}$ is a lower-triangular matrix with identity matrices on the diagonal. Thus, multiplying on the left by Ψ^{-1} is equivalent to multiplying on the left by $\mathbf{L}^{-T}\mathbf{D}^{-1}\mathbf{L}^{-1}$. This can be computed in two phases: First, multiplying on the left by $\mathbf{D}^{-1}\mathbf{L}^{-1}$ results in the forward pass shown in Algorithm 1. Second, multiplying on the left by $\mathbf{L}^{-T}$ results in the backward pass shown in Algorithm 2. Note that both algorithms can be implemented as in-place algorithms by overwriting the inputs with the outputs.

This smoothing algorithm has a few benefits compared to the previous methods [11, 12]. First, the algorithm needs to compute only one matrix inversion per time instance, whereas the parallel and the augmented variants require three and two inversions, respectively. Second, the augmented variant requires the Cholesky decomposition of $\langle \mathbf{A}^T\mathbf{A} \rangle$ and $\langle \mathbf{C}^T \text{diag}(\tau)\mathbf{C} \rangle$ for each time instance if $\mathbf{A}$, $\mathbf{C}$ or τ varies in time or if the data $\mathbf{Y}$ contains missing values. Third, the proposed Cholesky approach makes it straightforward to modify the algorithm if one changes the graph structure of the Markov random field to something else than a Markov chain. Fourth, if Ψ is modified directly, for instance, if optimizing the natural parameters, the covariance and the mean can be computed without needing to solve what would be the parameters of the corresponding Markov

---

Algorithm 2. Backward pass.

Input: $\{\tilde{\Sigma}_{n,n}\}_{n=0}^{N}, \{\tilde{\Sigma}_{n,n+1}\}_{n=0}^{N-1}, \{\tilde{\mu}_n\}_{n=0}^{N}$

$\quad \Sigma_{N,N} \leftarrow \tilde{\Sigma}_{N,N}$

$\quad \mu_N \leftarrow \tilde{\mu}_N$

$\quad$ **for** $n = N - 1 \rightarrow 0$ **do**

$\quad\quad \Sigma_{n,n+1} \leftarrow -\tilde{\Sigma}_{n,n+1}\Sigma_{n+1,n+1}$

$\quad\quad \Sigma_{n,n} \leftarrow \tilde{\Sigma}_{n,n} - \tilde{\Sigma}_{n,n+1}\Sigma_{n,n+1}^{\mathrm{T}}$

$\quad\quad \mu_n \leftarrow \tilde{\mu}_n - \tilde{\Sigma}_{n,n+1}\mu_{n+1}$

$\quad$ **end for**

Output: $\{\Sigma_{n,n}\}_{n=0}^{N}, \{\Sigma_{n,n+1}\}_{n=0}^{N-1}, \{\mu_n\}_{n=0}^{N}$

---

chain. However, whatever smoothing algorithm is used, significant speeding up can be obtained by using the parameter expansion discussed in the next section.

5   Speeding Up the Inference

The variational Bayesian EM algorithm may converge extremely slowly because it updates only one variable at a time resulting in zigzagging for strongly coupled variables. It may be possible to speed up the algorithm using parameter expansion which reduces the coupling between the variables [5, 6]. For instance, parameter expanded VB-EM has been used for factor analysis [7], canonical correlation analysis [8], and common spatial patterns [9].

In state-space models, the states $\mathbf{x}_n$ and the loadings $\mathbf{C}$ are coupled through a dot product $\mathbf{C}\mathbf{x}_n$, which is unaltered if the latent space is rotated arbitrarily:

$$\mathbf{y}_n = \mathbf{C}\mathbf{x}_n = \mathbf{C}\mathbf{R}^{-1}\mathbf{R}\mathbf{x}_n. \tag{23}$$

Thus, one intuitive transformation would be $\mathbf{C} \rightarrow \mathbf{C}\mathbf{R}^{-1}$ and $\mathbf{X} \rightarrow \mathbf{R}\mathbf{X}$. In order to keep the dynamics of the latent states unaffected by the transformation, the state dynamics matrix $\mathbf{A}$ must be transformed accordingly:

$$\mathbf{R}\mathbf{x}_n = \mathbf{R}\mathbf{A}\mathbf{R}^{-1}\mathbf{R}\mathbf{x}_{n-1}, \tag{24}$$

resulting in a transformation $\mathbf{A} \rightarrow \mathbf{R}\mathbf{A}\mathbf{R}^{-1}$.

The parameter expansion is performed by parameterizing the posterior distributions with $\mathbf{R}$ and maximizing the lower bound of the marginal log likelihood (9) with respect to $\mathbf{R}$. Thus, the method optimizes the posterior distributions of several variables jointly instead of one at a time. In general, the optimal value for the parameter $\mathbf{R}$ is found using numerical optimization methods, although for a simple factor analysis model, the solution can be found analytically [7]. The optimal transformation is guaranteed not to decrease the lower bound if the initial value $\mathbf{R} = \mathbf{I}$ recovers the original posterior unaffected. The optimization is computationally efficient because the lower bound terms affected by $\mathbf{R}$ are low-dimensional.

The rotation can be optimized using nonlinear conjugate gradient (CG) algorithm. It is sufficient to find only a rough estimate of the optimal rotation

R to speed up the algorithm significantly, thus 10 iterations of CG was used in this paper, and CG was run after each iteration of the VB-EM algorithm. The gradients required by CG are not given in order to keep the paper concise but the derivations are straightforward.

The work required deriving the cost function for **R** for different models may be reduced by deriving the cost function for small general blocks that appear in several models. These general results may be used directly if a similar transformation for a similar block appears in another model. In order to provide modular results that can be applied to other models, the following subsections consider the following transformations of small blocks: rotating a Gaussian with an ARD prior (**C** and γ), rotating a Gaussian with an ARD prior from left and right (**A** and α), and rotating a Gaussian Markov chain (**X**).

5.1   Rotation of a Gaussian Variable with an ARD Prior

Let us examine the rotation of **C** as $\mathbf{CR}^{-1}$ in our linear state-space model. This transformation corresponds to a trivial rotation of the posterior mean and covariance of **C**. However, recall that **C** has an ARD prior with hyperparameters γ as defined in (6). It would be possible to simply rotate **C** without changing γ but it is more efficient to also transform the hyperparameters γ. This allows $q(\mathbf{C})$ and $q(\gamma)$ to be optimized jointly. The transformation of γ is motivated by the VB-EM update equation (16).

The rotation of **C** can be seen as the following transformation of $q(\mathbf{C})$ and $q(\gamma)$:

$$q_*(\mathbf{C}) = \prod_{m=1}^{M} \mathcal{N}\left(\mathbf{c}_m \middle| \mathbf{R}^{\text{-T}}\boldsymbol{\mu}_c^{(m)}, \mathbf{R}^{\text{-T}}\boldsymbol{\Sigma}_c^{(m)}\mathbf{R}^{-1}\right), \tag{25}$$

$$q_*(\gamma) = \prod_{d=1}^{D} \mathcal{G}(\gamma_d | \bar{a}_\gamma^{(d)}, \beta_\gamma^{(d)}), \tag{26}$$

where $\boldsymbol{\mu}_c^{(m)}$, $\boldsymbol{\Sigma}_c^{(m)}$ and $\bar{a}_\gamma^{(d)}$ are the parameters of the original distributions defined in (12),

$$\beta_\gamma^{(d)} = b_\gamma + \frac{1}{2}\left[\mathbf{R}^{\text{-T}}\langle\mathbf{C}^{\text{T}}\mathbf{C}\rangle\mathbf{R}^{-1}\right]_{dd}, \tag{27}$$

$\langle\cdot\rangle$ is the expectation with respect to the original posterior distribution, and $[\cdot]_{ij}$ is the element on the i-th row and j-th column. Note that the original posterior distributions $q(\mathbf{C})$ and $q(\gamma)$ are recovered by setting $\mathbf{R} = \mathbf{I}$, thus the optimal transformation is guaranteed not to worsen the posterior approximation.

The transformation affects only a small number of lower bound terms in (9) making the optimization of the rotation efficient. The transformation of **C** affects the likelihood term $\langle\log p(\mathbf{Y}|\mathbf{C}, \mathbf{X}, \tau)\rangle_*$ but this effect is cancelled by the transformation of **X** and can thus be ignored. The remaining terms are affected as

$$\langle \log q(\mathbf{C}) \rangle_* = -M \log |\mathbf{R}^{-T}| + \text{const}, \tag{28}$$

$$\langle \log p(\mathbf{C}|\boldsymbol{\gamma}) \rangle_* = -\frac{1}{2} \text{tr} \left(\langle \mathbf{C}^T \mathbf{C} \rangle_* \langle \text{diag}(\boldsymbol{\gamma}) \rangle_* \right) + \frac{M}{2} \sum_{d=1}^{D} \langle \log \gamma_d \rangle_* + \text{const}, \tag{29}$$

$$\langle \log q(\boldsymbol{\gamma}) \rangle_* = \sum_{d=1}^{D} \langle \log \gamma_d \rangle_* + \text{const}, \tag{30}$$

$$\langle \log p(\boldsymbol{\gamma}) \rangle_* = (a_\gamma - 1) \sum_{d=1}^{D} \langle \log \gamma_d \rangle_* - b_\gamma \sum_{d=1}^{D} \langle \gamma_d \rangle_d + \text{const}, \tag{31}$$

where const is the part that is constant with respect to $\mathbf{R}$, $\langle \cdot \rangle_*$ is the expectation with respect to the transformed posterior distribution q_*, and

$$\langle \mathbf{C}^T \mathbf{C} \rangle_* = \mathbf{R}^{-T} \langle \mathbf{C}^T \mathbf{C} \rangle \mathbf{R}^{-1}. \tag{32}$$

The expectations $\langle \gamma_d \rangle_*$ and $\langle \log \gamma_d \rangle_*$ are computed as shown in (19). When optimizing the rotation $\mathbf{R}$, it is not necessary to compute the rotated covariance matrix nor the rotated mean of $\mathbf{C}$ because the cost function only requires $\langle \mathbf{C}^T \mathbf{C} \rangle_*$ which can be computed efficiently using (32). After the optimization, the parameters of the posterior distribution are transformed using the optimal rotation in (25).

5.2 Double Rotation of a Gaussian Variable with an ARD Prior

The state dynamics matrix $\mathbf{A}$ should be rotated as $\mathbf{R}\mathbf{A}\mathbf{R}^{-1}$. However, performing this transformation exactly would make the rows of $\mathbf{A}$ dependent in the posterior approximation causing the VB-EM algorithm to be computationally much more intensive. Thus, the posterior distribution is transformed in such a way that the rows remain independent but that the transformation resembles the "true" transformation. The idea is to use a transformation which gives true values for the relevant expectations $\langle \mathbf{A} \rangle$ and $\langle \mathbf{A}^T \mathbf{A} \rangle$ although the covariance of $\mathbf{A}$ is transformed "incorrectly". Note that the transformation of the posterior distribution is not really incorrect even if it does not correspond to $\mathbf{R}\mathbf{A}\mathbf{R}^{-1}$ because, in principle, the transformation can be chosen arbitrarily.

In addition to rotating $\mathbf{A}$, the ARD parameter $\boldsymbol{\alpha}$ in (5) is also transformed in order to improve the effect of the transformation. Thus, the transformation of $q(\mathbf{A})$ and $q(\boldsymbol{\alpha})$ is

$$q_*(\mathbf{A}) = \prod_{d=1}^{D} \mathcal{N} \left(\mathbf{a}_d \left| \sum_{j=1}^{D} r_{dj} \mathbf{R}^{-T} \boldsymbol{\mu}_a^{(d)}, \left(\sum_{i=1}^{D} r_{id} \right)^2 \mathbf{R}^{-T} \boldsymbol{\Sigma}_a^{(d)} \mathbf{R}^{-T} \right. \right) \tag{33}$$

$$q_*(\boldsymbol{\alpha}) = \prod_{d=1}^{D} \mathcal{G}(\alpha_d | \bar{a}_\alpha^{(d)}, \beta_\alpha^{(d)}) \tag{34}$$

where $\boldsymbol{\mu}_a^{(d)}$, $\boldsymbol{\Sigma}_a^{(d)}$ and $\bar{a}_\alpha^{(d)}$ are the parameters of the original distributions in (11), r_{ij} is the element $[\mathbf{R}]_{ij}$, and

$$\beta_\alpha^{(d)} = b_\alpha + \frac{1}{2}\left[\langle\mathbf{A}^{\mathrm{T}}\mathbf{A}\rangle_*\right]_{dd}. \tag{35}$$

This transformation differs from the exact transformation $\mathbf{R}\mathbf{A}\mathbf{R}^{-1}$ in that the cross-covariances between the rows of $\mathbf{A}$ are kept zero and the covariance of each row is not a weighted sum of the covariances of all rows but only a rotated and scaled version of the covariance of the same row. Although this transformation does not exactly correspond to $\mathbf{R}\mathbf{A}\mathbf{R}^{-1}$, it has the nice property that the expectations $\langle\mathbf{A}\rangle_*$ and $\langle\mathbf{A}^{\mathrm{T}}\mathbf{A}\rangle_*$ are transformed "correctly". Also, note that setting $\mathbf{R} = \mathbf{I}$ recovers the original posterior approximation unaffected.

The terms in the lower bound of the marginal log likelihood (9) are affected as

$$\langle\log q(\mathbf{A})\rangle_* = -D\log|\mathbf{R}^{-\mathrm{T}}| - D\sum_{j=1}^{D}\log\left|\sum_{i=1}^{D}r_{ij}\right| + \mathrm{const}, \tag{36}$$

$$\langle\log p(\mathbf{A}|\boldsymbol{\alpha})\rangle_* = -\frac{1}{2}\,\mathrm{tr}\left(\langle\mathbf{A}^{\mathrm{T}}\mathbf{A}\rangle_*\langle\mathrm{diag}(\boldsymbol{\alpha})\rangle_*\right) + \frac{D}{2}\sum_{d=1}^{D}\langle\log\alpha_d\rangle_* + \mathrm{const}, \tag{37}$$

$$\langle\log q(\boldsymbol{\alpha})\rangle_* = \sum_{d=1}^{D}\langle\log\alpha_d\rangle_* + \mathrm{const}, \tag{38}$$

$$\langle\log p(\boldsymbol{\alpha})\rangle_* = (a_\alpha - 1)\sum_{d=1}^{D}\langle\log\alpha_d\rangle_* - b_\alpha\sum_{d=1}^{D}\langle\alpha_d\rangle_* + \mathrm{const}. \tag{39}$$

In addition, the transformation of $q(\mathbf{A})$ affects the lower bound term $\langle\log p(\mathbf{X}|\mathbf{A})\rangle_*$ but that is examined in the next subsection. The transformed expectations of $\mathbf{A}$ are

$$\langle\mathbf{A}\rangle_* = \mathbf{R}\langle\mathbf{A}\rangle\mathbf{R}^{-1}, \tag{40}$$

$$\langle\mathbf{A}^{\mathrm{T}}\mathbf{A}\rangle_* = \mathbf{R}^{-\mathrm{T}}\left[\langle\mathbf{A}\rangle^{\mathrm{T}}\mathbf{R}^{\mathrm{T}}\mathbf{R}\langle\mathbf{A}\rangle + \sum_{d=1}^{D}\left(\sum_{i=1}^{D}r_{id}\right)^2\boldsymbol{\Sigma}_a^{(d)}\right]\mathbf{R}^{-1}. \tag{41}$$

The expectations $\langle\alpha_d\rangle_*$ and $\langle\log\alpha_d\rangle_*$ are computed as shown in (19).

5.3   Rotation of a Gaussian Markov Chain

The state variables $\mathbf{x}_n$ are rotated as $\mathbf{R}\mathbf{x}_n$. Because the states $\mathbf{x}_n$ are not independent in the posterior approximation, the rotation is written equivalently for all the states as $(\mathbf{I}\otimes\mathbf{R})[\mathbf{X}]_{:}$. This results in the transformed posterior

$$q_*(\mathbf{X}) = \mathcal{N}\left([\mathbf{X}]_{:}\,\big|\,(\mathbf{I}\otimes\mathbf{R})\boldsymbol{\mu}_x, (\mathbf{I}\otimes\mathbf{R})\boldsymbol{\Sigma}_x(\mathbf{I}\otimes\mathbf{R})^{\mathrm{T}}\right), \tag{42}$$

where $\boldsymbol{\mu}_x$ and $\boldsymbol{\Sigma}_x$ are the original posterior mean and covariance parameters in (10). Note that it is not necessary to compute nor store the full covariance matrix $\boldsymbol{\Sigma}_x$, because the cost function of the rotation requires only the cross-covariance $\text{cov}(\mathbf{x}_{n-1}, \mathbf{x}_n)$ between consecutive states and the covariance $\text{cov}(\mathbf{x}_n, \mathbf{x}_n)$.

The transformation affects only a few lower bound terms in (9). The effect on the likelihood term $\langle \log p(\mathbf{Y}|\mathbf{C}, \mathbf{X}, \boldsymbol{\tau}) \rangle_*$ is cancelled by the transformation of $\mathbf{C}$. Assuming that $q(\mathbf{A})$ is transformed as described in the previous subsection, the lower bound terms are affected as

$$\langle \log q(\mathbf{X}) \rangle_* = -(N+1) \log |\mathbf{R}| + \text{const}, \tag{43}$$

$$\langle \log p(\mathbf{X}|\mathbf{A}) \rangle_* = \text{tr} \left(-\frac{1}{2} \boldsymbol{\Lambda}_0 \langle \mathbf{x}_0 \mathbf{x}_0^{\mathrm{T}} \rangle_* + \boldsymbol{\Lambda}_0 \mathbf{m}_0 \langle \mathbf{x}_0 \rangle_*^{\mathrm{T}} + \sum_{n=1}^{N} \left[-\frac{1}{2} \langle \mathbf{x}_n \mathbf{x}_n^{\mathrm{T}} \rangle_* \right.\right.$$
$$\left.\left. -\frac{1}{2} \langle \mathbf{A}^{\mathrm{T}} \mathbf{A} \rangle_* \langle \mathbf{x}_{n-1} \mathbf{x}_{n-1}^{\mathrm{T}} \rangle_* + \langle \mathbf{A} \rangle_* \langle \mathbf{x}_{n-1} \mathbf{x}_n^{\mathrm{T}} \rangle_* \right] \right), \tag{44}$$

where $\langle \mathbf{A} \rangle_*$ and $\langle \mathbf{A}^{\mathrm{T}} \mathbf{A} \rangle_*$ are defined in (40) and (41), respectively, and

$$\langle \mathbf{x}_n \rangle_* = \mathbf{R} \langle \mathbf{x}_n \rangle, \tag{45}$$

$$\langle \mathbf{x}_n \mathbf{x}_n^{\mathrm{T}} \rangle_* = \mathbf{R} \langle \mathbf{x}_n \mathbf{x}_n^{\mathrm{T}} \rangle \mathbf{R}^{\mathrm{T}}, \tag{46}$$

$$\langle \mathbf{A} \rangle_* \langle \mathbf{x}_{n-1} \mathbf{x}_n^{\mathrm{T}} \rangle_* = \mathbf{R} \langle \mathbf{A} \rangle \langle \mathbf{x}_{n-1} \mathbf{x}_n^{\mathrm{T}} \rangle \mathbf{R}^{\mathrm{T}}, \tag{47}$$

$$\text{tr} \left(\langle \mathbf{A}^{\mathrm{T}} \mathbf{A} \rangle_* \langle \mathbf{x}_{n-1} \mathbf{x}_{n-1}^{\mathrm{T}} \rangle_* \right) = \text{tr} \left(\langle \mathbf{A} \rangle^{\mathrm{T}} \mathbf{R}^{\mathrm{T}} \mathbf{R} \langle \mathbf{A} \rangle \langle \mathbf{x}_{n-1} \mathbf{x}_{n-1}^{\mathrm{T}} \rangle \right)$$
$$+ \sum_{d=1}^{D} \left(\sum_{k=1}^{D} r_{kd} \right)^2 \text{tr} \left(\boldsymbol{\Sigma}_{\mathbf{A}}^{(d)} \langle \mathbf{x}_{n-1} \mathbf{x}_{n-1}^{\mathrm{T}} \rangle \right). \tag{48}$$

Note that the sum over n in $\langle \log p(\mathbf{X}|\mathbf{A}) \rangle_*$ can be computed independently of $\mathbf{R}$ before starting the optimization of $\mathbf{R}$ in order to reduce the computational cost.

6   Experiments

6.1   Artificial Data

The method was tested on an artificial dataset, which was generated using the model with known parameter values. We generated $N = 400$ latent states $\mathbf{x}_n$ by using the following state dynamics matrix:

$$\mathbf{A} = \begin{bmatrix} \cos(\omega) & -\sin(\omega) & 0 & 0 \\ \sin(\omega) & \cos(\omega) & 0 & 0 \\ 0 & 0 & 1 & 0 \\ 0 & 0 & 0 & 0 \end{bmatrix}, \tag{49}$$

where $\omega = 0.3$. Thus, the four latent signals in $\mathbf{X}$ are as follows: the first and the second signals are noisy oscillators, the third signal is Gaussian random walk, and the fourth signal is Gaussian white noise. The loading matrix $\mathbf{C}$ for projecting the observations with dimensionality $M = 30$ was sampled from the Gaussian distribution with zero mean and unit variance for each element. The variance of the observation noise was set to $\tau^{-1} = 9$.

The dataset was used to estimate the parameters with the variational Bayesian learning method. The learning was performed both with and without the rotations in order to compare the effect of the rotations on the performance. Both methods used the same model and initialization. The dimensionality of the latent states was set to $D = 8$, which is larger than the true dimensionality, to let the ARD prior prune out any irrelevant dimensions. The hyperparameters were given broad priors by setting $a_\alpha = b_\alpha = a_\gamma = b_\gamma = a_\tau = b_\tau = 10^{-5}$, $\mu_0 = \mathbf{0}$ and $\mathbf{\Lambda}_o = 10^{-3} \cdot \mathbf{I}$.

The approximate posterior distributions of the variables were initialized by using the VB-EM update formulas (13)–(17) but ignoring the variational messages from the child nodes, that is, taking into account only the parent nodes. This means roughly that the variables were initialized from the prior. For instance, the mean of $\boldsymbol{\alpha}$, $\boldsymbol{\gamma}$ and $\boldsymbol{\tau}$ were set to ones, the mean of $\mathbf{A}$ to zeros and the covariance of $\mathbf{A}$ to the identity matrix. In order to get some initial latent space spanned, the mean of the loading matrix $\mathbf{C}$ was initialized randomly from the standard Gaussian distribution and the covariance was set to zero. The latent states $\mathbf{X}$ were not initialized because they were updated first in the VB-EM algorithm.

The performance of the methods was measured by monitoring the lower bound of the marginal log likelihood (9) and root-mean-square error (RMSE) on the training and test sets. The test set was created by removing training data y_{mn} randomly with probability 0.8, thus approximately 20% of the data was used for training.

Figure 1 shows the performance of both methods as a function of iterations. The number of iterations is used for simplicity as the computational cost of the rotations is negligible. The standard learning method converges in approximately 10000 iterations whereas the method with rotations converges in 10–20 iterations based on the lower bound of the marginal log likelihood in Fig. 1a. The rotations do not only affect the lower bound but also the reconstruction of the data. The standard method overfits at the beginning of the learning phase as can be seen from Fig. 1b and the predictions are improved very slowly as the test error shows in Fig. 1c. In comparison, the method with rotations finds the solution orders of magnitude faster.

6.2   Weather Data

The methods were tested on a real-world weather dataset provided by the Helsinki Testbed project of the Finnish Meteorological Institute (FMI). From the large dataset, we used temperature measurements in Southern Finland over a period of almost two years with an interval of ten minutes resulting in $N = 89202$ time instances.[1] Measurements from some weather stations were badly corrupted

---

[1] The data is available at http://users.ics.aalto.fi/jluttine/ecml2013/ under the FMI Open Data License.

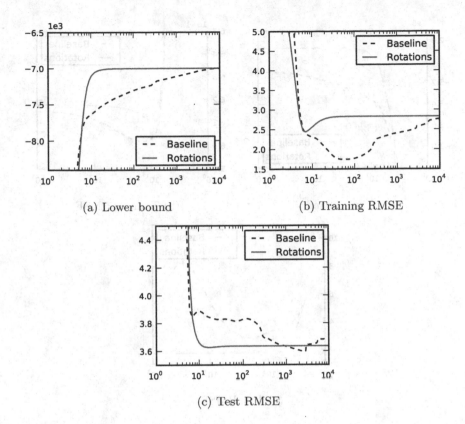

(a) Lower bound   (b) Training RMSE

(c) Test RMSE

Fig. 1. Results of the artificial experiment. (a) The marginal log likelihood lower bound, (b) training error, and (c) testing error shown as a function of iterations for the standard learning method (baseline) and the proposed method using rotations. Note that the x-axis has a logarithmic scale.

as discussed in [15] and were therefore discarded. Thus, $M = 66$ stations remained for the analysis. From the remaining data, approximately 35% of the measurements were missing.

The standard method and the proposed method using rotations were used to estimate the linear state-space model from the data. The model used latent space dimensionality $D = 10$ and broad hyperpriors as in the artificial experiment. The initialization was done similarly as in the artificial experiment. Test data was formed by removing training data randomly with probability 0.2 and completely for periods of one day at ten day intervals resulting in a large number of short gaps in the training data.

Figure 2 shows the lower bound of the marginal log likelihood, the training error and test error for both methods. Based on the lower bound in Fig. 2a, the standard method has not converged in 1000 iterations and the progress is extremely slow, whereas the method with rotations converges in 20–30 iterations.

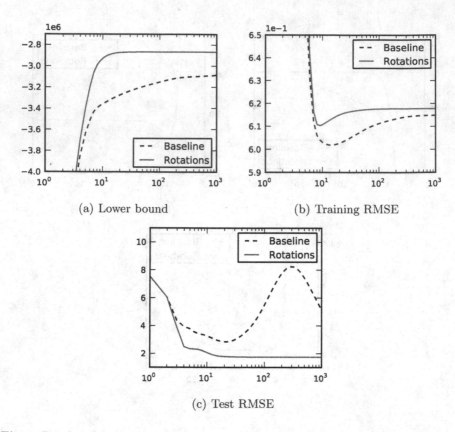

(a) Lower bound   (b) Training RMSE

(c) Test RMSE

Fig. 2. Results of the weather experiment. (a) The marginal log likelihood lower bound, (b) training error, and (c) testing error shown as a function of iterations. Note that the x-axis has a logarithmic scale.

The computational cost of the rotations is again negligible and can be ignored for simplicity. Similarly to the artificial experiment, the standard method overfits at the beginning as can be seen in Fig. 2b. The performance difference of the learning methods can be seen clearly from the test error in Fig. 2c as the test error for the standard method is significantly larger and decreasing very slowly. From these measures it is evident that the standard learning method has not yet converged in 1000 iterations and it might require several thousands iterations more to reach convergence.

7   Conclusions

The paper presented a parameter expansion for improving the speed of the variational Bayesian inference of linear state-space models. The expansion was based on optimizing the rotation of the latent space, which corresponds to a parameterized joint optimization of multiple variables. The transformations improved the speed of the inference by orders of magnitude compared to the standard VB-EM algorithm as shown in the experiments for artificial and real-world data. Thus, the proposed parameter expansion should be a standard technique if using VB inference for variants of linear state-space models in problems that are not very small.

The paper also gave a new perspective on estimating the posterior distribution of the states in the VB-EM algorithm. The states were estimated by using a smoothing algorithm based on the block LDL decomposition instead of the standard filters and smoothers used in previous papers. This approach reduced the number of required matrix inversions and avoided a model augmentation. In addition, as the algorithm is based on the LDL decomposition, one can utilize existing sparse LDL algorithms in order to generalize the smoothing algorithm, for instance, to other Gaussian Markov random fields with different graph structure.

A Python implementation of the presented method is published as a part of the Bayesian Python (BayesPy) package under the GNU General Public License.[2] In addition, the data and the scripts for running the experiments shown in the paper are available under open licenses.[3]

Acknowledgements. This work was supported by the Helsinki Doctoral Programme in Computer Science (Hecse) and the Academy of Finland (Finnish Centre of Excellence in Computational Inference Research COIN, 251170). The author would like to thank the Finnish Meteorological Institute and the Helsinki Testbed project for providing the dataset and sharing it under the FMI Open Data License.

References

1. Bar-Shalom, Y., Li, X.R., Kirubarajan, T.: Estimation with Applications to Tracking and Navigation. Wiley-Interscience (2001)
2. Shumway, R.H., Stoffer, D.S.: Time Series Analysis and its Applications. Springer (2000)
3. Beal, M.J., Ghahramani, Z.: The variational Bayesian EM algorithm for incomplete data: with application to scoring graphical model structures. Bayesian Statistics 7, 453–464 (2003)
4. Bishop, C.M.: Pattern Recognition and Machine Learning. Information Science and Statistics, 2nd edn. Springer, New York (2006)
5. Liu, C., Rubin, D.B., Wu, Y.N.: Parameter expansion to accelerate EM: the PX-EM algorithm. Biometrika 85, 755–770 (1998)
6. Qi, Y.A., Jaakkola, T.S.: Parameter expanded variational Bayesian methods. In: [16], pp. 1097–1104
7. Luttinen, J., Ilin, A.: Transformations in variational Bayesian factor analysis to speed up learning. Neurocomputing 73, 1093–1102 (2010)
8. Klami, A., Virtanen, S., Kaski, S.: Bayesian canonical correlation analysis. Journal of Machine Learning Research 14, 899–937 (2013)

---

[2] BayesPy is available at https://github.com/jluttine/bayespy
[3] The material is available at http://users.ics.aalto.fi/jluttine/ecml2013/

9. Kang, H., Choi, S.: Probabilistic models for common spatial patterns: Parameter-expanded EM and variational Bayes. In: Proceedings of the Twenty-Sixth AAAI Conference on Artificial Intelligence (2012)
10. Rauch, H.E., Tung, F., Striebel, C.T.: Maximum likelihood estimates of linear dynamic systems. AIAA Journal 3(8), 1445–1450 (1965)
11. Beal, M.J.: Variational algorithms for approximate Bayesian inference. PhD thesis, Gatsby Computational Neuroscience Unit, University College London (2003)
12. Barber, D., Chiappa, S.: Unified inference for variational Bayesian linear Gaussian state-space models. In: [16]
13. Eubank, R.L., Wang, S.: The equivalence between the Cholesky decomposition and the Kalman filter. The American Statistician 56(1), 39–43 (2002)
14. Kalman, R.E., Bucy, R.S.: New results in linear filtering and prediction theory. Journal of Basic Engineering 85, 95–108 (1961)
15. Luttinen, J., Ilin, A., Karhunen, J.: Bayesian robust PCA of incomplete data. Neural Processing Letters 36(2), 189–202 (2012)
16. Schölkopf, B., Platt, J., Hoffman, T. (eds.): Advances in Neural Information Processing Systems 19. MIT Press, Cambridge (2007)

Inhomogeneous Parsimonious Markov Models

Ralf Eggeling[1,*], André Gohr[1,*], Pierre-Yves Bourguignon[2],
Edgar Wingender[3], and Ivo Grosse[1,4]

[1] Institute of Computer Science, Martin Luther University,
06099 Halle, Germany
[2] Max Planck Institute for Mathematics in the Sciences,
04103 Leipzig, Germany
[3] Institute of Bioinformatics, University Medical Center Göttingen,
37077 Göttingen, Germany
[4] German Center of Integrative Biodiversity Research (iDiv) Halle-Jena-Leipzig,
04103 Leipzig, Germany

Abstract. We introduce inhomogeneous parsimonious Markov models
for modeling statistical patterns in discrete sequences. These models are
based on parsimonious context trees, which are a generalization of context trees, and thus generalize variable order Markov models. We follow
a Bayesian approach, consisting of structure and parameter learning.
Structure learning is a challenging problem due to an overexponential
number of possible tree structures, so we describe an exact and efficient
dynamic programming algorithm for finding the optimal tree structures.

We apply model and learning algorithm to the problem of modeling binding sites of the human transcription factor C/EBP, and find an
increased prediction performance compared to fixed order and variable
order Markov models. We investigate the reason for this improvement
and find several instances of context-specific dependences that can be
captured by parsimonious context trees but not by traditional context
trees.

1 Introduction

Discrete sequential data as diverse as bit strings in computer science, DNA and
polypeptide molecules in bioinformatics, or alphabetic strings in linguistics are
omnipresent in todays science and technology. Despite highly diverse applications, the characterization of ensembles of sequences based on a finite sample is
a common and fundamental statistical challenge raised in these different fields.
Examples are data compression [1, 2], the prediction of functional sites in biological macromolecules [3–6], or the study of the structure of languages [7, 8].

While reducing a sequence to a set of independent letters may yield satisfactory results for certain tasks and certain data sets [9], one can easily name a
wealth of other settings where this is unlikely to be the case. Examples are written texts, where the occurrence of a letter at a certain position is significantly

---

* Contributed equally.

H. Blockeel et al. (Eds.): ECML PKDD 2013, Part I, LNAI 8188, pp. 321–336, 2013.
© Springer-Verlag Berlin Heidelberg 2013

constrained by the language [10], or DNA sequences, where the occurrence of a base at a certain position of a functional site on a chromosome influences its activity [11, 12]. Hence, there is a wealth of applications where the characterization of finite ensembles of sequences bearing statistical dependences is needed.

Inferring the probability distribution of the sequences from a finite ensemble of sequences becomes challenging already for moderate sequence lengths. In such situations, trading simplifications of the model against statistical strength has been shown to be potentially beneficial. For each model class, the joint probability of a sequence can be decomposed into a product of conditional probabilities of single symbols given all predecessors. While the class of Markov models of order d is based on the simplification that all conditional dependences except for those given the d previous symbols are dropped [2], the richer class of variable order Markov models (VOMMs) [13] makes this order context dependent. Most other approaches proposed to date also share the feature of dropping certain entries from the conditional probability distributions in a Markovian manner.

Bourguignon and Robelin [14] propose an alternative approach to the reduction of the dimension of the space of conditional distributions, where conditional independence assumptions are formed with respect to a partition of the conditions, i.e., a partition of context words, by means of a parsimonious context tree (PCT). Particular choices for the partition of the context words may result in conditional independence assumptions that coincide with those formed by a regular Markov model, as well as those formed by a variable order Markov model. The parallel with the VOMM is actually much further reaching, since PCTs can be understood as a generalization of the context trees that are used by VOMMs. However, parsimonious Markov models that use parsimonious context trees are in general not representable in a sheer Markovian manner, i.e., by dropping entries in the conditions.

Here, we aim at exploring the merits of this form of parsimony for modeling discrete sequential data of fixed length. We introduce inhomogeneous parsimonious Markov models (PMMs) based on a sequence of parsimonious context trees and follow a Bayesian approach for structure and parameter learning. Whereas parameter learning is straightforward, structure learning is challenging due to an overexponential number of possible tree structures. However, this optimization problem can be solved by an efficient dynamic programming algorithm, which generalizes the context tree maximization algorithm [1]. We apply inhomogeneous PMMs to the prediction of binding sites of the human transcription factor C/EBP [15], and investigate if the richer expressiveness of inhomogeneous PMMs might possibly lead to an improved prediction compared to inhomogeneous VOMMs.

2   Theory

In this section, we introduce inhomogeneous parsimonious Markov models in a Bayesian framework by defining likelihood and prior. We subsequently describe structure and parameter learning, and finally discuss the relation to variable order Markov models and further special cases.

We denote a single symbol by $x \in \mathcal{A}$, a sequence of length L by $\vec{x} = (x_1, \ldots, x_L)$, and a data set of N sequences of fixed length L by $\mathbf{x} = (\vec{x}_1, \ldots, \vec{x}_N)$. Further, we denote the power set of $\mathcal{A}$ by $\mathcal{P}(\mathcal{A})$, and $\mathcal{P}_{\geq 1}(\mathcal{A}) = \mathcal{P}(\mathcal{A}) \setminus \emptyset$. We call each element in $\mathcal{A}^d$ *context word* of length d.

2.1   Model

Similar to context trees, which are used by variable order Markov models for reducing and representing their parameter space, parsimonious context trees as proposed by Bourguignon and Robelin [14] are the central data structure of inhomogeneous PMMs. A PCT τ of depth d for alphabet $\mathcal{A}$ is a rooted, balanced tree. Each node of a PCT is labeled by a non-empty subset of $\mathcal{A}$, except for the root, which is labeled by the empty subset. The set of labels of all children of an arbitrary inner node forms a partition of $\mathcal{A}$.

It follows that the cross product of the symbol sets encountered along each path from a leaf to the root defines a non-empty subset of $\mathcal{A}^d$, which we call *context*. Hence, a context is a set of context words, and the set of the contexts of all leaves of a PCT forms a partition of $\mathcal{A}^d$. Thus, the PCT is a data structure that represents a partition of the whole set of context words. For example, the PCT of depth two for the four-letter DNA alphabet $\mathcal{A} = \{A, C, G, T\}$ shown in Figure 1 encodes the contexts $\{A\} \times \{A\}, \{C, G\} \times \{A\}, \{T\} \times \{A\}, \{A, G\} \times \{C, G, T\}$, and $\{C, T\} \times \{C, G, T\}$. A PCT of depth d interpolates between two extreme cases: a *minimal* tree with only one leaf, which represents the union of all context words into one set, and a *maximal* tree with $|\mathcal{A}|^d$ leaves, each of which represents a single context word.

An inhomogeneous PMM of order d for sequences of length L is based on exactly L PCTs, which we denote by $\vec{\tau} = (\tau_1, \ldots, \tau_L)$. For the ease of presentation, we exclude the first d PCTs, which have an increasing depth of $0, \ldots, d-1$, from the following discussion. Since there is a bijective mapping from the leaves of a PCT to the corresponding contexts, we can perceive a PCT as a set of contexts as well as a set of leaf nodes. Hence, we denote a single context by c, the number of context words represented by a context by $|c|$, and the set of all contexts in a PCT by τ itself.

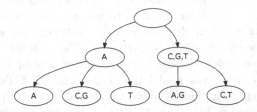

Fig. 1. Example PCT of depth 2 over DNA alphabet. It encodes the partition of all 16 possible context words into subsets $\{AA\}, \{CA, GA\}, \{TA\}, \{AC, AG, AT, GC, GG, GT\}$, $\{CC, CG, CT, TC, TG, TT\}$.

We denote the conditional probability of observing a symbol $a \in \mathcal{A}$ given that the concatenation of the preceding d symbols is in c by θ_{ca}. We denote the model parameters of a single position by $\Theta = \left(\tau, (\vec{\theta}_c)_{c \in \tau}\right)$ and all model parameters by $\vec{\Theta} = (\Theta_1, \ldots, \Theta_L)$. We now define the likelihood of an inhomogeneous PMM by

$$P(\mathbf{x}|\vec{\Theta}) = \prod_{\ell=1}^{L} \prod_{c \in \tau_\ell} \prod_{a \in \mathcal{A}} (\theta_{\ell ca})^{N_{\ell ca}}, \tag{1}$$

where $N_{\ell ca}$ is the number of occurrences of symbol a at position ℓ in all sequences of $\mathbf{x}$ where the concatenation of the symbols from position $\ell - d$ to position $\ell - 1$ is in c.

The likelihood of an inhomogeneous PMM is similar to that of a fixed order inhomogeneous Markov model since it is a product over all possible observations a for all possible contexts c at all possible positions ℓ. However, in contrast to fixed order inhomogeneous Markov models, where each c is a single context word, we here allow arbitrary sets of context words defined by the PCT τ_ℓ.

2.2 Prior

Assuming local and global parameter independence [16], we define the prior of an inhomogeneous PMM by

$$P(\vec{\Theta}) = P(\vec{\tau}) \prod_{\ell=1}^{L} \prod_{c \in \tau_\ell} P(\vec{\theta}_{\ell c}), \tag{2}$$

where $P(\vec{\tau})$ is the prior probability of all PCTs $\vec{\tau}$ (and could thus be referred to as structure prior) and $P(\vec{\theta}_{\ell c})$ is the prior over the probability parameters of one particular context c at position ℓ. We specify the structure prior by

$$P(\vec{\tau}) \propto \prod_{\ell=1}^{L} \kappa^{|\tau_\ell|}, \tag{3}$$

where $|\tau_\ell|$ denotes the number of leaves of τ_ℓ. It depends on one scalar hyperparameter $\kappa \in (0, \infty)$, which can be used to influence the number of leaves and thus the complexity of the model, interpolating between the two extreme cases: When $\kappa \to +\infty$, the model that has maximal PCTs at all positions, and is thus equivalent to a fixed order Markov model, receives a prior probability of one. Conversely, when $\kappa \to 0$, the model that has minimal PCTs at all positions, and is thus equivalent to an independence model, receives full prior support. For the local parameter priors $P(\vec{\theta}_{\ell c})$ we choose Dirichlet distributions with hyperparameters $\vec{\alpha}_{\ell c}$. In this work, we further restrict the parameter priors to symmetric Dirichlet distributions. Following the equivalent sample size concept [16], we obtain a natural computation of the pseudocounts from the equivalent sample size η that is inspired by Bayesian networks, namely $\alpha_{\ell ca} = \frac{\eta |c|}{|\mathcal{A}|^{d+1}}$.

2.3   Learning

In resonance with learning many other probabilistic graphical models, learning inhomogeneous PMMs consists of structure and parameter learning with the former being the more challenging task.

In order to learn the structure of the model, we intend to find the parsimonious context trees that maximize $P(\vec{\tau}|\mathbf{x})$. Since $P(\mathbf{x})$ is constant w.r.t. the tree structures, it is sufficient to maximize

$$P(\vec{\tau}, \mathbf{x}) = \int P(\mathbf{x}|\vec{\Theta})P(\vec{\Theta})d\vec{\Theta}_{\vec{\tau}}, \qquad (4)$$

where $\vec{\Theta}$ denotes all parameters of the model, and $\vec{\Theta}_{\vec{\tau}}$ denotes here the conditional probability parameters within $\vec{\Theta}$ for given PCT structures $\vec{\tau}$. Due to global parameter independence (outer product of Eq. 2), we can decompose the structure learning problem into finding the optimal PCT for each position separately. Due to local parameter independence (inner product of Eq. 2), we can decompose the score of a PCT into a product of scores of its leaves. Solving the remaining integral, we obtain the optimization problem

$$\forall_{\ell=1}^{L} : \hat{\tau}_{\ell} := \underset{\tau_{\ell}}{\operatorname{argmax}} \prod_{c \in \tau_{\ell}} \kappa \frac{\mathcal{B}(\vec{N}_{\ell c} + \vec{\alpha}_{\ell c})}{\mathcal{B}(\vec{\alpha}_{\ell c})}, \qquad (5)$$

where $\mathcal{B}$ denotes the multinomial beta function. Hence, the target function is a product over local marginal likelihoods for all contexts multiplied by the structure prior hyperparameters for each context.

While the score for a given PCT can be computed easily, finding the optimal out of an overexponential number of possible PCTs (with respect to model order and alphabet size) without computing the score for every single PCT explicitly is challenging. This problem can be solved by a dynamic programming (DP) algorithm similar to the context tree maximization algorithm [1]. The algorithm runs on a data structure that we call the extended PCT of depth d and that we denote by $\mathcal{T}_d^{\mathcal{A}}$. In contrast to a PCT, the children of a node of an extended PCT do not form a partition of alphabet $\mathcal{A}$, but rather encompass all elements of $\mathcal{P}_{\geq 1}(\mathcal{A})$ (Figure 2). The leaves of an extended PCT are thus all possible leaves (identified by their label concatenation up to the root) that may occur in any PCT of same depth and alphabet.

Let $\mathcal{N}(\mathcal{T})$ denote the set of nodes of an extended PCT $\mathcal{T}$, n one element of $\mathcal{N}(\mathcal{T})$, and $r(\mathcal{T})$ the root of $\mathcal{T}$. Each node can be uniquely identified by the label concatenation on the path from that node up to the root of the extended PCT. Let $s(n)$ denote the score of the optimal PCT subtree rooted at n. Let $\mathcal{C}(n)$ denote the set of all children of n in the extended PCT. Let $\mathcal{V}(\mathcal{C}(n))$ denote the set of all valid child combinations, i.e., all subsets of children whose labels form a partition of $\mathcal{A}$. Let further $\mathcal{L}(\mathcal{T})$ denote the leaves and $\mathcal{I}(\mathcal{T})$ the remaining inner nodes of $\mathcal{N}(\mathcal{T})$. Using this notation, we specify the dynamic programming approach in Algorithm 1, which consists of a single function for computing the optimal PCT subtree rooted at an arbitrary node of the extended PCT.

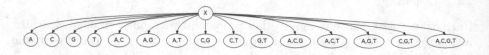

Fig. 2. Here, we show an arbitrary inner node (labeled by X) and its children in the extended PCT over the DNA alphabet. The labels of all children form $\mathcal{P}_{\geq 1}(\mathcal{A})$.

Algorithm 1. Dynamic programming for finding optimal PCT subtrees

---

findOptimalSubtree(n)

  if $n \in \mathcal{L}(\mathcal{T}_d^{\mathcal{A}})$ **then**

    $s(n) := \kappa \frac{\mathcal{B}(\bar{N}_{\ell n} + \bar{\alpha}_{\ell n})}{\mathcal{B}(\bar{\alpha}_{\ell n})}$

  end if

  if $n \in \mathcal{I}(\mathcal{T}_d^{\mathcal{A}})$ **then**

    for all $m \in \mathcal{C}(n)$ **do**

      findOptimalSubtree(m)

    end for

    for all $v \in \mathcal{V}(\mathcal{C}(n))$ **do**

      $s(v) := \prod\limits_{m \in v} s(m)$

    end for

    $v^* := \underset{v \in \mathcal{V}(\mathcal{C}(n))}{\text{argmax}} \ s(v)$

    $s(n) := s(v^*)$

    for all $m \in \mathcal{C}(n) \setminus v^*$ **do**

      remove m and subtree below

    end for

  end if

Applying this function to the root of the extended PCT, i.e., calling the function findOptimalSubtree($r(\mathcal{T}_d^{\mathcal{A}})$), yields the optimal PCT. The algorithm can be intuitively described as bottom-up reduction of the extended PCT towards a valid PCT by selecting at each inner node the locally optimal PCT subtree. The correctness of the algorithm follows from the property that the score of a PCT is a product of leaf scores (Eq. 5), which further implies that the score of a PCT subtree rooted at node n depends (apart from its own structure) only on the nodes on the path from n up to the root, but is independent of the structure of the PCT subtrees rooted at siblings of n.

The complexity of the DP algorithm is given by the size of the extended PCT, which must be completely traversed, multiplied by the number of valid child combinations, for which a score must be computed in each inner node of the extended PCT. Whereas the former is exponential with the base being the number of possible subsets of $\mathcal{A}$, the latter is equivalent to the Bell number $B_{|\mathcal{A}|}$. Hence, we obtain a time complexity of roughly $\mathcal{O}\left(B_{|\mathcal{A}|}\left(2^{|\mathcal{A}|} - 1\right)^d\right)$ for learning one PCT, stating that the complexity grows exponentially with model order d and overexponentially with alphabet size $\mathcal{A}$. Structure learning for an inhomogeneous PMM is linear in the sequence length as the DP algorithm is called $L - 1$ times, once for each PCT of non-zero depth.

Having determined optimal PCTs, we estimate their conditional probability parameters according to the posterior mean [17]. It is in general defined by $\hat{\theta} = \int_\theta \theta P(\theta|\mathbf{x})d\theta$ and yields for inhomogeneous PMMs

$$\forall_{\ell=1}^L \forall_{c\in\mathcal{T}_\ell} \forall_{a\in\mathcal{A}} : \hat{\theta}_{\ell ca} := \frac{N_{\ell ca} + \alpha_{\ell ca}}{N_{\ell c\cdot} + \alpha_{\ell c\cdot}}. \tag{6}$$

A common task is prediction, i.e., computing the probability of a data point $\vec{x}_{N+1}$ after having observed N data points $(\vec{x}_1, \ldots, \vec{x}_N)$. In a Bayesian setting, this is done by integrating over the space of parameters, which is in resonance with structure learning, where the target function is the probability of the model structure given data, obtained by integrating over the space of parameters. Here, we obtain for inhomogeneous PMMs

$$P(\vec{x}_{N+1}|\mathbf{x}, \vec{\tau}) = \int P(\vec{x}_{N+1}|\vec{\Theta}) \prod_{\ell=1}^L P(\vec{\theta}^{\tau_\ell}|\mathbf{x})d\vec{\Theta}_{\vec{\tau}}, \tag{7}$$

which is equivalent to computing $P(\vec{x}_{N+1}|\vec{\tau}, \hat{\Theta}_{1,\tau_1}, \ldots, \hat{\Theta}_{L,\tau_L})$, where $\hat{\Theta}_{\ell,\tau_\ell}$ is the posterior mean of the parameters (Eq. 6) of the PCT at position ℓ [16].

2.4   Special Cases

Context trees (CTs), which are used by variable order Markov models [13], are special cases of PCTs. Hence, inhomogeneous VOMMs are special cases of inhomogeneous PMMs. The differences between CTs and PCTs arise from a different concept of tree-building. Whereas the idea of building CTs is to *prune* a maximal tree by removing unimportant subtrees, the idea of PCTs is to *fuse* nodes if subtrees and corresponding conditional probability distributions are not sufficiently different. Since removing nodes can be also expressed by fusing them into one pseudo-node [18], CTs are special cases of PCTs (Figure 3). The opposite does not hold, though. There are many PCTs that represent a set of contexts that cannot be represented by CTs, since the notion of pruning yields several limitations of the possible CT structures that are relaxed by PCTs. Two structural features distinguish PCTs from CTs. First, an inner node in a PCT may have an arbitrary number of fused children as long as their labels form a partition of $\mathcal{A}$, whereas a CT allows at most one fused child (the pseudo-node). Second, a PCT allows arbitrary subtrees below a fused node, whereas a CT allows only a completely fused node as single child of a fused parent, which is equivalent to removing the entire subtree below the first occurrence of a fused node.

PCTs are more expressive than CTs, but this comes at the cost of a larger time complexity for structure learning, which limits the straightforward applicability of PMMs to problems with comparatively small alphabets. Even though there are plenty of such applications, with the most well known example being DNA and

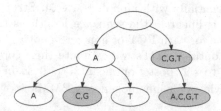

Fig. 3. Example CT of depth 2 over DNA alphabet. Pruned contexts are here shown as pseudo-nodes (displayed in gray) in order to achieve depth two for all possible contexts and thus allow a visualization of the CT in PCT-style.

RNA sequence analysis, it might be desirable to benefit from more expressive tree structures also for problems where the alphabet size becomes a limiting factor.

The DP algorithm offers the possibility to reduce the allowed tree structures (and thus the space that must be searched for the optimal structure) by redefining $\mathcal{V}(\mathcal{C}(n))$, the set of allowed child combinations of an inner node n. In PCTs this is the set of all partitions of the alphabet, which yields the Bell number factor in time complexity. Restricting $\mathcal{V}(\mathcal{C}(n))$ to all partitions that include one fused node at maximum, is one of the two necessary restrictions for obtaining a CT. Enforcing it, but allowing a fused node to have more than one child, yields a data structure that lies in between CTs and PCTs in terms of complexity. Conversely, restricting $\mathcal{V}(\mathcal{C}(n))$ to only one choice – the partition that lumps all symbols together into one node – if n is already a fused node, represents the second necessary restriction for obtaining CTs, which could also be solely enforced.

Besides these two options, which are inspired by the special case CT, there are further possible modifications such as restricting the maximal number of children of n to a value smaller than $|\mathcal{A}|$ or restricting $\mathcal{V}(\mathcal{C}(n))$ based on the label and/or the location of n in the extended PCT. Hence, a plethora of model classes of almost arbitrary complexity could be defined and learned by slight modifications of Algorithm 1.

3   Experiments

In the experimental part of this work, we apply inhomogeneous PMMs to the prediction of DNA binding sites of the eukaryotic transcription factor C/EBP [15] and compare it with inhomogeneous VOMMs, both implemented within the open source Java library Jstacs [19]. For the sake of convenience, we drop the explicit reference to the inhomogeneity in the following discussion. The C/EBP data set consists of $N = 96$ DNA binding sites from human and mouse, retrieved from the TRANSFAC® database [20]. These binding sites are aligned sequences of fixed length $L = 12$ over the DNA alphabet $\mathcal{A} = \{\texttt{A,C,G,T}\}$.

3.1 Comparing Prediction Performance

The Bayesian learning approach for PMMs and VOMMs described above allows influencing the complexity of the model via the structure prior (Eq. 3). Since it is not immediately clear, which value of hyperparameter κ translates to which model complexity, specifying the structure prior manually is not a trivial task. While a uniform prior over all structures, which we obtain by setting $\kappa = 1$, may appear as a reasonable option in the absence of a priori knowledge, it might yield tree structures that are not optimal for prediction and related tasks.

Hence, in a first study we investigate the performance of third-order PMMs and VOMMs for different model complexities. Even though statistical models are often used for classification purposes (e.g. positive vs. negative sites), we here focus on prediction as the main challenge of many classification approaches. Evaluation by prediction has the advantage of not requiring the choice of a negative data set and a corresponding statistical model, which both may influence results heavily.

Since the C/EBP data set is rather small, we perform a leave-one-out cross validation (CV). In the i-th step, we remove the i-th sequence from the data set, learn a model (using $\eta = 1$ for the parameter prior) on the remaining 95 sequences and compute the predictive probability of the i-th sequence. We repeat this procedure for $i = 1, \ldots, 96$, compute the average number of leaves of the models, and compute the arithmetic mean of the 96 logarithmic predictive probabilities, as well as the corresponding standard error.

In Figure 4 we plot, for both model classes, the mean log predictive probability against the average model complexity, quantified by the number of leaves of all context trees in the model, which is proportional to the total number of parameters. We choose values of κ that cover the whole range of model complexity,

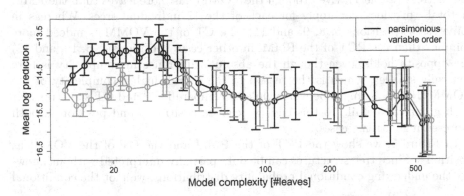

Fig. 4. We compare the prediction performance of third-order PMMs with third-order VOMMs. For both model classes, we plot the mean logarithmic prediction resulting from a leave-one-out cross validation experiment on the C/EBP data set against different model complexities (proportional to the number of parameters) obtained by varying the structure prior hyperparameter κ. Error bars depict double standard error.

interpolating from the minimal model with only 12 leaves (independence model) to the maximal model with 597 leaves (third-order Markov model).

We observe that for low model complexities of less than 50 leaves PMMs yield a substantially higher prediction than VOMMs. For high model complexities both approaches show a similar prediction, lower than the prediction achieved by a simple independence model, which indicates that overfitting occurs. These results are interesting in three aspects. First, PMMs are capable of utilizing statistical dependences in the data for improving prediction if the structure prior is chosen well. Second, a uniform structure prior corresponds here to a model structure of approximately 110 leaves, which confirms that using it is not an optimal choice. Third, VOMMs are barely capable of benefiting from statistical dependences no matter how κ is chosen. This observation raises the question why PMMs are capable of finding a good compromise between modeling dependences and avoiding overfitting whereas VOMMs are not.

3.2 Comparing Tree Structures

In a second study, we attempt to answer that question by comparing the learned model structures of PMM and VOMM. We choose for both model classes the values of κ that yield the highest mean log prediction in the leave-one-out CV experiment of Figure 4. For the PMM, this is $\kappa = e^{-2.5}$ with an average number of leaves of 32.6 and a mean log prediction of -13.6, while for the VOMM this is $\kappa = e^{-1.8}$ with an average number of leaves of 42.8 and a mean log prediction of -14.5. We use these structure priors to learn two models on the complete C/EBP data set of 96 sequences and scrutinize the resulting models in the following.

The resulting PMM and VOMM have 32 and 43 leaves respectively, which is in resonance with the average number of leaves of the leave-one-out CV experiment. First, we analyze how the total numbers of leaves of both models distribute over the 12 trees (Table 1). Even though the VOMM has more leaves than the PMM in total, this does not apply for each of the 12 individual trees. Whereas in some cases (positions 5, 8, 9, and 11), the CT of the VOMM is indeed more complex than the PCT of the PMM, in other cases (positions 4, 6, 10, and 12) the opposite holds, even though the absolute difference in complexity is here generally smaller, which is the reason of the overall higher complexity of the VOMM. Hence, it might be worthwhile to compare PCT and CT structures for both groups in detail. To this end, we choose position 5 and position 4, both representing extreme cases.

In Figure 5, we show the PCT of the PMM and the CT of the VOMM at position 5. Since tree structures can be only partially interpreted without knowing the underlying conditional probability distributions, we plot the conditional

Table 1. Numbers of leaves for all trees of best third-order PMM and VOMM

Position	1	2	3	4	5	6	7	8	9	10	11	12	Σ
PMM	1	1	4	3	2	3	1	2	6	5	3	2	32
VOMM	1	1	4	1	12	2	1	3	8	2	7	1	43

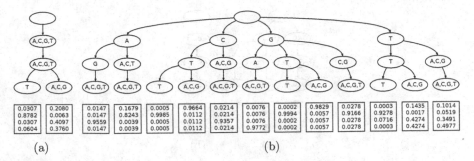

(a)   (b)

Fig. 5. We compare the PCT and the CT at position 5. We choose for both the PMM and the VOMM the optimal structure prior hyperparameter κ with respect to the leave-one-out experiment of Figure 4. Next, we learn both models using their respective optimal structure prior on the complete data set of 96 sequences and depict both the PCT of the PMM (a) and the CT of the VOMM (b) at position 5.

probabilities for each context, estimated according to Eq. 6, in rectangular boxes below the corresponding leaf node in lexicographical order of the observations.

The PCT in Figure 5(a) has only two leaves, so it partitions all context words into only two sets. The first and the second layer of the tree are completely fused, so the first and second predecessor symbol does not influence the probability distribution at position 5. At the third layer, however, the context words are partitioned into two subsets according to the observed symbol at the third predecessor (position 2). Observing a T at position 2 yields a high conditional probability of 0.8782 for finding a C at position 5, whereas any other symbol at position 2 yields a low conditional probability of 0.063 for a C at the fifth position. Conversely, for the second context, the conditional probability of finding A, G, and T is highly increased. This shows that there is a strong statistical dependence among positions 5 and 2, and a PCT is capable of exploiting it with only two parameters sets, which can be estimated comparatively robustly from 96 data points (partitioned into two sets of sizes 67 and 29 respectively).

The CT in Figure 5(b) has twelve leaves, but many of the contexts represent only few occurrences of context words in the data set. For example, the first, fourth, and ninth leaf represent only a single sequence in the data set each. Hence, the reliability of the corresponding parameter estimates is highly questionable. The reason why such a context tree is learned despite the indication of overfitting is the strong statistical dependence among positions 5 and 2. Leaves number two, three, seven, and ten represent most of the context words that are combined in the first leaf of the PCT in Figure 5(a). But since a CT does not allow a split in the tree structure below a fused node, the only possibility to learn this third-order dependence is a broad tree with many dispensable parameter sets.

We conclude that one reason for the inability of the VOMM to effectively capture dependences in this data set is its structural limitation of not being capable of "skipping" a position, which may lead to strong overfitting if skipping positions were actually required.

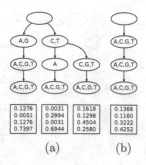

(a)   (b)

Fig. 6. We compare learned PCT (a) and CT (b) at position 4 of the C/EBP data set. The experimental setup is identical to that of Figure 5.

In Figure 6, we display the PCT and CT at position 4. The CT of Figure 6(b) is completely pruned, resulting in a minimal tree corresponding to full statistical independence. At this position the VOMM does not suffer from overfitting, but it may neglect existing dependences.

The PCT at the same position has three leaves, resulting in three different parameter sets. Each leaf represents a substantial amount of sequences from the data set (24, 10, 62) so that the parameter estimates may not be completely unrealiable. We observe that the first leaf yields a high conditional probability of 0.7397 for a T, given the symbol of the preceding position being either A or G. The second and third leaf represent the other contexts that have either C or T at the previous position and differ in the second predecessor. The second leaf represents the subset of context words that have an A at position 2. The corresponding conditional probability of a T is 0.6944, whereas A and C rarely occur. However, if the symbol at the second predecessor is not A, then G has the highest probability at position 4 (third leaf).

This implies that a certain amount of statistical dependences exists among the fourth position of the C/EBP data set and its predecessors, and that these dependences can be modeled – at least to some degree – by a PCT. A PCT is capable of splitting the contexts at any layer so that there is more than one fused child node per parent. This feature may be required to properly represent statistical dependences at position 4. Apparently a CT is not capable of representing these splits, so it here neglects statistical dependences completely. This indicates that VOMMs are not necessarily always overfitted compared to PMMs, but also the opposite, underfitting due to structural limitations, may occur.

We may conclude that, compared to the third-order PMM, the third-order VOMM is both over- and underfitted. The PMM is capable of using the full potential of the inhomogeneity of the model better than a VOMM, since it yields – on average over all positions – a better tradeoff between capturing dependences and reducing the parameter space.

3.3   Model Validation

The previous two experiments show that a PMM is capable of modeling dependences in a small real-world data set and how it finds a reasonable balance in avoiding both over- and underfitting due to its structural flexibility. However, as we have seen in Figure 4, the prediction performance depends on the choice of the structure prior, for which real a priori knowledge is rarely available.

Hence, we must devise a method that can automatically provide us with an adequate choice for κ in order to validate the model class against other alternatives. To this end, we perform in each step of the CV described in Section 3.1 another internal CV on the 95 training sequences. We then choose in the i-th step that κ that yields the highest mean log prediction in the CV on the 95 training data sequences, learn a model on that data set, and compute the predictive probability of the i-th sequence. Finally, we average all logarithmic predictive probabilities and use that single number for evaluating the performance of the model class.

In Figure 7, we compare PMMs, VOMMs, and inhomogeneous Markov models of orders 1-3. In addition, we consider the independence model, which neglects all dependences. Despite its simplicity, it is the most popular choice for modeling DNA binding sites in bioinformatics and in that field known as position weight matrix model [21, 22]. For the independence model and the fixed order Markov models, there is no internal cross validation.

We find that the independence model yields a mean log prediction value of -14.91. A first-order Markov model improves it to a value of -14.56, showing that taking into account first-order dependences is reasonable and beneficial. Second- and third-order Markov models yield a lower prediction than the

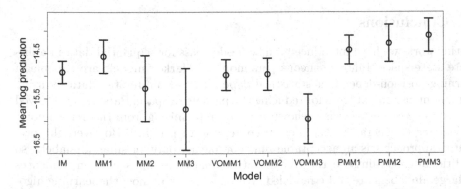

Fig. 7. We show the prediction performance of the independence model (IM), fixed order MMs, VOMMs, and PMMs of orders 1-3. The experimental setup is identical to that of Figure 4. For the parsimonious and variable order models, we perform an additional internal cross validation on the $N - 1$ training sequences for determining the optimal structure prior hyperparameter κ.

independence model. This is not surprising since we expect overfitting for complex models when sample size is as small as 96 data points.

First- and second-order VOMM yield a prediction accuracy that is comparable with that of an independence model. Despite reducing the parameter space, they are – at least on this data set – not capable of utilizing statistical dependences effectively. The third-order VOMM yields an even lower prediction, comparable to a third-order Markov model, indicating that overfitting occurs. This also shows that the internal CV fails in this case: At more than one position it selects in some iterations very complex and thus poorly generalizing tree structures, comparable to that in Figure 5(b).

The first-order PMM yields a mean log prediction value of -14.42, which is comparable to that of the first-order Markov model. Apparently, overfitting is not a serious problem for the first-order Markov model, so the potential reduction of the parameter space yields only a small improvement. However, in contrast to fixed order MMs, PMMs of second- and third-order continue to increase the prediction performance. The overall best prediction is achieved by a third-order PMM with a mean log prediction value of -14.1, which is slightly lower than the best prediction in Figure 4, but close to the average prediction within the range of reasonable complexities (15 to 40 leaves).

We summarize that PMMs yield a higher prediction of C/EBP binding sites than the independence model, than fixed order Markov models, and than variable order Markov models. Among the three PMMs, the third-order PMM yields the overall highest prediction. Hence, PMMs are capable of exploiting dependences in the small data set of only 96 sequences effectively, whereas the effectivity of VOMMs is harmed by their structural limitations. This makes it tempting to speculate that PMMs might be a useful model class for other types of sequential data as well, especially when certain dependences among non-neighboring positions exist and when the sample size is comparatively small.

4   Conclusions

In this work, we have introduced a new model class for sequential data in a discrete state space. Inhomogeneous parsimonious Markov models are capable of learning position-dependent statistical dependences from limited data by using parsimonious context trees for reducing the parameter space. Parsimony achieved by grouping context words is shown here to be promising from theoretical point of view as it generalizes the idea of context word pruning. However, the presented approach has an acceptable time complexity only for small alphabets, so additional constraints on the tree structures must be imposed when sequences of large alphabets are to be modeled. We have discussed how the learning algorithm can be adapted to incorporate these constraints, admitting an acceptable time complexity while retaining specific merits of parsimonious context trees.

Predicting functional DNA sequences is an important application where this model class can be used in a straightforward manner. In a case study on binding sites of the human transcription factor C/EBP, we have observed that inhomogeneous parsimonious Markov models yield more accurate predictions than the corresponding variable order Markov models. Scrutinizing the structural differences between the best models of both model classes, we found that strong third-order dependences but comparatively weak first- and second-order dependences exist at several positions. These are features that a parsimonious context tree can take into account with very few parameters, whereas a traditional context tree is limited by its structural constraints, either requiring substantially more parameters, yielding unreliable parameter estimates, or neglecting those dependences completely. We conclude that inhomogeneous parsimonious Markov models are a promising alternative to inhomogeneous Markov models and inhomogeneous variable order Markov models. The adaptation to different applications might possibly require additional algorithmic work, but taking such challenges might be worth the effort.

Acknowledgments. This work was funded by *Reisestipendium des allg. Stiftungsfonds der MLU Halle–Wittenberg*, DFG (grant no. GR 3526/1-2), and *CNRS/MPG GDRE in Systems Biology*.

References

1. Volf, P., Willems, F.: Context maximizing: Finding MDL decision trees. In: 15th Symp. Inform. Theory Benelux, pp. 192–200 (May 1994)
2. Cover, T., Thomas, J.: Elements of Information Theory, 2nd edn. Wiley Interscience (2006)
3. Ding, Y.: Statistical and Bayesian approaches to RNA secondary structure prediction. RNA 12(3), 323–331 (2006)
4. Xu, X., Ji, Y., Stormo, G.D.: RNA sampler: a new sampling based algorithm for common RNA secondary structure prediction and structural alignment. Bioinformatics 23(15), 1883–1891 (2007)
5. Busch, J.R., Ferrari, P.A., Flesia, A.G., Fraiman, R., Grynberg, S.P., Leonardi, F.: Testing statistical hypothesis on random trees and applications to the protein classification problem. The Annals of Applied Statistics 3(2), 542–563 (2009)
6. Won, K.-J., Ren, B., Wang, W.: Genome-wide prediction of transcription factor binding sites using an integrated model. Genome Biology 11(1), R7 (2010)
7. Ramus, F., Nespor, M., Mehler, J.: Correlates of linguistic rhythm in the speech signal. Cognition 73, 265–292 (1999)
8. Kolmogorov, A., Rychkova, N.: Analysis of russian verse rhythm, and probability theory. Theory Probab. Appl. 44, 375–385 (2000)
9. Rissanen, J., Langdon, G.: Arithmetic coding. IBM Journal of Research and Development 23, 149–162 (1979)
10. Galves, A., Galves, C., Garcia, J., Garcia, N., Leonardi, F.: Context tree selection and linguistic rhythm retrieval from written texts. Ann. Appl. Stat. 6(1), 186–209 (2012)
11. Stormo, G.D.: DNA binding sites: representation and discovery. Bioinformatics 16(1), 16–23 (2000)
12. Bejerano, G., Yona, G.: Variations on probabilistic suffix trees: statistical modeling and prediction of protein families. Bioinformatics 17(1), 23–43 (2001)
13. Rissanen, J.: A universal data compression system. IEEE Trans. Inform. Theory 29(5), 656–664 (1983)

14. Bourguignon, P., Robelin, D.: Modèles de Markov parcimonieux. In: Proceedings of JOBIM (2004)
15. Ramji, D., Foka, P.: CCAAT/enhancer-binding proteins: structure, function and regulation. Biochem. J. 365, 561–575 (2002)
16. Heckerman, G., Geiger, D., Chickering, D.: Learning Bayesian networks: The combination of knowledge and statistical data. Machine Learning 20, 197–243 (1995)
17. Jaynes, E.T.: Probability Theory: The Logic of Science. Cambridge University Press (2003)
18. Bühlmann, P., Wyner, A.: Variable length Markov chains. Annals of Statistics 27, 480–513 (1999)
19. Grau, J., Keilwagen, J., Gohr, A., Haldemann, B., Posch, S., Grosse, I.: Jstacs: A Java Framework for Statistical Analysis and Classification of Biological Sequences. Journal of Machine Learning Research 13, 1967–1971 (2012)
20. Matys, V., Fricke, E., Geffers, R., Gossling, E., Haubrock, M., Hehl, R., Hornischer, K., Karas, D., Kel, A., Kel-Margoulis, O., Kloos, D., Land, S., Lewicki-Potapov, B., Michael, H., Münch, R., Reuter, I., Rotert, S., Saxel, H., Scheer, M., Thiele, S., Wingender, E.: TRANSFAC: transcriptional regulation, from patterns to profiles. Nucleic Acids Research 33, 374–378 (2003)
21. Stormo, G., Schneider, T., Gold, L.: Characterization of translational initiation sites in E.coli. Nucleic Acids Research 10(2), 2971–2996 (1982)
22. Staden, R.: Computer methods to locate signals in nucleic acid sequences. Nucleic Acids Research 12, 505–519 (1984)

Explaining Interval Sequences by Randomization

Andreas Henelius, Jussi Korpela, and Kai Puolamäki

Finnish Institute of Occupational Health,
Topeliuksenkatu 41 a A, FI-00250 Helsinki, Finland
{andreas.henelius,jussi.korpela,kai.puolamaki}@ttl.fi

Abstract. Sequences of events are an ubiquitous form of data. In this paper, we show that it is feasible to present an event sequence as an interval sequence. We show how sequences can be efficiently randomized, how to choose a correct null model and how to use randomizations to derive confidence intervals. Using these techniques, we gain knowledge of the temporal structure of the sequence. Time and Fourier space representations, autocorrelations and arbitrary features can be used as constraints in investigating the data. The methods presented are applied to two real-life datasets; a medical heart interbeat interval dataset and a word dataset from a book. We find that the interval sequence representation and randomization methods provide a powerful way to explore interval sequences and explain their structure.

1 Introduction

Time series are sequences of consecutive, time-stamped events. The events can have properties, such as values of measurements at the particular time instances. A single event can have multiple properties, in which case one ends up with a multidimensional time series. In this paper we, however, ignore the properties of the events and study only the fundamental temporal structure of the time series, which can be represented as a sequence of intervals.

Interval sequences are ubiquitous. They can be analyzed and compared by numerous methods, and many application areas such as medical signal processing have established conventions on how to study them. The structure of event sequences can be described by complex models. However, before addressing more complex properties of the event sequence, the first question is whether it is meaningful to look for complex structures. Can the structure observed in the interval sequence be explained by a random occurrence? If not, then what constitutes a good description?

Randomization methods provide a means of studying non-random structures. These techniques have a long tradition in statistics and are used increasingly in data analysis as well. To use randomization methods one first needs to define the null distribution from which random samples are drawn. If the observed event sequence differs, in terms of one or more test statistics, from the random samples, we can conclude that there are non-random structures.

The null distribution encodes our prior information and assumptions about the data as constraints. The choice of constraints is, however, far from trivial and

H. Blockeel et al. (Eds.): ECML PKDD 2013, Part I, LNAI 8188, pp. 337–352, 2013.

no clear guidelines exist. A suitable set of constraints depends on the research question and hence there is no universally appropriate null model. A natural choice is to select constraints that explain those aspects of the data we assume known and wish to account for. This makes previously unknown patterns stand out. The randomization methodology can therefore be used as a probabilistically robust means of detecting statistically significant patterns [14,5,6,20].

1.1   Structure and Contributions of This Paper

In this paper we show how an event sequence can be represented as a sequence of events in the time and Fourier domains, and in the autocorrelation space. We present the main theoretic properties of these representations in Section 2. We introduce fast and convenient randomization methods in which various properties such as Fourier amplitudes or phases, autocorrelation coefficients, or arbitrary statistics of the event sequence are kept fixed. We demonstrate how these randomization techniques can be used to determine if the observed features of the event sequence are just random artifacts, and show how to detect the features explaining the event sequence. In Section 3 we demonstrate our approach by applying the methods to important real-life data consisting of heart rate variability data and the occurrence of words in natural language.

Summarizing, the main contributions of this paper are:

- Interval sequence representation and its theoretic properties.
- Efficient randomization techniques for interval sequences.
- Using randomization to derive confidence limits and to explain non-random features of the data.
- Application of the proposed methods in two real-life applications.

2   Methods and Theory

2.1   Definitions

Assume that we have a sequence of $N+1$ events that occur at times $t_0, t_1, \ldots, t_N$, where $t_0 \leq t_1 \leq \ldots \leq t_N$. In this paper, we consider a sequence of N intervals S, defined by

$$S = (x_0, x_1, \ldots, x_{N-1}),$$

where $x_n = t_{n+1} - t_n$. In most of the numerical formulæ we use the logarithmic interval sequence $S_z = (z_0, z_1, \ldots, z_{N-1})$, where $z_n = \log x_n$. The logarithmic scale is more appropriate for our two applications: doubling and halving the interval both cause equal absolute changes in the value of the logarithm of the interval sequence, whereas without use of the logarithm, long intervals would receive much larger weight in the analysis. Furthermore, logarithms of intervals can take any value, including negative, which is numerically convenient.

For convenience and where appropriate, we extend the interval sequence by assuming that it is cyclic with a cycle of length N, i.e., $x_{n+N} = x_n$ and $z_{n+N} =$

z_n for all n. We denote the mean by $\bar{z}$ and the variance by σ_z^2, defined by $\bar{z} = \sum_{n=0}^{N-1} z_n/N$ and $\sigma_z^2 = \sum_{n=0}^{N-1} (z_n - \bar{z})^2/N$, respectively.

Fourier Representation. The Fourier representation of the data is defined by the sine and cosine series,

$$z_n = a_0 + \sum_{k=1}^{K} a_k \cos \frac{2\pi kn}{N} + \sum_{k=1}^{K} b_k \sin \frac{2\pi kn}{N}$$

$$= a_0 + \sum_{k=1}^{K} c_k \cos \left(\frac{2\pi kn}{N} - \varphi_k \right), \tag{1}$$

where $K = \lfloor N/2 \rfloor$. The data can be parametrized either by the parameters $(a_0, \{a_k\}, \{b_k\})$ or $(a_0, \{c_k\}, \{\varphi_k\})$, where $k \in \{1, \dots, K\}$ The Fourier amplitudes satisfy $c_k = \sqrt{a_k^2 + b_k^2}$ and the Fourier phases satisfy $\varphi_k \in [0, 2\pi)$. The Fourier parameters and the inverse transformation can be computed in $O(N \log N)$ time by a Fast Fourier Transform (FFT).

Autocorrelation. We use the autocorrelation function r_l with lag l, defined by

$$r_l = \frac{1}{N} \sum_{n=0}^{N-1} \frac{(z_n - \bar{z})(z_{n+l} - \bar{z})}{\sigma_z^2}. \tag{2}$$

A value of the autocorrelation function for a single lag can be computed in $O(N)$ time, and the values of the autocorrelation function for all lags can be computed in $O(N \log N)$ time by using the fast Fourier transformation. Notice that due to the cyclicity assumption, the lags satisfy $r_l = r_{N-l}$; therefore, it is sufficient to consider lags in $l \in \{1, \dots, \lceil N/2 \rceil\}$ only.

2.2 Randomization Methods

We define several distributions of interval sequences, and the respective randomization methods. Each of the distributions preserves some aspect of the original sequence.

Interval Randomization. The INTERVAL distribution is a uniform distribution over all permutations of sequence S. A sample S^* from the INTERVAL distribution can be drawn by permuting S uniformly in random.

Fixed Subsequence Randomization. The SUBSEQUENCE distribution is a uniform distribution over all permutations of the sequence S where a given subsequence $G_x \subseteq \{0, \dots, N-1\}$ of the intervals is kept fixed. A sample S^* is obtained by permuting all intervals in S that are not in G_x uniformly in random.

Fixed Fourier Parameters Randomization. The FOURIER distribution is a distribution of interval sequences in which given subsets of Fourier amplitudes and phases have been fixed. The FOURIER distribution is obtained by fixing a subset $G_c \subseteq \{1, \ldots, K\}$ of Fourier amplitudes c_k where $k \in G_c$, and a subset $G_\varphi \subseteq \{1, \ldots, K\}$ of Fourier phases φ_k where $k \in G_\varphi$. A sample S^* from FOURIER is obtained by first taking a sample S'^* from the INTERVAL distribution, and then replacing the Fourier amplitudes c_k and Fourier phases φ_k not in G_c and G_φ, respectively, by the respective Fourier parameters of the sample S'^*. The sample S^* is then obtained by applying the inverse Fourier transformation of Equation (1) to the randomized Fourier parameters.

Uniform Randomization. As a comparison to the INTERVAL distribution, we define the UNIFORM distribution to a uniform distribution over all sequences of N intervals in which the duration is fixed to $t_N - t_0$.

Fixed Distance Function Randomization. Finally, we define a randomization method which approximately preserves any arbitrary constraint. We define the constraint by a *distance function* $d(S')$ which is a non-negative function of permutations of the original interval sequence and zero for the original non-permuted sequence, $d(S) = 0$. We define a distribution DISTANCE by

$$f(S') \propto e^{-d(S')}, \tag{3}$$

where S' is a permutation of the original interval sequence S. A sample from the distribution f is likely to include intervals which are close to the original interval sequence in terms of the distance function. A sample S^* from DISTANCE can be obtained via Markov chain Monte Carlo (MCMC) integration, described in more detail in Section 2.4.

We use the distribution DISTANCE to sample intervals preserving the autocorrelation function at lags given in $G_r \subseteq \{1, \ldots, \lceil N/2 \rceil\}$. We use the distance function $d(S') = \lambda \sum_{l \in G_r} |r'_l - r_l|$, where $\lambda > 0$ is a parameter describing the accuracy to which we want to preserve the autocorrelations and r'_l is the value of autocorrelations for the resampled sequence. There is a critical value of λ in Equation (3) corresponding to the phase transition in statistical physics: the threshold value is recognized from the fact that when λ exceeds the threshold most of the probability mass of f is close to the original interval sequence (i.e., the expected value of the distance function is small). For all datasets considered in this paper a sufficiently high value is $\lambda = 10^4$, which is used in all experiments. Notice that the MCMC method could also be used to preserve the Fourier parameters, but it would be much slower than using the earlier introduced FOURIER randomization.

Time Complexity of the Randomizations. The time complexity of the UNIFORM, INTERVAL and SUBSEQUENCE randomizations is $O(N)$, and of FOURIER $O(N \log N)$. MCMC DISTANCE randomization is in practice always much slower, but its time complexity cannot be given for a general case because the number of MCMC iterations needed depends on the original sequence and the distance function. The time required by one MCMC iteration is typically dominated by the time complexity of the distance function. However, typical wall-clock running times to produce 1000 MCMC samples from the word and interbeat interval datasets used here are on the order of 1–3 and 10–20 minutes, respectively, using a non-optimized R [27] implementation and a standard desktop computer.

2.3   Properties of Fourier Parameters

In this section we show some properties of the Fourier parameters under the INTERVAL distribution: (i) the Fourier amplitudes are uncorrelated and their variance is proportional to σ_z^2, (ii) the phases φ_k approximately obey the uniform distribution on $[0, 2\pi)$.

Theorem 1. *The Fourier parameters satisfy the following properties under the* INTERVAL *distribution:*

- *The coefficient a_0 is the mean of the intervals in S_z.*
- *The expectations $E(a_k)$ and $E(b_k)$ vanish for every k.*
- *The variances are $E(a_k a_l) = E(b_k b_l) = 2\delta_{kl}\sigma_z^2(N-2)/(N(N-1))$ for every k and l, where δ_{kl} is the Kronecker delta.*
- *The cross-correlations $E(a_k b_l)$ vanish for every k and l.*
- *For all k, the phases $\varphi_k + 2\pi l/N$ (mod 2π) are equally probable for all values of $l \in \{0, \ldots, N-1\}$.*

We omit the proof for brevity.

2.4   MCMC and Parallel Tempering

We use MCMC integration with parallel tempering [9,22] to draw samples from the distribution f defined by Equation (3). Instead of drawing samples directly from f the samples are drawn from a product distribution F

$$F(\{S'_\alpha\}_{\alpha \in \Lambda}) \propto \prod_{\alpha \in \Lambda} f(S'_\alpha)^\alpha, \tag{4}$$

where $\{0, 1\} \subseteq \Lambda \subseteq [0, 1]$ is a finite set and S'_α is a permutation of the original sequence. At each MCMC iteration, the value of S'_1 gives a sample from f.

The distribution for specific values of α in Λ are called "chains". We perform sampling using the Metropolis-Hastings algorithm in which the proposal distributions include changes into one chain (within-chain jumps) and swapping chains with adjacent values of α (chain swaps). Here we use the following proposal distributions for within-chain jumps, repeated 10 times per MCMC iteration: (i) permuting the interval sequence in random, (ii) permuting a randomly chosen subsequence of the sequence in random, (iii) reversing a randomly chosen subsequence, (iv) swapping randomly chosen intervals, and (v) swapping adjacent items.

The idea is that the chain mixes well for low values of α ("high temperatures"). Indeed, for the chain with $\alpha = 0$ each consecutive state is a random permutation of the interval sequence. If the set Λ is chosen suitably then there is a sufficient number of chain swaps, which brings states from the well-mixed high temperature chains into the target distribution for which $\alpha = 1$; we have verified that the chains mix sufficiently with the high temperatures and hence include a sufficient number of practically independent samples for use in the computation of confidence intervals. See [22] for a more detailed discussion on parallel tempering.

2.5   Confidence Intervals and Hypothesis Testing

Confidence intervals cannot in general be determined analytically for non-trivial features of interest and must hence be obtained by simulation. Confidence intervals of level α for any feature of interest (e.g., Fourier amplitudes, Fourier phases or the autocorrelation structure) can be computed by calculating the value of the feature for a set of simulated samples obtained from a chosen null distribution. The confidence intervals here are defined to be the $\alpha/2$ and $1 - \alpha/2$ quantiles, where we always use $\alpha = 0.05$ in confidence levels and as a limit of significance. The parameters of interest can also be averaged in bins, in which case the confidence intervals will be narrower.[1] In this paper, we use the term *significant feature* to denote features that are outside the confidence intervals of some chosen null distribution. We further define a *non-random feature* as a feature that lies outside the confidence intervals calculated using the INTERVAL distribution.

The Fourier phases are approximately uniformly distributed under INTERVAL randomization (see Theorem 1). Due to the cyclic nature of the phases, confidence intervals for the phases cannot be computed in a meaningful way. Instead, the null hypothesis that the phases are uniformly distributed on the interval $[0, 2\pi)$ is tested by the Kolmogorov-Smirnov test.

3   Experiments

3.1   Datasets

The application of the randomization methods presented in this paper are illustrated using one artificial and two real-life datasets.

Toy Dataset. The toy dataset consists of two sequences: (1) The AR sequence is an autoregressive sequence of order 1 obeying $z_{n+1} \sim N(z_n, \varepsilon)$ and (2) the periodic sequence obeying $z_n \sim N(\cos{(2\pi k_{\text{toy}} n/N)}, \varepsilon)$ is a cosine embedded in noise. $N(\mu, \sigma)$ denotes a normal distribution with mean μ and standard deviation σ; here we have used $k_{\text{toy}} = 7$ and $\varepsilon = 0.7$.

IBI Dataset. The signals in the IBI dataset are interval sequences representing the time between two successive heartbeats, forming an interbeat-interval (IBI)

---

[1] In this paper, we always use bins of width one for the word data and bins of width 10 for the heartbeat data.

series. It has been shown that the IBI series have different time domain (e.g. [23]) and frequency domain (e.g. [1]) properties for normal subjects and for subjects with heart failure. The IBI dataset was hence formed from two different datasets from the PhysioBank biomedical signal archive [11]: (1) The normal rhythm dataset[2] contains recordings from 54 subjects in normal sinus rhythm, (2) The heart failure dataset[3]) contains recordings from 29 subjects with congestive heart failure. The first 1500 intervals were chosen for analysis, which translates to about 25 minutes of data at a heart rate of 60 beats per minute.

Word Dataset. The word dataset is composed of words from the book *Pride and Prejudice* by Jane Austen, publicly available from Project Gutenberg[4]. The interval sequence in this case represents the number of words between successive occurrences of a particular word.[5] A previously reported [19] representative set of words was chosen for analysis, forming the (1) bursty and (2) non-bursty datasets, each containing 12 words. In addition, the words were also divided into frequency classes of low, medium and high, corresponding to frequencies of roughly 40, 200 and 1200, respectively.

3.2 Null Model Selection and Confidence Intervals

Here we demonstrate the advantages of the INTERVAL distribution over the UNIFORM distribution. In Fig. 1, both UNIFORM and INTERVAL randomizations are shown for the word *met*. We notice that for the UNIFORM randomization many of the Fourier amplitudes are significant, whereas for the INTERVAL randomization all Fourier parameters are consistent with the random data. This shows that there is structure in *met* not present in the UNIFORM distribution, but explained by the INTERVAL distribution. Observing non-random features under the INTERVAL distribution is always due to the ordering of the intervals, which is not the case if the UNIFORM distribution is used, as shown by the example in Fig. 1. See [21] for further discussion on the unsuitability of the uniform distribution in the analysis of interval sequences.

3.3 Investigating the Structure of the Datasets

In this section, we present an overview of the datasets and illustrate their general properties using examples. We determine non-random features of the sequences by calculating confidence intervals in accordance with Section 3.2, using the INTERVAL distribution as the null model.

The Toy Dataset. The Toy dataset is shown in Fig. 2. For AR, most of the Fourier coefficients are non-random, as the complex structure of the sequence

---

[2] http://www.physionet.org/physiobank/database/nsr2db/
[3] http://www.physionet.org/physiobank/database/chf2db/
[4] http://www.gutenberg.org
[5] More specifically, the interval is one plus the number of words between successive occurrences of a word, i.e., adjacent words have an interval value of one.

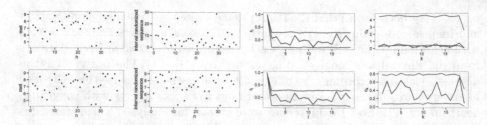

Fig. 1. UNIFORM (top) and INTERVAL (bottom) randomizations of the word *met*. Shown are, from the left: (1) the original sequence, (2) a realization from the random distribution, (3) the autocorrelation function, and (4) the Fourier amplitudes. Confidence intervals are shown in blue. Values outside confidence intervals are shown in red.

cannot be easily captured by a low number of Fourier amplitudes. For `periodic`, the Fourier coefficient for $k = 7$ is clearly non-random, corresponding to the number of periods in the sequence. The other Fourier amplitudes are mostly non-random, except for some high frequencies corresponding to the noise. INTERVAL randomization does not explain the autocorrelation structure of either sequence.

The IBI Dataset. Example sequences from the IBI dataset are shown in Fig. 3. Sequences from both the `normal rhythm` and `heart failure` datasets exhibit clear temporal structures, caused e.g. by different activities undertaken by the subject. This leads to segments with varying IBI distributions within one record. The temporal structure of the IBI sequences is usually characterized by a slow global trend containing segments with more rapid local variation.

The Fourier amplitudes for records with a strong temporal structure are only partially explained under the INTERVAL distribution. For such records, most of the Fourier amplitudes are non-random (see records `chf201` and `nsr033` in Fig. 3).

The non-random low-order Fourier amplitudes probably reflect the global trend, whereas the higher-order non-random Fourier amplitudes likely reflect short-range temporal variation. In contrast, some records with a weak global

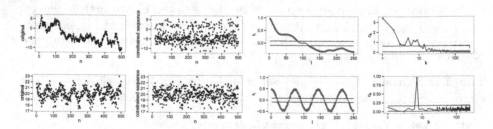

Fig. 2. The sequences in the Toy dataset. The `AR` (top row) and the `periodic` sequences (bottom row). Subplots follow the same order as in Fig. 1. The confidence intervals are based on INTERVAL randomization.

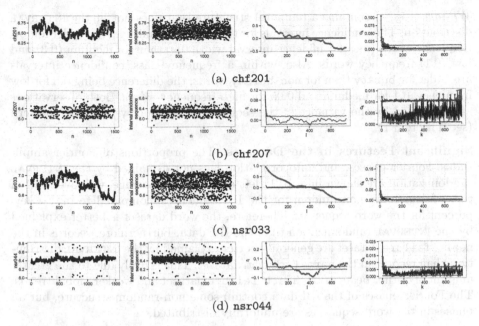

Fig. 3. Example IBI sequences. Subfigures and confidence intervals as in Fig. 2.

trend (nsr044 in Fig. 3) or a high degree of outliers (chf207 in Fig. 3) are better explained by the INTERVAL distribution, and for these records only a few Fourier coefficients are non-random.

The significant temporal structure of the sequences is also strongly reflected in the autocorrelation, which for most records is non-random.

The Word Dataset. Three example sequences from the word dataset are shown in Figs. 1 and 4. Only a few Fourier amplitudes or autocorrelation lags are non-random, usually marking visible temporal patterns in the data. The words *met* (Fig. 1) and *soon* (Fig. 4) are explained by the INTERVAL distribution. The word

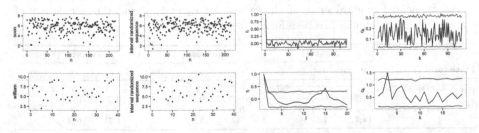

Fig. 4. Examples of word sequences. *Soon* (top) is a frequent word mostly (but not completely) explained by the INTERVAL distribution, *William* (bottom) contains a clear temporal structure. Subfigures and confidence intervals as in Fig. 2.

William (Fig. 4) contains a temporal structure that does not fit the INTERVAL distribution. The confidence intervals for the Fourier amplitudes are wider for low frequency words (mean confidence interval width 0.9) than for medium (0.3) and low (0.1) frequency words. Also, within a frequency class, confidence intervals are wider for bursty than for non-bursty words, the difference being 0.6 for low frequency, 0.1 for medium and 0.05 for high frequency words. Both observations are explained by the variance of the sequence, which follows a similar pattern (see Theorem 1).

Significant Features in the Datasets. The proportions of Fourier amplitudes, Fourier phases, and autocorrelation lags not explained by the INTERVAL randomization are shown in Tab. 1 (left column). On average, well over half of these features are non-random for the IBI sequences, compared to only a few percent for the word sequences. Therefore, the word dataset is better explained by the INTERVAL randomization than the IBI data. Furthermore, records in the heart failure dataset are generally better explained by the INTERVAL distribution than records in the normal rhythm dataset. This is likely due to the greater amount of outlier beats in the heart failure dataset and weaker global trends. The Fourier phases of the IBI data contain some non-random structure, but all phases in the word sequences are uniformly distributed.

3.4 Constrained Randomizations

We construct constrained randomizations by fixing a specific set of features in the FOURIER or DISTANCE randomizations. If the data are explained by the constrained null hypothesis, we can conclude that we have successfully located the features explaining the non-random characteristics of the data.

Connection between Fourier Amplitudes and Autocorrelation. In this section, both the AR and periodic sequences are randomized by fixing the most significant feature (with respect to INTERVAL randomization, see Fig. 2). The fixing of features is performed separately for autocorrelations and Fourier amplitudes. The results are shown in Fig. 5.

Table 1. Percentages of non-random features in the different datasets. The values represent mean (standard error of the mean) for c_k and r_l, and the percentage of sequences with non-uniform phases for φ_k (see Section 2.5).

Dataset		INTERVAL			FOURIER (c_k)			DISTANCE (r_l)		
		c_k	r_l	φ_k	c_k	r_l	φ_k	c_k	r_l	φ_k
IBI	normal rhythm	89 (1.5)	94 (0.9)	69	0.2 (0.07)	2 (0.5)	0	5 (0.6)	3 (0.4)	5
	heart failure	66 (4.7)	81 (4.5)	30	0.8 (0.2)	6 (1.3)	0	7 (1.7)	2 (0.4)	5
Word	bursty	6 (1.0)	10 (1.7)	0	0.4 (0.2)	2 (0.4)	0	9 (2.6)	8 (2)	6
	non-bursty	4 (0.9)	3 (0.9)	0	0.5 (0.1)	1 (0.4)	0	8 (0.9)	5 (0.9)	4

For AR, fixing the autocorrelation r_1 produces a sequence that retains the local temporal structure of the original sequence. The Fourier amplitudes are almost explained, but the autocorrelation function matches the original only for short lags. In contrast, fixing a single non-random feature in the Fourier domain performs much worse in explaining the data.

For periodic, Fourier amplitude randomization yields signals resembling the original. The majority of the Fourier amplitudes are explained, and the confidence intervals for the autocorrelations follow the course of the original autocorrelation function, albeit not perfectly. In contrast, fixing a single autocorrelation lag for periodic does not explain the features of the signal at all.

Fixing Fourier Amplitudes and Autocorrelation Structure. Constrained randomization of Fourier amplitudes and autocorrelation lags was applied to both the IBI and word datasets, keeping the non-random Fourier amplitudes and autocorrelation lags constant. The percentage of Fourier amplitudes and autocorrelation lags that remain significant under the randomizations are shown in Tab. 1. For the constrained Fourier amplitude randomization (middle column) the percentages are low, indicating that the data are well explained. Only the autocorrelations of the heart failure dataset shows a slightly higher percentage of significant features. For the constrained autocorrelation randomization

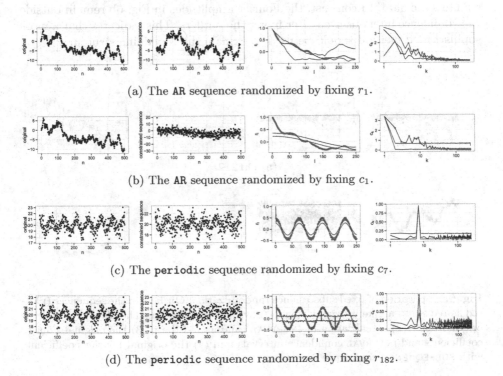

(a) The AR sequence randomized by fixing r_1.

(b) The AR sequence randomized by fixing c_1.

(c) The periodic sequence randomized by fixing c_7.

(d) The periodic sequence randomized by fixing r_{182}.

Fig. 5. Constrained randomizations of the toy data. Subplots are as in Fig. 1.

(rightmost column), the Fourier amplitudes of the word data and `heart failure` IBI data have above 5% of significant features, indicating that the randomization does not fully explain the Fourier amplitudes. There is also unexplained autocorrelation structure in the word data.

Since 95% confidence intervals were used, one must note in the interpretation of Tab. 1, that if the randomization explains the data, at most 5% of features should remain significant. In practice, this value is lower as a large portion of the features is kept fixed, especially for the IBI data. Also, the autocorrelation lags are not independent, causing them to fit the simple quantile based confidence intervals better than Fourier amplitudes.

Fixed Subsequence Randomization. Outliers in the data can significantly affect the structure and interpretation of the data. In order to investigate the structure of the data, outliers can be considered subsequences and kept fixed in the SUBSEQUENCE randomization.

In Fig. 6, outliers detected using a commonly used algorithm by [37] were kept fixed while the rest of the data were randomized using the SUBSEQUENCE method. In Fig. 6a several of the Fourier coefficients are outside the confidence intervals calculated using INTERVAL randomization, i.e., the structure of the data is not modeled by the INTERVAL distribution. However, fixing the outliers and calculating the confidence intervals using the SUBSEQUENCE distribution explains the data. In contrast, the Fourier amplitudes in Fig. 6b remain outside the confidence intervals even after fixing the outliers. This indicates that a more sophisticated method should be used to explain the remaining structure.

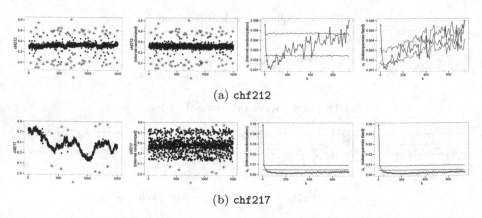

(a) chf212

(b) chf217

Fig. 6. Application of fixed subsequence randomization. Outliers in the sequences (plotted as red stars) are kept fixed during INTERVAL randomization. The plots show (1) the original data, (2) a realization of INTERVAL randomized data, (3) the original Fourier coefficients and INTERVAL confidence intervals and (4) the original Fourier coefficients with SUBSEQUENCE confidence intervals.

3.5 Application to Hypothesis Testing

Constrained realizations obtained e.g. by fixing non-random Fourier amplitudes can be used in statistical hypothesis testing. As an example of this, the significance of the pNN50-value commonly used in heart rate variability analysis [23] was calculated[6] for the records in the IBI datasets. The results are shown in Tab. 2. There are clearly differences between the choices of constraints.

On one hand, the INTERVAL randomization appears to provide realizations that are consistently too extreme for hypothesis testing, at least if the objective is to study the differences between normal rhythm and heart failure. On the other hand, the number of significant p-values with the Fourier amplitude constraint is much smaller for `normal rhythm` than for `heart failure`, suggesting that part of the IBI signal measured by the pNN50 statistic and not explained by the amplitudes, is related to the heart failure condition. Therefore, the null hypothesis with the amplitude constraint might be suitable for modeling healthy individuals.

Table 2. Percentages (%) of significant pNN50 -values for the datasets. From the left: by constraining Fourier coefficients, phases, autocorrelation lags, and using INTERVAL randomization.

	Randomization method			
Dataset	FOURIER (c_k)	FOURIER (φ_k)	DISTANCE (r_l)	INTERVAL
`normal rhythm`	22.2	75.9	100.0	100.0
`heart failure`	58.6	79.3	86.2	96.5

4 Related Work

Randomization testing in statistical analysis has a long history; see, e.g., [12,36] for a review. Randomization methods are useful in hypothesis testing and defining confidence bounds when sampling from the null hypothesis is easier than to define the null hypothesis analytically. Randomization methods have been devised for various kinds of data structures, such as binary matrices [10], graphs [13,38], gene periodicity (e.g., [15]), and real matrices [24].

Time series randomization has been studied, e.g., in [3,16,25,34,2,30,35]. Some of the prior randomization methods work in the Fourier space (see, e.g., [26] for use of phase-randomization in hypothesis testing) or in the wavelet space, see [17] for a review. However, usually the time series has not been represented as (equally-spaced) sequence of intervals, but as an event sequence with variable event interval (see, e.g., [4,31]).

In the field of data analysis, a recently promoted approach [14,20,5,6] to the use of randomization is to interpret the patterns as constraints to the null hypothesis. The use of surrogate data [32] in the hypothesis testing of data

---

[6] Here we consider all interbeat intervals, not just normal-to-normal (NN) intervals.

structures is a common technique, and has been applied in the generation of constrained realizations for hypothesis testing regarding the properties of a time series, e.g., by [33,29,28].

Randomization techniques have been applied in the analysis of heart rate variability (HRV), e.g., by [18,8], who used Fourier phase randomization for generating surrogate data for hypothesis testing. Time-varying surrogates were used by [7] for studying non-linearity in interbeat interval (IBI) series.

5   Conclusions

We have shown that interval sequences form a natural representation for event sequences, and offer a principled and robust basis to sequence randomization. We have investigated the problem of interpreting commonly used Fourier parameters and autocorrelation structures. We find that the interpretation depends on the null hypothesis used; for example, a naïve use of the UNIFORM distribution may lead to false conclusions regarding the temporal structure of sequences.

Furthermore, we have provided computationally efficient randomization methods for studying Fourier parameters, and an MCMC based method for studying autocorrelation structures and arbitrary constraints. The randomization methods allow the user to efficiently test different null hypotheses by fixing chosen subsets of parameters. This makes it possible to infer possible causes for the observed significant patterns.

In this paper, we have shown how the proposed randomization methods can be used in hypothesis testing, and examined the role of the null hypothesis. There is no universally suitable null hypothesis. The null hypothesis should encompass our best understanding of the features of the data and hence depends on the research question.

With the help of the randomization methods presented here, simple and understandable explanations for the structure of the data can be found efficiently and in a statistically robust way. If there are structures left unexplained by the proposed methods, more complex constraints or models of different types can be used to further investigate and explain the remaining patterns in the data.

Acknowledgements. This study was supported by the SalWe Research Programme for Mind and Body (Tekes – the Finnish Funding Agency for Technology and Innovation grant 1104/10).

References

1. Bigger, J.T., Fleiss, J.L., Steinman, R.C., Rolnitzky, L.M., Schneider, W.J., Stein, P.K.: RR variability in healthy, middle-aged persons compared with patients with chronic coronary heart disease or recent acute myocardial infarction. Circulation 91(7), 1936–1943 (1995)

2. Bullmore, E., Long, C., Suckling, J., Fadili, J., Calvert, G., Zelaya, F., Carpenter, T.A., Brammer, M.: Colored noise and computational inference in neurophysiological (fMRI) time series analysis: Resampling methods in time and wavelet domains. Human Brain Mapping 12(2), 61–78 (2001)
3. Carlstein, E.G.: Resampling techniques for stationary time-series: some recent developments. University of North Carolina at Chapel Hill (1990)
4. Clifford, G.D., Azuaje, F., McSharry, P., et al. (eds.): Advanced Methods and Tools for ECG Data Analysis. Artech House, London (2006)
5. De Bie, T.: An information theoretic framework for data mining. In: Proceedings of the 17th ACM SIGKDD International Conference on Knowledge Discovery and Data Mining, KDD 2011, pp. 564–572. ACM, New York (2011)
6. De Bie, T.: Maximum entropy models and subjective interestingness: an application to tiles in binary databases. Data Mining and Knowledge Discovery 23(3), 407–446 (2011)
7. Faes, L., Zhao, H., Chon, K., Nollo, G.: Time-varying surrogate data to assess nonlinearity in nonstationary time series: Application to heart rate variability. IEEE Transactions on Biomedical Engineering 56(3), 685–695 (2009)
8. Garde, S., Regalado, M.G., Schechtman, V.L., Khoo, M.C.: Nonlinear dynamics of heart rate variability in cocaine-exposed neonates during sleep. American Journal of Physiology-Heart and Circulatory Physiology 280(6), H2920–H2928 (2001)
9. Geyer, C.J.: Markov chain Monte Carlo Maximum Likelihood. In: Computing Science and Statistics: The 23rd Symposium on the Interface, pp. 156–163. Interface Foundation, Fairfax (1991)
10. Gionis, A., Mannila, H., Mielikäinen, T., Tsaparas, P.: Assessing data mining results via swap randomization. ACM Trans. Knowl. Discov. Data 1(3) (December 2007)
11. Goldberger, A.L., Amaral, L.A.N., Glass, L., Hausdorff, J.M., Ivanov, P.C., Mark, R.G., Mietus, J.E., Moody, G.B., Peng, C.K., Stanley, H.E.: PhysioBank, PhysioToolkit, and PhysioNet: Components of a new research resource for complex physiologic signals. Circulation 101(23), e215–e220 (2000)
12. Good, P.I.: Permutation Tests: A Practical Guide to Resampling Methods for Testing Hypotheses. Springer (2000)
13. Hanhijärvi, S., Garriga, G.C., Puolamäki, K.: Randomization techniques for graphs. In: Proceedings of the 9th SIAM International Conference on Data Mining (SDM 2009), pp. 780–791 (2009)
14. Hanhijärvi, S., Ojala, M., Vuokko, N., Puolamäki, K., Tatti, N., Mannila, H.: Tell me something I don't know: randomization strategies for iterative data mining. In: Proceedings of the 15th ACM SIGKDD International Conference on Knowledge Discovery and Data Mining, KDD 2009, pp. 379–388. ACM, New York (2009)
15. Kallio, A., Vuokko, N., Ojala, M., Haiminen, N., Mannila, H.: Randomization techniques for assessing the significance of gene periodicity results. BMC Bioinformatics 12(1), 330 (2011)
16. Kreiss, J.P., Franke, J.: Bootstrapping stationary autoregressive moving-average models. Journal of Time Series Analysis 13(4), 297–317 (1992)
17. Laird, A.R., Rogers, B.P., Meyerand, M.E.: Comparison of fourier and wavelet resampling methods. Magnetic Resonance in Medicine 51(2), 418–422 (2004)
18. Li, C., Ding, G.H., Wu, G.Q., Poon, C.S.: Band-phase-randomized surrogate data reveal high-frequency chaos in heart rate variability. In: 2010 Annual International Conference of the IEEE Engineering in Medicine and Biology Society (EMBC), pp. 2806–2809 (2010)

19. Lijffijt, J., Papapetrou, P., Puolamäki, K.: Size matters: Finding the most informative set of window lengths. In: Flach, P.A., De Bie, T., Cristianini, N. (eds.) ECML PKDD 2012, Part II. LNCS, vol. 7524, pp. 451–466. Springer, Heidelberg (2012)
20. Lijffijt, J., Papapetrou, P., Puolamäki, K.: A statistical significance testing approach to mining the most informative set of patterns. Data Mining and Knowledge Discovery (December 2012) (to appear) (published online before print)
21. Lijffijt, J., Papapetrou, P., Puolamäki, K., Mannila, H.: Analyzing word frequencies in large text corpora using inter-arrival times and bootstrapping. In: Gunopulos, D., Hofmann, T., Malerba, D., Vazirgiannis, M. (eds.) ECML PKDD 2011, Part II. LNCS, vol. 6912, pp. 341–357. Springer, Heidelberg (2011)
22. Liu, J.: Monte Carlo Strategies in Scientific Computing. Series in Statistics. Springer (2008)
23. Mietus, J., Peng, C., Henry, I., Goldsmith, R., Goldberger, A.: The pnnx files: re-examining a widely used heart rate variability measure. Heart 88(4), 378–380 (2002)
24. Ojala, M., Vuokko, N., Kallio, A., Haiminen, N., Mannila, H.: Randomization methods for assessing data analysis results on real-valued matrices. Statistical Analysis and Data Mining 2(4), 209–230 (2009)
25. Politis, D.N.: The impact of bootstrap methods on time series analysis. Statistical Science 18(2), 219–230 (2003)
26. Prichard, D., Theiler, J.: Generating surrogate data for time series with several simultaneously masured variables. Physical Review Letters 73(7), 951–954 (1994)
27. R Core Team: R: A Language and Environment for Statistical Computing. R Foundation for Statistical Computing, Vienna, Austria (2013) ISBN 3-900051-07-0, http://www.R-project.org/
28. Schreiber, T.: Constrained randomization of time series data. Physical Review Letters 80(10), 2105–2108 (1998)
29. Schreiber, T., Schmitz, A.: Improved Surrogate Data for Nonlinearity Tests. Physical Review Letters 77(4), 635–638 (1996)
30. Schreiber, T., Schmitz, A.: Surrogate time series. Physica D: Nonlinear Phenomena 142(3-4), 346–382 (2000)
31. Sörnmo, L., Laguna, P.: Bioelectrical Signal Processing in Cardiac and Neurological Applications. Academic Press (2005)
32. Theiler, J., Eubank, S., Longtin, A., Galdrikian, B., Doyne Farmer, J.: Testing for nonlinearity in time series: the method of surrogate data. Physica D: Nonlinear Phenomena 58(1), 77–94 (1992)
33. Theiler, J., Prichard, D.: Constrained-realization Monte-Carlo method for hypothesis testing. Physica D: Nonlinear Phenomena 94(4), 221–235 (1996)
34. Vinod, H.D.: Maximum entropy ensembles for time series inference in economics. Journal of Asian Economics 17(6), 955–978 (2006)
35. Vuokko, N., Kaski, P.: Significance of patterns in time series collections. In: Proceedings of the Eleventh SIAM International Conference on Data Mining, Mesa, AZ, April 28-30, pp. 676–686. SIAM, Philadelphia (2011)
36. Westfall, P.H., Young, S.: Resampling-Based Multiple Testing: Examples and Methods for P-Value Adjustment. A Wiley-Interscience publication, Wiley (1993)
37. Xu, X., Schuckers, S.: Automatic detection of artifacts in heart period data. Journal of Electrocardiology 34(4), 205–210 (2001)
38. Ying, X., Wu, X.: Graph generation with prescribed feature constraints. In: Proceedings of the 9th SIAM International Conference on Data Mining (SDM 2009), pp. 966–977 (2009)

Itemset Based Sequence Classification

Cheng Zhou[1,2], Boris Cule[1], and Bart Goethals[1]

[1] University of Antwerp, Belgium
[2] National University of Defense Technology, China
Cheng.Zhou@student.ua.ac.be

Abstract. Sequence classification is an important task in data mining. We address the problem of sequence classification using rules composed of interesting itemsets found in a dataset of labelled sequences and accompanying class labels. We measure the interestingness of an itemset in a given class of sequences by combining the cohesion and the support of the itemset. We use the discovered itemsets to generate confident classification rules, and present two different ways of building a classifier. The first classifier is based on the CBA (Classification based on associations) method, but we use a new ranking strategy for the generated rules, achieving better results. The second classifier ranks the rules by first measuring their value specific to the new data object. Experimental results show that our classifiers outperform existing comparable classifiers in terms of accuracy and stability, while maintaining a computational advantage over sequential pattern based classification.

1 Introduction

Many real world datasets, such as collections of texts, videos, speech signals, biological structures and web usage logs, are composed of sequential events or elements. Because of a wide range of applications, sequence classification has been an important problem in statistical machine learning and data mining.

The sequence classification task can be defined as assigning class labels to new sequences based on the knowledge gained in the training stage. There exist a number of studies integrating pattern mining techniques and classification, such as classification based on association rules [10], sequential pattern based sequence classifiers [8,13], and many others. These combined methods can output good results as well as provide users with information useful for understanding the characteristics of the dataset.

In this paper, we propose to utilise a novel itemset mining technique [4] in order to obtain an accurate sequence classifier. An itemset in a sequence should be evaluated based on how close to each other its items occur (cohesion) and how often the itemset itself occurs (support). We therefore propose a new method called sequence classification based on interesting itemsets (SCII), that greatly improves on the accuracy obtained by other classifiers based on itemsets, as they typically do not take cohesion into account. Moreover, we also achieve a reduction in complexity compared to classifiers based on sequential patterns, as

H. Blockeel et al. (Eds.): ECML PKDD 2013, Part I, LNAI 8188, pp. 353–368, 2013.
© Springer-Verlag Berlin Heidelberg 2013

we generate and evaluate fewer patterns, and, yet, using cohesion, we still take the location of items within the sequences into account.

The main contribution of this paper consists of two SCII classifiers, both based on frequent cohesive itemsets. By using cohesion, we incorporate the sequential nature of the data into the method, while, by using itemsets, we avoid the complexity of mining sequential patterns. The two classifiers differ in how the rules are selected and ranked within the classifier, and we experimentally demonstrate that both give satisfactory results.

The rest of the paper is organised as follows. Section 2 gives a review of the related work. In Sections 3 and 4, we formally describe the sequence classification problem setting and present our two approaches for generating rules and building classifiers, respectively. We end the paper with an experimental evaluation in Section 5 and a summary of our conclusions in Section 6.

2   Related Work

The existing sequence classification techniques deploy a number of different approaches, ranging from decision trees, Naïve Bayes, Neural Networks, K-Nearest Neighbors (KNN), Hidden Markov Model (HMM) and, lately, Support Vector Machines (SVMs) [7].

In this section, we give an overview of pattern-based classification methods. Most such work can be divided into the domains of classification based on association rules and classification based on sequential patterns. The main idea behind the first approach is to discover association rules that always have a class label as their consequent. The next step is to use these patterns to build a classifier, and new data records are then classified in the appropriate classes. The idea of classification based on association rules (CBA) was first proposed by Liu et al. [10]. In another work, Li et al. [9] proposed CMAR, where they tackled the problem of overfitting inherent in CBA. In CMAR, multiple rules are employed instead of just a single rule. Additionally, the ranking of the rule set in CMAR is based on the weighted Chi-square of each rule replacing the confidence and support of each rule in CBA. Yin and Han [15] proposed CPAR which is much more time-efficient in both rule generation and prediction but its accuracy is as high as that of CBA and CMAR.

The concept of sequential pattern mining was first described by Agrawal and Srikant [2], and further sequential pattern mining methods, such as Generalized Sequential Patterns (GSP) [12], SPADE [16], PrefixSpan [11], and SPAM [3], have been developed since. A number of sequence classifiers have been based on these methods.

Lesh et al. [8] combined sequential pattern mining and a traditional Naïve Bayes classification method to classify sequence datasets. They introduced the FeatureMine algorithm which leveraged existing sequence mining techniques to efficiently select features from a sequence dataset. The experimental results showed that BayesFM (combination of Naïve Bayes and FeatureMine) is better than Naïve Bayes only. Although pruning is used in their algorithm, there was still a great number of sequential patterns used as classification features. As a result, the algorithm could not effectively select discriminative features from a large feature space.

Tseng and Lee [14] proposed the Classify-By-Sequence (CBS) algorithm for classifying large sequence datasets. The main methodology of the CBS method is mining classifiable sequential patterns (CSPs) from the sequences and then assigning a score to the new data object for each class by using a scoring function, which is based on the length of the matched CSPs. They presented two approaches, CBS_ALL and CBS_CLASS. In CBS_ALL, a conventional sequential pattern mining algorithm is used on the whole dataset. In CBS_CLASS, the database is divided into a number of sub-databases according to the class label of each instance. Sequential pattern mining was then implemented on each sub-database. Experimental results showed that CBS_CLASS outperforms CBS_ALL. Later, they improved the CBS_CLASS algorithm by removing the CSPs found in all classes [13]. Furthermore, they proposed a number of alternative scoring functions and tested their performances. The results showed that the length of a CSP is the best attribute for classification scoring.

Exarchos et al. [6] proposed a two-stage methodology for sequence classification based on sequential pattern mining and optimization. In the first stage, sequential pattern mining is used, and a sequence classification model is built based on the extracted sequential patterns. Then, weights are applied to both sequential patterns and classes. In the second stage, the weights are tuned with an optimization technique to achieve optimal classification accuracy. However, the optimization is very time consuming, and the accuracy of the algorithm is similar to FeatureMine.

Additionally, several sequence classification methods have been proposed for application in specific domains. Exarchos et al. [5] utilised sequential pattern mining for protein fold recognition, while Zhao et al. [17] used a sequence classification method for debt detection in the domain of social security.

The main bottleneck problem for sequential pattern based sequence classification being used in the real world is efficiency. Mining frequent sequential patterns in a dense dataset with a large average sequence length is time and memory consuming. None of the above sequence classification algorithms solve this problem well.

3   Problem Statement

In this paper, we consider multiple event sequences where an event e is a pair (i, t) consisting of an item $i \in I$ and a time stamp $t \in \mathbb{N}$, where I is the set of all possible items and $\mathbb{N}$ is the set of natural numbers. We assume that two events can never occur at the same time. For easier readibility, in our examples, we assume that the time stamps in a sequence are consecutive natural numbers. We therefore denote a *sequence* of events by $s = e_1, \cdots, e_l$, where l is the length of the sequence, and $1, \cdots, l$ are the time stamps.

Let L be a finite set of class labels. A sequence database SDB is a set of *data objects* (s, L_k), such that s is a sequence and $L_k \in L$ is a *class label*

($k = 1, 2, \cdots, m$, where m is the number of classes). The set of all sequences in SDB is denoted by S. We denote the set of sequences carrying class label L_k by S_k.

The patterns considered in this paper are itemsets, or sets of items coming from the set I. The support of an itemset is typically defined as the number of different sequences in which the itemset occurs, regardless of how many times the itemset occurs in any single sequence. To determine the interestingness of an itemset, however, it is not enough to know how many times the itemset occurs. We should also take into account how close the items making up the itemset occur to each other. To do this, we will define interesting itemsets in terms of both support and cohesion. Our goal is to first mine interesting itemsets in each class of sequences, and then use them to build a sequence classifier, i.e., a function from sequences S to class labels L.

We base our work on an earlier work on discovering interesting itemsets in a sequence database [4], and we begin by adapting some of the necessary definitions from that paper to our setting. The interestingness of an itemset depends on two factors: its support and its cohesion. Support measures in how many sequences the itemset appears, while cohesion measures how close the items making up the itemset are to each other on average.

For a given itemset X, we denote the set of sequences that contain all items of X as $N(X) = \{s \in S | \forall i \in X, \exists (i, t) \in s\}$. We denote the set of sequences that contain all items of X labelled by class label L_k as $N_k(X) = \{s \in S_k | \forall i \in X, \exists (i, t) \in s\}$. The *support* of X in a given class of sequences S_k can now be defined as $F_k(X) = \frac{|N_k(X)|}{|S_k|}$.

We begin by defining the length of the shortest interval containing an itemset X in a sequence $s \in N(X)$ as $W(X, s) = \min\{t_2 - t_1 + 1 | t_1 \leq t_2 \text{ and } \forall i \in X, \exists (i, t) \in s, \text{ where } t_1 \leq t \leq t_2\}$. In order to calculate the cohesion of an itemset within class k, we now compute the average length of such shortest intervals in $N_k(X)$: $\overline{W_k}(X) = \frac{\sum_{s \in N_k(X)} W(X, s)}{|N_k(X)|}$. It is clear that $\overline{W_k}(X)$ is greater than or equal to the number of items in X, denoted as $|X|$. Furthermore, for a fully cohesive itemset, $\overline{W_k}(X) = |X|$. Therefore, we define cohesion of X in $N_k(X)$ as $C_k(X) = \frac{|X|}{\overline{W_k}(X)}$. Note that all itemsets containing just one item are fully cohesive, that is $C_k(X) = 1$ if $|X| = 1$. The *cohesion* of X in a single sequence s is defined as $C(X, s) = \frac{|X|}{W(X, s)}$.

In a given class of sequences S_k, we can now define the *interestingness* of an itemset X as $I_k(X) = F_k(X)C_k(X)$. Given an interestingness threshold min_int, an itemset X is considered interesting if $I_k(X) \geq min\_int$. If desired, minimum support and minimum cohesion can also be used as separate thresholds.

Once we have discovered all interesting itemsets in each class of sequences, the next step is to identify the classification rules we will use to build a classifier.

We define $r_{km} : p_m \Rightarrow L_k$ as a rule where p_m is an interesting itemset in S_k and L_k is a class label. p_m is the *antecedent* of the rule and L_k is the *consequent* of the rule. We further define the interestingness, support, cohesion and size of r_{km}

to be equal to the interestingness, support, cohesion and size of p_m, respectively. The *confidence* of a rule can now be defined as:

$$conf(p_m \Rightarrow L_k) = \frac{|N_k(p_m)|}{|N(p_m)|} \tag{1}$$

A rule $p_m \Rightarrow L_k$ is considered confident if its confidence exceeds a given threshold *min_conf*.

If all items in the antecedent of the rule can be found in the sequence of a given data object, we say that the rule *matches* the data object. We say that a rule *correctly classifies* or *covers* a data object in *SDB* if the rule matches the sequence part of the data object and the rule's consequent equals the class label part of the data object.

In practice, most datasets used in the sequence classification task can be divided into two main cases. In the first case, the class of a sequence is determined by certain items that co-occur within it, though not always in the same order. In this case, a classifier based on sequential patterns will not work well, as the correct rule will not be discovered, and, with a low enough threshold, the rules that are discovered will be far too specific. For an itemset of size n, there are $n!$ orders in which this itemset could appear in a sequence, and therefore $n!$ rules that could be discovered (none of them very frequent). Our method, however, will find the correct rule. In the other case, the class of a sequence is determined by items that occur in the sequence always in exactly the same order. At first glance, a classifier based on sequential patterns should outperform our method in this situation. However, we, too, will discover the same itemset (and rule), only not in a sequential form. Due to a simpler candidate generation process, we will even do so quicker. Moreover, we will do better in the presence of noise, in cases when the itemset sometimes occurs in an order different from the norm. This robustness of our method means that we can handle cases where small deviations in the sequential patterns that determine the class of the sequences occur. For example, if a class is determined by occurrences of sequential pattern abc, but this pattern sometimes occurs in a different form, such as acb or bac, our method will not suffer, as we only discover itemset $\{a, b, c\}$. This means that, on top of the reduced complexity, our method often gives a higher accuracy than classifiers based on sequential patterns, as real-life data is often noisy and sequential classification rules sometimes prove to be too specific. On the other hand, in cases where two classes are determined by exactly the same items, but in different order, our classifier will struggle. For example, if class A is determined by the occurrence of abc and class B by the occurrence of cba, we will not be able to tell the difference. However, such cases are rarely encountered in practice.

4 Generating Rules and Building Classifiers

Our algorithm, SCII (Sequence Classification Based on Interesting Itemsets), consists of two stages, a rule generator (SCII_RG), which is based on the Apriori algorithm [1], and two different classifier builders, SCII_CBA and SCII_MATCH. This section discusses SCII_RG, SCII_CBA and SCII_MATCH.

4.1   Generating the Complete Set of Interesting Itemsets

The SCII_RG algorithm generates all interesting itemsets in two steps. Due to the fact that the cohesion and interestingness measures introduced in Section 3, are not anti-monotonic, we prune the search space based on frequency alone. In the first step, we use an Apriori-like algorithm to find the frequent itemsets. In the second step, we determine which of the frequent itemsets are actually interesting. An optional parameter, max_size, can be used to limit the output only to interesting itemsets with a size smaller than or equal to max_size.

Let n-$itemset$ denote an itemset of size n. Let A_n denote the set of frequent n-$itemsets$. Let C_n be the set of candidate n-$itemsets$ and T_n be the set of interesting n-$itemsets$. The algorithm for generating the complete set of interesting itemsets in a given class of sequences is shown in Algorithm 1.

Algorithm 1. GENERATINGITEMSETS. An algorithm for generating all interesting itemsets in S_k.

---

    input   : S_k, minimum support threshold min_sup, minimum interestingness threshold min_int, max size constraint max_size

    output : all interesting itemsets P_k

1 $C_1 = \{i | i \in I_k\}$, I_k is the set of all the items which occur in S_k;

2 $A_1 = \{f | f \in C_1, F_k(f) \geq min\_sup\}$;

3 $T_1 = \{f | f \in A_1, F_k(f) \geq min\_int\}$;

4 $n = 2$;

5 **while** $A_{n-1} \neq \emptyset$ and $n \leq max\_size$ **do**

6      $T_n = \emptyset$;

7      $C_n = \text{candidateGen}(A_{n-1})$;

8      $A_n = \{f | f \in C_n, F_k(f) \geq min\_sup\}$;

9      $T_n = \{f | f \in A_n, I_k(f) \geq min\_int\}$;

10     $n{+}{+}$;

11 $P_k = \bigcup\limits_{i=1}^{n-1} T_i$;

12 **return** P_k;

Lines 1-2 count the supports of all the items to determine the frequent items. Lines 3 stores the interesting items in T_1 (note that the interestingness of a single item is equal to its support). Lines 4-12 discover all interesting itemsets of different sizes n ($n \geq max\_size \geq 2$). First, the already discovered frequent itemsets of size $n - 1$ (A_{n-1}) are used to generate the candidate itemsets C_n using the candidateGen function (line 7). The candidateGen function is similar to the function Apriori-gen in the Apriori algorithm [1]. In line 8, we store the frequent itemsets from C_n into A_n. Line 9 stores the interesting itemsets (as defined in Section 3) from A_n into T_n. The final set of all interesting itemsets in S_k is stored in P_k and produced as output.

The time cost of generating candidates is equal to that of Apriori. We will now analyse the time needed to evaluate each candidate. We denote the time needed for computing the interestingness of a frequent itemset f with $T_{I_k(f)}$. To get $I_k(f)$, we first need to find a minimal interval $W(f, s)$ of an itemset f in a sequence $s \in S_k$, whereby the crucial step is the computation of the candidate intervals $W(f, t_i)$ for the time stamps t_i at which an item of f occurs. In our implementation, we keep the set of candidate intervals associated with f in a list. To find the candidate interval around position t_i containing all items of f, we start by looking for the nearest occurrences of items of f both left and right of position t_i. We then start reading from the side on which the furthest element is closest to t_i and continue by removing one item at a time and adding the same item from the other side. This process can stop when the interval on the other side has grown sufficiently to make it impossible to improve on the minimal interval we have found so far. When we have found this minimal interval, we compare it to the smallest interval found so far in s, and we update this value if the new interval is smaller. This process can stop if we get a minimal interval which equals to $|f|$, and then $W(f, s) = |f|$. Otherwise, $W(f, s)$ equals to the smallest value in the list of candidate intervals.

Theoretically, in the worst case, the number of candidate intervals that need to be found can be equal to the length of sequence s, $|s|$. To find a candidate interval, we might need to read the whole sequence both to the left and to the right of the item. Therefore, the time to get a $W(f, t_i)$ is $O(|s|)$. So, $T_{I_k(f)}$ is $O(|s|^2)$. However, this worst case only materialises if we are computing $I_k(f)$ when f is composed of all items that appear in s, and even then only if item appearing at each end of s do not appear anywhere else.

4.2   Pruning the Rules

Once we have found all interesting itemsets in a given class, all confident rules can be found in a trivial step. However, the number of interesting itemsets is typically very large, which leads to a large amount of rules. Reducing the number of rules is crucial to eliminate noise which could affect the accuracy of the classifier, and to improve the runtime of the algorithm.

We therefore try to find a subset of rules of high quality to build an efficient and effective classifier. To do so, we use the idea introduced in CMAR [9], and prune unnecessary rules by the database coverage method.

Before using the database coverage method, we must first define a total order on the generated rules R including all the rules from every class. This is used in selecting the rules for our classifier.

Definition 1. *Given two rules in R, r_i and r_j, $r_i \succ r_j$ (also called r_i precedes r_j or r_i has a higher precedence than r_j) if:*

1. the confidence of r_i is greater than that of r_j, or

2. their confidences are the same, but the interestingness of r_i is greater than that of r_j, or

3. both the confidences and interestingnesses of r_i and r_j are the same, but the size of r_i is greater than that of r_j

4. all of the three parameters are the same, r_i is generated earlier than r_j.

We apply the database coverage method to get the most significant subset of rules. The main idea of the method is that if a rule matches a data object that has already been matched by a high enough number of higher ranked rules (this number is defined by a user chosen parameter δ, or the coverage threshold), this rule would contribute nothing to the classifier (with respect to this data object). The algorithm for getting this subset is described in Algorithm 2. The algorithm has 2 main steps. First, we sort the set of confident rules R according to definition 1 (line 1). This makes it faster to get good rules for classifying. Then, in lines 2-13, we prune the rules using the database coverage method. For each rule r in sorted R, we go through the dataset D to find all the data objects correctly classified by r and increase the cover counts of those data objects (lines 3-7). We mark r if it correctly classifies a data object (line 8). If the cover count of a data object passes the coverage threshold, its id will be stored into $temp$ (line 9). Finally, if r is marked, we store it into PR and remove those data objects whose ids are in $temp$ (lines 10-13). Line 14 returns the new set of rules PR. In the worst case, to check whether a data object is correctly classified by r, we might need to read the whole sequence part s of the data object, resulting in a time complexity of $O(|s|)$.

Algorithm 2. PRUNINGRULES. An algorithm for finding the most significant subset among the generated rules.

---

input   : training dataset D, a set of confident rules R, coverage threshold δ
output : a new set of rules PR

1 sort R according to Definition 1;
2 **foreach** data object d in D **do** $d.cover\_count = 0$;
3 **foreach** rule r in sorted R **do**
4 $\quad$ $temp = \emptyset$;
5 $\quad$ **foreach** data object d in D **do**
6 $\quad\quad$ **if** rule r correctly classifies data object d **then**
7 $\quad\quad\quad$ $d.cover\_count + +$;
8 $\quad\quad\quad$ mark r;
9 $\quad\quad\quad$ **if** $d.cover\_count >= \delta$ **then** store $d.id$ in $temp$;
10 $\quad$ **if** r is marked **then**
11 $\quad\quad$ select r and store it into PR;
12 $\quad\quad$ **foreach** data object d in D **do**
13 $\quad\quad\quad$ **if** $d.id \in temp$ **then** delete d from D;

14 **return** PR;

4.3   Building the Classifiers

Based on the generated rules, we now propose two different ways to build a classifier, SCII_CBA and SCII_MATCH.

SCII_CBA. We build a classifier using the rules we discovered after pruning based on the CBA-CB algorithm (the classifier builder part of CBA [10]). In other words, we use rules generated in section 4.2 instead of using all the rules before pruning as the input for CBA-CB. We can also skip the step of sorting rules in CBA-CB because we have already sorted them in the pruning phase. However, the total order for generated rules defined in CBA-CB is different from that given in Definition 1. We use interestingness instead of support and, if the confidence and the interestingness of two rules are equal, we consider the larger rule to be more valuable than the smaller rule.

After building the classifier using the CBA-CB method, the classifier is of the following format: $< r_1, r_2, \ldots, r_n, default\_class >$, where $r_i \in R$, $r_a \succ r_b$ if $a < b$, and $default\_class$ is the default class produced by CBA-CB. When classifying a new data object, the first rule that matches the data object will classify it. This means that the new data object will be classified into a class which the consequent of this rule stands for. If there is no rule that matches the data object, it is classified into the default class.

SCII_MATCH. Rather than ranking the rules using their confidence, interestingness and size, we now propose incorporating the cohesion of the antecedent of a rule in the new data object into the measure of the appropriateness of the rule for classifying the object. Obviously, we cannot entirely ignore the confidence of the rules. Therefore, we will first find all rules that match the new object, and then compute the product of the rule's confidence and the antecedent's cohesion in the new data object. We then use this new measure to rank the rules, and classify the object using the highest ranked rule.

Considering there may not exist a rule matching the given data object, we must also add a default rule, of the form $null \Rightarrow L_d$, to the classifier. If there is no rule that matches the given data object, the default rule will be used to classify the data object. To find the default rule, we first delete the data objects matched by the rules in PR. Then we count how many times each class label appears in the remainder of the dataset. Finally we set the label that appears the most times as the default class label L_d. If multiple class labels appear the most times, we choose the first one as the default class label. So the default rule $default\_r$ is $null \Rightarrow L_d$. In the worst case, to check whether a data object is matched by rule r in PR, we might need to read the whole sequence part s of the data object. Since we need to do this for all data objects and all rules, the time complexity of finding the default rule is $O(|PR| \sum_{s \in D} |s|)$.

The classifier is thus composed of PR and the default rule $default\_r$. We now show how we select a rule for classifying the sequence in a new data object. The algorithm for finding the rule used to classify a new data object is shown in Algorithm 3.

First, we find all the rules that match the given data object d and store them into MR (lines 1). Then, we handle three different cases:

1. (lines 2-7): If the size of MR is greater than 1, we go through MR to compute the cohesion of each rule in MR with respect to the given data object.

Algorithm 3. CLASSIFYINGRULE. An algorithm for finding the rule used to classify a new sequence.

---

    input   : PR and $default\_r$, a new unclassified data object $d = (s, L_?)$
    output : the classifying rule r_c
1 $MR = \{r \in PR \,|\, r$ matches $d\}$;
2 **if** $MR.size > 1$ **then**
3   **foreach** rule $r : p \Rightarrow L_k$ in MR **do**
4   **if** $r.length > 1$ **then** $r.measure = r.confidence * C(p, d.s)$;
5   **else** $r.measure = r.confidence$;
6   sort rules in MR in descending order by $r.measure$;
7   **return** the first rule in sorted MR;
8 **else**
9   **if** $MR.size == 1$ **then return** the only rule in MR;
10   **else return** $default\_r$;

Let us go back to the cohesion defined in section 3. We use the antecedent of a rule to take the place of itemset X to compute the cohesion of a rule. We then compute the value of every rule in MR (the product of the rule's confidence and the antecedent's cohesion), and sort the rules according to their value (the higher the value, the higher the precedence). We then utilize the first rule in the sorted MR to classify the given data object.

2. (line 9): If the size of MR is 1, then we classify the sequence using the only rule in MR.

3. (line 10): If there is no rule in MR, then we use the default rule to classify the given data object.

The only time-consuming part of Algorithm 3 is the computation of $C(p, d.s)$. The time complexity of this computation has already been analysed at the end of Section 4.1.

4.4 Example

To illustrate how our methods work, we will use a toy example. Consider the training dataset consisting of the data objects (sequences and class labels) given in Table 1. We can see that itemset $abcd$ exists in all sequences regardless of class. It is therefore hard to distinguish the sequences from different classes using the traditional frequent itemset methods. We now explain how our approach works.

Using the definitions given in Section 3 and Algorithm 1, assume $min\_sup = min\_int = \frac{2}{3}$, $max\_size = 4$ and make the sequences of class 1 as input S_1 in Algorithm 1. First, we discover frequent itemsets in S_1, which turn out to be itemset $abcd$ and all its subsets. Then we generate interesting itemsets from frequent itemsets, and find itemsets ab, a, b, c and d, whose interestingness is equal to 1. Meanwhile, in the sequences of class 2, S_2, itemsets bcd, cd a, b, c and d are interesting. If we now set $min\_conf = 0.5$, we get the confident rules sorted using Definition 1, as shown in Table 2.

Table 1. An example of a sequence dataset

ID	Sequence	Class Label	ID	Sequence	Class Label
1	$c\,c\,x\,y\,a\,b\,d$	class1	5	$a\,d\,z\,z\,c\,d\,b$	class2
2	$a\,b\,e\,e\,x\,x\,e\,c\,f\,d$	class1	6	$b\,x\,y\,d\,d\,c\,d\,d\,d\,x\,a$	class2
3	$c\,g\,h\,a\,b\,d\,d$	class1	7	$b\,d\,c\,c\,c\,c\,a\,y$	class2
4	$d\,d\,e\,c\,f\,b\,a$	class1	8	$a\,x\,x\,c\,d\,b$	class2

Table 2. Sorted rules from the example

Rule	Cohesion	Confidence	Rule	Cohesion	Confidence
$a\,b \Rightarrow Class1$	1.0	0.5	$c \Rightarrow Class2$	1.0	0.5
$c\,d \Rightarrow Class2$	1.0	0.5	$a \Rightarrow Class2$	1.0	0.5
$c \Rightarrow Class1$	1.0	0.5	$b \Rightarrow Class2$	1.0	0.5
$a \Rightarrow Class1$	1.0	0.5	$d \Rightarrow Class2$	1.0	0.5
$b \Rightarrow Class1$	1.0	0.5	$c\,b\,d \Rightarrow Class2$	0.8	0.5
$d \Rightarrow Class1$	1.0	0.5	$b\,d \Rightarrow Class2$	0.8	0.5

Given a new input sequence $s_9 = a\,x\,b\,y\,c\,d\,z$, we can see that it is not easy to choose the correct classification rule, as all rules match the input sequence, and only the last two score lower than the rest. The first two rules are ranked higher due to the size of the antecedent, but the CBA, CMAR and SCII_CBA methods would have no means to distinguish between the two rules, and would classify s_9 into class 1, simply because rule $a\,b \Rightarrow Class1$ was generated before rule $c\,d \Rightarrow Class2$. Using the SCII_MATCH method, however, we would re-rank the rules taking the cohesion of the antecedent in s_9 into account. In the end, rule $c\,d \Rightarrow Class2$ is chosen, as $C(cd, s_9) = 1$, while $C(ab, s_9) = \frac{2}{3}$. The cohesion of all antecedents of size 1 in s_9 would also be equal to 1, but rule $c\,d \Rightarrow Class2$ would rank higher due to its size. We see that the SCII_MATCH method classifies the new sequence correctly, while other methods fail to do so.

5   Experiments

We compared our classifiers SCII_CBA and SCII_MATCH with five classifiers: CBA, CMAR, BayesFM [8] and CBS [13]. The CBS paper proposes a number of different scoring functions [13], and we chose the length policy as it gave the best results. For better comparison, we also added a *max_size* constraint into the pattern mining stage of CBS. Our methods, BayesFM and CBS are implemented in Java of Eclipse IDE, while CBA and CMAR are implemented in LUCS-KDD Software Library[1]. We use SPADE [16] to mine subsequential patterns for BayesFM and CBS and we transform the sequence dataset into a transaction dataset for CBA and CMAR. All experiments are performed on a

---

[1] http://cgi.csc.liv.ac.uk/~frans/KDD/Software/

laptop computer with Intel i7 (2 CPUs 2.7GHz), 4GB memory and Windows 7 Professional.

In order to evaluate the proposed methods, we used four real-life datasets. Three of these datasets were formed by making a selection from the Reuters-21578 dataset[2], consisting of news stories, assembled and indexed with categories by Reuters Ltd personnel. We consider the words appearing in the texts as items and treat each paragraph as a sequence. We formed the *Reuters1* dataset using the two biggest classes in the Reuters-21578 dataset, "acq" (1596 paragraphs) and "earn" (2840 paragraphs). *Reuters2* consists of the four biggest classes in Reuters-21578, "acq", "earn", "crude" (253 paragraphs) and "trade" (251 paragraphs), and is therefore an imbalanced dataset. *Reuters3* is a balanced dataset obtained from *Reuters2* by keeping only the first 253 paragraphs in the top two classes. *Reuters1* consists of 4436 sequences composed of 11947 different items, *Reuters2* of 4940 sequences containing 13532 distinct items, and *Reuters3* of 1010 sequences composed of 6380 different items.

Our fourth dataset is a protein dataset obtained from PhosphoELM[3]. The data consists of different combinations of amino acids for each kind of protein. We chose two of the biggest protein groups (PKA with 362 combinations and SRC with 304 combinations) to form the *Protein* dataset. We treat each combination of amino acids as a sequence and consider each amino acid as an item. Each sequence is labelled by the protein group it belongs to. This dataset consists of 666 sequences containing 20 different items. All the reported accuracies in all of experiments were obtained using 10-fold cross-validation.

5.1   Analysis of the Predictive Accuracy

Table 3 reports the accuracy results of all six classifiers. In the experiments, we set min_conf to 0.6 and min_sup to 0.1 for all of the classifiers, while varying the max_size threshold. Additionally, we set min_int to 0.05 for the SCII classifiers. For the SCII methods and CMAR, the database coverage threshold was set to 3. The best result for each dataset is highlighted in bold. As shown in Table 3, the SCII algorithms generally outperform other classifiers.

To further explore the performance of the six classifiers, we conducted an analysis of the predictive accuracy under different support, confidence, and interestingness thresholds, respectively. We first experimented on *Reuters1* and *Protein*, with min_conf fixed at 0.6, max_size set to 3 and min_int for SCII classifiers set to 0.05. We can see in Fig. 1 that the SCII_C classifier is not sensitive to the minimum support thresholds as minimum interestingness threshold is the main parameter deciding the output rules. As the number of output rules drops, SCII_M begins to suffer, as it picks just the highest ranked rules to classify a new object. SCII_C compensates by using a combination of rules, and the accuracy therefore does not suddenly drop once some rules drop out of the classifier.

We then compared the predictive accuracy of the classifiers using different minimum confidence thresholds on the *Protein* dataset. We compare just four

---

[2] http://web.ist.utl.pt/~acardoso/datasets/r8-train-stemmed.txt
[3] http://phospho.elm.eu.org/

Table 3. Comparison of Predictive Accuracy (%)

Dataset	*max_size*	SCII_C	SCII_M	CBA	CMAR	BayesFM	CBS
Reuters1	2	**92.74**	92.54	67.05	65.76	92.38	89.27
	3	**92.74**	92.54	66.63	65.65	92.88	88.12
	4	**92.74**	92.54	66.63	65.54	92.79	88.25
	5	**92.74**	92.54	66.63	65.54	92.76	88.82
	∞	**92.74**	92.54	66.63	65.54	92.72	88.88
Protein	2	86.63	**87.56**	81.34	81.91	52.66	54.92
	3	87.69	**88.59**	86.55	80.88	72.97	74.94
	4	**91.01**	90.71	87.93	78.64	85.14	86.81
	5	90.71	**91.73**	89.14	78.94	85.14	86.96
	∞	90.56	**91.86**	89.14	78.94	85.14	86.96
Reuters2	2	**90.40**	90.16	57.69	57.45	83.68	78.87
	3	**90.51**	90.22	57.65	57.17	83.22	75.99
	4	**90.67**	90.28	57.65	57.09	82.94	74.31
	5	**90.75**	90.26	57.65	57.09	82.87	72.96
	∞	**90.61**	90.45	57.65	57.09	82.82	72.11
Reuters3	2	92.48	**92.97**	78.61	62.28	78.71	87.82
	3	92.28	**92.87**	78.71	62.38	74.16	88.22
	4	**92.87**	92.67	78.71	62.38	72.57	87.92
	5	92.77	**92.97**	78.71	62.38	72.18	87.52
	∞	**93.07**	92.77	78.71	62.38	71.98	86.73

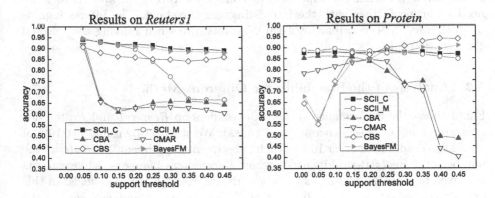

Fig. 1. The impact of varying the support threshold on various classifiers

classifiers of the classifiers, as BayesFM and CBS do not use a confidence threshold. Here, *min_sup* is fixed at 0.1, *max_size* is set to 3 and *min_int* for the SCII classifiers is set to 0.05. From Fig. 2, we can see that the SCII classifiers are not sensitive to the minimum confidence threshold at all. When the confidence threshold is not lower than 0.8, the accuracies of CBA and CMAR decline sharply. It shows the performance of CBA and CMAR is strongly related to the number of produced rules.

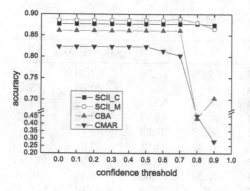 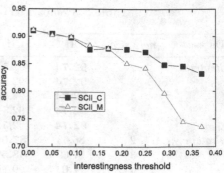

Fig. 2. The impact of varying the confidence threshold on various classifiers

Fig. 3. The impact of varying the interestingness threshold on the SCII classifiers

Fig. 3 shows the accuracy of the SCII classifiers on the *Reuters2* dataset with different minimum interestingness thresholds. Here, min_sup is fixed at 0.1, min_conf at 0.6, and max_size is set to 3. We can see that the accuracies of both SCII_C and SCII_M decrease when the minimum interestingness threshold increases. When the minimum interestingness threshold is greater than 0.2, fewer rules are discovered, and the accuracy of SCII_M, once again, declines faster than that of SCII_C. We can conclude that the selection of good classification rules is already done using the support and confidence threshold, and there is no need to prune further with the interestingness threshold. The interestingness of an itemset, however, remains a valuable measure when ranking the selected classification rules.

5.2   Analysis of the Scalability for Different Methods

Fig. 4 shows the performance of the six classifiers on *Reuters1* and *Protein* for a varying number of sequences (*#sequences*). We start off by using just 10% of the dataset, adding another 10% in each subsequent experiment. In *Reuters1* the number of items (*#items*) increases when *#sequences* increases, while *#items* is a fixed number in *Protein*, as there are always exactly 20 amino acids. In this experiment we set $min\_int = 0.01$ for SCII, $max\_size = 3$ and $min\_conf = 0.6$ for all methods.

The first two plots in Fig. 4 show the effect of an increasing dataset size on the run-times of all six algorithms. We began with a small subset of *Reuters1*, adding further sequences until we reached the full dataset. We plot the runtimes compared to the number of sequences, and the number of different items encountered in the sequences. For all six algorithms, the run-times grew similarly, with the classifiers based on sequential patterns the slowest, and the classifiers that took no sequential information into account the fastest.

The last two plots show the run-times of the algorithms on the *Protein* dataset. Here, too, we kept increasing the dataset size, but the number of items was

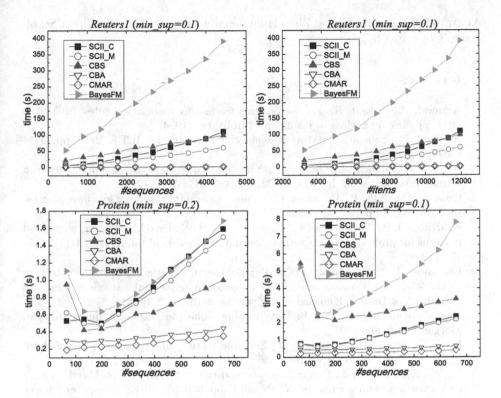

Fig. 4. Scalability analysis

always equal to 20. We performed the experiments with two different support thresholds, and it can be noted that classifiers based on sequential patterns are much more sensitive to the lowering of the threshold than our two classifiers. Once again, as expected, CMAR and CBA were fastest, but as was already seen in Table 3, their accuracy was unsatisfactory.

6   Conclusions

In this paper, we introduce a sequence classification method based on interesting itemsets named SCII with two variations. Through experimental evaluation, we confirm that the SCII methods provide higher classification accuracy compared to existing methods. The experimental results show that SCII is not sensitive to the setting of a minimum support threshold or a minimum confidence threshold. In addition, the SCII method is scalable as the runtime is proportional to the dataset size and the number of items in the dataset. Therefore, we can conclude that SCII is an effective and stable method for classifying sequence data. What is more, the output rules of SCII are easily readable and understandably represent the features of datasets.

Acknowledgement. Cheng Zhou is financially supported by the China Scholarship Council (CSC).

References

1. Agrawal, R., Srikant, R.: Fast algorithms for mining association rules. In: VLDB 1994, pp. 487–499. Morgan Kaufmann Publishers (1994)
2. Agrawal, R., Srikant, R.: Mining sequential patterns. In: ICDE 1995, pp. 3–14 (1995)
3. Ayres, J., Flannick, J., Gehrke, J., Yiu, T.: Sequential pattern mining using a bitmap representation. In: KDD 2002, pp. 429–435. ACM (2002)
4. Cule, B., Goethals, B., Robardet, C.: A new constraint for mining sets in sequences. In: SDM 2009, pp. 317–328 (2009)
5. Exarchos, T.P., Papaloukas, C., Lampros, C., Fotiadis, D.I.: Mining sequential patterns for protein fold recognition. Journal of Biomedical Informatics 41(1), 165–179 (2008)
6. Exarchos, T.P., Tsipouras, M.G., Papaloukas, C., Fotiadis, D.I.: A two-stage methodology for sequence classification based on sequential pattern mining and optimization. Data & Knowledge Engineering 66(3), 467–487 (2008)
7. Han, J., Kamber, M., Pei, J.: Data mining: concepts and techniques, 3rd edn. Morgan Kaufmann (2011)
8. Lesh, N., Zaki, M.J., Ogihara, M.: Scalable feature mining for sequential data. IEEE Intelligent Systems 15(2), 48–56 (2000)
9. Li, W., Han, J., Pei, J.: Cmar: Accurate and efficient classification based on multiple class-association rules. In: ICDM 2001, pp. 369–376. IEEE Computer Society (2001)
10. Liu, B., Hsu, W., Ma, Y.: Integrating classification and association rule mining. In: KDD 1998, pp. 80–86 (1998)
11. Pei, J., Han, J., Mortazavi-Asl, B., Wang, J., Pinto, H., Chen, Q., Dayal, U., Hsu, M.C.: Mining sequential patterns by pattern-growth: The prefixspan approach. IEEE Transactions on Knowledge and Data Engineering 16(11), 1424–1440 (2004)
12. Srikant, R., Agrawal, R.: Mining sequential patterns: Generalizations and performance improvements. In: Apers, P.M.G., Bouzeghoub, M., Gardarin, G. (eds.) EDBT 1996. LNCS, vol. 1057, pp. 3–17. Springer, Heidelberg (1996)
13. Tseng, V.S., Lee, C.H.: Effective temporal data classification by integrating sequential pattern mining and probabilistic induction. Expert Systems with Applications 36(5), 9524–9532 (2009)
14. Tseng, V.S.M., Lee, C.H.: Cbs: A new classification method by using sequential patterns. In: SDM 2005, pp. 596–600 (2005)
15. Yin, X., Han, J.: Cpar: Classification based on predictive association rules. In: SDM 2003, pp. 331–335 (2003)
16. Zaki, M.J.: Spade: An efficient algorithm for mining frequent sequences. Machine Learning 42(1-2), 31–60 (2001)
17. Zhao, Y., Zhang, H., Wu, S., Pei, J., Cao, L., Zhang, C., Bohlscheid, H.: Debt detection in social security by sequence classification using both positive and negative patterns. In: Buntine, W., Grobelnik, M., Mladenić, D., Shawe-Taylor, J. (eds.) ECML PKDD 2009, Part II. LNCS, vol. 5782, pp. 648–663. Springer, Heidelberg (2009)

A Relevance Criterion for Sequential Patterns

Henrik Grosskreutz, Bastian Lang, and Daniel Trabold

Fraunhofer IAIS, Schloss Birlinghoven, 53754 St. Augustin, Germany
{henrik.grosskreutz,bastian.lang,daniel.trabold}@iais.fraunhofer.de

Abstract. The theory of relevance is an approach for redundancy avoidance in labeled itemset mining. In this paper, we adapt this theory to the setting of sequential patterns. While in the itemset setting it is suggestive to use the closed patterns as representatives for the relevant patterns, we argue that due to different properties of the space of sequential patterns, it is preferable to use the minimal generator sequences as representatives, instead of the closed sequences. Thereafter, we show that we can efficiently compute the relevant sequences via the minimal generators in the negatives. Unlike existing iterative or post-processing approaches for pattern subset selection, our approach thus results both in a reduction of the set of patterns and in a reduction of the search space – and hence in lower computational costs.

1 Introduction

Sequential pattern mining is concerned with finding frequent subsequences in sequence databases [1]. These subsequences, or sequential patterns, have many real-world applications. For example, they can be used to characterize sequences of credit card transactions having high fraud probability, or DNA subsequences having particular properties.

Like most pattern mining tasks, sequential pattern mining suffers from the problem that it mostly comes up with huge amounts of patterns. This problem is well-know in the pattern mining community, and various approaches have been proposed to cope with this issue (e.g. [2–5]).

In this paper, we take one particular approach, namely the *theory of relevance* [6, 7], and adapt it to the case of sequential data. Originally, the theory of relevance was developed in the setting of *labeled itemset data*, and assumes that one is interested in characterizing a particular *target class*. The basic idea is related to the concept of Pareto domination: remove all itemsets which are *dominated* by some other itemset, meaning that the dominated itemset is strictly inferior in characterizing the target class. More precisely, an itemset is considered as dominated if there is another, dominating itemset which supports at least all *positives* (i.e. target-class sequences) supported by the dominated itemset, but no additional *negative* (i.e. non target-class sequences).

The theory of relevance not only reduces the size of the resulting set of itemsets, but also allows for efficient algorithms. Unlike iterative or post-processing approaches, the relevant itemsets can be collected by traversing, once only, a

H. Blockeel et al. (Eds.): ECML PKDD 2013, Part I, LNAI 8188, pp. 369–384, 2013.

small subset of all itemsets. The foundation for these algorithms is a set of properties that relate the relevant itemsets to the itemsets that are closed in a particular subset of the data, namely the positives.

The adaptation of the theory of relevance to sequential data raises interesting challenges. These are due to the different characteristics of the space of sequential patterns, compared with the space of itemsets. One well-known difference, illustrated in Figure 1, is that the closed sequences are no *unique* representatives of their equivalence class. This is unlike in the itemset setting, where the closed itemsets are used as unique representatives for the relevant itemsets [8]. The different characteristics of sequential patterns makes the use of closed (sequential) patterns much less suggestive for this new setting. As we will show, there are other important differences (for example, the relevance of a sequence cannot be checked by considering its generalizations, unlike in the setting of itemsets). Altogether, we make the following contributions:

label	sequence
+	abc
+	acb
-	b
-	c

Fig. 1. In this sequence database, the sequences a, ab and ac have the same support set: while the first sequence is a (minimal) generator, the other two sequences are closed. Applying the idea of domination to this example, we see that all sequences but the three above-mentioned are dominated: for example, the sequence b is dominated by a, because b supports a superset of the negatives supported by a but the same set of positives. In our approach, only the minimal generator a will be kept.

- We show that if the concept of domination is transfered from itemsets to sequences, several important properties no longer hold. As a consequence, the standard algorithmic approach cannot be applied to find the relevant sequences (Section 4.2);
- We propose to use generators as representatives for the relevant sequences, instead of closed sequences. Besides the obvious advantage of shorter descriptions, this allows dealing efficiently with maximum pattern length constraints (Section 4.3);
- We show that our new definition of relevance has the consequence that the relevant sequences are a subset of the minimal generator sequences in the negatives (Section 5.1);
- Subsequently, we describe how this connection can be turned into an efficient algorithm (Section 5.2);
- Finally, we experimentally investigate the impact of our new relevance criterion on the number of patterns and the computational costs (Section 6).

The rest of this paper is structured as follows: After discussing related work in Section 2 and introducing our notation in Section 3, we present the contributions listed above, before we conclude in Section 7.

2   Related Work

Sequential pattern mining was first considered by Agrawal and Srikant [1]. The notions closed pattern, minimal generator pattern etc. have been transfered from itemset data to the setting of sequences, and different algorithms have been proposed to find the closed sequences [9, 10], respectively the minimal generator sequences [11, 12].

While the use of closed patterns resp. minimal generators reduces the number of patterns, the outcome can still be huge. A variety of pattern selection approaches have been proposed to cope with this issue. Most of these approaches are either post-processing or iterative solutions. The post-processing approaches expect, as input, a set of patterns, from which they choose a subset, typically in a greedy fashion [2, 4]. The iterative approaches, on the other hand, run a new search in every iteration; The different runs assess the pattern quality differently, taking into account the set of patterns already collected [5, 13–15]. As all these approaches rely, somehow, on an underlying mining algorithm, they are no alternative to our approach but could, instead, be combined with our approach: that is, for labeled sequential data they could rely on our algorithm to enumerate the candidate patterns.

Other approaches exist that reduce the set of patters by relying strongly on the properties and operations that can be performed on itemset data [3, 16]. As these operations are not directly applicable to sequences, there is no easy way to transfer these approaches to the setting of sequential patterns (Note that the theory of relevance considered here falls into this category of approaches).

Different approaches have been proposed to define a closure operator in a sequential data setting. However, none of these approaches is directly applicable to our setting, as they all consider different patterns families, which are only connected via some post-processing to the (classical) sequential patterns we consider. In particular, Garriga has proposed a new closure operator, which however is not defined on individual sequential patterns, but on *sets* of sequential patterns. While this approach allows for advanced summarization [17], it relies on a classical closed sequential pattern miner to produce the patterns to be post-processed respectively summarized. Raïssi et al. [18] presents a similar approach, which also considers sets of sequences instead of individual sequences. Finally, Tatti et al. [19] proposed a new notion of closedness, called *i-closed*. Unlike us, they don't consider sequence databases but consider the episode mining setting (where frequency is defined in terms of sliding windows over a single sequence) and consider patterns taking the form of directed acyclic graphs. Above all, the computation of i-closed episodes is only the first step: the i-closed episodes are a *superset* of the classical closed episodes, from which the closed episodes must then be computed in a second step.

3   Preliminaries

In this section, we review the standard notions from itemset and sequential pattern mining, which will then be used in the remainder of this paper.

3.1   Itemsets, Closed Patterns and Minimal Generators

As the theory of relevance has been defined in the scope of itemsets, we will first review the notions from itemset and closed pattern mining [20].

Itemsets and Itemset Databases. An *itemset* over an alphabet Σ is a subset of Σ. A labeled dataset $\mathcal{DB}$ over an alphabet Σ is a collection of records (l, I), where l is a label and I is an itemset. Given a database $\mathcal{DB}$ and an itemset P, the *support set* of P in a dataset $\mathcal{DB}$, denoted by $\mathcal{DB}[P]$, is defined as the set of records $r = (l, I) \in \mathcal{DB}$ such that P is a subset of I. The *support* of an itemset is the size of its support set.

Positives, Negatives, True Positives etc. In the rest of this paper we assume a binary setting where the set of classes consists of "+" and "-". We call the subset of "+"-labeled records the *positives*. Similarly, we call the "-"-labeled records the *negatives*. The term *true positives*, denoted by $TP(\mathcal{DB}, P)$, refers to the support set of P in the positives. The *false positives*, $FP(\mathcal{DB}, P)$, are defined analogously on the negatives.

Equivalence Classes, Closed Itemsets and Generators. If two itemsets have the same support set, then the two are said to be *equivalent*. The space of itemsets can thus be partitioned into *equivalence classes*: all itemsets with the same support set belong to the same equivalence class. Within an equivalence class, there are two interesting subsets of itemsets: the *minimal generators* and the *closed* itemsets. The minimal generators are the minimal members of an equivalence class, meaning that any true generalization (i.e. sub-itemset) has a strictly higher support in the dataset. The closed itemsets are their counterpart: they are maximal members of the equivalence class, meaning that any true specialization (i.e. super-itemset) has strictly lower support.

3.2   The Theory of Relevance

We will now turn to the theory of relevance [6–8].

Domination and Relevance. The basic idea of the theory of relevance is to reduce the number of itemsets by removing itemsets that are irrelevant for the purpose of characterizing the *target class*, which by convention is the "+" class. An itemset is considered to be irrelevant if there is another itemset, called the *dominating* itemset, which allows characterizing the target class at least as good as the former (dominated) itemset. Formally:

Definition 1. *The itemset P dominates the itemset P_{irr} in dataset $\mathcal{DB}$ iff (i) $TP(\mathcal{DB}, P) \supseteq TP(\mathcal{DB}, P_{irr})$ and (ii) $FP(\mathcal{DB}, P) \subseteq FP(\mathcal{DB}, P_{irr})$.*

Note that it is possible that two itemsets dominate each other, however only in the case that they belong to the same equivalence class. Given that for itemsets there is exactly one closed itemset in every equivalence class, the closed itemsets can be used as unique representatives. Garriga et al. [8] thus define the relevant itemsets as follows:

Definition 2. *Itemset P is relevant in database DB iff (i) P is closed and (ii) there is no itemset having a different support set that dominates P (in DB).*

The Connection to Closed-on-the-Positives Garriga et al. [8], have shown that when searching for relevant itemsets, it is sufficient to consider only itemsets that are *closed on the positives*, that is, itemsets that are closed in the subset of the positively labeled records:

Proposition 1 ([8]). *The space of relevant itemsets consists of all itemsets P_{rel} satisfying the following:*

- *P_{rel} is closed on the positives, and*
- *there is no generalization $P \subsetneq P_{rel}$ closed on the positives such that $|FP(DB, P)| = |FP(DB, P_{rel})|$.*

The above proposition provides an elegant way to compute the relevant itemsets, sketched in Algorithm 1.

Algorithm 1. CPOS Relevant Itemset Miner

Input   : an itemset database DB
Output : the relevant itemsets in DB

1: collect all closed-on-the-positive itemsets (using some closed itemset mining algorithm, e.g. [21]).
2: remove all itemsets having a (closed-on-the-positives) generalization with the same negative support.

3.3   Sequences and Sequence Databases

We will now review the most important notions from sequence mining [1].

Sequences and Sequence Databases. A *sequence* over a set of items Σ is a sequence of items $i_1, \ldots, i_l$, $i_i \in \Sigma$. The *length* of the sequence is the number of items in the sequence. A sequence $S_a = a_1, \ldots, a_n$ is said to be *contained* in another $S_b = b_1, \ldots b_m$, denoted by $S_a \sqsubseteq S_b$, if $\exists i_1, \ldots i_n$ such that $1 \le i_1 < \cdots < i_n \le m$ and $a_1 = b_{i_1}, \ldots, a_n = b_{i_n}$. We also call S_a a *generalization* of S_b.

A *sequence database* SDB is a collection of labeled sequences. A *labeled sequence* is a tuple (l, S), where S is a sequence and l a label – i.e, "+" or "-". Again, we call the subset of "+"-labeled sequences the *positives*. Similarly we call the "-"-labeled sequences the *negatives*.

The *support set* of a sequence S in a sequence database SDB, denoted by $SDB[S]$, consists of all labeled sequences in SDB that contain S. Here, a labeled sequence (l, S) contains a sequence S_a iff $S_a \sqsubseteq S$. Again, the term *true positives*, denoted by $TP(SDB, S)$, refers to the support set of S in SDB's positives. $FP(SDB, S)$ is defined analogously on the negatives. Finally, the *support* denotes the size of the support set.

Patterns, Closed Patterns and Minimal Generators. In the rest of this paper, we will use the general term *pattern* to refer to either an itemset or a sequence. In general, patterns have a support set (wrt. a given database), and moreover there is a partial generalization order between patterns (defined via the subset relation for itemsets, resp. the *contained* relation for sequences).

Based upon these generalized definitions of support set and generalization, the terms *closed* and *minimal generator* from Section 3.1 can be carried over to sequences, and can hence be applied to both types of patterns.

4   Relevant Sequences

We will now adapt the definition of relevance to sequential data. While is is straightforward to transfer the concept of domination to sequential patterns, defining relevance will raise subtle issues.

4.1   Domination between Sequences

Unlike the original definition (Definition 1), our definition of domination between sequential patterns explicitly distinguishes between *weak* and *strong* domination. This will be useful in situations where two different patterns dominate each other circularly (in the original definition).

Definition 3. *The sequence S_d weakly dominates the sequence S iff*

- $TP(\mathcal{SDB}, S_d) \supseteq TP(\mathcal{SDB}, S)$, *and*
- $FP(\mathcal{SDB}, S_d) \subseteq FP(\mathcal{SDB}, S)$.

Moreover, S_d strongly dominates S iff S_d weakly dominates S and $\mathcal{SDB}[S_d] \neq \mathcal{SDB}[S]$.

4.2   Relevant Sequences: Problems and Differences to the Itemset Setting

While we directly carried over the definition of *domination* to sequential patterns, our proposed definition of *relevant sequences* will differ from the definition used in the setting of itemsets. This is due do the fact that several properties that hold in the space of itemsets do not transfer to the space of sequences.

One main issue is the choice of representatives for the patterns that are not strongly dominated. In the itemset setting, Garriga et al. chose to use the closed itemsets as representatives. In the itemset setting, this is very suggestive: it provides unique representatives and allows for efficient computation. In this section, we will argue that in the sequential setting, the use of closed patterns as representatives is much less appealing. Beside the issue that there can be several equivalent closed sequences (as illustrated in the introduction), the use of closed patterns as representatives results in the following issues:

1. The computational approach proposed by Garriga et al. is not applicable, because Proposition 1 does not carry over to sequential patterns;
2. The use of a length limit is problematic, resulting in counter-intuitive results and/or excessive computational costs. In practice, however, specifying a limit for the length of the patterns to be considered is very useful: it allows reducing the computational costs to a reasonable amount of time, and is often more suitable than using a minimum support threshold.

We will now discuss these issues in detail.

Garriga's Computational Approach Is Not Applicable to Sequences.
Proposition 1 is the foundation for many fast relevant itemset mining algorithms [8, 22]. We will now show that it does not carry over to sequential patterns:

Proposition 2. *There is a sequence database such that a closed-on-the-positives sequence pattern S exists which is strongly dominated, yet not dominated by any of its generalizations.*

The correctness of the above proposition is shown by the example in Table 1. Here, the sequence c is closed on the positives. It is, however, dominated, namely by ab. Yet, c is not dominated by any generalization of itself.

Table 1. Example: the closed sequence c is strongly dominated (e.g. by a), yet it is not dominated by any generalization

label	sequence
+	cab
+	abc
-	c
-	d

The above proposition shows that it is not sufficient to consider *generalizations* to verify the relevance of a sequence. While the above example alone shows that Proposition 1 does not hold, we could still hope that testing for relevance is possible by comparing only other patterns with same negative support (as in Proposition 1). However, this also does not carry over:

Proposition 3. *There is a sequence database such that a closed-on-the-positives sequence S exists which is strongly dominated, yet it is not strongly dominated by any sequence having the same negative support.*

Again, this is illustrated by Example 1. c is closed on the positives and is strongly dominated. However, all strongly dominating sequences (a, b, and ab) have a different negative support.

The above two propositions show that the second step of Algorithm 1 cannot be adapted to the sequential pattern setting: neither can relevance be tested by considering only generalizations; nor is it possible to consider only patterns with same negative support.

Problems with Length Limits. We will now turn to the issues that arise if a length limit is introduced and closed patterns are used as representatives for the relevant sequences (Issue 2). Here, instead of considering the space of all sequences, we are only concerned with the space of sequences satisfying the length limit. We wish to remove all sequences that are strongly dominated, and to keep only a set of representatives for the remaining sequences.

Again, the example from Table 1 illustrates the problems that arise if closed patterns are used as representatives: assume that we are searching for relevant sequences with a maximum length limit of 1. Then:

- c is dominated, namely by the patterns a, b and ab. It should thus not be in the result set, because it is dominated by some pattern satisfying the length limit.
- c is, however, not dominated by any closed pattern satisfying the length limit (a and b are not closed). Checking domination would hence require a computationally much more expensive approach, for example considering *all* sequences up to the length limit, not only closed sequences.
- a should not be in the result because it is not closed; same for b. However, ab, which is closed and lies in the same equivalence class as the earlier two sequences, has a too long description. The result is that there is no representative in the result set for this equivalence class. This is somewhat counter-intuitive.

While it might be possible to ensure efficient computation by using a different, computationally-motivated definition of relevance wrt. a length limit, this is likely to result in awkward and unintuitive results.

4.3   The Relevant Sequences

As we have seen in the previous section, the closed sequences are not a particularly suggestive set of representatives for the relevant sequences. Therefore, we propose to use a different set of patterns as representatives: the *minimal generator sequences*:

Definition 4. *Given a sequence database SDB and a length limit L, the set of relevant minimal generator sequences (wrt. SDB and L) consists of all sequences S that satisfy the following:*

1. *S is a minimal generator in SDB of length $\leq L$,*
2. *S is not strongly dominated (in SDB) by any other sequence of length $\leq L$.*

In the following, the database and length limit will be clear from the context, so they will not be explicitly listed. Moreover, if no length limit is given, this is handled as if $L = \infty$. Finally, we will use the expression *relevant sequence* to refer to an element of the set of relevant minimal generator sequences.

Using minimal generators as representatives has several advantages. First, it produces shorter descriptions, which can be an important advantage if the patterns are to be read and interpreted by human experts; second, it allows for efficient computation via the minimal-generators-in-the-negatives, as we will describe in Section 5; and finally it allows for maximum length constraints with clear and simple semantics:

Proposition 4. *Let SDB be a sequence database, L a positive integer and S some sequence of length $\leq L$. Then, there is a relevant sequence S^* in SDB such that S^* is of length $\leq L$ and S^* weakly dominates S.*

Hence, for every sequence S satisfying the length limit there is a relevant sequence as good as S in characterizing the target class.

Proof. Let S_G be (one of the) minimal generator of S. Obviously, S_g weakly dominates S and satisfies the length limit. If S is a relevant sequence, then $S^* = S_g$ and we're finished. Else, S_g must be strongly dominated. As strong domination is transitive, non-reflexive and the set of minimal generators satisfying the length constraint is finite, there must be (at least one) minimal generator S^* that (i) satisfies the length constraint, (ii) strongly dominates S_g and (iii) is not dominated by any other minimal generator of length $\leq L$.

It remains to show that this pattern S^* dominates S_g and that it is a relevant sequence. The first fact follows by transitivity of weak domination. Concerning the second fact, S^* is a minimal generator and satisfies the length constraint by construction. It remains to show that it is not strongly dominated, which we show by contradiction. Assume it is dominated by a sequence of length $\leq L$, then it would also be dominated by the minimal generators of that pattern. Contradiction with (iii) above. $\qquad\square$

5   Computing the Relevant Sequences

We will now present a new approach that allows computing the relevant sequences much more efficiently than by simply checking, for every pair of patterns, the dominance criterion from Definition 3.

5.1   Relevant Sequences and Minimal Generators in the Negatives

This approach is based on the observation that the set of patterns not-strongly-dominated is not only related to the closed-on-the-positives (as investigated by Garriga et al.), but also to their counterpart: namely to the *minimal generators* in the *negatives*.

Proposition 5. *Let SDB be a sequence database and S some relevant sequence in SDB. Then S is a minimal generator in SDB's negatives.*

Proof. By contradiction. Assume that S is a relevant sequence but is no minimal generator in the negatives. The latter implies that there is a generalization S' of S with same support in the negatives. Thus, we have that $FP(SDB, S) \supseteq$

$FP(\mathcal{SDB}, S')$. Just as in the case of classical itemsets, for sequential patterns we have the property that the support is anti-monotonic. That is, the support set of S' in the positives is a superset of the support set of S in the positives. Thus, we also have $TP(\mathcal{SDB}, S) \subseteq TP(\mathcal{SDB}, S')$. The above implies that S' weakly dominates S. Moreover, by the assumption that S is a relevant sequence together with Definition 4, we have that S is a minimal generator, hence S and S' have different support sets. Hence, S' strongly dominates S – which contradicts the assumption that S is relevant.   □

Please note that in the above proposition, unlike in the work of Garriga et al. we consider a different pattern type (minimal generators instead of closed patterns) but also a different subset of the data (negatives instead of positives).

label	sequence
+	ab
+	abc
-	abc
-	ac
-	c

(a) dataset

seq.	dominated	closed	gen	g-neg
a	yes, by b	y	y	y
b	no	-	y	y
c	yes, by b	y	y	-
ab	no	y	-	-
ac	yes, by b	y	y	-
bc	yes, by b	-	y	-
abc	yes, by b	y	-	-

(b) sequential patterns

Fig. 2. Subfigure 2(b) considers all sequential patterns occurring in the dataset in Subfigure 2(a). As the 2nd column shows, all sequences but b and ab are strongly dominated. These two patterns belong to the same equivalence class, with b being a minimal generator (column "gen") and ab a closed sequence (column "closed"). As we opted for the minimal generators as representatives, we want to come up with b. While this result can be computed using the minimal generators as candidates (column "gen"), using the minimal generators in the negatives (column "g-neg") is more efficient as this yields a smaller candidate set.

We will now illustrate the above proposition and its implications using the example in Figure 2. In this database, only the sequences b and ab are potentially useful in characterizing the target class. All other sequential patterns are strongly dominated, and should thus be removed as irrelevant. The two un-dominated patterns b and ab are equivalent, hence we would be happy with just one of these two as representative. More precisely, according to our new approach, we would select b as representative, which is the (only) minimal generator. Now Proposition 5 shows that to compute this result, it is sufficient to consider the set of minimal generators in the negatives (i.e. a and b) as candidates, instead of considering the whole set of minimal generators (which comprises 5 sequences).

5.2   Our Algorithm

The new relation stated in Proposition 5 suggests the following approach: first compute the minimal generators in the negatives (using some standard

minimal generator sequence miner, e.g. [11, 12]) and then remove the dominated generators. So far, we have not considered the second step – the removal
of the dominated generators. The following proposition shows that it is possible to decide whether a pattern is strongly dominated solely by considering the
minimal-generators-in-the-negatives:

Proposition 6. *Let SDB be a sequence database, L a positive integer, and S_{irr}
some minimal generator of length $\leq L$ that is not relevant (wrt. SDB and L).
Then, there is a minimal-generator-in-negatives S_g of length $\leq L$ strongly dominating S_{irr}.*

Proof. Let S_d be one of the sequences strongly dominating S_{irr} and satisfying the
length limit. By Proposition 4, there is a relevant sequence S_g weakly dominating
S_d and satisfying the length limit. By Proposition 5, S_g is a minimal generator
in the negatives. Moreover, by transitivity S_g strongly dominates S_{irr}, which
completes the proof.   □

Algorithm 2. Relevant Sequence Miner

Input   : a sequence database SDB and optionally a length limit L
Output : the relevant sequences

  1: Calculate the set $\mathcal{G}_{\mathcal{N}}$ of sequences that are minimal generator in the negatives and
     have length $\leq L$;
  2: Group the candidate patterns $\mathcal{G}_{\mathcal{N}}$ into sets having same extension in SDB. Let $\mathcal{G}_{\mathcal{N}}^{\equiv}$
     denote the resulting set of equivalence classes
  3: sort the set $\mathcal{G}_{\mathcal{N}}^{\equiv}$ by (i) descending positive support and (ii) in case of ties ascending
     negative support
  4: let $\mathcal{R}$ be an empty set of equivalence classes
  5: **for** every class e in $\mathcal{G}_{\mathcal{N}}^{\equiv}$ **do**
  6:     **if** e is not dominated by any class in $\mathcal{R}$ **then**
  7:         add e to $\mathcal{R}$
  8:     **end if**
  9: **end for**
 10: **return** The set of minimal generators in $\mathcal{R}$

The above Proposition, together with Proposition 5, is the foundation for
our algorithmic approach, sketched in Algorithm 2. In Line 1, the algorithm
can make use of any minimal generator sequence miner, e.g. [11, 12] to compute the minimal-generators-in-the-negatives. The rest of the pseudo-code takes
care of filtering strongly dominated sequences from this candidate set. Instead
of the naive approach – comparing every pair of candidates, which would result in a quadratic number of comparisons – we use a slightly more efficient
solution. First, we group the candidate patterns (i.e. the minimal-generators-in-
the-negatives) into equivalence classes (Line 2). The reason is that the definition
of domination immediately carries over from patterns to equivalence classes and

it is thus sufficient to consider those instead of the individual patterns. The second improvement is that we sort the candidates, resp. the equivalence classes, by descending positive support and then, in case of ties, by ascending negative support. This step, done in Line 3, ensures that a pattern can only be dominated by a predecessor in the sorted list. As a consequence, during the following iteration over the candidates one only has to compare a candidate with the predecessors in the sorted list *which have been verified to be relevant*. Hence, the number of comparisons per candidate is limited by the number of relevant patterns.

5.3   Analysis of the New Algorithm

The correctness of our algorithm follows directly from Propositions 5 and 6. We will now turn to its complexity: Let n denote the number of items, m the number of sequences in the database, L the length limit, l the maximum length of the sequences, $|\mathcal{G}_\mathcal{N}|$ the number of minimal generators in the negatives, $|\mathcal{G}_{\overline{\mathcal{N}}}^{\equiv}|$ the equivalence classes including a minimal generator in the negatives, and $|\mathcal{R}^{\equiv}|$ the number of relevant equivalence classes. Then

1. The runtime of the first step in Algorithm 2 – computing the minimal generators – is $O(n^L \cdot m \cdot l)$.
2. The grouping of the candidate sequences can be done using a hash function mapping the support set to an integer. The runtime is then $O(|\mathcal{G}_\mathcal{N}| \cdot m \cdot l)$.
3. The runtime for sorting is $O(|\mathcal{G}_{\overline{\mathcal{N}}}^{\equiv}| \log(|\mathcal{G}_{\overline{\mathcal{N}}}^{\equiv}|))$.
4. The loop is executed $|\mathcal{G}_{\overline{\mathcal{N}}}^{\equiv}|$ times. The condition in the if requires to check each equivalence class against at most $|\mathcal{R}^{\equiv}|$ relevant patterns. Every comparison can be done in $O(m)$, assuming that hash-sets are used to check for inclusion of a record. The total runtime is thus $O(|\mathcal{G}_{\overline{\mathcal{N}}}^{\equiv}| \cdot |\mathcal{R}^{\equiv}| \cdot m)$.

Overall, the runtime for the computation of Algorithm 2 is hence

$$O(n^L \cdot m \cdot l + |\mathcal{G}_\mathcal{N}| \cdot m \cdot l + |\mathcal{G}_{\overline{\mathcal{N}}}^{\equiv}| \log(|\mathcal{G}_{\overline{\mathcal{N}}}^{\equiv}|) + |\mathcal{G}_{\overline{\mathcal{N}}}^{\equiv}| \cdot |\mathcal{R}^{\equiv}| \cdot m).$$

As the number of minimal generators is typically much smaller than n^L, the overall runtime is typically dominated by the first summand – that is, the runtime is dominated by the first step which computes the minimal generators in the negatives. This will be confirmed by experiments presented in Section 6.

6   Experimental Evaluation

In this section, we experimentally evaluate the impact of our approach. As several investigations have demonstrated that removing dominated patterns is beneficial for classification purposes [6, 8, 23], we only investigate the effect on the size of the result pattern set and on the computational costs.

6.1   Implementation and Setup

Our approach requires, as a building block, a minimal generator sequence miner. To this end, we have used a (slightly modified) reimplementation of FEAT [11], a state-of-the-art generator sequence mining algorithm. In particular, our implementation allows for *maximum length constraints*. This can easily be realized by stopping the recursive traversal of the candidate space if a pattern violates the length constraint.

We used five sequence datasets in our evaluation. The datasets 'hill-valley', 'libras', 'person-activity', and 'promoter' are publicly available datasets from the UCI repository [24]. The last dataset, 'wlan', is from an ongoing project and cannot be made publicly available. Table 3 shows all datasets together with their most important statistics.

dataset	# seq.	# pos.	# items	max. length
hill-valley	606	301	5	100
libras	360	192	979	89
person	273	198	116	8610
promoter	106	53	4	57
wlan	206	166	15	2920

Fig. 3. Datasets

6.2   Results

We will now show how the concept of relevance affects the number of patterns obtained. Figures 4(a) to 4(e) show, for different datasets and length limits, the number of minimal generators ("Gen"), the number of minimal generators in the negatives ("G-neg") and of relevant sequences ("Rel"). In the experiments, we also used a minimum support of 10%.

Reduction of the Pattern Set. The figures show, first, that the concept of relevance dramatically reduces the number of patterns. At higher length limits, the reduction from all generator sequences ("Gen") to the relevant sequences ("Rel") amounts to several orders of magnitude. The results are similar if the size of the outcome is controlled using a support threshold instead of a length limit. We show a corresponding plot for the 'hill-valley' dataset in Figure 4(f), where we additionally used a length limit of 10 (the result for other datasets are similar and omitted for space reasons). This demonstrates the main benefit of our relevance criterion for sequential patterns: it tremendously reduces the number of sequential patterns.

Computational Speedup. The second observation is that the computation via the minimal-generators-in-the-negatives reduces the computational costs. Again, this can be seen in Figure 4, which shows the reduction from generators ("Gen") to generators-in-the-negatives ("G-neg"). These numbers are less implementation-dependent than the runtime, and hence a more convenient assessment of the computational costs (for comparison, we also show the runtimes

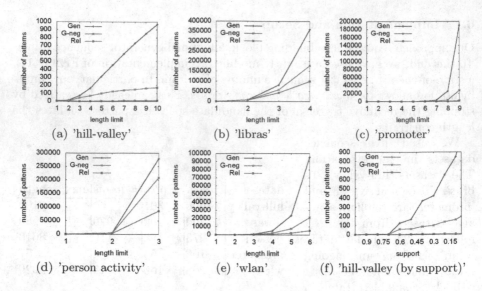

(a) 'hill-valley'   (b) 'libras'   (c) 'promoter'

(d) 'person activity'   (e) 'wlan'   (f) 'hill-valley (by support)'

Fig. 4. The relevance constraint tremendously reduces the number of patterns. Moreover, the generators-in-the-negatives approach significantly reduces the candidate set.

on a Core 2 Duo E8400 for the hill-valley dataset in Figure 5(a)). The experiments show that while the reduction varies between the datasets, it can amount to an order of magnitude.

We also compared the runtimes for the first and second step of our algorithm, namely computing the generators-in-the-negatives and removing the dominated candidates. The result is shown in Figure 5(b). It shows that for all datasets, the computational costs are dominated by the candidate mining step. While the table shows the values for a maximum length of 2 and a support threshold of 30%, the results are similar for other settings.

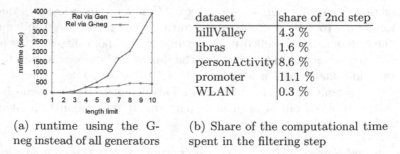

dataset	share of 2nd step
hillValley	4.3 %
libras	1.6 %
personActivity	8.6 %
promoter	11.1 %
WLAN	0.3 %

(a) runtime using the G-neg instead of all generators

(b) Share of the computational time spent in the filtering step

Fig. 5. Runtime figures showing (a) that the generators-in-the-negatives approach reduces the computation time, and (b) that the overall costs are dominated by the candidate mining step

7   Conclusions

In this paper, we have adapted the idea of relevance [6] to sequential data. We have shown that several important properties do not carry over from itemset to sequence data. This makes the use of the closed-on-the-positives as representatives less appealing, which motivated our proposal to use, instead, the minimal generator sequences as representatives. Besides coming up with shorter descriptions, this has the important advantage that it allows for a meaningful maximum pattern length constraint, which can be very useful in practical applications.

Subsequently, we have presented a computational approach for mining the relevant sequences. Our approach is based on the relation between relevant sequences and *minimal generators* in the *negatives*. This relation is kind of the counterpart to the relation between relevant itemsets and *closed itemsets* in the *positives*, discovered by Garriga et al. [8].

In the experimental section, we have shown that the concept of relevance results in a tremendous reduction of the number of patterns, and that the generators-in-the-negatives approach reduces the computational costs. Our approach thus improves upon the use of all sequence generators in a similar way as Garriga's approach exceeds over the use of all closed itemsets. For sequence data, computing the minimal generators is, in general, not more demanding than computing closed patterns. Thus, our algorithm would also be a good choice as underlying miner in post-processing [2, 4] or iterative approaches [5, 13–15].

There are several lines in which our research can be extended. For one, it could be adapted to sequences of itemsets, as opposed to the sequences of items considered here. For another, the relation between minimal generators in the negatives and relevant patterns also holds in the case of itemsets. It might be exploited to design algorithms for mining relevant minimal-length itemsets. This would result in shorter itemsets, which is an important advantage if the itemsets are to be read and interpreted by human experts. Another interesting question would be whether the approaches proposing closure operators on patterns taking the form of *sets* of sequences [17, 18] could be combined with the concept of relevance. Finally, it would be interesting to investigate whether the notion of relevance can be further relaxed, following the ideas of [25, 26].

Acknowledgments. This publication has been produced in the context of the EU Collaborative Projects P-Medicine and EURECA, which are co-funded by the European Commission under the contracts ICT-2009-6-270089 respectively ICT-2011-288048.

References

1. Agrawal, R., Srikant, R.: Mining sequential patterns. In: Yu, P.S., Chen, A.L.P. (eds.) ICDE, pp. 3–14. IEEE Computer Society (1995)
2. Knobbe, A.J., Ho, E.K.Y.: Pattern teams. In: Fürnkranz, J., Scheffer, T., Spiliopoulou, M. (eds.) PKDD 2006. LNCS (LNAI), vol. 4213, pp. 577–584. Springer, Heidelberg (2006)
3. Webb, G.I.: Discovering significant patterns. Mach. Learn. 68, 1–33 (2007)

4. Bringmann, B., Zimmermann, A.: One in a million: picking the right patterns. Knowl. Inf. Syst. 18(1), 61–81 (2009)
5. Mampaey, M., Tatti, N., Vreeken, J.: Tell me what i need to know: succinctly summarizing data with itemsets. In: KDD (2011)
6. Lavrac, N., Gamberger, D., Jovanoski, V.: A study of relevance for learning in deductive databases. J. Log. Program. 40(2-3), 215–249 (1999)
7. Lavrač, N., Gamberger, D.: Relevancy in constraint-based subgroup discovery. In: Boulicaut, J.-F., De Raedt, L., Mannila, H. (eds.) Constraint-Based Mining. LNCS (LNAI), vol. 3848, pp. 243–266. Springer, Heidelberg (2006)
8. Garriga, G.C., Kralj, P., Lavrač, N.: Closed sets for labeled data. J. Mach. Learn. Res. 9, 559–580 (2008)
9. Yan, X., Han, J., Afshar, R.: Clospan: Mining closed sequential patterns in large databases. In: SDM (2003)
10. Wang, J., Han, J.: Bide: Efficient mining of frequent closed sequences. In: ICDE, pp. 79–90 (2004)
11. Gao, C., Wang, J., He, Y., Zhou, L.: Efficient mining of frequent sequence generators. In: WWW, pp. 1051–1052 (2008)
12. Lo, D., Khoo, S.C., Li, J.: Mining and ranking generators of sequential patterns. In: SDM, pp. 553–564. SIAM (2008)
13. Lavrac, N., Kavsek, B., Flach, P., Todorovski, L.: Subgroup discovery with CN2-SD. Journal of Machine Learning Research 5, 153–188 (2004)
14. Cheng, H., Yan, X., Han, J., Yu, P.S.: Direct discriminative pattern mining for effective classification. In: ICDE (2008)
15. Zimmermann, A., Bringmann, B., Rückert, U.: Fast, effective molecular feature mining by local optimization. In: Balcázar, J.L., Bonchi, F., Gionis, A., Sebag, M. (eds.) ECML PKDD 2010, Part III. LNCS, vol. 6323, pp. 563–578. Springer, Heidelberg (2010)
16. Grosskreutz, H.: Class relevant pattern mining in output-polynomial time. In: SDM (2012)
17. Casas-Garriga, G.: Summarizing sequential data with closed partial orders. In: SDM (2005)
18. Raïssi, C., Calders, T., Poncelet, P.: Mining conjunctive sequential patterns. Data Min. Knowl. Discov. 17(1), 77–93 (2008)
19. Tatti, N., Cule, B.: Mining closed strict episodes. In: IEEE International Conference on Data Mining, pp. 501–510 (2010)
20. Bastide, Y., Pasquier, N., Taouil, R., Gerd, S., Lakhal, L.: Mining minimal non-redundant association rules using frequent closed itemsets. In: Proceedings of the First International Conference on Computational Logic (2000)
21. Uno, T., Asai, T., Uchida, Y., Arimura, H.: An efficient algorithm for enumerating closed patterns in transaction databases. In: Suzuki, E., Arikawa, S. (eds.) DS 2004. LNCS (LNAI), vol. 3245, pp. 16–31. Springer, Heidelberg (2004)
22. Grosskreutz, H., Paurat, D.: Fast and memory-efficient discovery of the top-k relevant subgroups in a reduced candidate space. In: Gunopulos, D., Hofmann, T., Malerba, D., Vazirgiannis, M. (eds.) ECML PKDD 2011, Part I. LNCS, vol. 6911, pp. 533–548. Springer, Heidelberg (2011)
23. Lemmerich, F., Atzmueller, M.: Fast discovery of relevant subgroup patterns. In: FLAIRS (2010)
24. Asuncion, A., Newman, D.: UCI machine learning repository (2007)
25. Lemmerich, F., Atzmueller, M.: Incorporating Exceptions: Efficient Mining of epsilon-Relevant Subgroup Patterns. In: Proc. LeGo 2009: From Local Patterns to Global Models, Workshop at ECMLPKDD 2009 (2009)
26. Grosskreutz, H., Paurat, D., Rüping, S.: An enhanced relevance criterion for more concise supervised pattern discovery. In: KDD 2012 (2012)

A Fast and Simple Method for Mining
Subsequences with Surprising Event Counts

Jefrey Lijffijt

Helsinki Institute for Information Technology HIIT,
Department of Information and Computer Science,
Aalto University, Finland
jefrey.lijffijt@aalto.fi

Abstract. We consider the problem of mining subsequences with surprising event counts. When mining patterns, we often test a very large number of potentially present patterns, leading to a high likelihood of finding *spurious* results. Typically, this problem grows as the size of the data increases. Existing methods for statistical testing are not usable for mining patterns in *big data*, because they are either computationally too demanding, or fail to take into account the dependency structure between patterns, leading to true findings going unnoticed. We propose a new method to compute the significance of event frequencies in subsequences of a long data sequence. The method is based on analyzing the joint distribution of the patterns, omitting the need for randomization. We argue that computing the p-values exactly is computationally costly, but that an upper bound is easy to compute. We investigate the tightness of the upper bound and compare the power of the test with the alternative of post-hoc correction. We demonstrate the utility of the method on two types of data: text and DNA. We show that the proposed method is easy to implement and can be computed quickly. Moreover, we conclude that the upper bound is sufficiently tight and that meaningful results can be obtained in practice.

Keywords: Big data, pattern mining, multiple hypothesis testing, event sequence, frequency of occurrence.

1 Introduction

The amount of collected data is growing rapidly. As a result, the focus in data mining research is more than ever on faster and simpler methods, where fast currently means linear or sublinear in the size of the data. However, *big data* presents more challenges. For example, when mining *patterns*—local structure, as opposed to global structure [15]—the number of patterns potentially present in the data is often exponential in the size of the data. Testing more patterns is nice, because it increases the likelihood of finding interesting results. However, testing more patterns is also dangerous, as it increases the likelihood of finding *spurious* results, i.e., patterns caused by randomness.

H. Blockeel et al. (Eds.): ECML PKDD 2013, Part I, LNAI 8188, pp. 385–400, 2013.
© Springer-Verlag Berlin Heidelberg 2013

Several methods have been developed in the past decade for testing the statistical significance of various types of patterns, and a few studies investigated post-hoc corrections to avoid finding many spurious patterns. Unfortunately, none of the proposed methods is usable for big data, because they rely either on randomization, Bonferroni-style post-hoc correction, or both.

Randomization testing is computationally expensive; a single randomization has a computational cost linear in the size of the data or higher, and thousands or millions of randomizations may be required for sufficient resolution. Bonferroni-style post-hoc correction is also problematic, because the studied patterns (which each correspond to a hypothesis test) are typically dependent, in which case the p-values become conservative, i.e., many true findings will go unnoticed. The problem is worse for large data, as the conservativeness depends on the number of patterns, which may be exponential in the size of the data [6].

We propose a new method for mining subsequences with surprising event counts that does not suffer from these problems. We formulate a statistical test that includes a correction for testing multiple hypotheses, i.e., the p-value for an observation will depend on the observation itself, as well as on the size of the data. This allows us to avoid using a conservative post-hoc correction. Although the method is not directly applicable to other data or pattern types, it may act as a model for methods on other data.

The method provides strong control over the *family-wise error rate* (FWER), that is, the probability that any of the significant results is a false positive. Put less formally, we ask the question "what is the probability that *any of the considered patterns* would have a statistic equal to or higher than the observed statistic?", where the *statistic* can be any interestingness measure: support, lift, WRAcc, etc. We illustrate FWER control in the following example.

Assume that the interestingness measure, and thus the test statistic, is the support of a pattern, and that the data is a transaction database in tabular form. For simplicity assume that all items have equal support. The probability that the statistic of a specific pattern P is significantly high can be assessed by, for example, using swap randomization [5] to generate randomized samples[1] and then computing how often we observe a similar or higher statistic for pattern P in the randomized samples. The obtained p-value corresponds to the question "what is the probability that *this specific pattern* has a test statistic equal to or higher than the observed statistic?".

Now assume that we repeat this procedure for all itemsets of some fixed size. Because we are testing many hypotheses, we are liable to finding many small p-values. To prevent this, we can instead compare the observed statistic with the maximum observed statistic over all itemsets of that size in each randomization. In that case, the p-values correspond to the question "what is the probability that *any of the considered patterns* would have a statistic equal to or higher than the observed statistic?", which is the same as FWER control. Significance

---

[1] Which randomization method to use depends on the assumptions that one wants to make.

testing with FWER control using randomization for mining frequent itemsets has been studied extensively by Hanhijärvi [7].

As stated earlier, randomization is unpractical for large data, and the method proposed in this paper is based on computing the p-values analytically. This means that we have to analyze the joint distribution of the statistics of all potential patterns. We discuss a specific type of data and patterns. We show that, although exact p-values are computationally costly to obtain, an upper bound can be computed efficiently. We show empirically that the upper bound is sufficiently tight.

The data that we consider are event sequences, and the aim is to find subsequences of a fixed length where a certain event is significantly frequent or infrequent. This is essentially a *subgroup discovery* problem: the target is a specific event, the descriptions or patterns are subsequences, and the aim is to find all descriptions where the target is exceptionally frequent or infrequent. This problem setting has many applications. For example, biologists are interested in detecting *isochores* and *CpG sites* in DNA sequences, which are regions that are especially rich or poor in CG content and rich in the dinucleotide CpG respectively [2], and another example is that in text analysis it is useful to identify text fragments where a certain word is under or overused.

Summary of contributions. We propose a new method to test the significance of event frequencies in subsequences that provides p-values under control of the family-wise error rate. That is, the p-value corresponds to the probability of observing the observed statistic or higher in *any* of the subsequences of a given length in a single long sequence. We show that computing the p-values exactly is computationally costly, but that an upper bound can be computed fast. We investigate the tightness of the upper bound and compare the power of the test against using a generic post-hoc correction. We demonstrate the utility of the method by applying the method to two types of data: text and DNA. We show that the proposed method is easy to implement and can be computed quickly. Moreover, we conclude that the upper bound is sufficiently tight and that meaningful results can be obtained in practice.

Outline. The method is introduced in Section 2. Results from the experiments on the tightness of the upper bound, comparison with the generic post-hoc correction, and the experiments on the two data sets are presented in Section 3. Related work is discussed in Section 4 and conclusions are given in Section 5.

2   Method

2.1   Notation

Given a finite set of *event labels* L, an *event sequence* S is defined as $S = (s_1, \ldots, s_n), \forall i \in \{1, \ldots, n\} : s_i \in L$, where n is the *length* of the sequence. We denote a *subsequence* of S as $S_{i,m} = (s_i, \ldots, s_{i+m-1})$, where m is the *length* of the subsequence. The *count* of an event $a \in L$ in subsequence $S_{i,m}$ is given by $\sigma(S_{i,m}, a) = \sum_{k=i}^{i+m-1} \mathbf{1}_{\{a\}}(s_k)$, where $\mathbf{1}_A(s_k)$ is the indicator function that equals 1 if $s_k \in A$ and 0 otherwise. The *frequency* of an event $a \in L$ in subsequence $S_{i,m}$ is $\zeta(S_{i,m}, a) = \sigma(S_{i,m}, a)/m$. The count and frequency of an event in a sequence S are defined as $\sigma(S, a) = \sigma(S_{1,n}, a)$ and $\zeta(S, a) = \zeta(S_{1,n}, a)$.

2.2 Background

Our aim is to test the hypothesis that an event is *significantly* frequent or infrequent in a given subsequence. To determine if an observed frequency is significant, we use the notion of *p-values*. Denote Z a random variable that represents the count of an event under the null hypothesis. The p-value for an observed count k is the probability of observing that count or higher, under the null hypothesis:

$$p_H = Pr(Z \geq k)$$

The observed count is *significantly high* if the probability of a observing that count or higher under the null hypothesis is less than or equal to the pre-specified threshold α:

$$p_H \leq \alpha$$

Vice versa, the observed count is *significantly low* if the probability of observing that count or lower is less than or equal to α:

$$p_L = Pr(Z \leq k) \leq \alpha$$

The null hypothesis that we are interested in is that the data has no structure, i.e., that all events in the sequence are i.i.d. samples:

Definition 1 (Null Hypothesis). *The null hypothesis is that the sequence is generated by a sequence of random variables $X_1, \ldots, X_n$, where each random variable X_i is defined by an independent Bernoulli distribution: $X_i \in \{0, 1\}$, and $Pr(X_i = 1) = p$.*

We assume that the parameter p, which represents the expected frequency of an event, is fixed. The parameter p can be, for example, estimated from the sequence S, in which case the method will find regions in the sequence where the event frequency is significantly high (or low) with respect to the rest of the sequence. Alternatively, p can be based on background knowledge, for example an estimate derived from a database of sequences.

Furthermore, we assume that we are going to test subsequences of a fixed length m, which is a parameter defined beforehand by the user, and we assume that the user chooses a priori the significance threshold α.

2.3 Computing P-Values When Testing One Subsequence

Given a sequence of independent random variables $X_1, \ldots, X_n$, each following a Bernoulli distribution with parameter p, define $Z_{i,m}$ as

$$Z_{i,m} = \sum_{j=i}^{i+m-1} X_j.$$

Because $Z_{i,m}$ is the sum of m independent and identically distributed Bernoulli variables, the probability distribution for $Z_{i,m}$ is a binomial distribution:

$$Pr(Z_{i,m} = k) = Bin\,(k; m, p) = \binom{m}{k} p^k (1-p)^{m-k}.$$

We find that, as expected, the distribution is independent of the location i.

We can now define the one-tailed p-value under the null hypothesis for a single subsequence at a random location. For the high frequency direction, the one-tailed p-value is given by

$$
\begin{aligned}
p_H &= Pr(\sigma(S_{i,m}, a) \geq k) \\
&= Pr(Z_{i,m} \geq k) \\
&= \sum_{j=k}^{m} \binom{m}{j} p^j (1-p)^{m-j},
\end{aligned}
\tag{1}
$$

while the one-tailed p-value in the low frequency direction is given by

$$
\begin{aligned}
p_L &= Pr(\sigma(S_{i,m}, a) \leq k) \\
&= Pr(Z_{i,m} \leq k) \\
&= \sum_{j=0}^{m} \binom{m}{j} p^j (1-p)^{m-j}.
\end{aligned}
\tag{2}
$$

As can be seen, the p-values correspond to the cumulative distribution function of the binomial distribution. These tests are also known as the *binomial test*. Many statistical software packages contain a function for computing its value.

2.4   Computing P-Values When Testing All Subsequences

When testing a single subsequence at a random location, the probability of rejecting the null hypothesis while it is actually true—a *false positive* or *type I error*—is exactly α, and thus the result is easy to interpret. However, if we test the significance of the event frequency in multiple subsequences, or in a subsequence at an optimized location, we increase the probability of false positives.

Let us assume that we test the observed counts for all subsequences of a given length, using a sliding window with step size one. In that case, the probability under the null hypothesis of observing a certain count or higher in at least one subsequence of length m is

$$Pr(\bigcup_{i=1,\dots,n-m+1} Z_{i,m} \geq k).
\tag{3}$$

When we test the event frequency in all subsequences, it seems reasonable to use this probability as a p-value. This is also theoretically justified: the probability expressed in Eq. (3) is equal to the probability of obtaining at least one false positive, thus, using this as the p-value corresponds to strong control of the family-wise error rate [18].

Thus, we redefine the one-tailed p-value, in the high direction, as

$$p_H = Pr(\bigcup_{i=1,\ldots,n-m+1} Z_{i,m} \geq k).$$

The p-value can be decomposed as

$$p_H = Pr(Z_{1,m} \geq k) + Pr(Z_{2,m} \geq k \cap \bigcap_{i=1} Z_{i,m} < k) + \ldots$$
$$+ Pr(Z_{n-m+1,m} \geq k \cap \bigcap_{i=1,\ldots,n-m} Z_{i,m} < k), \tag{4}$$

which highlights that the p-value equals the standard case (Eq. (1)) *plus* a correction term.

This correction term is in general difficult to compute exactly. A straightforward approach would be to define a column vector v with a probability for each possible initial state, and a transition matrix W that specifies the transition probabilities between the states, and use one sink state for all subsequences with at least k ones. Then the exact p-value is given by computing $W^{n-m} \cdot v$. However, the matrix W will have $O(2^{2m})$ entries, so this approach works only when the length of the subsequences, m, is very small.

The main result of this paper is that we can instead obtain an upper bound that is very easy to compute. Let us define the following approximation:

$$\tilde{p}_H = Pr(Z_{1,m} \geq k) + (n - m) \cdot Pr(Z_{2,m} \geq k \cap Z_{1,m} < k).$$

Theorem 1. $\tilde{p}_H$ *is an upper bound on the exact p-value* p_H, *i.e.,* $\tilde{p}_H \geq p_H$.

Proof. Notice that for the correction terms of p_H it holds that

$$Pr(Z_{2,m} \geq k \cap \bigcap_{i=1} Z_{i,m} < k) \geq Pr(Z_{3,m} \geq k \cap \bigcap_{i=1,2} Z_{i,m} < k)$$
$$\geq Pr(Z_{4,m} \geq k \cap \bigcap_{i=1,2,3} Z_{i,m} < k) \tag{5}$$
$$\geq \ldots$$
$$\geq Pr(Z_{n-m+1,m} \geq k \cap \bigcap_{i=1,\ldots,n-m} Z_{i,m} < k).$$

Combining Eqs. (4) and (5) gives

$$p_H = Pr(Z_{1,m} \geq k) + Pr(Z_{2,m} \geq k \cap \bigcap_{i=1} Z_{i,m} < k) + \ldots$$
$$+ Pr(Z_{n-m+1,m} \geq k \cap \bigcap_{i=1,\ldots,n-m} Z_{i,m} < k)$$
$$\leq Pr(Z_{1,m} \geq k) + (n - m) \cdot Pr(Z_{2,m} \geq k \cap Z_{1,m} < k).$$

Thus, $\tilde{p}_H$ is an upper bound on the exact p-value p_H. ☐

Notice that the first term of $\tilde{p}_H$ can be computed using Eq. (1), while the second term can be rewritten as follows:

$$Pr(Z_{2,m} \geq k \cap Z_{1,m} < k)$$
$$= Pr(Z_{1,1} = 0 \cap Z_{2,m-1} = k - 1 \cap Z_{m+1,1} = 1)$$
$$= Pr(Z_{1,1} = 0) \cdot Pr(Z_{2,m-1} = k - 1) \cdot Pr(Z_{m+1,1} = 1)$$
$$= (1 - p) \cdot Bin\,(k - 1; m - 1, p) \cdot p.$$

Thus, the upper bound $\tilde{p}_H$ is easy to compute.

We propose to use the upper bound $\tilde{p}_H$ as a statistical test. This test may be conservative, but that only means that results may be statistically more significant. As the exact p-value p_H is difficult to compute, we cannot analyze directly how tight the upper bound is. In Section 3.1 we study empirically how tight the approximation is, and in Section 3.2 we compare the power of this test to the alternative of combining the binomial test with a general post-hoc correction.

To complete the story, we obtain an upper bound to the one-tailed p-value in the low direction analogously to the previous case. For brevity we just list the result. Define

$$\tilde{p}_L = Pr(Z_{1,m} \leq k) + (n - m) \cdot Pr(Z_{2,m} \leq k \cap Z_{1,m} > k).$$

Theorem 2. $\tilde{p}_L$ is an upper bound on the exact p-value p_L, i.e., $\tilde{p}_L \geq p_L$.

Proof. Analogous to Theorem 1. □

The correction term can be computed using

$$Pr(Z_{2,m} \leq k \cap Z_{1,m} > k) = p \cdot Bin\,(k; m - 1, p) \cdot (1 - p).$$

2.5 A Generalization for Sliding Windows with Constant Step Size

If we use a sliding window with step size larger than one, we test fewer hypotheses, but the dependency between the consecutive subsequences will also change. The upper bound from Section 2.4 is also an upper bound when using a larger step size, but a tighter bound can be obtained relatively easily.

Let r be the user-defined step size. The p-value in the high direction is

$$p_H = Pr(\bigcup_{i=1,1+r,1+2r,\ldots,1+\lfloor \frac{n-m}{r} \rfloor r} Z_{i,m} \geq k)$$

Since there are $1 + \lfloor \frac{n-m}{r} \rfloor$ subsequences, we define $\tilde{p}_H$ as

$$\tilde{p}_H = Pr(Z_{1,m} \geq k) + \left\lfloor \frac{n - m}{r} \right\rfloor \cdot Pr(Z_{1+r,m} \geq k \cap Z_{1,m} < k).$$

Theorem 3. $\tilde{p}_H$ is an upper bound on the exact p-value p_H, i.e., $\tilde{p}_H \geq p_H$.

Proof. p_H can be decomposed as

$$p_H = Pr(Z_{1,m} \geq k) + Pr(Z_{1+r,m} \geq k \cap \bigcap_{i=1} Z_{i,m} < k) + \ldots$$

$$+ Pr(Z_{1+\lfloor \frac{n-m}{r} \rfloor r,m} \geq k \cap \bigcap_{i=1,1+r,1+2r,\ldots,1+(\lfloor \frac{n-m}{r} \rfloor -1)r} Z_{i,m} < k). \quad (6)$$

Also, it holds that

$$Pr(Z_{1+r,m} \geq k \cap \bigcap_{i=1} Z_{i,m} < k) \geq$$

$$Pr(Z_{1+2r,m} \geq k \cap \bigcap_{i=1,1+r} Z_{i,m} < k) \geq \quad (7)$$

$$\ldots$$

Combining Eqs. (6) and (7) gives

$$p_H \leq Pr(Z_{1,m} \geq k) + \left\lfloor \frac{n-m}{r} \right\rfloor \cdot Pr(Z_{1+r,m} \geq k \cap Z_{1,m} < k).$$

Thus, $\tilde{p}_H$ is an upper bound on the exact p-value p_H.   □

In this setting, the correction term is more involved. For convenience, we split the correction term into three parts: the overlap between the two subsequences, $Z_{1+r,m-r}$, and the two non-overlapping parts, $Z_{1,r}$ and $Z_{1+m,r}$. We have that

$$Z_{1+r,m} \geq k \Rightarrow Z_{1+r,m-r} + Z_{1+m,r} \geq k, \text{ and}$$

$$Z_{1,m} < k \Rightarrow Z_{1,r} + Z_{1+r,m-r} < k.$$

Both right hand sides are satisfied simultaneously if and only if

$$Z_{1+m,r} \geq k - Z_{1+r,m-r}, \quad Z_{1+r,m-r} \geq k - Z_{1+m,r},$$

$$Z_{1,r} < k - Z_{1+r,m-r}, \quad Z_{1+r,m-r} < k - Z_{1,r}. \quad (8)$$

Since $Z_{1+m,r}$ and $Z_{1,r}$ are both by definition between 0 and r, we have that

$$k - r \leq Z_{1+r,m-r} < k. \quad (9)$$

We can rewrite the correction term to an explicit sum using Eqs. (8) and (9):

$$Pr(Z_{1+r,m} \geq k \cap Z_{1,m} < k)$$

$$= \sum_{j=max(0,k-r)}^{k-1} Pr(Z_{1+r,m-r} = j \cap Z_{1+m,r} \geq k - j \cap Z_{1,r} < k - j)$$

$$= \sum_{j=max(0,k-r)}^{k-1} Pr(Z_{1+r,m-r} = j) \cdot Pr(Z_{1+m,r} \geq k - j) \cdot Pr(Z_{1,r} < k - j)$$

$$= \sum_{j=max(0,k-r)}^{k-1} \left(Bin(j; m-r, p) \cdot \sum_{l=k-j}^{r} Bin(l; r, p) \cdot \sum_{l=0}^{k-j-1} Bin(l; r, p) \right).$$

One may verify that the result for $r = 1$ is the same as in Section 2.4. The binomial pmf and cmf can be computed in constant time [14], thus the computational complexity of the correction term is $O(min(k, r))$ and independent of the size of the full sequence. An upper bound $\tilde{p}_L$ can be derived analogously.

3 Experiments

We studied the power of the test on synthetic data and compared the power of the test with the alternative of post-hoc correction, results of which are discussed in Sections 3.1 and 3.2. We also investigated the practical utility of the test on two types of data: an English novel and a part of the human reference genome. The findings of these experiments are presented in Sections 3.3 and 3.4.

3.1 Tightness of the Upper Bound

Since the proposed test provides strong control over the family-wise error rate, we know that the probability of observing one or more false positives is at most α. Unfortunately, this provides no information on the *power* of the test, i.e., the probability of rejecting a false null hypothesis. Ideally, we would study the probability or rate of false negatives directly. But that is not possible, unless we specify an alternative hypothesis; there is no general false negative rate. Instead, we use the fact that there is a trade-off between the probability false positives and the probability of false negatives.

By definition we have that the probability of false negatives is minimized when the probability of false positives is maximized. Thus, preferably, the probability of observing one or more false positives should be as close to α as possible. To study how close the probability of encountering one or more false positives is in practice, we designed the following experiment.

The tightness of the upper bound may depend both on the length of sliding window, as well as the event probability. Thus, we tried various window lengths ($m \in \{100, 1000, 10000\}$) and event probabilities ($p \in \{0.001, 0.01, 0.1\}$). For each combination, we generated 1,000 sequences of length $n = 9,999 + m$ (such that there are 10,000 p-values per sequence) and computed the p-values $\tilde{p}_H$ for all subsequences using a sliding window with step size 1.

The quantity of interest is the minimal p-value per sequence, because if the minimal p-value in a sequence is below the threshold α, then we have at least one false positive. Ideally, the distribution of minimal p-values over the sequences is uniform, which means that for any value α, the probability of observing one or more p-values below α is exactly α itself. This ensures that the probability of false positives is maximal (while providing FWER control), and that the probability of false negatives is minimal. Note that this holds by definition for the exact p-values under the null hypothesis, but the upper bound that we propose to use instead may have a higher probability of false negatives.

The results of the experiment are presented in Figure 1. We find that the p-values are reasonably close to the optimal distribution and that they are further

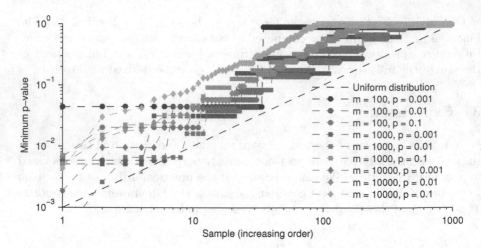

Fig. 1. The distribution of minimal p-values over 1,000 synthetic sequences for the proposed method, using various window lengths m and event probabilities p, compared to the ideal distribution. We find that the p-values are reasonably close to the uniform distribution and that they are further from uniform when the expected number of events $(= m \cdot p)$ is higher.

from the optimal distribution when the expected event count $(= m \cdot p)$ is larger. The largest observed effect is approximately 1 order of magnitude ($m = 10,000$, $p = 0.1$), indicating that the p-values are 1 order of magnitude too high in that case. Note that the results for very low expected counts (e.g., $m = 100$, $p = 0.001$) may appear more conservative, but they are skewed mostly because there are very few distinct p-values: the highest number of events observed in any subsequence is 3 ($\tilde{p}_H = 0.0437$), and for $k \in \{0, 1\}$, we have $\tilde{p}_H = 1$.

We expect that p-value estimates that are conservative by one order of magnitude will not be a problem in most practical settings; much larger differences in the choice of α can be observed in the literature: from $\alpha = 0.1$ to $\alpha = 0.00001$. Also, because the p-values are controlled for family-wise error rate, use of a 'large' α, such as 0.05, still guarantees that obtaining any false-positive results has very low probability.

3.2   Comparison to Hochberg's Step-Up Procedure

An alternative approach to obtaining p-values for the tested hypotheses under strong control of the family-wise error rate is to use the binomial test (Eqs. (1) and (2)) with post-hoc correction. The correction with largest power that we are aware of that provides strong control for the family-wise error rate, and which is applicable in this setting, and that does not require specifying the dependency structure of the p-values, is Hochberg's step-up procedure [8]. Hochberg's procedure is valid for independent and positively dependent p-values [17]. The latter is the case here, as the p-values for overlapping windows have positive correlation.

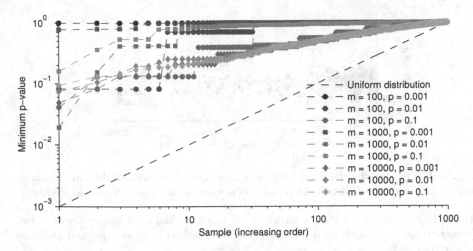

Fig. 2. The distribution of minimal p-values for the binomial test with Hochberg's post-hoc correction, on the data from Figure 1. We find that the p-values are far from the uniform distribution, for any combination of parameters, while the distribution is more uniform when the expected number of events $(= m \cdot p)$ is larger.

We computed the p-values for the binomial test for each sequence generated in the previous experiment (Section 3.1), using a sliding window of the same length, and adjusted these using Hochberg's procedure. Thus, the p-values are directly comparable to those in the previous experiment. We computed the minimal p-value per sequence, and compared the results with those from the upper-bound method.

The distribution of minimal p-values is shown in Figure 2. We observe that p-values from the method with post-hoc correction are far from uniform, for any combination of parameters, while the distribution becomes more uniform as the expected number of events per subsequence increases. The proposed method outperforms the post-hoc approach for any combination of parameters, although we cannot be certain that this holds for much larger expected number of events.

3.3   Bursty and Non-bursty Words in an English Novel

The prime motivation for this work comes from the domain of text analysis. Church and Gale [3] and Katz [9] both studied *burstiness* of words in the context of probabilistic modeling of word counts, and the concept is related to relevance measures in information retrieval, such as inverse document frequency [19]. More recently, using a quantification of burstiness based on the inter-arrival time distributions of words, burstiness of words has been related to semantic categories [1], statistical tests for comparing corpora that take into account burstiness have been proposed [13], and the impact of burstiness on choosing appropriate window lengths for sequence analysis has been studied [12].

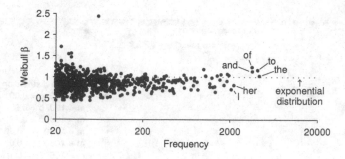

Fig. 3. The relationship between burstiness, measured using the Weibull distribution, and frequency of words. Each dot represents a word in the novel *Pride and Prejudice*.

Table 1. We studied the local behavior of the five least and most bursty words in two frequency bins to investigate the suitability of our method to locate over and underuse of words in text.

Frequency	Low [$\sigma = 40$–50]	High [$\sigma = 300$–600]
Non-bursty	hardly, help, perfectly, point, scarcely	an, elizabeth, more, there, when
Bursty	marry, pride, read, rosings, william	are, me, their, will, your

For the purpose of text analysis, it is useful to know if there are fragments in a text where a certain word is over or underused and to locate such fragments. We investigated the suitability of the proposed method to this task. As an experiment, we downloaded the book Pride & Prejudice by Jane Austen, which is freely available via Project Gutenberg[2]. We computed the frequency and the maximum-likelihood estimates for the Weibull distribution [1,13] for all words, and then selected the five most and least bursty words in two frequency bins, see Table 1. An overview of the relation between the frequency and burstiness of words is given in Figure 3.

For each of the selected words, we tracked the frequency throughout the book using a sliding window of length 5,000 and step size 1. The book contains $n = 121{,}892$ words, thus there are 116,893 windows. We chose a window length of 5,000 to ensure that low event counts could also be significant; for example, for a window length of 2,000 and event probability $p = 1/300$, we have that the p-value for $k = 0$ is $\tilde{p}_L = 0.4833$. Thus, an event count of zero is not significant, even for fairly frequent words. With a window length of 5,000, event counts of 3 and less are significant at $\alpha = 0.05$ ($\tilde{p}_L = 0.0164$).

We computed the significance of the observed frequencies, for both the high and low direction. Because the results are for illustrative purposes, we did not apply any additional correction for testing multiple sets of hypothesis. Figure 4 shows the results for three words. The word *an* is frequent and non-bursty, and no parts of the book show significant under or overuse of the word. For the pronoun *me*, which is frequent and bursty, we observe two areas of overuse, and

---

[2] http://www.gutenberg.org/

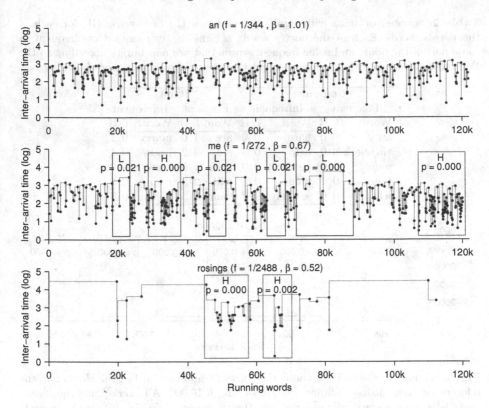

Fig. 4. Significant over and underuse of three words in the novel *Pride and Prejudice*, compared to the average frequency in the book. Each blue dot corresponds to an occurrence of the word in the text. To aid the visualization of the results, all overlapping significant subsequences have been merged together. We observe that for *an*, no parts of the book show significant under or overuse of the word, while for the pronoun *me*, two areas show significant overuse, and four areas show underuse of the word. Finally, the family name *rosings* is used mainly in two parts of the book.

four areas of underuse, compared to the average frequency. Finally, the family name *rosings*, which is infrequent and bursty, is used a lot in two text fragments and occurs a few times in other parts of the book.

A full overview of results is given in Table 2. As expected, we find that each of the bursty words is significantly over or underrepresented in at least one fragment of the book. Surprising is that some frequent words that are non-bursty according to the Weibull distribution estimate are also under or overused in one or more fragments. This indicates that there is local structure that is not captured by the Weibull measure of word burstiness. The results from the proposed method are confirmed by visual inspection of the data and we conclude that the method has a clear potential to find novel and interesting patterns.

Table 2. Number of areas with significant underuse (L) or overuse (H) for each of the twenty words. Each of the bursty words is significantly more or less frequent in some part of the book, and some frequent words that are non-bursty according to the Weibull distribution estimate are also under or overused in one or more book parts.

| | | | | | Non-bursty | | | | | | | Bursty | | | |
| Frequent | | | Infrequent | | | Frequent | | | Infrequent | | |
Word	L	H	Word	L	H	Word	L	H	Word	L	H
an	0	0	hardly	0	0	are	1	0	marry	0	1
elizabeth	2	0	help	0	0	me	4	2	pride	0	1
more	0	0	perfectly	0	0	their	1	0	read	0	2
there	0	1	point	0	0	will	2	3	rosings	0	2
when	0	0	scarcely	0	0	your	2	3	william	0	1

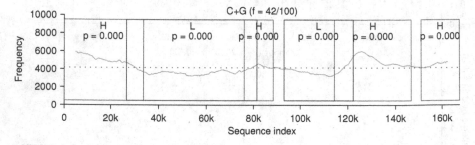

Fig. 5. Analysis of the GC content at the start of Chromosome 1 of the Homo Sapiens reference genome, using a sliding window of length 10,000. All overlapping significant parts have been merged. We observe that the frequency of GC is quite volatile: parts where the content is significantly high overlap with parts where the content is significantly low. We also observe that the test is sufficiently powerful, there are many significant results, even though we are testing a total of 225,270,622 hypotheses.

3.4   Variation in GC and TA Content in DNA

Variation of GC content in DNA sequences is used to define *isochores*, which in turn are used to identify gene structure [2]. We tested if we could find significant variation in GC and TA content in chromosome 1 from the Homo Sapiens reference genome, which we downloaded from the NCBI repository[3]. We computed the frequency of C+G using a sliding window of length 10,000 and step size 1. Chromosome 1 of the reference genome (build 37, patch 9) contains 225,280,621 fixed nucleotides, thus the number of tested hypotheses is in this case very large.

Analysis of the first consecutive fixed part can be found in Figure 5. We observe that the test is sufficiently powerful, because several parts of the sequence are identified as having significantly high or low GC content. We find that the GC content is quite volatile: the parts where the content is significantly low and high overlap each other. We conclude again that the proposed method has potential for finding novel and interesting patterns in the data.

---

[3] http://www.ncbi.nlm.nih.gov

4   Related Work

The popularity of significance testing methods in data mining has increased considerably over the past decade. Gionis et al. [5] introduced swap randomization for mining significant patterns while maintaining row and column margins, while De Bie [4] proposed a maximum-entropy approach that can also take into account other types of constraints. Webb [20] and Hanhijärvi [7] studied the problem of multiple testing for mining patterns. These studies are all restricted to mining itemsets or tiles. A generic approach to mining structure in data using statistical testing has been presented by Lijffijt et al. [11].

There are only a few studies on statistical testing approaches for mining sequential data. Most related is the statistical test proposed by Kifer et al. [10] for detecting change points in streams. However, they rule out the possibility of controlling the family-wise error rate, as they consider only streams of infinite length. Another drawback of that method is that the critical points cannot be computed analytically, but require randomization.

Complementary to this work are the randomization-based statistical tests for comparing event counts between databases of sequences put forward by Lijffijt et al. [13]. Segmentation methods may provide an alternative to modeling frequency variation, although the focus is then on global modeling, while the aim here is to find local structure. Mannila and Salmenkivi [16] study efficient methods for sequence segmentation, while the approach by Lijffijt et al. [11] can be used to assess the significance of such a segmentation.

5   Conclusions

We have introduced a novel statistical test for assessing the significance of event frequencies in subsequences when using a sliding window. The test provides strong control of the family-wise error rate and takes into account the dependency structure of overlapping subsequences. We have shown that, although exact p-values under the null hypothesis are difficult to compute, an easy-to-compute upper bound can be used instead. We have shown empirically that the upper bound is sufficiently tight and that the test offers increased power compared to combining the binomial test with a generic post-hoc correction.

We have also investigated the utility and practicality of the test on linguistic and biological sequences and found several novel and interesting patterns. We have shown that meaningful results can be obtained, and that the method remains sufficiently powerful even when testing a very large number of hypotheses. We conclude that the proposed method is simple, fast and powerful and that it can produce meaningful results on various types of data.

Acknowledgements. The author has received support from the Finnish Doctoral Programme in Computational Sciences (FICS) and the Academy of Finland's Centre of Excellence in Algorithmic Data Analysis (ALGODAN). I thank Heikki Mannila and Petteri Kaski for useful discussion and feedback.

References

1. Altmann, E.G., Pierrehumbert, J.B., Motter, A.E.: Beyond word frequency: Bursts, lulls, and scaling in the temporal distributions of words. PLoS One 4(11), e7678 (2009)
2. Bernardi, G.: Isochores and the evolutionary genomics of vertebrates. Gene 241(1), 3–17 (2000)
3. Church, K.W., Gale, W.A.: Poisson mixtures. Nat. Lang. Eng. 1(2), 163–190 (1995)
4. De Bie, T.: Maximum entropy models and subjective interestingness: an application to tiles in binary databases. Data Min. Know. Disc. 23(3), 407–446 (2011)
5. Gionis, A., Mannila, H., Mielikäinen, T., Tsaparas, P.: Assessing data mining results via swap randomization. ACM TKDD 1(3), 14 (2007)
6. Han, J., Pei, J., Yin, Y., Mao, R.: Mining frequent patterns without candidate generation: A frequent-pattern tree approach. Data Min. Know. Disc. 8(1), 53–87 (2004)
7. Hanhijärvi, S.: Multiple hypothesis testing in pattern discovery. In: Elomaa, T., Hollmén, J., Mannila, H. (eds.) DS 2011. LNCS, vol. 6926, pp. 122–134. Springer, Heidelberg (2011)
8. Hochberg, Y.: A sharper Bonferroni procedure for multiple tests of significance. Biometrika 75(4), 800–802 (1988)
9. Katz, S.M.: Distribution of content words and phrases in text and language modelling. Nat. Lang. Eng. 2(1), 15–59 (1996)
10. Kifer, D., Ben-David, S., Gehrke, J.: Detecting change in data streams. In: Nascimento, M.A., Özsu, M.T., Kossmann, D., Miller, R.J., Blakeley, J.A., Schiefer, K.B. (eds.) Proc. of VLDB, pp. 180–191. VLDB Endowment (2004)
11. Lijffijt, J., Papapetrou, P., Puolamäki, K.: A statistical significance testing approach to mining the most informative set of patterns. Data Min. Know. Disc. (in press)
12. Lijffijt, J., Papapetrou, P., Puolamäki, K.: Size matters: Finding the most informative set of window lengths. In: Flach, P.A., De Bie, T., Cristianini, N. (eds.) ECML PKDD 2012, Part II. LNCS, vol. 7524, pp. 451–466. Springer, Heidelberg (2012)
13. Lijffijt, J., Papapetrou, P., Puolamäki, K., Mannila, H.: Analyzing word frequencies in large text corpora using inter-arrival times and bootstrapping. In: Gunopulos, D., Hofmann, T., Malerba, D., Vazirgiannis, M. (eds.) ECML PKDD 2011, Part II. LNCS, vol. 6912, pp. 341–357. Springer, Heidelberg (2011)
14. Loader, C.: Fast and accurate computation of binomial probabilities (2000) (unpublished manuscript)
15. Mannila, H.: Local and global methods in data mining: Basic techniques and open problems. In: Widmayer, P., Triguero, F., Morales, R., Hennessy, M., Eidenbenz, S., Conejo, R. (eds.) ICALP 2002. LNCS, vol. 2380, pp. 57–68. Springer, Heidelberg (2002)
16. Mannila, H., Salmenkivi, M.: Finding simple intensity descriptions from event sequence data. In: Proc. of ACM SIGKDD, pp. 341–346. ACM, New York (2001)
17. Sarkar, S.K., Chang, C.K.: The Simes method for multiple hypothesis testing with positively dependent test statistics. J. Am. Stat. Ass. 92(440), 1601–1608 (1997)
18. Shaffer, J.P.: Multiple hypothesis testing. Ann. Rev. Psych. 46, 561–584 (1995)
19. Spärck Jones, K.: A statistical interpretation of term specificity and its application in retrieval. J. Doc. 28(1), 11–21 (1972)
20. Webb, G.I.: Layered critical values: A powerful direct-adjustment approach to discovering significant patterns. Mach. Learn. 71(2-3), 307–323 (2008)

Relevant Subsequence Detection
with Sparse Dictionary Learning

Sam Blasiak, Huzefa Rangwala, and Kathryn B. Laskey

George Mason University, Fairfax, VA 22030, USA
sblasiak@masonlive.gmu.edu, rangwala@cs.gmu.edu, klaskey@gmu.edu

Abstract. Sparse Dictionary Learning has recently become popular for discovering latent components that can be used to reconstruct elements in a dataset. Analysis of sequence data could also benefit from this type of decomposition, but sequence datasets are not natively accepted by the Sparse Dictionary Learning model. A strategy for making sequence data more manageable is to extract all subsequences of a fixed length from the original sequence dataset. This subsequence representation can then be input to a Sparse Dictionary Learner. This strategy can be problematic because self-similar patterns within sequences are over-represented. In this work, we propose an alternative for applying Sparse Dictionary Learning to sequence datasets. We call this alternative Relevant Subsequence Dictionary Learning (RS-DL). Our method involves constructing separate dictionaries for each sequence in a dataset from shared sets of relevant subsequence patterns. Through experiments, we show that decompositions of sequence data induced by our RS-DL model can be effective both for discovering repeated patterns meaningful to humans and for extracting features useful for sequence classification.

1 Introduction

Sparse Dictionary Learning has recently become popular for discovering latent components that can be used to reconstruct elements in a dataset. It has seen particular success in computer vision where it has been incorporated into solutions for problems in image reconstruction, in-painting, and classification [1–6].

Sparse Dictionary Learning's success in computer vision makes it a promising candidate as an algorithm for discovering patterns in sequence data. Sequence data, however, is not natively accepted by the Sparse Dictionary Learning model: sequences can be of variable length and patterns within sequences are not associated with a fixed set of indices. These patterns can occur at any point within a sequence and can be repeated multiple times. A strategy for making sequence data more manageable is to extract all subsequences of length K from the original dataset and use these as input to a Sparse Dictionary Learning algorithm. This strategy poses a problem, however, because self-similar patterns within sequences are over-represented.

In this work, we propose an alternative to this standard subsequence dataset approach, which we call Relevant Subsequence Dictionary Learning (RS-DL).

H. Blockeel et al. (Eds.): ECML PKDD 2013, Part I, LNAI 8188, pp. 401–416, 2013.

Our method involves constructing separate dictionaries for each sequence in a dataset from shared "relevant subsequence patterns." This structured dictionary can be used to pick out shared information from sets of sequences and can be learned using standard optimization methods. An important contribution of our work is in showing how to efficiently run the LARS algorithm given our relevant subsequence dictionary structure.

To show the utility of the RS-DL model, we run experiments on several types of sequence data. Running our algorithm on synthetic sets of sequences with discrete-valued elements, continuous electrocardiogram data, and text datasets, we show that our RS-DL model is effective for discovering repeated patterns meaningful to humans (also called motifs). We also show that RS-DL is effective for classification. To do so, we use RS-DL to extract features from time-series data and show that these features can reduce classification error compared to standard methods.

2 Background: Sparse Dictionary Learning

Sparse Dictionary Learning is a method of decomposing a dataset into the product of a dictionary matrix and a sparse vector of coefficients. Here we represent the N dataset vectors as $x_{1:N}$, with the nth vector given by $x_n \in \mathbb{R}^d$, the dictionary matrix as $W \in \mathbb{R}^{d \times C}$, and the set of N sparse vectors of coefficients as $\alpha_{1:N}$, $\alpha_n \in \mathbb{R}^C$. The number of dictionary columns, C, is chosen beforehand. The Sparse Dictionary Learning objective is typically defined as follows:

$$f(x_n; W) = \min_{\alpha_n} \underbrace{\frac{1}{2} ||x_n - W\alpha_n||_2^2}_{\text{loss}} + \underbrace{\lambda \psi(\alpha_n)}_{\text{sparsity-inducing term}} \tag{1}$$

where ψ is a regularization function, typically an L_1 norm.

There has been a significant amount of research to develop efficient algorithms for solving the Sparse Dictionary Learning problem [3]. These algorithms typically consist of repeating two optimization steps. In the first step, a linear regression problem with the sparsity-inducing regularization term is solved to compute $\alpha_n = \min_{\alpha_n} ||x_n - W\alpha_n||_2^2 + \lambda \psi(\alpha_n)$ given the current value of the dictionary, W, for each example in the dataset. Common algorithms to perform this task include pursuit algorithms [7], Least Angle Regression (LARS) [8], coordinate-wise descent methods [9], and proximal methods [10].

In the second step, the value of the dictionary, W, is updated given the current minimum values of α_n. As with methods for optimizing with respect to the α's, it is possible to use any of a number of different methods to minimize with respect to the dictionary. These methods include K-SVD [7] (which also updates the α terms), stochastic gradient methods [6], and solutions of the dual problem (for a constrained dictionary) [5], among others.

Sparse Dictionary Learning is similar to other decomposition techniques like Principal Component Analysis (PCA). PCA decomposes elements of a dataset into linear combinations of vectors from an orthogonal basis. Sparse Dictionary

Learning differs from PCA in two important respects. First, dictionary columns are non-orthogonal, and second, the sparsity inducing regularization term forces only a small number of columns to be used for reconstruction. These characteristics can be advantageous compared to PCA because the sparsity inducing term allows the dictionary to include more columns that the dimensionality of the vector being reconstructed [3]. This "overcomplete" representation allows a large number of patterns to be found in the data but only a small number of these patterns are used to reconstruct each data element.

3   Relevant Subsequence Dictionary Learning

We propose an approach, which we call Relevant Subsequence Dictionary Learning (RS-DL)[1], to extend Sparse Dictionary Learning to the domain of sequences. Sequences differ from more-standard vector representations in that they can vary in length across a single dataset, and patterns within sequences can occur at any position rather than being associated with a fixed set of indices. To account for these characteristics of sequence data, RS-DL constructs dictionaries from C different subsequence dictionary components, each of length K. We refer to these constituent components as "relevant subsequence patterns" and indicate these patterns by the two-dimensional array, $\mathbf{v}$, of size $C \times K$, where $v_{c,k}$ is a value associated with the kth position in the cth relevant subsequence pattern.

Unlike standard Sparse Dictionary Learning, RS-DL constructs a separate dictionary, W_n, for each sequence, x_n, in a dataset by positioning relevant subsequence parameters, $v_{c,:}$, so that they cover all possible subsequence starting positions. Positions in dictionary columns that are not given by relevant subsequence parameters are set to zero.

Figure 1 shows how the array of constituent relevant subsequence patterns, $\mathbf{v}$, is used to construct W_n, the dictionary associated with sequence x_n. Table 1 gives descriptions of all parameters in the RS-DL formulation. After building the dictionaries, W_n, we are left with an objective very similar to that of standard Sparse Dictionary Learning:

$$f(x_{1:N}; \mathbf{v}) = \sum_{n=1}^{N} \min_{\alpha_n} \frac{1}{2}||x_n - W_n\alpha_n||_2^2 + \lambda|\alpha_n|_1 \qquad (2)$$

To optimize with respect to this objective, we employ a stochastic gradient descent procedure where sequences are received by the learner in random order. The learner alternatively solves first for α_n, then takes a gradient step with respect to the array, $\mathbf{v}$, in a similar manner to existing Sparse Dictionary Learning optimization algorithms. For the optimization step with respect to α_n, we apply a variation of the Least Angle Regression (LARS) [8] algorithm. The LARS algorithm requires computing a number of matrix products involving W_n. However, computing these matrix products directly would be inefficient, as each W_n matrix is of size $O(T_n) \times O(CT_n)$, where T_n is the length of the nth sequence.

---

[1] We have made code available at http://cs.gmu.edu/~sblasiak/RS-DL.tar.gz

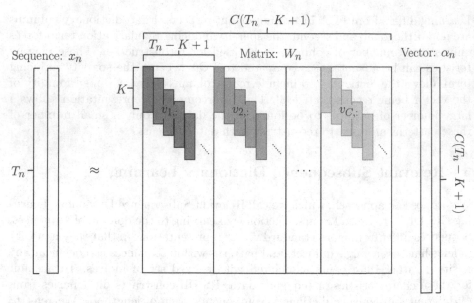

Fig. 1. The figure above illustrates the Relevant Subsequence Dictionary Learning setup. The matrix W_n is constructed from the weights $v_{c,k}$ in C blocks so that the relevant subsequence patterns given by each $v_{c,:}$ are arranged to create dictionary elements (columns of W_n) that cover every K length subsequence of the sequence x_n (illustrated in blue). White areas of the W_n matrix are set to zero. The vector α_n is L_1-regularized to select a small number of dictionary columns associated with positioned relevant subsequences patterns. The α_n-weighted sum of these positioned relevant subsequences patterns approximates x_n.

To improve performance, we can take advantage of the sparse construction of each W_n, allowing these products to be computed more quickly, as we describe in the next section.

After computing each new value of α_n, the RS-DL algorithm takes a single stochastic gradient step in $\mathbf{v}$: $\mathbf{v}^{i+1} \leftarrow \mathbf{v}^i - \left(\frac{\gamma}{i+1}\right) \frac{\partial \frac{1}{2}\|x_n - W_n^i \alpha_n\|_2^2}{\partial \mathbf{v}}$, where γ is a learning rate term. We found empirically that, for RS-DL, this single stochastic gradient step is often faster than solving for $\mathbf{v}$ after accumulating information from a batch of α_n's as in Mairal et. al. [3].

3.1   Efficiently Running the LARS Algorithm with RS-DL

RS-DL involves constructing dictionaries, W_n of size $O(T_n) \times O(CT_n)$, many of whose entries are set to zero. If not carefully handled, this large, sparse matrix can cause the RS-DL training algorithm to operate inefficiently. The LARS algorithm constitutes a major substep in RS-DL training and requires a number of computations involving W_n. Efficiency of these computations can be considerably improved by taking W_n's sparse construction from elements of the array $\mathbf{v}$ into account. Three LARS computations involving the dictionaries, W_n are (i) the matrix-matrix product $(W_n)_{\mathcal{A}}^{\top}(W_n)_{\mathcal{A}}$, (ii) the matrix-vector product

Table 1. Relevant Subsequence Dictionary Learning parameters

Parameter	Definition
M	The size of the alphabet for discrete sequences. We omit the M parameter when dealing with continuous-valued sequences.
$\mathbf{x}_{1:N}$	A set of N observed sequences. Individual sequences, x_n, can be of variable length. Discrete sequences are expanded to M concatenated sequences of T_n indicator variables.
T_n	The length of the nth sequence.
$\alpha_{1:N}$	A set of N vectors. Each α_n vector is of length $C(T_n - K + 1)$.
W_n	A weight matrix of size $T_n \times C(T_n - K + 1)$ created from elements of the array $\mathbf{v}$.
$\mathbf{v}$	An array of values used to construct dictionary elements. $v_{c,k}$ is associated with the kth position in the cth relevant subsequence pattern.
λ	The L_1 regularization parameter associated with each α_n.

$(W_n)_{\mathcal{A}} \omega_{\mathcal{A}}$, and (iii) the matrix-vector product $W_n^{\top} u$, where $\mathcal{A}$ indicates an active set of columns (the number of non-zero components of α), $(W_n)_{\mathcal{A}}$ indicates a matrix constructed from this active set, $\omega_{\mathcal{A}}$ is a vector of length $|\mathcal{A}|$, and u is a vector of length T_n. Below, $t(i)$ indicates the index of the start of the subsequence associated with the ith column of W_n (see Figure 1), $c(i)$ indicates the relevant subsequence position associated with the ith column of W_n, and $\text{sgn}(i)$ indicates the sign of the correlation between the ith matrix column, $(W_n)_i^{\top}$, and the current residual: $\text{sgn}\left((W_n)_i^{\top}\left(x_n - (W_n)_{\mathcal{A}}^{\top}\alpha_n\right)\right)$.

The matrix-matrix product, $X = (W_n)_{\mathcal{A}}^{\top}(W_n)_{\mathcal{A}}$, can be computed as follows:

$$X_{ij} = \begin{cases} \sum_{k=0}^{\max(K-t(j)+t(i),0)} \text{sgn}(i)\, v_{c(i),k}\text{sgn}(j)\, v_{c(j),t(j)-t(i)+k} & t(j) \geq t(i) \\ \sum_{k=0}^{\max(K-t(i)+t(j),0)} \text{sgn}(i)\, v_{c(i),t(i)-t(j)+k}\text{sgn}(j)\, v_{c(j),k} & t(j) < t(i) \end{cases} \tag{3}$$

This matrix-matrix product has an overall complexity of $O(|\mathcal{A}|^2 K)$. However, the full product does not need to be computed at each LARS iteration. Rather, as additional columns are added to the active set, we update a stored Cholesky decomposition of $(W_n)_{\mathcal{A}}^{\top}(W_n)_{\mathcal{A}}$, at a cost of $O(|\mathcal{A}|K)$ for each update (updates involve computing a single column of the product in Equation 3), plus $O(|\mathcal{A}|^2)$ for a back-substitution operation.

We compute the matrix-vector product, $\mathbf{x} = (W_n)_{\mathcal{A}} \omega_{\mathcal{A}}$, incrementally as the weighted sum of components of $\mathbf{v}$:

$$\mathbf{x}^{(1:i)} = \sum_{k=1}^{K} \mathbf{x}_{t(i)+k}^{(1:i-1)} + \text{sgn}(i)\,(\omega_{\mathcal{A}})_i\, v_{c(i),k} \tag{4}$$

where $\mathbf{x}^{(1:i)}$ indicates the sum up to the ith term, and $i \in [1 \ldots |\mathcal{A}|]$. This matrix-vector product has an overall complexity of $O(|\mathcal{A}|K)$.

Finally, we compute the matrix-vector product, $\mathbf{x} = W_n^{\top} u$, as follows:

$$x_i = \sum_{k=1}^{K} v_{c(i),k} u_{t(i)+k} \tag{5}$$

with an overall complexity of $O(CT_n K)$.

For each LARS iteration, we must also compute CT_n correlations between each column of the matrix, W_n, and the current residual at a cost of K each.

These are computed in the same way as the matrix-vector product in Equation 5. In most of out experiments, we restrict $|\mathcal{A}|$ to values less than or equal to C. Thus, each LARS iteration has a complexity of $O(CT_nK)$ when $|\mathcal{A}|$ is small, which can be a significant reduction from $O(CT_n^2)$. However, with no restrictions on the size of the active set, $|\mathcal{A}|$ can potentially grow to CT_n. In this case, complexity is eventually dominated by back-substitution operations involving the incrementally-updated Cholesky decomposition of $(W_n)_{\mathcal{A}}^{\top}(W_n)_{\mathcal{A}}$ at a cost of up to $O(C^2T_n^2)$ per iteration.

3.2 Modifications to RS-DL

The procedure for constructing the RS-DL dictionary (Figure 1) is applicable only to sequences with continuous-valued elements. To allow RS-DL to find decompositions of sequences of discrete symbols, we first transform each original sequence into M separate binary sequences, where elements of the mth binary sequence indicate if the symbols in the original sequence are equal to the mth symbol in the alphabet. These M binary sequences are then concatenated to obtain the input sequence to RS-DL. Dictionary construction must also be modified for discrete sequences. In this case, $\mathbf{v}$ becomes a three-dimensional constituent array, where $v_{c,k,m}$ is associated with the mth symbol of the kth position in the cth relevant subsequence. Separate dictionaries are constructed for each of the M possible symbols using $v_{c,:,m}$ for the cth relevant subsequence pattern associated with the mth constructed binary sequence. These M dictionaries are then stacked vertically to create a composite dictionary.

It is also possible to use RS-DL to find decompositions of multi-variate sequences. To do so, we rearrange each multivariate sequence as a concatenation of univariate sequences. We then create a stack of M dictionaries as we did to create the dictionary for discrete sequences.

Another modification of the basic RS-DL algorithm includes appending a column to the dictionary whose entries are set to a constant value. This addition has the effect of including a bias term whose magnitude varies depending on the associated α_n term. This bias term is useful for modeling time series datasets where the amplitudes of major trends that occur in individual sequences are offset by varying amounts. We employ this bias term in all experiments conducted on time series sequences. A similar strategy can also be employed to capture linear trends.

Finally, we can modify the LARS algorithm so that, rather than finding an L_1-regularized solution for α, it finds solutions with one or fewer non-zero α terms associated with each of the C relevant subsequence patterns. Although the L_1 regularization is no longer enforced in this case, sparsity is maintained in a similar manner to the L_0-regularized[2] version of LARS[8].

---

[2] The "L_0 norm" [7] is a pseudo-norm that counts the number of non-zero components in a vector, i.e., $\|\mathbf{x}\|_0 = \sum_i \mathbb{I}(x_i \neq 0)$.

4   Relationship to Hidden Markov Models

A form of the Factorial Hidden Markov Model, which we describe later in this section, shares characteristics of RS-DL. To understand Factorial Hidden Markov Models, one must first understand the basic Hidden Markov Model (HMM), which defines a probability distribution over sequences. The HMM assumes that each symbol in a sequence is generated from a mixture distribution. Mixture components are indexed by "hidden states" in the HMM. The Markov property holds over these hidden states, meaning that the value of the hidden state indexing the observation at time point t depends only on the value of the hidden state associated with time point $t - 1$.

The Profile HMM (pHMM) [11] is an HMM with specific restrictions on transitions and emissions. In Profile HMMs, hidden states are divided into three types: *Match* states, which describe important sequence elements, *Insert* states, which model noise, and *Delete* states, which do not emit a symbol and allow the model to skip a *Match* or *Insert* state. Emission distributions from the pHMM's *Match* hidden states capture an archetypal sequence or sequence fragment, and the likelihood of an observed sequence under a pHMM can be viewed as a measure of distance to the archetypal sequence encoded in the model. Blasiak et. al. [12] defined a simplified version of the pHMM, called the Simplified Local pHMM (SL-pHMM), which generates observed sequences using a contiguous sequence of *Match* states surrounded by *Insert* states. This structure simplifies the pHMM in a convenient way, as the only information needed to encode the model's entire hidden state configuration is the position of the first *Match* state.

The Factorial HMM [13] extends the basic HMM by postulating that the distribution over sequence elements depends on hidden states from multiple, parallel HMMs. If SL-pHMM factors are used, then the resulting Factorial SL-pHMM, with Gaussian emission distributions, operates very similarly to RS-DL.

Figure 2 shows an example configuration of *Match* hidden states in a Factorial SL-pHMM. This hidden state configuration leads to the same additive composition of parameters used to represent symbols of an observed sequence in RS-DL. The primary differences between RS-DL and the Factorial SL-pHMM lie in how the parameters of each model are constrained. In the Factorial SL-pHMM, the model's hidden states can be encoded in a vector, $\alpha_n^{(FHMM)}$ of length $C(T_n - K + 1)$, where C indicates the number of factors in the model, T_n is the length of the sequence, and K is the number of *Match* states in the SL-pHMM. Because the hidden state sequence in the SL-pHMM only allows a single start position for each chain of *Match* states, encoding the positions of initial *Match* hidden states requires that $\alpha_n^{(FHMM)}$ be constrained as $\alpha_{n,i}^{(FHMM)} \in \{0, 1\}$ and $\sum_{t=0}^{T_n - K} \alpha_{n,c(T_n - K + 1) + t}^{(FHMM)} = 1 \quad \forall c \in [1 \dots C]$. In contrast, the α_n-vectors in RS-DL are not explicitly constrained but are instead subject to L_1 regularization. Substituting an L_1 regularizer in RS-DL for the binary constraint in the Factorial SL-pHMM is advantageous, because it converts the combinatorial optimization problem associated with the MAP solution over hidden state configurations of the Factorial SL-pHMM into one that is more-easily solvable.

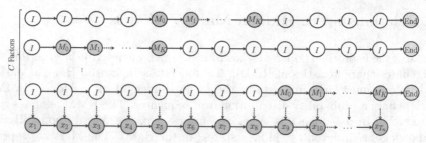

Fig. 2. The diagram above illustrates example hidden state assignments for the Factorial SL-pHMM. Sequences of SL-pHMM *Match* states are indicated by blue nodes with the text "M_k," indicating the kth *Match* state. *Insert* states are indicated by white-colored nodes with the text "I." SL-pHMM transition probabilities are defined so that only a single sequence of *Match* states per individual SL-pHMM can occur. For a Factorial SL-pHMM with Gaussian emission distributions, hidden states are associated with different weights which are summed over the C constituent SL-pHMMs (vertically in the diagram) to obtain the mean parameter used to generate the appropriate observed sequence element (in gray).

5   Experiments

We evaluate Relevant Subsequence Dictionary Learning using two types of measurements. First, we expect RS-DL to find meaningful subsequences within a dataset. This task is also referred to as "motif finding" [14, 15] (other authors [16] use a different definition of the term "motif"). We quantitatively assess motif finding on a synthetic dataset consisting of discrete sequences where the ground truth motif positions are known. We also qualitatively assess motif finding results on sets of both time-series and text sequences to verify that RS-DL can pick out portions of a sequence meaningful to humans. In the next sections we make a distinction between the terms "relevant subsequence pattern" and "motif". We use "relevant subsequence pattern" to indicate the pattern encoded in RS-DL parameters, and "motif" to denote subsequences selected from a dataset because of their association with a particular relevant subsequence pattern.

We also test RS-DL in sequence classification. We hypothesize that if RS-DL can discover informative subsequences with no access to label information, then these subsequence features will be effective for classification. In these experiments, RS-DL features are input to a one-nearest-neighbor classifier to isolate the effect of different feature representations.

5.1   Datasets

We employ four types of datasets to evaluate our algorithm. To evaluate the ability of RS-DL to discover known motifs, we generated a synthetic dataset of discrete-valued sequences containing three predefined subsequences. We also assessed motif finding ability using a set of continuous-valued ECG sequences[3]

---

[3] http://www.cs.ucr.edu/~eamonn/discords/ECG_data.zip

and the Associated Press (AP) dataset[4], consisting of English language text. We assessed classification ability using only continuous-valued sequences. These included both a synthetic dataset, which we call the "Bumps" dataset, and datasets from the University of California Riverside (UCR) Time Series Classification Database [17].

5.2   Finding Motifs in Synthetic Sequences

To verify basic motif finding abilities of RS-DL, we constructed a synthetic dataset, allowing us to control the location and frequency of motifs. The synthetic dataset consisted of 20 sequences, generated to contain up to three non-overlapping motifs. These motifs consisted of 5 repetitions of "a," "r," or "n," with a 10% chance at each motif position for a motif character to be replaced by a character generated uniformly from the full sequence alphabet of 20 possible characters. Non-motif sequence elements were chosen uniformly at random from the full 20-characters alphabet. Sequence lengths were generated uniformly at random from a range of 25 to 75.

To explore the behavior of the RS-DL model, we ran a number of experiments, varying the values of K, the length of the relevant subsequence pattern, from 3 to 7, and the values of λ, the L_1-regularization parameter, from 0.4 to 0.8 in steps of 0.05. We configured the algorithm to use at most one of each relevant subsequence pattern to reconstruct each sequence.

Figure 3a shows graphs of the average precision and recall associated with motifs recovered by the RS-DL algorithm over 20 trials for each configuration of K and λ. We counted a ground truth motif as "discovered" if its start position was within $\lfloor K/2 \rfloor$ of the motif returned by the RS-DL algorithm. To verify the upper limit of algorithm performance and to confirm the trend that ground truth motifs were associated with larger values of α than false positive motifs, we counted motifs as "not found" if their associated α values were below 0.25. Motifs were extracted by taking the subsequence of length K at the position of an associated non-zero component of the α vector.

Figure 3b, shows the output of the run of the RS-DL algorithm with the lowest mean-squared error (MSE) out of 20 random initializations. Columns in the figure display both values of α and motifs selected from the dataset sequence. Different dataset sequences are associated with different rows in the figure. In this run, low values of α are consistently associated with incorrectly discovered motifs (in red), and, out of four possible relevant subsequence patterns given by the model, only three are used, which is consistent with the ground truth.

Figure 3a shows that both precision and recall tend to increase as the value of K increases. In addition, the figure shows that if we set λ to a value that is too high, both precision and recall are degraded. This behavior, when varying λ, occurs because at high λ levels, the model becomes too sparse, reducing the number of motifs returned. In this case, we do not see a corresponding increase in precision because sparsity is only enforced in the number of relevant

---

[4] http://www.cs.princeton.edu/~blei/lda-c/ap.tgz

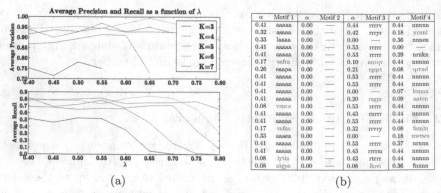

α	Motif 1	α	Motif 2	α	Motif 3	α	Motif 4
0.41	aaaaa	0.00	—	0.44	rrrrv	0.44	nnnnn
0.32	asnaa	0.00	—	0.42	rrryr	0.18	yrnnl
0.33	laaaa	0.00	—	0.00	—	0.36	nnnen
0.41	aaaaa	0.00	—	0.53	rrrrr	0.00	—
0.41	aaaaa	0.00	—	0.53	rrrrr	0.29	nrnkn
0.17	safrn	0.00	—	0.10	anrqv	0.44	nnnnn
0.26	eaapa	0.00	—	0.21	rgqrt	0.08	qrtnd
0.41	aaaaa	0.00	—	0.53	rrrrr	0.44	nnnnn
0.41	aaaaa	0.00	—	0.53	rrrrr	0.44	nnnnn
0.41	aaaaa	0.00	—	0.00	—	0.07	lenma
0.41	aaaaa	0.00	—	0.20	rsgpr	0.09	aaten
0.08	vsnca	0.00	—	0.53	rrrrr	0.44	nnnnn
0.41	aaaaa	0.00	—	0.43	rnrrr	0.44	nnnnn
0.41	aaaaa	0.00	—	0.53	rrrrr	0.44	nnnnn
0.17	vafsa	0.00	—	0.32	rrvry	0.08	famln
0.33	aaaea	0.00	—	0.00	—	0.18	nwtwn
0.41	aaaaa	0.00	—	0.53	rrrrr	0.37	nrnnn
0.41	aaaaa	0.00	—	0.43	rrrrm	0.44	nnnnn
0.08	lytia	0.00	—	0.43	rtrrr	0.44	nnnnn
0.08	uigpa	0.00	—	0.08	lkrei	0.36	fnnnn

(a)   (b)

Fig. 3. (a) The graphs in the left-hand figure depict precision and recall over 20 runs of the RS-DL model on a synthetic dataset. As the L_1-regularization term, λ, increases, fewer motifs are returned, leading to a drop in both recall and precision. As the length of the relevant subsequence patterns increase, precision and recall tend to increase. **(b)** Recovered α coefficients (left side of Figure b) and associated subsequences (right) from a low-error run (the run with the smallest MSE out of 20 random initializations) of the RS-DL algorithm on the synthetic dataset. The low-error run gave a precision of 0.7 (with an α cutoff of 0) and a recall of 1.0. The number of relevant subsequences patterns, C, in the model was set to four, while the number of ground truth motifs was three. Consistent with the ground truth, the model only used three relevant subsequences patterns to reconstruct the data. Incorrectly discovered motifs are depicted in red.

subsequences patterns used to reconstruct a sequence. Also from Figure 3a, the best precision scores were near 1.0, occurring with $K = 6$ and $\lambda = 0.65$ and filtering motifs with α coefficients less that 0.25. This result contrasts with top precision values of 0.5 (not shown in the figure) when the α filtering level is set to zero. The reason for this trend is illustrated in Figure 3b, where motifs associated with small α coefficients also tend to be less correlated with core relevant subsequence patterns.

To avoid low recall solutions it is possible to rerun the model for a number of trials with initial relevant subsequence patterns, $\mathbf{v}$, drawn from a standard Normal distribution. Because the RS-DL problem is non-convex, the optimization algorithm will converge to different areas in the parameter space depending on initial parameter settings. We found that, with our synthetic dataset, low-MSE runs consistently produced recall values of 1.0 (see Figure 3b). Selecting a low-MSE run also allows us to better take advantage of RS-DL's sparsity. For instance, if we set C, the number of relevant subsequence patterns, to 4, larger than the 3 ground truth motifs in our dataset, then low-MSE solutions return only 3 discovered motifs (higher error solutions find a fourth motif with noisy parameters).

5.3   Motifs in Time-Series Data

To show that RS-DL can pick out patterns meaningful to humans in continuous-valued sequences, we trained it on a single sequence of electrocardiogram (ECG) data, containing 3750 datapoints. The ECG sequence consists of a recording of electrical signals from the human heart measured at the surface of the skin. A plot of the signal (Figure 4) contains repeated patterns easily identifiable to humans. The ECG sequence also contains an anomalous motif, which, like the main set of patterns in the sequence, is easily identified by humans. We ran the RS-DL algorithm on the sequence with the L_1 regularization term, λ, set to .1 and the length of the relevant subsequence pattern, K, set to 150, and C, the number of relevant subsequence patterns, set to 15. In Figure 4, we plotted the relevant subsequence patterns learned by RS-DL associated with the largest 50 regression coefficients, α. Each pattern in the plot (top three graphs) consists of 150 values of the relevant subsequence pattern given by the constituent vector, $\mathbf{v}$, multiplied by its corresponding α coefficient. Summing over all of these plotted subsequences gives the approximate sequence reconstructed by RS-DL (bottom plot in green, offset by -1). The original sequence is also shown in the figure (bottom plot in blue). The MSE between reconstructed sequence and the original sequence was 0.98. As expected, the figure demonstrates how the relevant subsequence patterns in the upper graphs are strongly correlated with the human-perceptible patterns from the original sequence in the bottom graph. Another interesting property of the RS-DL decomposition shown in Figure 4 relates to the sparsity of the model. Only a three (6, 10, and 11) out of fifteen possible relevant subsequence patterns account for the main patterns in the sequence while additional patterns are responsible for increasingly fine-grained approximations. This type of behavior is similar to commonly-used orthonormal bases, such as the DCT basis, which consist of low frequency components that capture major trends, while high-frequency basis elements capture finer-grained variations. Another characteristic of the RS-DL solution is that the anomalous portion of the sequence is associated with a different relevant subsequence pattern (Relevant Subsequence Pattern 10) than the common ECG pattern. This characteristic shows how RS-DL can be used not only to find positions of recurrent patterns but also to distinguish between pattern types.

5.4   Motifs in Text Data

As an additional test of RS-DL's motif-finding ability, we trained the model on the Associated Press (AP) corpus. We preprocessed the corpus by removing words that occurred more than 500 times or in fewer than three documents. We then removed documents containing fewer than 10 words. The processed corpus size was 2213 documents. Finally, to make processing the text dataset tractable, rather than representing each word as a large binary vector (which would typically have a length of at least 10,000), we used the "word embedding" representation from Collobert et. al. [18]. These word embeddings are vectors in $\mathbb{R}^{50}$ and were constructed so that the Euclidean distance between a pair of vectors is be small if the meanings of the associated words are similar.

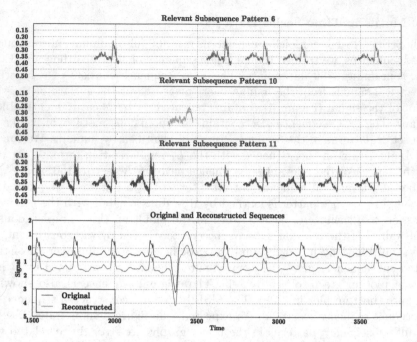

Fig. 4. A plot of the relevant subsequences patterns (upper plots) associated with the largest 50 coefficients of the vector α that were learned by RS-DL to approximate the ECG sequence (bottom plot, blue line). Only 3 out of the 15 possible relevant subsequence patterns appear in this set of 50. Relevant subsequence patterns learned by RS-DL are strongly associated with human-identifiable patterns in the sequence. The figure also shows that the approximation learned by RS-DL (bottom plot, green line) is very similar to the original sequence with an MSE of 0.98. The RS-DL approximation is offset by -1 on the y-axis to aid in presentation.

Figure 5 shows the top 15 examples, as ordered by the absolute value of the associated α coefficient, of the top four relevant subsequence patterns (out of $C = 10$ total possible relevant subsequence patterns) learned from a run of the RS-DL algorithm. Unlike text processing methods that treat words independently, RS-DL preserves the order of words within each document (minus words removed in the document preprocessing step). As the columns of five-word groups in the figure show, RS-DL, in minimizing reconstruction error over sequences of word embeddings, is capable of finding and grouping together meaningful sequences of words within the text. In the figure, all columns of discovered motifs share internally consistent semantic themes. Moreover, these themes tend to center around phrases containing important nouns. For instance, Motif 1 includes organization-related phrases like "product safety commission defended", "public health system plagued", and "environmental protection agency banned". Motifs 2 and 4 contain phrases including a person and occupation description such as "defense attorney thomas e. wilson", "district attorney william h. ryan", and "secretary james a. baker" in Motif 2 and "attorney michael rosen," "education secretary william bennett," and "assistant district attorney ted stein." Motif 3 is

Motif 1, max $|\alpha| = 5.357$					Motif 2, max $|\alpha| = 1.659$				
product	safety	commission	defended	record	defense	attorney	thomas	e.	wilson
community	service	act	provides	community	district	attorney	william	h.	ryan
standards	computer	industry	announced	technology	coalition	secretary	james	a.	baker
intelligence	support	ship	passed	black	assistant	secretary	john	h.	kelly
flight	training	manuals	brought	date	treasury	secretary	james	a.	baker
public	service	employees	received	raise	treasury	secretary	james	a.	baker
public	health	system	plagued	months	district	judge	william	t.	hart
foreign	debt	economy	warned	loss	navy	secretary	james	h.	webb
center	health	promotion	issued	warning	washington	secretary	james	a.	baker
korea	security	force	placed	armed	letter	secretary	james	a.	baker
u.n.	security	council	passed	use	washington	secretary	james	a.	baker
emergency	fund	funds	provided	embassies	date	secretary	james	a.	baker
environmental	protection	agency	banned	chemical	announced	secretary	james	a.	baker
forces	law	order	opened	fire	campaign	chairman	james	a.	baker
justice	information	organization	conducted	survey	assistant	attorney	stephen	a.	mansfield

Motif 3, max $|\alpha| = 1.635$					Motif 4, max $|\alpha| = 1.628$				
workers	discrimination	foreign	workers	impact	education	university	attorney	michael	rosen
authority	conduct	foreign	policy	direct	debt	education	secretary	william	bennett
detainees	paying	legal	fees	work	hearing	assistant	attorney	john	carroll
assets	include	private	loans	trade	assistant	district	attorney	ted	stein
workers	give	local	unions	grievances	assistant	immigration	commissioner	james	kennedy
program	financed	foreign	aid	assets	white	assistant	attorney	richard	roberts
bills	raise	environmental	protection	agency	questioning	assistant	attorney	john	carroll
planning	aid	foreign	organizations	perform	controversy	press	secretary	david	beckwith
penalties	sought	corporate	executives	release	talks	treasury	secretary	james	brady
problem	finding	foreign	aid	money	deputy	assistant	attorney	john	martin
policies	managing	public	affairs	front	phone	treasury	secretary	james	baker
law	banned	agricultural	products	coal	hearing	district	judge	edward	davis
proposal	noted	natural	resources	issue	list	assistant	attorney	william	reynolds
money	available	public	services	whose	ruling	district	judge	edward	davis
reasons	tough	economic	conditions	particular	acting	prime	minister	ben	jones

Fig. 5. The figure above shows motifs discovered by the RS-DL model in the Associated Press corpus. It lists the top 15 motifs by α coefficient of the top four (out of ten possible) relevant subsequence patterns. Motifs found by RS-DL have, in general, captured sets of semantically coherent phrases. Motifs 1 and 2 contain phrases including organization and noun/concept phrases while Motifs 2 and 4 contain phrases including a person and occupation descriptions.

centered on organizations and concepts like "natural resources," "public services," and "tough economic conditions."

5.5 Classification Experiments

To assess whether features derived by RS-DL are useful for classification, we compared the performance of these features on both a synthetic dataset of our own design and five UCR Time Series datasets that satisfied the underlying assumptions of our model. Because RS-DL selects subsequences, we do not expect features from the algorithm to be effective for classification when discriminative information between sequence categories lies in global trends over an entire sequence or if the order of different patterns within a sequence is highly correlated with its category. Similarly, because RS-DL is a sparse regression algorithm, we expect relevant subsequence patterns to be matched to high-magnitude areas of dataset sequences. Therefore, if dataset sequences contain large-magnitude areas (e.g. spikes in an ECG sequence), but discriminative information found elsewhere in the sequence, we do not expect RS-DL features to be effective for classification.

Table 2. Classification results using RS-DL features on the UCR Time Series datasets. The "Sequence", "DTW", and "RS-DL" columns give error rates from the one-nearest-neighbor algorithm using the Euclidean distance between sequences, Dynamic Time Warping scores, and Euclidean distance between RS-DL features respectively. RS-DL features improved the classification error for all three datasets.

Dataset	# Categories	# Train	# Test	Sequence	DTW	RS-DL
Bumps (synthetic)	2	50	50	0.460	0.140	**0.056**
CBF	3	30	900	0.148	**0.030**	0.108
Coffee	2	28	28	0.250	0.214	**0.171**
DiatomSizeReduction	4	16	306	0.065	0.042	**0.028**
ECGFiveDays	2	23	861	0.203	0.249	**0.095**
TwoLeadECG	2	23	1139	0.253	0.073	**0.035**

With these assumptions in mind, we generated a set of continuous sequences, which we call the "Bumps" dataset[5] (see Figure 6c). Each sequence in this dataset contains two large magnitude bumps placed at random and without overlap. In the negative category one out of the two bumps in each sequence contains a divot. We also selected five datasets from the UCR Time Series database that conform to the underlying assumptions about RS-DL: CBF, Coffee, DiatomSizeReduction, ECGFiveDays, and TwoLeadECG. These datasets consist of sequences that contain large magnitude patterns occurring in all or nearly all sequences, satisfying the assumptions needed for RS-DL to extract useful features.

We ran RS-DL with randomly initialized v arrays for ten trials on all sequences in both the training and test sets, excluding label information, for each dataset. For all experiments, we set $C = 10$, K to 30% of the sequence length, and $\lambda = 3.0$. We also enabled the restriction on the LARS algorithm (see Section 3.2) where only a single relevant subsequence pattern of each type was used. For each sequence, we created feature vectors by concatenating the subsequences associated with each relevant subsequence pattern. Table 2 shows a comparison of classification errors using the one-nearest-neighbor algorithm on (i) features given by treating sequences as vectors in Euclidean space, (ii) Dynamic Time Warping (DTW)[6] [19] distances between sequences, and (iii) Euclidean distance between RS-DL feature vectors. As assessed by McNemar's test [20], RS-DL features reduce classification error over raw sequence vectors with p-values of less than 0.014 for all datasets. For all datasets except for the CBF dataset, RS-DL features improved on the classification error over DTW. Here, all results were significant with p-values of less than 0.01, except for the Coffee dataset, where RS-DL's improvement over DTW was significant with a p-value of 0.17.

Figure 6 shows examples of positive and negative category sequences from three of the classification datasets. In each case, RS-DL features lead to improved time-series category prediction by isolating large-magnitude trends in subsequences shared across the set of sequences (as shown in the upper portions of each plot in the figure). Constructing features from these isolated subsequences

---

[5] This dataset can be found at http://cs.gmu.edu/~sblasiak/RS-DL.tar.gz

[6] DTW scores were computed in R using http://dtw.r-forge.r-project.org/

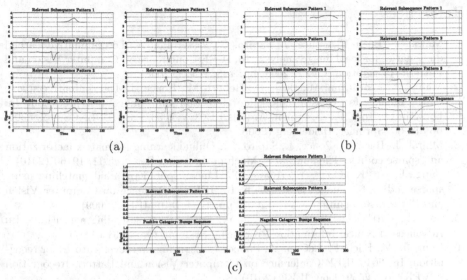

(a)   (b)

(c)

Fig. 6. Figures **a**, **b**, and **c** above show the top two (by α value) relevant subsequence patterns that approximate positive (bottom blue) and negative category (bottom red) sequences in the ECGFiveDays, TwoLeadECG, and our synthetically-generated Bumps datasets respectively. For each of these datasets, RS-DL features improve classification performance by picking out similarly shaped subsequences from different dataset categories. Classification performance improves because class distinctions occur in minor variations in the major trends captured by RS-DL. After processing by RS-DL, these minor variations can more easily be distinguished by standard classification algorithms.

aligns these major subsequence trends, allowing minor variations that occur between the positive and negative sequence categories to be more-easily distinguished. When variations in the general trend are highly correlated with a category label, then the feature isolation provided by RS-DL can lead to more accurate classification.

6   Conclusions

In this paper, we have presented Relevant Subsequence Dictionary Learning, a novel method for adapting Sparse Dictionary Learning to discover interesting subsequence patterns across sets of sequences. RS-DL is related to standard statistical models over sequences through a version of the Factorial HMM with specially formulated restrictions on transition probabilities. In a series of experiments, we have shown that RS-DL can discover useful information across a variety of sequence domains. In addition, as demonstrated on time-series data, sequence features extracted using RS-DL can improve sequence classification performance.

Acknowledgments. This work was supported by NSF grant IIS-0905117 and NSF Career Award IIS-1252318 to HR.

References

1. Mairal, J., Bach, F., Ponce, J., Sapiro, G., Zisserman, A.: Discriminative learned dictionaries for local image analysis. In: IEEE Conference on Computer Vision and Pattern Recognition, CVPR 2008, pp. 1–8. IEEE (2008)
2. Mairal, J., Leordeanu, M., Bach, F., Hebert, M., Ponce, J.: Discriminative sparse image models for class-specific edge detection and image interpretation. In: Forsyth, D., Torr, P., Zisserman, A. (eds.) ECCV 2008, Part III. LNCS, vol. 5304, pp. 43–56. Springer, Heidelberg (2008)
3. Mairal, J., Bach, F., Ponce, J., Sapiro, G.: Online learning for matrix factorization and sparse coding. The Journal of Machine Learning Research 11, 19–60 (2010)
4. Yang, J., Yu, K., Gong, Y., Huang, T.: Linear spatial pyramid matching using sparse coding for image classification. In: IEEE Conference on Computer Vision and Pattern Recognition, CVPR 2009, pp. 1794–1801. IEEE (2009)
5. Lee, H., Battle, A., Raina, R., Ng, A.Y.: Efficient sparse coding algorithms. In: Advances in Neural Information Processing Systems 19, p. 801 (2007)
6. Boureau, Y., Bach, F., LeCun, Y., Ponce, J.: Learning mid-level features for recognition. In: 2010 IEEE Conference on Computer Vision and Pattern Recognition (CVPR), pp. 2559–2566. IEEE (2010)
7. Aharon, M., Elad, M., Bruckstein, A.: K-svd: Design of dictionaries for sparse representation. IEEE Transactions on Signal Processing 54(11), 4311–4322 (2006)
8. Efron, B., Hastie, T., Johnstone, I., Tibshirani, R.: Least angle regression. The Annals of Statistics 32(2), 407–499 (2004)
9. Friedman, J., Hastie, T., Höfling, H., Tibshirani, R.: Pathwise coordinate optimization. The Annals of Applied Statistics 1(2), 302–332 (2007)
10. Beck, A., Teboulle, M.: A fast iterative shrinkage-thresholding algorithm for linear inverse problems. SIAM Journal on Imaging Sciences 2(1), 183–202 (2009)
11. Eddy, S.R.: Profile hidden markov models. Bioinformatics 14(9), 755 (1998)
12. Blasiak, S., Rangwala, H., Laskey, K.B.: A family of feed-forward models for protein sequence classification. In: Flach, P.A., De Bie, T., Cristianini, N. (eds.) ECML PKDD 2012, Part II. LNCS, vol. 7524, pp. 419–434. Springer, Heidelberg (2012)
13. Ghahramani, Z., Jordan, M.I.: Factorial hidden markov models. Machine Learning 29(2-3), 245–273 (1997)
14. Bailey, T.L., Elkan, C.: Fitting a mixture model by expectation maximization to discover motifs in biopolymers. Vectors 1, 2
15. Buhler, J., Tompa, M.: Finding motifs using random projections. Journal of Computational Biology 9(2), 225–242 (2002)
16. Mueen, A., Keogh, E., Zhu, Q., Cash, S., Westover, B.: Exact discovery of time series motifs. In: Proc. of 2009 SIAM International Conference on Data Mining: SDM, pp. 1–12 (2009)
17. Keogh, E., Xi, X., Wei, L., Ratanamahatana, C.A.: The ucr time series classification/clustering homepage (2011)
18. Collobert, R., Weston, J., Bottou, L., Karlen, M., Kavukcuoglu, K., Kuksa, P.: Natural language processing (almost) from scratch. The Journal of Machine Learning Research 12, 2493–2537 (2011)
19. Sakoe, H., Chiba, S.: Dynamic programming algorithm optimization for spoken word recognition. IEEE Transactions on Acoustics, Speech and Signal Processing 26(1), 43–49 (1978)
20. Dietterich, T.G.: Approximate statistical tests for comparing supervised classification learning algorithms. Neural Computation 10(7), 1895–1923 (1998)

Future Locations Prediction
with Uncertain Data*

Disheng Qiu[1], Paolo Papotti[2], and Lorenzo Blanco[1]

[1] Università Roma Tre, Roma, Italy
{disheng,blanco}@dia.uniroma3.it
[2] Qatar Computing Research Institute (QCRI), Doha, Qatar
ppapotti@qf.org.qa

Abstract. The ability to predict future movements for moving objects enables better decisions in terms of time, cost, and impact on the environment. Unfortunately, future location prediction is a challenging task. Existing works exploit techniques to predict a trip destination, but they are effective only when location data are precise (e.g., GPS data) and movements are observed over long periods of time (e.g., weeks). We introduce a data mining approach based on a Hidden Markov Model (HMM) that overcomes these limits and improves existing results in terms of precision of the prediction, for both the route (i.e., trajectory) and the final destination. The model is resistant to uncertain location data, as it works with data collected by using cell-towers to localize the users instead of GPS devices, and reaches good prediction results in shorter times (days instead of weeks in a representative real-world application). Finally, we introduce an enhanced version of the model that is orders of magnitude faster than the standard HMM implementation.

1 Introduction

The ability to predict future movements can enable novel applications in a wide range of scenarios. For example, in the context of Location-Based Services, it can be used to deliver advertising to customers approaching shops of interest.

Work has been done to predict future locations for moving objects, both in the short term [20] and in longer timeframes [11]. These works rely on two hypotheses: (i) moving objects follow some patterns in (most of) their movements, and (ii) such movements can be observed with a certain accuracy. The first argument relies on the intuition that most people drive among their points of interest along a usually small number of routes (e.g., from home to work, from work to the gym). The same observation holds for public transportation (such as buses or flights), and even animals in their migrations. The second argument is due to the increasing popularity of GPS devices in the last years. On one hand, such systems are becoming popular because of the large diffusion of smartphones equipped with GPS sensors. On the other hand, it is not reasonable to assume

---

* Supported in part by a Working Capital 2011 grant from Telecom Italia.

H. Blockeel et al. (Eds.): ECML PKDD 2013, Part I, LNAI 8188, pp. 417–432, 2013.

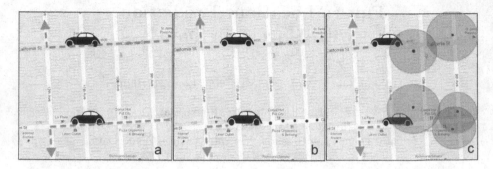

Fig. 1. Two cars running in different streets are going to turn in opposite directions. (a) shows their complete trajectories; (b) and (c) show how they are observed before they turn with GPS and cell-towers data, respectively.

the availability of these data on long periods. In fact, previous algorithms were designed with car navigation kits in mind, which do not have any constraints on power consumption. On the contrary, battery life is a primary issue for mobile users and GPS sensors are known for their high battery consumption. Existing technologies cannot be used for applications where movements of the users have to be continuously observed. For instance, consider the following application.

Carpooling is defined as a meeting of two or more people to share a car and travel together. Carpooling has a strong appeal for multiple reasons, but it experiences rather low levels of participation because of the complexity of social and work schedules, which make such arrangements hard to plan. Existing services (such as iCarpool and Avego) allow drivers and passengers to arrange occasional shared rides on short notice with smartphone applications. Such services facilitate one-time ride matches but users still have to define the routes and times of their trips, exactly as in traditional carpooling. This requirement is unacceptable in a context where users want more flexibility and short-term commitments.

This problem can be solved with methods to automatically predict the route and destination of the drivers, but existing solutions cannot be applied as they require the user to constantly keep the GPS sensor active. This is a tight requirement, as prediction techniques require to collect the routes of the users (to build the patterns repository) and to continuously verify their actual position.

Our approach tackles this problem by relying on location information inferred by the cell-towers the phone is connected to. Cell towers multilateration is the standard solution for location aware applications that need to run in background continuously: a standard network connectivity allows the smartphone OS to access a repository of GPS coordinates for the physical towers and to compute an estimated position for the user based on signal strength. Therefore network based localization comes at no additional cost and reduces battery consumption.

Consider Figure 1a, where two cars run on parallel streets before turning in opposite directions. Figure 1b shows how the cars are observed with GPS equipment, every dot is an *observation*. Given the current observations and an history of previous trips of the user, a system can match the current trip with the repository to find the most probable destination for it. This is possible when there is a sufficient number of previous trips and one of them matches to some degree with the actual one. But what happens with noisy observations, such as in Figure 1c? Inferring location data with cell-towers is a difficult task: real-world data shows that the approximation in urban areas varies between 400 and 1600 meters. Predictor systems have problems in the matching step when locations are inferred with cell-towers, thus the prediction can fail.

To overcome these issues, we introduce a new prediction model, based on the Hidden Markov Model (HMM) [18], that naturally supports the uncertainty of the observations. HMMs have been used before in similar setting for *destination prediction* [19], and for handling *uncertain location data* [15], but our formulation tackles the two problems in a unified, principled solution. In particular, we designed a *future location prediction system* with the following contributions:

I We introduce a HMM for future locations prediction that naturally supports imprecise data and efficiently predicts both future locations (i.e., final destinations) and their corresponding trajectories (i.e., travel routes).
II A direct implementation of the HMM leads to unacceptable execution times in a real-world application. We therefore present a refined version with lossless optimization and a faster one with approximate solutions.
III We conduct an experimental study on the models and compare them with a state-of-the-art solution in simulated and real environments. For the synthetic scenarios, we introduce a tool to generate routes for moving objects.

In the following, Section 2 discusses related work and Section 3 formally introduces the problem. Sections 4 and 5 present our algorithm and its optimization, respectively. Finally, Section 6 validates the proposed approach with extensive experiments and Section 7 discusses future directions of research.

2 Related Work

Many studies have faced the problem of predicting future locations and routes for moving objects [16]. Considering the adopted techniques, they can be divided in two main trends. The first one is about the prediction of paths in an euclidean space [11,20]: given the location and velocity of an object, the future location of the object is predicted with a function. The second trend is based on pattern matching: the algorithms compare the current location of the object with previously observed routes and return the best match as a prediction. Considering the domain of application, these can be further divided in three groups.

Prediction Based on GSM Network. Solutions for the prediction problem with wireless cellular communication networks differ from our approach as user history is not considered and they focus on optimizing network resources [7,17,21]. In [22], the authors build transaction rules based on neighbor cell towers with a support computed from the user history. The approach does not predict the final goal but only the most probable next cell tower.

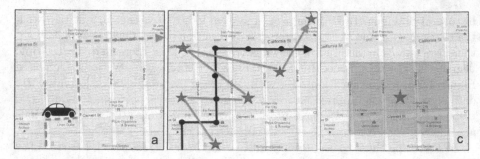

Fig. 2. A car running across an urban area: **a** shows its real trajectory; in **b** dots and stars show the trajectory with GPS and Network data, respectively; **c** shows one way to distribute the error rate for an observed block

Prediction Based on Movement. Proposals that focus on predicting the destination and routes require precise locations [5,9,19], whereas our approach can handle uncertain observations. The notable exception is CRPM [6], as it can tolerate different kinds of disturbance in trajectory data. CRPM relies on a frequency analysis and adopts different heuristics to manage imprecise data.

Prediction Based on Location. Other methods focus on predicting the destination without considering the current partial route [1,2,12,14]. They rely on different approaches, such as a prediction model leveraging the history of group of users [14], or a route recommendation system considering user interest and a database of geotagged photos [12].

HMMs have been successfully applied for many tasks related to sequence clustering [18], such as map matching with noisy observations [15], and to hidden intents recognition [1,19]. We also differ from [4], where a traditional Markov model is used to classify different observations to an internal taxonomy. The work proposes a one-to-one mapping to couple state-symbol, but due to the absence of hidden states, Baum-Welch and Viterbi algorithms are not applied. Our proposal is the first attempt to addresses the problem of predicting the destination and the route to reach it with a HMM formulation over noisy observations.

3   Preliminaries

Road Networks. Our model does not require road networks given as input, as it infers the underlying roads from the observations. We introduce a simple urban trip to describe our setting. Consider the real trajectory of a car in Figure 2a. As in [23], we divide the area of interest in blocks and use the centers of such areas as observations points. Blocks can have variable size and their dimension depends on the granularity required by the actual application. Once blocks are fixed, we associate the reported location with the center of the block that contains the location. We model a road network as a directed graph $G = (V, E)$, where V is the set of vertices, E is the set of edges. A vertex $v \in V$ represents a block in the road network, an edge $(v, v') \in E$ represents blocks connected by a road.

Objects of Interest. We assume a set of moving objects $d \in D$ on the edges E of the road network G. At a given time t, each object d has its current spatial location $d.b$ (i.e., a block) and the predicted route $d.r^*$. A predicted route is a sequence of predicted locations $d.b_1^*, \ldots, d.b_n^*$, where $d.b_n^*$ is the final *goal* (or *destination*) of the trip.[1] The real route of an object d is denoted by $d.r$ and it is a sequence of locations $d.b_1, \ldots, d.b_n$. A route between two goals is called a *trip* or *trajectory*. The size (cardinality) of the set of objects D is denoted as $|D|$. The shortest path between locations $d.b$ and $d.b'$ in G is defined as $||d.b, d.b'||$.

Problem Statement. The problem faced in this work is the following: *given a time t, a moving object d with route $d.r$, and the subroute of d until t, set up a probabilistic prediction model s.t. the most probable predicted route corresponds to the route $d.r^*$ that minimizes the distance between $d.r^*$, $d.r$.*

The effective route is not available in general and the goal of our work is to minimize the distance between $d.r^*$ and $d.r$ with efficient algorithms. In other words, the goal is to predict the future locations of the moving objects and find the most similar route w.r.t. their real movements.

4   A Model for Destination and Route Prediction

The problem above can be modeled as a Markov chain reproducing the state of a system with a random variable for the final destination that changes through time. A state coincides with an observation and the distribution for this variable depends only on the distribution of the previous observations. Given a Markov chain and a sequence of observations, it is possible to predict the resulting state distribution [2]. Unfortunately, such approach fails in practice: (i) it predicts only the final destination and not the route; (ii) it is designed to work on precise GPS data, thus it suffers the problems we discussed above when used with noisy location information. The first problem can be faced with a model where l is a known location and every state represents a transition from location l to l' with a pair $(l'|l)$ [23]. By considering a n order Markov Model, this approach predicts all the possible routes for the user by considering, for every transition, all the n transitions that it has seen before. Unfortunately, Markov chains are not flexible w.r.t. the order n (which is fixed and has to be chosen at design time), so it is not easy to find a compromise between the right complexity and a good prediction.

Given the limits of Markov models, we build our solution on top of a Hidden Markov Model (HMM), for which the modeled system is assumed to be a Markov process with unobserved (hidden) states. In our solution, a state represents a pair (block, goal) $s_i = (b|g)$, and each state can be associated to multiple observations O_{s_i} with a given distribution of probability, i.e., O_{s_i} is the set of blocks observable from s_i. If we ignore hidden states observations, a moving object in a trip towards a given goal g covers the states $S = \{(b_i|g), (b_j|g), \ldots, (b_k|g)\}$. Given the sequence of states S, multiple observed sequences are possible. In fact, there is no one-to-one mapping from the state to the observation. Given a sequence of states $S = \{s_1, \ldots, s_n\}$, the possible observation sequences are $b_1, b_2, \ldots, b_n$

---

[1] A goal is a location visited by the observed object for long periods of time.

with $b_i \in O_{s_i}$. In our setting $|O_{s_i}|$ grows if the measurement error of the block associated to the state is high and it decreases to 1 if measurement error is low (e.g., GPS observations). In Figure 2c, $|O_s| = 9$, because 9 blocks are covered by the measurement error of the observation.

Given an observation b, the possible related states are $s_i : b \in O_{s_i}$. Notice that this formulation models both noisy observations and user intents. In fact, given b_1 and b_2 close to observed b, possible states are $(b_1|g_1)$ or $(b_2|g_1)$, modeling uncertainty w.r.t. the current location, but also $(b_1|g_1)$ and $(b_1|g_2)$, modeling uncertainty w.r.t. the future location.

At prediction time, the HMM is used to calculate what is the most likely sequence of states associated to the sequence of blocks observed so far: the last state $(b_k|g)$ models that the location is b_k and the predicted goal g.

We build our Hidden Markov Model as follows:

- $A = \{a_{i,j}\}$ is the *state transition probability matrix*, where $a_{i,j}$ is the transition probability from s_i to s_j. If $s_i = (b_n|g)$ and $s_j = (b_m|g)$, $a_{i,j}$ is the probability to move from b_n to b_m with same goal g, when the object is in s_n.
- $B = \{s_j(o_k)\}$ is the *observation probability distribution matrix*. If $s_j = (b_n|g)$ and $o_k = b_m$, then $s_j(o_k)$ is the probability to see b_m, when in b_n with goal g.
- π is the *initial state distribution*.

HMM is a useful representation for our application. Consider again the example in Figure 2. Dots in Figure 2b show the trajectory with observed data as blocks (i.e., data observed with GPS), while stars show the trajectory with the blocks identified with the typical degraded data observed with systems based on cell-towers only. Data retrieved with cell-towers are approximate, but we can use a function to assign a certain probability to an observed block and spread the rest of the information over the 8 blocks around it as shown for the third observation in Figure 2c. As cell-towers localization services in smartphones give an estimation of the error of measurement, the number of blocks that have to be considered dynamically adapts to the quality of the observations.

Training. *Baum-Welch* algorithm [18] is used to train the model. Given a sequence of observations, it computes a maximum likelihood estimation, i.e., it produces A, B, and π such that the probability that the given sequence is observed is maximized. Unfortunately, it cannot be directly applied in our setting:

1. In our setting the system receives as input many sequences of observations (i.e., different trips), but the algorithm is designed to train a single sequence.
2. We want to train our model with a route to a destination. In the training, for each observation the corresponding state has two pieces of information: the current block b and the final destination g. The algorithm, given an observation for block b for which two states exist with distinct goals g and g', would train both states $(b|g)$ and $(b|g')$. This is motivated by the fact that no constraints are present in the HMM about the modeled goal.
3. The algorithm complexity is $O(\mathcal{N}^2\mathcal{T})$, where $\mathcal{N}$ is the number of states and $\mathcal{T}$ is the size of the observation sequence. Its performance is unacceptable in settings with a large number of states and long sequences of observations.

To overcome the above issues, we implemented an ad-hoc training. First, we split the states domain by considering a matrix for each destination. Instead of training a matrix A of size $\mathcal{N}=(\mathcal{GB})^2$, where $\mathcal{G}$ is the number of goals and $\mathcal{B}$ is the number of blocks, we separately train $\mathcal{G}$ matrices of size $\mathcal{B}^2$ (thus addressing issue 3). Second, we split the training set in groups of trips with the same destination; this allows us to train our model with multiple sequences (thus addressing issue 1) for which each goal is known and can be used to train the correct sequence of states (thus addressing issue 2).

From the algorithmic point of view, we initialize A with flat values, $a_{i,j}^0 = avg = \frac{1}{N}$. Given a sequence of trips $T = \{trip^1, \dots, trip^{|T|}\}$, for each $trip^t$ we train a $\mathcal{B}^2$ matrix $S^t = \{s_{i,j}^t\}$ associated to the observed destination g^t and we interpolate S with A. In this way we constrain the training process to consider one trip at a time and with the interpolation process we train the original matrix A. In particular, each value $a_{i,j}^t$ is updated as follows:

$$a_{i,j}^t = \begin{cases} K \cdot s_{i,j}^t + (1 - K) \cdot a_{i,j}^{t-1} \text{ , iff the goals of } s_i \text{ and } s_j \text{ are equal to } g^t \\ F \cdot avg + (1 - F) \cdot a_{i,j}^{t-1} \text{ , otherwise} \end{cases}$$

Where K is a parameter called "learning factor", F is the "forgetting factor", $s_{i,j}^t$ is the trained value in S^t, and $0 \leq s_{i,j}^t \leq 1$, $0 \leq F \leq 1$, $0 \leq K \leq 1$. Blocks of states $s_{i,j}^t$ have to match the blocks of states $a_{i,j}^t$. By tuning K and F, we can find a trade-off between the importance given to old trips and new ones.[2] We train our model by repeating this task for every trip in T.

Predicting Destinations and Routes. Given a sequence of observations O, we use the *Viterbi algorithm* [18] to efficiently compute the most likely states sequence associated to the observations. The algorithm matches the current observations O (unaware of the final destination) with all the previous trips and finds the one that maximizes the probability of the given sequence of observations. By applying Viterbi we obtain the sequence of states $S = \{s_1, s_2, \dots, s_n\}$ that maximizes the probability of observing O. S models the information about the final destination. In the actual implementation we consider only the last predicted state s_n and its goal.

Once we are able to recognize the current state of an object, we can find its destination and the route to reach it. We compute how the object reaches the predicted goal in s_n by using Dijkstra's algorithm in a connected graph where the nodes are the states of the model, and the costs on the edges are the probabilities from the matrix A. Our goal is to find the most likely path by maximizing the probability to get from the current state to the destination.

This approach can predict a route $d.r^*$ that is closer to $d.r$ than each of the previous observed trips. In fact, given a training set with a route observed multiple times, the noise in the observations is likely to be in different portions for each of the previous trips. But, when the shortest path is computed from A, the algorithm combines pieces matching the current route from multiple previous

---

[2] We experimentally identified $F = 0.05$ and $K = 0.9$ as the optimal parameters.

trips, thus automatically filtering noisy observations. Experiments show that this behavior improves the quality of the predicted path w.r.t. alternative approaches.

Dijkstra algorithm is one of the several algorithms that can be used to compute the route to the predicted goal given a trained matrix A, while the prediction is done by Viterbi algorithm. The role of the HMM is to feed the weights to the graph used by the search algorithm to efficiently find a clean $d.r^*$.

5   Revisiting the Model to Improve Performance

State Pruning. For a fixed $\mathcal{T}$, all algorithms are $O(\mathcal{N}^2)$ with $\mathcal{N} = \mathcal{G} * \mathcal{B}$, where $\mathcal{B}$ is the number of blocks and $\mathcal{G}$ the number of goals. In particular, the cardinality of A is $(\mathcal{G}\mathcal{B})^2$, thus the matrix grows by a factor $\mathcal{B}^2$ for each new goal. This easily leads to a very large number of states. A natural approach to the problem is to reduce the size of the matrices. In fact, reducing the matrix dimension has direct effects in response time in all stages. We reduce its size with two pruning techniques.

Lossless Pruning In the model, **all** possible pairs $(b|g)$ are considered as states, even if most of them are useless as many pairs are never observed in the user routes. For example, it is possible to remove all the blocks that are not crossed by an objects in its trips. In our experiments with a large urban area, this simple heuristic removes up to 85% of the blocks. Moreover, there are blocks crossed to reach multiple distinct goals, but there is a larger number of blocks crossed to always reach the same goal. In the original model, the set of all observed blocks L contains different noisy states such as follows: given f distinct goals and the user going to goal g_i through block b, state $(b|g_i)$ is generated together with $(b|g_1), \ldots, (b|g_{f-1})$. These $f - 1$ states would be part of the model matrix, even if b has never been observed with the user moving to a goal different from g_i. Therefore $f - 1$ states are pruned to reduce the size of the matrix.

Consider an incremental creation of the matrix A of dimension $\mathcal{N}$, by adding a new (i.e., never seen before) sequence $trip^{t+1}$ of size M to a trained model; the original model would generate a new matrix A of dimension:

$$(L + M)(\mathcal{G} + 1) = \mathcal{N} + (M\mathcal{G} + M + L)$$

On the contrary, the optimized model generates a new matrix of dimension $\mathcal{N} + M$, because a new goal is added only for the blocks that lead to it and the already trained blocks are not involved.

This optimized model is already orders of magnitude faster than the original model in both the training and the prediction, and the results of the prediction are of same or better quality. In the following, we will refer to this model implementing the lossless pruning as the **standard HMM**.

Approximate Pruning With network observations the accuracy of the measures varies between 400 and 1600 meters. With such uncertain signals we spread the information over a number of blocks that depends on the observation quality. This approach leads to a big A matrix: from GPS to network data there is at least a five times increase in the number of observed blocks.

A possible solution is to manage the blocks added for error compensation in the B matrix, without reporting all blocks in A. In the standard model, when we observe a block, **all** the blocks within its measurement error are "promoted" to states inside the HMM. Here instead, we do not create new states for each block reachable within the measurement error, but all the blocks added to distribute the error for an observation are stored in the emission matrix B.

We pushed this idea further, by doing the same pruning also for observed blocks. The basic intuition is the following: suppose that during a trip, observed through network data, we obtain in a sequence three blocks: b_1, b_2, b_3, with b_2 inside the error area of b_1. Given b_1, it is added to A, but b_2 is stored in the emission matrix B, while b_3 is promoted to A if it is out of the measurement error of b_1. In the approximate model only observed blocks out of the measurement error are promoted as states. In Section 6 we show how this aggressive pruning drastically reduces the execution times without reducing the prediction quality.

Notice that uncertainty is managed by the model and the pruning is applied only for efficiency purpose. The complexity of HMM algorithms is $O(\mathcal{N}^2\mathcal{T})$ and it is not affected by the dimension of matrix B. In fact, HMM is designed to support a large set of possible observations, and its growth does not increase the complexity of the system. Consider an observation sequence of 10 distinct blocks with an estimated error of 600 meters. If for each block we add the 8 blocks around it, adding this sequence adds $10*9$ new observations. By managing the 8 extra blocks as observations, we add only 10 new blocks and the remaining $10*8$ are managed without an overhead in the model. In the following, we will refer to this model implementing the approximate pruning as the **smart HMM**.

Split Training. Splitting if the training also brings a performance improvement. Instead of working with a matrix of size $\mathcal{N}^2 = (\mathcal{G}*\mathcal{B})^2$, we train $\mathcal{G}$ matrices of size $\mathcal{B}^2$ so that, given a trip to a goal, we train it only for the B matrix of its goal. Given $|T|$ sequences of observations, $\mathcal{G}$ goals, and $\mathcal{B}$ known symbols, the cost for a training of the original matrix is $|T| \cdot \mathcal{T}(\mathcal{G}\mathcal{B})^2$ (considering every observation of fixed length $\mathcal{T}$). The cost for the split training is instead $|T|\cdot\mathcal{T}(\mathcal{B})^2$ for the Baum-Welch execution and $|T| \cdot (\mathcal{G}\mathcal{B})^2$ for the interpolation step, with an important improvement w.r.t. the original training:

$$\frac{Original}{Split} = \frac{|T| \cdot \mathcal{T}\mathcal{G}^2\mathcal{B}^2}{|T| \cdot \mathcal{T}\mathcal{B}^2 + |T| \cdot \mathcal{G}^2\mathcal{B}^2} = \frac{\mathcal{T}\mathcal{G}^2}{\mathcal{T}+\mathcal{G}^2}$$

For 6 goals and $\mathcal{T} = 30$, the improvement is $\frac{1080}{66} = 16.36$.

Maximum Observation Length. Response time can be improved also by optimizing the observation sequence length $\mathcal{T}$. This applies in the Viterbi algorithm, as we do not need the complete sequence of states, but cannot be used with Baum-Welch, because in the model setup we need a training for the entire observation sequences. We experimentally observed that only significant changes to the size of the observations sequence affect the results. A short sequence is not sufficient to recognize a route, thus providing wrong predictions, while large sizes do not improve predictions quality while increasing execution times. In the experiments we set $m=10$ after verifying the prediction accuracy in real data.

6   Experiments

We implemented and tested alternative models to compare predictions quality and execution times. Algorithms have been implemented in Java and tested on an i5 2.4GHz CPU with 8GB of RAM.[3]

Metrics. We consider a training set of routes T, and an evaluation dataset E. Two kinds of dataset are used: *synthetic*, built with a route generator based on features obtained from real data, and *real-world*, made with a collection of real users movements in a large urban area. The real route $d.r$ is known for every object $d \in E$ and is used as ground truth for evaluation.

In the evaluation, we use as input a fragment i of the real route (e.g., the first 10% observations) to compute the predictions. For each object $d \in E$, we collect the predicted route $d.r^* = d.l_1^*, \ldots, d.l_n^*$, with the last location being the predicted goal. We test eight executions with an increasing percentage of input data from $d.r$ (from 10% to 80%).

We rely on two metrics for quality evaluation. In the first metric, we match the predicted goal $d.l_n^*$ with the effective goal in $d.r$. A match is valid if the predicted goal is in the area covered by the measurement error of the block containing the effective goal. We report the percentage of errors in the goal prediction in a set of executions as the Wrong Goal Percentage (**WG-P**). The second metric is based on our definition of distance and represents the quality of the predicted route. We define the distance between $d.r^*$ and $d.r$ by using the *Levenshtein* edit distance [13] with the following costs:
- Substitute: the cost to replace a wrong location $d.b_i^*$ with the correct one $d.b_i$ is computed as the distance between them: $||d.b_i^*, d.b_i||$.
- Add: the cost to add a location $d.b_i$ to $d.r^*$ is the distance between the location to be added and the last predicted location in the route: $||d.b_n^*, d.b_i||$.
- Delete: the cost to remove a location $d.b_j^*$ from $d.r^*$ is the distance between the location to be deleted and the closest location $d.b_j \in d.r$: $||d.b_j^*, d.b_j||$.

For settings with several goals, an incorrect goal prediction can be close to the correct one. We therefore distinguish two sets of routes: the ones with correct predicted goal are measured in the Correct Goal Route Distance (**CG-RD**); those with erroneous predictions are in the Wrong Goal Route Distance (**WG-RD**).

Synthetic Data. Despite there exist tools to generate routes [8,10], we created our own generator to control parameters that are peculiar to our setting, such as the probability of having alternatives route between the same pair of goals. This cannot be defined with existing tools where the route is the shortest path between two points. The generator takes as input the following parameters.

Goals. Goals can be randomly distributed or manually set, to verify ad-hoc settings. Goals act as both starting and ending points for a route.

---

[3] The scenario generator, the implementation of the models, and a sample of the real-world datasets can be downloaded at http://www.placemancy.com/public/code.zip

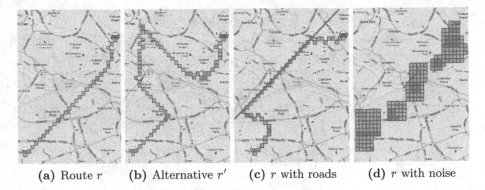

(a) Route r (b) Alternative r' (c) r with roads   (d) r with noise

Fig. 3. Synthetic routes generation

Roads. We initialize a weighted undirected graph, with a node for every block in the grid, and a weighted edge for every pair of adjacent blocks. The weight for an edge is in inverse proportion to the speed associated to the block transaction. Once goals are set, a route is the shortest path among them as in Figure 3a.

Alternative Routes. To model alternative paths to reach the same destination, different routes between a pair of goals are computed with two parameters: probability to have an alternative route and maximum number of alternatives.

Fast Roads. Users use fast roads when available, thus such roads are added to make the scenarios more realistic as in Figure 3c. This makes more challenging the prediction as users often drive the same fast roads to reach different goals. Weight over normal edges is three times the weight of edges for fast roads.

Noise. We add noise to the above routes to simulate real-world observations. We extracts two measurement error distributions (GPS and Network) from real data. An error probability distribution is the probability that a specified error rate is seen for a given block. Given a distribution, we associate to every block an error rate according to it, then we take a route as input and use the error rate associated to its blocks to "perturb" it. Given route r in Figure 3a, for each block $i \in r$ with error rate e_i, we remove from r the following $e_i - 1$ blocks. We then shift block i by a random number of blocks limited by e_i as in Figure 3d.

Results. We discuss results for the following models: standard HMM, smart HMM, and CRPM [23]. CRPM is one of the current state of the art solutions and it has been shown that it improves by 71% the goal prediction and by 30% the route distance w.r.t. a second order Markov model. We tested [19], but it failed in every test as it is not resistant to uncertain observations.

We consider 20 users and report the average results. We start with a *standard configuration* with a 80×80 grid, 5 goals distributed randomly, and 8 executions with increasing input. Once the parameters for the route generator are fixed, for the same scenario and for every user we create T (50 trips) and E (50 trips).

Table 1 shows a comparison of the standard configuration for Smart HMM and CRPM over three settings with different complexity.

Table 1. Results on synthetic scenarios

	Base		Fast Roads		Alternatives	
	Smart HMM	CRPM	Smart HMM	CRPM	Smart HMM	CRPM
WG-P	0.19	0.25	0.28	0.42	0.19	0.27
CG-RD	82	82	65	77	167	256
WG-RD	509	715	1131	1791	651	1064

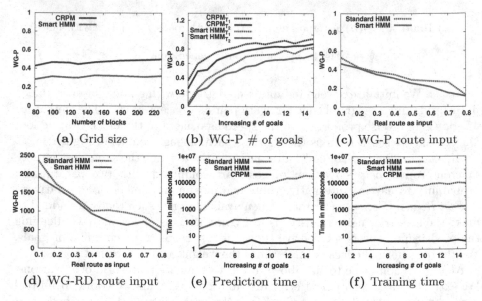

(a) Grid size   (b) WG-P # of goals   (c) WG-P route input

(d) WG-RD route input   (e) Prediction time   (f) Training time

Fig. 4. Synthetic experiments: (a) grid size, (b) number of goals with different training sizes, (c-d) route input, (e) prediction time, (f) training time

- In *Base*, fast roads and alternatives are not enabled. This simple scenario can be used as a measure of comparison for the other settings. HMM has a lower error rate in the prediction of the goals (WG-P). This is reflected in the lower values for WG-RD, while there is no difference in the easier case (CG-RD).
- With *Fast Roads* enabled, the goal prediction becomes more challenging: errors in WG-P increases by 68% for CRPM, while for Smart HMM is limited to 47%. Consequentially, the gap between HMM and CRPM increases in all measures.
- In *Alternatives*, each route has 3 possible alternatives. WG-P values are close to the *Base* setting for both models, but the predicted routes differ more from *d.r.* The distances increase more for CRPM than Smart HMM.

We now discuss how results are affected by varying the parameters fixed in the standard configuration. Finally, we report execution times.

Grid Size The standard configuration represents a city in a 80x80 blocks grid, with a 500 meters side for each block. An increase in the number of blocks does not shows remarkable differences in the goal error rate. Figure 4(a) reports that the two models show a 10% increase of the WG-P measure with a 230x230 grid w.r.t. the 80x80 version, while for other measures the results are stable.

Number of Goals and Training Size Figure 4(b) shows results with increasing number of randomly located goals for models with different training sets. We denote with T_1 a set of 20 trips and with T_2 a set of 70 trips, and with CRPM_{T_x} (Smart HMM_{T_x}) the model trained on T_x. Smart HMM outperforms CRPM in all scenarios, and, as expected, increasing the number of goals makes the prediction more difficult: while CG-RD is of course stable, results degrade for WG-P and WG-RD. The number of goals is important, but also their location play a role. In fact, WG-P has a lower growth for both models starting from 7 goals because of their positions: when there are many of them, even a wrong goal can lead to a route which is relatively close to the target $d.r$. Interestingly, goals number can be reduced by using clustering techniques [2]. With a clustering radius of 1.5 miles they reduce the number of goals for all users to less then 10. Given the size of our blocks, we implicitly consider a clustering radius of 500m from raw GPS data. As expected, models improve with bigger training sets. For complex settings, longer training is necessary to obtain acceptable results. Similar trends are observed with other measures.

Increasing Input Figures 4(c-d) show how an increasing percentage of the effective route $d.r$ given as input linearly increases the performances of the algorithms. This is important: when the user starts a trip the system may predict a wrong goal, but it promptly changes prediction along the route. HMMs outperform CRPM and similar trends are observed with CG-RD. The smart HMM obtains a slightly better error goal rate than the standard. Similarly, it has a lower average value in route distance. This indicates that, even in cases where it is not possible to predict the correct goal, the smart model can predict better routes and the pruning does not penalize the performances.

Execution Times The better prediction performance of the HMMs has a cost in term of computational time. Figures 4(e-f) show that CRPM is faster than HMMs in every test. However, the execution times for Smart HMM are acceptable and its execution times do not show a quadratic growth w.r.t. the number of blocks. Compared to the standard model in the worst setting (15 goals), the smart model reduces the prediction time from about 8 minutes to less than 0.2 second, and from about 1.5 minute to 2 seconds for the training.

Real-World Data. We collected real-world data from 10 volunteers with mobile applications for smartphones. Raw data trajectories were collected for 4 weeks by separately storing and cleaning [9] both GPS and network data. The datasets allow a comparison of the prediction models with precise and imprecise real data on the same routes; we report an example of the same trip observed by the two technologies in Figure 5a. Data has been pre-processed in order to identify goals and isolate the sequences of observations to be evaluated. For each learning system, we created one model with GPS data and one with network data. Once the four models were trained, we computed the routes to conduct a leave-one-out cross-validation. Examples of the predictions are in Figure 5(b-c).

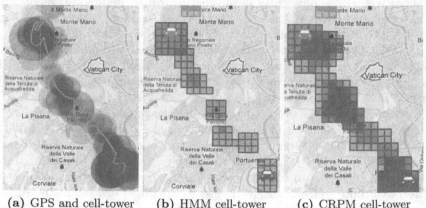

(a) GPS and cell-tower   (b) HMM cell-tower   (c) CRPM cell-tower

Fig. 5. In (a) the green line is the trajectory with GPS data, while the red circles represent the cell-towers data with the measurement error for each observation. In (b-c) green blocks are the input observations, while red are predictions done with cell-towers data (fewer blocks indicate a more precise prediction).

Table 2. Experimental results with both GPS and cell-towers real data

		HMM		CRPM	
		GPS	NETWORK	GPS	NETWORK
WG-P	Average all users	0.21	0.30	0.21	0.39
	Standard deviation	0.05	0.07	0.09	0.11
ALL-RD	Average all users	106.2	244.0	137.7	384.2
	Standard deviation	38	66	57	134

In Table 2 we report statistics about the observed distributions over the collected data. The standard deviation of the distributions is averaged over 10 users. GPS data are used as the ground truth. We denote with **ALL-RD** the route distance computed on all predictions. Training for Smart HMM required 259ms and a prediction required an average of 52ms. For goal prediction, Smart HMM outperforms CRPM on real network data by more than 23% (error goal rate of 0.30 compared to 0.39), while there is no significant difference on GPS data. For route distance on GPS data, Smart HMM returns routes closer to the correct one by 24% w.r.t to CRPM distance (distance of 106 compared to 137), while for the cell-tower data the difference in favor of HMM increases to 37% (244 compared to 384). In particular, Smart HMM outperforms CRPM for all our volunteers. The advantage in WG-P ranges from 15% to 27%, while for the ALL-RD it goes from 16% to 50%. HMM predictions are more stable than CRPM ones w.r.t. the standard deviation.

The differences in WG-P are due to a better modeling of the problem in the HMM. Even in cases when the WG-P values coincide, HMM obtains a better ALL-RD because the route is mined over all user trained paths (i.e., finding the shortest path inside A), while CRPM returns the single trained path that matches best. The execution times and the accuracy of predictions are in line with those obtained with the synthetic data generator. In particular, from the study of real data, an average user has less than 7 goals, and it is easy to verify that the numbers in Table 2 match those computed in Figure 4(b). Moreover, the number of goals for a user can be reduced by considering models with weekend days or weekdays only, or if we distinguish goals observed in the morning hours from the ones in the evening time. This leads to a solution with multiple HMMs for each user (e.g., one for the weekdays, one for the weekend, etc), such that the number of goals is always small.

Memory Consumption. HMM is known to be space consuming. We use a trained HMM for every user, thus it can result in a large amount of parameters that have to be maintained. For a space complexity analysis, we trained the two HMMs over an increasing number of users with a month of cell-tower observations. For each user, a new prediction model is created. Then, each trained model is added to a global list, making sure that the models are not deleted. In this setting, a laptop (8GB RAM) can manage up to 50 trained models with the Standard HMM and up to 2000 models with the Smart HMM. This is a consequence of the different number of states in the models: the number of states in the smart implementation is almost an order of magnitude smaller than the standard model. In fact, the trained matrix has a $|S| \times |S|$ dimension, with $|S|$ the number of the states, and occupies about 95% of the allocated memory.

7   Conclusions

We presented novel prediction models for moving objects designed for imprecise data. We first modeled the problem with a HMM and then refined it to improve quality of the results and execution times. Experiments show that our model needs a shorter training than existing methods and predicts more precise routes.

Our work can be easily apply to a large audience. Existing location aware applications, such as Foursquare, already collect users' data and have APIs to access them. Using such systems brings advantages w.r.t. ad-hoc solutions: (1) a large user base; (2) optimized software to save battery power, also using network-based localization, thus highlighting the necessity of supporting imprecise data; (3) historical observations that enable our model to make predictions from the installation; (4) privacy issues are managed by the application provider. We plan to implemented our solution on top of such location aware applications.

Our model is effective when users travel over routes known to the system. This does not apply for new routes. A promising approach to overcome this limitation is to automatically extract appointments and events from social media (such as Facebook and shared calendars [3]) to enable the prediction of new locations.

References

1. Alvarez-Garcia, J.A., Ortega, J.A., Gonzalez-Abril, L., Velasco, F.: Trip destination prediction based on past gps log using a HHM. ESA 37(12), 8166–8171 (2010)
2. Ashbrook, D., Starner, T.: Using gps to learn significant locations and predict movement across multiple users. Personal Ubiquitous Comput. 7, 275–286 (2003)
3. Blanco, L., Crescenzi, V., Merialdo, P., Papotti, P.: Flint: Google-basing the web. In: EDBT, pp. 720–724 (2008)
4. Bonchi, F., Castillo, C., Donato, D., Gionis, A.: Taxonomy-driven lumping for sequence mining. Data Mining and Knowledge Discovery 19, 227–244 (2009)
5. Burbey, I., Martin, T.L.: When will you be at the office? Predicting future locations and times. In: Gris, M., Yang, G. (eds.) MobiCASE 2010. LNICST, vol. 76, pp. 156–175. Springer, Heidelberg (2012)
6. Chen, L., Lv, M., Ye, Q., Chen, G., Woodward, J.: A personal route prediction system based on trajectory data mining. Inf. Sci. 181, 1264–1284 (2011)
7. Cheng, C., Jain, R., van den Berg, E.: Location prediction algorithms for mobile wireless systems. In: Wireless Internet Handbook, pp. 245–263 (2003)
8. Duntgen, C., Behr, T., Guting, R.H.: Berlinmod: a benchmark for moving object databases. The VLDB Journal 18(6), 1335–1368 (2009)
9. Froehlich, J., Krumm, J.: Route prediction from trip observations. In: Society of Automotive Engineers (SAE) 2008 World Congress (2008)
10. Giannotti, F., Mazzoni, A., Puntoni, S., Renso, C.: Synthetic generation of cellular network positioning data. In: GIS, pp. 12–20 (2005)
11. Jeung, H., Yiu, M.L., Zhou, X., Jensen, C.S.: Path prediction and predictive range querying in road network databases. The VLDB Journal 19, 585–602 (2010)
12. Kurashima, T., Iwata, T., Irie, G., Fujimura, K.: Travel route recommendation using geotags in photo sharing sites. In: CIKM, pp. 579–588 (2010)
13. Levenshtein, V.I.: Binary codes capable of correcting deletions, insertions, and reversals. Soviet Physics Doklady 10(8), 707–710 (1966)
14. Monreale, A., Pinelli, F., Trasarti, R., Giannotti, F.: Wherenext: a location predictor on trajectory pattern mining. In: KDD, pp. 637–646 (2009)
15. Newson, P., Krumm, J.: Hidden markov map matching through noise and sparseness. In: GIS, pp. 336–343 (2009)
16. Nizetic, I., Fertalj, K.: Automation of the moving objects movement prediction process independent of the application area. ComSIS 7(4), 931–945 (2010)
17. Paramvir, T.L., Liu, T., Bahl, P., Chlamtac, I.: Mobility modeling, location tracking, and trajectory prediction in wireless atm networks. IEEE Journal on Selected Areas in Communications 16, 922–936 (1998)
18. Rabiner, L.R.: A tutorial on hidden Markov models and selected applications in speech recognition. Proceedings of the IEEE 77(2), 257–286 (1989)
19. Simmons, R., Browning, B., Zhang, Y., Sadekar, V.: Learning to predict driver route and destination intent. In: IEEE ITSC, pp. 127–132 (2006)
20. Tao, Y., Faloutsos, C., Papadias, D., Liu, B.: Prediction and indexing of moving objects with unknown motion patterns. In: SIGMOD, pp. 611–622 (2004)
21. Xiao, Y.Y., Zhang, H., Wang, H.Y., Wang, F.: Location Prediction of Moving Objects with Uncertain Motion Patterns. DCDIS 14(S2), 503–507 (2007)
22. Yavas, G., Katsaros, D., Ulusoy, O., Manolopoulos, Y.: A data mining approach for location prediction in mobile environments. Data Knowl. Eng. 54(2), 121–146 (2005)
23. Ye, Q., Chen, L., Chen, G.: Predict personal continuous route. In: IEEE ITSC, pp. 587–592 (2008)

Modeling Short-Term Energy Load
with Continuous Conditional Random Fields

Hongyu Guo

National Research Council Canada,
1200 Montreal Road, Ottawa, ON., K1A 0R6, Canada
hongyu.guo@nrc-cnrc.gc.ca

Abstract. Short-term energy load forecasting, such as hourly predictions for the next n ($n \geq 2$) hours, will benefit from exploiting the relationships among the n estimated outputs. This paper treats such multi-steps ahead regression task as a sequence labeling (regression) problem, and adopts a Continuous Conditional Random Fields (CCRF) strategy. This discriminative approach intuitively integrates two layers: the first layer aims at the prior knowledge for the multiple outputs, and the second layer employs edge potential features to implicitly model the interplays of the n interconnected outputs. Consequently, the proposed CCRF makes predictions not only basing on observed features, but also considering the estimated values of related outputs, thus improving the overall predictive accuracy. In particular, we boost the CCRF's predictive performance with a multi-target function as its edge feature. These functions convert the relationship of related outputs with continuous values into a set of "sub-relationships", each providing more specific feature constraints for the interplays of the related outputs. We applied the proposed approach to two real-world energy load prediction systems: one for electricity demand and another for gas usage. Our experimental results show that the proposed strategy can meaningfully reduce the predictive error for the two systems, in terms of mean absolute percentage error and root mean square error, when compared with three benchmarking methods. Promisingly, the relative error reduction achieved by our CCRF model was up to 50%.

Keywords: Conditional Random Fields, Energy Demand Forecast.

1 Introduction

Commercial building owners are facing rapidly growing energy cost. For example, energy accounts for approximately 19% of total expenditures for a typical commercial building in the U.S.; in Canada, annual energy cost for commercial buildings is about 20 billion dollars. Particularly, these numbers are expected to double in the next 10 years[1]. Aiming at reducing this operational cost, buildings have started to respond to utility's Time of Use Pricing or Demand and

---

[1] http://www.esource.com, http://nrtee-trnee.ca/

H. Blockeel et al. (Eds.): ECML PKDD 2013, Part I, LNAI 8188, pp. 433–448, 2013.

Response signals. Such smart energy consumption, however, requires accurate short-term load predictions.

One of the main challenges for short-term energy load is to predict multiple time-ticks ahead, namely multiple target variables. Typically, these predicted outcomes are correlated. For instance, knowing the current hour's overall energy usage will help estimate the next hour's energy demand. To make use of the relationships among predicted outputs, this paper deploys the Conditional Random Fields (CRF) [8], a sequential labeling method. More specifically, we adopt the Continuous Conditional Random Fields (CCRF) [10]. As depicted on the left subfigure in Figure 1, our CCRF approach intuitively integrates two layers. The first layer consists of variable (node) features (filled squares in Figure 1), and aims at the prior knowledge for the multiple outputs. The second layer employs edge potential features (unfilled squares in Figure 1) to implicitly model the interplays of the interconnected outputs, aiming at improving the predictions from the first layer. Consequently, the proposed method makes predictions not only basing on observed features, but also considering the estimated values of related outputs, thus improving the overall predictive accuracy.

In addition to its capability of implicitly modeling the interplays between outputs through its edge potential functions, the CCRF strategy can include a large number of accurate regression algorithms or strong energy predictors as its node features, thus enhancing its prior knowledge on each individual output. Importantly, the proposed CCRF method has the form of a multivariate Gaussian distribution, resulting in not only efficient learning and inference through matrix computation, but also being able to provide energy load projects with smooth predicted confidence intervals, rather than only the forecasted load values, thus further benefiting the decision makings for energy load management.

In particular, we addressed the weak feature constraint problem in the CCRF with a novel edge function, thus boosting its predictive performance. Such weak constraint issue arises because CCRF takes aim at target outputs with continuous values. In detail, CRF's function constraints are weak for edge features with

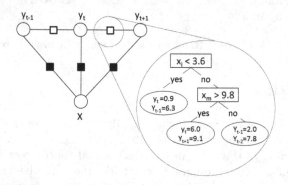

Fig. 1. A chain-structured Continuous CRF with PCTs trees (right subfigure) as edge potential functions (unfilled squares). Here the filled squares are the node features.

continuous values, compared to that of binary features, because of CRF's linear parameterization characteristics [13,14]. That is, for a binary feature, knowing the mean is equivalent to knowing its full probability distribution. On the contrary, knowing the mean may not tell too much about the distribution of a continuous variable. Since CRF strategies are devised to form models satisfying certain feature constraints [14], such weak feature constraints will limit the resultant CRF's predictive performance. Moreover, typical approaches of dividing continuous values into "bins" cannot be applied to our CCRF method here because for energy load forecasting, one has to be able to simultaneously "bin" multiple target variables that are unknown in inference time. To address the above concern for the CCRF model, we employ a multi-target function, namely the Predictive Clustering Trees (PCTs) strategy [1], as the CCRF's edge feature. The PCTs method first partitions instances with similar values for multiple related target variables, only based on their shared observation features, into disjoint regions. Next, it models a separate relationship among these target variables in each smaller region. In other words, the PCTs convert the relationship of the related target variables into a set of sub-relationships, each containing more specific constraints for the related target variables. As a result, it enables the CCRF to better capture the correlations between related outputs, thus boosting the CCRF's predictive performance.

We applied the proposed method to two real-world energy load forecasting systems: one for gas which is used to warm buildings in winter, and another for electricity for building cooling in summer. Also, we compared our approach with three benchmarking strategies: 1) a random forests method where each branch is a multi-objective decision tree for multiple target variables, 2) a collection of regression trees each targeting an individual target variable, and 3) a CCRF model with basic features. Our experimental results show that the proposed method can significantly reduce the predictive error, in terms of mean absolute percentage error and root mean square error, for the two energy systems, when compared with the three baseline algorithms.

This paper is organized as follows. Section 2 introduces the background. Next, a detailed discussion of the proposed algorithm is provided in Section 3. In Section 4, we describe the comparative evaluation. Section 5 presents the related work. Finally, Section 6 concludes the paper and outlines our future work.

2   Background

2.1   Conditional Random Fields

Conditional Random Fields (CRF) are undirected graphical models that define the conditional probability of the label sequence $Y = (y_1, y_2, \cdots, y_n)$, given a sequence of observations $X = (x_1, x_2, \cdots, x_r)$. That is, the discriminative strategy aims to model $P(Y|X)$. Specifically, benefiting from the Hammersley-Clifford theorem, the conditional probability can be formally written as:

$$P(Y|X) = \frac{1}{Z(X)} \prod_{c \in C} \Phi(y_c, x_c)$$

where C is the set of cliques[2] in the graph, Φ is a potential function defined on the cliques, and $Z(x)$ is the normalizing partition function which guarantees that the distribution sums to one.

One of the most popular CRFs is the linear chain CRF (depicted on the left of Figure 1), which imposes a first-order Markov assumption between labels Y. This assumption allows the CRF to be computed efficiently via dynamic programming. In addition, the clique potentials Φ in the linear chain CRF are often expressed in an exponential form, so that the formula results in a maximum entropy model. Formally, the linear-chain CRF is defined as a convenient log-linear form:

$$P(Y|X) = \frac{1}{Z(X)} \prod_{t=1}^{n} exp(v^T \cdot f(t, y_{t-1}, y_t, X)) \tag{1}$$

$$where, \ Z(X) = \sum_{Y} \prod_{t=1}^{n} exp(v^T \cdot f(t, y_{t-1}, y_t, X))$$

Here, $f(t, y_{t-1}, y_t, X)$ is a set of potential feature functions which aim to capture useful domain information; v is a set of weights, which are parameters to be determined when learning the model; and y_{t-1} and y_t are the label assignments of a pair of adjacent nodes in the graph.

2.2 Continuous Conditional Random Fields

The CRF strategy is originally introduced to cope with discrete outputs in labeling sequence data. To deal with regression problems, Continuous Conditional Random Fields (CCRF) has recently been presented by Qin et al. [10], aiming at document ranking. In CCRF, Equation 1 has the following form.

$$P(Y|X) = \frac{1}{Z(X, \alpha, \beta)} exp(\sum_{1}^{n} H(\alpha, y_i, X) + \sum_{i \sim j} G(\beta, y_i, y_j, X)) \tag{2}$$

where $i \sim j$ means y_i and y_j are related, and

$$Z(X, \alpha, \beta) = \int_{y} exp(\sum_{1}^{n} H(\alpha, y_i, X) + \sum_{i \sim j} G(\beta, y_i, y_j, X)) dy$$

Here, potential feature functions $H(y_i, X)$ and $G(y_i, y_j, X)$ intend to capture the interplays between inputs and outputs, and the relationships among related outputs, respectively. For descriptive purpose, we denote these potential functions as *variable (node) feature* and *edge feature*, respectively. Here, α and β represent the weights for these feature functions. Typically, the learning of the CCRF is to find weights α and β such that conditional log-likelihood of the

---

[2] A clique is a fully connected subgraph.

training data, i.e., $L(\alpha, \beta)$, is maximized, given training data $D = \{(X, Y)\}_1^L$ (L is the number of sample points in D):

$$(\widehat{\alpha}, \widehat{\beta}) = \underset{\alpha, \beta}{\operatorname{argmax}}(L(\alpha, \beta)), \quad \text{where } L(\alpha, \beta) = \sum_{l=1}^{L} log P(Y_l | X_l) \qquad (3)$$

After learning, the inference is commonly carried out through finding the most likely values for the $P(Y_l)$ vector, provided observation X_l:

$$\widehat{Y_l} = \underset{Y_l}{\operatorname{argmax}}(P(Y_l | X_l)) \qquad (4)$$

Promisingly, as shown by Radosavljevic et al. [12], if the potential feature functions in Equation 2 are quadratic functions of output variables Y, the CCRF will then have the form of a multivariate Gaussian distribution, resulting in a computationally tractable CCRF model. Our approach deploys such a Gaussian form CCRF model with newly designed edge and variable features. We will discuss our model and feature design in detail next.

3   CCRF for Energy Load

One of the core developments for a CCRF model is its edge and variable features.

3.1   Model Design

Our edge features are designed to capture the relationships between *two adjacent* target variables, and to ensure that the resultant CCRF has a multivariate Gaussian form. Our motivations are as follows. Our analysis on real-world short-term energy load data indicates that, for these data the adjacent target variables are highly correlated. As an example, Figure 2 pictures the partial autocorrelation graphs of two years' hourly gas demand and electricity load (we will discuss these two data sets in details in Section 4) in the left and right subfigures, respectively.

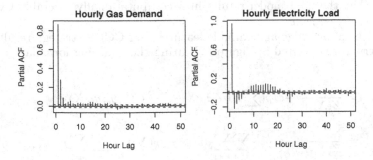

Fig. 2. Partial autocorrelation graphs of the gas demand and electricity load data

These two subfigures indicate that adjacent target variables show significant correlation, compared to the other related target variables. For example, as shown in Figure 2, for both the gas demand and electricity load data, the first lag bears a correlation value of over 0.8. In contrast, other lags have a correlation of less than 0.3. These results suggest that the energy loads of a pair of adjacent hours are highly correlated.

Aiming to capture the above mentioned correlation between *two adjacent* target variables, we deploy $G(\beta, y_i, y_j, X) = \sum_S \beta_s \omega_s (y_i - y_j)^2$ as our edge function form. Here y_i and y_j are the i-th and j-th target outputs, respectively. Since the correlation of energy usages of a pair of adjacent hours is much higher than other hours (as previous shown), in our design the i-th and j-th represent two adjacent hours. The ω_s is the s-th of a set of S indicator functions with values of either zero or one, indicating if the correlation between y_i and y_j should be measured or not. The β here represents the weights for these feature functions, and these weights will be learned by the CCRF during the training. In particular, the quadratic function forms here are specially designed to ensure that the CCRF results in a multivariate Gaussian form with efficient computation for the learning and inference, as will be further discussed later in this section.

In contrast to the edge potential feature which takes into account the interactions between predicted target variables, the variable potential feature of the CCRF, as described in Equation 2, aims at making good use of many efficient and accurate regression predictors. To this end, we consider variable features of the form $H(\alpha, y_i, X) = \sum_{k=1}^m \alpha_k (y_i - f_k(X))^2$. Here, y_i indicates the i-th target output, $f_k(X)$ is the k-th of m predictors for the target output y_i. This specific variable feature form is motivated by the following two reasons. First, with this particular form, the resultant CCRF strategy is able to include many efficient and accurate single-target regression models, such as Regression Trees or Support Vector Machines, or existing state-of-the-art energy load predictors as its features. One may include a large number of such predictors, namely with a large m, and the CCRF will automatically determine their relevance levels during training. For example, for target output y_i, we can have the output from a single-target Regression Trees and the prediction from a SVM as its two features; during the learning, the CCRF will determine their contribution to the final prediction of the y_i through their weights. Second, the quadratic form here ensures that the final model results in a computationally tractable CCRF, as will be discussed next.

With the above edge and variable features, our CCRF strategy results in the graph structure depicted in Figure 1, bearing the following formula.

$$P(Y|X) = \frac{1}{Z(X, \alpha, \beta)} exp(\sum_1^n H(\alpha, y_i, X) + \sum_{i \sim j} G(\beta, y_i, y_j, X))$$

$$= \frac{1}{Z(X, \alpha, \beta)} exp(-\sum_{i=1}^n \sum_{k=1}^m \alpha_k (y_i - f_k(X))^2 - \sum_{i \sim j} \sum_{s=1}^S \beta_s \omega_s (y_i - y_j)^2) \quad (5)$$

In this equation, we have n target outputs (i.e., $\{y_i\}_1^n$), m variable features (i.e., $\{f_k(X)\}_1^m$) for each target y_i, and S edge features (with s as index) for modeling the correlation between two outputs y_i and y_j (where indicator function ω_s indicates if the correlation between the i-th and j-th outputs will be taken into account or not). In our case, we use edge features to constrain the square of the distance between two outputs when the two outputs are adjacent. Note that we here assume that the neighboring information between two target outputs will be given.

Intuitively, the integration of the variable and edge feature, as described in Equation 5, forms a model with two layers. The variable features $\alpha_k(y_i - f_k(X))^2$ are predictors for individual target variables. That is, these variable features depend only on the inputs. Hypothetically, if the edge functions are disabled, the predictions of the CCRF model will be the outputs of these individual predictors. In this sense, we can consider the variable features as the *prior knowledge* for the multiple outputs. On the other hand, the edge potential functions $\beta_s\omega_s(y_i - y_j)^2$ involve multiple related target variables, constraining the relationships between related outputs. In fact, we can think of the edge features as representing a separate set of weights for each multi-targets output configuration. In other words, these weights serve as a second layer on top of the variable features. This second layer aims to fine-tune the predictions from the first layer, namely the prior knowledge provided by the variable features.

Promisingly, following the idea presented by Radosavljevic et al. [12], the above CCRF, namely Equation 5 can be further mapped to a multivariate Gaussian because of their quadratic forms for the edge and variable potential features:

$$P(Y|X) = \frac{1}{(2\pi)^{n/2}|\sum|^{1/2}} \cdot exp(-\frac{1}{2}(Y - \mu(X))^T \sum{}^{-1}(Y - \mu(X))) \qquad (6)$$

In this Gaussian mapping, the inverse of the covariance matrix Σ is the sum of two $n \times n$ matrices, namely $\Sigma^{-1} = 2(Q^1 + Q^2)$ with

$$Q_{ij}^1 = \begin{cases} \sum_{k=1}^m \alpha_k & \text{if } i = j \\ 0 & otherwise \end{cases} \quad \text{and} \quad Q_{ij}^2 = \begin{cases} \sum_{j=1}^n \sum_{s=1}^S \beta_s\omega_s & \text{if } i = j \\ -\sum_{s=1}^S \beta_s\omega_s & \text{if } i \neq j \end{cases}$$

Also the mean $\mu(X)$ is computed as $\Sigma\theta$. Here, θ is a n dimensional vector with values of

$$\theta_i = 2 \sum_{k=1}^m \alpha_k f_k(X)$$

Practically, this multivariate Gaussian form results in efficient computation for the learning and inference of the CCRF model, which is discussed next.

Training CCRF. In the training of a CRF model, feature function constraints require the expected value of each feature with respect to the model be the same as that with respect to the training data [14]. Following this line of research, with a multivariate Gaussian distribution that aims at maximizing log-likelihood, the learning of a CCRF as depicted in Equation 3 becomes a convex optimization problem. As a result, stochastic gradient ascent can be applied to learn the parameters.

Inference in CCRF. In inference, finding the most likely predictions Y, given observation X as depicted in Equation 4, boils down to finding the mean of the multivariate Gaussian distribution. Specifically, it is computed as following:

$$\widehat{Y} = \underset{Y}{\text{argmax}}(P(Y|X)) = \mu(X) = \Sigma\boldsymbol{\theta}$$

Furthermore, the 95%-confidence intervals of the estimated outputs can be obtained by $\widehat{Y} \pm 1.96 \times diag(\Sigma)$, due to the Gaussian distribution.

3.2   Cope with Weak Feature Constraint in CCRF

Recall from Section 3.1 that the edge features in our CCRF have the form of $(y_i - y_j)^2$. This particular function form aims to ensure that not only the correction between adjacent outputs are taken into account, but also the resultant CCRF has a multivariate Gaussian form with efficient computation for the learning and inference. This design, however, results in a weak feature constraint problem for the CCRF because now each edge function depends on multiple, continuous target variables. We detail this challenge as follows.

In a nutshell, CRF is a maximum entropy model with feature constraints that capture relevant aspects of the training data. That is, training a CRF amounts to forcing the expected value of each feature with respect to the model to be the same as that with respect to the training data. Consequently, the constraints with binary feature, for example, contain essential information about the data because knowing the mean of the binary feature is equivalent to knowing its full distribution. On the other hand, knowing the mean may not tell too much about the distribution of continuous variables because of CCRF's linear parameterization characteristics [13,14]. As an example, the mean value of the red curve distribution on the left subfigure in Figure 3 does not tell us too much about the distribution of the curve. As a result, the CCRF may learn less than it should from the training data.

To tackle this constraint weakness, one typically introduces the "Binning" technique. That is, one can divide the real value into a number of bins, and then each bin is represented by a binary value. However, in the CCRF, typical "Binning" techniques are difficult to apply to the edge functions because all the values for these features are predicted values of the target variables, and we do not know these values beforehand. That is, we do not know, for example, the values of y_i and y_j in inference time. To cope with unknown target variables, one may have to "Bin" these features using only the known input variables. Nevertheless, relying on only the observed inputs may not be enough to distinguish the interactions between the pair of unknown outputs. For example, a large y_i

value and a small y_j value may have the same result, as computed by $(y_i - y_j)^2$, as that of a small y_i and a large y_j value pair. These observations suggest that it will be beneficial to has a "Binning" technique that is able to *simultaneously* take the interactions of a pair of outputs and the observed inputs into account.

Following this line of thought, we propose to use the Predictive Clustering Trees (PCTs) [1]. The aim here is to use the PCTs to divide the relationships of related outputs into a set of "sub-relationships", each providing more specific feature constraints for the interplays of the related outputs. The PCTs strategy considers a decision tree as a hierarchy of clusters. The root node corresponds to one cluster containing all data, which is recursively partitioned into smaller clusters while moving down the tree. When dealing with multiple target attributes, the PCTs approach can be viewed as a tree where each leaf has multiple targets, compared to that of traditional decision tree which learns a scalar target. The PCTs method extends the notion of class variance towards the multi-dimensional regression case. That is, given a distance function, such as the sum of the variances of the target variables, for the multi-dimensional target space, the PCTs algorithm partitions the input space, namely X, into different disjoint regions, where each is a leaf and each groups instances with similar values for the target variables Y_s. When deployed for CCRF, each PCTs tree can be used to model the interactions between the related Y_s through its leaves. The graph structure in Figure 1 pictures our CCRF model, where each *unfilled square* describes an edge feature, and each is represented by a PCTs tree on the right of the figure.

We illustrate the above weak edge feature constraint and the proposed PCTs solution with Figure 3. In this figure, the red curve shows the distribution of the edge potential feature of $(y_i - y_j)^2$ in the gas demand (used for heating) data set. Here, y_i and y_j represent the energy loads of two neighboring hours, namely hours i and j, respectively. This distribution subsumes three sub-distributions, depicted by the blue, brown, and green curves, respectively. In detail, the blue curve pictures the distribution of $(y_i - y_j)^2$ where the hours i and j have similar

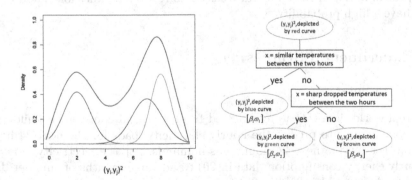

Fig. 3. Left subfigure: distribution of $(y_i - y_j)^2$ (red curve), and the subsumed three sub-distributions (blue, brown, and green curves); right subfigure: the PCT tree that shows what X values were used to convert the red curve into the three sub regions

temperature; the brown curve presents the same two hours with a dramatically increasing temperature; and the green curve shows the distribution of the same two hours where the temperature drops sharply. Intuitively, one can consider the red curve pictures a joint probability of $P((y_i - y_j)^2, X)$, and the other three curves show the conditional probability of $P((y_i - y_j)^2 | X)$ when X takes one of the three weather scenarios, namely, similar, sharply increasing, and dramatically dropping temperatures between two neighboring outputs.

As can be seen from this example, the edge feature [3] of $(y_i - y_j)^2$, as shown by the red curve, is not able to distinguish the three sub-relationships clustered by the blue, brown, and green curves. That is, the edge feature constraints represented by the red curve cannot distinguish between a similar, increasing, or reducing energy consumption trends. Such weak edge feature will limit the constraining power of the edge potential functions in the CCRF. It is worth to further noting that, if we do not *simultaneously* consider the input variables and the interplays between the target variables, as what the PCTs do, we may not be able to distinguish the brown and green curves since these two curves represent similar $(y_i - y_j)^2$ values.

Let us continue with the above example. Tackled by the PCTs, the original edge feature of $(y_i - y_j)^2$, as depicted by the red curve, will be replaced by three sub features, namely the distributions shown in the blue, brown, and green curves. In other words, *three* edge feature constraints, instead of *only one*, will be used by the $G(\beta, y_i, y_j, X)$ function, representing three different types of interplays between the (y_i, y_j) pair: one constraining a small change between y_i and y_j, another defining a sharp increase of energy consumption, and the other confining a quick drop in term of energy consumption.

Let's sum up the above example. The edge function with PCTs here can naturally model the multi-steps ahead energy consumptions: 1) if the temperature (which can be observed or forecasted) is sharply dropping, the constraint of a small y_i and a large y_j will have a high probability; 2) if the temperature is dramatically increasing, the constraint of a large y_i and a small y_j will have a high probability; 3) if the temperature is similar, similar values for y_i and y_j will then have a high probability.

4   Experimental Studies

4.1   Data Sets

Two real world data sets were collected from a typical commercial building in Ontario: one aims to predict the hourly electricity loads for the next 24 hours, and another for the next 24 hours' gas demands. For the electricity, one year of hourly energy consumption data in 2011 and three months of summer data, from March 1st to May 31rd in 2012, were collected; for gas, we have the whole

---

[3] Note that, as discussed in Section 3.1, the quadratic function forms here are specially designed to ensure that the CCRF results in a multivariate Gaussian form with efficient computation for the learning and inference.

year's data in 2011 and winter data from January to March in the year of 2012. In our experiments, for both the electricity and gas, we trained the model with the 2011 data and then tested the model using the data from 2012.

4.2 Features and Settings

In these two energy load forecasting systems, the proposed CCRF method deployed 23 edge features, as discussed in Section 3.2. Each such feature aims to capture the interplays of an adjacent pair of target variables, namely two consecutive hours of the 24 hours. The number of sub-regions generated for each of these edge features were controlled by the search depth of the PCTs trees. The larger this number, potentially more sub-regions or clusters will be created to group a pair of related target variables. In our experiments, we set this number to 3. In fact, we compared with different settings and the model was insensitive to this parameter.

Also, 24 variable features were used, each focusing on one target variable, namely an individual hour of the 24 output hours. To this end, we deploy Friedman's additive gradient boosted trees [5,6] as our CCRF model's variable features. Friedman's additive boosted trees can be considered as a regression version of the well-known Boosting methodology for classification problems. Promising results of applying this additive approach have been observed, in terms of improving the predictive accuracy for regression problems [5]. In our studies here, each such variable feature, namely each target y_i, is modeled using an additive gradient boosted strategy with the following parameters: a learning rate of 0.05, 100 iterations, and a regression tree as the base learner. The input features for the Friedman machine include past energy usages, temperatures, the day of the week, and the hour of the day.

In addition, to avoid overfitting in the training of the CCRF, penalized regularization terms $0.5\alpha^2$ and $0.5\beta^2$ were subtracted from the log-likelihood function depicted in Equation 3. Also, the number of iterations and learning rate for the gradient ascent in the CCRF learning were set to 100 and 0.0001, respectively.

4.3 Methodology

We compared our method with three benchmarking approaches. The first comparison algorithm is a state-of-the-art multi-target system, namely the ensembles of Multi-Objective Decision Trees (MODTs) [7]. We obtained the settings of the ensembles of MODTs from their authors. That is, in our experiments, a random forest strategy was applied to combine 100 individual multi-objective decision trees. The second benchmarking algorithm is a strategy that trains independent regression models for each target attribute and then combines the results [9,15]. In our studies, a collection of regression trees were used where each tree models a target variable. The last comparison approach we compared with is a CCRF model with basic features. That is, this CCRF strategy used 24 single-target regression trees as its variable features. Also, each of the 23 edge features captures the square of the distance between two adjacent target variables. The comparison here aims to evaluate the impact of the newly designed features, namely the predictive clustering approach, to the CCRF strategy.

We implemented the CCRF models in Java on a 2.93GHz PC with 64 bit Windows Vista installed. We measured the performance of the tested algorithms with the mean absolute percentage error (MAPE) and the root mean square error (RMSE). For descriptive purpose, we referred to the random forests approach with multi-objective decision trees, the method of learning a collection of regression trees, the basic CCRF algorithm, and the proposed CCRF strategy as MODTs, RTs, CCRFs_BASE, and CCRFs_EP, respectively.

4.4   Experimental Results

In this section we examine the predictive performance of the proposed method against both the electricity and gas data, in terms of MAPE and RMSE.

Electricity Usage. Our first experiment studies the performance of the tested methods on the electricity load data. We present the MAPE and RMSE obtained by the four tested approaches for each of the three months, namely March, April, and May, in Figure 4. In this figure, we depicted the MAPE and RMSE obtained on the left and right subfigures, respectively.

The MAPE results, as presented on the left subfigure of Figure 4 show that the CCRF method appears to consistently reduce the error rate for each of the three months, when compared to all the other three tested strategies, namely the collection of regression trees, the random forests with multi-objective decision trees, and the CCRF model with basic features. For example, when compared with the collection of regression trees method, namely the RTs approach, the CCRFs_EP model decreases the absolute MAPE for months March, April, and May with 0.51, 0.53, and 1.0, respectively. The relative average error reduced for these three tested months was 17.9% (drop from 3.80 to 3.12 as shown on the left of Figure 4). In terms of RMSE, for each of the three months, the error was

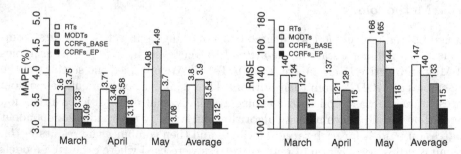

Fig. 4. MAPE and RMSE obtained by the four methods, against the electricity data in the months of March, April, and May in 2012

reduced by the CCRFs_EP method from 139.63, 136.91, and 165.86 to 112.11, 114.76, and 118.17, respectively. As depicted on the right of Figures 4, a relative average reduction was 22.0% (drop from 147.47 to 115.01).

When considering the comparison with the random forests of multi-objective decision trees, namely the MODTs method, the results as depicted in Figure 4 indicate that the CCRFs_EP model was also able to meaningfully reduce the error. As shown in Figure 4, for both MAPE and RMSE, the CCRFs_EP strategy was able to reduce the error for all the three months. On average, relative error reductions of 20.08% and 17.66% were achieved by the CCRFs_EP model over the MODTs strategy, in terms MAPE and RMSE, respectively.

Comparing to the CCRFs_BASE algorithm, the CCRFs_EP method also appears to consistently outperform the CCRFs_BASE strategy for each of the three months regardless the evaluation metrics used, namely no matter if the MAPE or RMSE was applied as the predictive performance metrics. As depicted in Figure 4, average relative error reductions of 11.87% and 13.75% were achieved by the CCRFs_EP model over the CCRFs_BASE approach, in terms MAPE and RMSE, respectively. These results suggest that the advanced potential feature functions as introduced in Section 3.2 enhanced the proposed CCRF model's predictive performance.

Gas Consumption. Our second experiment investigates the performance of the tested methods on the gas demand data. We present the MAPE and RMSE obtained by the four tested methods for each of the three months, namely January, February, and March, in Figure 5. In this figure we depicted the MAPE and RMSE obtained on the left and right subfigures, respectively.

The MAPE results, as presented on the left subfigure of Figure 5 show that the proposed CCRF_EP method appears to consistently reduce the error for all the three months, when compared to the RTs, MODTs, and CCRFs_BASE methods. For instance, when compared with the RTs algorithm, the results on the left subfigure of Figure 5 show that the CCRFs_EP model decreases the absolute MAPE for months January, February, and March with 2.61, 1.81, and

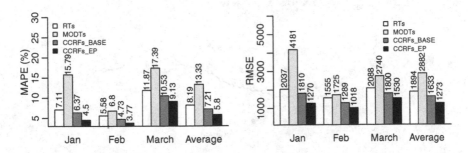

Fig. 5. MAPE and RMSE obtained by the four methods, against the gas data in the months of January, February, and March in 2012

2.74, respectively. The relative average error reduced for these three months was 29.15% (drop from 8.19 to 5.80 as indicated in Figure 5). In terms of RMSE, results on the right of Figure 5 demonstrate that, the CCRF_EP strategy outperformed the RTs algorithm for all the three tested months. A relative average reduction was 32.77% (drop from 1894 to 1272 as shown on the right of Figure 5).

When considering the comparison with the MODTs method, the results in Figure 5 indicate that the proposed CCRF model meaningfully reduce the error rate. For example, for both MAPE and RMSE, the CCRF_EP strategy was able to reduce the error for all the three months. As shown on the right of Figure 5, average relative error reductions of 56.47% and 55.82% were achieved by the CCRF_EP model over the random forests ensemble.

Comparing to the CCRFs_BASE, the CCRFs_EP method again appears to consistently outperform the CCRFs_BASE strategy for all the three months in terms of both the MAPE or RMSE. As depicted in Figure 5, average relative error reductions of 19.55% and 22.05% were achieved by the CCRFs_EP model over the CCRFs_BASE strategy, in terms MAPE and RMSE, respectively.

In summary, the experimental results on the six data sets indicate that, the proposed CCRF model consistently outperformed the other three tested methods in terms of MAPE and RMSE. Promisingly, the relative error reduction achieved by the proposed CCRF algorithm was at least 11.87%, and up to 56.47%.

In addition to its superior accuracy, the proposed CCRF has the form of a multivariate Gaussian. Therefore, it can provide projects with probability distributions rather than only the forecasted numbers. In Figure 6, we depicted a sample of the 24 predictions with their 95% confidence intervals from our gas forecasting system. The 24 hours ahead predictions, along with their confidence intervals, were generated for the date of April 1st, 2012, at mid night. In this figure, the dark curve in the middle shows the 24 predictions, and the two dot curves depict the two confidence interval bands. These smooth, uncertainty information could be beneficial for better decision makings in energy load management.

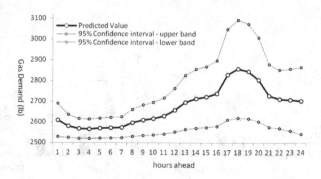

Fig. 6. Outputs with 95% confidence bands for the gas consumptions of April 1st, 2012

5   Related Work

Short-term energy load forecasting has been an active research area for decades, and a variety of machine learning techniques have been proposed to cope with this challenge, including regression algorithms, time series analysis strategies, Neural networks, and Support Vector Machines, amongst others. An informative review has been reported by Feinberg and Genethliou [4]. Comparing with the CCRF methods, many existing approaches either have difficulties to make use of different types of features (such as dependent features, categorical features etc.), to generate statistical information of the estimated values (e.g., the confidence intervals), or to explore the interrelationships among the multiple outputs (e.g., structured outputs).

Recent years, Conditional Random Fields has been devised to provide a probabilistic model to represent the conditional probability of a particular label sequence. This discriminative framework has been very successfully applied to many classification tasks, including text labeling [8], activity recognition [14], recommendation [16], and image recognition [11], amongst others. Also, within the CRF research community, issues related to the powerful and flexible CRF model have also been actively studied [2,17]. In contrast, only a few applications of applying this framework on regression tasks have been reported. These applications include document ranking [10], Aerosol optical depth estimation [12], and travel speed prediction [3]. To our best knowledge, this paper is the first to report an application of Conditional Random Fields on short-term energy load forecasting. Also, we focus on designing a CCRF with tractable computation cost for training and inferring, through the carefully designed potential feature functions. Most importantly, we cope with the weak feature constraint in a CCRF model, which, to our best knowledge, has not been addressed by any CCRF paper before.

6   Conclusions and Future Work

Embracing "smart energy consumption" to optimize energy usage in commercial buildings has provided a unique demand for modeling short-term energy load. We have devised a Continuous Conditional Random Fields strategy to cope with these structured outputs tasks. The CCRF can naturally model the multi-steps ahead energy load with its two layers design. In particular, we deployed a novel edge feature, namely a multi-target regression strategy, to enable the CCRF to better capture the interplays between correlated outputs with continuous values, thus boosting the CCRF model's accuracy. We evaluated the proposed method with two real-world energy load forecasting systems. When compared with three benchmarking strategies, our experimental studies show that the proposed approach can meaningfully reduce the predictive error for the two energy systems, in terms of mean absolute percentage errors and root mean square errors.

To our best knowledge, this is the first study on adopting a CRF to model multiple-steps-ahead energy loads. Furthermore, we introduced a novel multi-target edge function to address the weak feature constraint problem in the

CCRF, thus boosting its accuracy. Our future work will test our approach against more data sets with comprehensive statistical analysis. Also, we plan to further conduct comparison studies with other state-of-the-art energy predictors.

Acknowledgments. We wish to thank the anonymous reviewers for their insightful comments on our submission, which helped improve the paper quality.

References

1. Blockeel, H., Raedt, L.D., Ramon, J.: Top-down induction of clustering trees. In: ICML 1998, pp. 55–63. Morgan Kaufmann (1998)
2. Dietterich, T.G., Ashenfelter, A., Bulatov, Y.: Training conditional random fields via gradient tree boosting. In: ICML 2004, pp. 28–36 (2004)
3. Djuric, N., Radosavljevic, V., Coric, V., Vucetic, S.: Travel speed forecasting by means of continuous conditional random fields. Transportation Research Record: Journal of the Transportation Research Board 2263, 131–139 (2011)
4. Feinberg, E., Genethliou, D.: Load Forecasting. Springer, US (2005)
5. Friedman, J.H.: Stochastic gradient boosting. Computational Statistics and Data Analysis 38, 367–378 (1999)
6. Friedman, J.H.: Greedy function approximation: A gradient boosting machine. Annals of Statistics 29, 1189–1232 (2000)
7. Kocev, D., Vens, C., Struyf, J., Džeroski, S.: Ensembles of multi-objective decision trees. In: Kok, J.N., Koronacki, J., Lopez de Mantaras, R., Matwin, S., Mladenič, D., Skowron, A. (eds.) ECML 2007. LNCS (LNAI), vol. 4701, pp. 624–631. Springer, Heidelberg (2007)
8. Lafferty, J.D., McCallum, A., Pereira, F.C.N.: Conditional random fields: Probabilistic models for segmenting and labeling sequence data. In: ICML 2001, San Francisco, CA, USA, pp. 282–289 (2001)
9. Petrovskiy, M.: Paired comparisons method for solving multi-label learning problem. In: Sixth International Conference on Hybrid Intelligent Systems, HIS 2006 (December 2006)
10. Qin, T., Liu, T.-Y., Zhang, X.-D., Wang, D.-S., Li, H.: Global ranking using continuous conditional random fields. In: NIPS, pp. 1281–1288 (2008)
11. Quattoni, A., Collins, M., Darrell, T.: Conditional random fields for object recognition. In: NIPS, pp. 1097–1104. MIT Press (2004)
12. Radosavljevic, V., Vucetic, S., Obradovic, Z.: Continuous conditional random fields for regression in remote sensing. In: ECAI 2010, pp. 809–814 (2010)
13. Sutton, C., McCallum, A.: An introduction to conditional random fields for relational learning. In: Getoor, L., Taskar, B. (eds.) Introduction to Statistical Relational Learning. MIT Press (2007)
14. Vail, D.L., Veloso, M.M., Lafferty, J.D.: Conditional random fields for activity recognition. In: AAMAS 2007, pp. 235:1–235:8. ACM, New York (2007)
15. Ženko, B., Džeroski, S.: Learning classification rules for multiple target attributes. In: Washio, T., Suzuki, E., Ting, K.M., Inokuchi, A. (eds.) PAKDD 2008. LNCS (LNAI), vol. 5012, pp. 454–465. Springer, Heidelberg (2008)
16. Xin, X., King, I., Deng, H., Lyu, M.R.: A social recommendation framework based on multi-scale continuous conditional random fields. In: CIKM, pp. 1247–1256 (2009)
17. Yu, D., Deng, L., Acero, A.: Using continuous features in the maximum entropy model. Pattern Recogn. Lett. 30(14), 1295–1300 (2009)

Fault Tolerant Regression for Sensor Data

Indrė Žliobaitė and Jaakko Hollmén

Aalto University, Dept. of Information and Computer Science,
Helsinki Institute for Information Technology (HIIT), Finland
{indre.zliobaite,jaakko.hollmen}@aalto.fi

Abstract. Many systems rely on predictive models using sensor data, with sensors being prone to occasional failures. From the operational point of view predictions need to be tolerant to sensor failures such that the loss in accuracy due to temporary missing sensor readings would be minimal. In this paper, we theoretically and empirically analyze robustness of linear predictive models to temporary missing data. We demonstrate that if the input sensors are correlated the mean imputation of missing values may lead to a very rapid deterioration of the prediction accuracy. Based on the theoretical results we introduce a quantitative measure that allows to assess how robust is a given linear regression model to sensor failures. We propose a practical strategy for building and operating robust linear models in situations when temporal sensor failures are expected. Experiments on six sensory datasets and a case study in environmental monitoring with streaming data validate the theoretical results and confirm the effectiveness of the proposed strategy.

Keywords: missing data, data streams, linear models, sensor failure.

1 Introduction

The amount of sensors installed in the urban and natural environments is rapidly increasing. It is predicted that sensor data collected from satellites, mobile devices, outdoor and indoor cameras will become the largest information trove for our society in the coming years [3]. Predictive models using sensor readings as inputs are widely applied in real-time systems, such as production quality control, air pollution monitoring, detecting traffic jams or severe road conditions, route recognition, road navigation, cargo tracking and many more [6].

Physical sensors are exposed to various risks due to, for instance, severe environmental conditions or exposure to physical damage. Moreover, typically sensors rely on batteries, are installed in remote or hardly accessible locations, or are unaccessible during operation runtimes. Sensors may break causing a sudden failure until replaced. Sensors may get covered in snow or water causing a seasonal temporary disruption. Some sensors may lose sensitivity due to wear and tear. Under such circumstances it is very common to have time intervals when readings from some sensors are missing. At the same time a predictive model needs to operate continuously and deliver predictions in real time.

H. Blockeel et al. (Eds.): ECML PKDD 2013, Part I, LNAI 8188, pp. 449–464, 2013.

We study how to make predictive models robust to temporary sensor failures during real-time operation. We focus on linear regression models. We assume that the observed process is stationary and the predictive model remains fixed during real-time operation. Our goal is to maximize prediction accuracy not only when all the input sensors are available, but also when readings from several sensors are missing for continuous periods of time. The problem is more challenging than it may seem due to the temporal nature of the missing data and limited computational resources under the stream setting.

Discarding the incoming instances that contain missing values is not an option, since there will be no predictions for continuous periods of time. The iterative multiple imputation (MI) [14] carries high computational costs and is not considered for real-time operation. Deploying additional predictive models for filling in missing values given the available sensors is computationally impractical, since an exponential number of models would be required to cover all combinations of failing sensors. One could consider an adaptive regression, e.g. [17]. However, the time for learning a stable model before recovery or the next failure is very limited, while the previous model is in principal correct. This applies to persistent temporal failures as well as once-off outlier failures. Hence, adaptation is not considered when to cover for frequent temporary failures.

A simple replacement of the missing values by the sensor mean value is a popular and easy to implement strategy in industrial applications. Unfortunately, in the stream setting where values of the same sensors are missing for a continuous period of time, this strategy may lead to a drastic deterioration of the prediction accuracy, particularly if the input sensors are highly correlated and a regression model exploits that correlation. Therefore, if sensor failures are expected in real-time operation it is not enough to replace missing values by the mean; we also need to ensure that the predictive model is robust to temporary missing data.

This paper presents a theoretical analysis of the predictive performance under sensor failures and formulates robustness criteria for real-time operation. We introduce the *deterioration index* that allows to assess robustness of a given linear model to partial loss of input data. We propose a practical strategy for building robust linear models that is based on a de-correlating transformation and a subsequent regularization of the model parameters in the transformed space. Experimental validation on six sensor datasets and a case study in environmental monitoring domain confirms the effectiveness of the proposed strategy.

Our study contributes a theoretically supported methodology for diagnosing robustness of linear regression models to loss of input data. This methodology makes it possible to *assess* the robustness of alternative models prior to deployment in real-time operation. The second contribution is a practical strategy for *optimizing* linear regression models such that they are robust to sensor failures.

The paper is organized as follows. Section 2 outlines the setting. In Section 3 we theoretically analyze how sensor failures affect the prediction accuracy and develop an index for diagnosing the performance. Section 4 gives practical recommendations. Experimental analysis is reported in Section 5 and the case study in Section 6. Section 7 discusses related work and Section 8 concludes the study.

2   Background and Problem Setting

We start by formalizing the problem and presenting a recap on linear models.

Setting. Suppose we have r sensors generating multidimensional streaming data vectors $\mathbf{x} \in \Re^r$ (e.g. weather observation sensors). Our task is to predict the target variable $y \in \Re^1$ (e.g. solar radiation) using these sensor readings as inputs. Data arrives in real time, the predictions need to be available in real time. The expected loss in accuracy due to sensor failures should be minimum. We assume that the observed process is stationary and the predictive model remains fixed during real-time operation. To keep the focus, we do not explicitly model potential spatial or temporal correlation between sensors.

Prerequisites. Without loss of generality we assume that the data (including the target variable) is standardized to zero mean and unit variance. To keep the focus we also assume that we know when a sensor fails (we do not need to detect it). We also assume that when a sensor fails, the missing values are automatically replaced with a constant value, say *zero* or the *median* value [2].

2.1   Linear Regression

In this study, we consider linear regression models for prediction [9], which assume that the relationship between r input sensors $\mathbf{x} = (x_1, \ldots, x_r)$ and the target variable y is linear. The model takes the form

$$y = b_1 x_1 + b_2 x_2 + \ldots + b_r x_r + \epsilon = \mathbf{x}\beta + \epsilon, \tag{1}$$

where ϵ is the error variable and the vector $\beta = (b_1, b_2, \ldots, b_r)^T$ contains the parameters of the linear model (regression coefficients). Since the data is assumed to have been standardized, the bias term in the regression model is omitted.

There are different ways to estimate the regression parameters [8,9]. Ordinary least squares (OLS) is a simple and probably the most common estimator. It minimizes the sum of squared residuals giving the following solution

$$\hat{\beta}_{OLS} = \arg\min_{\beta} \left((\mathbf{X}\beta - \mathbf{y})^T (\mathbf{X}\beta - \mathbf{y}) \right) = (\mathbf{X}^T \mathbf{X})^{-1} \mathbf{X}^T \mathbf{y}, \tag{2}$$

where $\mathbf{X}_{n \times r}$ is a sample data matrix containing n records from r sensors, and $\mathbf{y}_{n \times 1}$ is a vector of the corresponding n target values. Having estimated a regression model $\hat{\beta}$ the predictions for on new data $\mathbf{x}_{new}$ can be made as $\hat{y} = \mathbf{x}_{new}\hat{\beta}$.

If the input data is correlated, regularization is often used for estimating the regression parameters. The Ridge regression (RR) [9, 10] regularizes the regression coefficients by imposing a penalty on their magnitude. The RR solution is

$$\hat{\beta}_{RR} = \arg\min_{\beta} \left((\mathbf{X}\beta - \mathbf{y})^T (\mathbf{X}\beta - \mathbf{y}) + \lambda \beta^T \beta \right) = (\mathbf{X}^T \mathbf{X} + \lambda \mathbf{I})^{-1} \mathbf{X}^T \mathbf{y}, \tag{3}$$

where $\lambda > 0$ controls the amount of shrinkage: the larger the value of λ, the greater the amount of shrinkage. $\mathbf{I}_{r \times r}$ is the identity matrix.

2.2   Prediction Error

The mean squared error (MSE) is a popular measure to quantify the discrepancy between the true target y and the prediction $\hat{y}$. For a test dataset it is computed as $MSE = \sum_{l=1}^{n} (\hat{y}^{(l)} - y^{(l)})^2/n = \hat{\mathbf{y}}^T \mathbf{y}/n$, where n is the number of samples. We use MSE since it punishes large deviations from the true values, that is relevant to industrial applications. Also, MSE has interesting analytical properties. We can decompose the expected mean squared error as

$$E[MSE] = E\Big[\frac{1}{n}\sum_{l=1}^{n}(\hat{y}^{[l]} - y^{[l]})^2\Big] = E[(\hat{y} - y)^2] = E[(\hat{y})^2 - 2\hat{y}y + y^2]$$

$$= E[\hat{y}^2] - 2E[\hat{y}y] + E[y^2] = Var[\hat{y}] - 2\,Cov[\hat{y}, y] + Var[y] \qquad (4)$$

The last equation follows from $Var[z] = E[z^2] - (E[z])^2$ and $Cov[x, z] = E[xz] - E[x]E[z]$. $E[y] = 0$ and $E[\hat{y}] = 0$, since the data has been standardized.

Let the prediction be $\hat{y} = \mathbf{x}\beta$. Then the variance of this prediction is

$$Var[\hat{y}] = Var\Big[\sum_{i=i}^{r} b_i x_i\Big] = \sum_{i=1}^{r}\sum_{j=i}^{r} b_i b_j\,Cov[x_i, x_j] = \beta^T \mathbf{\Sigma}\beta, \qquad (5)$$

where $\mathbf{\Sigma} = \mathbf{X}^T\mathbf{X}/(n-1)$ is the covariance matrix of the input data.

The covariance of the prediction is

$$Cov[\hat{y}, y] = E[\hat{y}y] = E\Big[y\sum_{i=1}^{r} b_i x_i\Big] = \mathbf{y}^T\mathbf{X}\beta/(n-1). \qquad (6)$$

In real-time predictive systems if a sensor fails, typically, a constant value is displayed. For convenience but without loss of generality we assume that the missing values are replaced by the mean (*zero*, since the data is standardized) as they arrive. Detecting sensor failures is beyond the scope of this work. We assume that the data collection system can signal sensor failures automatically. If this is not the case one can set up a simple rule based detector, such as: if the value is constant for a period of time declare sensor failure.

3   Theoretical Analysis of the Effect of Sensor Failures

Let us consider theoretically the prediction error of a linear model when the input sensors start to fail. Surprisingly, the jump in error can be nonlinear in the number of sensors failed and highly depends on the correlation of the inputs.

Denote by MSE_m the mean square error after m sensors have failed. Let MSE_0 be the error when all the sensors are working. Correspondingly, Var_m and Cov_m denote the variance and the covariance after m sensors have failed. Note that cross-validtion MSE_0 is often the only consideration when assessing the performance of a model or deploying in practice.

Proposition 1. *Suppose sensors fail independently with the uniform prior probability. With m failed sensors the expected error of a linear model is*

$$E[MSE_m] = \frac{r-m}{r} E[MSE_0] + \frac{m}{r} - \frac{(r-m)m}{r(r-1)} \beta^T (\Sigma - \mathbf{I})\beta,$$

where r is the number of input sensors, β is a vector of the regression coefficients, Σ is the covariance matrix of the input data.

MSE can be decomposed into variance of the prediction, covariance of the prediction and the target and the variance of the target, as given in Eq.(4). For proving Proposition 1 we will analyze each component separately.

Proposition 2. *Suppose sensors fail independently with the uniform prior probability. With m failed sensors the expected variance of the prediction $\hat{y} = \mathbf{x}\beta$ is*

$$Var_m[\hat{y}] = \frac{r-m}{r} Var_0[\hat{y}] - \frac{(r-m)m}{r(r-1)} \beta^T (\Sigma - \mathbf{I})\beta.$$

Proof (of Proposition 2). We decompose the variance from Eq. (5) into $Var_0[\hat{y}] = \beta^T \Sigma \beta = \beta^T \beta + \beta^T (\Sigma - \mathbf{I})\beta$. The first part $\beta^T \beta = \sum_{i=1}^{r} b_i Var[x_i]$ describes the variance when the inputs are linearly independent. The second part $\beta^T (\Sigma - \mathbf{I})\beta = 2 \sum_{i=1}^{r-1} \sum_{j=i+1}^{r} b_i b_j Cov(x_i, x_j)$ is due to correlation of the inputs.

Consider the first component $\beta^T \beta$. If a sensor fails, the individual variance becomes zero and the term vanishes. The total variance decreases by $\frac{1}{r}b_1^2 + \frac{1}{r}b_2^2 + \cdots + \frac{1}{r}b_p^2 = \frac{1}{r}\beta^T \beta$. Likewise, if m sensors fail, the variance decreases by $\frac{m}{r}\beta^T \beta$.

Now consider the second component $\beta^T (\Sigma - \mathbf{I})\beta$. Since $(\Sigma - \mathbf{I})$ has zeros on the diagonal, the component is a weighted sum of $r(r-1)$ covariances from the covariance matrix. If one sensor (say, sensor i) fails, all the covariances of other sensors with x_i will become zero and all the terms containing covariance with x_i will vanish. The sum will lose $2(r-1)$ elements (such is the amount of elements with $Cov[x_i, \ldots]$), the total loss will be $\frac{2(r-1)}{r(r-1)}\beta^T (\Sigma - \mathbf{I})\beta$.

However, if two sensors fail then only $2(r-1) + 2(r-2)$ elements will be lost from the sum. If sensors i and j fail, there will be $2(r-1)$ lost containing covariance with x_i, but only $2(r-2)$ more terms lost containing covariance with x_j, as the term $Cov(x_i, x_j)$ has already been lost earlier. Hence, if m sensors fail then $2(r-1) + 2(r-2) + \cdots + 2(r-m) = (2r-1-m)m$ elements will be lost and the collinearity component will decrease by $\frac{(2r-1-m)m}{r(r-1)}\beta^T (\Sigma - \mathbf{I})\beta$.

Plugging the terms into $Var[\hat{y}]$ expression gives
$Var_m[\hat{y}] = \beta^T \beta - \frac{m}{r}\beta^T \beta + \beta^T (\Sigma - \mathbf{I})\beta - \frac{(2r-1-m)m}{r(r-1)}\beta^T (\Sigma - \mathbf{I})\beta = \frac{r-m}{r}\beta^T \beta + \frac{r-m}{r}\beta^T (\Sigma - \mathbf{I})\beta - \frac{(r-m)m}{r(r-1)}\beta^T (\Sigma - \mathbf{I})\beta = \frac{r-m}{r} Var_0[\hat{y}] - \frac{(r-m)m}{r(r-1)}\beta^T (\Sigma - \mathbf{I})\beta.$ $\square$

Proposition 3. *Suppose sensors fail independently with the uniform prior probability. With m failed sensors the expected covariance of the prediction $\hat{y} = X\beta$ is*

$$Cov_m[\hat{y}, y] = \frac{r-m}{r} Cov_0[\hat{y}, y].$$

Proof (of Proposition 3). The covariance is $Cov_0[\hat{y}, y] = E\left[y \sum_{i=1}^{r} b_i x_i\right]$. If a sensor fails, the expected value of that sensor becomes zero, the term becomes independent from y and vanishes. If one sensor fails, the expectation decreases by $\frac{1}{r} y b_1 x_1 + \frac{1}{r} y b_2 x_2 + \cdots + \frac{1}{r} y b_r x_r = \frac{1}{r} y \sum_{i=1}^{r} b_i x_i$. If m sensors fail the expectation decreases by $\frac{m}{r} y \sum_{i=1}^{r} b_i x_i$. Plugging that into the covariance expression gives $Cov_m[\hat{y}, y] = E\left[y \sum_{i=1}^{r} b_i x_i - \frac{m}{r} y \sum_{i=1}^{r} b_i x_i\right] = E\left[\frac{r-m}{r} y \sum_{i=1}^{r} b_i x_i\right] = \frac{r-m}{r} Cov_0[\hat{y}, y]$. Note that the effect of sensor failure on $Cov[\hat{y}, y]$ is the same no matter whether the input data is correlated. □

Given Proposition 2 and Proposition 3 we can now prove Proposition 1.

Proof (of Proposition 1). Following Eq. (4) we decompose the error with m failed sensors into $E[MSE_m] = Var_m[\hat{y}] - 2 Cov_m[\hat{y}, y] + Var_m[y]$. Failing input sensors do not affect the variance of the true target, thus $Var_m[y] = Var_0[y] = 1$. From Propositions 2, 3 and Eq. (4) we get $E[MSE_m] = \frac{r-m}{r} Var_0[\hat{y}] - 2 \frac{r-m}{r} cov_0[\hat{y}, y] + Var_0[y] - \frac{(r-m)m}{r(r-1)} \beta^T(\Sigma - \mathbf{I})\beta = \frac{r-m}{r} E[MSE_0] + \frac{m}{r} - \frac{m(r-m)}{r(r-1)} \beta^T(\Sigma - \mathbf{I})\beta$. □

For constructing fault tolerant models we will need the next proposition.

Proposition 4. *Given a $r \times r$ covariance matrix Σ and a vector $\beta \in \Re^r$ with at least one non-zero element, the term $\beta^T(\Sigma - \mathbf{I})\beta$ is bounded by*

$$-\beta^T\beta \leq \beta^T(\Sigma - \mathbf{I})\beta \leq (r-1)\beta^T\beta.$$

Proof (of Proposition 4). The Rayleigh quotient of the covariance matrix is defined as $\frac{\beta^T \Sigma \beta}{\beta^T \beta}$, for non-zero $\beta \in \Re^r$ and is bounded by the maximum and the minimum eigenvalues of Σ: $\ell_{min} \leq \beta^T \Sigma \beta / \beta^T \beta \leq \ell_{max}$, where ℓ are eigenvalues, and takes the extreme values when β is equal to the corresponding eigenvectors.

Since Σ is a covariance matrix, all eigenvalues are non-negative and their sum is equal to the sum of the trace. As the data is standardized the sum of eigenvalues is r, hence the maximum eigenvalue does not exceed r: $0 \leq \beta^T \Sigma \beta / \beta^T \beta \leq r$. Algebraic manipulations give the bound $-\beta^T\beta \leq \beta^T(\Sigma - \mathbf{I})\beta \leq (r-1)\beta^T\beta$. □

Our analysis relies on theoretical variance and covariance of the prediction. Potentially it could be extended to higher order regression models (e.g. quadratic), that would require much more involved theoretical analysis due to interaction terms. Alternatively, one could obtain non linear prediction models by using the same linear regression with non-linear input features.

Proposition 1 has an important implication. If input data is uncorrelated then *MSE* is increasing linearly with the number of sensors failed. If some sensor fails, the predictive information is lost, there is no source for replacement.

On the other hand, if input data is correlated, *MSE* changes quadratically in the number of sensors lost. From Proposition 4 we see that this quadratic term can be positive or negative depending on the regression model (β). The good news is that a well chosen β may reduce the loss in accuracy to sub-linear. The next section considers strategies for building regression models such that the expected *MSE*, when sensors are failing, is minimized.

Comparison of three regression models

	∂	MSE_m
$\beta_1 = (1,0,0,0)^T$	0	$0.25m$
$\beta_2 = (\frac{1}{4}, \frac{1}{4}, \frac{1}{4}, \frac{1}{4})^T$	-0.75	$(0.25m)^2$
$\beta_3 = (2, -\frac{3}{2}, 1, -\frac{1}{2})^T$	6.5	$1.7m - 0.4m^2$

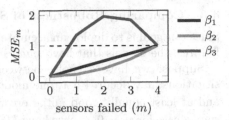

Fig. 1. Three regression models and the deterioration of the prediction accuracy as a function of number of failed sensors (m)

4 How to Build Fault Tolerant Regression Models

Based on the theoretical results next we propose a quantitative measure for assessing robustness of regression models and present practical guidelines on how to build fault tolerant regression models.

4.1 Deterioration Index

In Proposition 1 the term $-\frac{(r-m)m}{r(r-1)}\beta^T(\Sigma - \mathbf{I})\beta$ decides whether MSE increases linearly, quadratically or sub-linearly due to sensor failures. Hence, for diagnosing model robustness to sensor failures we define a *deterioration index* as

$$\partial = -\beta^T(\Sigma - \mathbf{I})\beta.$$

When input data is uncorrelated, i.e. $\Sigma = \mathbf{I}$, then $\partial = 0$. When input data is correlated ∂ may be positive or negative (see Proposition 4). If $\partial > 0$ then MSE deteriorates quadratically, if $\partial < 0$ then MSE deteriorates sub-linearly. Thus, the lower the *deterioration index* the more robust the predictive model is.

4.2 An Illustrative Example

For illustrative purposes, let us consider a small regression problem where four input sensors are perfectly correlated with each other and the target variable: $x_1 \sim \mathcal{N}(0,1)$, $x_1 = x_2 = x_3 = x_4 = y$. Note that $\Sigma = \mathbf{1}_{4\times4}$. Figure 1 gives three regression models that would give perfect predictions if all sensors are working, i.e. $MSE_0 = 0$ and their respective MSE_m after m sensors fave failed (from Proposition 1). Figure 1 plots the expected errors when sensors start to fail.

Model β_1 utilizes only one input sensor and the deterioration of MSE_m is linear to the number of sensors failed (m). Can we do better? In fact we can do better with model β_2, which makes use of the redundancy in sensors. As a result, the loss in accuracy is lower. Model β_3 represents a really bad case of overfitting with the regression, although this model can predict perfectly well, the weights grow unnecessary high. In such a case, if a sensors start failing, the variance of the prediction grows really high and so does the MSE_m. We can observe that even if a single sensor has failed $MSE_1 > 1$ making the predictions worse than a naive baseline that always predicts constant value ($MSE_{constant} = 1$).

4.3 Comparing Robustness of Several Regression Models

Since our goal is to deploy an accurate model that would also be robust to sensor failures, the models that have *small* initial MSE_0 and *small* $\mathfrak{d}$ are preferred.

Suppose we have a choice between two regression models A and B. Two situations may occur. First, one model, say A, has a better *deterioration index* and at least equally good initial error: $MSE_0^{(A)} \leq MSE_0^{(B)}$ and $\mathfrak{d}^{(A)} < \mathfrak{d}^{(B)}$. In such a case for $m = 0 \ldots r$ we have $MSE_m^{(A)} \leq MSE_m^{(B)}$ (results from Proposition 1); hence, model A is preferable. Example in Section 4.2 showed such a situation.

The other situation is more tricky. One model, say A, may have a better *deterioration index*, but worse initial error: $MSE_0^{(A)} > MSE_0^{(B)}$ and $\mathfrak{d}^{(A)} < \mathfrak{d}^{(B)}$. In this case model A is preferable if we expect less than $m^\star$ sensors to fail, and otherwise model B is preferable. We can find $m^\star$ from Proposition 1 by assigning $E[MSE_m^{(A)}] = E[MSE_m^{(B)}]$ and solving for m. The number of sensors to fail is

$$m^\star = (r - 1)(MSE_0^{(A)} - MSE_0^{(B)})/(\mathfrak{d}^{(B)} - \mathfrak{d}^{(A)}). \tag{7}$$

4.4 Building Fault Tolerant Regression Models

A regression model β obtained using the ordinary least squares procedure OLS minimizes MSE_0. The index $\mathfrak{d}$ takes its minimum when β is equal to the eigenvector with the maximum eigenvalue (Proposition 4). Unfortunately, such β does not guarantee correct predictions, since eigenvectors are obtained not taking into account the target variable. Hence, for making fault tolerant models we need an optimization criteria that would minimize $\mathfrak{d}$ and MSE_0 at the same time.

For a predictive model $\hat{y} = \mathbf{x}\beta$ the *deterioration index* can be decomposed into $\mathfrak{d} = -\beta^T(\mathbf{\Sigma} - \mathbf{I})\beta = -Var[\hat{y}] + \beta^T\beta$. We can rewrite Eq. (4) as $Var[\hat{y}] = MSE_0 + 2Cov[\hat{y}, y] - Var[y] \to 1$. Accurate prediction requires $Cov[\hat{y}, y] \to 1$ and $Var[y]$ is fixed. Thus, we cannot vary $Var[\hat{y}]$ without affecting the error.

However, we could vary $\beta^T\beta$ to a certain extent with little impact to MSE_0, for instance, as in the toy example in Section 4.2. Hence, $\mathfrak{d}$ will be minimized when $\beta^T\beta$ shrinks. To achieve that we recommend using regularization for building regression models, such as the Ridge regression (Section 2.1).

In addition, we recommend reducing the dimensionality rotating the input data towards the first k principal components. Let $\mathbf{X} = \mathbf{U}\mathbf{D}\mathbf{V}^T$ be the singular value decomposition of the training data. Let the rotation matrix $\mathbf{R}_{p \times k}$ be composed of the eigenvectors in $\mathbf{V}$ that correspond to the largest eigenvalues recorded in the diagonal of $\mathbf{D}$. Then new k-dimensional input data is $\mathbf{X}^\star = \mathbf{X}\mathbf{R}$. Let $\beta^\star$ be a vector of regression coefficients in the transformed k-dimensional space. The model $\beta = \mathbf{R}\beta^\star$ would give the same predictions in the original.

In order to minimize $\mathfrak{d}$ we need to minimize $\beta^T\beta$ in the original space, but at the same time we need to find the optimal $\beta^\star$ in the transformed space. Hence, our optimization criteria is $\hat{\beta}^\star = \arg\min_{\beta^\star} \left((\mathbf{X}^\star\beta^\star - y)^T(\mathbf{X}^\star\beta^\star - y) + \lambda\beta^T\beta\right)$. Since $\mathbf{R}$ is orthogonal, thus $\beta^T\beta = \beta^{\star T}\mathbf{R}^T\mathbf{R}\beta^\star = \beta^{\star T}\beta^\star$. Therefore, optimizing the above criteria is equivalent to the Ridge regression in the $\mathbf{X}^\star$ space.

Given these considerations, our recommendation for building fault tolerant models is to apply PCA and then train the Ridge regression in the new space.

5   Experimental Analysis

Next we experimentally analyze the robustness of selected regression techniques against sensor failure on a synthetic benchmark and real sensor datasets.

5.1   Datasets

Chemi dataset from the INFER project[1] describes a chemical production process via 70 real valued sensor variables sampled once per hour over a two years period (17562 instances). The goal is to predict concentration of a product.

ChemiR extends the Chemi data, an additional variable indicates the concentration of the product an hour ago (71 sensors, 17562 instances).

Catalyst dataset[2] is a chemical modeling dataset. Given 13 input variables the goal is to predict catalyst activity. The dataset contains 8687 instances.

Wine dataset from the UCI repository[3] presents 10 chemical measurements as inputs and 4897 instances. The goal is to predict a wine quality score.

CPU dataset from the DELVE repository[4] collects computer systems activity measures described by 19 real valued attributes. The goal is to predict the portion of time that cpus run in user mode. The dataset contains 8192 instances.

Gaussian is a synthetic dataset in 30-dimensional space. Input data is sampled from $\mathcal{N}(\mathbf{0}, \Sigma)$, a random covariance matrix is generated as $\Sigma = s^T s$, where $s \sim \mathcal{U}(-1, 1)$. The target variable is set to $y = \sum_{i=1}^{30} x_i + u$, where $u \sim \mathcal{N}(0, 6)$.

5.2   Regression Models

We test the following regression models.

ALL uses all r sensors as inputs. A regression model is built using the ordinary least squares (OLS) optimization approach.

rALL uses all r sensors and the Ridge regression (RR) optimization.

SEL builds OLS regression on k sensors that have the largest absolute correlation with the target variable (measured on the training data).

PCA rotates the input data using principal component analysis (PCA) and builds the OLS regression on k new attributes with the largest eigenvalues.

rPCA extracts attributes using PCA, but then RR is used instead of OLS.

sPCA rotates the data using PCA, selects k new attributes that are the most correlated with the target. A regression model is built using OLS.

PLS regression is very popular in chemometrics [18]. It is similar to PCA, but instead of maximizing the variance with the rotation, a covariance between the inputs and the target is maximized. We keep k new attributes.

---

[1] Source: `http://infer.eu/`

[2] Source: `http://www.nisis.risk-technologies.com/`

[3] Source: `http://archive.ics.uci.edu/ml/`

[4] Source: `http://www.cs.toronto.edu/~delve/`

Table 1. Testing MSE_0 and *deterioration index* on the six sensory datasets

		rPCA	PCA	rALL	SEL	sPCA	PLS	ALL
Chemi	MSE_0	0.47	0.47	0.38	0.52	0.42	0.41	0.41
	ȼ	-0.25	-0.25	1.03	0.67	6.42	6.54	7.44
ChemiR	MSE_0	0.42	0.43	0.36	0.35	0.37	0.35	0.34
	ȼ	-0.34	-0.33	0.09	0.53	4.90	3.83	5.57
Catalyst	MSE_0	0.51	0.51	0.45	0.82	0.47	0.44	0.43
	ȼ	-0.11	-0.08	0.21	0.69	0.63	0.82	1.71
Wine	MSE_0	0.83	0.83	0.73	0.76	0.74	0.73	0.73
	ȼ	-0.07	-0.07	0.04	0.12	0.04	0.08	0.18
CPU	MSE_0	0.31	0.31	0.28	0.30	0.29	0.28	0.28
	ȼ	-0.31	-0.32	-0.25	-0.26	-0.16	-0.14	-0.11
Gaussian	MSE_0	0.23	0.23	0.12	0.25	0.14	0.12	0.12
	ȼ	-0.32	-0.33	-0.13	-0.16	-0.10	0	7.58

5.3 Experimental Protocol and Parameters

Each dataset is split into training and testing at random (equal sizes). Some data may have temporal dependencies, hence some predictive information (such as autocorrelation) cannot be utilized, that applies to all the tested models in the same way, while random splits allow multiple tests. We repeat every experiment 100 times and report averaged results. The input data and the target variable is standardized, the mean and the variance for standardization is calculated on the training data. The regression models are trained on the training part and the reported errors and sample covariances are estimated on the testing part. The regression coefficients are always reported in the original (not transformed) feature space. For feature selection SEL, PCA and PLS models we set the number of components to be a half of original number of features: $k = r/2$. The regularization parameter in the Ridge regression experiments is fixed to 200.

5.4 Robustness versus Accuracy

Table 1 reports the testing errors and the *deterioration index* (ȼ) on the six sensory datasets. The models are grouped according to the potential deterioration of their accuracies. We can distinguish three groups of models.

The first group contains PCA and regularized rPCA. These models consistently achieve a very good *deterioration index* (below zero), that guarantees preservation of prediction accuracy. The initial MSE_0 of PCA and rPCA is typically larger than the peer approaches, that is the price to pay for robustness. The superior performance of rPCA and PCA is consistent across the six datasets.

The second group contains regularized rALL and SEL, which have varying *deterioration index* , but typically not too high. rALL maintains a reasonable accuracy (typically better than the first group); however, the accuracy of SEL varies a lot, due to varying predictive power of the individual sensors (depends on the prediction task at hand). The third group contains sPCA, PLS and ALL,

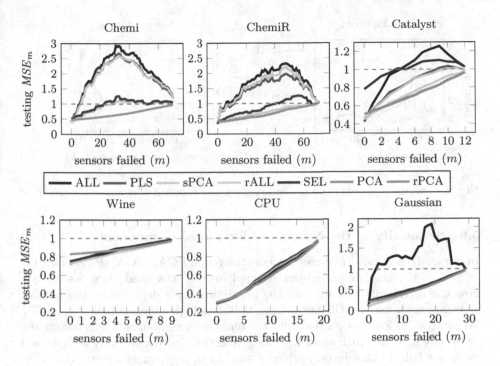

Fig. 2. Empirical MSE_m versus the number of sensors failed

these models mostly show a very high *deterioration index* , especially in Chemi, ChemiR and Catalyst datasets, where the inputs are highly correlated.

Thus, we recommend using rPCA (or PCA) if sensors are expected to fail often and predictions are needed continuously. If failures are rare we recommend rALL that is less robust but more accurate with all the sensors working.

5.5   Empirical Analysis of Deterioration of Accuracy

Next we investigate how the error depends on the number of sensors that have failed. Figure 2 shows testing MSE_m as a function of the failed sensors. We chose sensors to fail uniformly at random, we report the results over 100 runs.

Advantages of PCA and rPCA are prominent in Chemi and ChemiR, where the dimensionality is large and the input data is strongly correlated. All the models perform similarly (nearly linear loss) in Wine and CPUact, where the input data is not much correlated. In Catalyst PCA and rPCA have an advantage as expected based on *deterioration index* . On this dataset SEL has notably worse performance. As the overall number of features is low (12), quite a lot of initial accuracy is lost by dropping half of the features. Gaussian data strictly follows the normal distribution, and the contribution of each sensor to the target variable is uniformly distributed. ALL performs notably badly, but we see that any regularization attempt (all the other methods) leads to a good performance.

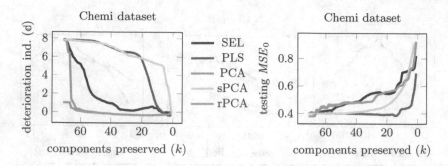

Fig. 3. Deterioration index and prediction error as a function of components preserved

5.6 Sensitivity to the Number of Extracted Components

In our analysis number of components extracted in PCA, rPCA, sPCA and PLS, as well as the number of features selected in SEL was fixed. Next we analyze how the *deterioration index* and the prediction error depends on the number of components on the Chemi dataset that has high dimensionality ($r = 70$). We analyze all five models that take the number of components as a parameter: PCA, rPCA, sPCA, SEL and PLS. Our goal is to assess the stability with respect to sensor failures at arbitrary selected number of components. Note, that if all the components are preserved ($k = r$) then PCA, sPCA, PLS and SEL are equivalent to ALL, and the regularized rPCA is equivalent to rALL.

Figure 3 shows the *deterioration index* and the prediction error as a function of extracted components (over 100 runs). The regularized rPCA performs much better đ than PCA when nearly all of the components are retained. PCA and rPCA demonstrates superior đ across all k in line with the previous experiments. SEL demonstrates a mediocre đ and sPCA together with PLS keeps a dangerously high đ until the majority of the components are discarded (k is below 10). We see from the right plot in Figure 3 that, unfortunately, at such a low k a lot of prediction accuracy is lost, *MSE* is nearly twice as large as with all the sensors.

Overall, we see a tendency to achieve a better *deterioration index* at an expense of a lower initial prediction accuracy. The regularized rPCA demonstrates the most stable performance and superior results throughout all the range of k.

5.7 Worse Than Blind Guessing

Blind guessing is a naive prediction, that does not use any input data and always predicts the average of the target variable. Next we analyse how many sensors can fail before predictions become worse than blind guessing. Table 2 reports empirical results on the six datasets averaged over 100 runs.

We see that in Gaussian the problem of sensor failure is very serious, it is enough for two sensors out of 30 to fail and the predictive model is useless. In case of Chemi and ChemiR it is enough for 6-7 sensors to fail out of 70-71 to make ALL or even PLS useless. PLS is a very popular state of the art technique

Table 2. No. of sensors to fail before the prediction becomes worse than blind guessing

	rPCA	PCA	rALL	SEL	sPCA	PLS	ALL	sensors
Chemi	-	-	29	23	7	7	6	out of **70**
ChemiR	-	-	70	41	10	9	6	out of **71**
Catalyst	-	-	-	4	10	9	5	out of **12**
Wine	-	-	-	-	-	-	-	out of **9**
CPU	-	-	-	-	-	-	-	out of **19**
Gaussian	-	-	-	-	-	-	3	out of **30**

often used in chemometrics applications [13,18], such as Chemi. From Figure 2 we can see that if just *one* sensor is lost, the error of PLS or ALL in Chemi already jumps up by nearly 40%. We see that the mean imputation of missing sensor values is a serious problem for these regression models.

The experimental results suggest that careful regularization measures are needed for ensuring that predictive models stay functional during real-time operation. The experimental results confirm our theoretical findings and the indications of the *deterioration index* that the proposed rPCA and PCA can effectively prevent rapid boosts in errors due to sensor failures. If a user does not have the capacity to determine the optimal k, as a rule of thumb from our practical experience we recommend using $k = r/2$, where r is the number of input sensors.

6   Case Study in Environmental Monitoring

To validate our findings we perform a case study in environmental monitoring where sensor failures are happening frequently. The task is to predict the *level of solar radiation* from meteorological sensor data (such as temperature, precipitation, wind speed). We use a data stream recorded at SMEAR II station in Finland [12]. This station can measure solar radiation; hence, the ground truth is available for us. In general, measuring solar radiation is delicate and expensive. Not many stations can afford to measure solar radiation and would be interested in predicting it from other data that can be collected much cheaper and easier.

We use data over a five years period (2007-2012), recorded every 30 min. from 39 meteorological sensors at one station. The data coming from the station has about 7% of missing values. There is no single sensor that would provide non interrupted readings over those five years; for any sensor from 1% up to 30% values are missing. The solar radiation (target variable) is available 99% of the times, we eliminate from the experiment the instances having no target value.

Our goal is to verify if the proposed *deterioration index* can effectively diagnose the performance of regression models and test the performance of our regression models with naturally occurring missing data. We use the first two years of data as a training set and the remaining three years as a testing set. From the training set we eliminate all the instances that contain missing values (−25% of train data). We standardize the training set (zero mean, unit variance). Then we standardize the testing set using the mean and the variance values obtained from the training set. After standardization we replace all the missing values in the testing set by zeros and test the regression models.

Table 3. Accuracy and stability on the environmental monitoring data

	rPCA	PCA	rALL	SEL	sPCA	PLS	ALL
MSE on all test data	0.37	0.37	0.51	15.67	8.55	0.93	7.39
MSE on non-missing test data	0.33	0.33	0.29	0.35	0.29	0.28	0.28
MSE on missing test data	0.39	0.39	0.63	23.88	12.99	1.27	11.21
MSE on train data (cross-validation)	0.40	0.39	0.37	0.34	0.31	0.32	0.29
d̶ on train data	-0.22	-0.20	1.61	574.58	253.86	8.66	302.84

Table 3 reports the testing results of the regression models ALL, rALL, SEL, PCA, rPCA, sPCA and PLS ($k = 20$, which is half of the input sensors following the rule of thumb suggested in Section 5.6). The recommended rPCA and PCA demonstrate an outstanding performance ($MSE = 0.37$), followed by rALL (0.51). The performance of PLS (0.91) is more than twice worse as of PCA and rPCA. ALL, sPCA and SEL perform much worse by a large margin.

Next we split of the test data into non-missing (35%) and missing data (65%) parts and inspect the errors separately. We see that the performance data of all the models is very similar when there is no missing data. However, the non-regularized models (ALL, SEL, sPCA and PLS) fail badly when there is missing data, except for PCA, which is consistent with our theoretical findings. Moreover, we can see from the last part of the table that if we selected a model for deployment based on cross-validation *MSE*, we would probably deploy ALL. It would perform on non-missing data well, but the performance would deteriorate very drastically when sensors started to fail. Finally, we can see that the proposed *deterioration index* computed on the training data indicates very well the future robustness of the model. Hence, after seeing a comparable cross-validation performance of all models we would deploy rPCA that gives the minimum d̶.

The results support our recommendation to use PCA and rPCA when temporal sensor failures are expected. The case study also confirms the effectiveness of the *deterioration index* in diagnosing robustness of predictive models.

7   Related Work

Our study is closely connected with handling missing data research, see e.g. [1,2, 4,14]. The main techniques are: imputation procedures where missing values are filled in and the resulting compete data is analyzed, reweighing procedures where instances with missing data are discarded or assigned low weights, and model-based procedures that define models for partially missing data. We investigate what happens *after* missing values are imputed during real-time operation using a very popular and practical mean value imputation. In our setting discarding streaming data is not suitable, since there would be continuous periods when we have no input data and thus no predictions. Model-based procedures could handle one-two missing sensors; however, when many sensors may fail, such a procedure is computationally impractical and likely infeasible, as we would need to keep an exponential number of models to account for all possible situations.

Handling missing values in regression is reviewed in [14]. The majority of research focuses on training regression models from data with partially missing values. In our setting discarding some *training* data with missing values is not a problem, since the volumes of data are typically large. The problem arises during real-time operation. We not only need to input missing values, but also make the regression models fault tolerant. Hence, our work solves a different problem and is not directly comparable with missing value imputation techniques.

Topic-wise our work relates to fault tolerant control that is widely researched and applied in industrial and aerospace systems [16, 19]. The main focus is on detecting the actual fault, not operating with a fault present. In our setting there is no fault in the system, just sensors fails, our model needs to remain accurate.

Redundancy in engineering duplicates critical components of a system to increase reliability, see e.g. [5]. A common computational approach is to to use an average the redundant sensors to reduce the impact of possible sensor failure. In fact, this is the effect we are aiming to achieve by minimizing the *deterioration index*. The main difference from our setting is in availability of backup sensors, it is even possible to install duplicate sensors on demand. In our setting; however, the data is given as is and we aim at exploiting it in the best way.

Robust statistics aims at producing models that are robust to *outliers* or other small departures from model assumptions, see e.g. [11]. The main idea is to modify loss functions so that they do not increase so rapidly, to reduce the impact of outliers. In our setting there are no large deviations in the input data due to sensor failure, in fact the opposite, the variance of a failed sensor goes to zero. Hence, robust statistics approaches target a different problem.

Our theoretical analysis of the mean squared error resembles bias variance analysis (see e.g. [7]) in the way we decompose *MSE* into components. Regarding the connection of the bias-variance decomposition to the Ridge regression solution, we well know that enforcing strong regularisation is likely to decrease variance and to increase bias. Further investigation is left for future work.

Finally, the setting relates to concept drift [20] and transfer learning [15] settings in a sense that the training and the testing data distributions are different. However, in our setting there is no model adaptation during real-time operation.

8   Conclusion

Systems relying on predictive models should be robust with regard to missing input values, due to transient failures in the sensors, for instance. We focused on linear models for predictions, and theoretically analyzed the criteria for linear regression to be robust to sensor failures. Based on this analysis we introduced the *deterioration index* measure that allows to quantify how robust is a given linear regression model to sensor failure. We also proposed a practical strategy for building robust linear models. Our experiments with real data confirmed the theoretical results and demonstrated the effectiveness of the proposed strategy.

The current work assumes that input sensors fail with the uniform prior probability, but does not quantify any distribution on how many are likely to fail, or how the failures form correlated patterns among the sensors. These questions would make an interesting follow up investigation. Mapping the findings of the current study to predictive models in the evolving data stream setting offers another important avenue for future research.

Acknowledgments. We thank the INFER project for Chemi dataset. This work has been supported by the Academy of Finland grant 118653 (ALGODAN).

References

1. Alippi, C., Boracchi, G., Roveri, M.: On-line reconstruction of missing data in sensor/actuator networks by exploiting temporal and spatial redundancy. In: Proc. of the 2012 Int. Joint Conf. on Neural Networks, IJCNN, pp. 1–8 (2012)
2. Allison, P.: Missing data. Sage, Thousand Oaks (2001)
3. Brobst, S.: Sensor data is data analytics' future goldmine (2010), http://www.zdnet.com
4. Ciampi, A., Appice, A., Guccione, P., Malerba, D.: Integrating trend clusters for spatio-temporal interpolation of missing sensor data. In: Di Martino, S., Peron, A., Tezuka, T. (eds.) W2GIS 2012. LNCS, vol. 7236, pp. 203–220. Springer, Heidelberg (2012)
5. Frank, P.: Fault diagnosis in dynamic systems using analytical and knowledge-based redundancy: a survey and some new results. Automatica 26(3), 459–474 (1990)
6. Gama, J., Gaber, M. (eds.): Learning from Data Streams: Processing Techniques in Sensor Networks. Springer (2007)
7. Geman, S., Bienenstock, E., Doursat, R.: Neural networks and the bias/variance dilemma. Neural Comput. 4(1), 1–58 (1992)
8. Golub, G., Van Loan, C.: Matrix Computations. Johns Hopkins Un. Press (1996)
9. Hastie, T., Tibshirani, R., Friedman, J.: The elements of statistical learning: data mining, inference, and prediction. Springer (2001)
10. Hoerl, A., Kennard, R.: Ridge regression: Biased estimation for nonorthogonal problems. Technometrics 42(1), 55–67 (1970)
11. Huber, P.: Robust statistics. Wiley (1981)
12. Junninen, H., Lauri, A., Keronen, P., Aalto, P., Hiltunen, V., Hari, P., Kulmala, M.: Smart-SMEAR: on-line data exploration and visualization tool for smear stations. Boreal Env. Res., 447–457 (2009)
13. Kadlec, P., Gabrys, B., Strandt, S.: Data-driven soft sensors in the process industry. Computers & Chemical Engineering 33(4), 795–814 (2009)
14. Little, R.: Regression with missing X's: A review. Journal of the American Statistical Association 87(420), 1227–1237 (1992)
15. Pan, S., Yang, Q.: A survey on transfer learning. IEEE Trans. on Knowl. and Data Eng. 22(10), 1345–1359 (2010)
16. Patton, R.: Fault-tolerant control: the 1997 situation. In: Proc. of the 3rd IFAC Symp. on Fault Detection, Superv. and Safety for Tech. Proc., pp. 1033–1055 (1997)
17. Qin, J.: Recursive PLS algorithms for adaptive data modeling. Computers & Chemical Engineering 22(4-5), 503–514 (1998)
18. Wold, S., Sjostroma, M., Eriksson, L.: PLS-regression: a basic tool of chemometrics. Chemometrics and Intelligent Laboratory Systems 58(2), 109–130 (2001)
19. Zhang, Y., Jiang, J.: Bibliographical review on reconfigurable fault-tolerant control systems. Annual Reviews in Control 32(2), 229–252 (2008)
20. Zliobaite, I.: Learning under concept drift: an overview. CoRR, 1010.4784 (2010)

Pitfalls in Benchmarking Data Stream Classification and How to Avoid Them

Albert Bifet[1], Jesse Read[2], Indrė Žliobaitė[3], Bernhard Pfahringer[4], and Geoff Holmes[4]

[1] Yahoo! Research, Spain
abifet@yahoo-inc.com
[2] Universidad Carlos III, Spain
jesse@tsc.uc3m.es
[3] Dept. of Information and Computer Science, Aalto University and Helsinki Institute
for Information Technology (HIIT), Finland
indre.zliobaite@aalto.fi
[4] University of Waikato, New Zealand
{bernhard,geoff}@waikato.ac.nz

Abstract. Data stream classification plays an important role in modern data analysis, where data arrives in a stream and needs to be mined in real time. In the data stream setting the underlying distribution from which this data comes may be changing and evolving, and so classifiers that can update themselves during operation are becoming the state-of-the-art. In this paper we show that data streams may have an important temporal component, which currently is not considered in the evaluation and benchmarking of data stream classifiers. We demonstrate how a naive classifier considering the temporal component only outperforms a lot of current state-of-the-art classifiers on real data streams that have temporal dependence, i.e. data is autocorrelated. We propose to evaluate data stream classifiers taking into account temporal dependence, and introduce a new evaluation measure, which provides a more accurate gauge of data stream classifier performance. In response to the temporal dependence issue we propose a generic wrapper for data stream classifiers, which incorporates the temporal component into the attribute space.

Keywords: data streams, evaluation, temporal dependence.

1   Introduction

Data streams refer to a type of data, that is generated in real-time, arrives continuously as a stream and may be evolving over time. This temporal property of data stream mining is important, as it distinguishes it from non-streaming data mining, thus it requires different classification techniques and a different evaluation methodology. The standard assumptions in classification (such as IID) have been challenged during the last decade [14]. It has been observed, for instance, that frequently data is not distributed identically over time, the distributions may evolve (concept drift), thus classifiers need to adapt.

Although there is much research in the data stream literature on detecting concept drift and adapting to it over time [10,17,21], most work on stream classification assumes

H. Blockeel et al. (Eds.): ECML PKDD 2013, Part I, LNAI 8188, pp. 465–479, 2013.

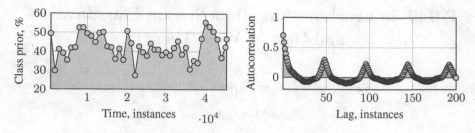

Fig. 1. Characteristics of the Electricity Dataset

that data is distributed not identically, but still *independently*. Except for our brief technical report [24], we are not aware of any work in data stream classification discussing what effects a *temporal dependence* can have on evaluation. In this paper we argue that the current evaluation practice of data stream classifiers may mislead us to draw wrong conclusions about the performance of classifiers.

We start by discussing an example of how researchers evaluate a data stream classifier using a real dataset representing a data stream. The Electricity dataset due to [15] is a popular benchmark for testing adaptive classifiers. It has been used in over 40 concept drift experiments[1], for instance, [10,17,6,21]. The Electricity Dataset was collected from the Australian New South Wales Electricity Market. The dataset contains 45,312 instances which record electricity prices at 30 minute intervals. The class label identifies the change of the price (UP or DOWN) related to a moving average of the last 24 hours. The data is subject to concept drift due to changing consumption habits, unexpected events and seasonality.

Two observations can be made about this dataset. Firstly, the data is not independently distributed over time, it has a temporal dependence. If the price goes UP now, it is more likely than by chance to go UP again, and vice versa. Secondly, the prior distribution of classes in this data stream is evolving. Figure 1 plots the class distribution of this dataset over a sliding window of 1000 instances and the autocorrelation function of the target label. We can see that data is heavily autocorrelated with very clear cyclical peaks at every 48 instances (24 hours), due to electricity consumption habits.

Let us test two state-of-the-art data stream classifiers on this dataset. We test an incremental Naive Bayes classifier, and an incremental (streaming) decision tree learner. As a streaming decision tree, we use VFDT [16] with functional leaves, using Naive Bayes classifiers at the leaves.

In addition, let us consider two naive baseline classifiers that do not use any input attributes and classify only using past label information: a moving majority class classifier (over a window of 1000) and a No-Change classifier that uses temporal dependence information by predicting that the next class label will be the same as the last seen class label. It can be compared to a naive weather forecasting rule: the weather tomorrow will be the same as today.

We use prequential evaluation [11] over a sliding window of 1000 instances. The prequential error is computed over a stream of n instances as an accumulated loss L between the predictions $\hat{y}_t$ and the true values y_t:

---

[1] Google scholar, 2013 March.

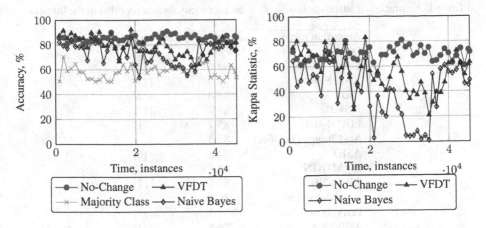

Fig. 2. Accuracy and Kappa Statistic on the Electricity Market Dataset

$$p_0 = \sum_{t=1}^{n} L(\hat{y}_t, y_t).$$

Since the class distribution is unbalanced, it is important to use a performance measure that takes class imbalance into account. We use the Kappa Statistic due to Cohen [7]. Other measures, such as, for instance, the *Matthews correlation coefficient* [19], could be used as well. The Kappa Statistic κ is defined as

$$\kappa = \frac{p_0 - p_c}{1 - p_c},$$

where p_0 is the classifier's prequential accuracy, and p_c is the probability that a chance classifier - one that assigns the same number of examples to each class as the classifier under consideration—makes a correct prediction. If the tested classifier is always correct then $\kappa = 1$. If its predictions coincide with the correct ones as often as those of a chance classifier, then $\kappa = 0$.

Figure 2 shows the evolving accuracy (left) of the two state-of-the-art stream classifiers and the two naive baselines, and the evolution of the Kappa Statistic (right). We see that the state-of-the-art classifiers seem to be performing very well if compared to the majority class baseline. Kappa Statistic results are good enough at least for the decision tree. Following the current evaluation practice for data stream classifiers we would recommend this classifier for this type of data. However, the No-Change classifier performs much better. Note that the No-Change classifier completely ignores the input attribute space, and uses nothing but the value of the previous class label.

We retrospectively surveyed accuracies of 16 new stream classifiers reported in the literature that were tested on the Electricity dataset. Table 1 shows a list of the results reported using this dataset, sorted according to the reported accuracy. Only 6 out of 16 reported accuracies outperformed the No-Change classifier. This suggests that current evaluation practice needs to be revised.

This paper makes a threefold contribution. First, in Section 2, we explain what is happening when data contains temporal dependence and why it is important to take

Table 1. Accuracies of adaptive classifiers on the Electricity dataset reported in the literature

Algorithm name	Accuracy (%)	Reference
DDM	89.6*	[10]
Learn++.CDS	88.5	[8]
KNN-SPRT	88.0	[21]
GRI	88.0	[22]
FISH3	86.2	[23]
EDDM-IB1	85.7	[1]
No-Change classifier	**85.3**	
ASHT	84.8	[6]
bagADWIN	82.8	[6]
DWM-NB	80.8	[17]
Local detection	80.4	[9]
Perceptron	79.1	[5]
ADWIN	76.6	[2]
Prop. method	76.1	[18]
Cont. λ-perc.	74.1	[20]
CALDS	72.5	[12]
TA-SVM	68.9	[13]

* tested on a subset

into account when evaluating stream classifiers. Second, in Section 3, we propose a new measure to evaluate data stream classifiers taking into account possible temporal dependence. Third, in Section 5, we propose a generic wrapper classifier that enables conventional stream classifiers to take into account temporal dependence. In Section 4 we perform experimental analysis of the new measure. Section 6 concludes the study.

2   Why the Current Evaluation Procedures May Be Misleading

We have seen that a naive No-Change classifier can obtain very good results on the Kappa Statistic measure by using temporal information from the data. This is a surprising result since we would expect that a trivial classifier ignoring the input space entirely should perform worse than a well-trained intelligent classifier. Thus, we start by theoretically analyzing the conditions under which the No-Change classifier outperforms the majority class classifier. Next we discuss the limitations of the Kappa Statistic for measuring classification performance on data streams.

Consider a binary classification problem with fixed prior probabilities of the classes $P(c_1)$ and $P(c_2)$. Without loss of generality assume $P(c_1) \geq P(c_2)$. The expected accuracy of the majority class classifier would be $p_{maj} = P(c_1)$. The expected accuracy of the No-Change classifier would be the probability that two labels in a row are the same $p_e = P(c_1)P(c_1|c_1) + P(c_2)P(c_2|c_2)$, where $P(c_1|c_1)$ is the probability of observing class c_1 immediately after observing class c_1.

Note that if data is distributed independently, then $P(c_1|c_1) = P(c_1)$ and $P(c_2|c_2) = P(c_2)$. Then the accuracy of the No-Change classifier is $P(c_1)^2 + P(c_2)^2$. Using the fact that $P(c_1) + P(c_2) = 1$ it is easy to show that

$$P(c_1) \geq P(c_1)^2 + P(c_2)^2,$$

that is $p_{maj} \geq p_{nc}$. The accuracies are equal only if $P(c_1) = P(c_2)$, otherwise the majority classifier is more accurate. Thus, if data is distributed independently, then we can safely use the majority class classifier as a baseline.

However, if data is *not independently distributed*, then, following similar arguments it can be shown that if $P(c_2|c_2) > 0.5$ then

$$P(c_1) < P(c_1)P(c_1|c_1) + P(c_2)P(c_2|c_2).$$

That is $p_{maj} < p_e$, hence, the No-Change classifier will outperform the majority class classifier if the probability of seeing consecutive minority classes is larger than 0.5. This happens even in cases of equal prior probabilities of the classes.

Similar arguments are valid in multi-class classification cases as well. If we observe the majority class, then the No-Change classifier predicts the majority class, the majority classifier predicts the same. They will have the same accuracy on the next data instance. If, however, we observe a minority class, then the majority classifier still predicts the majority class, but the No-Change classifier predicts a minority class. The No-Change strategy would be more accurate if the probability of observing two instances of that minority class in a row is larger than $1/k$, where k is the number of classes.

Table 2 presents characteristics of four popular stream classification datasets. Electricity and Airlines are available from the MOA[2] repository, and KDD99 and Ozone are available from the UCI[3] repository. Electricity and Airlines represent slightly imbalanced binary classification tasks, we see by comparing the prior and conditional probabilities that data is not distributed independently. Electricity consumption is expected to have temporal dependence. The Airlines dataset records delays of flights, it is likely that e.g. during a storm period many delays would happen in a row. We see that as expected, the No-Change classifier achieves higher accuracy than the majority classifier. The KDD99 cup intrusion detection dataset contains more than 20 classes, we report on only the three largest classes. The problem of temporal dependence is particularly evident here. Inspecting the raw dataset confirms that there are time periods of intrusions rather than single instances of intrusions, thus the data is not distributed independently over time. We observe that the No-Change classifier achieves nearly perfect accuracy. Finally, the Ozone dataset is also not independently distributed. If ozone levels rise, they do not diminish immediately, thus we have several ozone instances in a row. However, the dataset is also very highly imbalanced. We see that the conditional probability of the minority class (ozone) is higher than the prior, but not high enough to give advantage to the No-Change classifier over the majority classifier. This confirms our theoretical results.

Thus, if we expect a data stream to contain temporal dependence, we need to make sure that any intelligent classifier is compared to the No-Change baseline in order to make meaningful conclusions about performance.

Next we highlight issues with the prequential accuracy in such situations, and then move on to the Kappa Statistic. The main reason why the prequential accuracy may

---

[2] http://moa.cms.waikato.ac.nz/datasets/
[3] http://archive.ics.uci.edu/ml/

Table 2. Characteristics of stream classification datasets

| Dataset | $P(c_1)$ | $P(c_2)$ | $P(c_3)$ | Majority acc. |
	$P(c_1\|c_1)$	$P(c_2\|c_2)$	$P(c_3\|c_3)$	No-Change acc.
Electricity	0.58	0.42	-	0.58
	0.87	0.83	-	**0.85**
Airlines	0.55	0.45	-	0.55
	0.62	0.53	-	**0.58**
KDD99	0.60	0.18	0.17	0.60
	0.99	0.99	0.99	**0.99**
Ozone	0.97	0.03	-	**0.97**
	0.97	0.11	-	0.94

mislead is because it assumes that the data is distributed *independently*. If a data stream contains the same number of instances for each class, accuracy is the right measure to use, and will be sufficient to detect if a method is performing well or not. Here, a random classifier will have a $1/k$ accuracy for a k class problem. Assuming that the accuracy of our classifier is doing better than $1/k$, we know that we are doing better than guessing the classes of the incoming instances at random.

We see that when a data stream has temporal dependence, using only the Kappa Statistic for evaluating stream classifiers may be misleading. The reason is that when the stream has a temporal dependence, by using the Kappa Statistic we are comparing the performance of our classifier with a random classifier. Thus, we can view the Kappa Statistic as a normalized measure of the prequential accuracy p_0:

$$p_0' = \frac{p_0 - \min p}{\max p - \min p}$$

In the Kappa Statistic, we consider that $\max p = 1$ and that $\min p = p_c$. This measure may be misleading because we assume that p_c is giving us the accuracy of the baseline naive classifier. Recall that p_c is the probability that a classifier that assigns the same number of examples to each class as the classifier under consideration, makes a correct prediction. However, we saw that the majority class classifier may not be the most accurate naive classifier when temporal dependence exists in the stream. No-Change may be a more accurate naive baseline, thus we need to take it into account within the evaluation measure.

3 New Evaluation for Stream Classifiers

In this section we present a new measure for evaluating classifiers. We start by more formally defining our problem. Consider a classifier h, a data set containing n examples and k classes, and a contingency table where cell C_{ij} contains the number of examples for which $h(x) = i$ and the class is j. If $h(x)$ correctly predicts all the data, then all non-zero counts will appear along the diagonal. If h misclassifies some examples, then some off-diagonal elements will be non-zero.

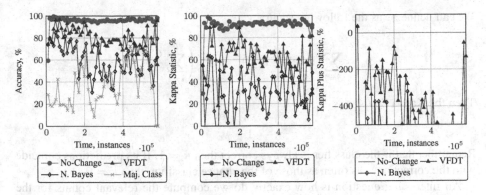

Fig. 3. Accuracy, κ and κ^+ on the Forest Covertype dataset

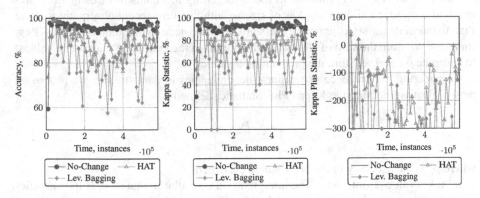

Fig. 4. Accuracy, κ and κ^+ on the Forest Covertype dataset

The classification accuracy is defined as

$$p_0 = \frac{\sum_{i=1}^{k} C_{ii}}{n}.$$

Let us define

$$\Pr[\text{class is } j] = \sum_{i=1}^{k} \frac{C_{ij}}{n}, \Pr[h(x) = i] = \sum_{j=1}^{k} \frac{C_{ij}}{n}.$$

Then the accuracy of a random classifier is

$$p_c = \sum_{j=1}^{k} \left(\Pr[\text{class is } j] \cdot \Pr[h(x) = j] \right)$$

$$= \sum_{j=1}^{k} \left(\sum_{i=1}^{k} \frac{C_{ij}}{n} \cdot \sum_{i=1}^{k} \frac{C_{ji}}{n} \right).$$

We can define p_e as the following accuracy:

$$p_e = \sum_{j=1}^{k} \left(\Pr[\text{class is } j] \right)^2 = \sum_{j=1}^{k} \left(\sum_{i=1}^{k} \frac{C_{ij}}{n} \right)^2 .$$

Then the Kappa Statistic is

$$\kappa = \frac{p_0 - p_c}{1 - p_c}.$$

Remember that if the classifier is always correct then $\kappa = 1$. If its predictions coincide with the correct ones as often as those of the chance classifier, then $\kappa = 0$.

An interesting question is how exactly do we compute the relevant counts for the contingency table: using all examples seen so far is not useful in time-changing data streams. Gama et al. [11] propose to use a forgetting mechanism for estimating prequential accuracy: a sliding window of size w with the most recent observations. Note that, to calculate the statistic for a k class problem, we need to maintain only $2k + 1$ estimators. We store the sum of all rows and columns in the confusion matrix ($2k$ values) to compute p_c, and we store the prequential accuracy p_0.

Considering the presence of temporal dependencies in data streams we propose a new evaluation measure the Kappa Plus Statistic, defined as

$$\kappa^+ = \frac{p_0 - p'_e}{1 - p'_e}$$

where p'_e is the accuracy of the No-Change classifier.

κ^+ takes values from 0 to 1. The interpretation is similar to that of κ. If the classifier is perfectly correct then $\kappa^+ = 1$. If the classifier is achieving the same accuracy as the No-Change classifier, then $\kappa^+ = 0$. Classifiers that outperform the No-Change classifier fall between 0 and 1. Sometimes it can happen that $\kappa^+ < 0$, which means that the classifier is performing worse than the No-Change baseline.

In fact, we can compute p'_e as the probability that for all classes, the class of the new instance i_{t+1} is equal to the last class seen in instance i_t. It is the sum for each class of the probability that the two instances in a row have the same class:

$$p'_e = \sum_{j=1}^{k} \left(\Pr[i_{t+1} \text{ class is } j \text{ and } i_t \text{ class is } j] \right) .$$

Two observations can be made about κ^+. First, when there is no temporal dependence, κ^+ is closely related to κ since

$$\Pr[i_{t+1} \text{ class is } j \text{ and } i_t \text{ class is } j] = Pr[i_t \text{ class is } j]^2$$

holds, and $p'_e = p_e$. It means that if there is no temporal dependence, then the probabilities of selecting a class will depend on the distributions of the classes, so does κ.

Second, if classes are balanced and there is no temporal dependence, then κ^+ is equal to κ and both are linearly related to the accuracy p_0:

$$\kappa^+ = \frac{n}{n-1} \cdot p_0 - \frac{1}{n-1}.$$

Therefore, using κ^+ instead of κ, we will be able to detect misleading classifier performance for data that is dependently distributed. For highly imbalanced, but independently distributed data, the majority class classifier may beat the No-Change classifier. κ^+ and κ measures can be seen as orthogonal, since they measure different aspects of the performance. Hence, for a thorough evaluation we recommend measuring both.

An interested practitioner can take a snapshot of a data stream and measure if there is a temporal dependency, e.g. by comparing the probabilities of observing the same labels in a row with the prior probabilities of the labels as reported in Table 2. However, even without checking whether there is a temporal dependency in the data a user can safely check both κ^+ and κ. If there is no temporal dependency, both measures will give the same result. In case there is a temporal dependency a good classifier should score high in both measures.

4 Experimental Analysis of the New Measure

The goal of this experimental analysis is to compare the informativeness of κ and κ^+ in evaluating stream classifiers. These experiments are meant to be merely a proof of concept, therefore we restrict the analysis to two data stream benchmark datasets. The first, the Electricity dataset was discussed in the introduction. The second, *Forest Covertype*, contains the forest cover type for 30×30 meter cells obtained from US Forest Service (USFS) Region 2 Resource Information System (RIS) data. It contains $581,012$ instances and 54 attributes, and has been used in several papers on data stream classification.

We run all experiments using the MOA software framework [3] that contains implementations of several state-of-the-art classifiers and evaluation methods and allows for easy reproducibility. The proposed κ^+ is not base classifier specific, hence we do not aim at exploring a wide range of classifiers. We select several representative data stream classifiers for experimental illustration.

Figure 3 shows accuracy of the three classifiers Naive Bayes, VFDT and No-Change using the prequential evaluation of a sliding window of 1000 instances, κ results and results for the new κ^+. We observe similar results to the Electricity Market dataset, and that for the No-Change classifier κ^+ is zero, and for Naive Bayes and VFDT, κ^+ is negative.

We also test two more powerful data stream classifiers:

- Hoeffding Adaptive Tree (HAT): which extends VFDT to cope with concept drift. [3].
- Leveraging Bagging: an adaptive ensemble that uses 10 VFDT decision trees [4].

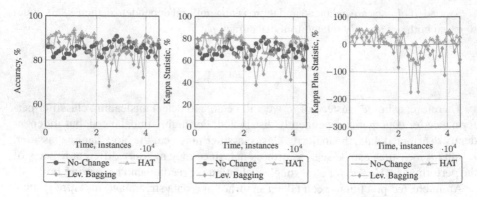

Fig. 5. Accuracy, κ and κ^+ on the Electricity Market dataset

For the Forest CoverType dataset, Figure 4 shows accuracy of the three classifiers HAT, Leveraging Bagging and No-Change using a prequential evaluation of a sliding window of 1000 instances. It also shows κ results and the new κ^+ results. We see how these classifiers improve the results over the previous classifiers, but still have negative κ^+ results, meaning that the No-Change classifier is still providing better results.

Finally, we test the two more powerful stream classifiers on the Electricity Market dataset. Figure 5 shows accuracy, κ and κ^+ for the three classifiers HAT, Leveraging Bagging and No-Change . κ^+ is positive for a long period of time, but still contains some negative results.

Our experimental analysis indicates that using the new κ^+ measure, we can easily detect when a classifier is doing worse than the simple No-Change strategy, by simply observing if negative values of this measure exist.

5   SWT: Temporally Augmented Classifier

Having identified the importance of temporal dependence in data stream classification we now propose a generic wrapper that can be used to wrap state-of-the-art classifiers so that temporal dependence is taken into account when training an intelligent model. We propose SWT, a simple meta strategy that builds meta instances by augmenting the original input attributes with the values of recent class labels from the past (in a sliding window). Any existing incremental data-stream classifier can be used as a base classifier with this strategy. The prediction becomes a function of the original input attributes and the recent class labels

$$Pr[\text{class is } c] \equiv h(x^t, c^{t-\ell}, \ldots, c^{t-1})$$

for the t-th test instance, where ℓ is the size of the sliding window over the most recent true labels. The larger ℓ, the longer temporal dependence is considered. h can be any of the classifiers we mentioned (e.g., HAT or Leveraging Bagging).

It is important to note that such a classifier relies on immediate arrival of the previous label after the prediction is casted. This assumption may be violated in real-world applications, i.e. true labels may arrive with a delay. In such a case it is still possible to use the proposed classifier with the true labels from more distant past. The utility of this approach will depend on the strength of the temporal correlation in the data.

We test this wrapper classifier experimentally using HAT and VFDT as the internal stream classifiers. In this proof of concept study we report experimental results using $\ell = 1$. Our goal is to compare the performance of an intelligent SWT, with that of the baseline No-Change classifier. Both strategies take into account temporal dependence. However, SWT, does so in an intelligent way considering it alongside a set of input attributes.

Figure 6 shows the SWT strategy applied to VFDT, Naive Bayes, Hoeffding Adaptive Tree, and Leveraging Bagging for the Electricity dataset. The results for the Forest Cover dataset are displayed in Figure 7. As a summary, Figure 8 (left and center) shows κ^+ on the Electricity and Forest Cover datasets. We see a positive κ^+ which means that the prediction is meaningful taking into account the temporal dependency in the data. Additional experiments reported in Figures 9, 10, 11 confirm that the results are stable under varying size of the sliding window (to $\ell > 1$) and varying feature space (i.e., $x^{t-\ell}, \ldots, x^{t-1}$). More importantly, we see a substantial improvement as compared to the state-of-the-art stream classifiers (Figures 3, 4, 5) that do not use the temporal dependency information.

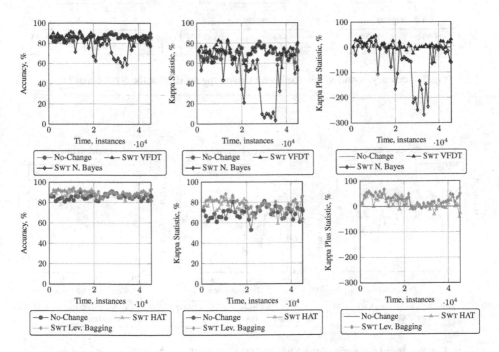

Fig. 6. Accuracy, κ and κ^+ on the Electricity Market dataset for the SWT classifiers

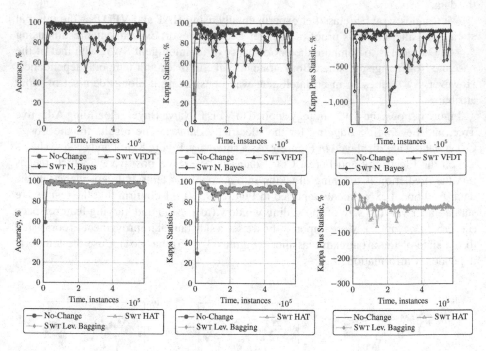

Fig. 7. Accuracy, κ and κ^+ on the Forest Covertype dataset for the SWT classifiers

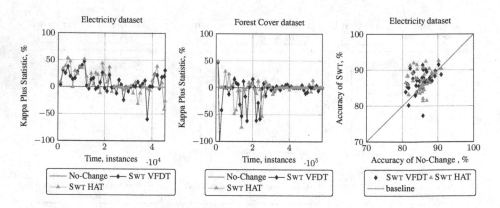

Fig. 8. κ^+ and accuracy of the SWT VFDT, SWT HAT, and No-Change classifiers

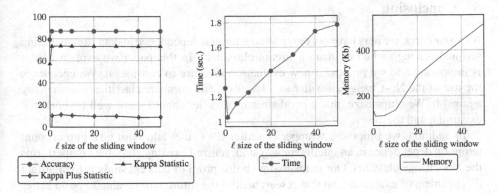

Fig. 9. Accuracy, κ, κ^+, time and memory of a VFDT on the Electricity Market dataset for the SWT classifiers varying the size of the sliding window parameter ℓ

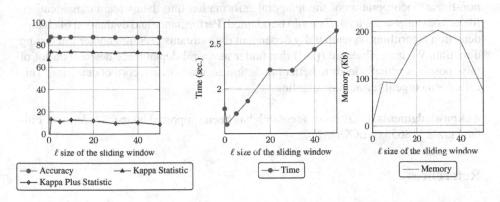

Fig. 10. Accuracy, κ, κ^+, time and memory of a Hoeffding Adaptive Tree on the Electricity Market dataset for the SWT classifiers varying the ℓ size of the sliding window parameter

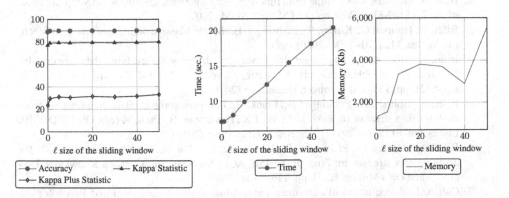

Fig. 11. Accuracy, κ, κ^+, time and memory of a Leveraging Bagging on the Electricity Market dataset for the SWT classifiers varying the ℓ size of the sliding window parameter

6   Conclusion

As researchers, we may have not considered temporal dependence in data stream mining seriously enough when evaluating stream classifiers. In this paper we explain why it is important, and we propose a new evaluation measure to consider it. We encourage the use of the No-Change classifier as a baseline, and compare classification accuracy against it. We emphasize, that a good stream classifier should score well on both: the existing κ and the new κ^+.

In addition, we propose a wrapper classifier SWT, that allows to take into account temporal dependence in an intelligent way and, reusing existing classifiers outperforms the No-Change classifier. Our main goal with this proof of concept study is to highlight this problem of evaluation, so that researchers in the future will be able to build better new classifiers taking into account temporal dependencies of streams.

This study opens several directions for future research. The wrapper classifier SWTis very basic and intended as a proof of concept. One can consider more advanced (e.g. non-linear) incorporation of the temporal information into data stream classification. Ideas from time series analysis could be adapted. Performance and evaluation of change detection algorithms on temporally dependent data streams present another interesting direction. We have observed ([24]) that under temporal dependence detecting a lot of false positives actually leads to better prediction accuracy than a correct detection. This calls for an urgent further investigation.

Acknowledgments. I. Žliobaitė's research has been supported by the Academy of Finland grant 118653 (ALGODAN).

References

1. Baena-Garcia, M., del Campo-Avila, J., Fidalgo, R., Bifet, A., Gavalda, R., Morales-Bueno, R.: Early drift detection method. In: Proc. of the 4th ECMLPKDD Int. Workshop on Knowledge Discovery from Data Streams, pp. 77–86 (2006)
2. Bifet, A., Gavalda, R.: Learning from time-changing data with adaptive windowing. In: Proc. of the 7th SIAM Int. Conf. on Data Mining, SDM (2007)
3. Bifet, A., Holmes, G., Kirkby, R., Pfahringer, B.: MOA: Massive online analysis. J. of Mach. Learn. Res. 11, 1601–1604 (2010)
4. Bifet, A., Holmes, G., Pfahringer, B.: Leveraging bagging for evolving data streams. In: Balcázar, J.L., Bonchi, F., Gionis, A., Sebag, M. (eds.) ECML PKDD 2010, Part I. LNCS, vol. 6321, pp. 135–150. Springer, Heidelberg (2010)
5. Bifet, A., Holmes, G., Pfahringer, B., Frank, E.: Fast perceptron decision tree learning from evolving data streams. In: Zaki, M.J., Yu, J.X., Ravindran, B., Pudi, V. (eds.) PAKDD 2010. LNCS, vol. 6119, pp. 299–310. Springer, Heidelberg (2010)
6. Bifet, A., Holmes, G., Pfahringer, B., Kirkby, R., Gavaldà, R.: New ensemble methods for evolving data streams. In: Proc. of the 15th ACM SIGKDD Int. Conf. on Knowledge Discovery and Data Mining, KDD, pp. 139–148 (2009)
7. Cohen, J.: A coefficient of agreement for nominal scales. Educational and Psychological Measurement 20(1), 37–46 (1960)
8. Ditzler, G., Polikar, R.: Incremental learning of concept drift from streaming imbalanced data. IEEE Transactions on Knowledge and Data Engineering (2013)

9. Gama, J., Castillo, G.: Learning with local drift detection. In: Li, X., Zaïane, O.R., Li, Z.-H. (eds.) ADMA 2006. LNCS (LNAI), vol. 4093, pp. 42–55. Springer, Heidelberg (2006)

10. Gama, J., Medas, P., Castillo, G., Rodrigues, P.: Learning with drift detection. In: Bazzan, A.L.C., Labidi, S. (eds.) SBIA 2004. LNCS (LNAI), vol. 3171, pp. 286–295. Springer, Heidelberg (2004)

11. Gama, J., Sebastião, R., Rodrigues, P.: On evaluating stream learning algorithms. Machine Learning 90(3), 317–346 (2013)

12. Gomes, J., Menasalvas, E., Sousa, P.: CALDS: context-aware learning from data streams. In: Proc. of the 1st Int. Workshop on Novel Data Stream Pattern Mining Techniques, StreamKDD, pp. 16–24 (2010)

13. Grinblat, G., Uzal, L., Ceccatto, H., Granitto, P.: Solving nonstationary classification problems with coupled support vector machines. IEEE Transactions on Neural Networks 22(1), 37–51 (2011)

14. Hand, D.: Classifier technology and the illusion of progress. Statist. Sc. 21(1), 1–14 (2006)

15. Harries, M.: SPLICE-2 comparative evaluation: Electricity pricing. Tech. report, University of New South Wales (1999)

16. Hulten, G., Spencer, L., Domingos, P.: Mining time-changing data streams. In: Proc. of the 7th ACM SIGKDD Int. Conf. on Knowl. Disc. and Data Mining, KDD, pp. 97–106 (2001)

17. Kolter, J., Maloof, M.: Dynamic weighted majority: An ensemble method for drifting concepts. J. of Mach. Learn. Res. 8, 2755–2790 (2007)

18. Martinez-Rego, D., Perez-Sanchez, B., Fontenla-Romero, O., Alonso-Betanzos, A.: A robust incremental learning method for non-stationary environments. Neurocomput. 74(11), 1800–1808 (2011)

19. Matthews, B.: Comparison of the predicted and observed secondary structure of T4 phage lysozyme. Biochimica et Biophysica Acta 405(2), 442–451 (1975)

20. Pavlidis, N., Tasoulis, D., Adams, N., Hand, D.: Lambda-perceptron: An adaptive classifier for data streams. Pattern Recogn. 44(1), 78–96 (2011)

21. Ross, G., Adams, N., Tasoulis, D., Hand, D.: Exponentially weighted moving average charts for detecting concept drift. Pattern Recogn. Lett. 33, 191–198 (2012)

22. Tomczak, J., Gonczarek, A.: Decision rules extraction from data stream in the presence of changing context for diabetes treatment. Knowl. Inf. Syst. 34(3), 521–546 (2013)

23. Zliobaite, I.: Combining similarity in time and space for training set formation under concept drift. Intell. Data Anal. 15(4), 589–611 (2011)

24. Zliobaite, I.: How good is the electricity benchmark for evaluating concept drift adaptation. CoRR, abs/1301.3524 (2013)

Adaptive Model Rules from Data Streams

Ezilda Almeida[1], Carlos Ferreira[1], and João Gama[1,2]

[1] LIAAD-INESC TEC, University of Porto
ezildacv@gmail.com, cgf@isep.ipp.pt
[2] Faculty of Economics, University Porto
jgama@fep.up.pt

Abstract. Decision rules are one of the most expressive languages for machine learning. In this paper we present Adaptive Model Rules (AMRules), the first streaming rule learning algorithm for regression problems. In AMRules the antecedent of a rule is a conjunction of conditions on the attribute values, and the consequent is a linear combination of attribute values. Each rule uses a Page-Hinkley test to detect changes in the process generating data and react to changes by pruning the rule set. In the experimental section we report the results of AM-Rules on benchmark regression problems, and compare the performance of our system with other streaming regression algorithms.

Keywords: Data Streams, Regression, Rule Learning.

1 Introduction

Regression analysis is a technique for estimating a functional relationship between a dependent variable and a set of independent variables. It has been widely studied in statistics, pattern recognition, machine learning and data mining. The most expressive data mining models for regression are model trees [18] and regression rules [19]. In [15], the authors present a large comparative study between several regression algorithms. Model trees and model rules are among the most performant ones. Trees and rules do automatic feature selection, being robust to outliers and irrelevant features; exhibit high degree of interpretability; and structural invariance to monotonic transformation of the independent variables. One important aspect of rules is modularity: each rule can be interpreted per si [6].

In the data stream computational model [7] examples are generated sequentially from time evolving distributions. Learning from data streams require incremental learning, using limited computational resources, and the ability to adapt to changes in the process generating data. In this paper we present the AMRules algorithm, the first one-pass algorithm for learning regression rule sets from time-evolving streams. It near follows FIMT [11], an algorithm to learn regression trees from data streams, and AVFDR [13], a one-pass algorithm for learning classification rules. AMRules can learn ordered or unordered rules. The antecedent of a rule is a set of literals (conditions based on the attribute values), and the consequent is a function that minimizes the mean square error of the target attribute computed from the set of examples covered by rule. This function might be either a constant, the mean of the target attribute, or a linear combination of

H. Blockeel et al. (Eds.): ECML PKDD 2013, Part I, LNAI 8188, pp. 480–492, 2013.

the attributes. Each rule is equipped with an online change detector. The change detector monitors the mean square error using the Page-Hinkley test, providing information about the dynamics of the process generating data.

The paper is organized as follows. The next Section presents the related work in learning regression trees and rules from data focusing on streaming algorithms. Section 3 describe in detail the AMRules algorithm. Section 4 presents the experimental evaluation using stationary and time-evolving streams. AMRules is compared against other regression systems including batch learners and streaming regression models. Last Section presents the lessons learned.

2 Related Work

In this section we analyze the related work in two dimensions: regression algorithms and incremental learning of regression algorithms.

In the field of machine learning, one of the most popular, and competitive, regression model is system M5, presented by [18]. It builds multivariate trees using linear models at the leaves. In the pruning phase for each leaf a linear model is built. Later, [5] have presented M5' a *rational reconstruction* of Quinlan's M5 algorithm. M5' first constructs a regression tree by recursively splitting the instance space using tests on single attributes that maximally reduce variance in the target variable. After the tree has been grown, a linear multiple regression model is built for every inner node, using the data associated with that node and all the attributes that participate in tests in the subtree rooted at that node. Then the linear regression models are simplified by dropping attributes if this results in a lower expected error on future data (more specifically, if the decrease in the number of parameters outweighs the increase in the observed training error). After this has been done, every subtree is considered for pruning. Pruning occurs if the estimated error for the linear model at the root of a subtree is smaller or equal to the expected error for the subtree. After pruning terminates, M5' applies a *smoothing* process that combines the model at a leaf with the models on the path to the root to form the final model that is placed at the leaf.

A widely used strategy consists of building rules from decision (or regression) trees, as it is done in [20]. Any tree can be easily transformed into a collection of rules. Each rule corresponds to the path from the root to a leaf, and there are as many rules as leaves. This process generates a set of rules with the same complexity as the decision tree. However, as pointed out by [22], a drawback of decision trees is that even a slight drift of the target function may trigger several changes in the model and severely compromise learning efficiency. Cubist [19] is a rule based model that is an extension of Quinlan's M5 model tree. A tree is grown where the terminal leaves contain linear regression models. These models are based on the predictors used in previous splits. Also, there are intermediate linear models at each level of the tree. A prediction is made using the linear regression model at the leaf of the tree, but it is *smoothed* by taking into account the prediction from the linear models in the previous nodes in the path, from the root to a leaf, followed by the test example. The tree is reduced to a set of rules, which initially are paths from the top of the tree to the bottom. Rules are eliminated via pruning of redundant conditions or conditions that do not decrease the error.

2.1   Regression Algorithms for Streaming Data

Many methods can be found in the literature for solving classification tasks on streams, but only few exists for regression tasks. One of the first incremental model trees, was presented by [17]. The authors present an incremental algorithm that scales linearly with the number of examples. They present an incremental node splitting rule, together with incremental methods for stopping the growth of the tree and pruning. The leaves contain linear models, trained using the RLS (Recursive Least Square) algorithm.

The authors of [11] propose an incremental algorithm FIMT for any-time model trees learning from evolving data streams with drift detection. It is based on the Hoeffding tree algorithm [4], but implements a different splitting criterion, using a standard deviation reduction (SDR) based measure more appropriate to regression problems. The FIMT algorithm is able to incrementally induce model trees by processing each example only once, in the order of their arrival. Splitting decisions are made using only a small sample of the data stream observed at each node, following the idea of Hoeffding trees. FIMT is able to detect and adapt to evolving dynamics. Change detection in the FIMT is carried out using the Page-Hinckley (PH) change detection test [14]. Adaptation in FIMT involves growing an alternate subtree from the node in which change was detected. When the performance of the alternate subtree improves over the original subtree, the latter is replaced by the former.

IBLStreams (Instance Based Learner on Streams) is an extension of MOA [2] that consists of an instance-based learning algorithm for classification and regression problems on data streams by [21]. IBLStreams optimizes the composition and size of the case base autonomously. When a new example (x_0, y_0) is available, the example is added to the case base. The algorithm checks whether other examples might be removed, either because they have become redundant or they are outliers. To this end, a set C of examples within a neighborhood of x_0 are considered as candidates. This neighborhood if given by the k_{cand} nearest neighbors of x_0, accordingly with a distance function D. The most recent examples are not removed due to the difficulty to distinguish potentially noisy data from the beginning of a concept change.

3   The AMRules Algorithm

In this section we present an incremental algorithm for learning model rules to address these issues, named Adaptive Model Rules from High-Speed Data Streams (AMRules). The pseudo code of the algorithm is given in Algorithm 1.

3.1   Learning a Rule Set

The algorithm begin with a empty rule set (RS), and a default rule $\{\} \rightarrow \mathcal{L}$. Every time a new training example is available the algorithm proceeds with checking whether for each rule from rule set (RS) the example is covered by any rule, that is if all the literals are true for the example. The target values of the examples covered by a rule are used to update the sufficient statistic of the rule. Before an example is covered by any rule change detection tests are updated with every example of this rule. We use

---

Algorithm 1. AMRules Algorithm

    Input: S: Stream of examples
    ordered-set: Boolean flag
    N_{min}: Minimum number of examples
    λ: Threshold
    α: the magnitude of changes that are allowed
    Result: RS Set of Decision Rules
    begin
        Let $RS \leftarrow \{\}$
        Let *defaultRule* $\mathcal{L} \leftarrow 0$
        foreach *example* $(\boldsymbol{x}, y_k) \in S$ **do**
            foreach *Rule* $r \in RS$ **do**
                if *r covers the example* **then**
                    Update change detection tests
                    Compute error $= x_t - \bar{x}_t - \alpha$
                    Call PHTest(error, λ)
                    if *Change is detected* **then**
                          └ Remove the rule
                else
                    if *Number of examples in* $\mathcal{L}_r > N_{min}$ **then**
                        $r \leftarrow ExpandRule(r)$
                        Update sufficient statistics of r
                if *ordered-set* **then**
                  └ BREAK
            if *none of the rules in RS triggers* **then**
                if *Number of examples in* $\mathcal{L}$ *mod* $N_{min} = 0$ **then**
                    $RS \leftarrow RS \cup ExpandRule(defaultRule)$
                    Update sufficient statistics of the defaultRule

---

the Page-Hinckley (PH) change detection test to monitor the online error of each rule. If a change is detected the rule is removed from the rule set (RS). Otherwise, the rule is expanded. The expansion of the rule is considered only after certain period (N_{min} number of example). The expansion of a rule is done with Algorithm 2.

The set of rules (RS) is learned in parallel, as described in Algorithm 1. We consider two cases: learning ordered or unordered set of rules. In the former, every example updates statistics of the first rule that covers it. In the latter every example updates statistics of all the rules that covers it. If an example is not covered by any rule, the default rule is updated.

3.2 Expansion of a Rule

Before discuss how rules are expanded, we will first discuss the evaluation measure used in the attribute selection process. [11] describe a standard deviation reduction measure

---

Algorithm 2. Expandrule: Expanding one Rule

Input:
   r: One Rule
   τ: Constant to solve ties
   δ : Confidence
Result: r' : Expanded Rule
begin
   Let X_a be the attribute with greater SDR
   Let X_b be the attribute with second greater SDR
   Compute $\epsilon = \sqrt{\frac{R^2 \ln(1/\delta)}{2n}}$ (Hoeffding bound)
   Compute $ratio = \frac{SDR(X_b)}{SDR(X_a)}$ (Ratio of the SDR values for the best two splits)
   Compute $UpperBound = ratio + \epsilon$
   if $UpperBound < 1 \vee \epsilon < \tau$ **then**
      Extend r with a new condition based on the best attribute
      Release sufficient statistics of $\mathcal{L}_r$
      $r \leftarrow r \cup \{X_a\}$
   return r

(SDR) for use in determining the merit of a given split. It can be efficiently computed in an incremental way. The formula for SDR measure of the split h_A is given below:

$$SDR(h_A) = sd(S) - \frac{N_{Left}}{N} sd(S_{Left}) - \frac{N_{Right}}{N} sd(S_{Right})$$

$$sd(S) = \sqrt{\frac{1}{N}(\sum_{i=1}^{N}(yi - \bar{y})^2)} =$$

$$= \sqrt{\frac{1}{N}(\sum_{i=1}^{N} yi^2 - \frac{1}{N}(\sum_{i=1}^{N} yi)^2)}$$

To make the actual decision regarding a spit, this SDR measure for the best two potential splits are compared, dividing the second-best value by the best one to generate a ratio $ratio$ in the range 0 to 1. Having a predefined range for the values of the random variables, the Hoeffding probability bound (ϵ) [10] can be used to obtain high confidence intervals for the true average of the sequence of random variables. The value of ϵ is calculated using the formula:

$$\epsilon = \sqrt{\frac{R^2 \ln(1/\delta)}{2n}}$$

where $R^2 = 1$ is the range of the random variable. The process to expand a rule by adding a new condition works as follows. For each attribute X_i, the value of the SDR is computed for each attribute value v_j. If the upper bound ($ratio + \epsilon$) of the sample average is below 1 then the true mean is also below 1. Therefore with confidence $1 - \epsilon$ the best attribute over a portion of the data is really the best attribute. In this case, the rule is expanded with condition $X_a \leq v_j$ or $X_a > v_j$. However, often two splits are extremely similar or even identical, in terms of their SDR values, and despite the ϵ intervals shrinking considerably as more examples are seen, it is still impossible to choose one split over the other. In these cases, a threshold (τ) on the error is used. If ϵ falls below this threshold and the splitting criterion is still not met, the split is made on the one with a higher SDR value and the rule is expanded. The pseudo-code of expanding a rule is presented in Algorithm 2.

3.3 Prediction Strategies

The set of rules learned by AMRules can be ordered or unordered. They employ different prediction strategies to achieve 'optimal' prediction. In the former, only the first rule that cover an example is used to predict the target example. In the latter, all rules covering the example are used for prediction and the final prediction is decided by aggregating predictions using the mean.

Each rule in AMrules implements 3 prediction strategies: i) the mean of the target attribute computed from the examples covered by the rule; ii) a linear combination of the independent attributes; iii) an adaptive strategy, that chooses between the first two strategies, the one with lower MSE in the previous examples.

Each rule in AMRules contains a linear model, trained using an incremental gradient descent method, from the examples covered by the rule. Initially, the weights are set to small random numbers in the range -1 to 1. When a new example arrives, the output is computed using the current weights. Each weight is then updated using the Delta rule: $w_i \leftarrow w_i + \eta(\hat{y} - y)x_i$, where $\hat{y}$ is the output, y the real value and η is the learning rate.

3.4 Change Detection

We use the Page-Hinckley (PH) change detection test to monitor the online error of each rule. Whenever a rule covers a labeled example, the rule makes a prediction and computes the loss function (MSE or MAD). We use the Page-Hinkley (PH) test [16] to monitor the evolution of the loss function. If the PH test signals a significant increase of the loss function, the rule is removed from the rule set (RS).

The PH test is a sequential analysis technique typically used for online change detection. The PH test is designed to detect a change in the average of a Gaussian signal [14]. This test considers a cumulative variable m_T, defined as the accumulated difference between the observed values and their mean till the current moment:

$$m_T = \sum_{t=1}^{T}(x_t - \bar{x}_T - \delta)$$

where $\bar{x}_T = 1/T \sum_{t=1}^{t} x_t$ and δ corresponds to the magnitude of changes that are allowed.

The minimum value of this variable is also computed: $M_T = min(m_t, t = 1 \ldots T)$. The test monitors the difference between M_T and m_T: $PH_T = m_T - M_T$. When this difference is greater than a given threshold (λ) we signal a change in the process generating examples. The threshold λ depends on the admissible false alarm rate. Increasing λ will entail fewer false alarms, but might miss or delay change detection.

4   Experimental Evaluation

The main goal of this experimental evaluation is to study the behavior of the proposed algorithm in terms of performance and learning times. We are interested in studying the following scenarios:

- How to grow the rule set?
  - Update only the first rule that covers training examples. In this case the rule set is ordered, and the corresponding prediction strategy uses only the first rule that covers test examples.
  - Update all the rules that covers training examples. In this case the rule set is unordered, and the corresponding prediction strategy uses a weighted sum of all rules that covers test examples.
- How does AMRules compares against others streaming algorithms?
- How does AMRules compares against others state-of-the-art regression algorithms?
- How does AMRules learned models evolve in time?

4.1   Experimental Setup

All our algorithms were implemented in java using the Massive Online Analysis (MOA) data stream software suite [2]. The performance of the algorithms is measured using the standard metrics for regression problems: Mean Absolute Error (MAE) and Root Mean Squared Error (RMSE) [24]. We used two evaluation methods. When no concept drift is assumed, the evaluation method we employ uses the traditional train and test scenario.

The experimental datasets include both artificial and real data, as well sets with continuous attributes. We use ten regression datasets from the UCI Machine Learning Repository [1] and other sources. The datasets used in our experimental work are briefly described here. **2dplanes** this is an artificial dataset described in [3]. **Ailerons** this dataset addresses a control problem, namely flying a F16 aircraft. **Puma8NH** and **Puma32H** is a family of datasets synthetically generated from a realistic simulation of the dynamics of a Unimation Puma 560 robot arm. **Pol** this is a commercial application described in [23]. The data describes a telecommunication problem. **Elevators** this dataset was obtained from the task of controlling a F16 aircraft. **Fried** is an artificial dataset used in Friedman (1991) and also described in [3]. **Bank8FM** a family of datasets synthetically generated from a simulation of how bank-customers choose their banks. **Kin8nm** this dataset is concerned with the forward kinematics of an 8 link robot arm. **Airlines** This dataset using the data from the 2009 Data Expo competition. The dataset consists of a huge amount of records, containing flight arrival and departure details for all the commercial flights within the USA, from October 1987 to April 2008. This is a very large dataset with nearly 120 million records (11.5 GB memory size) [11]. The Table 1 summarizes the number of instances and the number of attributes of each dataset.

All algorithms learn from the same training set and the performance is estimated from the same test set. In scenarios with concept drift, we use the prequential error estimates [8]. This method evaluates a model on a stream by testing then training with each example in the stream. For all the experiments, we set the input parameters to:

Table 1. Summary of datasets

Datasets	# Instances	# Attributes
2dplanes	40768	11
Ailerons	13750	41
Puma8NH	8192	9
Puma32H	8192	32
Pol	15000	49
Elevators	8752	19
Fried	40769	11
bank8FM	8192	9
kin8nm	8192	9
Airline	115Million	11

$N_{min} = 200$, $\tau = 0.05$ and $\delta = 0.01$. All of the results in the tables are averaged of ten-fold cross validation [12], except for the Airlines dataset. For all the simulations of evolving data and change detection we set the PH test parameters to $\lambda = 50$ and $\alpha = 0.005$.

4.2 Experimental Results

In this section, we empirically evaluate the adaptive model rules algorithm. The results come in four parts. In the first part, we compare the AMRules variants. In the second part we compare AMRules against others streaming algorithms. In the third part, we compare AMRules against others state-of-the-art regression algorithms. In the last part, we assess AMRules models in time-evolving data streams.

Comparison between AMRules Variants. In this section we focus in two strategies that we found potentially interesting. It is a combination of expanding only one rule, the rule that first triggered, with predicting strategy uses only the first rule that covers test examples. Obviously, for this approach it is necessary to use ordered rules ($AMRules^o$). The second setting employs unordered rule set, where all the covering rules expand and the corresponding prediction strategy uses a weighted sum of all rules that covers test examples ($AMRules^u$).

Ordered rule sets specializes one rule at time. As a result it often produces fewer rules than the unordered strategy. Ordered rules need to consider the previous rules and remaining combinations, which might not be easy to interpret in more complex sets. Unordered rule sets are more modular, because they can be interpreted alone.

Table 2 summarize the mean absolute error and the root mean squared error of these variants. Overall, the experimental results points out that unordered rule sets are more competitive than ordered rule sets in terms of MAE and RMSE.

Comparison with Other Streaming Algorithms. We compare the performance of our algorithm with three others streaming algorithms, FIMT and IBLStreams. FIMT is an incremental algorithm for learning model trees, described in [11]. IBLStreams is an extension of MOA that consists in an instance-based learning algorithm for classification and regression problems on data streams by [21].

Table 2. Comparative performance between AMRules variants. Results of 10-fold cross-validation except for the AIrlines dataset.

Datasets	Mean Absolute Error (variance)		Root Mean Squared Error (variance)	
	AMRules^o	AMRules^u	AMRules^o	AMRules^u
2dplanes	1.23E+00 (0.01)	**1.16E+00** (0.01)	1.67E+00 (0.02)	**1.52E+00** (0.01)
Ailerons	1.10E-04 (0.00)	**1.00E-04** (0.00)	1.90E-04 (0.00)	**1.70E-04** (0.00)
Puma8NH	3.21E+00 (0.04)	**2.66E+00** (0.02)	**4.14E+00** (0.05)	4.28E+00 (0.03)
Puma32H	**1.10E-02** (0.00)	1.20E-02 (0.00)	1.60E-02 (0.00)	**1.00E-04** (0.00)
Pol	**14.0E+00** (25.1)	15.6E+00 (3.70)	**2.30E+01** (44.50)	23.3E00 (4.08)
Elevators	3.50E-03 (0.00)	**1.90E-03** (0.00)	4.80E-03 (0.00)	**2.20E-03** (0.00)
Fried	2.08E+00 (0.01)	**1.13E+00** (0.01)	2.78E+00 (0.08)	**1.67E+00** (0.25)
bank8FM	4.31E-02 (0.00)	**4.30E-02** (0.00)	4.80E-02 (0.00)	**4.30E-02** (0.00)
kin8nm	1.60E-01 (0.00)	**1.50E-01** (0.00)	2.10E-01 (0.00)	**2.00E-01** (0.00)
Airlines	1.42E+01 (0.00)	**1.31E+01** (0.00)	2.44E+01 (0.00)	**2.25E+01** (0.00)
Average Rank	1.7	1.3	1.8	1.2

The performance measures for these algorithms are given in Table 3. The comparison of these streaming algorithms shows that AMRules get better results.

Table 3. Comparative performance between streaming algorithms. Results of 10-fold cross-validation except for the AIrlines dataset.

Datasets	Mean Absolute Error (variance)			Root Mean Squared Error (variance)		
	AMRules^u	FIMT	IBLStreams	AMRules^u	FIMT	IBLStreams
2dplanes	1.16E+00 (0.01)	**8.00E-01** (0.00)	1.03E+00 (0.00)	1.52E+00 (0.01)	**1.00E+00** (0.00)	1.30E+00 (0.00)
Ailerons	**1.00E-04** (0.00)	1.90E-04 (0.00)	3.20E-04 (0.00)	1.70E-04 (0.00)	**1.00E-09** (0.00)	3.00E-04 (0.00)
Puma8NH	**2.66E+00** (0.01)	3.26E+00 (0.03)	3.27E+00 (0.01)	**4.28E+00** (0.03)	12.0E+00 (0.63)	3.84E+00 (0.02)
Puma32H	1.20E-02 (0.00)	**7.90E-03** (0.00)	2.20E-02 (0.00)	**1.00E-04** (0.01)	1.20E-02 (0.00)	2.70E-02 (0.00)
Pol	**15.6E+00** (3.70)	38.2E+00 (0.17)	29.7E+00 (0.55)	**23.3E+00** (4.08)	1,75E+03 (1383)	50,7E+00 (0.71)
Elevators	**1.90E-03** (0.00)	3.50E-03 (0.00)	5.00E-03 (0.00)	2.20E-03 (0.00)	**3.00E-05** (0.00)	6.20E-03 (0.00)
Fried	**1.13E+00** (0.01)	1.72E+00 (0.00)	2.10E+00 (0.00)	**1.67E+00** (0.25)	4.79E+00 (0.01)	2.21E+00 (0.00)
bank8FM	4.30E-02 (0.00)	**3.30E-02** (0.00)	7.70E-02 (0.00)	4.30E-02 (0.00)	**2.20E-03** (0.00)	9.60E-02 (0.00)
kin8nm	**1.60E-01** (0.00)	**1.60E-01** (0.00)	9.50E-01 (0.00)	**2.00E-01** (0.00)	2.10E-01 (0.00)	1.20E-01 (0.00)
Airlines	**1.31E+01** (0.00)	1.39E+01 (0.00)	1.45E+01 (0.00)	**2.25E+01** (0.00)	2.30E+01 (0.00)	2.51E+01 (0.00)
Average Rank	1.5	1.9	2.6	1.8	2.0	2.3

Comparison with Other State-of-the-Art Regression Algorithms. Another experiment which involves adaptive model rules is showed in Table 4. We compared AMRules with other non-incremental regression algorithms from WEKA [9]. We use the standard method of ten-fold cross-validation, using the same folds for all the algorithms included.

The comparison of these algorithms show that AMRules has lower accuracy (MAE, RMSE) than M5Rules and better accuracy than the others methods. These results were somewhat expected, since these datasets are relatively small for the incremental algorithm.

Evaluation in Time-Evolving Data Streams. In this subsection we first study the evolution of the error measurements (MAE and RMSE) and evaluate the change detection method. After, we evaluate the streaming algorithms on non-stationary streaming real-world problems, using the Airline dataset from the DataExpo09 competition.

Table 4. Comparative performance between AMRules[u] and other regression algorithms. Results of 10-fold cross validation.

	Root Mean Squared Error (variance)			
Datasets	AMRules[u]	M5Rules	MLPerceptron	Linear Regression
2dplanes	1.52E+00 (0.01)	**9.8E-01** (0.01)	1.09E+00 (0.01)	2.37E+00 (0.00)
Ailerons	**1.70E-04** (0.00)	2.00E-04 (0.00)	1.71E-04 (0.00)	2.00E-04 (0.00)
Puma8NH	4.28E+00 (0.03)	**3.19E+00** (0.01)	4.14E+00 (0.20)	4.45E+00 (0.01)
Puma32H	**1.00E-04** (0.00)	8.60E-03 (0.00)	3.10E-02 (0.00)	2.60E-02 (0.00)
Pol	23.3E+00 (4.08)	**6.56E+00** (0.45)	20.1E+00 (15.1)	30.5E+00 (0.16)
Elevators	**2.20E-03** (0.00)	2.23E-03 (0.00)	2.23E-03 (0.00)	2.29E-03 (0.00)
Fried	1.67E+00 (0.25)	**1.60E+00** (0.00)	1.69E+00 (0.04)	2.62E+00 (0.00)
bank8FM	4.30E-02 (0.00)	**3.10E-02** (0.00)	3.40E-02 (0.00)	3.80E-02 (0.00)
kin8nm	2.00E-01 (0.00)	**1.70E-01** (0.00)	1.60E-01 (0.00)	2.00E-01 (0.00)
Average Rank	2.2	1.7	2.6	3.6

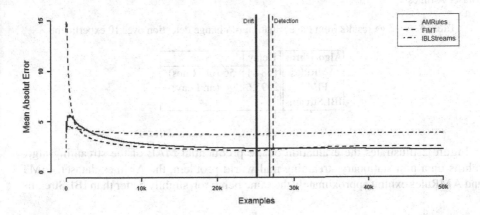

Fig. 1. Evolution of the prequential Mean Absolute Error of streaming algorithms using the dataset Fried

To simulate drift we use Fried dataset. The simulations allow us to control the relevant parameters and to assess the change detection algorithm. Figure 1 depict the prequential MAE curve of the streaming algorithms using the dataset Fried. The figure also illustrates the change point and the points where the change was detected. Only two of the algorithms detected the change (FIMT and AMRules). Table 5 shows the average results over 10 experiments using the Fried dataset. We measure the number of nodes, the number of rules and the Page-Hinckley test delay (number of examples monitored by PH test before the detection). The delay of the Page-Hinckley test is an indication of how fast the algorithm will be able to start the adaptation strategy. These two algorithms obtained similar results. The general conclusions are that FIMT and AMRules algorithms are robust and have better results than IBLStreams.

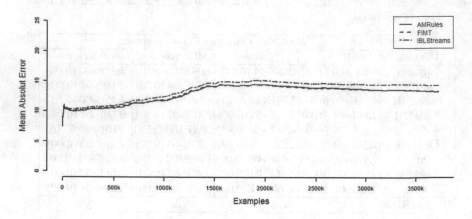

Fig. 2. Evolution of the prequential Mean Absolute Error of streaming algorithms using the dataset Airlines

Table 5. Average results from the evaluation of change detection over 10 experiments

Algorithms	Delay	Size
AMRules	1484.1	56 (nr. Rules)
FIMT	2095.6	290 (nr. Leaves)
IBLStreams	-	-

Figure 2 illustrates the evaluation of the prequential MAE of the streaming algorithms on a non-stationary streaming real-world problem, the Airlines dataset. FIMT and AMRules exhibit approximately the same behavior, slightly better than IBLStreams.

Learning Times. Table 6 reports the relative learning times required for the 10-cross-validation experiments. Being one-pass algorithms, both versions of AMRules are much

Table 6. Relative learning times of experiences reported in the paper. The reference algorithm is AMRules^o.

Datasets	Time(seconds)						
	AMRules^o	AMRules^u	M5Rules	MLPerceptron	LinRegression	IBLStreams	FIMT
2dplanes	1	1.5	317	577	1.0	4627	2.3
Airlerons	1	1.2	535	737	4.6	8845	1.5
Puma8NH	1	1.1	497	113	0.5	1700	2.0
Puma32H	1	1.4	801	280	2.0	1084	2.2
Pol	1	1.4	806	1260	1.4	12133	1.5
Elevators	1	1.5	678	341	0.9	16856	1.8
Fried	1	1.4	5912	359	0.7	154157	1.4
bank8FM	1	1.4	78	73	0.8	17189	0.4
kin8nm	1	1.1	146	75	0.7	29365	0.4
Airlines	1	8.3	-	-	-	203	10.6

faster than M5 Rules and multi-layer perceptron. AMRules^o generates fewer rules being slightly faster than AMRules^u.

5   Conclusions

Regression rules are one of the most expressive languages to represent generalizations from examples. Learning regression rules from data streams is an interesting research line that has not been widely explored by the stream mining community. To the best of our knowledge, in the literature there is no method that addresses this issue. In this paper, we present a new regression model rules algorithm for streaming and evolving data. The AMRules algorithm is a one-pass algorithm, able to adapt the current rule set to changes in the process generating examples. It is able to induce ordered and unordered rule sets, where the consequent of a rule contains a linear model trained with the perceptron rule.

The experimental results point out that, in comparison to ordered rule sets, unordered rule sets are more competitive in terms of performance (MAE and RMSE). AMRules is competitive against batch learners even for medium sized datasets.

Acknowledgments. The authors acknowledge the financial support given by the project FCT-KDUS PTDC/ EIA/ 098355/2008 FCOMP -01-0124-FEDER-010053, the ERDF through the COMPETE Programme and by National Funds through FCT within the project FCOMP - 01-0124-FEDER-022701.

References

1. Bache, K., Lichman, M.: UCI machine learning repository (2013)
2. Bifet, A., Holmes, G., Pfahringer, B., Kranen, P., Kremer, H., Jansen, T., Seidl, T.: MOA: Massive online analysis. Journal of Machine Learning Research (JMLR), 1601–1604 (2010)
3. Breiman, L., Friedman, J., Olshen, R., Stone, C.: Classification and Regression Trees. Wadsworth and Brooks, Monterey (1984)
4. Domingos, P., Hulten, G.: Mining High-Speed Data Streams. In: Parsa, I., Ramakrishnan, R., Stolfo, S. (eds.) Proceedings of the ACM Sixth International Conference on Knowledge Discovery and Data Mining, pp. 71–80. ACM Press, Boston (2000)
5. Frank, E., Wang, Y., Inglis, S., Holmes, G., Witten, I.H.: Using model trees for classification. Machine Learning 32(1), 63–76 (1998)
6. Fürnkranz, J., Gamberger, D., Lavrač, N.: Foundations of Rule Learning. Springer (2012)
7. Gama, J.: Knowledge Discovery from Data Streams. Chapman & Hall, CRC Press (2010)
8. Gama, J., Sebastião, R., Rodrigues, P.P.: On evaluating stream learning algorithms. Machine Learning 90(3), 317–346 (2013)
9. Hall, M., Frank, E., Holmes, G., Pfahringer, B., Reutemann, P., Witten, I.H.: The WEKA data mining software: an update. SIGKDD Explor. Newsl. 11, 10–18 (2009)
10. Hoeffding, W.: Probability inequalities for sums of bounded random variables. Journal of the American Statistical Association 58(301), 13–30 (1963)
11. Ikonomovska, E., Gama, J., Dzeroski, S.: Learning model trees from evolving data streams. Data Min. Knowl. Discov. 23(1), 128–168 (2011)

12. Kohavi, R.: A study of cross-validation and bootstrap for accuracy estimation and model selection, pp. 1137–1143 (1995)
13. Kosina, P., Gama, J.: Handling Time Changing Data with Adaptive Very Fast Decision Rules. In: Flach, P.A., De Bie, T., Cristianini, N. (eds.) ECML PKDD 2012, Part I. LNCS, vol. 7523, pp. 827–842. Springer, Heidelberg (2012)
14. Mouss, H., Mouss, D., Mouss, N., Sefouhi, L.: Test of Page-Hinkley, an approach for fault detection in an agro-alimentary production system. In: Proceedings of the Asian Control Conference, vol. 2, pp. 815–818 (2004)
15. Ould-Ahmed-Vall, E., Woodlee, J., Yount, C., Doshi, K.A., Abraham, S.: Using model trees for computer architecture performance analysis of software applications. In: IEEE International Symposium on Performance Analysis of Systems & Software, ISPASS 2007, pp. 116–125. IEEE (2007)
16. Page, E.S.: Continuous inspection schemes. Biometrika 41(1/2), 100–115 (1954)
17. Potts, D., Sammut, C.: Incremental learning of linear model trees. Machine Learning 61(1-3), 5–48 (2005)
18. Quinlan, J.R.: Learning with continuous classes. In: Australian Joint Conference for Artificial Intelligence, pp. 343–348. World Scientific (1992)
19. Quinlan, J.R.: Combining instance-based and model-based learning, pp. 236–243. Morgan Kaufmann (1993)
20. Quinlan, R.: C4.5: Programs for Machine Learning. Morgan Kaufmann Publishers, Inc., San Mateo (1993)
21. Shaker, A., Hüllermeier, E.: Iblstreams: a system for instance-based classification and regression on data streams. Evolving Systems 3, 235–249 (2012)
22. Wang, H., Fan, W., Yu, P.S., Han, J.: Mining concept-drifting data streams using ensemble classifiers. In: Proceedings of the ACM SIGKDD International Conference on Knowledge Discovery and Data Mining, pp. 226–235. ACM Press, Washington, D.C. (2003)
23. Weiss, S.M., Indurkhya, N.: Rule-based machine learning methods for functional prediction. Journal of Artificial Intelligence Research 3, 383–403 (1995)
24. Willmott, C.J., Matsuura, K.: Advantages of the mean absolute error (mae) over the mean square error (rmse) in assessing average model performance. Climate Research 30, 79–82 (2005)

Fast and Exact Mining of Probabilistic Data Streams

Reza Akbarinia and Florent Masseglia

Zenith team (INRIA-UM2), LIRMM, Montpellier, France
{reza.akbarinia,florent.masseglia}@inria.fr

Abstract. Discovering Probabilistic Frequent Itemsets (PFI) is very challenging since algorithms designed for deterministic data are not applicable in probabilistic data. The problem is even more difficult for probabilistic data streams where massive frequent updates need to be taken into account while respecting data stream constraints. In this paper, we propose FEMP (Fast and Exact Mining of Probabilistic data streams), the first solution for exact PFI mining in data streams with sliding windows. FEMP allows updating the frequentness probability of an itemset whenever a transaction is added or removed from the observation window. Using these update operations, we are able to extract PFI in sliding windows with very low response times. Furthermore, our method is exact, meaning that we are able to discover the exact probabilistic frequentness distribution function for any monitored itemset, at any time. We implemented FEMP and conducted an extensive experimental evaluation over synthetic and real-world data sets; the results illustrate its very good performance.

Keywords: Probabilistic Data Streams, Probabilistic Frequent Itemsets, Sliding Windows.

1 Introduction

Dealing with probabilistic data has gained increasing attention these past few years in both static and streaming data management and mining [3], [9], [2], [11], [10]. There are many possible reasons for probabilistic data, such as noise occurring when data are collected, noise injected for privacy reasons, semantics of the results of a search engine (often ambiguous), etc. Thus, many sensitive domains now involve massive probabilistic data. Example 1 illustrates a collection of probabilistic data, where each record is associated to a probability of occurrence.

Example 1. Let us consider animals' health monitoring in a zoo, and more particularly the health of Pandas, for which reproduction is an important issue. In our scenario, a set of body sensors gathers physiological data (blood pressure, temperature, etc.) and transforms it into possible activities thanks to a given model. For instance, the rule $\{pressure = [100..150], temperature = [80..90] \rightarrow$

H. Blockeel et al. (Eds.): ECML PKDD 2013, Part I, LNAI 8188, pp. 493–508, 2013.

sleeping, 75%} means that with a blood pressure between 100 and 150mmHg, and a body temperature between 80 and 90 Fahrenheit, the probability that a Panda is sleeping is 75%. Figure 1 illustrates the activities inferred for the pandas. We can observe, for instance, that Yuan Zi was eating at 9am, with a probability of 40%.

With the probabilistic approach illustrated by Example 1, there are two cases for each probabilistic record: either it really occurred in the real world or it did not. A reliable framework for handling such probabilistic data lies in the theory of "possible worlds" [6] where each unique combination of records' existence corresponds to a possible world. Unfortunately, there is a combinatorial explosion in the number of possible worlds (n records, each associated to 2 possible values of existence, leading to 2^n possible worlds). Therefore, in this context, frequent itemset mining [1] must be carefully adapted. Finding the number of occurrences of an itemset X in a database D (also called the support of X in D) is at the core of frequent itemset mining. In the literature, we find two main support measures for probabilistic data: Expected Support [5] (an approximate measure of support) and Probabilistic Support [3] (that is an exact measure of support in probabilistic data). We work with Probabilistic, and we propose a solution for Probabilistic Frequent Itemset (PFI) mining in data streams using this measure of interest.

Huan Huan				Yuan Zi			
e	h	activity	Prob.	e	h	activity	Prob.
1	8	sleeping	0.3	2	8	sleeping	0.9
3	9	eating	0.3	4	9	eating	0.4
5	10	sleeping	0.3	6	10	drinking	1
7	11	grooming	0.4	8	11	grooming	0.9
9	12	sleeping	0.3	10	12	marking	0.4
11	13	drinking	0.3	12	13	resting	0.2
13	14	courting	0.9	14	14	climbing	0.2
15	15	resting	0.2	16	15	courting	0.4
17	16	playing	0.4	18	16	playing	0.3
19	17	growling	0.2	20	17	growling	0.9

Fig. 1. Panda's activities inferred from body sensor data

There are several ways to observe a data stream, two important ones being batches and sliding windows [8]. Both techniques have pros and cons. Batches allow fast processing but the result is available only after the batch has been fulfilled (which is not compatible with real time constraints). Sliding windows allow maintaining the result any time the stream is updated, but they need more CPU. Today, existing methods for probabilistic data stream mining are batch-based and work with Expected Support [17], [11], [10]. Meanwhile, working with sliding windows is a major matter for numerous monitoring applications where handling "anytime queries" is crucial. Data stream mining over a sliding window requires to provide efficient solutions for updating probabilistic supports after adding/removing transactions and, in the probabilistic context, this is quite challenging.

In this paper, we introduce FEMP (Fast and Exact Mining of Probabilistic data streams), a framework adopting the exact approach while meeting the time limitations of data stream environments. *To the best of our knowledge, FEMP is the first solution for PFI mining in a sliding window over probabilistic data streams.*

Our contributions are i) a new model for probabilistic data streams, where an item may have multiple occurrences (each associated to a probability) for one transaction ii) a new approach for computing probabilistic support by recursion on the transactions (this approach allows to develop efficient algorithms for updating probabilistic support after any modification in the sliding window) and iii) new algorithms for probabilistic frequent itemset mining with sliding windows, where transactions are inserted or deleted. Our algorithms allow updating the new probabilistic support of any monitored itemset with a low complexity since they avoid scanning the whole sliding window from scratch.

We validated our solution through experimentation over synthetic and real-world data sets. The results show that it is able to discover and manage PFI in data streams with response times that are up to several orders of magnitude faster than baseline methods of the literature employed in a sliding window context.

2   Problem Definition

We now describe the problem we address with formal definitions of the probabilistic model we adopt, probabilistic itemset mining and probabilistic data streams.

2.1   Probabilistic Data

Let I be a set of literals. I is also called the *vocabulary*. An event e_i is a tuple $e_i =< Oid, ts, x, P >$ where i is the identifier of the event, Oid is an object identifier, ts is a timestamp, $x \in I$ is an item and P is an existential probability $P \in [0,1]$ denoting the probability that e_i occurs.

Example 2. Consider the data given by Figure 1, the first two events for Huan Huan are: $e_1 =< Huan\ Huan, 8,\ sleeping, 0.3 >$ and $e_3 =< Huan\ Huan, 9, eating, 0.3 >$.

Definition 1. *A probabilistic item x is an item that appears in an event, the probability of x is the probability of its event.*

Definition 2. *A probabilistic transaction t is a set of pairs (x, P) for an object such that x is a probabilistic item and P is the probability of the event of x. $P(x \in t)$ is the probability of existence of x in t. A probabilistic database is a set of probabilistic transactions.*

Panda	Id	Transaction
Huan Huan	t_1	(eating, 0.3); (sleeping, 0.3)
Yuan Zi	t_2	(eating, 0.4); (drinking, 1)

Fig. 2. The pandas' activities (probabilistic transactions) from 9am to 10am

Example 3. Figure 2 gives the probabilistic transaction database of Huan Huan and Yuan Zi for two hours, from 9am to 10am. We can observe that Yuan Zi's activities in this time window were: eating with a probability of 40% and drinking with a probability of 100%.

	Possible Worlds	Probability
w_1	{}; {drinking}	0.294
w_2	{eating}; {drinking}	0.126
w_3	{sleeping}; {drinking}	0.126
w_4	{eating, sleeping}; {drinking}	0.054
w_5	{}; {eating, drinking}	0.196
w_6	{eating}; {eating, drinking}	0.084
w_7	{sleeping}; {eating, drinking}	0.084
w_8	{eating, sleeping}; {eating, drinking}	0.036

Fig. 3. Possible worlds for the database illustrated in Figure 2

A probabilistic database can be treated as a set of deterministic databases, called possible worlds. The possible worlds are generated from the possible instances of transactions. Let w be a possible world, then the instance of a transaction t in w is denoted by t_w. Figure 3 shows the possible worlds for the database in Figure 2. In this database, the instance of transaction t_1 in w_3 is {*sleeping*}, and that of transaction t_2 is {*drinking*}. For each possible world w, there is probability $P(w)$ that is computed based on the probability of its transaction instances. The sum of the probabilities of all possible worlds of a database is equal to one. In the case of independence of events, the probability of a given world is computed as $P(w) = \prod_{t \in I} P(t_w)$, where $P(t_w)$ is the probability of t's instance in w. $P(t_w)$ is computed as follows: $P(t_w) = (\prod_{x \in t_w} (P(x \in t)) \times (\prod_{x \notin t_w} (1 - P(x \in t)))$.

Intuitively, we multiply the existential probability of t items that are present in t_w by the probability of absence of those that are not present in t_w.

Example 4. In the possible worlds shown in Figure 3, the probability of w_4 is equal to the occurrence of eating and sleeping for transaction t_1, drinking for t_2, and the non-occurrence of eating for t_2. Thus $P(w_4) = (0.3 \times 0.3 \times 1) \times (1 - 0.4) = 0.054$.

2.2   Probabilistic Frequent Itemsets

The problem of frequent itemset mining from a set of transactions T, as defined in [1], aims at extracting the itemsets that occur in a sufficient number of transactions in T. This is based on the number of transactions in T where an itemset X appears (i.e. the support of X in T). In the deterministic context, computing this support is straightforward (with a scan over T). In probabilistic databases, however, the support varies from one possible world to another. For this reason, the support of an itemset in a probabilistic database, introduced in [3], is given as a *probability distribution function*. In other words, each possible value $i \in \{0, \ldots, |T|\}$ for the support of X is associated to a probability that is the probability that X has this support in the probabilistic database. Definition 3 gives a more formal definition of this notion.

Definition 3. *Let W be the set of possible worlds and $S_{X,w}$ be the support of $X \in I$ in world $w \in W$. The probability $P_{X,T}(i)$ that X has support i in the set of probabilistic transactions T is given by: $P_{X,T}(i) = \sum_{w \in W, S_{X,w}=i} P(w)$. In other words, $P_{X,T}(i)$ is the cumulative probability of all possible worlds in which support of X is i. The probability distribution function $P_{X,T}(i)$ for $i \in [0..|T|]$ is called the probabilistic support of X.*

Example 5. In the possible worlds given by Figure 3, we have $P_{eating,T}(1) = P(w_2) + P(w_4) + P(w_5) + P(w_7) = 0.46$. In other words, the probability that exactly one Panda is eating between 9am and 10am is 46%.

Definition 4. *Given a support value i, the probability $P_{\geq X,T}(i)$ that an itemset X has at least i occurrences in T, is given by: $P_{\geq X,T}(i) = \sum_{j=i}^{|T|} P_{X,T}(j)$. Given minSup and minProb, a user minimum support and minimum probability, and T a set of probabilistic transactions, an itemset X is a probabilistic frequent itemset (PFI) iff $P_{\geq X,D}(minSup) \geq minProb$. $P_{\geq X,D}(minSup)$ is also called the frequentness probability of X.*

For example, the probability that "eating" has support of at least 1 is given by $P_{eating,T}(1) + P_{eating,T}(2) = 0.46 + 0.12 = 0.58$. In other words, the probability that at least one panda was eating between 9am and 10am is 58%.

2.3   Probabilistic Data Stream Mining

In many applications, the data production rate is so high that their analysis in real time with traditional methods is impossible. Sensor networks, Web usage data, scientific instruments or bio-informatics, to name a few, have added to this situation. Because of their rate, data streams should often be observed through a limited observation window and their analysis is highly constrained (*e.g.* "in real-time", "with ongoing queries", "with no access to outdated data", etc.). There are several models for this observation, including sliding windows [16]. Definition 5 gives a formal definition of this notion.

Definition 5. *An event data stream (or data stream) is an unbounded stream of ordered events. Given n, the maximum number of events to maintain in memory, a sliding window over a data stream contains the last n events from the stream.*

The problem of *probabilistic frequent itemset mining in a sliding window* is to extract the set of probabilistic frequent itemsets after each update. The updates occur when a new event is added to the stream and the oldest one is removed from the sliding window.

3   PFI Mining in Sliding Windows

We now introduce FEMP, our framework for PFI mining in probabilistic data streams with a sliding window SW. FEMP allows monitoring the probabilistic support of all the itemsets of SW in real time, as opposed to the batch model where these results are obtained only when a batch is complete. However, the main challenge in this approach consists in updating the probabilistic support of an itemset X when a transaction t is added to, or removed from the sliding window. In deterministic data, this operation is simple, we just check if $X \subseteq t$ and update its support consequently. In the context of possible worlds, there is no such straightforward approach, because the set of possible worlds changes completely, after adding/removing a transaction to/from the sliding window.

Before describing our solution, we mention that one of its requirements is to know $P(X \subseteq t)$, the probability that itemset X is included in transaction t. In the case of independent items, it can be computed as $P(X \subseteq t) = \prod_{i=1}^{|X|} P(x_i \in t)$. In the case where items of transaction t are dependent, for computing $P(X \subseteq t)$ we have to take into account the rules defined on the dependency of items. For example, if two items x_1 and x_2 have a mutual exclusion dependency, then the probability that $X = \{x_1, x_2, \dots\}$ is a subset of a transaction t is zero.

3.1   Sliding Window Model

Our sliding window model maintains a set of probabilistic transactions in memory. When the stream produces a new event $e_i = \,<Oid, time, x, P>$, the corresponding object in the model is either created or updated in the window. With streaming data, an item x may occur at several points in time and each occurrence is associated to a probability. Therefore, we must give a reliable probability

Panda	Sliding window of size 6, after e_8
Huan Huan	(eating, 0.3); (sleeping, 0.3); (grooming, 0.4)
Yuan Zi	(eating, 0.4); (drinking, 1); (grooming, 0.9)
	Sliding window of size 6, after e_9
Huan Huan	(eating, 0.3); (sleeping, 0.51); (grooming, 0.4)
Yuan Zi	(eating, 0.4); (drinking, 1); (grooming, 0.9)
	Sliding window of size 6, after e_{10}
Huan Huan	(sleeping, 0.51); (grooming, 0.4)
Yuan Zi	(eating, 0.4); (drinking, 1); (grooming, 0.9); (marking, 0.4)

Fig. 4. Sliding windows of size 6 from e_3 to e_{10}

of existence of x, by taking each probability of occurrence into account. To that end, we consider $P(x \in t)$ as the probability that at least one occurrence of x exists in t (*i.e.* 1 minus the probability that x does not exist in t).

Example 6. Consider the stream of events illustrated in Figure 1 and SW, the sliding window limited to the last 6 events. Figure 4 illustrates the content of SW from 11am (*i.e.* e_3 to e_8) to 12am (*i.e.* e_5 to e_{10}). In this example, when e_9 is added, we update the probability of *sleeping* for Huan Huan, but we do not need to remove any item from SW. Then, after e_{10}, we add *marking* to the probabilistic transaction of Yuan Zi and e_3, the oldest event, must be removed.

3.2   Computing Frequentness Probability

For computing the probability that an itemset X is frequent, we need to sum up the probabilities of all supports i for $i > minsup$. In other words, we have $P_{\geq X,T}(minsup) = \sum_{i=minsup}^{|T|} P_{X,T}(i)$, where $P_{X,T}(i)$ is the probability of support i for X in T. Notice that the sum of the probabilities in each row is equal to one. Therefore, we have: $P_{\geq X,T}(minsup) = (1 - \sum_{i=0}^{minsup-1} P_{X,T}(i))$. We use this equation for computing the frequentness probability of itemsets. To update the frequentness probabilities after inserting/deleting a transaction, we need to compute and update the probability of support i ($0 \leq i \leq minSup - 1$) for an itemset X after inserting/deleting a transaction to/from the sliding window. Our approach for computing the probabilistic support of itemsets uses a recursion on transactions.

3.3   Recursion on Transactions

Let X be an itemset, DB^n be a probabilistic database involving transactions $T = \{t_1, \ldots, t_n\}$, and $P_{X,T}(i)$ be the probability that the support of X, in the set of transactions T, is i. We develop an approach for computing $P_{X,T}(i)$ by doing recursion on the number of transactions.

Base. Let us first consider the recursion base. Consider DB^1 be a database that involves only transaction t_1. In this database, the support of X can be zero or one. The support of X in DB^1 is 1 with probability $P(X \subseteq t_1)$, and its support is 0 with probability $(1 - P(X \subseteq t_1))$. Thus, for the probabilistic support of X in DB^1, we have the following formula:

$$P_{X,\{t_1\}}(i) = \begin{cases} P(X \subseteq t_1) & \text{for i=1;} \\ (1 - P(X \subseteq t_1)) & \text{for i=0;} \\ 0 & \text{otherwise} \end{cases} \tag{1}$$

Recursion Step. Assume we have DB^{n-1}, a database involving the transactions $t_1, \ldots, t_{n-1}$. We construct DB^n by adding transaction t_n to DB^{n-1}. If $X \not\subseteq t_n$ then the probability of support i for X in DB^n is exactly the same as that in DB^{n-1}. If $X \subseteq t_n$ then two cases can lead to a support of i for X in DB^n:

1. $X \subseteq t_n$ in DB^n and the support of X in DB^{n-1} is equal to $i - 1$. Thus, we have:
$$P_{X,T}(i) = P_{X,T-\{t_n\}}(i - 1) \times (P(X \subseteq t_n)).$$
2. $X \not\subseteq t_n$ and the support of X in DB^{n-1} is equal to i. Thus, we have:
$$P_{X,T}(i) = P_{X,T-\{t_n\}}(i) \times (1 - P(X \subseteq t_n)).$$

Then, the probability of support i for X in a database containing $t_1, \ldots, t_n$ is computed based on theorem 1.

Theorem 1. *Given an itemset X and a set of transactions $T = \{t_1, \ldots, t_{n-1}, t_n\}$, the probabilistic support of X in T can be computed based on the probabilistic support in $T - \{t_n\}$ by using the following equation:*

$$P_{X,T}(i) = P_{X,T-\{t_n\}}(i - 1) \times (P(X \subseteq t_n))$$
$$+P_{X,T-\{t_n\}}(i) \times (1 - P(X \subseteq t_n)) \tag{2}$$

Proof. Implied by the above discussion.

3.4 Updating Probabilistic Support after Inserting a Transaction

To efficiently support data mining over probabilistic data streams, we need to update efficiently the probabilistic support of itemsets after each update. Here, we deal with the insertion of a new transaction to the sliding window. The case of transaction removal will be addressed in Section 3.5.

After inserting a new transaction to the sliding window, the probabilistic support can be updated by an algorithm that proceeds as follows (we removed the pseudo-code due to space restrictions). Let $P_{X,T}[0..|SW|]$ be an array such that $P_{X,T}[i]$ shows the probability of support i for itemset X in a set of transactions T. $|SW|$ is the maximum support of a transaction in the sliding window, i.e. the size of the window. Given $P_{X,T}$, we generate an array $P_{X,T+\{t\}}$ such that $P_{X,T+\{t\}}[i]$ shows the probability of support i for X in $T + \{t\}$. For filling the array $P_{X,T+\{t\}}$, our algorithm considers two main cases: either T is empty or T is not empty (so $P_{X,T}$ is available). In the first case, we have only one transaction in the sliding window. Thus, our algorithm initializes $P_{X,T+\{t\}}$ using the base of our recursive formula (described in Section 3.3) by setting $P_{X,T+\{t\}}[1] = P(X \subseteq t)$ and $P_{X,T+\{t\}}[0] = 1 - P(X \subseteq t)$. In the second case, i.e. where T is not empty, the algorithm computes the values of $P_{X,T+\{t\}}$ based on those in $P_{X,T}$ by using our recursive formula (i.e. Equation 2) as follows:
$$P_{X,T+\{t\}}[i] = (P_{X,T}[i - 1] \times P(X \subseteq t)) + (P_{X,T}[i] \times (1 - P(X \subseteq t)))$$

When $P(X \subseteq t) = 0$ we can simply ignore the transaction since it has no impact on the support, thus we have $P_{X,T+\{t\}} = P_{X,T}$. Recall that for computing the frequentness probability of itemsets, we need to know only the probability of supports between zero and $minSup - 1$. This is the reason why in our algorithm we fill the array only for the values that are lower than $minSup$. Example 7 illustrates our algorithm.

Example 7. Figure 5 shows the execution of our algorithm over the database shown in Figure 2, with X=*eating*. Recall that, in this database, we have: $P(X \subseteq t_1) = 0.3$ and $P(X \subseteq t_2) = 0.4$. Initially $T = \{\}$, then we add t_1 and afterwards t_2 to it. In the fist row, the algorithm sets the probabilistic supports for $T = \{t_1\}$. Thus, we have $P_{X,T+\{t_1\}}[1] = P(X \subseteq t_1) = 0.3$ and $P_{X,T+\{t_1\}}[0] = (1 - P(X \subseteq t_1)) = 1 - 0.3 = 0.7$. The probabilities in the second row are computed using our recursive formula. For example, $P_{X,\{t_1,t_2\}}[1] = (P_{X,\{t_1\}}[0] \times P(X \subseteq t_2)) + (P_{X,\{t_1\}}[1] \times (1 - P(X \subseteq t_2))) = (0.7 \times 0.4) + (0.3 \times 0.6) = 0.46$.

T			
$\{t_1, t_2\}$	0.42	0.46	0.12
$\{t_1\}$	0.7	0.3	
	0	1	2 possible supports

Fig. 5. Computing the probabilistic support of *eating* in the probabilistic database of Figure 2

The time complexity of our algorithm for updating the probabilistic support of an itemset X after inserting a new transaction to the sliding window is $O(minsup)$. Its space complexity is $O(|SW|)$ where $|SW|$ is the size of the sliding window, i.e. the maximum number of transactions in the window.

3.5   Updating Probabilistic Support after Deleting a Transaction

Assume we have the probabilistic support of an itemset X for a set of transactions T, then the question is: "how to compute the probabilistic support in $T - \{t\}$?" One might think that the probabilistic support i for X in $T - \{t\}$ (i.e. $P_{X,T-\{t\}}(i)$) could be computed as $P_{X,T}(i-1)/P(X \subseteq t)+P_{X,T}(i)/(1-P(X \subseteq t))$. Unfortunately, this formula will not work. For example, if we use it for computing $P_{eating,\{t_1\}}(1)$ after deleting transaction t_2 from the database used in Example 7, then we obtain $0.42 \times 0.4 + 0.46 \times 0.6 = 0.444$, whereas the value of $P_{eating,\{t_1\}}(1)$ is equal to 0.3 (see Figure 5). To solve the problem of updating the probabilistic support of X in $T - \{t\}$, we develop the following theorem:

Theorem 2. *Let X be an itemset, T a set of transactions, and $P_{X,T}$ an array denoting the probabilistic support of X in T. Assume we delete a transaction t from T. Let $P_{X,T-\{t\}}(i)$ be the probability for X to have support i in $T - \{t\}$, then $P_{X,T-\{t\}}(i)$ can be computed as:*

$$- \frac{P_{X,T}(i)-(P_{X,T-\{t\}}(i-1)\times P(X \subseteq t))}{1-P(X \subseteq t)} \text{ if } P(X \subseteq t) \neq 1$$
$$- P_{X,T}(i+1) \text{ otherwise}$$

Proof. In the case where $P(X \subseteq t) = 1$, it is obvious that by removing t from T, the support of X is reduced by one. Thus, the probability of support i in

$T - \{t\}$ is equal to the probability of support $i + 1$ in T. For the case where $P(X \subseteq t) \neq 1$, it is sufficient to show that:

$P_{X,T-\{t\}}(i) \times (1 - P(X \subseteq t)) = P_{X,T}(i) - (P_{X,T-\{t\}}(i-1) \times P(X \subseteq t)).$

For this, we expand the right side of this equation by using Equation 2 in Section 3.3. We replace $P_{X,T}(i)$ by its equivalent, that is:

$P_{X,T-\{t\}}(i-1) \times (P(X \subseteq t)) + P_{X,T-\{t\}}(i) \times (1 - P(X \subseteq t))$

Thus, we have:

$P_{X,T}(i) - (P_{X,T-\{t\}}(i-1) \times P(X \subseteq t))$
$= P_{X,T-\{t\}}(i-1) \times (P(X \subseteq t)) + P_{X,T-\{t\}}(i) \times (1 - P(X \subseteq t)) - (P_{X,T-\{t\}}(i-1) \times P(X \subseteq t))$
$= P_{X,T-\{t\}}(i) \times (1 - P(X \subseteq t))$ □

Theorem 2 suggests to compute $P_{X,T-\{t\}}(i)$ based on $P_{X,T}(i)$ and $P_{X,T-\{t\}}(i-1)$. To develop an algorithm based on this theorem, we need to compute $P_{X,T-\{t\}}(0)$ that is the probability of support 0 for X in $T - \{t\}$. This can be done as follows. We use the fact that when a transaction t is added to the sliding window, the probability of support 0 is multiplied by the probability of absence of t. Thus, when t is removed from T, to compute $P_{X,T-\{t\}}(0)$ we can divide $P_{X,T}(0)$ by $(1 - P(X \subseteq t))$, if $P(X \subseteq t) \neq 1$. In other words, we have:

$$P_{X,T-\{t\}}(0) = \frac{P_{X,T}(0)}{1 - P(X \subseteq t)}, for P(X \subseteq t) \neq 1 \qquad (3)$$

Equation 3 works iff $P(X \subseteq t) \neq 1$. In the case where $P(X \subseteq t) = 1$, we have $P_{X,T-\{t\}}(0) = P_{X,T}(1)$.

Based on Theorem 2 and Equation 3, we developed Algorithm 1 that updates the probabilistic support after removing a transaction from a sliding window. Recall that for finding frequent itemsets, we need only to compute the probabilistic supports for values that are lower than $minSup$. This is why the "for loop" in the algorithm (started at Line 10) is from 1 to $min\{minSup - 1, |T| - 1\}$. The time complexity of Algorithm 1 is $O(minsup)$, and its space complexity $O(|SW|)$ where $|SW|$ is the size of the sliding window.

4   Experiments

We evaluate the performance of FEMP by a thorough comparison to existing algorithms in the literature that use Probabilistic Support in exact [3] and approximate [14] mining. Since we do not find sliding window approaches in the literature, we implemented these algorithms as follows: each time an event is added or removed from the sliding window, the algorithm runs, from scratch, on the content of the updated sliding window. **PFIM** is the algorithm of [3] implemented with all the optimizations (including the 0-1 optimization). However, due to extremely high response time in batch mode, we implemented two other versions of this algorithm. In **PFIM-50%** the discovery is not performed for each event but for each two events (only 50% of the events are considered). In

Algorithm 1. Probabilistic support update after deleting a transaction.

Input: X: itemset; t: deleted transaction; T: set of transactions before delete; $P_{X,T}$: an array containing probabilistic support of X in T

Output: $P_{X,T-\{t\}}$: an array containing probabilistic supports for X in $T - \{t\}$

1: **if** $P(X \subseteq t) = 0$ **then**
2:     $P_{X,T-\{t\}} = P_{X,T}$
3: **else**
4:     **if** $P(X \subseteq t) < 1$ **then**
5:         $P_{X,T-\{t\}}[0] = \frac{P_{X,T}[0]}{1 - P(X \subseteq t)}$
6:     **else**
7:         $P_{X,T-\{t\}}[0] = P_{X,T}[1]$
8:     **end if**
9:     $k = min\{minSup - 1, |T| - 1\}$
10:    **for** $i = 1..k$ **do**
11:        **if** $P(X \subseteq t) = 1$ **then**
12:            $P_{X,T-\{t\}}[i] = P_{X,T}[i + 1]$
13:        **else**
14:            $P_{X,T-\{t\}}[i] = \frac{P_{X,T}[i] - (P_{X,T-\{t\}}[i-1] \times P(X \subseteq t))}{1 - P(X \subseteq t)}$
15:        **end if**
16:    **end for**
17: **end if**
18: **return** $P_{X,T-\{t\}}$

PFIM-25%, the discovery on the sliding window is performed each 4 events. Eventually, **Poisson** is the algorithm of [14] (that allows approximate PFI mining) running on the whole sliding window after each update. A brief discussion on these algorithms is given in Section 5.

We use two datasets for these experiments: a synthetic one (by the IBM[1] generator) and a real one (the "accident" dataset from the FMI repository[2]). The synthetic dataset contains 38 millions of events, 8 millions of transactions and 100 items. The accidents dataset contains 11 millions of events, 340K transactions and 468 items. We have added an existential probability $P \in]0..1]$ to each event in these datasets, with a uniform distribution. For both datasets, $minSup$ has been set to 30% of the window size and $minProb$ to 40%.

We implemented two versions of FEMP. The first one is "Dynamic-FEMP" (**d-FEMP** in our experiments). In this version, when a new candidate itemset is generated, it's frequentness probability will be checked over the next updates in the stream thanks to our algorithms presented in Section 3. This is the fastest approach but it implies a delay in the pattern discovery (similar to the delay described in [13]). The second version is "Exact-FEMP" (**e-FEMP** in our experiments). Here, each time a candidate itemset is generated it is immediately verified, from scratch, over all the transactions maintained in the current sliding window.Besides that, the probabilistic support of all existing itemsets is

---

[1] http://www.cs.loyola.edu/~cgiannel/assoc_gen.html
[2] http://fimi.ua.ac.be/data/

maintained at each update using our algorithms presented in Section 3. e-FEMP guarantees an exact PFI discovery at any point in the stream. However, this is done at the price of a higher time complexity compared to d-FEMP.

We present the results obtained by these approaches on a probabilistic data streams with sliding windows. Our goal is, on the one hand, to show that a sliding window approach in probabilistic data streams with probabilistic support is possible thanks to our algorithms and, on the other hand, to illustrate the behavior of our approach in a context where transactions can be added or removed from the observation window.

4.1   Feasibility

Figure 6 shows the time needed by each algorithm to extract the PFI in a growing sliding window SW. The size of SW grows from 0 to 5000 transactions for the synthetic dataset and from 0 to 10000 for the accident dataset. This corresponds to the initialization of the stream. We observe that the response time of d-FEMP increases barely since it needs very few calculations. e-FEMP increases more clearly, since it must scan SW each time a new candidate is proposed. Meanwhile, all the versions of PFIM and Poisson have much higher response times. d-FEMP needs 7.34s to fill SW for the accident dataset, where PFIM needs 618s. Furthermore, we can see that Poisson is faster than all versions of PFIM after a number of transactions, but not for the first ones. This is due to the large number of infrequent patterns extracted by Poisson, caused by the approximation of Expected Support. Actually, for the first hundreds of transactions, Poisson may extract up to 146 PFI while the real number of PFI is 36 at most. Such a large number of erroneous PFI is a cause of unnecessary computations and high response times.

Figure 7 shows the time needed by each algorithm to process 100 events, while the transaction data is fed in a pass-through fashion. Although probabilistic supports are maintained after each update in the cases of d-FEMP, e-FEMP,

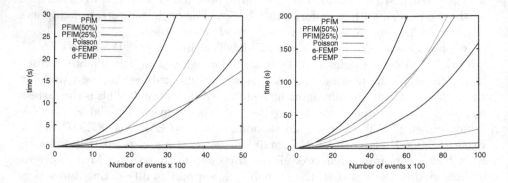

Fig. 6. Initialization time (filling SW) on synthetic (top) and accident (bottom) datasets

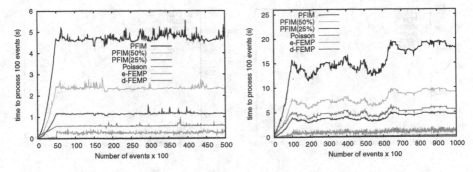

Fig. 7. Processing times for 100 events on synthetic (top) and accident (bottom) datasets

Poisson and PFIM, we report the time for 100 events because the response time of d-FEMP, for only one event, would always be 0s. That time is recorded as the number of processed events increases, from the 100th event to the 50000th one in the case of synthetic dataset (100000th for accident dataset). We observe that d-FEMP needs less than 0.05s to update the supports of the monitored itemsets in memory for each 100 updates to the stream. e-FEMP needs more time (up to 1s) since it has to scan SW when new candidate itemsets are generated. Depending on the dataset, Poisson is faster or slower that PFIM-25%. This is due to the difference in density between these datasets, where Poisson can extract itemsets that are not frequent (slowing down the extraction process). Over the synthetic dataset, the time needed by e-FEMP is 5 times faster than Poisson (while extracting exact probabilistic support, whereas Poisson gives an approximation with Expected Support) and up to 20 times faster than PFIM. We also observe that d-FEMP is very close to 0s. In fact, in our experimental data, d-FEMP appears to run up to two orders of magnitude faster than PFIM on the accident dataset to process 100 events. The global response time of d-FEMP, as the stream passes through, is several orders of magnitude lower than that of PFIM.

4.2 Scalability

Figure 8 shows the execution times of each algorithm for a full sliding window. More precisely, when a sliding window SW is full (after initialization), we measure the time needed to process $|SW|$ updates (one update is made of an event insertion and an event removal). This time is measured for an increasing size of SW. Our experiments clearly show that d-FEMP incurs very few overhead to the computations needed for maintaining the data structures.

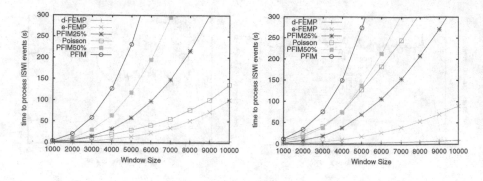

Fig. 8. Processing time for $|SW|$ updates with increasing size of SW on synthetic (top) and accident (bottom) datasets

5   Related Work

Expected Support. Probabilistic data mining is a recent research topic that is gaining increasing attention [9], [4], [12], [10], [15]. In [5], the problem of itemset mining from probabilistic data is introduced and the authors propose the notion of *Expected Support* as a first solution. Let $P(X \subseteq t)$ be the probability that itemset X is included in transaction t, the Expected Support $ES(X)$ of X in database D is given by: $ES(X) = \sum_{j=1}^{|D|} P(X, t_j)$. This support is then used as a frequency measure (compared to a user minimum threshold) in U-Apriori, a level-wise approach based on the Apriori principle for frequent itemset mining.

Probabilistic Support. In [3], the authors introduce the notion of *probabilistic support* which is an exact measure of an itemset support in the possible world model. They define the probability that an itemset X has support i as in Definition 4. The authors propose to compute the frequentness probability of an itemset X using a dynamic programming approach. However, their approach is incremental in the support (*i.e.* the transaction set is fixed and each iteration of their recursion allows computing the support probability of an itemset for an increasing support), thus not appropriate for a data stream environment.

Some approximation methods for the probabilistic support of an itemset have also been proposed. The idea of [14] is to approximate the support distribution function by means of a Poisson law. In [4], the authors propose another approximation of frequentness probability based on the central limit theorem. The main drawbacks of these approaches are to use Expected Support as a measure of probabilistic frequentness [14] and to work only on statistical independent data [14], [4].

Probabilistic Data Streams. Itemset mining in data streams is an important topic of knowledge discovery [13], [7]. Mainly, we find contributions on the extraction techniques and the data models, such as batches [7] or sliding windows [13], [16]. In [8], we find a comparative study of these models. In [17], the authors propose to extract frequent items in probabilistic data. Their approaches allow finding items (itemsets of only one item) in static data and likely frequent items in data streams. [11] proposes to extract frequent itemset from streaming probabilistic data by means of Expected Support and a batch model. In [10], we find a batch-based approach to extract frequent itemsets using Expected Support in probabilistic data streams with a technique inspired from [7].

Despite the interest of exact PFI mining with sliding windows [8], [16], we do not find any proposal in the literature for such an approach. As we discuss in Section 3, the main challenge in this context is to update the probabilistic support of an itemset when a transaction is added to or removed from the window. Our work is therefore motivated by the needs and challenges of providing an approach that is able to i) extract PFI from data streams; ii) use sliding windows and update the support of an itemset upon transaction insertion or removal; and iii) work with statistical dependent and independent data.

6 Conclusion

In this paper, we proposed FEMP, the first solution for exact PFI mining in data streams with sliding windows. FEMP allows efficient computation of the exact probabilistic support of itemsets whenever a transaction is added or removed from the observation window. Compared to non-incremental algorithms, that need to scan the whole sliding window after each update, our approach shows very low execution time. Through an extensive experimental evaluation on synthetic and real datasets, we observed that FEMP can be up to several orders of magnitude faster than traditional approaches adapted to sliding windows.

References

1. Agrawal, R., Imieliński, T., Swami, A.: Mining association rules between sets of items in large databases. SIGMOD Rec. 22, 207–216 (1993)
2. Akbarinia, R., Valduriez, P., Verger, G.: Efficient Evaluation of SUM Queries Over Probabilistic Data. IEEE Transactions on Knowledge and Data Engineering (2012)
3. Bernecker, T., Kriegel, H.-P., Renz, M., Verhein, F., Zuefle, A.: Probabilistic frequent itemset mining in uncertain databases. In: Proceedings of the 15th ACM SIGKDD International Conference on Knowledge Discovery and Data Mining, KDD 2009, pp. 119–128. ACM, New York (2009)
4. Calders, T., Garboni, C., Goethals, B.: Approximation of frequentness probability of itemsets in uncertain data. In: Proceedings of the 2010 IEEE International Conference on Data Mining, ICDM 2010, pp. 749–754. IEEE, Washington, DC (2010)
5. Chui, C.-K., Kao, B., Hung, E.: Mining frequent itemsets from uncertain data. In: Zhou, Z.-H., Li, H., Yang, Q. (eds.) PAKDD 2007. LNCS (LNAI), vol. 4426, pp. 47–58. Springer, Heidelberg (2007)
6. Dalvi, N., Suciu, D.: Efficient query evaluation on probabilistic databases. The VLDB Journal 16, 523–544 (2007)

7. Giannella, C., Han, J., Pei, J., Yan, X., Yu, P.: Mining Frequent Patterns in Data Streams at Multiple Time Granularities. In: Kargupta, H., Joshi, A., Sivakumar, K., Yesha, Y. (eds.) Next Generation Data Mining. AAAI/MIT (2003)

8. Kranen, P., Seidl, T.: Harnessing the strengths of anytime algorithms for constant data streams. Data Min. Knowl. Discov. 19, 245–260 (2009)

9. Leung, C.K.-S., Brajczuk, D.A.: Efficient algorithms for the mining of constrained frequent patterns from uncertain data. SIGKDD Explor. Newsl. 11, 123–130 (2010)

10. Leung, C.K.-S., Jiang, F.: Frequent itemset mining of uncertain data streams using the damped window model. In: Proceedings of the 2011 ACM Symposium on Applied Computing, SAC 2011, pp. 950–955. ACM, New York (2011)

11. Leung, C.-S., Hao, B.: Mining of frequent itemsets from streams of uncertain data. In: Proceedings of IEEE 25th International Conference on Data Engineering (ICDE), pp. 1663–1670 (2009)

12. Sun, L., Cheng, R., Cheung, D.W., Cheng, J.: Mining uncertain data with probabilistic guarantees. In: Proceedings of the 16th ACM SIGKDD International Conference on Knowledge Discovery and Data Mining, KDD 2010, pp. 273–282. ACM, New York (2010)

13. Teng, W.-G., Chen, M.-S., Yu, P.S.: A Regression-Based Temporal Pattern Mining Scheme for Data Streams. In: VLDB, pp. 93–104 (2003)

14. Wang, L., Cheng, R., Lee, S.D., Cheung, D.: Accelerating probabilistic frequent itemset mining: a model-based approach. In: Proceedings of the 19th ACM International Conference on Information and Knowledge Management, CIKM 2010, pp. 429–438. ACM, New York (2010)

15. Liu, Y.-H.: Mining frequent patterns from univariate uncertain data. Data and Knowledge Engineering 71(1), 47–68 (2012)

16. Zhang, C., Masseglia, F., Lechevallier, Y.: ABS: The anti bouncing model for usage data streams. In: Proceedings of the 2010 IEEE International Conference on Data Mining, ICDM 2010, pp. 1169–1174. IEEE Computer Society, Washington, DC (2010)

17. Zhang, Q., Li, F., Yi, K.: Finding frequent items in probabilistic data. In: Proceedings of the 2008 ACM SIGMOD International Conference on Management of Data, SIGMOD 2008, pp. 819–832. ACM, New York (2008)

Detecting Bicliques in GF[q]

Jan Ramon[1], Pauli Miettinen[2], and Jilles Vreeken[3]

[1] Department of Computer Science, KU Leuven, Belgium
Jan.Ramon@cs.kuleuven.be
[2] Max-Planck Institute for Informatics, Saarbrücken, Germany
pmiettin@mpi-inf.mpg.de
[3] Dept. of Mathematics and Computer Science, University of Antwerp, Belgium
Jilles.Vreeken@ua.ac.be

Abstract. We consider the problem of finding planted bicliques in random matrices over $GF[q]$. That is, our input matrix is a $GF[q]$-sum of an unknown biclique (rank-1 matrix) and a random matrix. We study different models for the random graphs and characterize the conditions when the planted biclique can be recovered. We also empirically show that a simple heuristic can reliably recover the planted bicliques when our theory predicts that they are recoverable.

Existing methods can detect bicliques of $O(\sqrt{N})$, while it is NP-hard to find the largest such clique. Real graphs, however, are typically extremely sparse and seldom contain such large bicliques. Further, the noise can destroy parts of the planted biclique. We investigate the practical problem of how small a biclique can be and how much noise there can be such that we can still approximately correctly identify the biclique. Our derivations show that with high probability planted bicliques of size logarithmic in the network size can be detected in data following the Erdős-Rényi model and two bipartite variants of the Barabási-Albert model.

1 Introduction

In this paper we study under what conditions we can recover a planted biclique from a graph that has been distorted with a noise. We consider the general setting of matrices under $GF[q]$, where the problem can be restated as finding the planted rank-1 matrix after noise has been applied. In addition to standard additive noise, we also allow destructive noise, that is, the noise can remove edges from the planted biclique. Therefore, we consider the planted biclique *recoverable* if it is the best rank-1 approximation of the noised matrix under $GF[q]$.

As tabular data essentially forms a bipartite graph, bicliques are meaningful for a wide variety of real data. Identifying bicliques, such as through factorization and bi-clustering, is an important topic in many fields, including machine learning, data mining, and social network analysis—each of these subfields naming bicliques differently, such as 'tiles', 'clusters', or 'communities'.

One of the main current challenges is the discovery of overlapping bicliques under noise. In particular, there is need for techniques that can model interactions where bicliques overlap. For example, say in our data we have records of

H. Blockeel et al. (Eds.): ECML PKDD 2013, Part I, LNAI 8188, pp. 509–524, 2013.

male conservatives, as well as of long-haired males, but none of long-haired male conservatives. Under $GF[2]$, where every subsequent factor can be seen to XOR the corresponding entries of a binary matrix, we only need two factors—one for conservatives, one for liberals, which corresponds to intuition. Methods unable to model interaction will need 3 factors, or have high errors. In bio-informatics, there are many examples of such complex interactions, such as inhibition and excitation in gene regulation as well as in protein-protein interaction [14,13]. By factorizing matrices in $GF[q]$ we can model arbitrary levels of interaction.

An important step towards factorizing data under $GF[q]$—i.e. discovering *all* important cliques in the data—is the reliable detection of *individual* planted bicliques. To this end, in this paper we study bounds on the dimensions of planted bicliques such that we can still reliably approximately identify these in quasi-polynomial time under different noise models. As many adversarial attacks exist that render exact solutions exponential, we focus on approximations—also, in practice, data analysts often do not require optimal results, but rather obtain good approximations in much less time.

Existing approaches aim at finding *complete* bicliques of size $O(\sqrt{N})$ [2], and it has been shown to be NP-hard to find the largest such biclique [6,11]. While most real-world graphs have very large number of vertices, they are, however, typically only very sparsely connected. Graphs that follow the popular Barabási-Albert model, for instance, only have a constant number of vertices with degrees of $O(\sqrt{N})$. Hence, finding a clique of size $O(\sqrt{N})$ can be trivially achieved by collecting those vertices with degree at least $O(\sqrt{N})$. As such, it is an interesting open question what the smallest size of a biclique and the largest amount of *destructive* noise are such that the biclique can still be approximately correctly discovered. In particular, we study Erdős-Rényi and Barabási-Albert background distributions for bipartite graphs.

Due to lack of space, we discuss the most fine-grained details of the proof of Lemma 6, i.e. Lemma 9, and Eq. 14, in the Appendix.[1]

2   Related Work

Finding large bicliques has many applications, and hence has received a lot of attention. Most research aims at finding *exact* bicliques, that is, complete bipartite subgraphs. One way of finding these is to remove edges from the graph until what is left is a complete bipartite subgraph. Hochbaum [6] showed that minimizing the number of edges to remove is NP-hard, though she also gave a 2-approximation algorithm for the problem. Later, Peeters [11] showed that finding the largest biclique is NP-hard in general.

Despite that finding the largest biclique is NP-hard, it is possible to recover a single planted biclique [2]. In particular, if the bipartite graph contains a biclique of $N+M$ nodes and an adversary adds up to $O(NM)$ edges, the planted biclique can still be recovered using nuclear norm minimization, provided that the added

---

[1] http://www.mpi-inf.mpg.de/~pmiettin/gf2bmf/appendix.pdf

edges do not have too many neighbors. Similarly, the random process can be characterized to add edges such that the biclique can still be found.

Our problem, however, is different as we do not aim to recover exact bicliques, but approximate *quasi-bicliques* (i.e. dense but not necessarily complete bipartite subgraphs). Compared to Ames et al. [2], we allow the noise to both *add* new edges as well as to *remove* edges from the planted bicliques.

An alternative approach is to consider the problem as rank-1 matrix factorization. If we work in $GF[2]$, any method discovering binary factor matrices works, including Boolean matrix factorization algorithms [10], binary matrix factorization algorithms [15], and PROXIMUS [7].

The problem of finding dense quasi-bicliques has been approached from different directions. In graph mining, a typical goal is to find all maximal quasi-bicliques satisfying a density condition. For example, Sim et al. [12] give an algorithm to mine all maximal quasi-bicliques where each vertex is connected to all but $\varepsilon \in \mathbb{N}$ vertices in the other side (for other algorithms, see [8]). Such algorithms can be used to find the quasi-biclique that best represents the data (in terms of error), but only by exhaustively iterating over density values.

There is existing research on finding rank-k approximations of given matrices under $GF[2]$. In fact, finding the rank of a matrix under $GF[2]$ is easy. This can be seen by noting that the problem is equivalent to the rank of the biadjacency matrix of the bipartite graph under the $GF[2]$, and therefore solvable in polynomial time using standard techniques. Finding the best $GF[2]$ rank-k factorization, however, is not so easy. In the Nearest Codeword Problem, we are given an N-by-M binary data matrix A and a binary N-by-k left factor matrix B, with the task to find the right factor matrix C such that we minimize $|A - B \oplus C|$. This problem is NP-hard to approximate within any constant factor, and there exists no polynomial-time algorithm for approximation within a factor of $2^{\log^{0.8-\varepsilon} N}$, unless $NP \subseteq DTIME(n^{poly(\log N)})$ [3]. There does exist, however, a polynomial-time randomized approximation algorithm with $O(k/\log N)$ approximation factor [4], and a deterministic approximation algorithm with the same factor and $N^{O(\log^* N)}$ running time [1].

3   Identifying Single Bicliques

We investigate bounds on discovering a single planted biclique under a given background distribution. As models for background noise we study resp. Erdős-Rényi graphs and the scale-free Barabási-Albert model.

3.1   A Generic Strategy

In the next sections, we will consider random graph models. In each of these cases, we assume that a 'planted biclique' is combined with 'noise' generated by the random graph model, and will consider the question of how easy it is to recognize the planted biclique. In the current section we present aspects common to the derivations for these random graph models.

In this section, we will denote the dimensions of the matrices by $N \times M$. We will use $\oplus$ and $\ominus$ to denote addition and subtraction in $GF[q]$ for vectors or matrices over $GF[q]$. Further, $x \equiv y$ denotes congruency in $GF[q]$, i.e. $x \equiv y$ (mod q). We also adopt the common notation $[n]$ for the set of all integers from 1 until n, i.e. $[n] = \{i \in \mathbb{N} \mid 1 \leq i \leq n\}$. We will use the indicator function $I(true) = 1$ and $I(false) = 0$.

We will often use vectors (Boolean or over $GF[q]$) to select a set of rows or columns, non-zero elements indicating a selected row or column, and similarly matrices to select a set of cells. Therefore, we define for two vectors a and b of the same dimensions that $a \setminus b$ is the vector for which $(a \setminus b)_i = I(a_i \neq 0 \wedge b_i = 0)$, and similarly $a \cap b$ and $a \cup b$ as the binary vectors for which $(a \cap b)_i = I(a_i \neq 0 \wedge b_i \neq 0)$ and $(a \cup b)_i = I(a_i \neq 0 \vee b_i \neq 0)$. We define the same operations for matrices, e.g. $(A \setminus B)_{i,j} = I(A_{i,j} \neq 0 \wedge B_{i,j} = 0)$. We will denoted with $|X|$ the number of non-zero elements of a vector or matrix X. We will denote the planted clique with uv where $u \in GF[q]^{N \times 1}$ and $v \in GF[q]^{1 \times M}$. We assume u and v fixed but unknown. We will denote approximations of u and v with x and y and express the quality of the approximation using a loss function

$$L(u, v, x, y) = \max(|u - x|, |v - y|) .\tag{1}$$

For sparse graphs, adding a planted clique to the graph usually increases the number of nonzero elements. We therefore adopt the following notations. Let A be a random graph according to some distribution $\mathcal{M}$. Let $B = A \oplus uv$ be the addition of the planted clique defined by u and v to this matrix. Let $x \in GF[q]^{m \times 1}$ and $y \in GF[q]^{1 \times n}$ be two vectors defining a biclique xy. We define the error of x and y wrt. identifying the biclique planted in B and characterized by u and v as

$$W'(x, y) = |B \ominus xy| = |\{(i, j) \mid B_{i,j} \not\equiv x_i y_j\}| = |\{(i, j) \mid A_{i,j} \oplus u_j v_j \not\equiv x_i y_j\}| ,$$

that is, $W'(x, y)$ counts the matrix cells which are nonzero after removing the hypothesized biclique xy (which we expect to be minimal if $xy = uv$). Furthermore, $W(x, y) = W'(x, y) - W'(u, v)$ characterizes whether xy yields better representation of B than uv ($W(x, y) < 0$) or vice versa ($W(x, y) > 0$). Clearly, $W(u, v) = 0$.

The set of elements where the approximated biclique xy differ from the planted biclique uv is denoted by $C_{\neq}(x, y)$, i.e.

$$C_{\neq}(x, y) = \{(i, j) \in [N] \times [M] \mid x_i y_j \neq u_i v_j\} .$$

If B is the matrix received on input, i.e. the matrix resulting from adding to a random graph a planted biclique, then we will denote with $\hat{u}$ and $\hat{v}$ the vectors minimizing $W'(\hat{u}, \hat{v}) = |B \ominus \hat{u}\hat{v}|$.

For each random graph model, our aim is to show that with reasonably high probability the planted biclique uv is well approximated by the biclique $\hat{u}\hat{v}$ minimizing $W'(\hat{u}, \hat{v})$. We will first show that the probability that $W(x, y) < 0$, with $xy \neq uv$, decreases exponentially with $|C_{\neq}(x, y)|$. Then, by the following lemma from such result we can derive that maximizing the objective function on the input will yield a good approximation of (u, v).

Lemma 1. *Let $\mathcal{M}$ be a distribution over $GF[q]^{N \times M}$, i.e. $N \times M$ matrices over $GF[q]$. Assume that there is an integer ζ and a constant c such that for any fixed $u \in GF[q]^{N \times 1}$ and $v \in GF[q]^{1 \times M}$ with $|u| \geq \zeta$ and $|v| \geq \zeta$, with probability at least $1 - \delta_1$ for a matrix A randomly drawn from $\mathcal{M}$, it holds for all $x \in GF[q]^{N \times 1}$ and $y \in GF[q]^{1 \times M}$ that*

$$P(W(x, y) \leq 0) \leq \exp(-|C_{\neq}(x, y)|c) . \tag{2}$$

Then, for all $\epsilon > 0$, u and v such that $|u| \geq \zeta$ and $|v| \geq \zeta$,

$$P_{A \sim \mathcal{M}}(L(u, v, \hat{u}, \hat{v}) \leq \epsilon) \geq 1 - \delta_1 - \delta_2$$

where $(\hat{u}, \hat{v}) = \arg\min_{(x,y)} |A \oplus uv \ominus xy|$ and

$$\delta_2 = T(\epsilon, |u|, |v|, |u|, |v|)T(\epsilon, N, M, |u|, |v|) \tag{3}$$

where

$$T(\epsilon, a, b, c, d) = \frac{\exp\left(\epsilon\left(\log\left(a + 1\right) + \log\left(b + 1\right) - \min\left(c, d\right)\right) c_{p,q}\right)}{1 - \exp\left(\left(\log(a + 1) + \log(b + 1) - \min(c, d)\right)c_{p,q}\right)} .$$

Proof. Equation (2) is a bound on the probability that for a given x and y, $W(x, y) < 0$. Several choices for x and y are possible. We will bound the probability that $\exists x, y : W(x, y) < 0$ by

$$P(\exists x, y : W(x, y) < 0) \leq \sum_{x,y} P(W(x, y) < 0)$$

For a given x and y we now bound $|C_{\neq}(x, y)|$. First, we define

$$C_{uv \setminus xy} = \{(i, j) \mid (uv \setminus xy)_{i,j} = 1\} \tag{4}$$

and

$$C_{xy \setminus uv} = \{(i, j) \mid (xy \setminus uv)_{i,j} = 1\} , \tag{5}$$

such that $|C_{\neq}(x, y)| = |C_{uv \setminus xy}| + |C_{xy \setminus uv}|$.

As we have both $|C_{uv \setminus xy}| \geq |u \setminus x| |v|$ and $|C_{uv \setminus xy}| \geq |v \setminus y| |u|$ it follows that

$$|C_{uv \setminus xy}(x, y)| \geq \max(|v \setminus y|, |u \setminus x|) \min(|u|, |v|) . \tag{6}$$

There are $\sum_{i=1}^{t} \binom{|u|}{i} \leq (|u| + 1)^t$ ways for choosing at most t rows out of the $|u|$ nonzero rows of u. Similarly, we have $\sum_{i=1}^{t} \binom{|v|}{i} \leq (|v| + 1)^t$ ways to choose at most t columns out of the $|v|$ nonzero columns of v. Hence, the number of ways to choose $u \setminus x$ and $v \setminus y$ such that both $|u \setminus x| \leq t$ and $|v \setminus y| \leq t$ hold is bounded by $(|u| + 1)^t(|v| + 1)^t$. Now, let us use $C_{uv \setminus}^{(t)} = \{(x, y) \mid \max(|u \setminus x|, |v \setminus y|) = t\}$

for the set of (x, y)'s such that for each the largest intersection with the rows or columns of (u, v) is t elements. We can now write

$$\sum_{t=s}^{\max(|u|,|v|)} \sum_{(x,y) \in C_{uv\backslash}^{(s)}} \exp\left(-\left|C_{uv\backslash xy}\right| c\right)$$

$$\leq \sum_{t=s}^{\max(|u|,|v|)} (|u| + 1)^t (|v| + 1)^t \exp\left(-t\min(|u|, |v|)c\right)$$

$$\leq \sum_{t=s}^{\max(|u|,|v|)} \exp\left(t(\log(|u| + 1) + \log(|v| + 1) - \min(|u|, |v|)c)\right)$$

$$= \frac{\exp\left(s(\log(|u| + 1) + \log(|v| + 1) - \min(|u|, |v|)c)\right)}{1 - \exp\left((\log(|u| + 1) + \log(|v| + 1) - \min(|u|, |v|)c)\right)}$$

$$= T(s, |u|, |v|, |u|, |v|)$$

Here, in the one-but last step we use the fact that $\sum_{i=0}^{\infty} x^i = 1/(1-x)$. Similarly, we define $C_{.\backslash uv}^{(t)} = \{(x, y) \mid \max(|x \backslash u|, |y \backslash v|) = t\}$ by which we have

$$\sum_{t=s}^{\max(N,M)} \sum_{(x,y) \in C_{.\backslash uv}^{(s)}} \exp\left(-\left|C_{xy\backslash uv}\right| c\right) \leq T(s, N, M, |x|, |y|)$$

Note that as u and v are fixed, we have only sets of size $|u|$ and $|v|$ to choose $u \backslash x$ and $v \backslash y$ from. Here however, x and y can be chosen from $N - |u|$ remaining rows and $M - |v|$ remaining columns, resp. Still, however, when $\log(N + 1) + \log(M + 1) < \min(|u|, |v|)$, $T(s, N, M, |u|, |v|) \leq 1$.

Finally, this allows us to combine these two inequalities into

$$P(W(x, y) < 0 \mid \max(|u - x|, |v - y|) \geq \epsilon) \leq T(\epsilon, |u|, |v|, |u|, |v|) T(\epsilon, N, M, |u|, |v|) .$$

This proves the lemma.   □

According to the above, if we know the dimensions of a biclique, we have a clear bound on its detectability. Further, it follows that when $|u| \ll |v|$ or $|x| \ll |y|$ the problem becomes much harder. This follows intuition as under an independence assumption large square blocks are much less probable than thin bicliques—as these could just as well be the result of few very high degree nodes.

3.2   Erdős-Rényi

The Erdős-Rényi (ER) model is one of the most well-studied models for graph generation. The general idea is that every edge is equally probable, regardless of other edges in the graph. That is, graphs of the same number of nodes and same total number of edges are all equally likely. For the case of factorizing data under $GF[q]$ with noise distributed according to ER, we have the following definition.

Definition 1. *With $\mathcal{M}^{ER}(p, q, N \times M)$ we will denote the model of sparse random matrices in $GF[q]^{N \times M}$ according to the Erdős-Rényi model, in particular if $A \in \mathcal{M}^{ER}(p, q, N \times M)$, for each $(i, j) \in [N] \times [M]$, A_{ij} is zero with probability $1 - p$ and non-zero with probability p. Non-zero elements are chosen randomly from $GF[q]$, i.e. each non-zero element of $GF[q]$ has probability $1/(q - 1)$.*

We will now show that the probability that *some* biclique yields lower error (i.e. residual) than the planted biclique uv decays exponentially with the difference between that biclique and the planted one.

Lemma 2. *Let $p < 1/2$. Let N, M and q be integers, $u, x \in GF[q]^{N \times 1}$ and $v, y \in GF[q]^{1 \times M}$. Then, there is a constant $c_{p,q}$ depending on p and q such that*

$$P_{A \sim \mathcal{M}^{ER}(p,q,N \times M)}(W(x, y) < 0) \leq \exp(c_{p,q}|C_{\neq}|) .$$

Proof. Let A be randomly drawn from $\mathcal{M}^{ER}(p, q, N \times M)$. As above, let $B = A \oplus uv$ be the matrix obtained by adding to A the biclique uv.

Let $C_{i,j} = u_i v_j \ominus x_i y_j$ be the difference between uv and xy, and let

$$D_{i,j} = I(A_{i,j} \oplus C_{i,j} \not\equiv 0) - I(A_{i,j} \not\equiv 0) ,$$

where $I(\cdot)$ is the indicator function. Now, $W(x, y) = \sum_{i,j} D_{i,j}$. If $C_{ij} \equiv 0$, then $D_{ij} = 0$, so let $C_{\neq}(x, y) = \{(i, j) \in [n] \times [m] \mid C_{ij} \not\equiv 0\}$ such that we have $W(x, y) = \sum_{(i,j) \in C_{\neq}} D_{i,j}$.

Following Section 3.1, we bound the probability that xy gives better representation of B than uv. That is, we bound $P(W(x, y) < 0)$. To that end, we define, for $z \in \{-1, 0, +1\}$, $W_z(x, y) = \{(i, j) \in C_{\neq} \mid D_{i,j} = z\}$, so we have

$$W(x, y) = |W_{+1}(x, y)| - |W_{-1}(x, y)| .$$

The three sets $W_z(x, y)$, $z \in \{-1, 0, 1\}$, partition the set $C_{\neq}(x, y)$. Let $X_{i,j}$ be a random variable defined as $X_{i,j} = I((i, j) \in W_{-1}(x, y) \cup W_0(x, y))$, so that $P(X_{i,j} = 1) = 1 - P((i, j) \in W_1(x, y))$ for all $(i, j) \in C_{\neq}(x, y)$, and $\sum_{i,j} X_{i,j} = |W_{-1}(x, y)| + |W_0(x, y)|$. We have for all $(i, j) \in C_{\neq}(x, y)$ that

$$P((i, j) \in W_{-1}(x, y)) = \frac{p}{q - 1} \quad \text{and} \quad P((i, j) \in W_0(x, y)) = \frac{p(q - 2)}{q - 1} ,$$

where on the second equation we use the fact that $C_{i,j} \not\equiv 0$. Therefore

$$P(X_{i,j} = 1) = P((i, j) \in W_{-1}(x, y)) + P((i, j) \in W_0(x, y)) = \frac{p}{q - 1} + \frac{p(q - 2)}{q - 1} = p$$

for $(i, j) \in C_{\neq}(x, y)$. Then, due to Chernoff's inequality, for any $\epsilon > 0$, we have

$$P\left(\frac{|W_{-1}(x, y)| + |W_0(x, y)|}{|C_{\neq}(x, y)|} \geq p + \epsilon\right) \leq \exp\left[-|C_{\neq}(x, y)|D_{KL}(p + \epsilon \parallel p)\right] ,$$

where

$$D_{KL}(p + \epsilon \parallel p) = (p + \epsilon) \log\left(\frac{p + \epsilon}{p}\right) + (1 - p - \epsilon) \log\left(\frac{1 - p - \epsilon}{1 - p}\right) , \quad (7)$$

In order for $W(x, y) < 0$ to be possible, we need to have $|W_{-1}(x, y)| + |W_0(x, y)| \geq |C_{\neq}(x, y)| - |W_1(x, y)|$. Hence,

$$P\left(\frac{|W_{-1}(x, y)| + |W_0(x, y)|}{|C_{\neq}(x, y)|} \geq 1 - \frac{|W_1(x, y)|}{|C_{\neq}(x, y)|}\right)$$

$$\leq \exp\left(-|C_{\neq}(x, y)|D_{KL}\left(1 - \frac{|W_1(x, y)|}{|C_{\neq}(x, y)|} \,\middle\|\, p\right)\right).$$

To have a chance to have $W(x, y) < 0$, we need at least $|W_{-1}(x, y)| + |W_0(x, y)| \geq |C_{\neq}(x, y)|/2$. Therefore, let $\epsilon = \frac{1}{2} - p$ to get

$$P\left(\frac{|W_{-1}(x, y)| + |W_0(x, y)|}{|C_{\neq}(x, y)|} \geq \frac{1}{2}\right) \leq \exp(-|C_{\neq}(x, y)|D_{KL}(1/2 \,\|\, p)), \qquad (8)$$

where

$$D_{KL}\left(\frac{1}{2} \,\middle\|\, p\right) = \frac{1}{2}\log\left(\frac{1}{2p}\right) + \frac{1}{2}\log\left(\frac{1}{2(1-p)}\right).$$

This already gives us a bound on $P(W(x, y) < 0)$:

$$P(W(x, y) < 0) \leq P(|W_1(x, y)| < |C_{\neq}(x, y)|/2) \leq \exp(-|C_{\neq}(x, y)|D_{KL}(1/2 \,\|\, p)).$$

In case $q > 2$, we can do better as we expect more (i, j)'s to land in $W_0(x, y)$ instead of $W_{-1}(x, y)$.

Suppose now $q > 2$. For a fixed value of $|W_{-1}(x, y)| + |W_0(x, y)|$, using $W_{-1}(x, y) \leq |C_{\neq}(x, y)|/2$ and Chernoff's inequality, we obtain

$$P\left(\frac{|W_{-1}(x, y)|}{|C_{\neq}(x, y)| - |W_1(x, y)|} > \frac{1}{q-1} + \left(\frac{|W_1(x, y)|}{|C_{\neq}(x, y)| - |W_1(x, y)|} - \frac{1}{q-1}\right)\right)$$

$$\leq \exp\left(-(|C_{\neq}(x, y)| - |W_1(x, y)|)D_{KL}\left(\frac{|W_1(x, y)|}{|C_{\neq}(x, y)| - |W_1(x, y)|} \,\middle\|\, \frac{1}{q-1}\right)\right)$$

$$\leq \exp\left(-\frac{|C_{\neq}(x, y)|}{2}D_{KL}\left(\frac{|W_1(x, y)|}{|C_{\neq}(x, y)| - |W_1(x, y)|} \,\middle\|\, \frac{1}{q-1}\right)\right).$$

The above equations imply that there exists some constant $c_{p,q}$ depending on p and q such that

$$P(W(x, y) < 0) \leq \exp(-|C_{\neq}(x, y)|c_{p,q}). \qquad (9)$$

This proves the lemma. □

The above lemma can be combined with Lemma 1 (substituting ζ with $\log(NM)$ and c with $c_{p,q}$) to show that one can retrieve a planted clique with high confidence and small error (according to the trade-off given by Equation 3), and in time quasipolynomial in N and M.

It should be noted that for clarity of explanation and space limitations we keep our derivation simple, but a constant factor can be gained by calculating more precise expressions for $c_{p,q}$ and performing less rough estimations in Lemma 1. Moreover, Lemma 1 does not properly take the value of q into account and doing so would yield another q-dependent factor.

Algorithm 1. Generating a bipartite Barabási-Albert graph

Require: density parameter s; seed $G^{(0)}$; $M, N \in \mathbb{N}$
Ensure: an $N \times M$ bipartite Barabási-Albert graph G sampled from $\mathcal{M}^{BA\text{-}gen}$.
1: **for** $i = 1 \ldots NM$ **do**
2:    $V_{row}^{(i)} \leftarrow V_{row}^{(i-1)}$; $V_{col}^{(i)} \leftarrow V_{col}^{(i-1)}$; $E^{(i)} \leftarrow E^{(i-1)}$
3:    **if** $|V_{row}^{(i)}|M < i$ **then**
4:        $v_{new} \leftarrow NewVertex()$
5:        $V_{row}^{(i)} \leftarrow V_{row}^{(i)} \cup \{v_{new}\}$
6:        Select a set A of s vertices from $V_{col}^{(i-1)}$ with probability proportional to their degree in $G^{(i-1)}$.
7:        $E^{(i)} \leftarrow E^{(i)} \cup (\{v_{new}\} \times A)$
8:    **if** $|V_{col}^{(i)}|N < i$ **then**
9:        $v_{new} \leftarrow NewVertex()$
10:        $V_{col}^{(i)} \leftarrow V_{col}^{(i)} \cup \{v_{new}\}$
11:        Select a set A of s vertices from $V_{row}^{(i-1)}$ with probability proportional to their degree in $G^{(i-1)}$.
12:        $E^{(i)} \leftarrow E^{(i)} \cup (A \times \{v_{new}\})$
13:    $V^{(i)} \leftarrow V_{row}^{(i)} \cup V_{col}^{(i)}$; $G^{(i)} \leftarrow (V^{(i)}, E^{(i)})$

3.3 Graphs Constructed by the Barabási-Albert Process

Next, we consider the background noise distributed according to the well-known Barabási-Albert (BA) model, of which the main intuition is also known as 'preferential attachment'. Nodes are added one at a time, and while edges are still selected independently, their probability depends on the degree of the target node. Instead of the ER model's Gaussian degree distribution, for BA we see a powerlaw—as we see for many real-world graphs [5].

For simplicity of the derivations, below we will assume that $q = 2$. If $q > 2$, a similar but more involved derivation is possible.

Definition 2 (bipartite Barabási-Albert graph). *Let $G^{(0)}$ be a small graph on a vertex set $V^{(0)} = V_{row}^{(0)} \cup V_{col}^{(0)}$ consisting of row vertices $V_{row}^{(0)}$ and column vertices $V_{col}^{(0)}$. Let N and M be integers. A bipartite Barabási-Albert $N \times M$ graph is generated from seed $G^{(0)}$ with density parameter s by following Algorithm 1. We denote the obtained probability distribution over $N \times M$ adjacency matrices with $\mathcal{M}^{BA\text{-}gen}(s, N \times M)$.*

Lemma 3. *Let $G = (V, E)$ be a bipartite Barabási-Albert $N \times M$ graph generated according to Definition 2, let V_{row} be its row vertices and V_{col} its column vertices. Let $X_{row} \subseteq V_{row}$ and $X_{col} \subseteq V_{col}$. Then,*

$$|E \cap X_{row} \times X_{col}| \leq s(|X_{col}| + |X_{row}| - (s+1)/2) .$$

Proof. The proof is straightforward from Definition 2: each vertex connects with s vertices when added, but can only connect to vertices added before. $\square$

Notice that analogue one can prove that a 'normal' (non-bipartite) Barabási-Albert graph can not contain an $(s + 2)$-clique. The probability distribution over graphs induced by the Barabási-Albert generative process is rather hard to analyse in detail, but we can provide the following non-probabilistic result:

Lemma 4. *Let A be drawn from $\mathcal{M}^{BA\text{-}gen}(s, N \times M)$. Let $B = A \oplus uv$ for some fixed u and v with $|u| > 4s$ and $|v| > 4s$. Then, $|B \cap uv| > |u||v|/2$.*

Proof. We know from Lemma 3 that in the area covered by uv in B, uv made all cells except at most $s(|u| + |v| - (s+1)/2)$ nonzero: $|B \cap uv| \geq |u||v| - s|u| - s|v| + s(s+1)/2$. If $|u| \geq 4s$ and $|v| \geq 4s$ we have $|u||v|/2 - s|u| - s|v| \geq 0$ from which $|B \cap uv| > |u||v|/2$ follows.   □

Lemma 5. *Let A be drawn from $\mathcal{M}^{BA\text{-}gen}(s, N \times M)$. Let $B = A \oplus uv$ for some fixed u and v. Then, for any x and y such that $|C_{\neq}(x,y)| > 2s|x| + 2s|y| - s(s + 1) + |u||v|$, it holds that $|B \cap xy| \leq |x||y|/2$.*

Proof. We know from Lemma 3 that in the area covered by xy, at most $s(|x| + |y| - (s+1)/2)$ cells are nonzero in $A \cap xy$. For $B \cap xy$, at most the area $(|x||y| + |u||v| - |C_{\neq}|)/2$ from overlap between xy and uv can be added. We get $|B \cap xy| \leq s|x| + s|y| - s(s+1)/2 + (|x||y| + |u||v| - |C_{\neq}|)/2$. From $2s|x| + 2s|y| - s(s+1) + |u||v| < |C_{\neq}|$ we can derive that $|B \cap xy| \leq |x||y|/2$.   □

Hence, to detect a planted biclique in Barabási-Albert data, one only needs to search for bicliques of size $4s$ and expand greedily.

3.4   Graphs with Barabási-Albert Degree Distribution

Here we consider random graphs with the same degree distribution as Barabási-Albert model but without following the generative process, of which we showed in the previous section that it prohibits the creation of large bicliques.

For simplicity, w.l.o.g. we assume $N = M$.

Definition 3. *With $\mathcal{M}^{BA\text{-}deg}(s, q, N \times M)$ we will denote the model of sparse random matrices in $GF[q]^{N \times M}$ according to the Barabási-Albert degree distribution, in particular if $A \in \mathcal{M}^{BA\text{-}deg}(s, q, N \times M)$, it is the result of the following random construction procedure:*

- *Consider the probability distribution P_{deg} over the set $\{s, s + 1, \ldots, N\}$ such that $P_{deg}(i) = i^{-3}/Z$ with $Z = \sum_{j=s}^{N-1} j^{-3}$*
- *For all $i \in [N]$, choose d_i^{row} according to distribution P_{deg}. For all $j \in [N]$, choose d_j^{col} according to P_{deg}. Repeat this step until $\sum_{i=1}^{N} d_i^{row} = \sum_{j=1}^{M} d_j^{col}$.*
- *Draw X uniformly from the set of all matrices of $GF[q]^{N \times M}$ such that for all $i \in [N]$, the number of nonzero elements of row i equals d_i^{row} and for all $j \in [M]$, the number of nonzero elements of column i equals d_j^{row}.*

In order to say something on the discernibility of cliques we need access to the connectivity within rows and columns in the form of degree lists.

Definition 4. *For an adjacency matrix $A \in GF[q]^{N \times M}$, let $f^{row(A)} \in \mathbb{Z}^N$ and $f^{col(A)} \in \mathbb{Z}^M$ such that*

$$f_i^{row(A)} = \sum_{j=1}^{N} I(A_{i,j} \neq 0) \quad , \quad and \quad f_j^{col(A)} = \sum_{i=1}^{M} I(A_{i,j} \neq 0) \, .$$

It is well-known that in Barabási-Albert graphs, the expected frequency of vertices with degree k is proportional to k^{-3}. Therefore, for sufficiently large N we can estimate the number of rows of with at least $c\sqrt{N}$ non-zero elements by

$$N \frac{\sum_{k=cN^{1/2}}^{N} k^{-3}}{\sum_{k=s}^{N}} \approx N \frac{\int_{k=cN^{1/2}}^{\infty} k^{-3} dk}{\int_{k=s}^{\infty} k^{-3} dk} = N \frac{(cN^{1/2})^{-2}/2}{s^{-2}/2} = \frac{s^2}{c^2}$$

which is a constant, not depending on N. The same holds for columns.

Lemma 6. *Let s be an integer. Let N and q be integers, $u, x \in GF[q]^{N \times 1}$ and $v, y \in GF[q]^{1 \times M}$ with $\log(N) \ll |u|$ and $\log(N) \ll |v|$. Then, there is a constant c_q^{BA} depending on q and δ_1 such that with probability at least $1 - \delta_1$*

$$P_{A \sim \mathcal{M}^{BA\text{-}deg}(s,q,N \times N)}(W(x,y) < 0) \leq \exp(c_q^{BA}|C_{\neq}|)$$

Proof. We will use notations similar to those used for the Erdős-Rényi case:

$$C_{i,j} = u_i v_j \ominus x_i y_j$$
$$D_{i,j} = I(A_{i,j} \oplus C_{i,j} \neq 0) - I(A_{i,j} \neq 0)$$
$$W'(x) = |B \bigcirc x^\top x| = |A \oplus uv \ominus xy|$$
$$W(x) = W'(x) - W'(u) = \sum_{i,j} D_{i,j}$$

We have $W(u) = 0$. Given a fixed degree list pair (f^{row}, f^{col}), we have (approximately, for sufficiently large N)

$$\mu_{i,j} = \mathbb{E}[D_{i,j}] = 1 - p_{i,j}q/(q-1) \quad \bullet$$

where $p_{i,j} = f_i^{row} f_j^{col} \left(\sum_{l=1}^{N} f_l^{row} \right)^{-1} \left(\sum_{l=1}^{N} f_l^{row} \right)^{-1}$, and

$$\sigma_{i,j}^2 = \mathbb{E}[(I(x \oplus C_{i,j} \neq 0) - I(x \neq 0) - \mu_{i,j})^2]$$
$$\leq p_{i,j}(1 - p_{i,j}) \frac{q+2}{q-1}$$
$$\leq p_{i,j} \frac{q+2}{q-1}.$$

We again define $C_{\neq}(x) = \{(i,j) \in [n] \times [m] \mid u_i v_j \neq x_i y_j\}$ and for $v \in \{-1, 0, +1\}$ we have $W_v(x) = \{(i,j) \mid D_{i,j} = v\}$. Moreover, let

$$\mu = \mathbb{E}[W(x)] = \sum_{(i,j) \in C_{\neq}(x)} \mu_{i,j} \, ,$$

and

$$\sigma^2 = \sum_{(i,j) \in C_{\neq}(x)} \sigma_{i,j}^2 .$$

From (14) we can see[2] that

$$\mathbb{E}[\mu] \leq |C_{\neq}| 2(s - 1/2)^2 / N(s - 1) .$$

By applying Chernoff's inequality, we then arrive at

$$P(W(x, y) < 0) = P(\mu - W(x, y) > \mu) \leq \exp\left(-\frac{\mu^2}{2(\sigma^2 + |C_{\neq}|)}\right) . \qquad (10)$$

Let $t = \lceil \Gamma^{-1}(2/\delta_1) \rceil - 1$ be such that $1/t! \leq \delta_1/2$. From Lemma 9 we know[5] that with probability $1 - \delta_1/2$ there is at most one i s.t. $f_i^{row} \geq s/\sqrt{N}$ and with probability $1 - \delta_1/2$ there is at most one j such that $f_j^{col} \geq s/\sqrt{N}$. Then, with probability at least $1 - \delta_1$,

$$\sum_{(i,j) \in C_{\neq}} p_{i,j} \leq \sum_{i \in u \setminus x} \sum_{j \in v} p_{i,j} + \sum_{i \in x \setminus u} \sum_{j \in y} p_{i,j} + \sum_{j \in v \setminus y} \sum_{i \in u} p_{i,j} + \sum_{j \in y \setminus v} \sum_{i \in x} p_{i,j}$$

$$\leq R(|u \setminus x|, |v|, t) + R(|x \setminus u|, |y|, t) + R(|v \setminus y|, |u|, t) + R(|y \setminus v|, |x|, t)$$

with

$$R(a, b, t) = R'(\min(a, t), \min(b, t), \max(a - t, 0), \max(b - t, 0))$$

and

$$R'(a_H, b_H, a_L, b_L) = R_1(a_H, a_L) R_1(b_H, b_L)$$

with $R_1(H, L) = H + \frac{s}{\sqrt{N}} L$. For $b \geq \eta \geq t$, $R(a, b, t) = R^*(\min(a, t), t, \max(a - t, 0), b - t)$ with

$$R^*(a_H, t, c_L, b - t) = R_1(a_H, a_L) \left(t + \frac{s}{\sqrt{N}}(b - t) \right)$$

$$\leq R_1(a_H, a_L) b \left(\frac{s}{\sqrt{N}} + (1 - \frac{s}{\sqrt{N}}) \frac{t}{b} \right)$$

$$\leq R_1(a_H, a_L) b R_C$$

with $R_C = \left(\frac{s}{\sqrt{N}} + \frac{t}{\eta} \right)$. Hence, as we assume $|u| \geq \eta$, $|v| \geq \eta$, $|x| \geq \eta$ and $|y| \geq \eta$, with probability at least $1 - \delta_1/2$ we have for every u, v, x and y

$$\sum_{(i,j) \in C_{\neq}} p_{i,j} \leq (R_1(|u \setminus x|)|v| + R_1(|v \setminus y|)|u| + R_1(|x \setminus u|)|y| + R_1(|y \setminus v|)|x|) R_C$$

---

[2] See appendix: http://www.mpi-inf.mpg.de/~pmiettin/gf2bmf/appendix.pdf

As
$$|C_{\neq}| \leq |u \setminus x||v| + |x \setminus u||y| + |v \setminus y||u| + |y \setminus v||x| \leq 2|C_{\neq}|$$
and $R_1(a) \leq a$,
$$\sum_{(i,j) \in C_{\neq}} p_{i,j} \leq 2R_C|C_{\neq}| \qquad (11)$$

and $\mu \geq |C_{\neq}|(1 - 2R_C)$. Combining Eq. (11) with the definition of σ^2 gives that $\sigma^2 \leq 2R_C|C_{\neq}|\frac{q+2}{q-1}$. Combining this with Eq. (10) results in

$$P(W(x,y) < 0) \leq \exp\left(-\frac{|C_{\neq}|(1 - 2R_C)^2}{2(1 + 2R_C(q+2)/(q-1))}\right). \qquad (12)$$

Setting $c_q^{BA} = (1 - 2R_C)^2/(2(1 + 2R_C(q+2)/(q-1)))$, we get

$$P(W(x,y) < 0) \leq \exp(-|C_{\neq}|c_q^{BA}).$$

This proves the lemma.   □

In practice, this means that as long as noise levels are not overly high, i.e. $s \ll N$, and the dimensions of the planted biclique are not overly small, i.e. $\log N \ll |u| \ll N^{-1/2}$, we can reliably identify the planted bicluster. We note that these assumptions are quite realistic under the BA model. More to the point, we find that a biclique uv is still discernible if $|u| > \log N$ and $|v| > \log N$.

4 Algorithm

In this section we describe a simple heuristic algorithm to recover the planted bicliques under $GF[2]$. We have already shown that the best biclique is the planted one (with high probability), and therefore we 'only' need to find the best biclique. Unfortunately, this problem is NP-hard (as finding the largest exact biclique is NP-hard [11]). Luckily, it seems that in practice a simple heuristic— which we present below—is able to recover the planted biclique very well.

Recall, that finding the best biclique in $GF[2]$ is equivalent to finding rank-1 binary matrix factorization that minimizes the Hamming distance. To compute find it, we used the Asso algorithm [10]. We note that our aim is not to perform a comparative study of different algorithms but to show that we can achieve the predicted performance using relatively simple, non-exhaustive method.

The crux of Asso is the use of pairwise *association confidences* for finding candidate column factors. Consider the N-by-M input matrix B. Asso will compute the association confidence between each row of B. The association confidence from row b_i to row b_j is defined as $conf(b_i \rightarrow b_j) = (\sum_{k=1}^m b_{ik}b_{jk})/(\sum_{k=1}^m b_{ik})$ and can be interpret as the (empirical) conditional probability that $b_{jk} = 1$ given that $b_{ik} = 1$. The intuition is that if rows i and j belong in the planted biclique, they should have relatively high confidence (each column of the biclique is 1 in both rows and each column not in the biclique is 0, save the effects of noise) and otherwise the confidence should be low. The `Asso` algorithm builds an N-by-N matrix D where $d_{ij} = conf(b_i \to b_j)$.

Matrix D is then round to binary matrix $\tilde{D}$ from some threshold τ. The *columns* of the binary matrix $\tilde{D}$ constitute the candidate columns of the biclique. The algorithm will then construct the optimum row for each of these columns, and select the best row-column pair (x, y) (measured by $A \ominus xy$). Computing the association accuracy takes $O(N^2 M)$ time (where we assume $N \le M$), there are N candidate vectors, and testing each of them takes $O(NM)$ time, giving the overall complexity for fixed τ as $O(N^2 M)$.

The last detail is how to select the rounding threshold τ. We can try every value of D, but that adds N^2 factor to the second term in the time complexity. To avoid quadratic running times, we opt to evaluate a fixed set of thresholds.

5   Experiments

In this section we experimentally evaluate the above theory. As we need to measure against a ground truth, we will experiment on synthetic data. For practical reasons we focus on $GF[2]$: in order to evaluate our bounds we require an algorithm to extract candidate bicliques from the data. While no polynomial time (approximate) biclique discovery algorithm exists for $GF[q]$ in general, we have seen in Section 4 that `Asso` [10] is relatively easy to adapt to $GF[2]$.

We implemented the $GF[2]$ version of `Asso` in Matlab/C, and provide for research purposes the source code together with the generators for bipartite Erdős-Rényi and Barabási-Albert graphs.[3]

As synthetic data, we consider square matrices over $GF[q]$ of dimensions $N = M = 1000$, to which we add noise. We consider the ER model as discussed in Section 3.2, and the probabilistic BA model from Section 3.4. We focus on this BA variant, as by allowing larger bicliques to be generated it corresponds to the hardest problem setting. In this matrix A we plant a square biclique uv (i.e. $|u| = |v|$), such that we obtain $B = A \oplus uv$. We run `Asso` on B for all values of $\tau \in \{0, 0.01, \ldots, 1.0\}$, and select the best candidate biclique. We report the $L(u, v, x, y)$ error between this candidate and the planted biclique.

We evaluate performance for different noise ratios, defined as $|A|/(NM)$, and for different biclique sizes. Figure 1 shows the results averaged over five independent runs. For Erdős-Rényi, we see that bicliques of 10×10 are easily discerned even for high noise ratios, despite that `Asso` is uninformed of the shape or size of the biclique. In accordance with theory, bicliques of 5×5 can still be detected reliably for lower noise levels, while those of 3×3 only barely so.

For Barabási-Albert (Figure 1, right), we also find that practice corresponds to theory. Per Section 3.4 bicliques need too have $|u| \gg \log N$ to be discernible; indeed, we here observe that for $|u| \ge 15$ the clique is discovered without error, while for smaller sized clusters error first rises and then stabilizes. By the scale-free property of the graph the noise ratio does not influence detection much.

---

[3] http://www.mpi-inf.mpg.de/~pmiettin/gf2bmf/

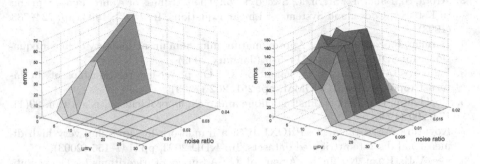

Fig. 1. Performance of `Asso` of finding a square planted biclique of dimensions ($|u| = |v|$), under different ratios of noise, resp. generated by the Erdős-Rényi model (left), and the Barabási-Albert model (right).

6 Conclusion

We consider the problem of finding planted bicliques in random matrices over $GF[q]$. More in particular, we investigated the size of the smallest biclique such that we can still approximately correctly identify it as the best rank-1 approximation against a background of either Erdős-Rényi or Barabási-Albert distributed noise. Whereas existing methods can only detect bicliques of $O(\sqrt{N})$ under non-destructive noise, we show that bicliques of resp. $n \geq 3$, and $n \gg \log N$ are discernible even under destructive noise. Experiments with the `Asso` algorithm confirm that we can identify planted bicliques under $GF[2]$ with high precision.

While the ER and BA models capture important graphs properties, they are stark simplifications. Studying whether similar derivations are possible for more realistic models, such as Kronecker Delta [9] will make for engaging future work.

The key extension of this work will be the development of theory for matrix factorization in $GF[q]$, by which we will be able to identify and analyse interactions between bicliques such as found in proteomics data [13].

Acknowledgements. Jan Ramon is supported by ERC Starting Grant 240186 "MiGraNT". Jilles Vreeken is supported by a Post-doctoral Fellowship of the Research Foundation – Flanders (FWO).

References

1. Alon, N., Panigrahy, R., Yekhanin, S.: Deterministic Approximation Algorithms for the Nearest Codeword Problem. In: Dinur, I., Jansen, K., Naor, J., Rolim, J. (eds.) APPROX 2009. LNCS, vol. 5687, pp. 339–351. Springer, Heidelberg (2009)
2. Ames, B.P.W., Vavasis, S.A.: Nuclear norm minimization for the planted clique and biclique problems. Math. Program. B 129(1), 69–89 (2011)

3. Arora, S., Babai, L., Stern, J., Sweedyk, Z.: The Hardness of Approximate Optima in Lattices, Codes, and Systems of Linear Equations. In: FOCS 1993, pp. 724–733 (1993)
4. Berman, P., Karpinski, M.: Approximating minimum unsatisfiability of linear equations. In: SODA 2002, pp. 514–516 (January 2002)
5. Faloutsos, M., Faloutsos, P., Faloutsos, C.: On power-law relationships of the internet topology. In: SIGCOMM, pp. 251–262 (1999)
6. Hochbaum, D.S.: Approximating clique and biclique problems. J. Algorithm 29(1), 174–200 (1998)
7. Koyutürk, M., Grama, A.: PROXIMUS: a framework for analyzing very high dimensional discrete-attributed datasets. In: KDD 2003, pp. 147–156 (2003)
8. Lee, V.E., Ruan, N., Jin, R., Aggarwal, C.: A Survey of Algorithms for Dense Subgraph Discovery. In: Aggarwal, C., Wang, H. (eds.) Managing and Mining Graph Data, pp. 303–336. Springer, New York (2010)
9. Leskovec, J., Chakrabarti, D., Kleinberg, J.M., Faloutsos, C., Ghahramani, Z.: Kronecker graphs: An approach to modeling networks. J. Mach. Learn. Res. 11, 985–1042 (2010)
10. Miettinen, P., Mielikäinen, T., Gionis, A., Das, G., Mannila, H.: The discrete basis problem. IEEE TKDE 20(10), 1348–1362 (2008)
11. Peeters, R.: The maximum edge biclique problem is NP-complete. Discrete Appl. Math. 131(3), 651–654 (2003)
12. Sim, K., Li, J., Gopalkrishnan, V., Liu, G.: Mining maximal quasi-bicliques: Novel algorithm and applications in the stock market and protein networks. Statistical Analysis and Data Mining 2(4), 255–273 (2009)
13. Wall, M.E.: Structure–function relations are subtle in genetic regulatory networks. Math. Bioscience 231(1), 61–68 (2011)
14. Yeger-Lotem, E., Sattath, S., Kashtan, N., Itzkovitz, S., Milo, R., Pinter, R.Y., Alon, U., Margalit, H.: Network motifs in integrated cellular networks of transcriptionregulation and proteinprotein interaction. PNAS 101(16), 5934–5939 (2004)
15. Zhang, Z.-Y., Li, T., Ding, C., Ren, X.-W., Zhang, X.-S.: Binary matrix factorization for analyzing gene expression data. Data Min. Knowl. Disc. 20(1), 28–52 (2010)

As Strong as the Weakest Link:
Mining Diverse Cliques in Weighted Graphs

Petko Bogdanov[1], Ben Baumer[2], Prithwish Basu[3], Amotz Bar-Noy[4],
and Ambuj K. Singh[1]

[1] University of California, Santa Barbara, CA 93106, USA
{petko,ambuj}@cs.ucsb.edu
[2] Smith College, Northampton, MA 01063, USA
bbaumer@smith.edu
[3] Raytheon BBN Technologies, 10 Moulton St., Cambridge, MA 02138, USA
pbasu@bbn.com
[4] The City University of New York, New York, NY 10016–4309, USA
amotz@sci.brooklyn.cuny.edu

Abstract. Mining for cliques in networks provides an essential tool for
the discovery of strong associations among entities. Applications vary,
from extracting core subgroups in team performance data arising in
sports, entertainment, research and business; to the discovery of functional complexes in high-throughput gene interaction data. A challenge
in all of these scenarios is the large size of real-world networks and the
computational complexity associated with clique enumeration. Furthermore, when mining for multiple cliques within the same network, the
results need to be diversified in order to extract meaningful information
that is both comprehensive and representative of the whole dataset.

We formalize the problem of *weighted diverse clique mining (mDkC)*
in large networks, incorporating both individual clique strength (measured by its weakest link) and diversity of the cliques in the result set.
We show that the problem is NP-hard due to the diversity requirement.
However, our formulation is sub-modular, and hence can be approximated within a constant factor from the optimal. We propose algorithms
for *mDkC* that exploit the edge weight distribution in the input network
and produce performance gains of more than *3* orders of magnitude compared to an exhaustive solution. One of our algorithms, *Diverse Cliques*
(DiCliQ), guarantees a constant factor approximation while the other,
Bottom Up Diverse Cliques (BUDiC), scales to large and dense networks
without compromising the solution quality. We evaluate both algorithms
on *5* real-world networks of different genres and demonstrate their utility
for discovery of gene complexes and effective collaboration subgroups in
sports and entertainment.

1 Introduction

It is often said that while the success of a sports or business team depends on
good individual performances, it depends even more on how individuals *gel* as

H. Blockeel et al. (Eds.): ECML PKDD 2013, Part I, LNAI 8188, pp. 525–540, 2013.
© Springer-Verlag Berlin Heidelberg 2013

a team – thus resulting in the idiom "There is no "I" in T-E-A-M". While this expression downplays the importance of individual performance, a team comprised of players with moderate individual talent but superior teamwork skills can outperform a dysfunctional team that emphasizes superlative individual performances. For example, the Detroit Pistons basketball team won the 2004 NBA championship with a collection of relatively unheralded players who were thought to collaborate so well together that "the whole was greater than the sum of its parts." Conversely, the team they defeated, the heavily-favored Los Angeles Lakers, featured four future Hall of Famers, none of whom appeared to collaborate particularly well together [14].[1] *Could the unexpected success of the Pistons or the demise of the Lakers have been predicted based on previous observations?*

While the importance of teamwork between elements in a group is easy to articulate, it is a non-trivial analytical task to isolate the core subgroups of entities that are responsible for the overall team performance. If the performance of the whole team can be measured, say, in terms of wins or losses, or revenue generated, the key problem is the discovery and identification of the *team cores* – subgroups within a team, whose inclusion results in higher-than-expected overall team performance, since their collaboration appears to motivate the success of the team as a whole. The discovery of core subgroups can illuminate distinctive individual characteristics, whose combination has a *super-additive* effect on the team [21]. This could provide important assistance to executives, who are ultimately judged by the success of the team, rather than the personal achievements of individual players. For example, sports executives could use team performance data from prior years to identify and acquire players exhibiting a combination of traits that lead to team success. Similarly, Hollywood studios can use data on prior collaborations among actors, directors, editors, cinematographers etc. while assembling a cast for an upcoming film, since successful past collaborations may portend similar success in the future.

"Teamwork" is not restricted to sports or business; it is also observed inside cells of living organisms – multiple proteins interact with each other to form a *multi-protein complex*, which is a cornerstone of many (if not most) biological processes, and together they form various types of molecular machinery that perform a vast array of biological functions [17]. The challenge here is to discover those complexes which are core groups of interacting genes within high-throughput pairwise interaction data [27]. The biological setting presents a distinctive challenge due to the difficulty in measuring the existence of a complex directly. The good news is that the strength of pairwise associations between genes can be tested efficiently via high-throughput methods employed to build functional interaction networks for analysis [28]. Hence, analytical techniques for mining strong gene subgroups can allow biologists to infer the existence of protein complexes that participate in the same cellular process and predict functional annotations for new gene sequences [11].

---

[1] Indeed, the open feud between Shaquille O'Neal and Kobe Bryant combusted after the season, resulting in the trade of O'Neal to Miami.

Our main goal is to extract an informative set of high-scoring[2] cliques that are representative of the entire network. To mitigate the presence of *free riders* (nodes attached by weak links to a strong clique), we define an intuitive score based on the *weakest link* in the clique. Alternative scoring schemes using sum or average are more forgiving to free riders since their link weights, albeit weak, would not drag the overall clique score down significantly. Similar scoring functions have been used in the Bioinformatics literature, namely Bandyopadhyay et al. [8] measure multi-way interaction strength as the minimum average weight of links adjacent to a node. A weakest link explanation of group success is also central to the *pooled interdependence theory* for business organizations as discussed by the classic text of Thompson [25].

Another challenge is to handle possible overlap among the best scoring cliques in order to represent all network locations of interest in the result set. Less overlap amounts to greater *diversity* among the reported cliques. Consideration of diversity is imperative in certain team sports such as ice hockey, in which a coach decides which *lines* (subgroups) of players play together on ice before being substituted by other *lines* Over the course of the season, the coach experiments with the makeup of these lines several times in order to figure out which players play well together. Diversity is also important in team formation for multiple tasks, where one aims to maximize the fitness of each team while simultaneously incorporating fairness by not overloading members with multiple tasks [6]. Thus, an analytical scheme that can mine a diverse set of subgroups is more useful than one which merely returns the top scoring ones without taking into account the overlap between them.

Our Contributions in this paper include the following:

Novelty: We formulate a novel *weighted diverse clique mining (mDkC)* problem that incorporates clique strength and diversity of result; and show that, although NP-hard, the formulation is sub-modular and allows accurate approximation schemes.

Scalability: We propose a $(1 - 1/e)$-approximation algorithm, DICLIQ, and a faster heuristic, BUDIC, for *mDkC*. Both achieve an improvement in running time of 3 orders of magnitude, when compared to an exhaustive search.

Quality: We demonstrate the utility of DICLIQ and BUDIC to identify team cores of significant performance and protein complexes in a gene interaction network.

2 Related Work

Clique mining work has focused on quasi (almost) cliques which allow a controlled number of missing edges [3,15,29]. While this relaxation is to accommo-

---

[2] The edge weights between entities (player/protein/stock symbol) are indicative of the strength of their pairwise relation. Particularly well-performing subgroups (or tightly interacting proteins) manifest in the resulting graph as "strong" cliques associating nodes with heavy edge weights.

date possible missing links and noise, all of the above methods operate in the scenario of unweighted and labeled graphs and optimize the frequency of clique occurrence as opposed to scores (labels are not unique and hence a quasi-clique may occur multiple times in the same data graph or in a database of small graphs). In contrast, we operate in settings of dense and weighted interactions exhibiting variance in the link strength. Weighted clique mining was considered by Bandyopadhyay et al. [8] who proposed a heuristic for the largest cardinality clique with average node-connectivity weight exceeding a user-defined threshold. Instead of using the link quality as a *constraint*, we incorporate it in the solution score. In addition, we are interested in finding a diverse set of (multiple) high scoring cliques as opposed to the single largest one.

Different subgraph "goodness" criteria have also been considered in graph mining, including diameter and spanning tree cost [16], Steiner tree and bottleneck cost [18]. Communities and modules have also been defined based on clique percolation (highly overlapping cliques) in CFinder [22] and applied to biological and social networks in a series of follow-up work. Such formulations allow sparse structures including nodes that do not interact directly. Instead, our methods are targeted to the discovery of "flat–organization" teams and all-to-all interactions in the case of gene complexes. As we show experimentally, our method (and formulation) outperforms CFinder by 20% when employed in gene complex discovery.

Alternative definitions of diversity have also been considered. Lappas et al. [16] investigated diversity of the node roles within single cliques. This within-clique diversity definition is targeted to cases where multiple nodes may have the same role (label) and can be considered in conjunction with ours. Anagnastopolus et al. [6] considered structures with fair assignment of tasks within the team and in follow-up work [7,18] combined communication cost with fair task assignment in online team formation. While these formulations target both overlap (task assignment) and structure, they incorporate one of the criteria (overlap or structure density) as a user-defined constraint and allow for sparse structures, i.e. they are suitable for hierarchical as opposed to "flat" teams/complexes.

3   Preliminaries

We model interaction strength between entities from sports, business, cinema and biology as a weighted undirected network $G = (V, E, w)$. Nodes V of the network graph correspond to agents or entities, while edge weights $w(u, v)$ between two nodes u and v reflect their connection strength: joint performance, interaction strength or similarity. For the rest of the presentation, without loss of generality, we assume edge weights are scaled to the interval $[0, 1]$.

Our goal is to extract a diverse set of groups, requiring that all internal pairwise connections are strong. In the graph setting, strong all-pair-connected groups map to cliques (complete graphs) of high weights on all edges. Finding the *Maximum Clique (*MAX CLIQUE*)* (the one of largest size in an unweighted graph) is NP-hard to solve and also approximate to within $n^{1-\epsilon}$ [13]. Introducing weights on edges preserves the same general complexity.

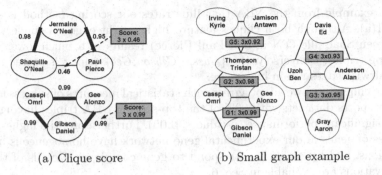

(a) Clique score   (b) Small graph example

Fig. 1. Comparison of two cliques based on the weakest link as their score (a). Example of five high-scoring triples from NBA (b). A diverse 2-set of cliques of size 3 minimizes the overlap in the resulting set, i.e. prefers $\{G1, G3\}$ over $\{G1, G2\}$ (slightly lower total scores but increased diversity).

4   Problem Definition

In this paper, we model the strength of a subgroup in terms of the pairwise interactions between individuals in that subgroup. Ideally, techniques that go beyond graph theory, e.g., hypergraphs or abstract simplicial complexes [9,19], are appropriate for modeling higher order interactions (triples, quadruples, etc.) directly. Such higher-order modeling approaches, however, are limited by both data scarcity and intractability. Enumerating and scoring all subgroups becomes computationally demanding for datasets of even hundreds of nodes. Moreover, in order to validate methods that score higher order groups directly, we need empirical data containing sufficiently many observations in which a particular group has interacted as a whole — a requirement that is hard to meet.

To address the above challenges, we require that all pairwise interactions (performances) are strong and relate the performance of a group to its weakest link, seeking to approximate the group behavior. Most professional sports teams, scientific collaborations, and protein networks are all examples of predominantly non-hierarchical ("flat") organizations that may collapse without strong pairwise links among all group members. This is true especially in a team sport such as basketball in which lack of good communication between any pair of players can easily threaten the success of the team (see [14]), especially since a highly competitive opponent is typically smart enough to exploit that vulnerability during the game. Thus, our clique scoring scheme is targeted to groups in which all pairs correlate/interact strongly, where the potential success of a group is limited by the worst pairwise connection therein.

We define the *score* of a group (network clique) in terms of its weakest link. That is, for any subset of nodes $C \subseteq V$, $s(C) = |C| \min_{u,v \in C} w(u,v)$. If C is not a clique (at least one missing edge), then the score is $s(C) = 0$. This scoring criterion prefers larger cliques, whose minimal edge score is high. It is also designed to eliminate the *free-rider* effect, i.e. inclusion of nodes that exhibit some weak links to others within the group.

An example from the NBA that illustrates our scoring method is shown in Fig. 1(a). Although both cliques feature high weight edges, a single weak link (between Shaquille O'Neal and Paul Pierce) results in a much lower score for the top triplet. Conversely, the {*Casspi, Gibson, Gee*} triplet is scored higher as all pairwise edges retain high score.

The clique score we adopt agrees with statistical measures of success for triples in our sports data sets with 90% of the top-scoring NBA triples having statistically significant performance (p-value ≤ 0.05). Furthermore, the highest scoring cliques of genes in our experimental gene network have homogeneous biological functions, and hence likely correspond to gene complexes. Details of the above observations are available in Sec. 6.

Based on our weakest link score definition, we formalize the problem of finding the best weighted fixed-size clique in a graph.

Definition 4.1. *The* maximum weighted k-clique *in a graph* $G = (V, E, w)$ *is the k-clique $C_k^*(G)$ of maximum score, i.e. $C_k^*(G) = \arg\max_{C \subseteq V, |C| = k} s(C)$. Given a weighted graph G,* MAXIMUM WEIGHTED k-CLIQUE *(Wk C) is the problem of finding the maximum weighted k-clique in G.*

WkC is NP-hard as it can be reduced from the MAX CLIQUE problem by restriction of the edge weights to 1. In our solutions, we will exploit the weight distribution of edges in a network in order to explore more promising cliques first and prune unpromising candidates for extension.

Beyond a single group, our goal is to report the best set of cliques, in order to represent all locations of interest in a large network. There are two main computational challenges to this end: (i) efficient discovery of good (high-scoring) cliques and (ii) ensuring informativeness of the result set by diversification.

To illustrate the intuition behind the importance of diversity in the resulting set, consider the NBA player example in Fig 1(b). There are five candidate cliques of size 3 and their scores are listed in shaded boxes in the corresponding triangles. Assuming that the top 2 highest scoring cliques are of interest, ignoring diversity amounts to reporting {$G1, G2$}. However, there is a lot of overlap (the duo *Casspi-Gee*) between {$G1, G2$}. Intuitively, reporting all super-cliques of *Casspi-Gee* is not representative of the overall network. Instead, we can diversify by pairing $G1$ with a slightly lower scoring clique of lower overlap such as $G3$. The idea of diversity has been considered in a number of other settings including information systems for web search [4,10,12], image retrieval [24,26], cheminformatics [5] and other domains.

Next, we formalize a joint diversity-score formulation and a corresponding solution with a good approximation guarantee.

Definition 4.2. *For a set $\mathcal{A}$ of cliques each of size k, we define their diversity score as*

$$ds(\mathcal{A}) = \alpha \frac{\sum_{C \in \mathcal{A}} s(C)}{k} + (1 - \alpha) \frac{|\bigcup_{C \in \mathcal{A}} C|}{k},$$

where $\alpha \in [0, 1]$, $\bigcup_{C \in \mathcal{A}} C$ is the union of nodes in the cliques of $\mathcal{A}$ and $ds(\emptyset) = 0$. Then given parameters k and α, the m DIVERSE k-CLIQUE *(mDkC) problem seeks the set $\mathcal{A}$ of size m that maximizes $ds(\mathcal{A})$.*

The above definition combines the average score of the answer set cliques and the diversity of their comprising nodes in a linear fashion. The trade-off between score and diversity can be controlled by the parameter α. Note that since both terms are bounded above by $|\mathcal{A}|$, $ds(\mathcal{A}) \in [0, |\mathcal{A}|]$.

If we set $m = \alpha = 1$, then $ds(\mathcal{A}) = s(C)$, and $mDkC$ is equivalent to WkC. Since, as we saw above, WkC is NP-hard, $mDkC$ is NP-hard. However, we prove the stronger statement that $mDkC$ is NP-hard even if an efficient heuristic for WkC (or an alternative high score structure) exists.[3] That is, we prove that the hardness of $mDkC$ comes not only from the hardness of WkC, but also from the difficulty of diversifying the result set.

Theorem 4.1. *For any scoring function $s()$ that maps a graph substructure to a non-negative real number, the decision problem corresponding to $mDkC$, namely: "Is there a set of m substructures $\mathcal{A}$, each of size k, such that $ds(\mathcal{A}) \geq B$ for some positive number B," is NP-complete.*

Proof. The proof is available in the Appendix [2]. □

While our focus is on cliques, the above theorem shows a more general result for arbitrary subsets of nodes in a graph given a scoring function for each subset. Hence, in different applications in which finding an optimal score substructure is computationally tractable, ensuring that the solution comprised of multiple substructures is diverse remains NP-hard.

Although the $mDkC$ problem is NP-hard (due to the NP-completeness of the decision version), we show that the diversity score function is monotonic and sub-modular. These properties allow a fixed-quality approximate solution based on a greedy scheme. Next, we formally show the monotonicity and sub-modularity of our diversity score formulation.

Theorem 4.2. *If k and α are fixed, the diversity score function $ds(\mathcal{A})$ is:*
- Monotonic, i.e. for any subset $\mathcal{A} \subseteq \mathcal{B}$, $ds(\mathcal{A}) \leq ds(\mathcal{B})$
- Sub-modular, i.e. for any sets $\mathcal{A}, \mathcal{B}$, $ds(\mathcal{A}) + ds(\mathcal{B}) \geq ds(\mathcal{A} \cup \mathcal{B}) + ds(\mathcal{A} \cap \mathcal{B})$.

Proof. The proof is available in the Appendix [2]. □

Due to the monotonicity and sub-modularity of the diversity score and based on the seminal result of Nemhauser et al [20], we can show the following corollary.

Corollary 4.1. *A GREEDY procedure for $mDkC$ that adds cliques in decreasing order of their diversity score improvement always achieves a solution within $1 - \frac{1}{e}$ from the optimal. Since $ds(\emptyset) = 0$, this the best possible approximation ratio for the problem.*

The $(1 - 1/e)$-approximation guarantee assumes that we can construct GREEDY and hence solve WkC optimally (when $\alpha = 1$, the first clique to be added is the WkC solution). This is by itself a hard problem as we argued above, however, we exploit the edge weights to provide scalable solutions for real-world datasets.

---

[3] Recall that by [13], the general clique problem cannot be approximated within $n^{1-\epsilon}$ for any given ϵ.

5   Weighted Diverse Clique Mining

In this section we propose two algorithms for the $mDkC$ problem: DICLIQ and BUDIC. Both adopt pruning of infeasible candidates based on partially explored cliques in order to reduce computation time. DICLIQ works by enumerating cliques within a thresholded version of the network: first high-scoring edges are considered and as the algorithm progresses lower-weight edges are included if needed. It provides a $(1 - 1/e)$-approximation guarantee as it implements a greedy strategy. For large and dense instances (exceeding $4,000$ nodes and $30,000$ edges) and for higher number of cliques and clique sizes, DICLIQ's running time worsens (requiring on the order of minutes to complete in our experimental datasets). To handle larger and denser instances, we develop a scalable heuristic BUDIC that achieves more than 90% of DICLIQ's diversity score (and at times even better scores than DICLIQ). BUDIC employs similar pruning, but avoids expensive enumeration of cliques by greedy expansion from a single edge.

5.1   Bounding the Diversity Score for Partial Cliques

We first show an upper bound for the contribution of a clique C when added to a set of cliques $\mathcal{A}$. If the newly added clique C is of the desired size k then its contribution to the overall score can be computed according to the definition of $ds()$. If, however, C is not a complete clique of the desired size $|C| < k$, one can bound the contribution of any of its super cliques (cliques that contains all nodes in C) of size k to the diversity score.

Theorem 5.1. *Let $C, |C| \leq k$ be a clique of size not exceeding k. The maximum improvement of ds score when adding any k super clique of C to a clique set $\mathcal{A}$ is bounded by:*

$$\delta(\mathcal{A}, C) = ds\left(\mathcal{A} \cup C\right) - ds(\mathcal{A}) = \alpha \min_{u,v \in C} w(u,v) + (1-\alpha)\frac{k - |(\cup_{B \in \mathcal{A}} B) \cap C|}{k},$$

where in the diversity part, the set $(\cup_{B \in \mathcal{A}} B) \cap C$ is the intersection of nodes included in $\mathcal{A}$ and nodes in C.

Proof. The proof is available in the Appendix [2].   □

The upper bound can be applied for incomplete cliques C of any size, even ones that are completely unobserved, i.e. $|C| = 0$. In the latter case, the score part increases by at most α (assuming the maximum possible edge weight is 1) and the diversity part increases by at most $(1 - \alpha)$. Equipped with the upper bound δ, we next define our edge thresholding algorithm DICLIQ that considers high-scoring cliques first and prunes infeasible candidates.

5.2   DiCliQ: Enumeration of Cliques with Thresholding

A naive *Baseline* heuristic for $mDkC$ (with the $(1-1/e)$-approximation) can (i) enumerate all possible cliques of the desired size k and then (ii) greedily (based

Algorithm 1. DiCliQ

Require: $G = (V, E, w), k, m, \alpha$, threshold schedule $T = \{T_i\}$
Ensure: A set of cliques $\mathcal{A} = \{C_i\}, |\mathcal{A}| = m, |C_i| = k$
1: $\mathcal{A} = \emptyset, l = 0$
2: **while** $|\mathcal{A}| < m$ AND $l < |T|$ **do**
3:      Obtain $G_l(V, E_l, w), e \in E_l \iff w(e) \geq T_l$
4:      Compute $\delta(\mathcal{A}, C_l^i), \forall |C_l^i| \leq k$ (incl. $C_l^i = \emptyset$)
5:      **while** $\max_{|C_l^i|=k} \delta(\mathcal{A}, C_l^i) \geq \max_{|C_l^j|<k} \delta(\mathcal{A}, C_l^j)$ **do**
6:          $\mathcal{A} = \mathcal{A} \bigcup argmax_{|C_l^i|=k} \delta(\mathcal{A}, C_l^i)$
7:          break if $|\mathcal{A}| = m$
8:          Update $\delta(\mathcal{A}, C_l^i)$ based on the new $\mathcal{A}$
9:      **end while**
10:     $l = l + 1$
11: **end while**
12: **return** $\mathcal{A}$

on best $ds()$ improvement) compile an m-size result-set. While such *Baseline* might be feasible for small sparse networks (up to $|V| = 500$) and small values of k, the clique enumeration step quickly becomes a bottleneck as the input size increases due to its combinatorial nature. It fails to complete in less than 4 hours in all but our smallest network from the NBA.

Different from *Baseline*, we observe that in order to maximize the diversity score, we can first consider only edges of high weights. Then, as needed, we can consider lower score edges completing cliques of small overlap with the partial result set. Following this intuition, DiCliQ enumerates cliques in a thresholded subgraph induced by the highest-score edges and gradually includes more edges on demand. This process is based on a decreasing schedule $T - \{T_l\}$ of edge weight thresholds. The best-scoring cliques are discovered first within a much smaller instance of the graph. In addition, DiCliQ employs the upper bound on the improvement of the ds score for candidate cliques in order to filter out infeasible candidates and guarantee that cliques are added to the result set in a greedy order ensuring a $(1 - 1/e)$-approximation.

DiCliQ is presented in Alg. 1. Apart from the input graph and parameters k, m and α, the algorithm also takes as an input a schedule $\{T_l\}$ of descending edge value thresholds. The result set $\mathcal{A}$ is first initialized as empty and the threshold level l to 0 (i.e. highest edge values). While a set $\mathcal{A}$ of size m is not obtained and we have not reached the last level of thresholding, the algorithm (i) filters the graph based on T_l (Line 3), (ii) enumerates and upper-bounds all cliques of size up to k (Line 4) from the filtered graph and (iii) attempts to add cliques to the result set if they are the best next cliques to add (Lines 5-9). Note, that on Line 4 an upper bound $\delta(\mathcal{A}, \emptyset) = \alpha T_l + 1 - \alpha$ on all yet unobserved cliques is also computed.

If the maximum improvement δ of a size-k clique C_l^i exceeds the upper bound on any incomplete clique, we add C_l^i to the solution (Line 6) and update the improvements of the cliques based on the new $\mathcal{A}$ (Line 8). Additions of cliques are performed until no complete clique exceeds the upper bound of an incomplete one, or until $|\mathcal{A}| = m$. After all possible additions are exhausted, if the result set does not contain m cliques, we lower the edge weight threshold (Line 10)

Algorithm 2. BUDiC

Require: $G = (V, E, w), k, m, \alpha$
Ensure: A set of cliques $\mathcal{A} = \{C_i\}, |\mathcal{A}| = m, |C_i| = k$
1: $\mathcal{A} = \emptyset$
2: **for** $i = 1 \rightarrow m$ **do**
3:     **for all** Edges $e \in E$ **do**
4:        $C_e = e$;
5:        **repeat**
6:           Grow C_e by $argmax_{v \in V} \delta(\mathcal{A}, C_e \bigcup v)$
7:        **until** $(|C_e| = k) \vee (C_e$ cannot be extended)
8:     **end for**
9:     **if** no C_e of size k are found **then**
10:        break
11:    **end if**
12:    $\mathcal{A} = \mathcal{A} \bigcup argmax_{|C_e|=k, e \in E} \delta(C_e)$
13: **end for**
14: **return** $\mathcal{A}$

---

and repeat Lines 3-5 for the new thresholded graph. Since we add cliques to the results set only if their score improvement exceeds the upper bound of all possible candidates (line 5) we ensure that the cliques are added in a greedy (descending score) order and hence DiCliQ implements a greedy strategy and obtains a $(1 - 1/e)$-approximation.

The thresholding scheme of DiCliQ is effective when the result set of m best cliques is completed before reaching the lowest threshold level, i.e. enumerating cliques in the whole graph. An important means to this end is choosing an appropriate schedule that reflects the distribution of edges. We divide the set of all edge weights into equi-size bins and adjust the threshold to incorporate one more of these bins at every iteration. Other schedules (exponentially increasing subsets of edges) are also possible, but were not more favorable in our experiments.

5.3   BUDiC: Scalable *Bottom-Up Diverse Clique* Heuristic

The bottleneck in DiCliQ is the enumeration and bounding of all cliques at a given edge weight level (Line 4, Alg. 1). This step is in general exponential and the algorithm is efficient only when the results set is computed at the first several thresholding levels. To scale to larger and denser graphs, while avoiding exhaustive enumeration of cliques, we employ a greedy *Bottom-up* scheme BUDiC.

The intuition behind BUDiC (Alg. 2) is that one can get good candidates for the result set by starting from a good edge and growing a clique, while avoiding overlap according to the diversity α. Cliques are added one at time in the outer loop (Lines 2-13). Good local cliques are grown greedily by nodes of best improvement (Line 3-8). If no clique of the desired size is found the main loop is terminated and an incomplete set of cliques is returned (Line 9-11). The best clique in each iteration is added to $\mathcal{A}$ (Line 12). Note, that BUDiC does not have the same approximation guarantee as DiCliQ because in the greedy expansion from an edge it does not consider all possible cliques.

The algorithm runs in polynomial time $O(k \cdot m \cdot n \cdot |E|)$ as every edge is grown to a clique of at most size k and this is repeated m times. The n term is due to the possibility of considering all graph nodes in Step 6 when the graph is complete. BUDiC is also suitable for parallel implementation, since Lines 3-8 can be executed on separate machines assuming the graph is partitioned with redundancy and distributed to all machines.

5.4   Discussion on Setting Parameters

While it is unlikely that a universally appropriate value of k exists, in certain applications there are domain-specific constraints that could be used. For example, in basketball, subgroups of sizes less than 5 are of interest, while in gene networks appropriate sizes are between 4 and 6 since many known yeast complexes contain 4.7 subunits on average [23]. When no prior domain knowledge is available, we envision varying k while tracking the relationship among consecutive result sets and concentrating on values for which the solution changes substantially (i.e. solution cliques of size k do not tend to include those of size $k-1$). A similar approach can be adopted to determine interesting values of the diversity weight α. We perform such analysis for α in the experimental section. The number of cliques in the resulting set (m) can be increased until the score contribution of adding additional cliques diminishes significantly relative to the average contributions of already included cliques.

6   Experimental Evaluation

We evaluate our algorithms on a variety of real world data from sports, cinema, biology and finance. Our goal in experimentation is to (i) assess the scalability of DiCliQ and BUDiC to large problem instances; (ii) demonstrate the quality of BUDiC compared to the $(1-1/e)$-approximation DiCliQ; and (iii) demonstrate the relevance of the mined diverse cliques to real world applications.

Data. We experiment with 5 publicly-available data sets including participation in teams sports (*NBA, MLB*), collaboration in movies (*IMDB*), a gene interaction network (*YeastNet*) and a correlation network of stock symbols (*Stocks*) (see Table 1). Edge strength in the sport/collaboration are based on the statistical significance of the performance of the pair of entities when in groups (sport team success or movie cast ratings). The edge weights in the gene network is based on strength of measured interaction of the genes, while the absolute Pearson's correlation serves as a weight in the stock network. The sizes of the datasets are listed in Table 1 (columns 2,3). We discuss in detail the sources and preprocessing of our datasets in the Appendix [2].

Scalability. All scalability measurements are on a Dell Desktop with $6GB$ RAM and Dual Core 4GHz processor. We measure the clock time of the exhaustive *Baseline*, DiCliQ, BUDiC and the iterative extension of [8] called *iMDV*. Note that both DiCliQ and *Baseline* implement a greedy strategy and hence obtain a constant $1-1/e$ factor approximate solution. An optimal solution for the problem would further require considering all possible (exponential) subsets of

Table 1. Summary of our datasets including time span, number of nodes V and edges E. The second part of the table lists the running time (in seconds) and quality (% of Baseline) on all datasets $\alpha = 0.5$, $m = 10, k = 5$. The quality of BUDiC and an iterative version of iMDV is measured as percentage of DiCliQ 's score. *Baseline* does not complete in $4h$ and its memory footprint exceeds $6GB$ causing an out-of-memory exception.

Source	$|V|$	$|E|$	Baseline Time	DiCliQ Time	iMDV Time	iMDV % sc.	BUDiC Time	BUDiC % sc.
NBA	532	5,945	23.00s	0.48s	0.018s	57	0.13s	**98**
MLB	1,569	40,126	>4h	3.00s	0.015s	52	0.24s	**95**
IMDB	25,141	417,705	>4h	107.00s	0.087s	54	0.17s	**94**
YeastNet	4,450	30,5416	>4h	5.8s	0.463s	73	0.77s	**99**
Stocks	1,194	32,406	>4h	12s	0.022s	44	0.4s	**99**

cliques and is not be feasible even for our smallest datasets. Details of competing techniques are available in the Appendix [2].

The right part of Table 1 shows the performance of competing techniques in all datasets. *Baseline* was able to complete only on our smallest dataset NBA and it was 40 times slower than DiCliQ and 2 orders of magnitude slower than BUDiC. On the rest of the datasets ($\alpha = 0.5$, $m = 10$ and $k = 5$) *Baseline* does not complete in 4 hours and runs out of memory, due to the exponential number of candidate cliques that it has to consider for inclusion in the result set. DiCliQ, BUDiC and *iMDV* have comparable running time on small datasets, while in denser and larger networks DiCliQ is 10 to 100 times slower. In terms of diverse clique score, our fast heuristic BUDiC dominates *iMDV* by $30 - 50\%$.

We present the scalability behavior of our techniques for varying clique size and number of cliques in the results set within YeastNet in Fig. 2. *Baseline* does not complete in 4 hours for $k = 5$ and any value of m. The reason for this long running time is that *Baseline* consumes all allocated memory ($6GB$) while enumerating all possible cliques. For smaller clique sizes it is 3 to 4 orders of magnitude slower than DiCliQ and BUDiC. When increasing k and m, DiCliQ slows down due to the need to lower its edge weight threshold and enumerate more cliques in progressively larger graphs. BUDiC's performance does not change significantly for these experimental settings, making it a good scalable method for higher k and m.

Quality of BUDiC. BUDiC reduces the computational time by up to 2 orders of magnitude, as expected due to its polynomial complexity. However, an immediate question is: *What is its quality?* We showed that BUDiC's quality on all datasets is above 95% (diversity score as a fraction of *Baseline*'s score) for one setting of parameters in Table 1. Next, we explore the quality dependence on the number of cliques m in the result set and on the value of α.

Fig. 2(c) summarizes the quality of BUDiC in the YeastNet and Stocks networks. We show its diversity score as a fraction of a Greedy solution (obtained by either *Baseline* or DiCliQ.) For high values of α (i.e. when the clique score matters more than diversity), BUDiC is able to find even better score solutions than *Baseline*. On average, it behaves similar to the greedy alternatives with

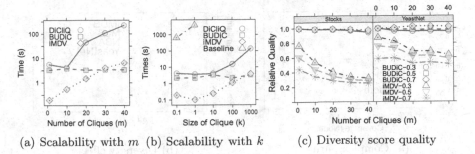

(a) Scalability with m (b) Scalability with k (c) Diversity score quality

Fig. 2. Scalability comparison of *Baseline*, DiCliQ and BUDiC for increasing number of cliques m in the result set for $k = 5, \alpha = 0.5$ (a) and increasing clique size k for $m = 5, \alpha = 0.5$ (b) in the *YeastNet* network. The Baseline approach does not complete in 4 hours for $k = 5, \alpha = 0.5$ and its memory footprint exceeds $6GB$. (c) Quality of BUDiC's diversity score as a fraction of the score obtained by a GREEDY heuristic (both *Baseline* and DiCliQ obtain the same score) in the *Stocks* (Left) and *YeastNet* (Right) networks($k = 5$).

$(1 - 1/e)$-approximation. We observe similar behavior on the rest of the datasets as well. We also explored qualitatively the mined clique sets and found that for m up to 20 the intersection of the cliques obtained by BUDiC and *Baseline* (and DiCliQ) remains above 80% as well (i.e. only 2-3 cliques differ in the result sets). Hence, BUDiC achieves tremendous savings in time at almost no cost in quality in the data we analyzed.

Gene Complexes and Influential Sub-groups. Next, we demonstrate the applicability of our formulation and methods for gene complex discovery and summarization of effective groups in sports. We label genes in YeastNet with known *process Gene Ontology (GO)*terms [1]. The GO labels are hierarchical with specificity increasing with the distance from the root. To account for varying specificity and hierarchy utilization, we only consider labels at level 4 and their descendants (i.e. 4 hops or more from the root). Annotations of higher specificity are mapped to their corresponding level 4 ancestors and the *YeastNet* network is filtered to include only genes that are annotated.

To evaluate the ability of BUDiC to identify meaningful gene complexes, we measure discovered groups' *purity* as the fraction of genes sharing the same label and compare to a recent overlapping community detection algorithm *CFinder* [22] and a *Random* subsets of genes as control (see details in the Appendix [2]). Figure 3(a) is a scatter plot of the average solution annotation purity versus coverage (the union size of nodes in the solution). Our diversity parameter α allows for control over the coverage/purity trade-off (labels of the BUDiC trace show the selected α). BUDiC's average group purity is 20% higher than that of CFinder [22] (at coverage 136 nodes) and 30% higher than the average random purity (at coverage 220 nodes). This separation demonstrates that our minimum edge weight formulation allows for discovery of biologically more relevant complexes, while allowing for diversity (overlap) control within the result set.

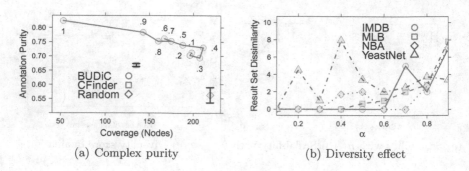

(a) Complex purity   (b) Diversity effect

Fig. 3. (a) Average group *GO biological process* purity (fraction of genes sharing labels) versus coverage (number of covered nodes) in *YeastNet*. Comparison among BUDiC (varying α labeling each point), CFinder ($\delta = 0.6, w = 0.15$) and *Random* grouping of genes ($k = 6, m = 37$); (b) Dissimilarity of consecutive solutions when increasing α. Peaks correspond to drastic changes in the result set ($m = 10, k = 4$).

In sports data, we compared the scores of DiCLiQ cliques and the significance scores of the corresponding groups of players (in the form of p-values). In the NBA data set, DiCLiQ retrieves 44% of the triples of lowest p-values (considering 1% of the lowest p-value triples). These high-performing triples are of paramount interest, since they represent the cliques that are likely driving team success. Furthermore, the average p-values of DiCLiQ's top cliques are comparable with the reference set of lowest p-value triples (0.010 versus 0.026) with 90% of DiCLiQ triples having p-value less than 0.05 (a common level of determining statistical significance in general scenarios). We discuss the mined subgroups and their relevance across the various datasets in the Appendix [2].

Effect of Diversity. By changing the value of the diversity parameter (α), we can alter the amount of overlap between the cliques returned by BUDIC. In Fig. 3(b), we show how the result sets change as a function of α. For any two consecutive values of α (e.g. 0.3 and 0.4), we obtain two result sets A and B. To measure their dissimilarity, we form the complete bipartite graph between the cliques in A and B, and assign weights to the edges based on the Jaccard similarity of the individuals cliques. Thus, for each clique $C_a \in A, C_b \in B$, the weight of the corresponding edge is given by $1 - Jaccard(C_a, C_b) = 1 - |C_a \cap C_b|/|C_a \cup C_b|$. The maximum weighted matching on this graph provides a dissimilarity score for A and B.

In Fig. 3(b), the dissimilarity between result sets in the NBA, for example, spikes at α between 0.3 and 0.6. The top 3 scoring quartets returned by BUDIC consist of only five distinct players, all playing for the Cleveland Cavaliers. However, by increasing diversity ($\alpha = 0.4$) we retain the first and third quartets only and bring in a different team quartet. Thus, α allows for application-specific control of the amount of diversity desired in the result sets. When exploring a new data set, appropriate α values can be chosen based on the tipping points of the solutions (spikes in Fig. 3(b)).

7   Conclusion

Mining strong subgroups in networks is an important, yet challenging computational problem. In this paper, we proposed a novel and flexible formulation $mDkC$, in which a *diverse* set of *strong* cliques is identified. We show that $mDkC$ is NP-hard, but due to its submodularity, allows a constant factor approximation. We develop scalable approximation schemes: DiCliQ with $(1 - 1/e)$-approximation guarantee and BUDiC that scales to large and dense networks. Both algorithms are more than 3 orders of magnitude faster compared to exhaustive counterparts, and BUDiC achieves 2 times higher scores than previous clique-mining heuristics. We demonstrate the utility of our algorithms for identifying interesting sets of high-performance collaborators in sports and entertainment, and complexes of similar biological function (30% improvement over earlier approaches) in gene networks. The developed algorithms thus present a useful tool for mining influential core subgroups in large networks from diverse sources.

Acknowledgements. This research was supported by the Army Research Laboratory under cooperative agreement W911NF-09-2-0053 (NS-CTA). The content of the information does not necessarily reflect the position or the policy of the Government, and no official endorsement should be inferred. The U.S. Government is authorized to reproduce and distribute reprints for Government purposes notwithstanding any copyright notice herein.

References

[1] The gene ontology, http://www.geneontology.org/
[2] Appendix (2013), http://cs.ucsb.edu/~dbl/papers/cliquesappendix.pdf
[3] Aggarwal, C.C., Li, Y., Yu, P.S., Jin, R.: On dense pattern mining in graph streams. In: PVLDB (2010)
[4] Agrawal, R., Gollapudi, S., Halverson, A., Ieong, S.: Diversifying search results. In: WSDM (2009)
[5] Akella, L., DeCaprio, D.: Cheminformatics approaches to analyze diversity in compound screening libraries. Current Opinion in Chemical Biology 14(3), 325–330 (2010)
[6] Anagnostopoulos, A., Becchetti, L., Castillo, C., Gionis, A., Leonardi, S.: Power in unity: forming teams in large-scale community systems. In: CIKM (2010)
[7] Anagnostopoulos, A., Becchetti, L., Castillo, C., Gionis, A., Leonardi, S.: Online team formation in social networks. In: WWW, pp. 839–848 (2012)
[8] Bandyopadhyay, S., Bhattacharyya, M.: Mining the largest dense vertexlet in a weighted scale-free graph. Fundam. Inform. 96(1-2), 1–25 (2009)
[9] Brinkmeier, M., Werner, J., Recknagel, S.: Communities in graphs and hypergraphs. In: CIKM (2007)
[10] Capannini, G., Nardini, F.M., Perego, R., Silvestri, F.: Efficient diversification of search results using query logs. In: WWW, pp. 17–18 (2011)
[11] Cui, G., Chen, Y., Huang, D.-S., Han, K.: An Algorithm for Finding Functional Modules and Protein Complexes in Protein-Protein Interaction Networks. J. of Biomedicine and Biotechnology (2008)

[12] Dubey, A., Chakrabarti, S., Bhattacharyya, C.: Diversity in ranking via resistive graph centers. In: KDD, pp. 78–86 (2011)

[13] Håstad, J.: Clique is hard to approximate within $n^{1-epsilon}$. In: FOCS (1996)

[14] Jackson, P., Arkush, M.: The last season: a team in search of its soul. Penguin (2004)

[15] Jiang, D., Pei, J.: Mining frequent cross-graph quasi-cliques. TKDD 2(4) (2009)

[16] Lappas, T., Liu, K., Terzi, E.: Finding a team of experts in social networks. In: KDD (2009)

[17] Li, X.-L., Tan, S.-H., Foo, C.-S., Ng, S.-K.: Interaction Graph Mining for Protein Complexes Using Local Clique Merging. Genome Informatics 16(2), 260–269 (2005)

[18] Majumder, A., Datta, S., Naidu, K.: Capacitated team formation problem on social networks. In: KDD (2012)

[19] Moore, T., Drost, R., Basu, P., Ramanathan, R., Swami, A.: Analyzing collaboration networks using simplicial complexes: A case study. In: INFOCOM WKSHPS, pp. 238–243 (2012)

[20] Nemhauser, G., Wolsey, L., Fisher, M.: An analysis of approximations for maximizing submodular set functions–i. Mathematical Programming 14(1), 265–294 (1978)

[21] Oliver, D., Fienen, M.: Importance of teammate fit: Frescoball example. J. of Quantitative Analysis in Sports 5(1) (January 2009)

[22] Palla, G., Derenyi, I., Farkas, I., Vicsek, T.: Uncovering the overlapping community structure of complex networks in nature and society. Nature 435, 814–818 (2005)

[23] Pu, S., Wong, J., Turner, B., Cho, E., Wodak, S.J.: Up-to-date catalogues of yeast protein complexes. Nucleic Acids Research 37(3), 825 (2009)

[24] Song, K., Tian, Y., Gao, W., Huang, T.: Diversifying the image retrieval results. ACM Multimedia (2006)

[25] Thompson, J.: Organizations in Action: Social Science Bases of Administrative Theory. Classics in Organization and Management Series. Trans. Publishers (1967)

[26] van Leuken, R.H., Pueyo, L.G., Olivares, X., van Zwol, R.: Visual diversification of image search results. In: WWW, pp. 341–350 (2009)

[27] Yang, Q., Lonardi, S.: A graph decomposition approach to the identification of network building modules. In: W. BIOKDD (2007)

[28] You, C.H., Holder, L.B., Cook, D.J.: Temporal and structural analysis of biological networks in combination with microarray data. In: CIBCB (2008)

[29] Zeng, Z., Wang, J., Zhou, L., Karypis, G.: Coherent closed quasi-clique discovery from large dense graph databases. In: KDD, pp. 797–802 (2006)

How Robust Is the Core of a Network?

Abhijin Adiga[1] and Anil Kumar S. Vullikanti[1,2]

[1] Virginia Bioinformatics Institute, Virginia Tech
[2] Department of Computer Science, Virginia Tech
abhijin@vbi.vt.edu, vsakumar@vt.edu

Abstract. The k-core is commonly used as a measure of importance and well connectedness for nodes in diverse applications in social networks and bioinformatics. Since network data is commonly noisy and incomplete, a fundamental issue is to understand how robust the core decomposition is to noise. Further, in many settings, such as online social media networks, usually only a sample of the network is available. Therefore, a related question is: How robust is the top core set under such sampling?

We find that, in general, the top core is quite sensitive to both noise and sampling; we quantify this in terms of the Jaccard similarity of the set of top core nodes between the original and perturbed/sampled graphs. Most importantly, we find that the overlap with the top core set varies *non-monotonically* with the extent of perturbations/sampling. We explain some of these empirical observations by rigorous analysis in simple network models. Our work has important implications for the use of the core decomposition and nodes in the top cores in network analysis applications, and suggests the need for a more careful characterization of the missing data and sensitivity to it.

1 Introduction

The k-core $C_k(G)$ of an undirected graph $G = (V, E)$ is defined as the maximal subgraph in which each node has degree at least k; the core number of a node is the largest k such that it belongs to the k-core (i.e., $v \in C_k(G)$). The set $S_k(G) = C_k(G) \setminus C_{k+1}(G)$, consisting of nodes with core-number k, is referred to as the k-shell; the core decomposition (i.e., the partitioning into shells) can be computed efficiently and combines local as well as global aspects of the network structure. This makes it a very popular measure (along with other graph properties, e.g., degree distribution and clustering coefficient) in a wide variety of applications, such as: the autonomous system level graph of the Internet [6,3], bioinformatics [18,26], social networks and epidemiology [20,17]; some of the key properties that have been identified include: the well-connectedness of the nodes with high core number and their significance in controlling cascades.

In most applications however, the networks are inferred by indirect measurements, e.g.: (i) the Internet router/AS level graphs constructed using traceroutes, e.g., [12], (ii) biological networks, which are inferred by experimental correlations, e.g., [18,26], (iii) networks based on Twitter data (related to which there is a

H. Blockeel et al. (Eds.): ECML PKDD 2013, Part I, LNAI 8188, pp. 541–556, 2013.

growing body of research, e.g., [17,4,16]), in which a limited 1% sample can be constructed by the APIs.[1] Therefore, networks studied in these applications are inherently noisy and incomplete; this raises a fundamental issue in the use of any graph property $\mathcal{P}(G)$ for graph G: How does the property, and conclusions based on it get affected by the uncertainty in G? Is there a smooth transition in the property with the uncertainty,[2] and is it possible to quantify the error in the observed measurement? An example of such an issue is the nature of degree distributions of the Internet router graph and its vulnerability: several papers, e.g., [12] observed that these are power laws. Achlioptas et al. [1] showed that there are significant sampling biases in the way traceroutes (which are used to infer the network) work; for a broad class of networks, they prove that such inference methods might incorrectly infer a power-law distribution (even when the underlying network is not).

Our work is motivated by these considerations of the sensitivity to noise and the adequacy of sampling. Specifically, we study how results about the core decomposition and top cores in the network, e.g., [6,3,18,26,20,17], are affected by the uncertainty, noise and small samples (as in the case of online social media networks). Such questions have been studied in the statistical physics literature, e.g., [10], who show that there is a threshold probability for random node deletions in infinite networks, above which the k-core disappears; it is not clear how relevant such results are to real world networks, which are finite and do not satisfy the symmetries needed in such results. Hamelin et al. [3] report robustness of their observations related to the shell structure in the Internet router graph, for specific sampling biases related to traceroute methods. We are not aware of any other empirical or analytical work on the sensitivity of the core decomposition.

Since there is very limited understanding of how noise should be modeled, we consider three different stochastic edge perturbation models, which are specified by how a pair u, v of nodes is picked: (i) uniformly at random (ERP, for Erdős-Rényi perturbations), (ii) in a biased manner, e.g., based on the degrees of u, v (CLP, for Chung-Lu or degree assortative perturbations), and (iii) by running a missing link prediction algorithm, such as [8] (LPP, for link prediction based perturbations); see Section 3 for complete definitions. We also study a model of stochastic node deletions. Let α denote the fraction of nodes/edges perturbed; typically we are interested in "small" α.

A complementary aspect (particularly relevant in the context of sampled data from social media such as Twitter) is the effect of sampling. We consider edge/node sampling with probability p (i.e., corresponding to deletion with probability $1-p$). We study the following question: can the properties of the core structure be identified by small edge/node samples, i.e., corresponding to small p? In our discussion below, we use G' to denote the graph resulting from perturbation/sampling of a

---

[1] Larger samples, e.g., 10% can be obtained form Twitter's commercial partners for a large fee.

[2] As observed in the case of centrality measures by [5], who claim that there is a gradual decrease in the accuracy of the centrality scores.

starting graph G. Let $k_{max}(G)$ denote the maximum core number in G. We study
the Jaccard similarity, $\eta_j(G, G')$, between the set of nodes in the top j-cores in G
and G'; we sometimes informally refer to η_j as the "similarity" between the top
cores. Our main results are summarized below.

1. Sensitivity to Noise. We consider perturbations with α ranging from less
than 1% to 10%. We find that η_j and the shell structure shows high sensitivity
to edge/node perturbations; however, the precise effects are very network and
noise model specific. Further, η_1 is quite sensitive in the CLP model for many
networks: perturbation with $\alpha < 5\%$ can alter η_1 by more than 20% in some
networks. More importantly, we find that in a large fraction of the networks,
η_j exhibits a *non-monotone* behavior as a function of α. This can be a serious issue in some applications where the core structure is used, and needs to
be examined critically. The sensitivity decreases as we increase j, but η_j varies
non-monotonically with j as well. In contrast, the top cores seem quite stable
to perturbations in the ERP model, which primarily affects the shell size distribution in some networks. The LPP model seems to affect both the low and high
cores. Further, node perturbations (modeled as random deletions) seem to have
a much higher impact than edge perturbations, in general. It is intriguing that
co-authorship and citation networks seem to be generally much more stable compared to influence and infrastructure networks. Further, we observe that sudden
changes in the similarity index are almost always accompanied with increase in
k_{max}.

   This motivates the CorePerturbation problem: given a graph G and a parameter k, what is the probability that a k-core forms in G after perturbation, if
it did not have a k-core initially. We prove that this problem is #P-hard, which
suggests rigorous quantification of the variation in the top core even in such
simple stochastic noise models is quite challenging. We attempt to further understand and explain the empirical observations analytically using simple mathematical models. We also prove that under some weak assumptions that usually
hold in social networks, the low core numbers can be altered quite significantly
in the ERP model.

2. Sensitivity to Sampling. We find most networks exhibit a high level of sensitivity to sampling, and η_j is a noisy and non-monotone function of p, especially
when p is close to 1; there is higher level of sensitivity to node sampling than
to edge sampling. For most of the networks we study, identifying a reasonably
large fraction (say 80%) of the top core set requires a fairly high sampling rate
p: higher than 0.6 in most networks, and as high as 0.8 in some. Specifically, in
the case of a Twitter "mentions" graph (see Section 3.2 for details), we find that
this entails a much higher level of sampling than what is supported by the public
API. Further, biased sampling based on edge weights can improve the similarity
index slightly. We analyze the effects of sampling to help explain some of these
results. We show that the maximum core number in G_p scales with the sampling
probability, and that non-monotonicity under sampling is an inherent aspect of

the Erdős-Rényi model. We also find that the top core can be very fragile, and can change completely even for very low sampling rate.

Organization. We briefly discuss the related work in Section 2. We introduce the main definitions, and summarize our data sets in Section 3. We discuss the sensitivity to noise and the effects of sampling in Sections 4 and 6, respectively. In Section 5, we discuss the CorePerturbation problem, and conclude in Section 7. Because of limited space, we present many of the details in the full version [2].

2   Related Work

Noise and sampling biases in networks are well recognized as fundamental issues in applications of complex networks, and many different models have been studied for it. A common approach in social networks, e.g., [9,5], is to examine stochastic node and edge deletions. There is a large body of work on predicting missing links in complex networks (based on expected clustering and other structural properties), e.g., [8], which could also be used as a possible candidate set. Since there is no clear understanding of noise/perturbations, we study three different models from the literature in this paper.

We briefly discuss a few of the results on understanding the impact of uncertainty on network properties. The impact of sampling bias on the properties of the Internet router graph [1] was already mentioned earlier. There has also been a lot of work in understanding the sensitivity of centrality to noise, e.g., [9,5]; it has been found that the impact on the centrality is variable and network dependent, but the general finding in [5] is that the accuracy of centrality measures varies smoothly and predictably with the noise. Morstatter et al. [22] study the effects of limited sampling in social media data by analyzing the differences in statistical measures, such as hashtag frequencies, and network measures, such as centrality.

The work by Flaxman and Frieze [14,15] is among the very few rigorous results on the impact of perturbations on network parameters— they rigorously analyze the impact of ERP on the diameter and expansion of the graph. The issue of noise has motivated a number of sampling based algorithmic techniques which are "robust" to uncertainty, in the form of "property testing" algorithms, e.g., [25] and "smoothed analysis", e.g., [27].

Finally, we briefly discuss some of the work on the core decomposition in graphs. As mentioned earlier, the core number and the top core set has been used in a number of applications, e.g., [6,3,18,26,20,17], in which the shell structure and the top core sets have been found to give useful insights. Conditions for existence of the k-core, and determining its size have been rigorously studied in different random graph models, e.g., [24,13]; the main result is that there is a sharp threshold for the sudden emergence of the k-core in these models. This has also been analyzed in the statistical physics literature, e.g., [10]; these papers also study the impact of node deletions on the core size in infinite graphs, and show a characterization in terms of the second moment of the degree distribution.

3   Definitions and Notations

The k-core $C_k(G)$ of an undirected graph $G = (V, E)$ is defined as the maximal subgraph of nodes in which each node has degree at least k; the core-number of a node v is the largest k such that $v \in C_k(G)$. The set $S_k(G) = C_k(G) \setminus C_{k+1}(G)$ is referred to as the kth-shell of G; we omit G, and refer to it by S_k, when the graph is clear from the context. The set $C_k(G)$, if it exists, can be obtained by repeatedly removing vertices of degree less than k until no further removal is possible. The maximum k such that $C_k(G) \neq \phi$ will be denoted by $k_{\max}(G)$; we use just $k_{\max}$, when there is no ambiguity. The core decomposition of a graph G corresponds to the partition $S_0, S_1, \ldots, S_{k_{\max}}$ of V. Let $s_i = |S_i|$. We use $\beta(G) = \langle s_1, s_2, \ldots, s_{k_{\max}} \rangle$ to denote the vector of shell size distribution in G. The *Jaccard index* is a measure of similarity between two sets and is defined as follows: For sets A and B, $JI(A, B) = \frac{|A \cap B|}{|A \cup B|}$. In our empirical analysis of networks we compare the top j cores of the unperturbed and the perturbed graphs using the Jaccard index. For this purpose we introduce the notation $\eta_j(G, G') := JI\left(\cup_{i \geq k_{\max}-j+1} C_i(G), \cup_{i \geq k_{\max}-j+1} C_i(G')\right)$. The *variation distance* between the core-number distributions of two graphs G and G' on the same vertex set V is defined as, $\delta(G, G') = \frac{1}{2|V|} \sum_i |s_i(G) - s_i(G')|$. We say that an event holds **whp** (with high probability) if it holds with probability tending to 1 as $n \to \infty$.

3.1   Noise Models

Since there is no clear understanding of how uncertainty/noise should be modeled, we introduce a generalized noise model for edge perturbations which captures most models in literature, and also enables us to control separately the extent of addition and deletion. Let G be the unperturbed graph. Let $\mathbb{G} = \mathbb{G}(n)$ denote a random graph model on n nodes which is specified by the probability $P_{\mathbb{G}}((u, v))$ of choosing the edge (u, v). We define a noise model $\mathcal{N}(G, \mathbb{G}, \epsilon_a, \epsilon_d)$ based on $\mathbb{G}$ as a random graph model where the edge probability between a pair u, v is given by

$$P_{\mathcal{N}}((u, v)) = \begin{cases} \epsilon_a P_{\mathbb{G}}((u, v)), & \text{if } (u, v) \notin E_G, \\ \epsilon_d P_{\mathbb{G}}((u, v)), & \text{if } (u, v) \in E_G, \end{cases} \tag{1}$$

where ϵ_a and ϵ_d denote the edge addition and deletion probabilities, respectively. The perturbed graph $G' = G \oplus R$ is obtained by XORing G with $R \in \mathcal{N}(G, \mathbb{G}, \epsilon_a, \epsilon_d)$, a sample from the noise model, i.e., if $(u, v) \in E_G$, then it is deleted with probability $\epsilon_d P_{\mathbb{G}}((u, v))$, but if $(u, v) \notin E_G$, (u, v) is added with probability $\epsilon_a P_{\mathbb{G}}((u, v))$. Depending on how we specify $P_{\mathbb{G}}$ and the parameters ϵ_a, ϵ_d, we get different models; we consider three specific models below.

Uniform Perturbation (ERP): In this model we set $\mathbb{G} = \mathcal{G}(n, 1/n)$, the Erdős-Rényi random graph model where each edge is chosen with probability $1/n$ independently, i.e., $P_{\mathbb{G}}((u, v)) = 1/n$. We use the following notation for this model:

$ERP(G, \epsilon_a, \epsilon_d) = \mathcal{N}(G, \mathcal{G}(n, 1/n), \epsilon_a, \epsilon_d)$. For example, $ERP(G, \epsilon, \epsilon)$ corresponds to adding an edge or removing an existing edge independently with probability ϵ/n, while $ERP(G, \epsilon, 0)$ corresponds to only adding edges. This is the simplest model, and has been studied in social network applications, e.g., [9,5].

Degree Assortative Perturbation (CLP): In this model, $\mathbb{G}$ corresponds to the Chung-Lu random graph model [7] for graphs with a given expected degree sequence. Each node u is associated with a weight w_u (which we take to be its degree), and edge is chosen independently with probability proportional to the product of the weights of its endpoints, i.e., $P_{\mathbb{G}}((u, v)) \propto w_u \cdot w_v = d(u) \cdot d(v)$. This model selects edges in a biased manner, and might be suitable in applications dealing with assortative graphs with correlations between degrees of the end points of edges, which has been observed in a number of networks, e.g., [23].

Link Prediction Based Model (LPP): Instead of the purely stochastic ERP and CLP models, we use the results of a missing link prediction algorithm to determine which edges to perturb. Here, we use the algorithm of Clauset, et al. [8], which has been used quite extensively in the social network literature; further, since it uses a hierarchical random graph model, it can be viewed as an instance of our generalized noise model. This model is based on the assumption that many real-life networks have a hierarchical structure, which can be represented by a binary tree with n leaves corresponding to the node (referred to as a "dendrogram"). Given such a dendrogram D, each internal node r is associated with a probability $p_r = \frac{E_r}{L_r R_r}$, where L_r and R_r are the number of leaves in the left and right subtrees of r respectively and E_r is the number of edges between L_r and R_r in G. The likelihood of D is defined as: $\mathcal{L}(D) = \Pi_r p_r^{E_r}(1 - p_r)^{L_r R_r - E_r}$. The algorithm of [8] specifies the probability, $P_D((u, v))$, of an edge between two vertices u, v, to be the value p_r, where r is the lowest common ancestor of u and v in D.

In the ERP model, the expected number of perturbed edges is $\approx n\epsilon/2$. For the purpose of fair comparison of noise models, we have normalized the weights of vertices in the CLP model such that the expected number of perturbed edges is again $\approx n\epsilon/2$. We use G_ϵ to denote the perturbed network. In the LPP model, we add edges as prescribed [8]; $n\epsilon/2$ edges are added in the decreasing order of their associated probabilities.

Additions vs Deletions: We find that, due to sparsity of the networks considered, perturbations involving edge additions/deletions do not alter the results by much, compared to perturbations involving just edge additions. Hence, unless explicitly specified, we only consider addition of edges. Also, henceforth, whenever we use the truncated notations ERP and CLP, we refer to $ERP(G, \epsilon, 0)$ and $CLP(G, \epsilon, 0)$, respectively.

Noise could also manifest in terms of missing nodes. We study a model of random node deletions with probability $1 - p$ (which corresponds to retaining nodes with probability p); we study the effect of this in the form of sampling in Section 6, instead of perturbations.

3.2 Data

In order to make our results as robust as possible, we analyze over 25 different real (from [21]) and random networks. We also used a Twitter mentions graph, constructed in the following manner: we consider a set of about 9 million tweets (corresponding to a 10% sample, obtained from a commercial source), and construct a graph on the Twitter users, in which an edge (u, v) denotes a mention of user v by user u (in the form of an "@v" in the tweet) or the other way around; this graph has over 2 million nodes and about 4.6 million edges. We then considered subgraphs constructed by sampling edges with probability $p \in [0.1, \ldots, 0.99]$; for $p \in [0.1, \ldots, 0.8]$, we use increments of 0.1, but for $p \in [0.8, 0.99]$, we use increments of 0.01, in order to increase the resolution. Finally, we also consider random graph models with Poisson and scale-free degree distributions. Table 1 contains a summary of the graphs analyzed.

Table 1. Real-world and synthetic graphs used in our experiments and their properties

Class	Network	N	E	$k_{\max}$	$|C_{k_{\max}}(G)|$
	As20000102	6474	12572	12	21
Autonomous Systems	Oregon1010331	10670	22002	17	32
	Oregon2010331	10900	31180	31	78
	Astroph	17903	196972	56	57
	Condmat	21363	91286	25	26
Co-authorship	Grqc	4158	13422	43	44
	Hepph	11204	117619	238	239
	Hepth	8638	24806	31	32
Citation	HepPh	34546	420877	30	40
	HepTh	27770	352285	37	52
Communication	Email-EuAll	265214	364481	37	292
	Email-Enron	33696	180811	43	275
Social	Epinion	75877	405739	67	486
	Slashdot0811	77360	469180	54	129
	Soc-Slashdot0902	82168	504230	55	134
	Twitter	22405	59898	20	177
	Wiki-Vote	7066	100736	53	336
	Twitter "mentions"	2616396	4677321	19	210
Internet peer-to-peer	Gnutella04	10876	39994	7	365
	Gnutella24	26518	65369	5	7480
Synthetic graphs	Regular ($d = 20$)	10000	100000	20	10000

4 Sensitivity of the Core Decomposition to Noise

We now study the effect of node/edge perturbations on the similarity index $\eta_j(G, G')$, and the changes in the shell size distribution $\beta(G)$ in terms of the variation distance, $\delta(G, G')$ (see Section 3 for definitions). We study these quantities on the networks mentioned in Section 3.2 and for the perturbation models discussed in Section 3.1. For the ERP and CLP models, we compute 100 to 1000 instances, for each choice of ϵ, over which $\eta_j(G, G')$ and $\delta(G, G')$ are averaged. The methodology for the LPP model is discussed later.

4.1   Sensitivity of the Top Cores

1. Sensitivity of the Top Core in the CLP Model: Figure 1 shows the variation in $\eta_1(G, G')$ for different networks in this model. The figure shows the variation with both ϵ and α (the fraction of edges added), the latter to account for the difference in the graph sizes. The most striking observation is the high sensitivity of η_1 and its highly non-monotonic variation in a large fraction of the networks. The specific points where significant jumps in η_1 happen correspond to the points where k_{max} changes in many cases, as shown in Figure 1(c). The specific behavior is highly variable and network dependent. For example, we note that while the top cores in collaboration and citation networks are, in general, highly resilient to perturbation, most social and peer to peer networks show great variation.

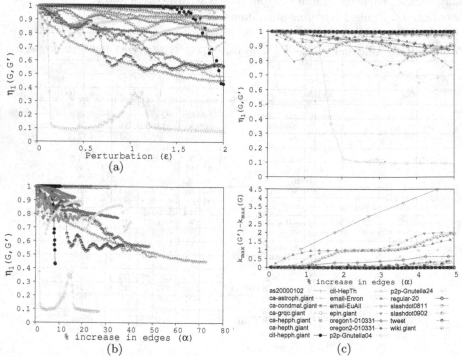

Fig. 1. Top core comparison for various networks under degree-weighted edge perturbation CLP: Here, (c) is a zoomed plot of (b). This is complemented by a plot of $k_{max}(G') - k_{max}(G)$ to depict the transition to a higher core and its effect on $\eta_1(G, G')$.

2. Sensitivity of the top core in the ERP Model: In contrast to the CLP model, we find that top cores are much more stable in the ERP model. The main reason for this stability is the fact that almost all networks considered here have very small fraction of nodes in the top core(s) (as shown in Figure 7 in the full version [2]), so that most of the edges in the ERP model are added to low core nodes

3. Explaining the Differences Between the CLP and ERP Models: We note that in the CLP model, the higher the degree of a node in the unperturbed graph, the greater is the number of edges incident with it after perturbation. This polarizing nature of the model needs to be taken into account to infer and quantify the stability of the top core. Figure 6 in the full version [2] shows scatter plots of core number vs. degree for some selected graphs. Even though it gives some idea about the behavior of the top core, we find it highly non-trivial to quantify the stability in any way. Later, in Section 5 we will be considering a theoretical formulation of this problem and showing that such a quantification of stability is in general hard.

4. Sensitivity of the Top 5 Cores: We extend our empirical analysis to $\eta_j(G, G')$ for $j = 1, \ldots, 5$ in Figure 2. Note that the non-monotonic behavior is mitigated in these plots, but η_j varies non-monotonically with j. However, as j is increased, the size of C_j can become very large, thus diminishing the main utility of the top cores in most applications.

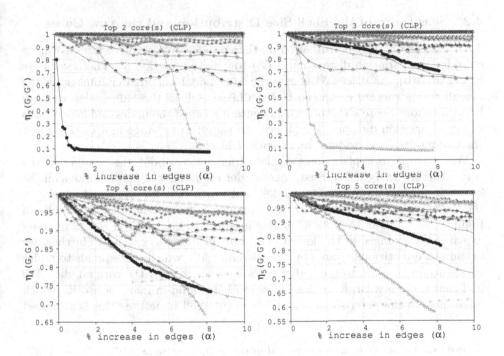

Fig. 2. Top 2–5 cores comparison ($\eta_j(G, G')$, $j = 2, \ldots, 5$) with respect to % increase in edges (α). The legends are the same as in Figure 1.

5. Sensitivity in the LPP Model: We considered the stability of the top cores in the LPP model by applying the link prediction algorithm given in [8]. We first generated the list of likelihood probabilities for each possible edge. For this purpose, we used the implementation of [11]. The edges were then added in the descending order of their probabilities. As shown in Figure 3(a) for a subset of graphs, the variation in η_1 is very network specific, and hard to characterize.

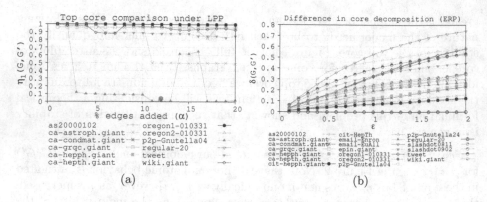

Fig. 3. Core stability with respect to LPP and ERP

4.2 Sensitivity of the Shell Size Distribution and the Low Cores

To study the effect of perturbation on the core decomposition of a network, we consider the variation distance $\delta(G, G')$ (defined in Section 3). The results are in Figure 3(b). As discussed above, the ERP model has greater impact on the overall core-structure compared to the CLP model; in the ERP model, changes happen to lower core structure which contain most of the nodes and hence leads to large variation distance. We observe no significant change in $\delta(G, G')$ under the CLP model, as is evident from Figure 7 in [2].

We attempt to explain some of the observations about the changes in the core structure analytically. First, we consider the impact of perturbations on the 2-core in any graph in the ERP model, and prove that for any constant $\epsilon > 0$, the 2-core always becomes of size $\Theta(n)$, which is consistent with the results in Figure 7 in [2]. Our results are similar in spirit to the work of [14]. This only explains the changes in the lowest core, and in order to extend it further, we examine a quantity motivated by the "corona" [10], which corresponds to nodes which need few edges to the higher cores in order to alter the core number. We find that there is a large fraction of nodes of this kind in many networks, which might help in characterizing the stability of the shell structure. This is discussed in [2].

Theorem 1. *Let G be any connected graph with n vertices and let $G_p = G \oplus R$ where, $R \sim \mathcal{G}(n, \epsilon/n)$ and ϵ is a constant. Then,* **whp** *G_p has a 2-core of size $\Theta(n)$.*

Proof. (Sketch) Consider a spanning tree T of G. We show that $T \oplus R$ itself has a $\Theta(n)$ sized 2-core. Let T^- denote the subgraph obtained by removing the edges common to T and R. Suppose e_d is the number of edges removed from T. Since each edge of T can be removed with probability ϵ/n, it can be verified that **whp** $O(\log n)$ edges are removed from T, so that T^- has $O(\log n)$ components **whp**.

Let I be a maximum independent set of T^-. We consider the graph induced by I in R and consider the edges of $R[I]$ (not belonging to T) one at a time. Let $e_i = (u, v)$ be the ith edge of $R[I]$ added to T^-. Let $T_i^- = T_{i-1}^- + e_i$ with $T_0^- = T^-$. If u and v belong to the same component of T_{i-1}^-, then, there is a path P in T_{i-1}^- with end points u and v and therefore, u and v both belong to the 2-core in $T \oplus R$. However, when u and v belong to different components, this does not happen. However, note that each time this happens, T_i^- has one less component compared to T_{i-1}^-. Since T_0^- has $O(\log n)$ components, there can be at most $O(\log n)$ such edges.

Consider vertices in $R[I]$ of degree at least 1. Note that $|I| = cn$ for some constant $c \geq 1/2$. It is easy to see that **whp** a constant fraction of these vertices in I have degree at least 1 in $R[I]$. Of these vertices, we will discard vertices which are end points of an edge e_i which is between two components of T_{i-1} for some i. From the previous discussion, there can be only $O(\log n)$ such edges. The rest of the vertices form a 2-core. Hence proved. □

5 The CorePerturbation Problem

From Section 4.1, it follows that the sudden and non-monotone changes in the similarity index correspond to an increase in the maximum core number. This motivates the COREPERTURBATION problem, which captures the probability of this change happening.

Definition 1. *The* CorePerturbation *problem* $(CP(G, E_A, p, k))$
Input: *A graph $G(V, E)$, an integer $k \geq 4$, edge probability p and a set of possible edges E_A (which are absent in G). Let G_p be the graph resulting from adding edges to G from E_A independently with probability p.*
Output: *Probability that G_p has a k-core.*

Theorem 2. $CP(G, E_A, p, k)$ *is #P-complete.*

The proof of Theorem 2 is in the full version [2]. The result also holds when $k_{\max}(G) = k - 1$, which implies that even in a very simple noise model, quantifying the precise effects of changes in the top core is very challenging. When this probability is not too small (e.g., larger than $1/n^c$ for some constant $c > 0$), it can be shown that a polynomial number of Monte-Carlo samples can give good estimates (within a multiplicative factor of $1 \pm \delta$, with any desired confidence, where $\delta > 0$ is a parameter).

6 Sensitivity of the Core Decomposition to Sampling

We now address the issue of sampling and focus on $\eta_k(G, G_p)$, where G_p denotes a node/edge sampled graph with probability p— our goal is to understand to what extent the core structure (especially the nodes in the top cores) can be identified from sparsely sampled data. As in the case of noise (Section 4), we find η_k is quite sensitive to sampling, and varies non-monotonically for many graphs. We attempt to explain these results rigorously in the following manner: (i) using the notion of edge density, we derive bounds on the maximum core in sampled graphs, which show that it scales with p, (ii) we analyze the sampling process in random graphs, and prove that the non-monotonicity in η_k is an inherent issue related to the core structure.

6.1 Variation in η_k

Figure 4(a) shows the variation in $\eta_1(G, G_p)$ for all networks, for an edge sampling probability $p \in [0.8, 1]$. We observe that η_1 is quite low in many networks; in order to identify at least 80% of the top core nodes (i.e., $\eta_1 \geq 0.8$), we need $p \geq 0.6$ in most networks. Figure 9 in the full version [2] shows additional results on the effect of edge sampling on η_k, for $k \in \{1, 2, 5, 10\}$. Like in the case of edge perturbations, we find η_k also exhibits non-monotonicity with respect to k for most networks. Further, it is interesting to note that the citation networks are very sensitive to sampling (and have η_1 below 0.6), but were found to be quite robust to edge perturbations (Section 4). However, collaboration networks seem to be robust to sampling as in the case of edge perturbations. We find that node sampling has a much higher impact than edge sampling; see Table 2 in the full

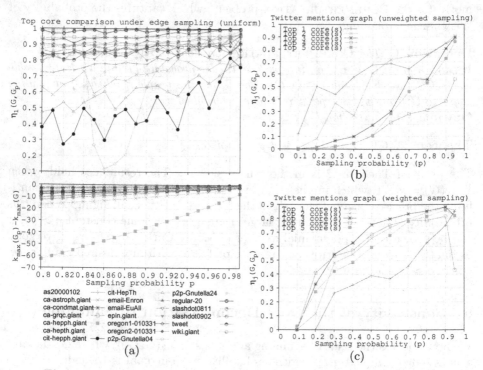

Fig. 4. Top core comparison for various networks under sampling edges

version [2] for details. For instance, with $p = 0.95$, we observe that η_1 is below 0.9 for all but four of the networks, and is below 0.62 in three networks.

Biased Sampling in Twitter Networks. Since sampling is an inherent aspect of the APIs provided by Twitter, we study its effects on the top cores. Our results for the Twitter mentions graph (see Section 3.2 for the details) are shown in Figures 4(b) and 4(c). The graph is weighted, in which the weight of an edge (u, v) corresponds to the number of mentions of u by v (or the converse). We observed that η_k is generally quite low, but is somewhat higher when edges are sampled with probability proportional to the edge weight (Figure 4(b)) instead of uniform (Figure 4(c)). Moreover, there is high non-monotonicity in both scenarios, suggesting that Twitter's public API is not adequate for identifying the core structure with high confidence (say 80% or more), and multiple calls to the API must be run to improve the accuracy. Table 3 in the Appendix of [2] gives additional details on the max core values in the sampled graphs.

Bounding the Max Core on Sampling. A first step towards understanding the effect of sampling is to determine $k_{\max}(G_p)$ in the sampled graph G_p. Table 3 in the full version [2] suggests that $k_{\max}$ scales with the sampling probability. This is examined in the following lemma, whose proof is discussed in the full version [2].

Lemma 1. *Consider a graph G such that $k_{\max}(G) \to \infty$ as $n \to \infty$. Let G_p denote the random subgraph of G obtained by retaining each edge of G with probability p, where p is a constant. Then, for any constant $\delta \in (0, 1)$, $k_{\max}(G_p) > (1 - \delta)k_{\max}(G)p/2$,* **whp**.

6.2 Core Structure in Random Graphs

In order to understand our empirical observations about the sensitivity of the core structure to noise and sampling, and especially the non-monotone behavior, we now study the effect of sampling in random graph models. We consider two random graph families: (a) the Erdős-Rényi random graphs and (b) Chung-Lu power law random graphs [7] with node weights picked from a power-law distribution (see Section 3 for a description of this model). Figure 5 shows the sensitivity of $\eta_k(G, G_p)$ to the sampling probability p. Figure 5(a) shows the results for a random graph from $\mathcal{G}(n, p)$ for $n = 10000$ and $p = 50/n$ and Figure 5(b) shows the results for a graph from the Chung-Lu model with power law exponent 2.5, $n = 10000$ and average degree 5. We observe non-monotone variation in η_k with p; this is more pronounced in the case of the Chung-Lu model, in which case η_k is quite low, which is consistent with the effect of perturbations on real networks in Section 4. Further, we observe that the variation in η_k is much smoother for $k > 1$, which is not the case of the networks in Section 4. This non-monotone variation of η_k with the sampling probability is explained to some extent through Lemma 2; by analyzing η_k in the Erdős-Rényi model, we show rigorously that this is an inherent aspect of most graphs.

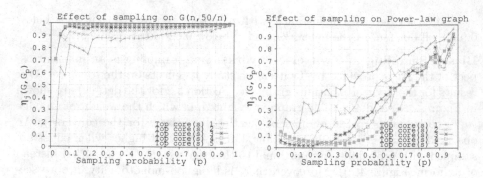

Fig. 5. Non-monotonicity of top-cores in random graphs: (a) Erdős-Rényi model $\mathcal{G}(n, c/n)$, with $n = 10000$ and $c = 50$; (b) Chung-Lu model with $n = 10000$ and average degree 5.

Lemma 2. *There exist constants c and pairs (p_1, p_2), where $0 < p_1 < p_2 < 1$ such that for $G \in \mathcal{G}(n, c/n)$, $\eta_1(G, G_{p_1}) > \eta_1(G, G_{p_2})$* **whp**.

This lemma relies on the result of [19]. Suppose $G \in \mathcal{G}(n, \lambda/n)$. Let $Po(\mu)$ denote a Poisson random variable with mean μ. For a positive integer j, let $\psi_j(\mu) := P(Po(\mu) \geq j)$ and let $\lambda_j := \min_{\mu > 0} \mu/\psi_{j-1}(\mu)$. Let, for $\lambda > \lambda_j$, $\mu_j(\lambda) > 0$ denote the largest solution to $\mu/\psi_{j-1}(\mu) = \lambda$. Pittel et al. [24] show that if $\lambda < \lambda_k$ and $k \geq 3$, then k-core $C_k(G)$ is empty **whp**, while if $\lambda > \lambda_k$, $|C_k(G)| = \psi_k(\mu_k(\lambda))n$, **whp**.

Proof. (of Lemma 2) First we note that for $G \in \mathcal{G}(n, c/n)$, the random graph sampled with probability p, G_p itself is a $\mathcal{G}(n, cp/n)$ random graph. We choose $c = 50$, for which **whp** $k_{\max}(G) = 38$ and $|C_{k_{\max}}(G)| \approx 0.91n$. We set $p_1 = 0.102$, such that cp_1 is slightly less than $\lambda_4 \approx 5.15$ (in the context of the result of [24]). For this p_1, $k_{\max}(G_{p_1}) = 3$ and $|C_{k_{\max}}(G_{p_1})| \geq 0.864n$ **whp**. We choose $p_2 = 0.198$, such that cp_2 is slightly more than $\lambda_7 \approx 9.88$. This means $k_{\max}(G_{p_2}) = 7$ and $|C_{k_{\max}}(G_{p_2})| \approx 0.694n$ **whp**. Now we show that $\eta_1(G, G_{p_1}) > \eta_1(G, G_{p_2})$ **whp**.

For any set U and subsets $A, B \subseteq U$, the following inequality follows trivially: $\frac{|A|+|B|-|U|}{|U|} \leq JI(A, B) \leq \frac{|B|}{|A|}$. We set $A = C_{k_{\max}}(G)$ and $U = V(G)$. Setting $B = C_{k_{\max}}(G_{p_1})$ and using the lower bound in the above inequality, the Jaccard Index at p_1 is $\geq 0.91 + 0.864 - 1 = 0.774$. Setting $B = C_{k_{\max}}(G_{p_2})$ and using the upper bound in the above inequality, the Jaccard Index at p_2 is $\leq 0.694/0.91 \approx 0.762$. Hence, proved. □

Remark 1. The proof of Lemma 2 is a rigorous analysis of the non-monotone behavior seen in Figure 5(a) in the interval $[0.1, 0.2]$. Similar pairs can be demonstrated for other values of c which correspond to $k_{\max} = 39, 40$ and so on.

7   Conclusions

Our results show that the top cores show significant sensitivity to perturbations, and can be recovered to a reasonable extent in sampled graphs, only if the sampling rate is sufficiently high. These results suggest that a careful sensitivity analysis is necessary when using the core structure, especially because of the non-monotone effects on the similarity index of the top cores. Our formulation of the CorePerturbation problem and its #P-hardness implies quantifying the effects of uncertainty can be a challenging problem even in very simplified noise models; developing efficient algorithms for this problem is an interesting open problem. Further, the non-monotone behavior in the similarity index of the top cores implies simple statistical tests that might try to improve the confidence by bounding the uncertainty might not work. The reduced non-monotonicity in η_k with k suggests considering the top few cores, instead of just the top core, as a way of dealing with these effects; however, as we observe, this would require considering a much larger set of nodes. The significant sensitivity to sampling also suggests the need for greater care in the use of networks inferred using small samples provided by public APIs of social media applications. We expect our approach to be useful in the analysis of the sensitivity of other network properties to noise and sampling.

Acknowledgments. We are grateful to the reviewers whose comments have helped improve the paper. This work has been partially supported by the following grants: DTRA Grant HDTRA1-11-1-0016, DTRA CNIMS Contract HDTRA1-11-D-0016-0010, NSF Career CNS 0845700, NSF ICES CCF-1216000, NSF NETSE Grant CNS-1011769 and DOE DE-SC0003957. Also supported by the Intelligence Advanced Research Projects Activity (IARPA) via Department of Interior National Business Center (DoI/NBC) contract number D12PC000337, the US Government is authorized to reproduce and distribute reprints for Governmental purposes notwithstanding any copyright annotation thereon.

Disclaimer: The views and conclusions contained herein are those of the authors and should not be interpreted as necessarily representing the official policies or endorsements, either expressed or implied, of IARPA, DoI/NBC, or the US Government.

References

1. Achlioptas, D., Clauset, A., Kempe, D., Moore, C.: On the bias of traceroute sampling. J. ACM 56(4), 21:1–21:28 (2009)
2. Adiga, A., Vullikanti, A.: How robust is the core of a network? http://ndssl.vbi.vt.edu/supplementary-info/vskumar/kcore.pdf
3. Alvarez-Hamelin, J.I., Dall'Asta, L., Barrat, A., Vespignani, A.: K-core decomposition of internet graphs: hierarchies, self-similarity and measurement biases. NHM 3(2), 371–393 (2008)
4. Bakshy, E., Hofman, J.M., Mason, W.A., Watts, D.J.: Everyone's an influencer: quantifying influence on Twitter. In: Proceedings of the Fourth ACM International Conference on Web Search and Data Mining, pp. 65–74. ACM (2011)
5. Borgatti, S., Carley, K., Krackhardt, D.: On the robustness of centrality measures under conditions of imperfect data. Social Networks 28, 124–136 (2006)

6. Carmi, S., Havlin, S., Kirkpatrick, S., Shavitt, Y., Shir, E.: A model of internet topology using k-shell decomposition. Proceedings of the National Academy of Sciences 104(27), 11150–11154 (2007)
7. Chung, F., Lu, L.: Connected components in random graphs with given expected degree sequences. Annals of Combinatorics 6, 125–145 (2002)
8. Clauset, A., Moore, C., Newman, M.: Hierarchical structure and the prediction of missing links in networks. Nature 453, 98–101 (2008)
9. Costenbader, E., Valente, T.: The stability of centrality measures when networks are sampled. Social Networks 25, 283–307 (2003)
10. Dorogovtsev, S.N., Goltsev, A.V., Mendes, J.F.F.: k-core architecture and k-core percolation on complex networks. Physica D: Nonlinear Phenomena 224(1), 7–19 (2006)
11. Dronen, N.: PyHRG (2013), https://github.com/ndronen/PyHRG
12. Faloutsos, M., Faloutsos, P., Faloutsos, C.: On power-law relationships of the internet topology. In: SIGCOMM, vol. 29, pp. 251–262 (1999)
13. Fernholz, D., Ramachandran, V.: Cores and connectivity in sparse random graphs. Technical report, UTCS TR04-13 (2004)
14. Flaxman, A.D., Frieze, A.M.: The Diameter of Randomly Perturbed Digraphs and Some Applications. In: Jansen, K., Khanna, S., Rolim, J.D.P., Ron, D. (eds.) RANDOM 2004 and APPROX 2004. LNCS, vol. 3122, pp. 345–356. Springer, Heidelberg (2004)
15. Flaxman, A.D.: Expansion and lack thereof in randomly perturbed graphs. Internet Mathematics 4(2-3), 131–147 (2007)
16. Galuba, W., Aberer, K., Chakraborty, D., Despotovic, Z., Kellerer, W.: Outtweeting the Twitterers-predicting information cascades in microblogs. In: Proceedings of the 3rd Conference on Online Social Networks, p. 3. USENIX Association (2010)
17. González-Bailón, S., Borge-Holthoefer, J., Rivero, A., Moreno, Y.: The dynamics of protest recruitment through an online network. Scientific Reports 1 (2011)
18. Hagmann, P., Cammoun, L., Gigandet, X., Meuli, R., Honey, C.J., Van Wedeen, J., Sporns, O.: Mapping the structural core of human cerebral cortex. PLoS Biology 6(7), e159 (2008)
19. Janson, S., Luczak, M.J.: A simple solution to the k-core problem. Random Structures & Algorithms 30(1-2), 50–62 (2007)
20. Kitsak, M., Gallos, L.K., Havlin, S., Liljeros, F., Muchnik, L., Stanley, E., Makse, H.A.: Identification of influential spreaders in complex networks. Nature Physics 6(11), 888–893 (2010)
21. Leskovec, J.: Stanford network analysis project (2011), http://snap.stanford.edu/index.html
22. Morstatter, F., Pfeffer, J., Liu, H., Carley, K.: Is the sample good enough? comparing data from Twitter's streaming API with Twitter's firehose. In: AAAI Conference on Weblogs and Social Media, ICWSM (2013)
23. Newman, M.: The structure and function of complex networks. SIAM Review 45(2), 167–256 (2003)
24. Pittel, B., Spencer, J., Wormald, N.C.: Sudden emergence of a giantk-core in a random graph. J. Comb. Theory, Ser. B 67(1), 111–151 (1996)
25. Ron, D.: Algorithmic and analysis techniques in property testing. Foundations and Trends in TCS 5(2), 73–205 (2010)
26. Schwab, D.J., Bruinsma, R.F., Feldman, J.L., Levine, A.J.: Rhythmogenic neuronal networks, emergent leaders, and k-cores. Physical Review E 82(5), 051911 (2010)
27. Spielman, D.: Smoothed analysis: An attempt to explain the behavior of algorithms in practice. Communications of the ACM, 76–84 (2009)

Community Distribution Outlier Detection in Heterogeneous Information Networks

Manish Gupta[1], Jing Gao[2], and Jiawei Han[3]

[1] Microsoft, India
gmanish@microsoft.com
[2] SUNY, Buffalo, NY
jing@buffalo.edu
[3] UIUC, IL
hanj@illinois.edu

Abstract. Heterogeneous networks are ubiquitous. For example, bibliographic data, social data, medical records, movie data and many more can be modeled as heterogeneous networks. Rich information associated with multi-typed nodes in heterogeneous networks motivates us to propose a new definition of outliers, which is different from those defined for homogeneous networks. In this paper, we propose the novel concept of *Community Distribution Outliers (CDOutliers)* for heterogeneous information networks, which are defined as objects whose community distribution does not follow any of the popular community distribution patterns. We extract such outliers using a type-aware joint analysis of multiple types of objects. Given community membership matrices for all types of objects, we follow an iterative two-stage approach which performs pattern discovery and outlier detection in a tightly integrated manner. We first propose a novel outlier-aware approach based on joint non-negative matrix factorization to discover popular community distribution patterns for all the object types in a holistic manner, and then detect outliers based on such patterns. Experimental results on both synthetic and real datasets show that the proposed approach is highly effective in discovering interesting community distribution outliers.

1 Introduction

Heterogeneous information networks are omnipresent. In such networks, the nodes are of different types and relationships between nodes are encoded using multi-typed edges. For example, bibliographic networks consist of authors, conferences, papers and title keywords. Edges in such a network represent relationships like "an author collaborated with another author", "an author published in a conference", and so on. Analysts often perform community detection on such networks with an aim of understanding the hidden structures more deeply. Although methods designed for homogeneous networks can be applied by extracting a set of homogeneous networks from the heterogeneous network, such a transformation causes inevitable information loss. For example, when converting bibliographic networks to co-authorship networks, some valuable connectivity information, e.g., paper title or conference an author published in, is lost. As objects of different types interact strongly with each other in the network, analysis on heterogeneous information networks at various levels must be conducted simultaneously from

H. Blockeel et al. (Eds.): ECML PKDD 2013, Part I, LNAI 8188, pp. 557–573, 2013.

multiple types of data. Such an analysis will help in exploiting the shared hidden structure of communities across object types, i.e., the common patterns across types that can explain the generation of these community distributions. For example, in bibliographic networks, when grouping authors based on their "research area" distributions, one must use the knowledge of the grouping of "research area" distributions for related conferences and keywords. This is because (1) the community space (research areas) is the same across different object types, and (2) all these objects interact strongly with each other in the network.

Although most of the objects in a heterogeneous network follow common community distribution patterns which can be uncovered by joint analysis of community membership of multiple heterogeneous object types, certain objects deviate significantly from these patterns. It is important to detect such outliers in heterogeneous information networks for de-noising data thereby improving the quality of the patterns and also for further analysis. Therefore, in this paper, we propose to detect such anomalous objects as *Community Distribution Outliers* (or *CDOutliers*) given the community distribution of each object of every type. In the following, we present a few *CDOutlier* examples and discuss the importance of identifying such outliers in real applications.

CDOutlier Examples. Consider a bibliographic network where the research area label associated with an author node depends on the community labels of the conferences where he publishes, keywords he uses in the title of the papers, and the other authors he collaborates with. There may exist some popular community distribution patterns extracted by analysis across various object types, which majority of the objects follow. For example, say there are four communities: data mining (DM), software engineering (SE), compilers (C) and machine learning (ML). Then popular distribution patterns could be (DM:1, SE:0, C:0, ML:0), (DM:0, SE:1, C:0, ML:0), (DM:0, SE:0, C:1, ML:0), (DM:0, SE:0, C:0, ML:1), and (DM:0.7, SE:0, C:0, ML:0.3). Then, an author who contributes to DM and C (with a distribution like (DM:0.5, SE:0, C:0.5, ML:0)) would be considered as a *CDOutlier*. Furthermore, there could be subtle patterns like (DM:0.8, Energy:0.2), i.e., 80% probability belonging to DM and 20% probability in Energy, which is followed by majority of the objects. If an author's community distribution is (DM:0.2, Energy:0.8), which deviates from the majority pattern, then he is considered as a *CDOutlier*. Similarly, one could compute outliers among other types of objects, such as conferences and title keywords, based on the popular distribution patterns derived by holistic analysis across all object types.

Besides these examples, applications of *CDOutliers* can be commonly observed in real-life scenarios, and we briefly mention a few here. (1) In the Delicious network, most users who tag pages about "Tech and Science" do not tag pages about "Arts and Design". A user doing so (user with unusual skill combinations) can be considered as a *CDOutlier*. (2) In the Youtube network, most of the users would be interested in videos of a particular category. However, certain users who act as middlemen in publishing and uploading videos may interact with videos of many different categories and would be detected as *CDOutliers*.

CDOutlier distributions should not be confused with "hub" distributions (i.e., distributions with high entropy) over communities. Certain "hub" distributions could be frequent patterns, but only those that are very rare should be labeled as *CDOutliers*. On the other hand, not all *CDOutlier* distributions have high entropy.

Brief Overview of CDOutlier Detection. Given the *soft* community distributions for each object of every type, one can compute distribution patterns. *CDOutliers* are objects that defy the trend, and the trend must be obtained from accurate pattern discovery. However, pattern discovery suffers from the presence of *CDOutliers* itself. Therefore, given community detection results, we design an iterative two-stage procedure to identify *CDOutliers*, which integrates community distribution pattern discovery and *CDOutlier* detection. First, we discover popular distribution patterns for all the object types together by performing a joint nonnegative matrix factorization (NMF) on the community distribution matrices, such that it ignores the outliers discovered in the previous iteration. At the second step, the outlierness score for an object is computed based on its distance from its nearest distribution pattern. The algorithm iterates until the set of outliers discovered do not change. Thus, distribution pattern discovery and outlier detection are improved through iterative update procedures, and upon convergence, meaningful outliers are output.

Summary. Our contributions are summarized as follows.

- We introduce the notion of identifying *CDOutliers* from heterogeneous networks based on the discovery of community distribution patterns.
- We propose a unified framework based on joint-NMF formulation, which integrates the discovery of distribution patterns across multiple object types and the detection of *CDOutliers* based on such patterns together.
- We show interesting and meaningful outliers detected from multiple real and synthetic datasets.

Our paper is organized as follows. In Sec. 2, we introduce the notion of distribution patterns and develop our method to extract heterogeneous community trends for objects of different types in the form of popular distribution patterns. In Sec. 3, we present discussions related to practical usage of the algorithm. We discuss datasets and results with detailed insights in Sec. 4. Finally, related work and conclusions are presented in Sec. 5 and 6 respectively.

2   CDOutlier Detection Approach

In this section, we will present our iterative two-stage approach for *CDOutlier* detection. Table 1 shows the important notations we will use in this paper. We denote an element (i, j) of a matrix A by $A_{(i,j)}$. More details about the notations will be found in the following problem definition.

2.1   Problem Definition

We start with introduction to a few basic concepts.

Community. Consider a heterogeneous network with K types of objects $\{\tau_1, \tau_2, \ldots, \tau_K\}$. A community is a probabilistic collection of similar objects, such that

Notation	Meaning
τ_k	k^{th} object type
k, l	Index for a type of objects
N_k	Number of objects of type k
K	Number of types of objects
C	Number of communities
C'	Number of distribution patterns
$T_k^{N_k \times C}$	Membership matrix for objects of type k
$W_k^{N_k \times C'}$	Distribution pattern indicator matrix for objects of type k
$H_k^{C' \times C}$	Distribution patterns matrix for objects of type k
O_k	Outlier objects set for type k
α	Regularization Parameter

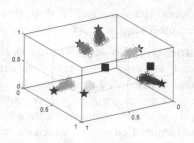

Fig. 1. Table of Notations

Fig. 2. Distribution Patterns in 3D Space

similarity between objects within the community is higher than the similarity between objects in different communities. For example, a research area is a community in a bibliographic network. For heterogeneous networks, one is often interested in identifying heterogeneous communities which contain objects of different types. We will use C to denote the number of communities.

Membership Matrix. Membership matrix T is a matrix such that the element T_{oi} corresponds to the probability with which an object o belongs to a community i. The rows of the matrix correspond to objects while the columns correspond to communities. Let $N_1, N_2, \ldots, N_K$ be the number of objects of each type. Let $T_1, T_2, \ldots, T_K$ denote the membership matrices for the objects of types $\tau_1, \tau_2, \ldots, \tau_K$ respectively. Thus, the membership matrix T_k is of size $N_k \times C$.

Distribution Patterns. The rows of a membership matrix can be grouped into clusters. To be able to capture inter-type interactions, such clusters should be obtained using a joint analysis of membership matrices of all types. The cluster centroid of each such cluster denotes a representative distribution in the community space. We call these cluster centroids as distribution patterns. For example, in Figure 2, we plot a membership matrix with C=3. Each axis represents probability of membership for the corresponding community. Different colors represent objects following different patterns. Black stars (★) are the representatives (cluster centroids) used to represent the distribution patterns.

Community Distribution Outlier. An object o in a heterogeneous network, is a *CD-Outlier* if its distance to the closest distribution pattern, which is obtained by a joint analysis of all the object types, is very high. For example, in Figure 2, the *CDOutlier* points are marked as black squares (■).

Communities and hence distribution patterns discovered from a heterogeneous network are very different from those obtained by processing a homogeneous projection of a heterogeneous network. Thus, *CDOutliers* are quite different from the community outliers obtained using homogeneous network analysis [6].

Community Distribution Outlier Detection Problem

Input: Community membership matrices $T_1, T_2, \ldots, T_K$ for the types $\tau_1, \tau_2, \ldots, \tau_K$.

Output: Top κ outlier objects of each type that deviate the most from distribution patterns for that type.

For example, for DBLP, the types are τ_1 =author, τ_2 =conference and τ_3 =keywords, and research areas are communities. T_1 will then be a matrix where each row denotes the probability with which an author belongs to various research areas. The expected output is top few authors (conferences, keywords) that deviate the most from the popular research area distribution patterns for the author (conference, keyword) type.

We will solve this problem using an iterative two-stage approach. In the first stage, distribution patterns are discovered ignoring the outliers detected in the previous iteration. In the second stage outliers are detected based on the patterns discovered at the first stage within the same iteration. The proposed pattern discovery step is a joint Non-negative Matrix Factorization (NMF) process, and thus we will first discuss basics about NMF in the next section. We then introduce the two stages in Sections 2.3 and 2.4, and finally present the complete algorithm.

2.2 Brief Overview of NMF

Given a non-negative matrix $T \in \mathbb{R}^{N \times C}$ (each element of T is ≥ 0), the basic NMF problem formulation aims to compute a factorization of T into two factors $W \in \mathbb{R}^{N \times C'}$ and $H \in \mathbb{R}^{C' \times C}$ such that $T \approx WH$. Both matrices W and H are constrained to have only non-negative elements in the decomposition.

It has been shown earlier ([4]) that NMF is equivalent to a relaxed form of *KMeans* [16] clustering. NMF can be considered as a form of clustering over the matrix T. Each row of H represents a cluster centroid (or a distribution pattern) in the C-dimensional space. Thus, H contains the information about the C' cluster centroids obtained by clustering T. Each element of row r of W represents the probability with which the object corresponding to row r belongs to the different clusters. Generally, the loss function used to represent the error between T and WH is the element-wise Euclidean distance. Thus the typical NMF can be expressed as the following optimization problem.

$$\min_{W,H} ||T - WH||^2 \tag{1}$$

subject to the constraints

$$W \geq 0, H \geq 0 \tag{2}$$

where $||A||$ is the sum of the square of each element in the matrix A.

2.3 Discovery of Distribution Patterns

In this sub-section, we will discuss how to learn distribution patterns from community membership matrices. These patterns will form the basis for outlier detection which we will discuss in Section 2.4.

For a homogeneous network, any clustering algorithm could be run over the community membership matrix to obtain distribution patterns. However, the case of heterogeneous networks is challenging. Each of the membership matrices T_k can be clustered individually (using the basic NMF) to obtain distribution patterns for that type. However, since all the membership matrices are defined for objects that are connected to each other, the hidden structures that can explain these objects' communities should be consistent across types. Also, the membership matrices represent objects in the same space of C components. Hence the clustering of matrix T_i should correspond to the clustering of matrix T_j for all $1 \leq i, j \leq K$. In other words, the divergence between any pair of clusterings should be low.

This intuition can be encoded in the form of an optimization problem, which conducts Non-negative Matrix Factorization (NMF) over multiple matrices together. In the proposed problem setting, each of the matrices in the set $T = \{T_1, T_2, \ldots, T_K\}$ needs to be factorized, and we expect them to share a lot of common factors or have factors which are quite similar to each other. We will factorize each matrix $T_k \in \mathbb{R}^{N_k \times C}$ into two factors $W_k \in \mathbb{R}^{N_k \times C'}$ and $H_k \in \mathbb{R}^{C' \times C}$. Also, we need to ensure that clustering across different types is somewhat related. We achieve this by introducing a new term $||H_k - H_l||^2$ to the basic NMF optimization objective function, and a parameter α which decides what degree of correspondence should be obtained across clusterings.

Problem Formulation. Based on the above discussion, the problem can be formulated as an optimization problem as follows. Let W and H represent the set of matrices $\{W_1, W_2, \ldots, W_K\}$ and $\{H_1, H_2, \ldots, H_K\}$ respectively.

$$\min_{W,H} \sum_{k=1}^{K} \{||T_k - W_k H_k||^2\} + \alpha \sum_{\substack{k=1 \\ l=1 \\ k<l}}^{K} \{||H_k - H_l||^2\} \tag{3}$$

subject to the constraints

$$W_k \geq 0 \quad \forall k = 1, 2, \ldots, K \tag{4}$$

$$H_k \geq 0 \quad \forall k = 1, 2, \ldots, K \tag{5}$$

For example, for DBLP, τ_1=author, T_1 is the research-area distribution matrix for the author type. Each row of H_1 represents a distribution pattern for the author type and each row of W_1 denotes the probability with which the author belongs to the C' author distribution patterns.

The objective function in Eq. 3 is quadratic with respect to W_k or H_k when the other variable matrices are fixed. Converting to Lagrangian form by introducing the Lagrangian multiplier matrix variables $P = \{P_1, P_2, \ldots, P_K\}$ and $Q = \{Q_1, Q_2, \ldots, Q_K\}$, we obtain the following.

$$\min_{W,H,P,Q} \sum_{k=1}^{K} \{||T_k - W_k H_k||^2\} + \alpha \sum_{\substack{k=1 \\ l=1 \\ k<l}}^{K} \{||H_k - H_l||^2\} + \sum_{k=1}^{K} \{tr(P_k W_k^T) + tr(Q_k H_k^T)\} \tag{6}$$

KKT optimality conditions require the following.

$$\frac{\partial \left[||T_k - W_k H_k||^2 + \alpha \sum_{\substack{l=1 \\ k \neq l}}^{K} ||H_l - H_k||^2 \right]}{\partial H_{k_{(i,j)}}} = Q_{k_{(i,j)}} \quad \forall k = 1, 2, \ldots, K \tag{7}$$

$$\frac{\partial \left[||T_k - W_k H_k||^2 \right]}{\partial W_{k_{(i,j)}}} = P_{k_{(i,j)}} \quad \forall k = 1, 2, \ldots, K \tag{8}$$

Also, the complementary slackness conditions can be expressed as follows.

$$Q_{k_{(i,j)}} \times H_{k_{(i,j)}} = 0 \quad \forall i, j, k \tag{9}$$

$$P_{k_{(i,j)}} \times W_{k_{(i,j)}} = 0 \quad \forall i, j, k \tag{10}$$

Substituting Eqs. 7 and 8 into Eqs. 9 and 10 respectively, we get the following.

$$\left[W_k^T W_k H_k - W_k^T T_k + \alpha \sum_{\substack{l=1 \\ k \neq l}}^{K} \left(I^{C' \times C'} H_k - I^{C' \times C'} H_l \right) \right]_{(i,j)} \times H_{k_{(i,j)}} = 0 \quad \forall i, j, k \tag{11}$$

$$\left[W_k H_k H_k^T - T_k H_k^T \right]_{(i,j)} \times W_{k_{(i,j)}} = 0 \quad \forall i, j, k \tag{12}$$

These set of equations can be solved using the following iterative equations.

$$W_k \leftarrow W_k \odot \frac{T_k H_k^T}{W_k H_k H_k^T} \quad \forall k = 1, 2, \ldots, K \tag{13}$$

$$H_k \leftarrow H_k \odot \frac{W_k^T T_k + \alpha \sum_{\substack{l=1 \\ k \neq l}}^{K} I^{C' \times C'} H_l}{W_k^T W_k H_k + \alpha \sum_{\substack{l=1 \\ k \neq l}}^{K} I^{C' \times C'} H_k} \quad \forall k = 1, 2, \ldots, K \tag{14}$$

Here $\odot$ denotes the Hadamard product (element-wise product) and $\frac{A}{B}$ denotes the element-wise division, i.e. $\left(\frac{A}{B} \right)_{i,j} = \frac{A_{ij}}{B_{ij}}$.

2.4 Community Distribution Outlier Detection

Using the joint-NMF formulation described in the previous sub-section, we obtain the matrices $\{W_k\}_{k=1}^{K}$. Each row of H_k is a distribution pattern (a cluster centroid) and each element (i, j) of W_k denotes the probability with which object i belongs to the distribution pattern j. We define the outlier score of an object as the distance of the object i of type T_k from the nearest cluster centroid. Thus, the outlier score for an object i, $OS(i)$ can be written as follows.

$$OS(i) = \underset{j}{\operatorname{argmin}} \ Dist(T_{k_{(i, \cdot)}}, H_{k_{(j, \cdot)}}) \tag{15}$$

An object which is far away from its nearest cluster centroid gets a high outlier score. Using this outlier definition, one can find outlier scores for all objects of all types. Top κ objects with highest outlier scores for each type can be marked as outliers.

Iterative Refinement. If the input data contains outliers, the distribution patterns will try to overfit to those outliers and hence will be distorted compared to the actual hidden structure of the clean data, so the distribution pattern discovery needs to be outlier-aware. Similarly, if the distribution patterns are accurate, outlier detection will be of a high quality. Therefore, we propose to perform the steps of pattern discovery and outlier detection iteratively until convergence. At each iteration, while performing pattern discovery we ignore the set of top-κ outliers from each type. For outlier detection, we use the patterns discovered during the same iteration, to compute outlier scores for all the objects of all types. Empirically we observed that such an iterative refinement always converges. However in case the algorithm oscillates (i.e., enters a loop where the set of outliers detected repeats), the algorithm can be terminated when the set of outliers detected after any iteration is the same as the one detected in any previous iteration.

We summarize the outlier detection algorithm in Algorithm 1. We initialize the set of outliers of each type to an empty set (Step 1). The set of outliers is updated iteratively and the algorithm terminates when the outliers detected across two consecutive iterations are the same. Within every iteration, we first obtain T_k for that iteration by removing the rows corresponding to the current outliers from the original membership matrix (Step 6). NMF is sensitive to initialization and hence we initialize W_k's and H_k's using clusters discovered by running *KMeans* [16] on T_k (Step 7). Steps 6 to 13 correspond to pattern discovery using joint-NMF. Steps 14 to 17 correspond to outlier detection based on the discovered patterns. Finally, the outlier objects are returned.

Algorithm 1. *CDOutlier* Detection Algorithm (CDODA)

Input: (1) Cluster membership matrices $T = \{T_1, T_2, \ldots, T_K\}$ corresponding to objects of types $\tau = \{\tau_1, \tau_2, \ldots, \tau_K\}$, (2) α, (3)κ.
Output: Top κ *CDOutlier* objects of each type ($\{O_1, O_2, \ldots, O_K\}$).
1: Initialize each element of $currOutliers = \{O_1, O_2, \ldots, O_K\}$ to ϕ.
2: Initialize each element of $prevOutliers = \{O'_1, O'_2, \ldots, O'_K\}$ to *null*.
3: $\{origT_k \leftarrow T_k\}_{k=1}^{K}$
4: **while** checkForChange($currOutliers, prevOutliers$) **do**
5:      $prevOutliers \leftarrow currOutliers$
6:      $\{T_k \leftarrow origT_k -$ rows corresponding to $O_k\}_{k=1}^{K}$ ▷ Pattern Discovery
7:      Initialize $\{W_k\}_{k=1}^{K}$ and $\{H_k\}_{k=1}^{K}$ using $\{KMeans(T_k)\}_{k=1}^{K}$.
8:      **while** NOT converged **do**
9:          **for** $k = 1$ to K **do**
10:             Update W_k using Eq. 13.
11:             Update H_k using Eq. 14.
12:         **end for**
13:     **end while**
14:     **for** $k = 1$ to K **do**   ▷ Outlier Detection
15:         Compute outlier scores for all objects of type τ_k.
16:         $O_k \leftarrow$ top κ objects of type τ_k with highest outlier scores.
17:     **end for**
18: **end while**

---

3   Discussions

In this section, we analyze the time complexity of the proposed *CDOutlier* detection method. We also discuss several important issues in implementing the method.

Initialization. The joint-NMF formulation will converge to a *local* optimum, and thus it could be sensitive to initialization. Therefore, it is very important to choose an appropriate initialization for the algorithm. To initialize the matrix H_k, we run *KMeans* [16] on the matrix T_k. W_k is then computed by finding the nearest cluster for each object and setting the corresponding entry in W_k to 1.

Computational Complexity. The time required for an update to a W_k or H_k matrix is $O(NKC'^2)$. Thus, the pattern discovery phase has a complexity of $O(K^2 INC'^2)$, where I is the number of iterations for joint-NMF and N is the average number of objects per type. The outlier detection phase consists of finding top κ outliers per type which can be done in $O(KN log(\kappa))$ time. Let the number of iterations for the external While loop (Steps 4 to 18) be I'. Thus, the overall complexity of the algorithm is $O(NI'K(KIC'^2 + log(\kappa)))$. Note that $I'K(KIC'^2 + log(\kappa))$ becomes a small constant when N is large. Thus the algorithm is linear in the number of objects.

Selecting Parameters (α and κ). α determines the amount of regularization applied when performing the joint-NMF. If we set α to 0, it is as good as performing NMF separately. A high value of α will favor a solution where there are many shared distribution patterns across various types, while a low value of α will try to fit the NMF for each of the types individually without trying to discover any shared distribution patterns. Hence, the setting of the parameter α is important and domain dependent. If we believe that the objects of different types interact a lot all across the network, we should use a higher value for α for better results. κ can be selected based on the percentage of outliers expected. Another way of principled thresholding is to set the variance level, for example, consider any point as an outlier if it is at least two standard deviations away from the nearest cluster centroid.

4 Experiments

Evaluation of outlier detection algorithms is quite difficult due to lack of ground truth. We generate multiple synthetic datasets by injecting outliers into normal datasets, and evaluate outlier detection accuracy of the proposed algorithms on the generated data. We also conduct case studies by applying the method to real data sets. We perform comprehensive analysis to justify that the top few outliers returned by the proposed algorithm are meaningful. The code and the data sets are available at: http://dais.cs.uiuc.edu/manish/CDOutlier/

4.1 Baselines

Community Distribution Outlier Detection Algorithm (*CDO*) is the proposed method. The baseline methods (*SI* and *Homo*) are explained as follows.

SingleIteration (SI). As described in Algorithm 1, *CDO* performs community pattern discovery and outlier detection iteratively until the set of top κ outliers for each type do not change. *SI* is a simpler version of *CDO*, which performs only one iteration. Thus the pattern discovery phase in *SI* suffers from the presence of *CDOutliers*. This baseline will help us evaluate the importance of ignoring the *CDOutlier* noise when computing the distribution patterns.

Homogeneous (Homo). *CDO* performs pattern discovery using joint-NMF across multiple types. In contrast to this, the baseline *Homo* treats all objects to be of the same type and then performs distribution pattern discovery using a single matrix NMF. This baseline will help us evaluate the importance of modeling heterogeneous data types rather than reducing them to homogeneous ones in heterogeneous information networks.

4.2 Synthetic Datasets

Dataset Generation

We generate our synthetic dataset as follows. The dataset is represented by the matrices T_k for $1 \leq k \leq K$. We start by generating H_k and W_k and then obtain $T_k = W_k H_k$. We first generate a single matrix $H^{C' \times C}$ which we consider as a template for generating the distribution patterns. It appears across different types in a slightly perturbed form. H is generated as follows. We first fix $C' = 2C$. Next, each cluster centroid (a row of H) could be an impulse probability distribution function at different dimensions or could have non-zero random probability value for 2 dimensions. Perturb H randomly such that all objects of the same type follow the same fixed perturbation to get $H_1, \ldots, H_K$ (Recall K=Number of types). Such a perturbation captures the fact that clusters across different types of objects deviate slightly from each other. Then $\{W_k\}_{k=1}^{K}$ are generated such that one element in every row is close to 1, and the remaining probability mass is distributed uniformly among other elements. These W_k's and H_k's could then be used to generate $\{T_k = W_k H_k\}_{k=1}^{K}$.

Outliers are injected as follows. First we set an outlierness factor Ψ and choose a random set of objects, R_k with $N_k \times \Psi$ objects of type k. For each object o in R_k, we choose either a pattern randomly from some other type $k' \neq k$ or a pattern quite different from any pattern in H_k's. We use this pattern to define the row in T_k corresponding to the object o, i.e., $T_{k_{(o,.)}}$. Note that patterns in different types are reasonably different from each other. Hence, such an object which follows a pattern from some other type, or a completely different pattern from H itself, can be considered as an outlier for type k.

Results on Synthetic Datasets

We generate a variety of synthetic datasets capturing different experimental settings. For each setting, we perform 20 experiments and report the average values. We fix the threshold for NMF objective function convergence to 0.01. We vary the number of objects as 1000, 2000 and 5000. We also study the accuracy with respect to variation in number of object types (2, 3, 4) and variation in the number of communities (4, 6, 8, 10). We also vary the percentage of injected outliers as 1%, 2% and 5%. We fixed α=0.5 for our experiments. Using these settings, we compare the actual outlier objects with the top outliers returned by various algorithms. For each algorithm, we show the accuracy with respect to matches in the set of detected outliers and the set of injected outliers, in Table 1 (False Positives(%)=100-accuracy). Results for $C = 6, 8$ are also similar and we omit them for lack of space. For each experimental setting, we show the best accuracy obtained in bold. Each of the accuracy values is obtained by averaging the accuracy across all types of objects for that experimental setting (across 20 runs). Average standard deviations are 3.07% for *CDO*, 3.48 % for *SI* and 2.19% for *Homo*.

As the table shows, the proposed algorithm outperforms both of the other algorithms for most of the settings by a wide margin. On an average across all experimental settings, CDO is 2.85% better than SI and 21.5% better than Homo. In general, the accuracy of the proposed algorithm decreases slightly as the amount of outlierness increases to 5%.

4.3 Running Time and Convergence

The experiments were run on a Linux machine with 4 Intel Xeon CPUs with 2.67GHz each. The code was implemented in Java. KMeans [16] implementation of Weka [11] was used for initialization of the H_k and W_k matrices. Figure 3 shows the execution time for CDO algorithm for different number of object types. Note that the algorithm is linear in the number of objects. These times are averaged across multiple runs of the algorithm across different settings for degree of outlierness and number of communities.

Figure 4 shows the decrease in the objective function value with respect to the number of iterations for different dataset sizes (for $K=3$ and $C=10$). The figure shows that the joint-NMF algorithm converges well. The average number of iterations for convergence of joint-NMF are 118, 173 and 242 for datasets of sizes 1000, 2000 and 5000 respectively.

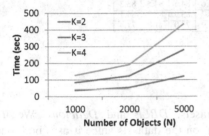

Fig. 3. Running Time (sec) for CDO (Scalability)

Fig. 4. Convergence of joint-NMF

On an average across all experimental settings, the proposed algorithm CDO takes the following number of external iterations (I') of pattern discovery and outlier detection: 6.21 for $N=1000$, 6.98 for $N=2000$ and 7.66 for $N=5000$.

Table 1. Synthetic Dataset Results (CDO=The Proposed Algorithm CDODA, SI= Single Iteration Baseline, Homo=Homogeneous (Single NMF) Baseline) for C=4 (left) and C=10 (right)

N	Ψ (%)	$K=2$ CDO	SI	Homo	$K=3$ CDO	SI	Homo	$K=4$ CDO	SI	Homo
	1	92	91.5	52	81.3	80	53.7	73.8	75	54.2
1000	2	94.2	85.8	60	83.3	83	57.3	76.1	75.4	56.4
	5	86.5	70.5	59.5	74.7	67.8	57.2	71	64.4	55.6
	1	95	91	56.5	81.2	81.3	54.8	73.1	74.5	52.1
2000	2	90.4	86.1	57.1	81.8	78.3	55.2	74.2	73.8	52.3
	5	91.7	72.8	58	73.4	65.4	57.2	74	67.7	55.4
	1	92.1	86.4	52.3	80.9	78.4	56.3	72.8	69.1	51.6
5000	2	95.4	94.4	56	79.9	77.2	54.6	74.6	74	53.8
	5	88.5	68	60.7	80.4	66.7	57.9	74.8	65.9	56.8

N	Ψ (%)	$K=2$ CDO	SI	Homo	$K=3$ CDO	SI	Homo	$K=4$ CDO	SI	Homo
	1	97	90.5	51	78	74.3	51.3	69.5	68.2	52.8
1000	2	81.8	81.2	55	67.3	66.8	56.8	65.9	65.6	59
	5	78.6	77.2	59.4	69.2	69.1	58.3	68.8	69	56
	1	79.2	78	55.5	72.7	71.5	58.2	71.9	72.2	56.6
2000	2	79	78.2	55.8	68.1	68.2	59.2	65.4	65.9	56.1
	5	74.4	72.4	61.5	73.1	73.4	58.4	66.4	67.2	56.2
	1	97.1	85.7	54.3	77.8	71.2	54.9	69.3	69	58.3
5000	2	75.8	74.4	57.1	68.9	69.3	56.9	69.3	70.8	57.3
	5	75	72.1	61.2	70.2	69.5	57.9	68.2	69.9	56.3

4.4 Regularization Parameter Sensitivity

The joint-NMF optimization problem (Eq. 3) includes a regularization parameter α. We study the sensitivity of the algorithm with respect to this parameter. Table 2 shows the accuracy of the proposed *CDO* algorithm for $K=3$ and $C=6$. Across different settings of the number of objects (N) and the degree of outlierness (Ψ), the table shows that the accuracy is not sensitive to the value of α. We observe that the algorithm provides good accuracy for any value of α between 0 and 1. Note that α decides how much importance the algorithm gives to the quality of object clustering within one type versus matches between clusters obtained across types. Thus, α should be decided for any dataset based on the size of the dataset and its inter-type cluster structure similarity.

Table 2. Regularization Parameter Sensitivity for $K=3, C=6$

N	Ψ (%)	α					
		0	0.2	0.5	0.8	1	10
1000	1	85.0	86.3	86.0	86.3	86.3	86.0
	2	81.5	83.5	83.3	82.8	82.7	82.2
	5	64.2	67.2	66.9	66.4	68.1	66.7
2000	1	82.1	85.5	85.8	85.5	85.3	83.5
	2	74.7	78.6	81.0	80.2	80.3	77.7
	5	62.3	70.6	70.5	70.5	70.4	69.9
5000	1	80.1	84.5	84.6	84.5	84.5	83.3
	2	79.6	82.3	82.7	83.9	83.9	84.1
	5	65.6	72.1	71.9	72.0	71.8	71.4

4.5 Real Datasets

Dataset Generation

We perform experiments using 2 real datasets: *DBLP* and *Delicious*. We use *NetClus* [20] to perform community detection on the datasets since it uses both data and link information for clustering and is specifically designed to handle heterogeneous networks. *NetClus* outputs the matrices $T_1, \ldots, T_K$ which we use as input for the proposed outlier detection algorithm. We found that the proposed method provides much more interesting top outliers compared to the *Homo* baseline and we provide case studies using *CDO* only, for lack of space.

DBLP: The *DBLP* network consists of papers, authors, keywords and conferences. We considered a temporal subset of DBLP[1] for 2001-2010. We removed authors with <10 papers during that time period. Our dataset consists of ~650K papers, ~480K authors, 3900 conferences, ~107K keywords and 14 research areas. We obtained a list of conferences from the Wikipedia Computer Science Conferences page[2] which labels conferences by research areas. By associating keywords from these conferences with research areas, we obtained term priors which were used as input for *NetClus*. We consider each research area as a community, and thus the number of communities is 14. We experimented with $C'=28$ (twice the number of communities), $\alpha=0.5$ and $\kappa = 1\%$.

---

[1] http://www.informatik.uni-trier.de/~ley/db/
[2] http://en.wikipedia.org/wiki/
List_of_computer_science_conferences

Delicious: The *Delicious* network consists of tagging events, users, URLs and tags. The dataset consists of all tagging events performed by a randomly chosen list of ~73K users from July 1 to July 28, 2010. The tagging events were obtained as RSS feeds[3] and were processed to obtain the desired network. Delicious provides a basic categorization on the home page[4]. We scrap category pages linked from home page to associate keywords with the categories. We consider these categories as communities and hence use the number of communities as 10 when running *NetClus* on the Delicious data. The categorized keywords are used to supply term priors for *NetClus*. Our Delicious dataset consists of ~73K users, ~1.3M tagging events, ~902K URLs, ~273K tags and 10 categories. We experimented with C'=20 (twice the number of communities), α=0.5 and κ =1%.

Results on Real Datasets

Running time for the algorithm is about 1.5 hours for both the datasets. Here, we will discuss case studies obtained from these datasets. We analyze the top 2 outliers of each type from the 2 datasets in terms of their community distribution. Objects that have very small frequency of occurrence may not have an appropriate community distribution. Hence, we analyze objects with at least 10 links in the network. Note that the outliers for each type have been obtained using a joint hidden structure analysis across multiple types, and hence are quite different from outliers obtained using homogeneous network analysis [6].

DBLP

In *DBLP*, we observe specialization in one of the 14 categories as clear patterns. Multiple types of objects share a few patterns, which combine several areas, for example, ("Databases":0.8, "Computational Biology":0.2). However, some of the other patterns with combinations of research areas are specific to particular types. For example, the pattern ("Software engineering":0.3, "Operating systems":0.6, "others":0.1) is observed for conferences but not for other types. Similarly, the pattern ("Concurrent Distributed and Parallel Computing":0.5, "Security and privacy":0.45, "others":0.05) is observed specifically for authors while ("Security and privacy":0.8,"Education":0.2) is observed specifically for title keywords. Thus some patterns are shared across types while others are slightly different. This stresses the need for a joint-NMF-based clustering.

Authors: Most of the authors publish frequently in such "commonly-paired" categories or in a single category of their expertise. However our top outliers show interesting combinations as follows. (Note that the community membership probabilities are shown in brackets and may not add up to 1; the residual is spread across other communities.) (1) Giuseppe de Giacomo: Algorithms and Theory (0.25), Databases (0.47), Artificial Intelligence (0.13), Human Computer Interaction (0.06). Note that the combination of Algorithms and Theory, Databases and Artificial Intelligence with small contributions to HCI is rare and hence interesting.

---

[3] http://feeds.delicious.com/v2/rss/
[4] http://delicious.com/

(2) Guang R Gao: Concurrent Distributed and Parallel Computing (0.41), Computer Architecture (0.3), Computational Biology (0.27). Similar to the case above, this combination of the research areas is quite rare.

Conferences: Among the top conference outliers are conferences that span across multiple streams of computer science. The top 2 conference outliers are as follows.
(1) From integrated publication and information systems to virtual information and knowledge environments[5]: Databases (0.5), Artificial Intelligence (0.09), Human Computer interaction (0.4). This conference is special because it celebrates an *occasion* (65th birthday of Erich J. Neuhold). From the name itself the reader can guess the wide nature of this conference.
(2) International Conference on Modelling and Simulation: Programming languages (0.18), Security and privacy (0.29), Databases (0.39), Computer Graphics (0.13). Again, this combination is quite rare.

Keywords: Finally, we also list the top 2 paper title keywords with high outlierness scores.
(1) military: Algorithms and theory (0.02), Security and Privacy (0.37), Databases (0.22), Computer Graphics (0.37). Lots of military sponsored research and paper motivations containing military scenarios results in such a diverse distribution for "military".
(2) inventory: Security and Privacy (0.29), Databases (0.31), Computer Graphics (0.34), Computational Biology (0.03). The nearest matching pattern for this one was (Databases: 0.8, Computational Biology: 0.2). But usually computer graphics and security and privacy are not associated with these.

Delicious

In *Delicious*, we observe specialization in one of the 10 categories as clear patterns, as expected. Different types of objects share a few patterns, which corresponds to combinations of categories, for example, "Education" and "Tech and Science". However, some of the other patterns with combinations of categories are specific to particular types. For example, "Arts and Design" and "Tech and Science" is observed for URLs but not for other types. Similarly, the pattern "Arts and Design" and "Entertainment" is observed specifically for users and "Lifestyle" and "Sports" is observed specifically for tags. Thus even in the Delicious dataset, some patterns are shared across types while others are slightly different.

Users: Most of the users (who tag a sizeable number of pages) tag pages related to a particular category only. However, there are some users who are experts across multiple categories. Sometimes their interests are quite diverse and do not follow patterns of other users. Here, we report top 2 users that the proposed algorithm reported as outliers, along with the probabilistic categories they belong to. Usually lifestyle and travel are highly correlated with food, unlike for the user "saassaga".
(1) saassaga: Arts and Design (0.25), Food (0.04), Lifestyle (0.35), Travel (0.34)
(2) lbbrad: Food (0.24), Lifestyle (0.37), News and Politics (0.37)

---

[5] http://dblp.dagstuhl.de/db/conf/birthday/neuhold2005.html

Tags: Top 2 tags detected as outliers by our algorithm along with the community distributions are as follows. It is interesting to note that people often mention "canoeing" as a sport that they perform often when they travel (e.g., on group outings).
(1) canoeing: Sports (0.62), Travel (0.38). Though there are other sports which people feel interesting in while traveling, canoeing seems to be a clear exception wrt number of travel pages it is mentioned on. The closest distribution pattern is (Sports: 1).
(2) rosary: Arts and Design (0.38), Education (0.02), Sports (0.6)

URLs: We find that not many web-pages belong to the Lifestyle and Travel categories together. As a result the pages that belong partially to the Travel and Lifestyle categories get marked as top outliers.

(1) `http://globetrooper.com/`: Lifestyle (0.35), Travel (0.38)
(2) `http://vandelaydesign.com/blog/galleries/travel-websites/`: Lifestyle (0.33), Travel (0.48)

In conclusion, our algorithm is effective at finding interesting outliers from real datasets.

5 Related Work

Outlier detection has been studied in the context of a large number of application domains [1,2,5,6,13,15]. Chandola et al. [3] and Hodge et al. [12] provide extensive overview of outlier detection techniques. Different from these studies, we perform community outlier detection for heterogeneous network data.

Individual, Global and Community Contexts. Outlier Detection can be performed at different levels of context. (1) Individual Context: For example, Type I and Type II Outliers [5] in time series are defined based on values observed for the same object across different time points. (2) Global Context: Stream Outliers [2], DB Outliers [13], Sub-Structure Outliers [18] are defined based on comparison with all the other objects in the dataset. (3) Community Context: Different from existing community outlier detection approaches (Community Outliers [6], CTOutliers [9], ECOutliers [10]), we model multiple data types in a *heterogeneous* network simultaneously to find outliers.

Homogeneous versus Heterogeneous Networks. Recently there has been work on outlier detection for homogeneous networks [2,6,7,10]. While previous work on outlier detection for heterogeneous networks [14,17] models the anomaly detection problem in heterogeneous networks as a tensor decomposition problem, we model the problem using a joint-NMF model to extract distribution patterns, which are further used to detect outliers. Also compared to our previous work (ABCOutliers [8]) which identified outlier cliques, this work focuses on finding outlier objects.

6 Conclusions

We introduced the notion of outliers with respect to latent communities for heterogeneous networks, i.e., *CDOutliers*. Such outliers represent objects that disobey the frequent community distribution patterns. The challenge in detecting such outliers is twofold: (1) correlation between patterns across different types of objects in the network should be considered; and (2) patterns need to be learned by ignoring the outliers, while outlier detection depends on effective discovery of patterns. To tackle such challenges, we proposed a joint-NMF optimization framework to learn distribution patterns across multiple object types, that uses a regularizer for distance between the cluster centroid matrices of different object types. We derive the update rules to learn the joint NMF model, which alternately updates the cluster membership and the cluster centroid matrices. Experiments on a series of synthetic data show the proposed algorithm's capability of detecting outliers under various levels of outlierness, data dimensionality, and number of types. Case studies on *DBLP* and *Delicious* datasets reveal some interesting and meaningful outliers. In the future, we plan to extend the framework to handle multiple temporal network snapshots in a stream scenario.

Acknowledgments. We thank the anonymous reviewers for their insightful comments. Research was sponsored in part by the U.S. Army Research Laboratory under Cooperative Agreements W911NF-11-2-0086 (Cyber-Security) and W911NF-09-2-0053 (NS-CTA), U.S. National Science Foundation grants IIS-0905215, CNS-0931975, CCF-0905014, IIS-1017362, DTRA, and NASA NRA-NNH10ZDA001N. The views and conclusions contained in this document are those of the authors and should not be interpreted as representing the official policies, either expressed or implied, of the Army Research Laboratory or the U.S. Government. The U.S. Government is authorized to reproduce and distribute reprints for Government purposes notwithstanding any copyright notation here on.

References

1. Aggarwal, C.C., Yu, P.S.: Outlier Detection for High Dimensional Data. SIGMOD Records 30, 37–46 (2001)
2. Aggarwal, C.C., Zhao, Y., Yu, P.S.: Outlier Detection in Graph Streams. In: ICDE, pp. 399–409 (2011)
3. Chandola, V., Banerjee, A., Kumar, V.: Anomaly Detection: A Survey. ACM Surveys 41(3) (2009)
4. Ding, C.H.Q., He, X.: On the Equivalence of Nonnegative Matrix Factorization and Spectral Clustering. In: SDM, pp. 606–610 (2005)
5. Fox, A.J.: Outliers in Time Series. Journal of the Royal Statistical Society 34(3), 350–363 (1972)
6. Gao, J., Liang, F., Fan, W., Wang, C., Sun, Y., Han, J.: On Community Outliers and their Efficient Detection in Information Networks. In: KDD, pp. 813–822 (2010)
7. Ghoting, A., Otey, M.E., Parthasarathy, S.: LOADED: Link-Based Outlier and Anomaly Detection in Evolving Data Sets. In: ICDM, pp. 387–390 (2004)
8. Gupta, M., Gao, J., Han, J.: On Detecting Association-Based Clique Outliers in Heterogeneous Information Networks. In: ASONAM (to appear, 2013)
9. Gupta, M., Gao, J., Sun, Y., Han, J.: Community Trend Outlier Detection Using Soft Temporal Pattern Mining. In: Flach, P.A., De Bie, T., Cristianini, N. (eds.) ECML PKDD 2012, Part II. LNCS, vol. 7524, pp. 692–708. Springer, Heidelberg (2012)
10. Gupta, M., Gao, J., Sun, Y., Han, J.: Integrating Community Matching and Outlier Detection for Mining Evolutionary Community Outliers. In: KDD, pp. 859–867 (2012)

11. Hall, M., Frank, E., Holmes, G., Pfahringer, B., Reutemann, P., Witten, I.H.: The WEKA Data Mining Software: An Update. SIGKDD Explorations 11(1), 10–18 (2009)
12. Hodge, V.J., Austin, J.: A Survey of Outlier Detection Methodologies. AI Review 22(2), 85–126 (2004)
13. Knorr, E.M., Ng, R.T., Tucakov, V.: Distance-Based Outliers: Algorithms and Applications. VLDBJ 8, 237–253 (2000)
14. Koutra, D., Papalexakis, E.E., Faloutsos, C.: TensorSplat: Spotting Latent Anomalies in Time. In: Panhellenic Conference on Informatics, pp. 144–149 (2012)
15. Kriegel, H.-P., Schubert, M., Zimek, A.: Angle-based Outlier Detection in High-Dimensional Data. In: KDD, pp. 444–452 (2008)
16. MacQueen, J.B.: Some Methods for Classification and Analysis of MultiVariate Observations. In: Berkeley Symposium on Mathematical Statistics and Probability, vol. 1, pp. 281–297 (1967)
17. Maruhashi, K., Guo, F., Faloutsos, C.: MultiAspectForensics: Pattern Mining on Large-Scale Heterogeneous Networks with Tensor Analysis. In: ASONAM, pp. 203–210 (2011)
18. Noble, C.C., Cook, D.J.: Graph-Based Anomaly Detection. In: KDD, pp. 631–636 (2003)
19. Sun, Y., Han, J., Yan, X., Yu, P.S.: Mining Knowledge from Interconnected Data: A Heterogeneous Information Network Analysis Approach. In: PVLDB (2012)
20. Sun, Y., Yu, Y., Han, J.: Ranking-based Clustering of Heterogeneous Information Networks with Star Network Schema. In: KDD, pp. 797–806 (2009)
21. Xu, X., Yuruk, N., Feng, Z., Schweiger, T.A.J.: SCAN: A Structural Clustering Algorithm for Networks. In: KDD, pp. 824–833 (2007)

Protein Function Prediction Using Dependence Maximization

Guoxian Yu[1], Carlotta Domeniconi[2], Huzefa Rangwala[2], and Guoji Zhang[3]

[1] School of Comp. & Inf. Sci., Southwest University, Chongqing, China
[2] Department of Comp. Sci., George Mason University, Fairfax, VA, USA
[3] School of Sci., South China University of Technology, Guangzhou, China
guoxian85@gmail.com, {carlotta,rangwala}@cs.gmu.edu, magjzh@scut.edu.cn

Abstract. Protein function prediction is one of the fundamental tasks in the post genomic era. The vast amount of available proteomic data makes it possible to computationally annotate proteins. Most computational approaches predict protein functions by using the labeled proteins and assuming that the annotation of labeled proteins is complete, and without any missing functions. However, partially annotated proteins are common in real-world scenarios, that is a protein may have some confirmed functions, and whether it has other functions is unknown.

In this paper, we make use of partially annotated proteomic data, and propose an approach called *Pro*tein Function Prediction using *D*ependency *M*aximization (ProDM). ProDM works by leveraging the correlation between different function labels, the 'guilt by association' rule between proteins, and maximizes the dependency between function labels and feature expression of proteins. ProDM can replenish the missing functions of partially annotated proteins (a seldom studied problem), and can predict functions for completely unlabeled proteins using partially annotated ones. An empirical study on publicly available protein-protein interaction (PPI) networks shows that, when the number of missing functions is large, ProDM performs significantly better than other related methods with respect to various evaluation criteria.

1 Introduction

Proteins are macromolecules that serve as the fundamental building blocks and functional components of a living cell. The knowledge of protein functions can promote the development of new drugs, better crops and synthetic biochemicals [14]. With the development of high-throughput biotechnologies, it is easy to collect various proteomic data, but the functions of these proteomic data cannot be determined at the same pace. The availability of vast amount of proteomic data enables researchers to computationally annotate proteins. Thus various computational models have been developed to reduce the cost associated with experimentally annotating proteins in the wet lab.

Numerous computational approaches have been proposed for protein function prediction. Some approaches assume that two proteins with similar sequences

H. Blockeel et al. (Eds.): ECML PKDD 2013, Part I, LNAI 8188, pp. 574–589, 2013.

should have similar functions. These methods use a kernel function (i.e., string kernel [12]) to measure the similarity between the sequences of a pair of proteins and predict their functions. A protein often interacts with other proteins to accomplish certain tasks. Some algorithms take advantage of this knowledge and use protein-protein interaction (PPI) networks to automatically make predictions [5,18,20,24]. Further, some approaches integrate multiple data types (i.e., PPI networks, protein sequences, and gene co-expression networks) for protein function prediction [13,21].

Proteins have multiple functions and each function can be viewed as a label. These function labels are typically correlated. Traditional protein function prediction approaches often formulate the problem as a multiple binary classification problem [12] and ignore the correlation between labels. To avoid this limitation, multi-label learning is widely used for protein function prediction [10,15,24]. Multi-label learning can make use of label correlations to boost the prediction accuracy and assign more than one function to a protein [20,22]. Other approaches train a binary classifier for each function label, and then organize these classifiers in a hierarchical (tree or direct acyclic graph) structure according to the Function Catalogue (FunCat) [16][1] or Gene Ontology [2][2] database [14]. In this paper, we focus on protein function prediction using multi-label learning and function correlation.

All these approaches assume that the available annotations for the labeled proteins are complete. In practice, we may just have a subset of the functions of a protein, and whether some functions are missing is unknown. In other words, proteins may not be completely annotated [4], i.e., function annotations may be only partial. This kind of multi-label learning problem is called *multi label weak-label learning* [19], a much less studied problem in the literature [4,22]. Unlike traditional multi-label learning methods [10,21,24], we study protein function prediction using incomplete annotations and propose a technique called *Protein Function Prediction using Dependency Maximization* (ProDM). ProDM can replenish the missing functions of partially annotated proteins and predict the function of completely unlabeled proteins using the partially annotated ones. Our empirical study on publicly available PPI datasets shows that ProDM performs better than other related approaches on these two prediction problems, and it is also computationally efficient.

2 Related Work

Various network-based methods have been proposed for protein function prediction [18]. Schwikowski et al. [17] make predictions for a protein based on the functions of its interacting proteins. They observed that the interacting proteins are likely to share similar functions, which is recognized as the 'guilt by association' rule. Chua et al. [6] found that indirectly interacting proteins share few functions, and extended the PPI network by integrating the level-1 (direct)

---

[1] http://mips.helmholtz-muenchen.de/proj/funcatDB/
[2] http://www.geneontology.org/

and level-2 (indirect) neighbors using different weights. These methods use a threshold on the predicted likelihood to attach more than one function to a protein. However, these methods do not take into account the correlation among functions.

More recently, multi-label learning approaches [23] have been introduced for protein function prediction. Pandey et al. [15] incorporated function correlations within a weighted k-nearest neighbor classifier, and observed that incorporating function correlations can boost the prediction accuracy. Jiang et al. [10] applied the learning with local and global consistency model [25] on a tensor graph to predict protein functions. Zhang et al. [24] included a function correlation term within the manifold regularization framework [3] to annotate proteins. Jiang et al. [9] conducted label propagation on a bi-relation graph to infer protein functions. To avoid the risk of overwriting functions during label propagation, Yu et al. [21] introduced a Transductive Multi-label Classifier (TMC) on a directed bi-relation graph to annotate proteins. Chi et al. [5] considered the fact that proteins' functions can influence the similarity between pairs of proteins and proposed an iterative model called Cosine Iterative Algorithm (CIA). In each iteration of CIA, the most confidently predicted function of an unlabeled protein is appended to the function set of this protein. Next, the pairwise similarity between training proteins and testing proteins is updated based on the similar functions within the two sets for each protein. CIA uses the updated similarity, function correlations, and PPI network structures to predict the functions on the unlabeled proteins in the following iteration.

All the above multi-label learning approaches focus on utilizing function correlation in various ways and assume that the function annotations on the training proteins are complete and accurate (without missing functions). However, due to various reasons (e.g., the evolving Gene Ontology scheme, or limitations of experimental methods), we may be aware of some of the functions of a protein, but don't know whether other functions are associated with the same protein. Namely, proteins are partially annotated. Learning from partially (or incomplete) labeled data is different from learning from partial labels [7]. In the latter case, one learns from a set of candidate labels of an instance, and assumes that only one label in this set is the ground-truth label. Learning from partially labeled data is also different from semi-supervised and supervised learning, as they both assume complete labels. In this paper, we study how to leverage partially annotated proteins, a less studied scenario in protein function prediction and multi-label learning literature [4,19,22].

Several multi-label weak-label learning approaches have been proposed. Sun et al. [19] introduced a method called WEak Label Learning (WELL). WELL is based on three assumptions: (i) the decision boundary for each label should go across low density regions; (ii) any given label should not be associated to the majority of samples; and (iii) there exists a group of low rank-based similarities, and the approximate similarity between samples with different labels can be computed based on these similarities. WELL uses convex optimization and quadratic programming to replenish the missing labels of a partially labeled sample. As such, WELL is computationally expensive. Buncak et al. [4] annotated unlabeled images using partially labeled images, and proposed a method called MLR-GL. MLR-GL optimizes the ranking loss and group Lasso in a convex optimization form. Yu et al. [22] proposed a method called *Pro*tein function prediction using *W*eak-label *L*earning (ProWL). ProWL can replenish the missing functions of partially annotated proteins, and can predict the functions of completely unlabeled proteins using the partially annotated ones. However, ProWL depends heavily on function correlations and performs the prediction for one function label at a time.

To alleviate these drawbacks associated with ProWL, we develop a new protein function prediction approach called *Pro*tein function prediction using *D*ependency *M*aximization (ProDM). ProDM uses function correlations, the 'guilt by association' rule [17], and maximizes the dependency between the features and function labels of proteins, to complete the prediction for all the function labels at one time. In our empirical study, we observe that ProDM performs better than the other competitive methods in replenishing the missing functions, and performs the best (or comparable to the best) in predicting function for completely unlabeled proteins.

3 Problem Formulation

For the task of *replenishing* missing functions, we have available n partially annotated proteins. The goal is to replenish the missing functions using such partially annotated proteins. For the task of *predicting* the functions of completely unlabeled proteins, we have a total of $n = l + u$ proteins, where the first l proteins are partially annotated and the last u proteins are completely unlabeled. The goal here is to use the l partially annotated proteins to annotate the u unlabeled ones.

Let $Y = [\mathbf{y}_1, \mathbf{y}_2, \ldots, \mathbf{y}_n]$ be the currently available function set, with $y_{ic} = 1$ if protein i has the c-th function, and $y_{ic} = 0$ otherwise. At first, we can define a function correlation matrix $M' \in \mathbb{R}^{C \times C}$ based on cosine similarity as follows:

$$M'_{st} = \frac{Y_{.s}^T Y_{.t}}{\|Y_{.s}\|\|Y_{.t}\|} \tag{1}$$

where M'_{st} is the correlation between functions s and t, and $Y_{.s}$ represents the s-th column of Y. There exists a number of ways (e.g., Jaccard coefficient [24] and Lin's similarity [15]) to define function correlation. Here, we use the cosine similarity for its simplicity and wide application [5,20,22]. If Y is represented in a probabilistic function assignment form, Eq. (1) can also be applied.

From Eq. (1), we can see that M'_{st} measures the fraction of times function s and t co-exist in a protein. We normalize M' as follows:

$$M_{st} = \frac{M'_{st}}{\sum_{c=1}^{C} M'_{sc}} \tag{2}$$

M_{st} can be viewed as the probability that a protein has function t given that it is annotated with function s.

Now, let's consider the scenario with incomplete annotations and extend the observed function set Y to $\tilde{Y} = YM$. Our motivation in using $\tilde{Y}$ is to append the missing functions using the currently known functions and their correlations. More specifically, suppose the currently confirmed functions Y_i for the i-th protein have a large correlation with the c-th function (which may be missing), then it is likely that this protein will also have function c. Based on this assumption, we define the first part of our objective function as follows:

$$\Psi_1(\mathbf{f}) = \frac{1}{2} \sum_{i=1}^{n} \sum_{c=1}^{C} (f_{ic} - \tilde{y}_{ic})^2 = \frac{1}{2} \sum_{i=1}^{n} \|F - \tilde{Y}\|_2^2 \tag{3}$$

where f_{ic} is the predicted likelihood of protein i with respect to the c-th function, $\tilde{y}_{ic}$ is the extended function annotation of protein i with respect to the c-th function, and $F = [\mathbf{f}_1, \mathbf{f}_2, \ldots, \mathbf{f}_n]$ is the prediction for the n proteins.

Since a protein has multiple functions, and the overlap between the function sets of two proteins can be used to measure their similarity, the larger the number of shared functions, the more similar the proteins are. This function induced similarity between proteins was used successfully in Chi et al. [5] and Wang et al. [20]. The function annotations of a protein can be used to enrich its feature representation. Thus, we define the function-based similarity matrix $W^f \in \mathbb{R}^{n \times n}$ between n proteins as follows:

$$W_{ij}^f = \frac{\mathbf{y}_i^T \mathbf{y}_j}{\|\mathbf{y}_i\| \|\mathbf{y}_j\|} \tag{4}$$

Note that W_{ij}^f measures the pairwise similarity (induced by the function sets of two proteins) between proteins i and j, whereas M_{st} in Eq. (2) describes the pairwise function correlations.

We now define a composite similarity W between pairwise proteins as:

$$W = W^p + \eta W^f \tag{5}$$

where $W^p \in \mathbb{R}^{n \times n}$ describes the feature induced similarity between pairs of proteins. Here W^p can be set based on the amino acid sequence similarity of a protein pair (i.e., string kernel [12] for protein sequence data), or by using the frequency of interactions found in multiple PPI studies (i.e., PPI networks in BioGrid[3]), or the weighted pairwise similarity based on reliability scores from all protein identifications by mass spectrometry (e.g., Krogan et al. [11][4]). η is a predefined parameter to balance the tradeoff between W^p and W^f. It is set to $\eta = \sum_{i=1,j=1}^{n,n} W_{ij}^p / \sum_{i=1,j=1}^{n,n} W_{ij}^f$.

The second part of our objective function leverages the knowledge that proteins with similar amino acid sequences are likely to have similar functions. In

---

[3] http://thebiogrid.org/

[4] http://www.nature.com/nature/journal/v440/n7084/suppinfo/nature04670.html

other words, we capture the 'guilt by association' rule [17], which states that interacting proteins are more likely to share similar functions. This rule is widely used in network-based protein function prediction approaches [5,17,18,22]. As in learning with local and global consistency [25], we include a smoothness term as the second part of our objective function:

$$\Psi_2(\mathbf{f}) = \frac{1}{2} \sum_{i,j=1}^{n} \| \frac{\mathbf{f}_i}{\sqrt{D_{ii}}} - \frac{\mathbf{f}_j}{\sqrt{D_{jj}}} \|^2 W_{ij}$$
$$= tr(F^T(I - D^{-\frac{1}{2}}WD^{-\frac{1}{2}})F)$$
$$= tr(F^T L F) \tag{6}$$

where D is a diagonal matrix with $D_{ii} = \sum_{j=1}^{n} W_{ij}$. I is an $n \times n$ identity matrix, $L = I - D^{-\frac{1}{2}}WD^{-\frac{1}{2}}$, and $tr(\cdot)$ is the matrix trace operation.

Here, we assume the function labels of a protein depend on the feature representation of this protein. We encode this assumption as the third part of our objective function. To capture the dependency between the function labels and the features of proteins we take advantage of the Hilbert-Schmidt Independence Criterion (HSIC) [8]. HSIC computes the squared norm of the cross-covariance operator over the feature and label domains in Hilbert Space to estimate the dependency. We choose HSIC because of its computational efficiency, simplicity and solid theoretical foundation. The empirical estimation of HSIC is given by:

$$HSIC(F, Y, p_{\mathbf{xy}}) = \frac{tr(KHSH)}{(n-1)^2} = \frac{tr(HKHS)}{(n-1)^2} \tag{7}$$

where $H, K, S \in \mathbb{R}^{n \times n}$, $K_{ij} = k(\mathbf{x}_i, \mathbf{x}_j)$ is used to measure the kernel induced similarity between two samples, $S_{ij} = s(\mathbf{f}_i, \mathbf{f}_j)$ is used to describe the label induced similarity between two samples, $H_{ij} = \delta_{ij} - \frac{1}{n}$, $\delta_{ij} = 1$ if $i = j$, otherwise $\delta_{ij} = 0$, $p_{\mathbf{xy}}$ is the joint distribution of $\mathbf{x}$ and $\mathbf{y}$. HSIC makes use of kernel matrices to estimate the dependency between labels and features of samples, thus it can also be applied in the case that there is no explicit feature representation for the n samples, as in the case of PPI network data. Although there are many other ways to initialize K and S, in this paper, we set $K = W$ and $S_{ij} = \mathbf{y}_i^T \mathbf{y}_j$ for its simplicity and its strong empirical performance. Alternative initializations of K and S will be investigated in our future study.

3.1 The Algorithm

By integrating the three objective functions introduced above, we obtain the overall objective function of ProDM:

$$\Psi(F) = tr(F^T L F) + \alpha \| F - \tilde{Y} \|_2^2 - \beta tr(HKHFF^T) + \gamma tr(F^T F) \tag{8}$$

where $\alpha > 0$ and $\beta > 0$ are used to balance the tradeoff between the three terms. Our motivation to minimize $\Psi(F)$ is three-fold: (i) two proteins with similar sequences (or frequently interacting) should have similar functions, which corresponds to the smoothness assumption in label propagation [25]; (ii) predictions in F should not change too much from the extended function labels $\tilde{Y}$; and (iii) the dependency between the function labels and the features of a protein should be maximized. In Eq. (8) we also add a term $tr(F^T F)$ (weighted by $\gamma > 0$) to enforce the sparsity of F, since each function is often associated with a relatively small number of proteins.

ProWL [22] makes use of function correlations and the 'guilt by association' rule to replenish the missing functions of partially annotated proteins. In addition, ProDM incorporates the assumption of dependency maximization. ProWL relies on the function correlation matrix M to extend the observed function annotations and to define the weight of each function label of a protein (see Eq. (3) in [22]). In contrast, ProDM exploits the function correlations to expand the incomplete function sets. As the number of missing functions increases, the function correlation matrix M becomes less reliable [22]. Therefore, when the number of missing functions is large, ProDM outperforms ProWL. In addition, ProWL predicts each function label separately and computes the inverse of a matrix for each label. ProDM, instead, predicts all C labels at once, and computes the inverse of a matrix only once. As a result, ProDM is faster than ProWL. These advantages of ProDM with respect to ProWL are corroborated in our experiments.

Eq. (8) can be solved by taking the derivative of $\Psi(F)$ with respect to F:

$$\frac{\partial \Psi(F)}{\partial F} = 2(LF + \alpha(F - \tilde{Y}) - \beta HKHF + \gamma F) \qquad (9)$$

By setting $\frac{\partial \Phi(F)}{\partial F} = 0$, we obtain:

$$F = \alpha(L + \alpha I - \beta HKH + \gamma I)^{-1}\tilde{Y} \qquad (10)$$

In Eq. (10), the complexity of the matrix multiplication HKH is $O(n^3)$ and the complexity of the matrix inverse operation is $O(n^3)$. Thus, the time complexity of ProDM is $O(n^3)$. In practice, though, L, H, and K are all sparse matrices, and Eq. (10) can be computed more efficiently. In particular, the complexity of sparse matrix multiplication is $O(nm_1)$, where m_1 is the number of nonzero elements in K. In addition, instead of computing the inverse of $(L + \alpha I - \beta HKH + \gamma I)$ in Eq. (10), we can use iterative solvers (i.e., Conjugate Gradient (CG)). CG is guaranteed to terminate in n iterations. In each iteration, the most time-consuming operation is the product between an $n \times n$ sparse matrix and a label vector (one column of $\tilde{Y}$). Thus, in practice, the time complexity of ProDM is $O(m_1 n + t m_2 n C)$, where C is the number of function labels, m_2 is the number of nonzero elements in $(L + \alpha I - \beta HKH + \gamma I)$, and t is the number of CG iterations. CG often terminates in no more than 20 iterations.

4   Experimental Setup

Datasets We investigate the performance of ProDM on replenishing missing functions and predicting protein functions on three different PPI benchmarks.

The first dataset, *Saccharomyces Cerevisiae* PPIs (ScPPI), is extracted from BioGrid[5]. We annotate these proteins according to FunCat [16] database and use the largest connected component of ScPPI for experiments, which includes 3041 proteins. FunCat organizes function labels in a tree structure. We filtered the function labels and used the 86 informative functions. Informative functions [10,24] are the ones that have at least 30 proteins as members and within the tree structure these functions do not have a particular descendent node with more than 30 proteins. The weight matrix W^p of ScPPI is specified by the number of PubMed IDs, where 0 means no interaction between two proteins, and $q > 0$ implies the interaction is supported by q distinct PubMed IDs. The second dataset, KroganPPI is obtained from the study of Krogan et al. [11][6]. We use its largest connected component for the experiments and annotate these proteins according to FunCat. After the preprocessing, KroganPPI contains 3642 proteins annotated with 90 informative functions. The weight matrix of W^p is specified by the provider. The third dataset, HumanPPI is obtained from the study of Mostafavi et al. [13][7]. HumanPPI is extracted from the multiple data types of Human Proteomic data. The proteins in HumanPPI are annotated according to the Gene Ontology [2]. Similarly to [10,13], we use the largest connected components of HumanPPI and the functions that have at least 30 annotated proteins. The weight matrix W^p of HumanPPI is specified by the provider. The characteristics of these processed datasets are listed in Table 1.

Table 1. Dataset Statistics (Avg±Std means average number of functions for each protein and its standard deviation)

Dataset	#Proteins	#Functions	Avg±Std
ScPPI	3041	86	2.49 ± 1.70
KroganPPI	3642	90	2.20 ± 1.60
HumanPPI	2950	200	3.80 ± 3.77

Comparative Methods. We compare the proposed method with: (i) ProWL [22], (ii) WELL [19][8], (iii) MLR-GL [4][9], (iv) TMC [21], and (v) CIA [5]. The first three approaches are multi-label learning models with partially labeled data, and the last two methods are recently proposed protein function prediction algorithms based on multi-label learning and PPI networks. WELL and MLR-GL need an input kernel matrix. We substitute the kernel matrix with W^p, which is semi-definite positive and can be viewed as a Mercer kernel [1]. WELL was proposed to replenish the missing functions of partially annotated proteins. We adopt it here to predict the functions of completely unlabeled proteins by including the unlabeled proteins in the input kernel matrix. MLR-GL is targeted

---

[5] http://thebiogrid.org/
[6] http://www.nature.com/nature/journal/v440/n7084/suppinfo/
   nature04670.html
[7] http://morrislab.med.utoronto.ca/~sara/SW/
[8] http://lamda.nju.edu.cn/code_WELL.ashx
[9] http://www.cse.msu.edu/~bucakser/MLR_GL.rar

at predicting the functions of completely unlabeled proteins using partially annotated proteins. We adapt it to replenish the missing functions of partially annotated proteins by using all the proteins as training and testing set. As was done for MLR-GL, we also adapt TMC to replenish the missing functions. Due to the iterative procedure of CIA, it cannot be easily adapted to replenish missing functions. The parameters of WELL, MLR-GL, ProWL, TMC, and CIA are set as the authors specified in their code, or reported in the papers. For ProDM, we search for optimal α values in the range $[0.5, 1]$ with step size 0.05, and β values in the range $[0.01, 0.1]$ with step size 0.01. In our experiments, we set α and β to 0.99 and 0.01, respectively, since we observed that the performance with respect to the various metrics does not change as we vary α and β around the fixed values. Similarly to ProWL, we set γ to 0.001.

Experimental Protocol In order to simulate the incomplete annotation scenario, we assume the annotations on the currently labeled proteins are complete and mask some of the ground truth functions. The masked functions are considered missing. For presentation, we define a term called *Incomplete Function* (IF) ratio, which measures the ratio between the number of missing functions and the number of ground truth functions. For example, if a protein has five functions (labels), and two of them are masked (two 1s are changed to two 0s), then the IF ratio is $2/5 = 40\%$.

Evaluation Criteria. Protein function prediction can be viewed as a multi-label learning problem and evaluated using multi-label learning metrics [10,22]. Various evaluation metrics have been developed for evaluating multi-label learning methods [23]. Here we use five metrics: *MicroF1*, *MacroF1*, *HammingLoss*, *RankingLoss* and adapted *AUC* [4]. These metrics were also used to evaluate WELL [19], MLR-GL [4], and ProWL [22]. In addition, we design *RAccuracy* to evaluate the performance of replenishing missing functions. Suppose the predicted function set of n proteins is F_p, the initial incomplete annotated function set is F_q, and the ground truth function set is Y. *RAccuracy* is defined as follows:

$$RAccuracy = \frac{|(Y - F_q) \cap F_p|}{|(Y - F_q)|}$$

where $|(Y - F_q)|$ measures how many functions are missing among n proteins and $|(Y - F_q) \cap F_p|$ counts how many missing functions are correctly replenished. To maintain consistency with other evaluation metrics, we report *1-HammLoss* and *1-RankLoss*. Thus, similarly to other metrics, the higher the values of *1-HammLoss* and *1-RankLoss*, the better the performance.

5   Experimental Analysis

5.1   Replenishing Missing Functions

We performed experiments to investigate the performance of ProDM on replenishing the missing functions of n partially labeled proteins. To this end, we consider all the proteins in each dataset as training and testing data. To perform comparisons against the other methods, we vary the IF ratio from 30% to 70%, with an interval of 20%. A few proteins in the PPI networks do not have any functions. To make use of the 'guilt by association' rule and keep the PPI network connected, we do not remove them and test the performance of replenishing missing functions on the proteins with annotations. We repeat the experiments 20 times with respect to each IF ratio. In each run, the missing functions are randomly masked for each protein according to the IF ratio. $F \in \mathbb{R}^{n \times C}$ in Eq. (10) is a predicted likelihood matrix. *MicroF1*, *MacroF1*, *1-HammLoss* and *RAccuracy* require F to be a binary indicator matrix. Here, we consider the functions corresponding to the r largest values of $\mathbf{f}_i$ as the functions of the i-th protein, where r is determined by the number of ground-truth functions of this protein. To simulate the incomplete annotation scenario, we assume the given functions of the i-th protein in a dataset are ground-truth functions, and mask some of them to generate the missing functions. The experimental results are reported in Tables 2-4. In these tables, best and comparable results are in **boldface** (statistical significance is examined via pairwise t-test at 95% significance level).

Table 2. Results of replenishing missing functions on **ScPPI**

Metric	IF Ratio	ProDM	ProWL	WELL	MLR-GL	TMC
MicroF1	30%	**93.88±0.12**	86.28±0.14	60.49±0.54	23.67±0.50	91.80±0.20
	50%	**79.09±0.28**	68.36±0.36	47.42±0.74	26.98±0.49	77.09±0.28
	70%	**71.67±0.51**	60.09±0.51	42.06±0.04	27.15±0.59	69.79±0.44
MacroF1	30%	**94.05±0.18**	86.28±0.18	55.35±0.52	24.06±0.79	90.98±0.24
	50%	**78.39±0.33**	67.81±0.36	43.80±0.55	27.45±0.72	74.72±0.35
	70%	**70.05±0.45**	59.45±0.62	38.25±0.87	27.98±0.72	67.34±0.52
1-HammLoss	30%	**99.65±0.01**	99.20±0.01	97.71±0.03	95.58±0.03	99.52±0.01
	50%	**98.79±0.02**	98.17±0.02	96.95±0.04	95.77±0.03	98.67±0.02
	70%	**98.36±0.03**	97.69±0.03	96.64±0.00	95.78±0.03	98.25±0.03
1-RankLoss	30%	**99.67±0.02**	95.16±0.02	94.78±0.07	44.38±0.39	99.65±0.02
	50%	96.80±0.12	91.95±0.24	90.41±0.24	41.43±0.66	**97.06±0.10**
	70%	**94.92±0.17**	88.03±0.24	89.01±0.26	38.06±0.77	94.52±0.29
AUC	30%	**98.79±0.05**	94.92±0.04	93.09±0.04	55.63±0.38	**98.77±0.04**
	50%	95.63±0.14	92.07±0.16	88.24±0.24	54.01±0.66	**95.97±0.10**
	70%	**93.09±0.22**	88.85±0.20	86.08±0.35	52.60±0.46	**93.04±0.29**
RAccuracy	30%	**49.24±1.28**	38.05±1.07	23.94±1.55	46.18±1.04	46.01±1.52
	50%	**46.57±0.71**	32.14±0.92	18.83±1.01	35.59±0.91	42.46±0.76
	70%	**44.18±1.03**	31.41±1.03	17.12±0.12	33.89±0.74	41.42±0.82

From these Tables (2-4), we can observe that ProDM performs much better than the competitive methods in replenishing the missing functions of proteins across all the metrics. Both ProDM and ProWL take advantage of function correlations and of the 'guilt by association' rule, but ProDM significantly outperforms ProWL. The difference in performance between ProDM and ProWL confirms our intuition that maximizing the dependency between functions and features of proteins is effective. The performance of WELL is not comparable to that of ProDM. The possible reason is that the assumptions used in WELL may be not suitable for the PPI network datasets. The performance of MLR-GL varies because it is targeted at predicting functions of unlabeled proteins using

Table 3. Results of replenishing missing functions on **KroganPPI**

Metric	IF Ratio	ProDM	ProWL	WELL	MLR-GL	TMC
MicroF1	30%	**95.51±0.13**	93.05±0.08	61.04±0.27	14.78±0.23	88.67±0.12
	50%	**79.46±0.22**	68.39±0.27	48.54±0.67	16.18±0.29	70.93±0.22
	70%	**70.23±0.35**	60.25±0.29	43.72±0.19	16.09±0.34	61.82±0.31
MacroF1	30%	**95.70±0.18**	94.57±0.15	58.24±0.20	13.71±0.28	88.41±0.12
	50%	**78.92±0.25**	71.51±0.32	52.09±1.08	15.12±0.34	69.20±0.33
	70%	**69.01±0.40**	62.30±0.46	48.79±0.52	14.92±0.35	60.20±0.44
1-HammLoss	30%	**99.78±0.01**	99.66±0.00	98.08±0.01	95.81±0.01	99.44±0.01
	50%	**98.99±0.01**	98.44±0.01	97.47±0.03	95.87±0.01	98.57±0.01
	70%	**98.53±0.02**	98.04±0.01	97.23±0.01	95.87±0.02	98.12±0.02
1-RankLoss	30%	**99.75±0.02**	99.61±0.02	96.50±0.03	39.88±0.37	99.52±0.02
	50%	**96.87±0.12**	94.55±0.12	91.60±0.09	39.99±0.27	96.20±0.16
	70%	**94.37±0.14**	91.02±0.25	89.89±0.06	38.48±0.39	93.28±0.19
AUC	30%	**98.87±0.04**	98.58±0.04	94.90±0.05	45.49±0.28	98.59±0.05
	50%	**95.47±0.12**	92.55±0.15	88.88±0.14	46.65±0.32	94.63±0.18
	70%	**91.91±0.16**	86.90±0.35	85.87±0.10	46.45±0.37	90.58±0.24
RAccuracy	30%	**44.97±1.63**	14.90±0.98	9.24±0.66	30.90±1.48	23.89±1.30
	50%	**42.20±0.63**	11.04±0.77	7.03±0.22	23.83±0.71	27.89±0.61
	70%	**36.25±0.75**	14.89±0.61	7.68±0.44	21.69±0.80	27.06±0.65

Table 4. Results of replenishing missing functions on **HumanPPI**

Metric	IF Ratio	ProDM	ProWL	WELL	MLR-GL	TMC
MicroF1	30%	**96.60±0.14**	95.12±0.14	86.21±0.10	15.76±0.30	91.90±0.15
	50%	**88.48±0.41**	77.18±0.24	64.93±0.26	16.36±0.21	77.98±0.27
	70%	**79.20±0.55**	61.91±0.30	51.91±0.46	16.10±0.29	69.05±0.31
MacroF1	30%	**96.21±0.16**	94.76±0.16	87.95±0.03	15.79±0.27	91.43±0.15
	50%	**87.49±0.46**	76.86±0.30	70.43±0.18	16.00±0.26	77.05±0.31
	70%	**77.58±0.53**	62.19±0.30	59.05±0.37	15.45±0.26	67.67±0.35
1-HammLoss	30%	**99.87±0.01**	99.81±0.01	99.48±0.00	96.80±0.01	99.69±0.01
	50%	**99.56±0.02**	99.13±0.01	98.67±0.01	96.82±0.01	99.16±0.01
	70%	**99.21±0.02**	98.55±0.01	98.17±0.02	96.82±0.01	98.83±0.01
1-RankLoss	30%	**99.81±0.02**	99.74±0.03	97.19±0.03	54.78±0.32	99.73±0.02
	50%	**98.73±0.07**	96.90±0.21	87.55±0.44	58.09±0.29	98.31±0.12
	70%	**97.50±0.15**	93.56±0.41	83.97±0.08	58.35±0.36	96.76±0.21
AUC	30%	**98.65±0.04**	98.52±0.05	93.51±0.13	54.32±0.22	98.44±0.04
	50%	**97.37±0.09**	95.86±0.15	83.05±0.20	55.90±0.21	96.82±0.10
	70%	**95.48±0.14**	91.31±0.28	76.12±0.48	55.69±0.26	94.64±0.18
RAccuracy	30%	**80.39±0.80**	71.86±0.79	20.50±0.59	30.92±1.09	53.35±0.83
	50%	**73.14±0.96**	46.78±0.55	18.23±0.62	23.92±0.56	48.66±0.63
	70%	**63.28±0.97**	32.76±0.53	15.09±0.82	21.41±0.45	45.36±0.55

partially annotated proteins, whereas here it is adapted for replenishing missing functions. TMC is introduced to predict functions for completely unlabeled proteins using completely labeled ones; TMC sometimes outperforms ProWL and WELL. This is because the missing functions can be appended in the bi-relation graph. In fact, TMC also makes use of function correlations and the 'guilt by association' rule, but it still loses to ProDM. The reason is that ProDM maximizes the dependency between proteins' functions and features. The margin in performance achieved by ProDM with respect to ProWL and TMC demonstrates the effectiveness of using *dependency maximization* in replenishing the missing functions of proteins.

We also observe that, as more functions are masked, ProWL downgrades much more rapidly than ProDM. As the IF ratio increases, the function correlation matrix M becomes less reliable. ProWL uses M to estimate the likelihood of missing functions and to weigh the loss function. ProDM only utilizes M to estimate the probability of missing functions and makes additional use of dependency maximization. Thus ProDM is less dependent on M. Taking *RAccuracy* on ScPPI as an example, ProDM on average is 33.55% better than ProWL, 49.60% better than WELL, 19.31% better than MLR-GL, and 8.21% better than TMC. These results confirm the effectiveness of ProDM in replenishing the missing functions. Overall, this experimental results confirm the advantages of combining the 'guilt by association' rule, function correlations, and dependency maximization.

5.2 Predicting Unlabeled Proteins

We conduct another set of experiments to study the performance of ProDM in predicting the function of completely unlabeled proteins using partially labeled ones. In this scenario, $l < n$ proteins are partially annotated and $n - l$ proteins are completely unlabeled. At first, we partition each dataset into a *training* set (accounting for 80% of all the proteins) with partial annotations and into a *testing* set (accounting for the remaining 20% of all the proteins) with no annotations. We run the experiments 20 times for each dataset. In each round, the dataset is randomly divided into training and testing datasets. We simulate the setting of missing functions (IF ratio=50%) in the training set as done in the experiments in Section 5.1, but r is determined as the average number of functions (round to the next integer) of all proteins. From Table 1: r is set to 3 for ScPPI and KroganPPI, and to 4 for HumanPPI. The results (average of 20 independent runs) are listed in Tables 5-7. Since *RAccuracy* is not suitable for the settings of predicting completely unlabeled proteins, the results for this metric are not reported.

Table 5. Prediction results on completely unlabeled proteins of **ScPPI**

Metric	ProDM	ProWL	WELL	MLR-GL	TMC	CIA
MicroF1	**32.78±1.37**	30.06±1.15	16.75±2.03	24.15±1.40	3.67±0.38	20.78±0.38
MacroF1	31.91±1.48	**31.33±1.74**	5.19±0.71	26.25±1.50	2.00±0.39	26.27±0.39
1-HammLoss	**95.73±0.10**	95.56±0.09	94.69±0.16	95.19±0.09	93.89±0.05	94.96±0.05
1-RankLoss	**73.13±2.72**	60.37±1.64	73.57±0.05	41.56±1.06	28.29±0.70	21.82±0.70
AUC	78.40±1.57	**78.63±0.74**	77.00±0.53	61.47±1.26	55.72±0.84	63.38±0.84

Table 6. Prediction results on completely unlabeled proteins of **KroganPPI**

Metric	ProDM	ProWL	WELL	MLR-GL	TMC	CIA
MicroF1	**22.55±1.35**	22.40±0.97	14.35±1.25	13.58±0.86	3.32±0.52	13.78±0.52
MacroF1	**18.26±1.53**	17.68±1.11	1.47±0.30	12.80±0.92	2.05±0.41	13.85±0.41
1-HammLoss	**96.40±0.08**	96.40±0.08	96.04±0.03	95.99±0.07	95.52±0.06	95.99±0.06
1-RankLoss	66.69±1.19	75.41±0.88	**75.43±0.22**	48.40±1.13	61.26±0.89	18.43±0.89
AUC	72.26±0.73	**74.78±0.73**	74.16±0.12	58.80±1.10	61.35±0.68	59.45±0.68

From Tables 5-7, we can observe that ProDM achieves the best (or comparable to the best) performance among all the comparing methods on various evaluation metrics. ProDM and ProWL have similar performance in the task of predicting the functions of completely unlabeled proteins. One possible reason is

Table 7. Prediction results on completely unlabeled proteins of **HumanPPI**

Metric	ProDM	ProWL	WELL	MLR-GL	TMC	CIA
MicroF1	**24.57±1.03**	23.18±1.24	16.43±1.78	12.87±0.76	1.91±0.28	12.86±0.28
MacroF1	**20.58±1.18**	19.32±0.90	15.55±1.30	11.95±0.78	1.61±0.26	9.90±0.26
1-HammLoss	**97.17±0.05**	97.11±0.09	96.85±0.10	96.73±0.05	96.33±0.07	96.72±0.07
1-RankLoss	**76.70±1.07**	76.64±2.01	62.98±1.82	67.89±1.44	50.93±0.77	33.87±0.77
AUC	**78.82±1.19**	77.41±0.92	62.30±1.38	66.23±0.85	51.78±1.21	67.08±1.21

that F is initially set to $\tilde{Y}$ and $\{\tilde{\mathbf{y}}_j\}_{j=l+1}^n$ are zero vectors. WELL works better than MLR-GL in replenishing the missing functions, and it loses to MLR-GL in predicting the functions of unlabeled proteins. One possible cause is that WELL is targeted at replenishing missing functions, and here it's adjusted to predict functions on completely unlabeled proteins. MLR-GL predicts protein functions under the assumption of partially annotated proteins, and it is outperformed by ProDM. MLR-GL optimizes the ranking loss and the group Lasso loss, whereas ProDM optimizes an objective function based on the function correlations, the 'guilt by association' rule, and the dependency between the function labels and the features of proteins. We can claim that ProDM is more faithful to the characteristics of proteomic data than MLR-GL. For the same reasons, ProDM often outperforms WELL, which takes advantage of low density separation and low-rank based similarity to capture function correlations and data distribution.

TMC sometimes performs similar to ProDM in the task of replenishing the missing functions. However, TMC is outperformed by other methods when making predictions for completely unlabeled proteins. A possible reason is that TMC assumes the training proteins are fully annotated, and the estimated function correlation matrix M may be unreliable when IF ratio is set to 50%. CIA also exploits function-based similarity and PPI networks to predict protein functions, but it's always outperformed by ProDM and by ProWL. There are two possible reasons. First, CIA does not account for the weights of interaction between two proteins. Second, CIA mainly relies on the function induced similarity W^f, and when training proteins are partially annotated, this similarity becomes less reliable. CIA performs better than TMC. One reason might be that CIA exploits a neighborhood count algorithm [17] to initialize the functions on unlabeled proteins in the kick-off step of CIA, whereas TMC does not. All these results show the effectiveness of ProDM in predicting unlabeled proteins by considering the partial annotations on proteins.

5.3   Component Analysis

To investigate the benefit of using the 'guilt by association' rule and of exploiting function correlations, we introduce two variants of ProDM, namely ProDM_nGBA and ProDM_nFC. ProDM_nGBA corresponds to *Pro*tein function prediction using *D*ependency *M*aximization with *no* '*G*uilt *B*y *A*ssociation' rule. Specifically, ProDM_nGBA is based on Eq. (8) without the first term; that is, ProDM_nGBA uses only the partial annotations and function correlations to replenish the missing functions. ProDM_nFC corresponds to *Pro*tein function prediction using *Dependency Maximization* with *no Function Correlation*. In ProDM_nFC, Y is used in Eq. (8) instead of $\tilde{Y}$. We increase the IF ratio from 10% to 90% at intervals of 10%, and record the results of ProDM, ProDM_nGBA and ProDM_nFC with respect to each IF ratio. For brevity, in Figure 1 we just report the results with respect to *MicroF1* and *AUC* on HumanPPI.

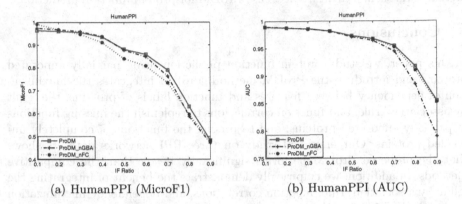

(a) HumanPPI (MicroF1)   (b) HumanPPI (AUC)

Fig. 1. The benefit of using both the 'guilt by association' rule and function correlations (ProDM_nFC is ProDM with no function correlation, and ProDM_nGBA is ProDM with no 'guilt by association' rule)

From Figure 1, we can observe that ProDM, ProDM_nGBA, and ProDM_nFC have similar performance when few functions are missing. This indicates that both the 'guilt by association' rule and function correlations can be utilized to replenish the missing functions. However, as the number of missing function increases, ProDM generally outperforms ProDM_nGBA and ProDM_nFC. The reason is that ProDM, unlike ProDM_nGBA and ProDM_nFC, makes use of *both* the 'guilt by association' rule and function correlations. This fact shows that it's important and reasonable to integrate these two components in replenishing missing functions.

5.4   Run Time Analysis

In Table 8 we record the average run time of each of the methods on the three datasets. The experiments are conducted on Windows 7 platform with Intel E31245 processor and 16GB memory. TMC assumes the training proteins are accurately annotated, and it takes much less time than the other methods. MLR-GL

Table 8. Runtime Analysis (seconds)

Dataset	ProDM	ProWL	WELL	MLR-GL	TMC
ScPPI	60.77	83.09	1687.09	22.66	2.29
KroganPPI	80.60	134.94	3780.24	32.40	3.62
HumanPPI	64.02	194.62	5445.97	50.68	3.49
Total	178.37	412.65	10913.30	105.74	9.40

relaxes the convex-concave optimization problem into a Second Order Cone Programming (SOCP) [4] problem, and it ranks 2nd (from fast to slow). ProDM takes less time than ProWL, since ProDM infers the functions of a protein in one step, whereas ProWL divides the prediction into C subproblems. WELL uses eigen-decomposition and convex optimization, and it costs much more than the other methods. As such, it is desirable to use ProDM for protein function prediction.

6   Conclusions

In this paper, we study protein function prediction using partially annotated proteins and introduce the ProDM method. ProDM integrates the maximization of dependency between features and function labels of proteins, the 'guilt by association' rule, and function correlations to replenish the missing functions of partially annotated proteins, and to predict the functions of completely un-labeled proteins. Our empirical study on three PPI networks datasets shows that the proposed ProDM performs significantly better than the competitive methods. In addition, we empirically demonstrate the benefit of integrating the 'guilt by association' rule, function correlations, and dependency maximization in protein function prediction.

Acknowledgements. The authors want to thank the anonymous reviewers for helpful comments. This work is partially supported by NSF IIS-0905117 and NSF Career Award IIS-1252318, NSFC (61003174, 61070090), Natural Science Foundation of Guangdong Province (S2012010009961), Specialized Research Fund for the Doctoral Program of Higher Education (20110172120027), and China Scholarship Council (CSC).

References

1. Aizerman, A., Braverman, E.M., Rozoner, L.I.: Theoretical foundations of the potential function method in pattern recognition learning. Automation and Remote Control 25, 821–837 (1964)
2. Ashburner, M., Ball, C.A., Blake, J.A., Botstein, D., Butler, H., Cherry, J.M., Davis, A.P., Dolinski, K., Dwight, S.S., Eppig, J.T., et al.: Gene ontology: tool for the unification of biology. Nature Genetics 25(1), 25–29 (2000)
3. Belkin, M., Niyogi, P., Sindhwani, V.: Manifold regularization: A geometric framework for learning from labeled and unlabeled examples. Journal of Machine Learning Research 7, 2399–2434 (2006)
4. Bucak, S.S., Jin, R., Jain, A.K.: Multi-label learning with incomplete class assignments. In: Proceedings of 24th IEEE Conference on Computer Vision and Pattern Recognition, pp. 2801–2808 (2011)
5. Chi, X., Hou, J.: An iterative approach of protein function prediction. BMC Bioinformatics 12(1), 437 (2011)
6. Chua, H.N., Sung, W., Wong, L.: Using indirect protein interactions for the prediction of gene ontology functions. BMC Bioinformatics 8(supp. 4), S8 (2007)
7. Cour, T., Sapp, B., Taskar, B.: Learning from partial labels. Journal of Machine Learning Research 12, 1501–1536 (2011)

8. Gretton, A., Bousquet, O., Smola, A.J., Schölkopf, B.: Measuring statistical dependence with hilbert-schmidt norms. In: Jain, S., Simon, H.U., Tomita, E. (eds.) ALT 2005. LNCS (LNAI), vol. 3734, pp. 63–77. Springer, Heidelberg (2005)

9. Jiang, J.Q.: Learning protein functions from bi-relational graph of proteins and function annotations. In: Przytycka, T.M., Sagot, M.-F. (eds.) WABI 2011. LNCS, vol. 6833, pp. 128–138. Springer, Heidelberg (2011)

10. Jiang, J.Q., McQuay, L.J.: Predicting protein function by multi-label correlated semi-supervised learning. IEEE/ACM Transactions on Computational Biology and Bioinformatics 9(4), 1059–1069 (2012)

11. Krogan, N.J., Cagney, G., Yu, H., Zhong, G., Guo, X., Ignatchenko, A., Li, J., Pu, S., Datta, N., Tikuisis, A.P., et al.: Global landscape of protein complexes in the yeast saccharomyces cerevisiae. Nature 440(7084), 637–643 (2006)

12. Leslie, C.S., Eskin, E., Cohen, A., Weston, J., Noble, W.S.: Mismatch string kernels for discriminative protein classification. Bioinformatics 20(4), 467–476 (2004)

13. Mostafavi, S., Morris, Q.: Fast integration of heterogeneous data sources for predicting gene function with limited annotation. Bioinformatics 26(14), 1759–1765 (2010)

14. Pandey, G., Kumar, V., Steinbach, M.: Computational approaches for protein function prediction. Technical Report TR 06-028, Department of Computer Science and Engineering, University of Minnesota, Twin Cities (2006)

15. Pandey, G., Myers, C., Kumar, V.: Incorporating functional inter-relationships into protein function prediction algorithms. BMC Bioinformatics 10(1), 142 (2009)

16. Ruepp, A., Zollner, A., Maier, D., Albermann, K., Hani, J., Mokrejs, M., Tetko, I., Güldener, U., Mannhaupt, G., Münsterkötter, M., et al.: The funcat, a functional annotation scheme for systematic classification of proteins from whole genomes. Nucleic Acids Research 32(18), 5539–5545 (2004)

17. Schwikowski, B., Uetz, P., Fields, S., et al.: A network of protein-protein interactions in yeast. Nature Biotechnology 18(12), 1257–1261 (2000)

18. Sharan, R., Ulitsky, I., Shamir, R.: Network-based prediction of protein function. Molecular Systems Biology 3(1) (2007)

19. Sun, Y., Zhang, Y., Zhou, Z.: Multi-label learning with weak label. In: Proceedings of 24th AAAI Conference on Artificial Intelligence (2010)

20. Wang, H., Huang, H., Ding, C.: Function-function correlated multi-label protein function prediction over interaction networks. In: Chor, B. (ed.) RECOMB 2012. LNCS, vol. 7262, pp. 302–313. Springer, Heidelberg (2012)

21. Yu, G., Domeniconi, C., Rangwala, H., Zhang, G., Yu, Z.: Transductive multi-label ensemble classification for protein function prediction. In: Proceedings of the 18th ACM SIGKDD International Conference on Knowledge Discovery and Data Mining, pp. 1077–1085 (2012)

22. Yu, G., Zhang, G., Rangwala, H., Domeniconi, C., Yu, Z.: Protein function prediction using weak-label learning. In: Proceedings of the ACM Conference on Bioinformatics, Computational Biology and Biomedicine, pp. 202–209 (2012)

23. Zhang, M., Zhou, Z.: A review on multi-label learning algorithms. IEEE Transctions on Knowledge and Data Engineering 99(99), 1 (2013)

24. Zhang, X., Dai, D.: A framework for incorporating functional interrelationships into protein function prediction algorithms. IEEE/ACM Transactions on Computational Biology and Bioinformatics 9(3), 740–753 (2012)

25. Zhou, D., Bousquet, O., Lal, T.N., Weston, J., Schölkopf, B.: Learning with local and global consistency. In: Proceedings of Advances in Neural Information Processing Systems, pp. 321–328 (2003)

Improving Relational Classification Using Link Prediction Techniques

Cristina Pérez-Solà[1] and Jordi Herrera-Joancomartí[1,2]

[1] Dept. d'Enginyeria de la Informació i les Comunicacions,
Universitat Autònoma de Barcelona,
08193 Bellaterra, Catalonia, Spain
{cperez,jherrera}@deic.uab.cat
[2] Internet Interdisciplinary Institute (IN3) - UOC

Abstract. In this paper, we address the problem of classifying entities belonging to networked datasets. We show that assortativity is positively correlated with classification performance and how we are able to improve classification accuracy by increasing the assortativity of the network. Our method to increase assortativity is based on modifying the weights of the edges using a scoring function. We evaluate the ability of different functions to serve for this purpose. Experimental results show that, for the appropriated functions, classification on networks with modified weights outperforms the classification using the original weights.

1   Introduction

Relational classification deals with the problem of classifying networked data, that is, data containing a set of entities that are interlinked with each other. Networked data can be found almost everywhere: from authorship networks, that link authors sharing a common paper, to the now very popular Online Social Networks, where users are mainly linked by friendship. Some traditional machine learning techniques, that deal with independent entities, have been adapted to handle networked datasets, and new algorithms have also been proposed to manage this kind of data. Classification is not an exception. In the last years, many algorithms have been proposed to take advantage of the linked nature of these datasets in order to perform classification [1–4].

In this paper, we build on these existing techniques and propose a method to increase assortativity mixing according to the node class labels. Prior works [5, 6] have suggested that assortativity with respect to class labels is an indicator of the level of performance that a relational classifier is able to achieve. So after proposing a method to increase assortativity, we will evaluate to what extents this statement is true. We will conduct a systematic analysis of the performances obtained when classifying different datasets with multiple configurations of the classifier, and we will show how these performances correlate with the assortativity obtained in both the original graphs and those modified to increase assortativity. Assortativity has been proposed as a metric to perform automatic edge selection [5] because preliminary results showed that choosing those edges

H. Blockeel et al. (Eds.): ECML PKDD 2013, Part I, LNAI 8188, pp. 590–605, 2013.
© Springer-Verlag Berlin Heidelberg 2013

for which higher assortativity was obtained resulted in higher classification performance. However, this preliminary study already showed that the procedure does not always lead to the best possible performance. It is thus interesting to evaluate to what extent assortativity is positively correlated with classification performance.

The contribution of this paper is threefold. First, we propose a method to increase both node and edge assortativity by modifying the weights of the edges. This method is based on the usage of scoring functions. We investigate several scoring functions abilities to increase assortativity for different datasets. Second, we evaluate the correlation between the level of assortativity found in a graph and the obtained performance when trying to classify nodes of that graph. We evaluate correlation for datasets modeling different entities and relationships and for multiple relational classifiers. Third, we compare the classification results of the increased assortativity graphs with the original graphs and analyze the performance improvement.

The rest of the paper is organized as follows. Section 2 describes our proposal for increasing assortativity by modifying the weights of the edges of the graph. Then, Section 3 presents the experimental results supporting our claims. First, Section 3.1 presents the datasets used in the experiments. Then, Section 3.2 demonstrates how the proposed method is able to increase assortativity. In Section 3.3, we define the classification problem that we are facing in order to show, in Section 3.4, how assortativity is positively correlated with classification performance. After that, Section 3.5 demonstrates the effects of using the proposed method on classification performance. Finally, Section 4 reviews the related work and Section 5 presents the conclusions and lines for further work.

2   Modifying Edges' Weight to Increase Assortativity

This section describes the proposed procedure for increasing assortativity. After defining the notation and the concept of assortativity, we present the set of scoring functions that we use to test our technique. Then, we show how to compute the new weights taking into account the results of the scoring functions.

2.1   Notation

Given a graph $G = (V, E_w)$, the set of vertexes V represents the entities in the networked dataset and the set of edges E_w represents the relationships between those entities. Since we are dealing with weighted graphs, edges are pairs of vertexes with an associated weight $e = (v_i, v_j, w_{ij})$ s.t $(v_i, v_j) \in V \times V$ and $w_{ij} \in \mathbb{R}$. Because we are dealing with undirected graphs, symmetry is assumed, $e = (v_i, v_j, w_{ij}) = (v_j, v_i, w_{ji})$. Let us denote by $\Gamma(v_i)$ the set of adjacent nodes of v_i, that is, $\Gamma(v_i) = \{v_j \in V$ s.t. $\exists e = (v_i, v_j, w_{ij}) \in E$ with $w_{ij} \neq 0\}$. Finally, we will use the words entities, nodes, or vertexes interchangeably through the rest of this paper, as we will do with edges, relationships, and links.

Classification is one of the basic techniques in data mining processes. Classification problems consist on assigning labels to entities for which the label is initially unknown. Given a set of labeled samples, the goal is to assign labels to the rest of the samples in the dataset. More formally, we denote by $\mathfrak{C} = \{\mathfrak{c}_k,$ for $k = 1, \cdots, m\}$ the set of all possible categories an entity can be labeled with. Then, there exist a set of nodes $V_l \subset V$ for which the mapping $A : V_l \to \mathfrak{C}$ is known before classification takes place, and a set of nodes $V_{nl} = V \setminus V_l$ for which the mapping is unknown.[1] The goal of the classification process is to discover this latter mapping, or a probability distribution over it. Notice that with this definition, the only uncertainty introduced is the class membership of the nodes in V_{nl}.

2.2 Assortativity

Assortativity mixing is the tendency for entities in a network to be connected to other entities that are like them in some way [7]. This phenomenon has been much studied for social networks, where users show a preference to link, follow, or listen to other users who are like them. When dealing with social networks, assortativity is usually known as homophily. Assortativity (or dissortativity, the tendency of nodes to be linked to other nodes that are not like them) has been reported in many kinds of networks. For instance, degree dissortativity has been observed in protein networks, neural networks, and metabolic networks [7].

Assortativity mixing can be computed according to an enumerative characteristic or a scalar characteristic. In the latter case, degree assortativity is of special interest because of its consequences on the structure of the network. In this paper, we are interested on the first alternative, assortativity according to an enumerative characteristic, where assortativity will be related to the class label of the nodes for which the classification will take place. From now on, we will refer to the assortativity regarding the class labels as merely assortativity.

The first hypothesis that we want to test is if it is possible to increase the assortativity of a graph with respect to the class labels assigned to its nodes without knowing these class labels. That is, given a graph $G = (V, E_w)$ for which all class labels are unknown, we want to see if it is possible to design a process that results in a new graph $G' = (V, E'_w)$ that presents higher assortativity than G. This scenario is even more restrictive than the usual within-network node classification scenario, where some of the labels will be known in advance. Note that although the described process does not need any class label, we make use of these class labels to evaluate its performance (i.e. to compute assortativity).

In order to compute edge assortativity [7] for a given graph $G = (V, E)$ for which the mapping $A : V \to \mathfrak{C}$ is known for all V, an edge assortativity matrix e of size $|\mathfrak{C}| \times |\mathfrak{C}|$ is constructed. Each cell e_{ij} contains the fraction of all edges that link nodes of class $\mathfrak{c}_i$ to nodes of class $\mathfrak{c}_j$, normalized such that $\sum_{\forall i,j} e_{ij} = 1$. Values a_i and b_i are defined as the fraction of each type of end of an edge that is attached to vertexes of type $\mathfrak{c}_i$: $a_i = \sum_{\forall j} e_{ij}$ and $b_i = \sum_{\forall i} e_{ij}$. The (edge) assortativity coefficient A_E is then defined as:

---

[1] Note that l stands for *labeled* and nl stands for *not labeled*.

$$A_E = \frac{\sum_{\forall i} e_{ii} - \sum_{\forall i} a_i b_i}{1 - \sum_{\forall i} a_i b_i}$$

Because A_E measures assortativity across edges and not across nodes, a node assortativity metric, A_N, is defined in [5]. A_N is computed in the same way, now using the node assortativity matrix e^* instead of the edge assortativity matrix e. There are also weighted versions of these metrics that take into account not only if there exists an edge between two nodes but also the weight of that edge. Through the rest of the paper, we make use of these weighted versions.

2.3 Scoring Functions

In order to increase both node and edge assortativity in a graph, our proposal is to modify the weights of the edges of the graph, so that the new weight is able to better quantify the strength of the relationship that the edges represent. So we need to find functions that quantify this strength. We make use of functions that receive as input an unweighted unlabeled graph $G = (V, E)$ and return a symmetric score, $s(v_i, v_j) = s(v_j, v_i)$, for every pair of nodes in V, such that it quantifies, somehow, the strength of the relationship between nodes v_j and v_i. Surely, strength is a very general word and, as a consequence, many functions meet the requirements to be used as scoring functions.

The set of scoring functions chosen to test our hypothesis was inspired from those used to solve the link prediction problem in Online Social Networks (OSN). OSN are very dynamic by nature. Over time, new members join the network and new relationships are created both between new and old members. The link prediction problem for OSN consists on inferring which new links are more likely to appear in the future in a network given only its current state [8]. One of the approaches that has been followed to deal with this problem is to define functions that evaluate how likely it is, for a given pair of nodes, to create a new link. After applying these functions to every pair of nodes in the network, the algorithm predicts that those pairs of nodes for which the function returns higher values are the ones who are going to create a new link in the near future. The used functions try to evaluate the proximity or similarity of the nodes, with the idea in mind that two nodes that are proximal are more likely to create a connection in the future than two distant nodes. Depending on which metric is used to define proximity, many link prediction models are created.

The set of metrics that are used to define proximity in the link prediction problem meets all the requirements for our scoring functions. What follows is a short summary of the metrics we have chosen to experiment with.

Number of Common Neighbors (CN): Proximity is usually understood in terms of describing the common neighborhood. The most direct metric to measure the common neighborhood is the number of common neighbors, that is, the cardinal of the intersection between each of the nodes' neighbors sets:

$$score_{CN}(v_i, v_j) = |\Gamma(v_i) \cap \Gamma(v_j)|$$

This measure captures how many neighbors two nodes have in common, but it does not take into account how many non shared neighbors do these nodes have. In order to also include this information, Jaccard Index is defined.

Jaccard Index (JI): JI is defined as the size of the intersection between the two nodes neighborhoods divided by the size of the union of the neighborhoods:

$$score_{JI}(v_i, v_j) = \frac{|\Gamma(v_i) \cap \Gamma(v_j)|}{|\Gamma(v_i) \cup \Gamma(v_j)|}$$

In a similar fashion, we could want to give higher score to nodes that share low degree neighbors. Intuitively, it is more difficult that these low degree nodes have the two evaluated nodes as neighbors than it is for higher degree nodes.

Adamic-Adar (AA): The adaptation to the link prediction model for the Adamic-Adar metric [9] would take into account the degree of the shared neighbors:

$$score_{AA}(v_i, v_j) = \sum_{v_k \in \Gamma(v_i) \cap \Gamma(v_j)} \frac{1}{\log(|\Gamma(v_k)|)}$$

However, other studies point metrics that do not follow this line of thought. Instead of rewarding connections between low degree nodes, some models assume that high degree nodes tend to create more new links.

Preferential Attachment (PA): The preferential attachment model postulates that the probability that a node v_i creates a new link in the network is proportional to the current degree of v_i. Then, the probability that a new link between two nodes is formed depends on the current degrees of these two nodes:

$$score_{PA}(v_i, v_j) = |\Gamma(v_i)||\Gamma(v_j)|$$

Apart from looking at the degree of the neighbors, we can also take into account the density of the common neighbors subgraph.

Clustering Coefficient (CC): The CC of the common neighborhood captures the number of links existing between the common neighbors, taking into account how many of those links could exist:

$$score_{CC}(v_i, v_j) = \frac{2 |\{e = (v_k, v_l) \in E \text{ s.t. } v_k, v_l \in \Gamma(v_i) \cap \Gamma(v_j)\}|}{|\Gamma(v_i) \cap \Gamma(v_j)|(|\Gamma(v_i) \cap \Gamma(v_j)| - 1)}$$

Note that all the proposed metrics are based on analyzing the common neighborhood that any two nodes may share. Apart from these metrics, other topological measures have been proposed to be used in link prediction. These measures take into account distances between nodes, paths among them, or similarity. A review of some of these metrics can be found in [8].

2.4 Modifying Edges' Weight

Once we have a set of functions evaluating the strength of a relationship, we need to define how to modify the original graph, which already has weights, so that it includes the results of the scoring functions. We propose to modify each weight by directly multiplying it by the result of the scoring function:

$$w'_{ij} = score_{func}(v_i, v_j) * w_{ij}$$

By doing so, we attain two different goals. On one hand, we ensure that no new edges are created. Recall that the scoring function is defined for every pair of nodes of the graph, whether they share a link or not. By multiplying the result of the scoring function by the original weight, we guarantee that all nodes that do not share a link in the original graph (and thus have $w = 0$) will not share a link on the modified graph. On the other hand, we allow all scoring functions to eliminate non-relevant edges by assigning them a score of 0.

3 Experimental Results

This section describes the methodology used to evaluate the proposed techniques as well as the results of the experiments performed in order to do this evaluation.

3.1 Datasets

This paper's experiments are based on several relational datasets which have already been used in the past by the relational learning community. This allows us to compare our results directly with those found on prior studies while providing a set of diverse graphs coming from different environments to prove our claims.

Table 1. Original datasets

Dataset	$|\mathcal{C}|$	Edge set	$|V|$	$|E|$
WebKB Cornell	7	Cocitations	351	26832
WebKB Cornell	7	Links	351	1393
WebKB Texas	7	Cocitations	338	32988
WebKB Texas	7	Links	338	1002
WebKB Washington	7	Cocitations	434	30462
WebKB Washington	7	Links	434	1941
WebKB Wisconsin	7	Cocitations	354	33250
WebKB Wisconsin	7	Links	354	1155
IMDb	2	All	1441	48419
IMDb	2	Prodco	1441	20317
Industry	12	Pr	2189	13062
Industry	12	Yh	1798	14165
Cora	7	All	4240	71824
Cora	7	Cite	4240	22516

All the experiments described in this paper are made using essentially 4 different datasets. For each of the datasets, various graphs can be created attending on the kind of relationships taken into account to define the edges or the source of information used to create the graph. This results in a total of 14 different graphs to experiment with. Table 1 presents a short summary of the key properties of each dataset. Note that these datasets are of very different nature and that the differences between graphs constructed using different edges or different datasets are strongly pronounced. The fact that our assumptions hold for most of the presented datasets is thus a good indicator of the soundness of the presented techniques. The original datasets used in this paper can be found in [10] together with a more detailed description of their content.

3.2   Assortativity Measurements

Table 2 shows the obtained edge assortativity (A_E) values for the original graph as well as for the graphs modified using the scoring functions (in bold type those of which assortativity improves w.r.t. the original graph). The first thing to notice is that original graphs present very different edge assortativity values, and even one of the graphs presents a negative value, although it is close to 0. So we are dealing with graphs that do not show any kind of assortativity nor dissortativity together with graphs that show very high assortativity (for instance, cora$_{cite}$ presents a value of 0.74). When analyzing the success of the different scoring functions in increasing edge assortativity, we can observe that using the Jaccard Index (JI) leads to an increase on A_E for all graphs. Then, there is a set of three graphs (Cornell$_{cocite}$, Washington$_{cocite}$, and Cora$_{all}$) for which none of the other scoring functions are able to increase A_E. Apart from Jaccard Index, both

Table 2. Edge assortativity

Graph	Original	AA	CC	CN	JI	PA
Cornell$_{cocite}$	0.22701	0.19305	0.21925	0.18095	**0.24969**	0.13277
Cornell$_{link}$	0.05404	**0.05860**	**0.09348**	**0.11501**	**0.12689**	−0.25756
Texas$_{cocite}$	0.46064	**0.47667**	0.45137	0.45240	**0.61685**	0.29227
Texas$_{link}$	−0.03256	**0.25315**	**0.29175**	**0.29279**	**0.50357**	−0.22091
Washignton$_{cocite}$	0.30070	0.27731	0.29330	0.25166	**0.36886**	0.19694
Washington$_{link}$	0.08401	**0.19725**	0.05016	**0.15769**	**0.43920**	−0.29734
Wisconsin$_{cocite}$	0.57683	**0.65363**	**0.58620**	**0.64662**	**0.74448**	0.44479
Wisconsin$_{link}$	0.16045	**0.45262**	**0.38430**	**0.50690**	**0.54182**	**0.21701**
IMDb$_{all}$	0.30519	**0.39482**	**0.33020**	**0.38908**	**0.44831**	0.24412
IMDb$_{prodco}$	0.50085	**0.52631**	0.49038	**0.53462**	**0.50723**	**0.52579**
Industry$_{pr}$	0.44210	**0.54537**	**0.47248**	**0.54325**	**0.53394**	**0.48832**
Industry$_{yh}$	0.44061	**0.47978**	0.41910	**0.45753**	**0.51627**	0.38919
Cora$_{cite}$	0.73664	**0.81468**	**0.81058**	**0.80629**	**0.84720**	0.65804
Cora$_{all}$	0.65627	0.65103	0.64375	0.64624	**0.67744**	0.58648

the Adamic-Adar metric (AA) and the size of the common neighborhood (CN) are also quite successful, with 11 out of 14 and 10 out of 14 graphs showing an increase on assortativity, respectively. Finally, Clustering Coefficient leads to an increase of A_E on just half of the graphs, while Preferential Attachment is able to do so for only 3 graphs.

The magnitude of the assortativity growth also differs depending on the used scoring function. While JI usually leads to the biggest growth, that is not true for all the cases. For instance, both CN and AA are able to surpass JI for the IMDB$_{prodco}$ and Industry$_{pr}$ graphs.

Table 3. Node assortativity

Graph	Original	AA	CC	CN	JI	PA
Cornell$_{cocite}$	0.15595	**0.17092**	0.15571	**0.16393**	**0.20798**	0.12103
Cornell$_{link1}$	0.03999	**0.08155**	0.03542	**0.12417**	**0.12070**	−0.12177
Texas$_{cocite}$	0.39393	**0.44062**	0.37213	**0.42317**	**0.55223**	0.28926
Texas$_{link1}$	0.04574	**0.24626**	**0.21532**	**0.29777**	**0.48132**	−0.12948
Washington$_{cocite}$	0.16165	**0.19945**	0.14190	0.17674	**0.21560**	0.15828
Washington$_{link}$	0.02381	**0.13928**	**0.03976**	**0.10134**	**0.36904**	−0.14483
Wisconsin$_{cocite}$	0.45537	**0.55342**	**0.46367**	**0.55227**	**0.60544**	0.39855
Wisconsin$_{link}$	0.19886	**0.41702**	**0.32907**	**0.47819**	**0.50172**	**0.20973**
IMDb$_{all}$	0.29626	**0.38699**	**0.32643**	**0.38093**	**0.44384**	0.23210
IMDb$_{prodco}$	0.50011	**0.52696**	0.49147	**0.53516**	**0.50827**	**0.52552**
Industry$_{pr}$	0.38282	0.38206	0.35325	0.37290	0.38263	0.38222
Industry$_{yh}$	0.38541	0.35086	0.37761	0.32570	**0.42881**	0.24248
Cora$_{cite}$	0.72968	**0.81079**	**0.81299**	**0.80202**	**0.84906**	0.65219
Cora$_{all}$	0.64420	0.63709	0.63393	0.63092	**0.67066**	0.55912

Table 3 shows the obtained values for node assortativity (A_N). In this case, there is a graph (Industry$_{pr}$) for which none of the modified graphs are able to surpass the original graph assortativity. Nonetheless, A_N does not decrease substantially for any of the modified graphs, so no negative consequences will appear by using the modifications. Leaving aside this graph, results for A_N are similar than those showed for A_E. Graphs modified using JI exhibit higher A_N than the original ones for all datasets, and both the AA and the CN are able to increase A_N for most of the graphs (11 out of 14 and 10 out of 14, respectively). Graphs modified using CC and PA do not show an increase on A_N for most of the graphs.

We have shown that it is possible to increase both edge and node assortativity without knowing the node class labels. Using Jaccard Index as a scoring function results in a general increase on (node and edge) assortativity. The usage of the CN and AA as scoring functions also leads to an increase on assortativity for most of the graphs, although this increase can not be observed for all them. In these cases where assortativity does not increase, it is worth to note that the magnitude of the decrease is small. The graphs modified using CC as scoring function do not show a significant increase in assortativity, so this metric does not seem to be a good alternative to use with general graphs. Lastly, the use of PA as a scoring function must be discarded, as it does not show any improvement over the non-modified graph.

The poor performance of PA in increasing assortativity may be explained by the fact that preferential attachment is a model of network growth, that is, it explains how likely it is for a node to get new links, but, unlike the other scoring functions, it does not quantify the strength of the created link in any manner. On the contrary, the relationships involving very high degree nodes (which get high scores when using PA), will most likely be very weak connections. Note that all the other scoring functions, although they can be used to predict the creation of non existing links, also quantify, in some way, the strength of the relationship between any two nodes.

3.3   Classification Algorithms

We use the Netkit toolkit [5] as the relational classification framework. By using Netkit, we are able to systematically test different classifiers and compare the results. Classifiers in Netkit are comprised by a local classifier (LC), a relational classifier (RL), and a collective inference procedure (CI). Each of the different modules can be instantiated with many components. In our experiments, we allow the LC to be instantiated with either classpriors (cp) or uniform (unif); the RL component can be instantiated with Weighted-Vote Relational Neighbor Classifier (wvrn), its Probabilistic version (prn), the Class-distribution Relational Neighbor Classifier[2] (cdrn-norm-cos), and Network-Only Bayes Classifier (no-bayes); the IC module can be specified with Relaxation Labeling (relaxLabel), Iterative Classification (it), or without any inference method (null).[3] This give us $2 \times 4 \times 3 = 24$ different full classifiers. For the rest of the paper, we will use the term *full classifier* (*fc*) to refer to a specific instantiation of the three modules (LC-RC-CI).

In order to measure classification accuracy or performance of each classifier we use the percentage of (initially unlabeled) nodes in the test set that the classifier is able to correctly classify. Since our datasets contain the class labels for all nodes, we are able to compute this accuracy by taking the labels as the ground of truth.

3.4   Correlation between Assortativity and Performance

Once we have shown in Section 3.2 that it is possible to increase assortativity, we have to analyze if this increase in assortativity leads to an increase on classification performance. Intuitively, this is almost tautological for some relational

---

[2] With Normalized values of neighbor-class and using the cosine distance metric.
[3] Readers can refer to the original Netkit paper [5] for a full explanation of these modules.

classifiers [5], but the relation is not so obvious for some other classifiers. In order to test our second hypothesis, namely, that assortativity is positively correlated with classification performance, we compute assortativity as in Section 3.2 and classification performances as described in Section 3.3.

We are interested in analyzing the correlation between assortativity and classification performance. We expect that when assortativity increases, classification performance also increases. So we want to discover if the function that describes the relationship between these two variables is monotonically increasing. However, we are not concerned on finding the exact function that describes this relationship.

Spearman's rank correlation coefficient is a measure of statistical dependence between two variables that assesses how well this relationship can be described using a monotonic function [11]. The Spearman's coefficient can take values between -1 and 1, with -1 describing a perfect decreasing monotonic function and 1 characterizing a perfect increasing monotonic function.[4] So we can use the Spearman's rank correlation coefficient to asses whether assortativity is positively correlated with performance.

In the interest of comparing classification performance between different datasets, we use the notion of relative error reduction as defined in [5]:

$$ER_{REL}(fc, D, r) = \frac{base\_error(D) - error(fc, D, r)}{base\_error(D)}$$

The base error for a given dataset D is the error committed when predicting that all samples belong to the most prevalent class. The error for a given dataset D, a full classifier fc, and a labeled ratio r is the error committed when trying to classify the $1 - r\%$ remaining samples with the specific configuration described by fc. Note that although the error reduction metric is not bounded, its value is inside the $[0, 1]$ interval when $base\_error(D) \geq error(fc, D, r)$, which is the most common scenario.

Although classification performance increased with the labeled set ratio (as we will see in Section 3.5), no significant differences where observed on the correlation between performance and assortativity for different r values. Table 4 shows the Spearman's rank correlation coefficient between the node and edge assortativity of each of the graphs and the error reduction achieved when classifying those graphs with the different full classifiers. Results presented on the table correspond to the experiments with r set to 35%. Each of the values represents the correlation between the 84 graphs[5] assortativity values and the $100-$run mean performance obtained when classifying those graphs. As it was expected, we found a positive correlation between both edge and node assortativity for all full classifiers, with the Spearman's rank coefficient ranging between 0.44 and 0.73 for node assortativity and between 0.44 and 0.71 for edge assortativity.

---

[4] When data does not contain repeated values.
[5] Notice that the total number of graphs tested comes from the 14 original graphs plus the ones obtained using each of the 5 scoring functions.

Table 4. Spearman's rank correlation coefficient between error reduction and assortativity (r=0.35)

Full classifier	A_N	A_E	Full classifier	A_N	A_E
cprior-wvrn-it	0.6264	0.6315	unif-wvrn-it	0.5986	0.6049
cprior-prn-it	0.4481	0.4474	unif-prn-it	0.4448	0.4417
cprior-nobayes-it	0.4949	0.5054	unif-nobayes-it	0.5002	0.5121
cprior-cdrn-norm-it	0.7362	0.7175	unif-cdrn-norm-it	0.6819	0.6676
cprior-wvrn-relaxLabel	0.5213	0.5171	unif-wvrn-relaxLabel	0.5355	0.5415
cprior-prn-relaxLabel	0.4534	0.4831	unif-prn-relaxLabel	0.4423	0.4757
cprior-nobayes-relaxLabel	0.5100	0.5357	unif-nobayes-relaxLabel	0.5205	0.5471
cprior-cdrn-norm-relaxLabl	0.4863	0.5015	unif-cdrn-norm-relaxLabl	0.4894	0.4848
cprior-wvrn-null	0.5318	0.5304	unif-wvrn-null	0.5390	0.5491
cprior-prn-null	0.4627	0.4893	unif-prn-null	0.4669	0.5016
cprior-nobayes-null	0.5103	0.5342	unif-nobayes-null	0.5431	0.5644
cprior-cdrn-norm-null	0.4963	0.5063	unif-cdrn-norm-null	0.4956	0.4903

The Spearman's rank correlation coefficient is positive and greater than 0.44 for all the classifiers, which denotes that there exists a positive correlation between both node and edge assortativity and classification performance. The strength of this correlation varies depending on the specific classifier configuration. However, the values are quite high considering that different datasets are compared together. Although relative error reduction is used instead of classification accuracy, which already tries to compensate the differences between base errors on the different datasets, the different nature of the used graphs introduces additional complexity. When evaluating the different datasets independently[6], we found that the correlation was almost perfect for some datasets and worse for some other datasets. For instance, Cornell$_{cocite}$, Texas$_{cocite}$, and IMDb$_{all}$ showed a correlation of 0.9429 (node assortativity and relative error reduction for the cp-wvrn-it configuration), while other datasets such as the four university ones with *link* edges showed very low correlation, or even a negative one.

3.5   Increasing Classification Performance

Once we have showed that we are able to increase assortativity using our scoring functions and that assortativity is positively correlated with performance, we want to observe the results of our third hypothesis, namely, that using scoring functions to correct weights can improve relational classification. In order to evaluate the degree in which using scoring functions improves networked classification, we use the 24 different full classifiers with all the available graphs. Since we have 14 original graphs and 5 different variations of each of these graphs can be obtained by using the different scoring functions, all the experiments are done with 14×6 graphs. For each graph and classifier, we repeat the process of selecting new train and test sets 100 times and define the performance of the full classifier with respect to a given graph and a labeled ratio r as the mean of these 100 different runs. We repeated the process for different labeled ratios (train set sizes): 20%, 35%, 50%, and 65%.

---

[6] We omit these individual results due to space constraints.

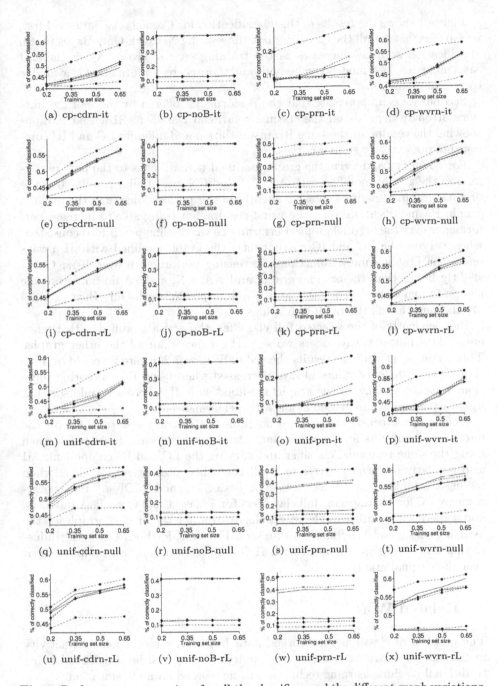

Fig. 1. Performances comparison for all the classifiers and the different graph variations for the Cornell$_{cocite}$ dataset: Original ($^-$), AA ($^+$), CC ($^{-x}$), CN ($^+$), JI ($^+$), PA ($^+$)

Figure 1 shows the results of this classification for Cornell$_{cocite}$ dataset. First, we can see that for all the classifiers but those using Network-Only Bayes (noB), classification accuracy increases as the training set size grows. As the labeled ratio increases, best models can be built and more correct information is available to do the predictions.

Second, we can appreciate that the performance offered by the scoring functions strongly depends on the specific relational classifier (RL) used. Graphs showing the results for the same RL instantiation and different LC and IC components present very similar curves.

For cdrn-cos and wvrn, the graph modified with JI leads to the best performance; the graphs modified with AA, CC and CN give similar results than the original graph, sometimes showing slightly better performance than the original graph; PA modifications offer the worst results, not being able to increase performance over the original graph. Performance when using prn is also consistent when varying the LC and IC components: the graph modified with JI always offers the best accuracy, sometimes increasing performance over 10%; CC is slightly better than the original graph; and AA, CN, and PA do not overcome the performance achieved with the original graph. Network-Only Bayes results are the same for all the LC-IC variations.

Independently of the selected full classifier, the graph modified with Preferential Attachment always offers worse performance than all the other graphs. This is consistent with the results showed in Section 3.2, where we could observe that the assortativity values always decreased when using PA as scoring function. This is also true for the graphs modified with JI, where we could see that assortativity always increased along with performance.

Due to space constrains, we are not able to include the results for all the datasets. The results for the other datasets showed the same consistency when using the same relational classifier and varying the LC and IC components. JI also regularly performed better than all the other alternatives for all fc when testing Washington$_{cocite}$, Wisconsin$_{cocite}$, Texas$_{cocite}$, and IMDb$_{all}$.[7] JI was overcome by CC for some specific full classifiers for Texas$_{link}$ and Cornell$_{link}$ datasets, and sometimes for other modified graphs or even for the original graph for the Washington$_{link}$ and Wisconsin$_{link}$ datasets. However, for both Cora and Industry datasets, the graph modified with JI did not show a significant improvement over the original graph.

4   Related Work

The problem of classifying networked data has been a recent focus of activity in the machine learning research community, with special interest on adapting traditional machine learning techniques to networked data classification.

---

[7] The exceptions were 2 out of 24 fc in Texas$_{cocite}$ for which the original graph performed better than JI, and some specific r values and classifiers in IMDb$_{all}$.

In [5], the authors present a relational classifier toolkit. Beyond the actual toolkit itself and by describing each of its modules, the authors review different algorithms that can be used to classify networked data.

Many algorithms for relational classifiers have been proposed in the past.

In [1] the authors present the Relational Neighbor (RL) classifier based on the principle of homophily, where the probability of a sample belonging to a given class is proportional to the number of neighbors of that sample belonging to the same class.

The Weighted Vote Relational Classifier (WVRN) estimates class-membership of a node as the weighted mean of the class-membership probabilities of the neighbors of that node.

The Class-Distribution Relational Neighbor classifier (CDRN) is presented in [5], where the probability of class membership of a node is estimated by the similarity of its class vector with the class reference vector. The class vector of a node is defined as the vector of summed linkage weights to the various classes and the class reference vector for a given class is the average of the class vectors for nodes known to be of that class.

Network Only Bayes classifier (nBC) [2] uses naive Bayes classification based on the classes of the nodes' neighbors to classify hyperlinked documents.

In [4] the Network-Only Link-Based classification (nLB) is presented, which uses regularized logistic regression models to classify networked data.

Since in relational classification problems entities are interlinked, the predicted class of a specific node may have consequences on the prediction of another node's class. For this reason, the method of independently classifying entities, which may be of use in traditional machine learning approaches, may not be the best way to deal with interlinked data. The process of simultaneously classifying a set of linked entities is known as collective inference. It has been shown that collective inference improves classification accuracy [12].

Collective inference may improve probabilistic inference in networked data [12]. Many CI methods are used in relational learning: Gibbs sampling [13], relaxation labeling [2], and iterative classification [4, 14] are the most used.

Relational classification has been applied to email classification [15], with a dataset of mails being linked only by parent-children relationships; to topic classification of hypertext documents [2]; to predict movie success with IMDb data, linking movies with a shared production company [1, 5]; to sub-topic prediction in machine learning papers [5]; to age, gender, and location prediction of bloggers [16]; and many other networked data classification problems.

5   Conclusions

We have showed that it is possible to increase the assortativity of a graph according to the node class labels with a very simple technique based on the usage of scoring functions. We have evaluated different scoring functions and demonstrated that using Jaccard Index (JI) always results in an increase on edge assortativity and, on all datasets but one, also in node assortativity. The usage of Common Neighbors (CN) and Adamic-Adar (AA) also leads to an increase on both node and edge assortativity for most of the tested datasets.

Although we have showed that there is a positive correlation between an increase on assortativity and an increase on classification performance, this correlation is not perfect (which supports preliminary tests done in [5]). Note that while we are dealing with a single assortativity value for each graph, many variables are involved in the performance obtained when classifying: from the specific configuration that the classifier adopts to the effect of choosing a concrete split of the training and test samples. So each assortativity value is compared against multiple classification performance results obtained when using different full classifiers.

Regarding the performance improvements achieved when using the modified graphs, the experiments showed that using Jaccard Index to modify the weights of the edges results in a general improvement of classification performance, although not for absolutely all the possible classifier configurations and datasets. The performance improvement when using CC, AA, and CN as scoring functions strongly depended on the selected dataset and, in a lesser extent, on the relational classifier instantiated. This opens an interesting line for future work: trying to identify the set of graph properties that determine which classifier (and scoring function) will lead to the best performance results.

Moreover, in this paper we were focused on evaluating different scoring functions and their effect on assortativity and performance. However, no specific efforts were devoted to construct the best possible scoring function. In this sense, a combination of the scoring functions that offered the best results may lead to a higher increase on classification performance. Trying to find the best possible scoring function is left as future work.

The techniques described in this paper can be applied to directed graphs following the naive procedure of computing the scoring functions over the underlying undirected graph obtained when omitting the direction of the edges. Since afterwards the results of the scoring functions are multiplied by the original weight in order to compute the new weight, edges between the same nodes differing only on the direction could be able to obtain different modified weights. Although the procedure seems feasible, it will be interesting to think about other techniques improving this naive approach.

Acknowledgments. This work was partially supported by the Spanish MCYT and the FEDER funds under grants TIN2011-27076-C03 "CO- PRIVACY", TSI2007-65406-C03-03 "E-AEGIS", CONSOLIDER CSD2007-00004 "ARES", TIN2010-15764 "N-KHRONOUS", and grant FPU-AP2010-0078.

References

1. Macskassy, S., Provost, F.: A simple relational classifier. In: Proceedings of the 2nd Workshop on Multi-Relational Data Mining, KDD 2003, pp. 64–76 (2003)
2. Chakrabarti, S., Dom, B., Indyk, P.: Enhanced hypertext categorization using hyperlinks. In: Proceedings of the ACM SIGMOD International Conference on Management of Data, vol. 27, pp. 307–318. ACM Press, New York (1998)
3. Perlich, C., Provost, F.: Distribution-based aggregation for relational learning with identifier attributes. Machine Learning 62(1-2), 65–105 (2006)
4. Lu, Q., Getoor, L.: Link-based classification using labeled and unlabeled data. In: Proceedings of the ICML 2003 Workshop on the Continuum from Labeled to Unlabeled Data (2003)
5. Macskassy, S.A., Provost, F.: Classification in networked data: A toolkit and a univariate case study. Journal of Machine Learning Research 8, 935–983 (2007)
6. Bilgic, M., Getoor, L.: Effective label acquisition for collective classification. In: Proceedings of the International Conference on Knowledge Discovery and Data Mining, pp. 43–51 (2008)
7. Newman, M.E.J.: Mixing patterns in networks. Physical Review E 67, 026126 (2003)
8. Liben, D., Kleinberg, J.: The link prediction problem for social networks. In: Proceedings of the International Conference on Information and Knowledge Management, pp. 556–559 (2003)
9. Adamic, L., Adar, E.: Friends and neighbors on the Web. Social Networks 25(3), 211–230 (2003)
10. Macskassy, S., Provost, F.: NetKit-SRL - network learning toolkit for statistical relational learning
11. Spearman, C.: The proof and measurement of association between two things. The American Journal of Psychology 15(1), 72–101 (1904)
12. Jensen, D., Neville, J., Gallagher, B.: Why collective inference improves relational classification. In: Proceedings of the International Conference on Knowledge Discovery and Data Mining, pp. 593–598 (2004)
13. Geman, S., Geman, D.: Stochastic relaxation, gibbs distributions, and the bayesian restoration of images. IEEE Transactions on Pattern Analysis and Machine Intelligence PAMI-6(6), 721–741 (1984)
14. Neville, J., Jensen, D.: Iterative classification in relational data. In: AAAI-2000 Workshop on Learning Statistical Models from Relational Data (2000)
15. Carvalho, V., Cohen, W.: On the collective classification of email speech acts. In: Proceedings of the International Conference on Research and Development in Information Retrieval, pp. 345–352 (2005)
16. Bhagat, S., Cormode, G., Rozenbaum, I.: Applying link-based classification to label blogs. In: Zhang, H., Spiliopoulou, M., Mobasher, B., Giles, C.L., McCallum, A., Nasraoui, O., Srivastava, J., Yen, J. (eds.) WebKDD/SNA-KDD 2007. LNCS, vol. 5439, pp. 97–117. Springer, Heidelberg (2009)

A Fast Approximation of the Weisfeiler-Lehman Graph Kernel for RDF Data

Gerben K.D. de Vries

System and Network Engineering Group, Informatics Institute,
University of Amsterdam, The Netherlands
g.k.d.devries@uva.nl

Abstract. In this paper we introduce an approximation of the Weisfeiler-Lehman graph kernel algorithm aimed at improving the computation time of the kernel when applied to Resource Description Framework (RDF) data. Typically, applying graph kernels to RDF is done by extracting subgraphs from a large RDF graph and computing the kernel on this set of subgraphs. In contrast, our algorithm computes the Weisfeiler-Lehman kernel directly on the large RDF graph, but still retains the subgraph information. We show that this algorithm is faster than the regular Weisfeiler-Lehman kernel for RDF data and has at least the same performance. Furthermore, we show that our method has similar or better performance, and is faster, than other recently introduced graph kernels for RDF.

Keywords: Resource Description Framework (RDF), Graph Kernels, Weisfeiler-Lehman.

1 Introduction

Machine learning techniques have been widely used to populate the semantic web, i.e. to create linked data. In contrast, there has been relatively little research into learning directly from the semantic web. However, the amount of linked data available is becoming larger and larger and provides interesting opportunities for data-mining and machine learning.

Kernel methods [1,2] are popular machine learning techniques for handling structured data. To deal with data structured as graphs, graph kernels, such as described in [3] and [4], have been developed.

The representation/storage format of the semantic web is the Resource Description Framework (RDF). The RDF format essentially represents a graph. Therefore, learning from RDF can potentially be accomplished using graph kernel methods on RDF. Research on this is in its infancy and, to the best of our knowledge, there currently exists one paper [5] on this topic. In [5] the authors introduce two types of graph kernels, designed for RDF, and compare these to general graph kernels in two tasks. The authors conclude that the introduced kernels work better or just as well as the general graph kernels. For the application of most of the graph kernels, instances are represented as (small) subgraphs extracted from a larger RDF graph.

H. Blockeel et al. (Eds.): ECML PKDD 2013, Part I, LNAI 8188, pp. 606–621, 2013.

Graph kernel computation is in general slow, since it is often based on computing some form of expensive (iso)morphism between graphs. In this paper we present an approximation of the Weisfeiler-Lehman graph kernel [4] to speed up the computation of the kernel on RDF data. This approximation exploits the fact that the subgraph instances for RDF learning tasks are usually extracted from the same large RDF graph. We test this kernel on a number of learning tasks with RDF data and compare its performance to the graph kernels designed for RDF described in [5].

Kernel methods have not been widely applied to RDF data. Earlier attempts have been [6] and [7]. In [6] kernels are manually designed by selecting task relevant properties and relations for the instances, and then incorporating these in the kernel measure. Kernels are built from RDF using inductive logic programming rules in [7]. The approach to learning from RDF using graph kernels is more generally applicable than both these methods. Other attempts at learning from semantic web data are based on description logic [8,9]. These approaches are applicable to a smaller part of the semantic web, since not everything on the semantic web is nicely formalized as description logic ontologies, whereas nearly everything on the semantic web is available as RDF. Other specifically tailored approaches for data-mining form the semantic web are, for instance, using statistical relational learning [10].

The rest of this paper is structured as follows. Section 2 introduces our adaption of the Weisfeiler-Lehman graph kernel for RDF data. We present our experiments with this kernel in Sect. 3. Finally, we end with conclusions and suggestions for future work.

2 Weisfeiler-Lehman Graph Kernel for RDF

In the following section we first briefly introduce RDF data. Then we define the regular Weisfeiler-Lehman kernel for graphs. Finally we introduce our adaption of this kernel to speed up computation on RDF data.

2.1 The Resource Description Framework

The Resource Description Framework (RDF) is the foundation for knowledge representation on the semantic web. It is based on the idea of making statements about resources in a subject-predicate-object form. Such expressions are dubbed *triples*. The RDF specification[1] defines a number of classes, for the subjects and objects, and properties, for the predicates. Moreover, users can, and should add their own classes and objects.

For example, suppose that we have an ontology about fruits, called fruit. Then the RDF triple fruit:Pear-rdfs:SubClassOf-fruit:Fruit expresses the fact the the class of Pear is a sub-class of the class of Fruit. And the triple fruit:elstar-rdf:type-fruit:Apple expresses the fact that an elstar is an instance of the class Apple. The

---

[1] http://www.w3.org/standards/semanticweb/

colon notation is used to indicate from which ontology a class or property is used, i.e. the class Pear comes from the example fruit ontology that we created ourselves, but the property type is defined by the RDF standard. These notations are shorthands for full-fledged Universal Resource Identifiers (URI) that uniquely identify the specific ontology used and where to find it, thereby forming the backbone of the semantic web.

RDF resources, i.e. the uri:value type of statements, can occur as subject, predicate and object in a triple. This means that they can be both vertex and edge in a graph at the same time. Therefore, formally, RDF represents a hypergraph. However, in practice interpreting RDF as an easier to handle directed multigraph does not lead to problems for applying graph kernels.

RDF is used as a representation scheme for more expressive knowledge representation formalisms such as the Web Ontology Language (OWL) and RDF Schema (RDFS). Therefore, RDF triple-stores, such as SESAME[2], often include a reasoning engine which allows the automatic derivation of new triples, using these more expressive formalisms.

In machine learning/data-mining for RDF, arguably, the most straightforward way to represent instances is as a set of RDF triples, or an RDF graph. For example, for each fruit that is of rdf:type fruit:Apple or fruit:Pear we can collect the RDF triples that describe properties of this fruit and then these sets of triples are our instances, which we can use to train a classifier for apples and pears.

2.2   Regular Weisfeiler-Lehman Graph Kernel

The Weisfeiler-Lehman Subtree graph kernel, from now on the Weisfeiler-Lehman kernel, is a state-of-the-art, efficient kernel for graph comparison introduced in [11] and elaborated upon in [4]. The kernel computes the number of subtrees shared between two graphs by using the Weisfeiler-Lehman test of graph isomorphism. The rewriting procedure underlying the Weisfeiler-Lehman kernel is given in Algorithm 1, which is taken from [4]. The idea of the rewriting process is that for each vertex we create a multiset label based on the labels of the neighbors of that vertex. This multiset is sorted and together with the original label concatenated into a string, which is the new label. For each unique string a new (shorter) label is introduced and this replaces the original vertex label. Note that the sorting of the strings in step 3 is not necessary, but provides a simple way to create the label dictionary f. The rewriting process can be efficiently implemented using counting sort, for which details are given in [4].

Using the rewriting techniques of Algorithm 1 it is straightforward to define a kernel, given in in Definition 1.

Definition 1. Let $G_n = (V, E, l_n)$ and $G'_n = (V', E', l_n)$ be the n-th iteration rewriting of the graphs G and G' using Algorithm 1 and h the number of iterations. Then the Weisfeiler-Lehman kernel is defined as:

---

[2] http://www.openrdf.org/

Algorithm 1. Weisfeiler-Lehman Relabeling

Input graphs $G = (V, E, \ell), G' = (V', E', \ell)$ and number of iterations h
Output label functions l_0 to l_h
Comments $M_n(v)$ are sets of labels for a vertex v and $N(v)$ is the neighborhood of v

- for $n = 0$ to h
    1. Multiset-label determination
        - for each $v \in V$
            - if $n = 0$, $M_n(v) = l_0(v) = \ell(v)$
            - if $n > 0$, $M_n(v) = \{l_{n-1}(u) | u \in N(v)\}$
    2. Sorting each multiset
        - for each $M_n(v)$, sort the elements in $M_n(v)$, in ascending order and concatenate them into a string $s_n(v)$
        - for each $s_n(v)$, if $n > 0$, add $l_{n-1}(v)$ as a prefix to $s_n(v)$
    3. Label compression
        - for each $s_n(v)$
            - sort all strings $s_n(v)$ together in ascending order
            - map each string $s_n(v)$ to a new compressed label, using a function $f : \Sigma^* \to \Sigma$, such that $f(s_n(v)) = f(s_n(v'))$ iff $s_n(v) = s_n(v')$
    4. Relabeling
        - for each $s_n(v)$, set $l_n(v) = f(s_n(v))$

$$k_{\mathrm{WL}}^h(G, G') = \sum_{n-0}^{h} k_\delta(G_n, G_n') \ , \tag{1}$$

where

$$k_\delta((V, E, l), (V', E', l')) = \sum_{v \in V} \sum_{v' \in V'} \delta(l(v), l'(v')) \ . \tag{2}$$

Here δ is the Dirac kernel, which tests for equality, it is 1 if its arguments are equal, and 0 otherwise.

Essentially this kernel counts the common vertex labels in each of the iterations of the graph rewriting process.

Instead of computing the Weisfeiler-Lehman relabeling on pairs of graphs, as in Algorithm 1, it can just as easily be computed on a set of graphs. Furthermore, the label dictionary f can be used to construct a feature vector for each graph. Then the kernel can be computed by taking the dot product of these feature vectors, which speeds up the computation of the kernel. See [4] for more details.

2.3   Fast Weisfeiler-Lehman for RDF

Since graph kernels compute a similarity between graphs, the most immediate approach to apply graph kernels to RDF is to extract subgraphs for the instances that we are interested in and to compute the kernel on these subgraphs. This approach is followed in [5] for most of the kernels. The intuition is that these subgraphs contain properties of the instances and that they say something about the position of the instances in the larger graph. However, the subgraphs are derived from the same underlying RDF graph and since they are instances of a similar concept they often have a number of vertices and edges in common. Potentially it can be more efficient to do the kernel computation directly on the larger underlying RDF graph.

In this section we introduce an approximation of the Weisfeiler-Lehman kernel designed for RDF data. We could just apply the Weisfeiler-Lehman relabeling as we have defined it above to the underlying RDF graph and then count the resulting labels in the neighborhood up to a certain depth for each instance vertex. However, by the nature of the relabeling process this means that vertices/edges on the border of the neighborhood are influenced by vertex/edges outside of the neighborhood. This means that the subgraph perspective for each instance is essentially gone, and that interesting information about the position of the instance in the graph is lost. We deal with this problem by tracking for each vertex and edge in the graph at what depth it occurs in the subgraphs of the instances.

First we define, in Definition 2, the type of graph that we apply the relabeling process to.

Definition 2. *A Weisfeiler-Lehman RDF graph is a graph $G = (V, E, l)$, where V is a set of vertices, E a set of directed edges, and $l : (V \cup E) \times \mathbb{N} \to \Sigma$ a labeling function from vertices V or edges E and a depth index $j \in \mathbb{N}$ to a set of labels Σ.*

This graph is a directed multigraph with a special labeling function, which gives the label for a vertex or edge given an index j. This index j indicates the depth at which this vertex or edge was encountered in the extraction of the subgraph.

We will also need our variant of neighborhoods of vertices and edges, as given in Definition 3.

Definition 3. *The neighborhood $N(v) = \{(v', v) \in E\}$ of a vertex is the set of edges going to the vertex v and the neighborhood $N((v, v')) = \{v\}$ of an edge is the vertex that the edge comes from.*

The graph extraction algorithm, given in Algorithm 2, creates an RDF graph, as given in Definition 2, for a set of instances I. For each instance i a subgraph up to depth d is extracted from the RDF dataset and this subgraph is added to the total graph G that the algorithm is building. Thus, vertices and edges are only added if they have not been added to the graph already (which is recorded using the *vMap* and *eMap* datastructures). For each vertex and edge encountered during the extraction process a label is saved (in ℓ) for the depth j at which the vertex or edge is encountered. For example, if a vertex v would only occur at depths 1 and 2 in all of the extracted subgraphs, then we would have $\ell(v, 1) = o$ and $\ell(v, 2) = o$. Next to the graph G we also construct mappings $\mathcal{V}_i$ and $\mathcal{E}_i$ for each instance i, which records which vertices and edges belong to the subgraph of instance i and at which depth.

Algorithm 2. Graph Creation from RDF

Input a set of RDF triples R, a set of instances I and extraction depth d
Output a Weisfeiler-Lehman RDF graph $G = (V, E, \ell)$, mappings $\mathcal{V}_i$ from vertices to integers and $\mathcal{E}_i$ from edges to integers for each instance i.

1. Initialization
   - for each $i \in I$:
     - add a vertex v to V, set $\ell(v, d) = \epsilon$ and set $vMap(i) = v$
2. Subgraph Extraction
   - for each $i \in I$:
     - $searchFront = \{i\}$
     - for $j = d - 1$ to 0
       - $newSearchFront = \emptyset$
       - for each $r \in searchFront$:
         - $triples = \{(r, p, o) \in R\}$
         - for each $(s, p, o) \in triples$:
           - add o to $newSearchFront$
           - if $vMap(o)$ is undefined, add vertex v to V and set $vMap(o) = v$
           - set $\ell(vMap(o), j) = o$
           - if $\mathcal{V}_i(vMap(o))$ is undefined, set $\mathcal{V}_i(vMap(o)) = j$
           - if $eMap(s, p, o)$ is undefined, add edge e to E and set $eMap(s, p, o) = e$
           - set $\ell(eMap(s, p, o), j) = p$
           - if $\mathcal{E}_i(eMap(s, p, o))$ is undefined, set $\mathcal{E}_i(eMap(s, p, o)) = j$
       - $searchFront = newSearchFront$

Algorithm 3 describes the Weisfeiler-Lehman relabeling for graphs constructed using Algorithm 2. There are two main differences compared to the standard Weisfeiler-Lehman algorithm of Algorithm 1. The first difference is the extension to directed edges with labels, which is relatively straightforward. The second difference is in the construction of the multisets M_n, which are now constructed for a vertex/edge with a depth index j. For the vertices these multisets are constructed using the labels of the edges at depth $j - 1$, and for the edges with labels of the vertices of depth j. Furthermore, a vertex label at depth 0 is never rewritten.

Definition 4. *Let G be a Weisfeiler-Lehman RDF graph, created using Algorithm 2 and rewritten for h iterations using Algorithm 3, and l_0 to l_h the resulting label functions. Then we compute a kernel between two instances $i, i' \in I$, as:*

$$k_{\mathrm{WLRDF}}^h(i, i') = \sum_{n=0}^{h} \frac{n+1}{h+1} k_{\delta,\mathrm{RDF}}^n((\mathcal{V}_i, \mathcal{E}_i), (\mathcal{V}_{i'}, \mathcal{E}_{i'})), \qquad (3)$$

where

Algorithm 3. Weisfeiler-Lehman Relabeling for RDF

Input a Weisfeiler-Lehman RDF graph $G = (V, E, \ell)$, subgraph depth d and number
of iterations h

Output label functions l_0 to l_h and label dictionary f

- for $n = 0$ to h
  1. Multiset-label determination
     - for each $v \in V$ and $e \in E$ and $j = 0$ to d
       - if $n = 0$ and $\ell(v, j)$ is defined, set $M_n(v, j) = l_0(v, j) = \ell(v, j)$
       - if $n = 0$ and $\ell(e, j)$ is defined, set $M_n(e, j) = l_0(e, j) = \ell(e, j)$.
       - if $n > 0$ and $\ell(v, j)$ is defined, set $M_n(v, j) = \{l_{n-1}(u, j) | u \in N(v)\}$
       - if $n > 0$ and $\ell(e, j)$ is defined, set $M_n(e, j) = \{l_{n-1}(u, j + 1) | u \in N(e)\}$
  2. Sorting each multiset
     - for each $M_n(v, j)$ and $M_n(e, j)$, sort the elements in $M_n(v, j)$, resp. $M_n(e, j)$,
       in ascending order and concatenate them into a string $s_n(v, j)$, resp.
       $s_n(e, j)$
     - for each $s_n(v, j)$ and $s_n(e, j)$, if $n > 0$, add $l_{n-1}(v, j)$, resp. $l_{n-1}(e, j)$, as a
       prefix to $s_n(v, j)$, resp. $s_n(e, j)$
  3. Label compression
     - for each $s_n(v, j)$ and $s_n(e, j)$, map $s_n(v, j)$, resp. $s_n(e, j)$, to a new com-
       pressed label, using a function $f : \Sigma^* \to \Sigma$, such that $f(s_n(v, j)) =$
       $f(s_n(v', j))$ iff $s_n(v, j) = s_n(v', j)$, resp. $f(s_n(e, j)) = f(s_n(e', j))$ iff
       $s_n(e, j) = s_n(e', j)$
  4. Relabeling
     - for each $s_n(v, j)$ and $s_n(e, j)$, set $l_n(v, j) = f(s_n(v, j))$ and $l_n(e, j) =$
       $f(s_n(e, j))$

$$k_{\delta, \text{RDF}}^n((\mathcal{V}_i, \mathcal{E}_i), (\mathcal{V}_{i'}, \mathcal{E}_{i'})) = \sum_{(v,d) \in \mathcal{V}_i} \sum_{(v',d') \in \mathcal{V}_{i'}} \delta(l_n(v, d), l_n(v', d'))$$

$$+ \sum_{(e,d) \in \mathcal{E}_i} \sum_{(e',d') \in \mathcal{E}_{i'}} \delta(l_n(e, d), l_n(e', d')) \ . \quad (4)$$

*Here δ is the Dirac kernel, which tests for equality, it is 1 if its arguments are
equal, and 0 otherwise.*

This kernel is very similar to the regular definition of the Weisfeiler-Lehman
Subtree kernel. Ones difference is that instances are not represented by their
graphs but by the two maps $\mathcal{V}_i$ and $\mathcal{E}_i$. Furthermore, there is an added part to
sum over all the edges. Like the regular Weisfeiler-Lehman Subtree kernel, this
kernel is an instance of a convolution kernel [12]. We have added the factor $\frac{n+1}{h+1}$
to put more weight on higher iterations, to weigh more complicated structural
similarity more heavily.

The resulting kernel k_{WLRDF} is an approximation of k_{WL} (provided that we
add edge relabeling to the algorithm). Differences occur when there are cycles
in the subgraph. For instance, let vertex v_1 have a label o_1 at depth 3 and v_1
has an edge to v_2. Vertex v_2 has a label o_2 at depth 2 and has an edge back to

v_1. Therefore v_1 will have a label o_3 at depth 1. During the relabeling, the label o_3 will (eventually) be combined with label o_2 and label o_2 will be combined with o_1 (not o_3, which seems more intuitive). In the regular Weisfeiler-Lehman variant, there would be no labels at different depths, so o_3 would be combined with o_2 and vice versa.

In [4] it is shown that the runtime for the relabeling algorithm on a set of graphs is $\mathcal{O}(Nhm)$, where N is the number of graphs, h is the number of iterations and m is the number of vertices (and edges) per graph. For our relabeling method we do not have N graphs, but we do introduce depth d labels per vertex/edge. Our larger graph has a number of vertices and edges k. Since our algorithm is essentially regular Weisfeiler-Lehman with the addition of multiple labels per vertex/edge, the runtime complexity for our relabeling algorithm is $\mathcal{O}(hkd)$. As for the regular Weisfeiler-Lehman Subtree kernel we can create feature vectors for each instance, using the label dictionary f. Hence, in situations with $kd < Nm$, our algorithm will be faster. This scenario is typical for the RDF use-case, where the subgraphs for each instance share a (large) number of vertices and edges, which means $Nm \gg k$ given large enough N, and therefore $kd < Nm$ if $d \ll N$. The same bounds hold for the space complexity as for the runtime complexity. Therefore, in situations with $kd < Nm$, our algorithm requires less memory than regular Weisfeiler-Lehman on a set of graphs.

3 Experiments

In this section we present a number of experiments with the Weisfeiler-Lehman for RDF (WL RDF) kernel presented above. The goal of these experiments is to compare the prediction performance of this kernel to the regular Weisfeiler-Lehman (WL) kernel, adapted to handle edge labels and using the same iteration weighting. For comparison we use three prediction tasks using RDF data. Since the WL RDF kernel is intended to be a faster variant of the WL kernel we also compare the runtimes.

Furthermore, we compare the WL RDF kernel with the kernels designed for RDF in [5]. These kernels are the efficient Intersection SubTree (IST) and Intersection Partial SubTree (IPST) kernels and the inefficient Intersection Graph Walk (IGW) and Intersection Graph Path (IGP) kernels. The intersection subtree kernels are based on counting the number of (partial) subtrees in the intersection tree of two graphs. The intersection graph kernels count the number of paths/walks in the intersection graph of two graphs.

Like we did for the WL RDF kernel, [5] compute the IST and IPST kernels directly on the RDF graph. For the WL, IGW and IGP kernels subgraphs have to be extracted. For each kernel we test 3 extraction depths (1,2,3) and we also test with and without RDFS inferencing by the triple-store. RDFS inferencing potentially derives new triples based on logical relations between the concepts defined in the RDF. We test these different settings to see the influence of larger subgraphs on the prediction performance.

All of the kernels and experiments where implemented in Java and the code is available online.[3] The Java version of the LibSVM [13] support vector machine library was used for prediction with the kernels and the SESAME[4] triple-store was used to handle the RDF data and do the RDFS inferencing. The experiments where run on an AMD X6 1090T CPU with 16 GB RAM.

3.1 Affiliation Prediction

For our first experiment we repeat the affiliation prediction task introduced in [6] and repeated in [5]. This experiment uses data of the semantic portal of the AIFB research institute modeled in the SWRC ontology [14], which models key concepts in a research community. The data contains 178 persons that belong to 1 of 5 research institutes. Furthermore it contains information about publications, students, etc. One institute contains only 4 members, which we ignore. The goal of the prediction task is to predict the affiliation for the remaining 174 instances. Since we know the affiliations, for training purposes the affiliation relation (and its inverse the employs relation) are removed from the RDF for each instance. Also we set the label of the root vertex for each instance to an identical special root label (like in [5]), since the original URI is unique for each instance.

For the Weisfeiler-Lehman kernels we test the h settings: $0, 2, 4, 6$. The two intersection graph kernels have a maximum path length parameter, which we also call h, for which we test $1, 2$.[5] All the four kernels from [5] have a discount factor parameter λ and are tested with the setting reported to give the best results.

For each kernel we use the C-SVC support vector machine algorithm from LibSVM to train a classifier to predict the affiliation. Per kernel we do a 10-fold cross-validation which is repeated 10 times. Within each fold the C parameter is optimized from the range: $\{10^{-3}, 10^{-2}, 0.1, 1, 10, 10^2, 10^3\}$ by doing 10-fold cross-validation. We also weigh the different classes with the inverse of their frequency. All kernels are normalized.

Table 1 presents the average accuracy and F1[6] scores. The best scores, and the scores that have no significant difference with these scores under a Student t-test with $p < 0.05$, are indicated using a bold type face.

The best performance is achieved by our Weisfeiler-Lehman RDF kernel variant, showing slightly better scores than regular Weisfeiler-Lehman in the '3,f' setting. The performance of the intersection graph kernels comes close to the WL kernels, but the intersection tree kernels clearly show worse performance. Increasing extraction depth increases performance for all the tested kernels but the intersection trees. Adding inferencing only benefits the WL-kernels.

The Weisfeiler-Lehman kernel under the $h = 0$ setting can be considered as a baseline method, because it is essentially a 'bag-of-labels' kernel, where

---

[3] https://github.com/Data2Semantics/d2s-tools
[4] http://www.openrdf.org/
[5] Higher settings take a very large amount of computation time and/or run out of memory.
[6] This is the average of the F1 scores for each class.

Table 1. Results for the affiliation prediction experiments. 1,2,3 indicate the subgraph depth and 'f' indicates that inferencing was applied.

	acc.	F1	acc.	F1	acc.	F1	acc.	F1
Weisfeiler-Lehman RDF								
	$h = 0$		$h = 2$		$h = 4$		$h = 6$	
1	0.84	0.67	0.84	0.67	0.84	0.67	0.84	0.67
2	0.83	0.66	0.87	0.72	0.86	0.70	0.86	0.70
3	0.85	0.71	0.89	0.79	0.89	0.77	0.88	0.76
1,f	0.79	0.59	0.79	0.59	0.79	0.59	0.79	0.59
2,f	0.57	0.35	0.84	0.66	0.81	0.62	0.81	0.61
3,f	0.73	0.56	**0.91**	**0.81**	0.90	**0.80**	0.90	0.79
IntersectionSubTree, $\lambda = 1$					**IntersectionPartialSubTree, $\lambda = 0.01$**			
1	0.83	0.64			0.81	0.61		
2	0.82	0.61			0.81	0.61		
3	0.82	0.61			0.79	0.60		
1,f	0.81	0.61			0.79	0.58		
2,f	0.79	0.58			0.78	0.58		
3,f	0.81	0.61			0.78	0.58		
Weisfeiler-Lehman								
	$h = 0$		$h = 2$		$h = 4$		$h = 6$	
1	0.83	0.66	0.84	0.67	0.84	0.67	0.84	0.67
2	0.87	0.74	0.84	0.67	0.78	0.54	0.75	0.48
3	0.86	0.72	0.88	0.77	0.88	0.75	0.86	0.71
1,f	0.79	0.60	0.79	0.60	0.79	0.60	0.79	0.60
2,f	0.58	0.36	0.83	0.64	0.79	0.57	0.73	0.47
3,f	0.73	0.56	0.89	0.78	0.89	0.77	0.87	0.72
IntersectionGraphPath, $\lambda = 1$					**IntersectionGraphWalk, $\lambda = 1$**			
	$h = 1$		$h = 2$		$h = 1$		$h = 2$	
1	0.84	0.65	0.84	0.65	0.84	0.64	0.84	0.64
2	0.82	0.61	0.80	0.59	0.82	0.61	0.80	0.59
3	0.88	0.76	0.90	0.77	0.89	0.76	0.89	0.77
1,f	0.81	0.61	0.81	0.61	0.81	0.61	0.81	0.61
2,f	0.79	0.58	0.75	0.51	0.79	0.58	0.71	0.48
3,f	0.88	0.76	0.88	0.76	0.88	0.76	0.88	0.75

no rewriting is performed. We can see that already quite good performance is achieved using this baseline.

3.2 Lithogenesis Prediction

We perform the next prediction experiment on the RDF dataset from the British Geological Survey[7], which contains information about geological measurements in Britain. This dataset was chosen because it is at least a factor 10 larger than the affiliation prediction set and has some potential nice properties to predict. The things that are measured by this survey are called 'Named Rock Units', which have a number of different properties. One of these is the lithogenesis property, for which the two largest classes have 93 and 53 instances. In this experiment we try to predict for these 146 instances which of these two classes it belongs too. Again we remove triples related to these properties from the dataset and set the labels of the root vertices to the same special root label.

---

[7] http://data.bgs.ac.uk/

The setup for this experiment is the same as for the affiliation prediction task. The results are presented in Table 2.

The results of this experiment are similar to the results for affiliation prediction. Again the best scores are achieved by the WL RDF kernel under the '3,f' setting. However, the intersection graph path kernel achieves a similar accuracy score and the intersection subtree kernel scores are closer to WL RDF kernel. Increasing the subgraph depth improves the performance for all kernels. The performance of the 'bag-of-labels' baseline is similar to the affiliation prediction task.

3.3   Runtimes

To test the differences in runtimes between the kernels we measure the runtimes for the computation of each of the kernels on the two datasets above under the highest extraction setting (depth 3 and inferencing on). We measure these runtimes for different fractions of the dataset, from 0.1 to 1. The computation times of the two intersection tree kernels are nearly identical, so we only include

Table 2. Results for the lithogenesis prediction experiments. 1,2,3 indicate the subgraph depth and 'f' indicates that inferencing was applied.

	acc.	F1	acc.	F1	acc.	F1	acc.	F1
Weisfeiler-Lehman RDF								
	$h = 0$		$h = 2$		$h = 4$		$h = 6$	
1	0.79	0.62	0.79	0.62	0.79	0.62	0.79	0.62
2	0.87	0.75	0.88	0.77	0.88	0.76	0.88	0.77
3	0.86	0.73	0.87	0.75	0.88	0.76	0.88	0.77
1,f	0.78	0.61	0.78	0.61	0.78	0.61	0.78	0.61
2,f	0.82	0.66	0.88	0.76	0.87	0.75	0.87	0.75
3,f	0.88	0.75	0.89	0.78	**0.91**	**0.82**	**0.91**	**0.82**
IntersectionSubTree, $\lambda = 1$				IntersectionPartialSubTree, $\lambda = 0.01$				
1	0.79	0.63			0.81	0.65		
2	0.85	0.71			0.82	0.66		
3	0.86	0.73			0.82	0.67		
1,f	0.78	0.60			0.79	0.62		
2,f	0.84	0.70			0.80	0.63		
3,f	0.85	0.72			0.80	0.64		
Weisfeiler-Lehman								
	$h = 0$		$h = 2$		$h = 4$		$h = 6$	
1	0.79	0.62	0.79	0.62	0.79	0.63	0.79	0.63
2	0.88	0.77	0.86	0.73	0.85	0.70	0.84	0.69
3	0.86	0.73	0.87	0.75	0.87	0.75	0.88	0.76
1,f	0.78	0.61	0.78	0.61	0.78	0.61	0.78	0.61
2,f	0.82	0.65	0.85	0.72	0.85	0.71	0.85	0.71
3,f	0.88	0.75	0.88	0.77	0.88	0.76	0.88	0.77
IntersectionGraphPath, $\lambda = 1$				IntersectionGraphWalk, $\lambda = 1$				
	$h = 1$		$h = 2$		$h = 1$		$h = 2$	
1	0.81	0.66	0.82	0.67	0.79	0.62	0.80	0.63
2	0.86	0.72	0.86	0.72	0.86	0.73	0.86	0.72
3	0.88	0.76	0.88	0.76	0.88	0.76	0.87	0.75
1,f	0.80	0.64	0.81	0.64	0.77	0.60	0.77	0.60
2,f	0.85	0.71	0.85	0.72	0.84	0.70	0.85	0.71
3,f	**0.90**	0.80	0.90	0.79	0.90	0.79	0.89	0.78

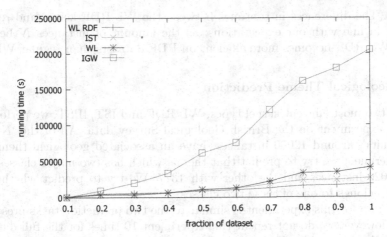

Fig. 1. Runtimes for the kernels: Weisfeiler-Lehman for RDF (WL RDF), Intersection SubTree (IST), Weisfeiler-Lehman (WL) and Intersection Graph Walk (IGW), for different fractions of the affiliation prediction dataset.

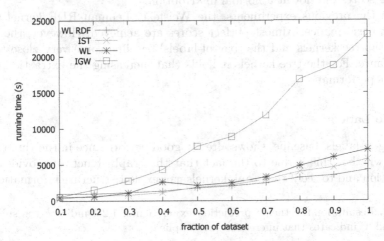

Fig. 2. Runtimes for the kernels: Weisfeiler-Lehman for RDF (WL RDF), Intersection SubTree (IST), Weisfeiler-Lehman (WL) and Intersection Graph Walk (IGW), for different fractions of the lithogenesis prediction dataset.

the IST kernel. The same is true for the two intersection graph kernels, so we only include the IGW kernel. The intersection subtree kernel is implemented as described in [5]. For the regular WL kernel and IGW kernel, the extraction of the subgraphs was also included, however this was only a small factor in the overall computation time.

Figure 1 presents the results for the four kernels on the affiliation prediction dataset. The results for the lithogenesis dataset are presented in Fig. 2.

Both figures show similar results: the IGW kernel is the slowest by a large margin and as the datasets grow larger, the WL RDF kernel becomes more efficient. The differences in runtimes between the WL RDF kernel and regular WL are in line with our expectations: as the amount of instances N become larger, WL RDF becomes more efficient on RDF datasets than regular WL.

3.4 Geological Theme Prediction

For the two most efficient kernel types, WL RDF and IST/IPST, we performed another experiment on the British Geological Survey data. All of the 'Named Rock Units', around 12000 instances, have an associated geological theme. In this experiment we try to predict that theme, which has two major classes, one with 10020 instances and the other with 1377. We try to predict whether an instance belongs to one of these two classes.

The setup for this experiment is similar to the two prediction tasks presented above. However, we do not repeat the experiment 10 times for the full dataset, but we take 10 random 10% subsets of the full dataset. The results are presented in Table 3. Again bold type face indicates the best scores. This time a MannWhitney U test with $p < 0.05$ was used as a significance test, since the resulting scores did not fit a normal distribution.

As in the previous experiments, the Weisfeiler-Lehman RDF kernel shows the best performance. Almost perfect scores are achieved. However, the intersection subtree kernels and the 'bag-of-labels' baseline come very close to this performance. For the tree kernels it holds that increasing the extraction depth increases performance.

3.5 No Labels

The 'bag-of-labels' baseline, shows already good performance in the three tasks. To test whether this is due to the fact that the graph structure provides little information and to see if the graph kernels can exploit structure information, we

Table 3. Results for the theme prediction experiments. 1,2,3 indicate the subgraph depth and 'f' indicates that inferencing was applied.

	acc.	F1	acc.	F1	acc.	F1	acc.	F1
Weisfeiler-Lehman RDF								
	$h = 0$		$h = 2$		$h = 4$		$h = 6$	
1	0.90	0.69	0.93	0.79	0.96	0.85	0.96	0.84
2	0.94	0.78	0.97	0.89	0.95	0.85	0.97	0.91
3	0.98	0.89	**1.0**	**0.98**	**1.0**	**0.98**	**0.99**	**0.98**
1,f	0.88	0.66	0.94	0.79	0.96	0.84	0.96	0.85
2,f	0.88	0.65	0.92	0.79	0.98	0.91	0.95	0.88
3,f	0.74	0.44	**1.0**	**0.98**	**0.99**	**0.98**	**1.0**	**0.98**
IntersectionSubTree, $\lambda = 1$				IntersectionPartialSubTree, $\lambda = 0.01$				
1	0.93	0.79			0.88	0.68		
2	0.96	0.89			0.97	0.88		
3	0.98	0.93			0.97	0.88		
1,f	0.94	0.84			0.94	0.78		
2,f	0.97	0.91			0.97	0.86		
3,f	0.99	0.96			0.98	0.92		

Table 4. Results for the affiliation prediction experiments with vertex and edge labels removed. 1,2,3 indicate the subgraph depth and 'f' indicates that inferencing was applied.

	acc.	F1	acc.	F1	acc.	F1	acc.	F1
	Weisfeiler-Lehman RDF							
	$h = 0$		$h = 2$		$h = 4$		$h = 6$	
1	0.14	0.07	0.61	0.40	0.63	0.41	0.63	0.41
2	0.48	0.23	0.61	0.40	0.62	0.42	0.64	0.44
3	0.43	0.24	0.87	0.75	**0.89**	**0.77**	0.88	0.75
1,f	0.11	0.06	0.51	0.31	0.51	0.31	0.51	0.31
2,f	0.26	0.12	0.57	0.36	0.60	0.39	0.61	0.40
3,f	0.18	0.09	0.85	0.70	0.87	0.74	0.88	0.75
	IntersectionSubTree, $\lambda = 1$				IntersectionPartialSubTree, $\lambda = 0.01$			
1	0.11	0.06			0.11	0.06		
2	0.22	0.09			0.11	0.06		
3	0.23	0.10			0.16	0.07		
1,f	0.11	0.06			0.11	0.06		
2,f	0.23	0.10			0.12	0.06		
3,f	0.23	0.10			0.25	0.09		
	Weisfeiler-Lehman							
	$h = 0$		$h = 2$		$h = 4$		$h = 6$	
1	0.14	0.07	0.50	0.24	0.51	0.25	0.50	0.24
2	0.48	0.23	0.47	0.25	0.46	0.24	0.46	0.24
3	0.43	0.24	0.59	0.37	0.72	0.51	0.72	0.49
1,f	0.11	0.06	0.32	0.17	0.33	0.17	0.33	0.17
2,f	0.29	0.15	0.53	0.33	0.51	0.29	0.52	0.29
3,f	0.16	0.08	0.57	0.37	0.73	0.54	0.71	0.50
	IntersectionGraphPath, $\lambda = 1$				IntersectionGraphWalk, $\lambda = 1$			
	$h = 1$		$h = 2$		$h = 1$		$h = 2$	
1	0.11	0.06	0.11	0.06	0.45	0.26	0.46	0.26
2	0.48	0.20	0.44	0.24	0.47	0.26	0.36	0.20
3	0.48	0.20	0.45	0.24	0.55	0.34	0.52	0.31
1,f	0.11	0.06	0.11	0.06	0.47	0.27	0.46	0.26
2,f	0.42	0.18	0.45	0.26	0.47	0.27	0.38	0.19
3,f	0.33	0.16	0.45	0.26	0.51	0.31	0.33	0.16

repeat the affiliation prediction experiment. This time we remove all the label information after the creation of the graph/subgraphs. Each label is replaced by the same dummy label. The rest of the experimental setup is identical to the affiliation prediction experiment. Results are given in Table 4.

Again, the Weisfeiler-Lehman RDF kernel shows the best performance. What is even more striking is that a performance close to the performance on labeled graphs can be achieved. The regular WL kernel also does reasonably well, however, the intersection graph and the intersection subtree kernel show very poor performance. These kernels are clearly not designed for unlabeled graphs and do not exploit graph structure as the WL kernels do.

4   Conclusions and Future Work

We presented an approximation of the Weisfeiler-Lehman Subtree kernel which speeds up computation on RDF graphs by computing the kernel on the underlying RDF graph instead of extracted subgraphs. The kernel shows performance that is better than the regular Weisfeiler-Lehman kernel applied to RDF. Also it is increasingly more efficient as the number of instances grows. This efficiency is achieved by exploiting the fact that the RDF instance subgraphs share vertices and edges in the underlying large RDF graph.

Furthermore, the presented kernel is faster and shows better classification performance than the intersection subtree and intersection graph kernels, which were recently introduced specifically for RDF data. When we remove the label data, then the Weisfeiler-Lehman for RDF still has relatively good classification performance on just the graph structure, whereas the intersection subtree and intersection graph kernels cannot use the structure information very well.

The performance difference between the presented approximation of the WL Subtree kernel and the regular version requires further investigation. However, we have observed that the computed feature vectors for WL RDF are shorter than for regular WL. Thus, our approximation probably leads to more shared features between instances, which can result in better generalization and less overfitting.

The presented kernel can be used in any machine learning situation where subgraphs derive from an underlying larger graph. It is particularly well-suited to RDF because the extracted subgraphs share a large number of vertices and edges. As future work it would be interesting to apply the kernel to other similar situations.

Another direction for future research is the application of the presented kernel to extremely large RDF datasets with 100 millions of triples or more, since it has the potential to scale well. For datasets with more instances, the computed feature vectors can be used directly with large scale linear SVM methods, such as LibLINEAR [15], which is not possible with the intersection subtree and intersection graph kernels. We also wish to investigate extensions of the Weisfeiler-Lehman kernel using label comparisons other the than the Dirac-kernel.

Acknowledgments. This publication was supported by the Dutch national program COMMIT. We express our gratitude to the anonymous reviewers for their helpful comments. We thank the authors of [5] for the AIFB dataset.

References

1. Shawe-Taylor, J., Cristianini, N.: Kernel Methods for Pattern Analysis. Cambridge University Press, New York (2004)
2. Schölkopf, B., Smola, A.J.: Learning with Kernels: Support Vector Machines, Regularization, Optimization, and Beyond. MIT Press, Cambridge (2001)
3. Vishwanathan, S.V.N., Schraudolph, N.N., Kondor, R.I., Borgwardt, K.M.: Graph kernels. Journal of Machine Learning Research 11, 1201–1242 (2010)
4. Shervashidze, N., Schweitzer, P., van Leeuwen, E.J., Mehlhorn, K., Borgwardt, K.M.: Weisfeiler-lehman graph kernels. J. Mach. Learn. Res. 12, 2539–2561 (2011)
5. Lösch, U., Bloehdorn, S., Rettinger, A.: Graph kernels for RDF data. In: Simperl, E., Cimiano, P., Polleres, A., Corcho, O., Presutti, V. (eds.) ESWC 2012. LNCS, vol. 7295, pp. 134–148. Springer, Heidelberg (2012)

6. Bloehdorn, S., Sure, Y.: Kernel methods for mining instance data in ontologies. In: Aberer, K., et al. (eds.) ISWC/ASWC 2007. LNCS, vol. 4825, pp. 58–71. Springer, Heidelberg (2007)
7. Bicer, V., Tran, T., Gossen, A.: Relational kernel machines for learning from graph-structured RDF data. In: Antoniou, G., Grobelnik, M., Simperl, E., Parsia, B., Plexousakis, D., De Leenheer, P., Pan, J. (eds.) ESWC 2011, Part I. LNCS, vol. 6643, pp. 47–62. Springer, Heidelberg (2011)
8. Fanizzi, N., d'Amato, C.: A declarative kernel for $\mathcal{ALC}$ concept descriptions. In: Esposito, F., Raś, Z.W., Malerba, D., Semeraro, G. (eds.) ISMIS 2006. LNCS (LNAI), vol. 4203, pp. 322–331. Springer, Heidelberg (2006)
9. Fanizzi, N., d'Amato, C., Esposito, F.: Statistical learning for inductive query answering on OWL ontologies. In: Sheth, A.P., Staab, S., Dean, M., Paolucci, M., Maynard, D., Finin, T., Thirunarayan, K. (eds.) ISWC 2008. LNCS, vol. 5318, pp. 195–212. Springer, Heidelberg (2008)
10. Rettinger, A., Nickles, M., Tresp, V.: Statistical relational learning with formal ontologies. In: Buntine, W., Grobelnik, M., Mladenić, D., Shawe-Taylor, J. (eds.) ECML PKDD 2009, Part II. LNCS, vol. 5782, pp. 286–301. Springer, Heidelberg (2009)
11. Shervashidze, N., Borgwardt, K.M.: Fast subtree kernels on graphs. In: Bengio, Y., Schuurmans, D., Lafferty, J.D., Williams, C.K.I., Culotta, A. (eds.) NIPS, pp. 1660–1668. Curran Associates, Inc. (2009)
12. Haussler, D.: Convolution kernels on discrete structures. Technical Report UCS-CRL-99-10, University of California at Santa Cruz, Santa Cruz, CA, USA (1999)
13. Chang, C.C., Lin, C.J.: LIBSVM: A library for support vector machines. ACM Transactions on Intelligent Systems and Technology 2, 27:1–27:27 (2011), Software available at http://www.csie.ntu.edu.tw/~cjlin/libsvm
14. Sure, Y., Bloehdorn, S., Haase, P., Hartmann, J., Oberle, D.: The SWRC ontology - semantic web for research communities. In: Bento, C., Cardoso, A., Dias, G. (eds.) EPIA 2005. LNCS (LNAI), vol. 3808, pp. 218–231. Springer, Heidelberg (2005)
15. Fan, R.E., Chang, K.W., Hsieh, C.J., Wang, X.R., Lin, C.J.: LIBLINEAR: A library for large linear classification. Journal of Machine Learning Research 9, 1871–1874 (2008)

Efficient Frequent Connected Induced Subgraph Mining in Graphs of Bounded Tree-Width

Tamás Horváth[1,2], Keisuke Otaki[3], and Jan Ramon[4]

[1] Dept. of Computer Science III, University of Bonn, Germany
[2] Fraunhofer IAIS, Schloss Birlinghoven, Sankt Augustin, Germany
tamas.horvath@iais.fraunhofer.de
[3] Dept. of Intelligence Science and Technology, Graduate School of Informatics,
Kyoto University, Kyoto, Japan
ootaki@iip.ist.i.kyoto-u.ac.jp
[4] Dept. of Computer Science, Katholieke Universiteit Leuven, Belgium
jan.ramon@cs.kuleuven.be

Abstract. We study the frequent connected induced subgraph mining problem, i.e., the problem of listing all connected graphs that are induced subgraph isomorphic to a given number of transaction graphs. We first show that this problem cannot be solved for arbitrary transaction graphs in output polynomial time (if $P \neq NP$) and then prove that for graphs of bounded tree-width, frequent connected induced subgraph mining is possible in incremental polynomial time by levelwise search. Our algorithm is an adaptation of the technique developed for frequent connected subgraph mining in bounded tree-width graphs. While the adaptation is relatively natural for many steps of the original algorithm, we need entirely different combinatorial arguments to show the correctness and efficiency of the new algorithm. Since induced subgraph isomorphism between bounded tree-width graphs is NP-complete, the positive result of this paper provides another example of efficient pattern mining with respect to computationally intractable pattern matching operators.

1 Introduction

Over the past 15 years substantial research efforts have been devoted toward designing effective frequent graph mining algorithms. Despite the numerous studies in this field of research, the theoretical aspects of the topic are still not well understood. The importance of a better understanding of the complexity aspects of the various graph mining problem settings appears somewhat neglected, which has as negative side effect that most algorithms are limited to some ten thousands transaction graphs only. In this work we study the *frequent connected induced subgraph mining* (FCISM) problem, which is the problem of listing all pairwise non-isomorphic connected graphs that are induced subgraph isomorphic to at least t transaction graphs for some parameter $t \in \mathbb{N}$. This problem, as we show, cannot be solved in output polynomial time for arbitrary transaction graphs. For forests, however, it can be solved in incremental polynomial time.

H. Blockeel et al. (Eds.): ECML PKDD 2013, Part I, LNAI 8188, pp. 622–637, 2013.
© Springer-Verlag Berlin Heidelberg 2013

As the main result for this work, we generalize the positive result on forests by showing that the FCISM problem can be solved in incremental polynomial time for graphs of *bounded tree-width*. Regarding the practical aspects of our result, we mention e.g. the ZINC dataset containing about 16.5 millions molecular graphs: 99.99% of these graphs have tree-width at most 3. Regarding its theoretical aspects, we note that induced subgraph isomorphism is one of the "persistent" problems that remain NP-complete even for graphs of tree-width 2 [8]. Thus, our positive result provides an example of the case that *efficient pattern mining is possible even for computationally intractable pattern matching operators*. To the best of our knowledge, there is only one further such example [5].

Finally, regarding the algorithmic aspects of our result, we note that the paradigm we followed here, and which is used also in [5] for frequent connected subgraph mining in graphs of bounded tree-width, appears sufficiently general for the design of graph mining algorithms for further pattern matching operators. This paradigm consists of the following main steps: (1) Give a *generic* levelwise search algorithm and, in addition to some natural conditions, (2) prove the existence of an *efficiently* computable pattern refinement operator that is *complete* with respect to the pattern matching operator, and (3) show that the otherwise exponential-time dynamic-programming algorithm [8] deciding the underlying pattern matching works in time *polynomial* in the size of the set of patterns generated by the algorithm so far. When comparing the (sub)steps of this paradigm for ordinary subgraph isomorphism [5] and for induced subgraph isomorphism, on the one hand one can notice a number of steps that are (almost) the same for the two problems. On the other hand, however, there are some crucial steps that require entirely different techniques. Thus, for example, the pattern refinement operator and the combinatorial characterization of the necessary information needed to calculate by the pattern matching algorithm become much more complicated for induced subgraph isomorphism, as we will show in Section 4.

The rest of the paper is organized as follows. In the next section we collect all necessary notions. In Section 3 we give a generic levelwise search algorithm and formulate five conditions for the efficiency of this algorithm. In Section 4 we prove that all these conditions are fulfilled by the class of bounded tree-width graphs. Finally, in Section 5 we conclude and mention some problems.

2   Preliminaries

In this section we collect and fix all necessary notions and notations used in the paper. Most of the definitions and notations are taken from [5].

Graphs. An *undirected graph* is a pair (V, E), where V is a finite set of *vertices* and $E \subseteq \{e \subseteq V : |e| = 2\}$ is a set of *edges*. We consider simple graphs, i.e., which do not contain loops or parallel edges. A *labeled undirected graph* is a triple (V, E, λ), where (V, E) is an undirected graph and λ is the labeling function $\lambda : V \cup E \to \mathbb{N}$. The set of vertices, the set of edges, and the labeling function of a graph G are denoted by $V(G)$, $E(G)$, and λ_G, respectively. Unless otherwise stated, by graphs in this paper we always mean *labeled undirected* graphs.

A *subgraph* of G is a graph G' with $V(G') \subseteq V(G)$, $E(G') \subseteq E(G)$, and $\lambda_{G'}(x) = \lambda_G(x)$ for all $x \in V(G') \cup E(G')$; G' is an *induced subgraph* of G if it is a subgraph of G satisfying $\{u, v\} \in E(G')$ iff $\{u, v\} \in E(G)$ for all $u, v \in V(G')$. For $S \subseteq V(G)$, $G[S]$ denotes the induced subgraph of G with vertex set S. For $v \in V(G)$, $G \ominus v$ denotes $G[V(G) \setminus \{v\}]$.

A *path* connecting two vertices v_1, v_k of a graph G, denoted by P_{v_1, v_k}, is a sequence $\{v_1, v_2\}, \{v_2, v_3\}, \ldots, \{v_{k-1}, v_k\} \in E(G)$ such that the v_i's are pairwise distinct. A graph is *connected* if there is a path between any pair of its vertices. A *connected component* of a graph G is a maximal subgraph of G that is connected. The set of all connected components of a graph G is denoted by $\mathcal{C}(G)$.

Graph Morphisms. Two graphs G_1 and G_2 are *isomorphic*, denoted $G_1 \simeq G_2$, if there is a *bijection* $\varphi : V(G_1) \to V(G_2)$ satisfying (i) $\{u, v\} \in E(G_1)$ iff $\{\varphi(u), \varphi(v)\} \in E(G_2)$ for every $u, v \in V(G_1)$, (ii) $\lambda_{G_1}(u) = \lambda_{G_2}(\varphi(u))$ for every $u \in V(G_1)$, and (iii) $\lambda_{G_1}(\{u, v\}) = \lambda_{G_2}(\{\varphi(u), \varphi(v)\})$ for every $\{u, v\} \in E(G_1)$. For G_1 and G_2 we say that G_1 is *subgraph isomorphic* to G_2, denoted $G_1 \preccurlyeq G_2$, if G_1 is isomorphic to a subgraph of G_2; it is *induced subgraph isomorphic* to G_2, denoted $G_1 \preccurlyeq_i G_2$, if it is isomorphic to an induced subgraph of G_2. In what follows, two graphs are regarded the same graph if they are isomorphic.

Tree-Width. A central notion to this work is *tree-width*, which was reintroduced in algorithmic graph theory in [10]. A *tree-decomposition* of a graph G is a pair $(T, \mathcal{X})$, where T is a rooted tree and $\mathcal{X} = (X_z)_{z \in V(T)}$ is a family of subsets of $V(G)$ satisfying (i) $\cup_{z \in V(T)} X_z = V(G)$, (ii) for every $\{u, v\} \in E(G)$, there is a $z \in V(T)$ such that $u, v \in X_z$, and (iii) $X_{z_1} \cap X_{z_3} \subseteq X_{z_2}$ for every $z_1, z_2, z_3 \in V(T)$ such that z_2 is on the path connecting z_1 with z_3 in T. The set X_z associated with a node z of T is called the *bag* of z. The nodes of T will often be referred to as the nodes of the tree-decomposition. The tree-width of $(T, \mathcal{X})$ is $\max_{z \in V(T)} |X_z| - 1$, and the *tree-width* of G, denoted $\mathrm{tw}(G)$, is the minimum tree-width over all tree-decompositions of G. By graphs of bounded tree-width we mean graphs of tree-width at most k, where k is some constant. The following notation will be used many times in what follows. Let G be a graph, $(T, \mathcal{X})$ be a tree-decomposition of G, and $z \in V(T)$. Then $G_{[z]}$ denotes the induced subgraph of G defined by the union of the bags of z's descendants, where z is considered also as a descendant of itself.

We will use a special kind of tree-decomposition. More precisely, a *nice* tree-decomposition of G, denoted $NTD(G)$, is a tree-decomposition $(T, \mathcal{X})$, where T is a rooted binary tree composed of three types of nodes: (i) a *leaf* node has no children, (ii) a *separator* node z has a single child z' with $X_z \subseteq X_{z'}$, and (iii) a *join* node z has two children z_1 and z_2 with $X_z = X_{z_1} \cup X_{z_2}$. It follows from [3] that for graphs of tree-width at most k, where k is some constant, a nice tree-decomposition of tree-width at most k always exists and can be constructed in linear time.

Tree-width is a useful parameter of graphs in algorithmic graph theory, as many NP-hard problems can be solved in polynomial time for graphs of bounded tree-width. However, subgraph isomorphism and induced subgraph isomorphism remain NP-complete even for graphs of tree-width 2 (see, e.g., [8,11]).

Listing Algorithms. A common feature of many listing problems is that the size of the output can be exponential in that of the input. Clearly, for such cases there exists no algorithm enumerating the output in time polynomial in the size of the input. Thus, the size of the output must also be taken into account. The following listing complexity classes are usually distinguished in the literature (see, e.g., [6]): For some input $\mathcal{I}$, let $\mathcal{O}$ be the output set of some finite cardinality N. Then the elements of $\mathcal{O}$, say $o_1, \ldots, o_N$, are listed with

polynomial delay if the time before printing o_1, the time between printing o_i and o_{i+1} for every $i = 1, \ldots, N-1$, and the time between printing o_N and the termination is bounded by a polynomial of the size of $\mathcal{I}$,

incremental polynomial time if o_1 is printed with polynomial delay, the time between printing o_i and o_{i+1} for every $i = 1, \ldots, N-1$ (resp. the time between printing o_N and the termination) is bounded by a polynomial of the combined size of $\mathcal{I}$ and the set $\{o_1, \ldots, o_i\}$ (resp. $\mathcal{O}$),

output polynomial time (or *polynomial total time*) if $\mathcal{O}$ is printed in time polynomial in the combined size of $\mathcal{I}$ and the *entire* output $\mathcal{O}$.

Clearly, polynomial delay implies incremental polynomial time, which, in turn, implies output polynomial time. Furthermore, in contrast to incremental polynomial time, the delay of an output polynomial time algorithm may be exponential in the size of the input even before printing the first element of the output.

3 Frequent Connected Induced Subgraph Mining

In this section we first define the frequent connected induced subgraph mining problem and show in Theorem 1 that it is computationally intractable. We then give a generic levelwise search algorithm [7] for mining frequent connected induced subgraphs and provide sufficient conditions in Theorem 2 for the efficiency of this algorithm. As a corollary of Theorem 2, we get that the frequent connected induced subgraph mining problem can be solved in incremental polynomial time for *forest* [1] transaction graphs. In the next section we generalize the positive result on forests to graphs of bounded tree-width. We start by defining the pattern mining problem we are interested in.

THE FREQUENT CONNECTED INDUCED SUBGRAPH MINING (FCISM) PROB-
LEM: *Given* a class $\mathcal{G}$ of graphs, a transaction database (i.e., multiset) DB of graphs from $\mathcal{G}$, and an integer frequency threshold $t > 0$, *list* the set $\mathcal{O}$ of all *distinct* frequent connected *induced* subgraphs, that is, all connected graphs that are induced subgraph isomorphic to at least t graphs in DB.

Note that we do not distinguish between isomorphic graphs and hence each isomorphism type (i.e., equivalence class under isomorphism) of $\mathcal{O}$ is a singleton. The *parameter* of the above problem is the size of DB. Clearly, the size of $\mathcal{O}$ can be exponential in that of DB. Thus, in general, the set of all frequent

---

[1] In this paper by forest we mean a set of disjoint unrooted trees.

Algorithm 1. FCISM

Require: transaction database DB of graphs and integer $t > 0$
Ensure: all frequent connected induced subgraphs

1: let $S_1 \subseteq \mathcal{G}$ be the set of frequent graphs consisting of a single labeled vertex
2: **for** $(l := 1;\ S_l \neq \emptyset;\ l := l + 1)$ **do**
3:      $C_{l+1} := S_{l+1} := \emptyset$
4:      **forall** $P \in S_l$ **do**
5:          **forall** $H \in \rho(P) \cap \mathcal{G}$ satisfying (i) $H \notin C_{l+1}$ and (ii) $\rho^{-1}(H) \subseteq S_l$ **do**
6:              add H to C_{l+1}
7:              **if** $|\{G \in DB : H \preccurlyeq_i G\}| \geq t$ **then**
8:                  **print** H and add it to S_{l+1}

---

connected induced subgraphs cannot be computed in time polynomial only in the size of DB. The following simple polynomial reduction shows that even output polynomial time enumeration is unlikely.

Theorem 1. *Unless* $P = NP$, *the FCISM problem cannot be solved in output polynomial time.*

Proof. We prove the claim by a reduction from the NP-complete k-CLIQUE problem. For an unlabeled graph G with n vertices, let DB consist of G and the clique K_n with n vertices. For DB and $t = 2$, the number of frequent connected induced subgraphs is at most n (i.e., all cliques up to size n). Thus, if the FCISM problem could be solved in output polynomial time, we could decide the k-CLIQUE problem in polynomial time by listing first the set $\mathcal{O}$ of all 2-frequent connected induced subgraphs and checking then whether $|\mathcal{O}| \geq k$ or not.   □

3.1 A Generic Levelwise Search Mining Algorithm

Our goal in this paper is to show that the FCISM problem can be solved in incremental polynomial time for graphs of *bounded tree-width*. To prove this result, we start by giving a generic algorithm, called FCISM, that lists frequent connected induced subgraphs with levelwise search (see Algorithm 1). The algorithm assumes the transaction graphs to be elements of some graph class $\mathcal{G}$ that is closed under taking subgraphs. Thus, as we are interested in mining frequent connected induced subgraphs, all patterns belong to $\mathcal{G}$ as well.

One of the basic features of the levelwise search algorithms is that the underlying pattern language $\mathcal{L}$ is associated with some, usually naturally defined partial order. Following the common pattern mining terminology (see, e.g., [7]), for a partially ordered pattern language $(\mathcal{L}, \leq)$ we say that a pattern $P_1 \in \mathcal{L}$ is a *generalization* of a pattern $P_2 \in \mathcal{L}$ (or P_2 is a *specialization* of P_1) if $P_1 \leq P_2$; P_1 is a *proper generalization* of P_2 (or P_2 is a *proper specialization* of P_1), denoted by $P_1 < P_2$, if $P_1 \leq P_2$ and $P_1 \neq P_2$. Furthermore P_1 is a *direct generalization* of P_2 (or P_2 is a *direct specialization* of P_1) if $P_1 < P_2$ and there is no $P_3 \in \mathcal{L}$ with $P_1 < P_3 < P_2$.

In case of the FCISM problem, the underlying pattern language $\mathcal{L}$ is the set of all finite connected (labeled) graphs of $\mathcal{G}$, associated with the following natural generalization relation $\leq$ defined as follows: For any $P_1, P_2 \in \mathcal{L}$, $P_1 \leq P_2$ if and only if $P_1 \preccurlyeq_i P_2$. The proofs of the two claims in the proposition below are straightforward.

Proposition 1. *Let $\mathcal{L}$ and $\leq$ be as defined above. Then $(\mathcal{L}, \leq)$ is a partially ordered set. Furthermore, for any $P_1, P_2 \in \mathcal{L}$ it holds that P_1 is a direct generalization of P_2 iff*

$$P_1 < P_2 \text{ and } |V(P_1)| = |V(P_2)| - 1 . \tag{1}$$

In the main loop of Algorithm 1 (lines 4–8), the set S_{l+1} of frequent connected induced subgraphs containing $l+1$ *vertices* are calculated from those containing l *vertices*, in accordance with condition (1). In particular, for each frequent pattern $P \in S_l$, we first compute a set $\rho(P) \cap \mathcal{G}$ of graphs, where $\rho(P)$ is a subset of the set of *direct* specializations of P. Clearly, the graphs in $\rho(P)$ are all connected by the choice of $\mathcal{L}$. Notice that we cannot define $\rho(P)$ as the set of *all* direct specializations of P, as this set can be of exponential cardinality. In Theorem 2 below we will provide sufficient conditions for ρ needed for efficient pattern enumeration.

For each direct specialization $H \in \rho(P) \cap \mathcal{G}$, we check whether it has already been generated during the current iteration (see condition (i) in line 5). If not, we also check for each connected direct generalization of H, denoted by $\rho^{-1}(H)$ in the algorithm, whether it is frequent (condition (ii) in line 5). Here we utilize that frequency is an anti-monotonic interestingness predicate for $(\mathcal{L}, \leq)$. In what follows, candidate patterns generated by Algorithm 1 that satisfy conditions (i) and (ii) in line 5 will be referred to as *strong candidates*. If H is a strong candidate, we add it to the set C_{l+1} of candidate patterns consisting of $l+1$ *vertices* and compute its support count (lines 6–7). If H is frequent, i.e., it is induced subgraph isomorphic to at least t transaction graphs in DB, we add it to the set S_{l+1} of frequent connected graphs containing $l+1$ *vertices*.

By Theorem 1 above, the FCISM problem cannot be solved in output polynomial time for the general problem setting. If, however, the class $\mathcal{G}$ of transaction graphs and the *refinement operator* ρ satisfy the conditions of Theorem 2 below, the FCISM problem can be solved in incremental polynomial time. To state the theorem, we recall some basic notions for refinement operators (see, e.g., [9]). A downward refinement operator Ξ for a poset $(\mathcal{L}, \leq)$ is a function $\Xi : \mathcal{L} \to 2^{\mathcal{L}}$ with $\Xi(P) \subseteq \{P' : P \leq P'\}$ for all $P \in \mathcal{L}$. That is, $\Xi(P)$ is a subset of the set of specializations of P. For Ξ, we define the n-th power $\Xi^n : \mathcal{L} \to 2^{\mathcal{L}}$ recursively by

$$\Xi^n(P) = \begin{cases} \Xi(P) & \text{if } n = 1 \\ \Xi(\Xi^{n-1}(P)) & \text{o/w} \end{cases}$$

for all $n \in \mathbb{N}$. Finally, we say that Ξ is *complete*, if for all $P \in \mathcal{L}$, there is some $n \in \mathbb{N}$ with $P \in \Xi^n(\bot)$, where $\bot$ denotes the empty graph. Using the above notions, we can formulate the following generic theorem:

Theorem 2. *Let $\mathcal{G}$ be a class of the transaction graphs, $\mathcal{L}$ be the set of connected graphs in $\mathcal{G}$, and $\rho : \mathcal{L} \to 2^{\mathcal{L}}$ be a downward refinement operator. If ρ and $\mathcal{G}$ satisfy the conditions below then Algorithm 1 solves the FCISM problem in incremental polynomial time and in incremental polynomial space.*

  (i) *$\mathcal{G}$ is closed under taking subgraphs.*
 (ii) *The membership problem in $\mathcal{G}$ can be decided in polynomial time.*
(iii) *ρ is complete and $\rho(P)$ can be computed in time polynomial in the combined size of the input and the set of frequent patterns listed so far by Algorithm 1.*
 (iv) *Isomorphism can be decided in polynomial time for $\mathcal{G}$.*
  (v) *For every $H, G \in \mathcal{G}$ such that H is connected, it can be decided in time polynomial in the combined size of the input and the set of frequent patterns listed so far by Algorithm 1 whether $H \preccurlyeq_i G$.*

Proof. The proof follows directly from the remarks and concepts above.

The following positive result on forests can immediately be obtained by applying the theorem above (it follows also e.g. from [5]):

Corollary 1. *The FCISM problem can be solved in incremental polynomial time for forest transaction graphs.*

4   Mining Graphs of Bounded Tree-Width

In this section we generalize the positive result of Corollary 1 to graphs of bounded tree-width and prove the main result of this paper:

Theorem 3. *The FCISM problem can be solved in incremental polynomial time for graphs of bounded tree-width.*

Before proving this result, we first note that the class of bounded tree-width graphs is not only of theoretic interest, but also of practical relevance. As an example, consider the ZINC dataset[2] consisting of more than 16 million chemical compounds. Regarding the distribution of the molecular graphs with respect to their tree-width, 99.99% of the 16,501,334 molecular graphs in this dataset have tree-width at most 3 and $99,31\%$ only tree-width at most 2.

To prove Theorem 3, it suffices to show that all conditions of Theorem 2 hold for bounded tree-width graphs. The proof of the claims in the theorem below is shown in [5] for the positive result on frequent connected subgraph mining in graphs of bounded tree-width; notice that the conditions considered in the theorem are all independent of the underlying pattern matching operator.

Theorem 4. *Conditions (i), (ii), and (iv) of Theorem 2 hold for the class of bounded tree-width graphs.*

Thus, only conditions (iii) and (v) have to be proven. We first show (iii).

---
[2] We used a commercial version of the ZINC dataset for the tree-width statistics.

Theorem 5. *For the class of bounded tree-width graphs there exists a refinement operator ρ satisfying condition (iii) of Theorem 2.*

Proof. For a connected pattern P with $\text{tw}(P) \leq k$, define the refinement $\rho(P)$ of P as follows: A connected graph P' with $\text{tw}(P') \leq k$ is in $\rho(P)$ iff P' has a vertex v with degree at most k such that $P \simeq P' \ominus v$. Notice that this definition is unique, as isomorphic graphs are not distinguished from each other by definition. Clearly, $\rho(P)$ is a subset of the set of direct specializations of P. Utilizing condition (i) of Theorem 2 and the basic fact that every graph of tree-width at most k has a vertex of degree at most k,[3] the completeness of ρ follows directly by induction on the number of vertices.

We now turn to the complexity of computing $\rho(P)$ and show the stronger property that $\rho(P)$ can actually be computed in time polynomial in the size of DB. Since each new vertex v is connected to P by at least one and at most k vertices, for the cardinality of $\rho(P)$ we have

$$|\rho(P)| \leq \sum_{i=1}^{k} |\Lambda|^{i+1} \binom{n}{i} < |\Lambda|^{k+1}(n+1)^k \ ,$$

where Λ is the set of vertex and edge labels used in DB and n is the number of vertices of P. Since k is a constant, $|\rho(P)|$ is polynomial in the size of DB, and hence, as condition (iv) of Theorem 2 holds by Theorem 4, $\rho(P)$ can be computed in time polynomial in the size of DB, as claimed.   □

It remains to show for the proof of Theorem 3 that condition (v) also holds. In Section 4.1 we first recall from [4] a dynamic programming algorithm deciding induced subgraph isomorphism for a restricted class of bounded tree-width graphs. Given a connected graph H and a transaction graph G, both of bounded tree-width, this algorithm decides $H \preccurlyeq_i G$ by computing recursively a certain set of tuples representing *partial* induced subgraph isomorphisms between H and G. The problem is, however, that for arbitrary graphs of bounded tree-width, the number of such partial solutions can be exponential in the size of H. Using the paradigm developed in [5] for frequent connected subgraph mining in graphs of bounded tree-width, we will show that $H \preccurlyeq_i G$ can be decided by computing only a polynomial number of *new* partial solutions and efficiently recovering all missing partials solutions from those calculated for the already generated frequent patterns.

4.1   A Dynamic Programming Algorithm

To make the paper as self-contained as possible, in this section we recall the dynamic programming algorithm from [4] that decides induced subgraph isomorphism for a restricted class of bounded tree-width graphs. The algorithm is

---

[3] This fact holds trivially if the graph has at most $k + 1$ vertices; o/w it has a tree-decomposition of tree-width at most k with a leaf z having a parent z' such that $X_z \not\subseteq X_{z'}$. But then there is a $v \in X_z$ that is not in the bag of any other node in the tree-decomposition and thus, v can be adjacent only to the vertices in $X_z \setminus \{v\}$.

based on an efficient algorithm [8] deciding various morphisms between bounded tree-width *and* bounded degree graphs, which, in turn, follows a generic dynamic programming approach designed in [2]. In order to be consistent with [5] on frequent connected subgraph mining in graphs of bounded tree-width, we naturally adapt the notions and notations from Section 4.1 of [5] from subgraph isomorphism to induced subgraph isomorphism.

In what follows, let H and G denote connected graphs with $\mathrm{tw}(H), \mathrm{tw}(G) \leq k$. In fact, as one can easily see, the results of this section hold also for the case that G is not connected. Given H and G, the algorithm in [4] decides whether $H \preccurlyeq_i G$ by computing a nice tree-decomposition $NTD(G)$ of G, traversing $NTD(G)$ in a postorder manner, calculating for each node in the tree-decomposition a set of tuples, called *characteristics*, and by testing whether the root of $NTD(G)$ has a characteristic satisfying a certain condition formulated in Lemma 1 below. More precisely, an *iso-quadruple* of H relative to a node z of $NTD(G)$ is a quadruple $(S, \mathcal{D}, K, \psi)$, where (i) $S \subseteq V(H)$ with $|S| \leq k+1$, (ii) $\mathcal{D} \subseteq \mathcal{C}(H[V(H) \setminus S])$, (iii) $K = H[S \cup V(\mathcal{D})]$, and (iv) $\psi : S \to X_z$ is an *induced* subgraph isomorphism from $H[S]$ to $G[X_z]$. Notice that K is redundant; it is used for keeping the explanation as simple as possible. The set of all iso-quadruples of H relative to a node z of $NTD(G)$ is denoted by $\Gamma(H, z)$.

For a node z in $NTD(G)$, an iso-quadruple $(S, \mathcal{D}, K, \psi) \in \Gamma(H, z)$ is a *z-characteristic* of H if there exists an induced subgraph isomorphism φ from K to $G_{[z]}$ satisfying (i) $\varphi(u) = \psi(u)$ for all $u \in S$ and (ii) $\varphi(v) \notin X_z$ for all $v \in V(\mathcal{D})$. These conditions imply that $\varphi(u) \in X_z$ for all $u \in S$. The set of all z-characteristics of H relative to z is denoted by $\Gamma_{\mathrm{ch}}(H, z)$. Clearly, $\Gamma_{\mathrm{ch}}(H, z) \subseteq \Gamma(H, z)$. The following lemma from [4] provides a characterization of induced subgraph isomorphism in terms of r-characteristics for the root r of $NTD(G)$.

Lemma 1. *Let r be the root of a nice tree-decomposition $NTD(G)$ of G. Then $H \preccurlyeq_i G$ iff there exists $(S, \mathcal{D}, K, \psi) \in \Gamma_{\mathrm{ch}}(H, r)$ with $K = H$.*

Thus, by the lemma above, we need to calculate the characteristics of the root of $NTD(G)$. Lemma 2 below from [4] shows how to compute the set of characteristics for leafs, and how for internal (i.e., separator or join) nodes from the sets of characteristics of their children. This enables the computation of the characteristics for all nodes of $NTD(G)$ by a postorder traversal of $NTD(G)$.

Lemma 2. *Let G, H be connected graphs of bounded tree-width and z be a node in $NTD(G)$. For all $(S, \mathcal{D}, K, \psi) \in \Gamma(H, z)$ it holds that $(S, \mathcal{D}, K, \psi) \in \Gamma_{\mathrm{ch}}(H, z)$ iff one of the following conditions holds:*

LEAF: *z has no children and $\mathcal{D} = \emptyset$.*

SEPARATOR: *z has a single child z' and $\exists (S', \mathcal{D}', K', \psi') \in \Gamma_{\mathrm{ch}}(H, z')$ with*
  (S.a) $S = \{v \in S' : \psi'(v) \in X_z\}$,
  (S.b) $\mathcal{D}' = \{D' \in \mathcal{C}(H[V(H) \setminus S']) : D'$ *is a subgraph of some $D \in \mathcal{D}\}$,*
  (S.c) $\psi(v) = \psi'(v)$ *for every $v \in S$.*

JOIN: *z has two children z_1, z_2 and there exist $(S_1, \mathcal{D}_1, K_1, \psi_1) \in \Gamma_{\mathrm{ch}}(H, z_1)$ and $(S_2, \mathcal{D}_2, K_2, \psi_2) \in \Gamma_{\mathrm{ch}}(H, z_2)$ satisfying*

(J.a) $S_i = \{v \in S : \psi(v) \in X_{z_i}\}$ *for* $i = 1, 2$,
(J.b) $\mathcal{D}_1$ *and* $\mathcal{D}_2$ *form a binary partition of the connected components of* $\mathcal{D}$,
(J.c) $\psi_i(v) = \psi(v)$ *for every* $v \in S_i$ *and for* $i = 1, 2$.

As mentioned, Lemma 2 provides a polynomial time algorithm for deciding induced subgraph isomorphism for restricted subclasses of bounded tree-width graphs (e.g., when the degree is also bounded [8] or when the graphs have log-bounded fragmentation [4]). Clearly, the algorithm is exponential for arbitrary bounded tree-width graphs; this follows directly from the negative result in [8].

It is important to stress that almost the same notions and conditions are used for the frequent *connected subgraph* mining (FCSM) problem (cf. Section 4.1 in [5]), where, in contrast to the FCISM problem, ordinary subgraph isomorphism is the underlying pattern matching operator. The only difference is in the definition of iso-quadruples, in particular, in the definition of ψ, in accordance with the semantic difference between the FCSM and FCISM problems. However, as it turns out in Section 4.2 below, we need a different combinatorial arguments to show the positive result for the FCISM problem.

4.2   Feasible Iso-Quadruples

Like in the FCSM problem, the main source of computational intractability of the algorithm based on Lemma 2 is the possibly exponential number of iso-quadruples needed to test. Using the paradigm developed for the FCSM problem [5], in this section we show that for each node of $NTD(G)$, it suffices to check only a polynomial number of iso-quadruples, as we can utilize the characteristics of the frequent patterns computed earlier by Algorithm 1. In order to show this result, we recall some necessary notions from [5]. As for the case of the FCSM problem, for all transaction graphs we fix a nice tree-decomposition computed in a preprocessing step for the entire mining process.

Let H_1, H_2, and G be connected graphs of bounded tree-width, $NTD(G)$ be some fixed nice tree-decomposition of G, and z be a node in $NTD(G)$. For any two $\xi_1 = (S_1, \mathcal{D}_1, K_1, \psi_1) \in \Gamma(H_1, z)$ and $\xi_2 = (S_2, \mathcal{D}_2, K_2, \psi_2) \in \Gamma(H_2, z)$, ξ_1 is *equivalent* to ξ_2, denoted $\xi_1 \equiv \xi_2$, if there is an isomorphism π between K_1 and K_2 such that π is a bijection between S_1 and S_2 and $\psi_1(v) = \psi_2(\pi(v))$ for every $v \in S_1$. The lemma below shows that it suffices to store only one representative z-characteristic for each equivalence class of the set of z-characteristics and that equivalence between iso-quadruples can be decided in polynomial time. The proof is similar to that of the corresponding lemma in [5].

Lemma 3. *Let G, H_1, and H_2 be connected graphs of tree-width at most k, z be a node in $NTD(G)$, and $\xi_i = (S_i, \mathcal{D}_i, K_i, \psi_i) \in \Gamma(H_i, z)$ $(i = 1, 2)$. Then*

(i) $\xi_1 \in \Gamma_{ch}(H_1, z)$ iff $\xi_2 \in \Gamma_{ch}(H_2, z)$ whenever $\xi_1 \equiv \xi_2$ and
(ii) $\xi_1 \equiv \xi_2$ can be decided in time $O\left(n^{k+4.5}\right)$.

For a strong candidate pattern H generated by Algorithm 1 (i.e., which satisfies both conditions in line 5), let $\mathcal{F}_H$ denote the set of patterns consisting of H

and all frequent patterns listed before H. For a transaction graph G and node z of $NTD(G)$, an iso-quadruple $\xi \in \Gamma(H, z)$ of a strong candidate pattern H is *redundant* if there are $P \in \mathcal{F}_H \setminus \{H\}$ and $\xi' \in \Gamma(P, z)$ with $\xi \equiv \xi'$; otherwise, ξ is *non-redundant*. Finally, $\Gamma_{\mathrm{nr}}(H, z)$ and $\Gamma_{\mathrm{nr,ch}}(H, z)$ denote the set of non-redundant iso-quadruples of H relative to a node z in $NTD(G)$ and the set of non-redundant z-characteristics of H, respectively.

Proposition 2 below implies that for a strong candidate pattern H and $\xi \in \Gamma(H, z)$, it has to be tested whether ξ is a z-characteristic of $NTD(G)$ only when ξ is non-redundant; otherwise, it suffices to check whether ξ is equivalent to a non-redundant z-characteristic for some frequent pattern $P \in \mathcal{F}_H \setminus \{H\}$ (the proof is similar to that of the corresponding claim in [5]).

Proposition 2. *Let H be a strong candidate pattern, G be a transaction graph, both of bounded tree-width, and let $\xi \in \Gamma(H, z)$ for some node z in $NTD(G)$. Then $\xi \in \Gamma_{\mathrm{ch}}(H, z)$ iff there exists $\xi' \in \bigcup_{P \in \mathcal{F}_H} \Gamma_{\mathrm{nr,ch}}(P, z)$ with $\xi \equiv \xi'$.*

Thus, induced subgraph isomorphism can be decided by using the non-redundant z-characteristics of the frequent patterns only. Instead of non-redundant iso-quadruples, as we will show below, we can use an efficiently computable superset of them, the set of *feasible* iso-quadruples. We first state a lemma that provides a necessary condition of non-redundancy.

Lemma 4. *Let H, G, and z be as in Proposition 2 and let $\xi \in \Gamma_{\mathrm{nr}}(H, z)$ with $\xi = (S, \mathcal{D}, K, \psi)$. Then, for all vertices $v \in V(H) \setminus V(K)$ it holds that*

(i) the degree of v in H is at least 2 and
(ii) v is a cut vertex in H.

Proof. The proof of (i) applies a similar argument used for ordinary subgraph isomorphism [5]. In particular, suppose for contradiction that $V(H) \setminus V(K)$ has a vertex v with degree 1 in H. Since, by assumption, H contains at least one edge and is connected, it has no isolated vertices. Let H' be the graph obtained from H by removing v and the (only) edge adjacent to it. Clearly, H' is a connected induced subgraph of H. Since H is a strong candidate pattern, H' is a frequent connected induced subgraph and has therefore already been generated by Algorithm 1. Furthermore, K is an induced subgraph of H' implying $\xi \in \Gamma(H', z)$. But ξ is then redundant for H, contradicting the assumption.

To prove (ii), suppose there is a non-cut vertex $v \in V(H) \setminus V(K)$ of H. Let $H' = H \ominus v$. Since H is connected and v is a non-cut vertex of H, H' is connected. Similarly to the case above, it holds that H' contains K as an induced subgraph because all edges that have been removed are outside of $E(K)$. Thus, $\Gamma(H', z)$ has an element equivalent to ξ, contradicting that ξ is non-redundant. $\square$

We now show that for any $S \subseteq V(H)$ of constant size, only a constant number of connected components in $H[V(H) \setminus S]$ can fulfill the two conditions of Lemma 4. Although the statement formulated below is similar to the corresponding claim stated for the case of ordinary subgraph isomorphism in [5], the arguments used in the proofs are entirely different, due to the difference between ordinary and induced subgraph isomorphism.

Lemma 5. *Let H be a strong candidate pattern generated by Algorithm 1, $S \subseteq V(H)$ with $|S| \leq k + 1$, and $\mathcal{C}_A$ be the set of connected components C from $\mathcal{C}(H[V(H) \setminus S])$ such that for all $v \in C$, v satisfies both conditions of Lemma 4. Then*

$$|\mathcal{C}_A| \leq \binom{k + 1}{2} .$$

To show the claim above, we first prove two technical lemmas.

Lemma 6. *Let H, S, and $\mathcal{C}_A$ be as defined in Lemma 5. Then for all $C \in \mathcal{C}_A$, C is connected to S by at least two edges ending in different vertices in S.*

Proof. The claim is straightforward if $|V(C)| = 1$; H has no parallel edges by construction and the only vertex of C for this case must be connected to S by at least two edges, as it is a cut vertex in H.

The proof of the case $|V(C)| > 1$ utilizes the fact that every connected graph has at least two non-cut vertices. More precisely, let u be a non-cut vertex of C. Since u is a cut vertex in H by condition (i) of Lemma 4, it must be the case that u is connected to at least one vertex in S. Thus, C is connected to S by at least two edges, as it has at least two non-cut vertices. Suppose that all non-cut vertices of C are adjacent to the same vertex in S, say w. Let $u, v \in V(C)$ be different non-cut vertices of C. Since, on the one hand, u is a non-cut vertex of C, and, on the other hand, it is a cut vertex in H by the condition of the lemma, there are two vertices $x, y \in V(H)$ that are disconnected by u (i.e., x and y belong to different connected components of $H \ominus u$). Since u is a non-cut vertex of C, at most one of x and y can belong to C. It can easily be seen for this case that in fact, *exactly* one of x and y, say x, belongs to C. Furthermore, $\{u, w\}$ must be an edge on the path connecting x and y in H, i.e., x and y are connected by a path of the form $P_{x,u} + \{u, w\} + P_{w,y}$, where $P_{x,u}$ is a path in C. Since $C \ominus u$ remains connected, there is a path $P_{x,v}$ connecting x and v in $C \ominus u$. Thus, the path $P_{x,v} + \{v, w\} + P_{w,y}$ connects x and y in $H \ominus u$, contradicting that u disconnects x and y. Hence, all connected components in $\mathcal{C}_A$ are connected to at least two different vertices in S, as stated. □

The second lemma states that each connected component of $\mathcal{C}_A$ "connects" such two vertices of S that are not "connected" by any other component of $\mathcal{C}_A$.

Lemma 7. *Let H, S, and $\mathcal{C}_A$ be as defined in Lemma 5. Then for all connected components $C \in \mathcal{C}_A$, there exist $u', v' \in S$ such that*

(i) $u' \neq v'$ and $\{u, u'\}, \{v, v'\} \in E(H)$ for some $u, v \in V(C)$, and
(ii) for all $C' \in \mathcal{C}_A \setminus \{C\}$, at least one of u' and v' is not adjacent to C'.

Proof. By Lemma 6, for all $C \in \mathcal{C}_A$ there are $u', v' \in S$ satisfying (i). Thus, to show the claim above, suppose for contradiction that there exists a connected component $C \in \mathcal{C}_A$ with the following property: for all $u', v' \in S$ satisfying (i) for C, there is a C' such that *both* u' and v' are adjacent to C'. Let u be the only vertex of C if $|V(C)| = 1$; otherwise let u be a non-cut vertex of C. Since $C \in \mathcal{C}_A$, u is a cut vertex in H by condition and hence, there are $x, y \in V(H)$

such that u disconnects x and y in H. Depending on the number of vertices of C and on the membership of x and y in C, we distinguish the following cases by noting that the case $x, y \in V(C)$ cannot occur by the choice of u:

Case 1. Suppose $|V(C)| = 1$. Then x and y must be connected in H by a path of the form $P_{x,v} + \{v, u\} + \{u, w\} + P_{w,y}$ for some $v, w \in S$ with $v \neq w$, where the length of $P_{x,v}$ and $P_{w,y}$ can be zero. By assumption, v and w are adjacent to some $C' \in \mathcal{C}_A$ and thus, there is a path $P_{v,w}$ in H that does not contain u. But x and y are then connected in H by the path $P_{x,v} + P_{v,w} + P_{w,y}$, contradicting that u disconnects x and y.

Case 2. Suppose $x \in V(C)$. Then x cannot be a non-cut vertex of C, as in this case x must be adjacent to a vertex $x' \in S$, which, in turn, is not adjacent to u. It can be shown in a way similar to the proof of Case 1, that for this case there is a path in H connecting x and y that does not contain u; a contradiction. Thus, as C has at least two non-cut vertices if $|V(C)| > 1$, there is a non-cut vertex $v \in V(C)$ with $v \neq u$. Since u disconnects x and y in H, there is a path of the form $P_{x,u} + \{u, u'\} + P_{u',y}$, where $P_{x,u}$ is a path in C, $\{u, u'\}$ is an edge of H with $u' \in S$, and $P_{u',y}$ is a path connecting u' and y in H. Let $v' \in S$ be a vertex adjacent to v. Notice that $v' \neq u'$, as otherwise the path $P_{x,v} + \{v, u'\} + P_{u',y}$ connects x and y in $H \ominus u$, contradicting that u disconnects x and y; clearly, a path $P_{x,v}$ connecting x and v in $C \ominus u$ always exists, as u is a non-cut vertex of C. Thus, u', v' fulfill condition (i) and hence, u' and v' are connected by a path $P_{u',v'}$ via some connected component $C' \in \mathcal{C}_A$ by assumption. But then x and y are connected by the path $P_{x,v} + \{v, v'\} + P_{v',u'} + P_{u',y}$ in $H \ominus u$, a contradiction.

Case 3. The case of $x, y \notin V(C)$ can be shown in a way similar to the cases above.   □

The proof of Lemma 5 follows directly from Lemma 7 and from $|S| \leq k + 1$. Following the paradigm designed for the FCSM problem in [5], we define *feasible* iso-quadruples, a superset of non-redundant iso-quadruples, and formulate in Theorem 6 the main result of this section, which states that feasible iso-quadruples can be used correctly to decide induced subgraph isomorphism and that the number of feasible iso-quadruples of a strong candidate pattern is polynomial in the pattern's size. More precisely, for a strong candidate pattern H generated by Algorithm 1 and for a node z in $NTD(G)$ of a transaction graph of bounded tree-width, an iso-quadruple $\xi \in \Gamma(H, z)$ is called *feasible* if it satisfies the conditions of Lemma 4. The set of feasible iso-quadruples relative to z and the set of feasible z-characteristics are denoted by $\Gamma_{\mathrm{f}}(H, z)$ and $\Gamma_{\mathrm{f,ch}}(H, z)$, respectively.

Theorem 6. *Let H be a strong candidate pattern generated by Algorithm 1 and z be a node of $NTD(G)$ for some transaction graph G with $\mathrm{tw}(G) \leq k$. Then*

(i) $\Gamma_{\mathrm{nr}}(H, z) \subseteq \Gamma_{\mathrm{f}}(H, z)$,

(ii) for all $\xi \in \Gamma(H, z)$, $\xi \in \Gamma_{\mathrm{ch}}(H, z)$ iff there exist a $\xi' \in \bigcup_{P \in \mathcal{F}_H} \Gamma_{\mathrm{f,ch}}(P, z)$ with $\xi \equiv \xi'$,

(iii) $|\Gamma_f(H, z)| = O\left(|V(H)|^{k+1}\right)$, *and*

(iv) $\Gamma_f(H, z)$ *can be computed in time polynomial in the size of* H.

Proof. The proof of (i) is immediate from the definitions and from Lemma 4. Claim (ii) follows from Proposition 2 and from the fact that $\bigcup_{P \in \mathcal{F}_H} \Gamma_{\mathrm{nr,ch}}(P, z)$ and $\bigcup_{P \in \mathcal{F}_H} \Gamma_{f,\mathrm{ch}}(P, z)$ are equal up to equivalence. To show (iii), let $S \subseteq V(H)$ with $|S| \le k + 1$ and $\mathcal{C}_A$ be the set of connected components as defined in Lemma 5. By definition, for every $\xi = (S, \mathcal{D}, K, \psi) \in \Gamma_f(H, z)$, $\mathcal{D}$ contains all connected components in $\mathcal{C}(H[V(H) \setminus S])$ that are not in $\mathcal{C}_A$. For a fixed subset $S \subseteq V(H)$ with $|S| \le k + 1$ and for a fixed injective function ψ mapping S to the bag X_z of z, the number of possible feasible quadruples is bounded by $2^{|\mathcal{C}_A|}$, which, in turn, is bounded by $2^{\binom{k+1}{2}}$ by Lemma 5. The number of induced subgraph isomorphisms from $H[S]$ to $G[X_z]$ is at most the number of injective functions from S to the bag X_z of z, which is bounded by $(k + 1)!$. Since S can be chosen in at most $|V(H)|^{k+1}$ different ways, we have

$$|\Gamma_f(H, z)| \le 2^{\binom{k+1}{2}} \cdot (k + 1)! \cdot |V(H)|^{k+1} \ ,$$

from which we get (iii) by noting that k is a constant. Finally, (iv) holds along the lines in the proof of (iii) above by noting that all cut vertices of H can be found in time $O(V(H) + E(H))$ and it can be decided whether an injective function $\psi : S \to X_z$ is an induced subgraph isomorphism from $H[S]$ to $G[X_z]$ in constant time, as $|S|, |X_z| \le k + 1$. □

4.3 Deciding Induced Subgraph Isomorphism

In this section we show how to utilize feasible characteristics efficiently for deciding induced subgraph isomorphism. Let H be a strong candidate pattern generated by Algorithm 1 and G be a transaction graph, both of tree-width at most k. Furthermore, let $NTD(G)$ be a nice tree-decomposition of G and r the root of $NTD(G)$. By Lemma 1 and Theorem 6, $H \preccurlyeq_i G$ iff there is a feasible r-characteristic $(S, \mathcal{D}, K, \psi) \in \Gamma_{f,\mathrm{ch}}(H, r)$ with $K = H$. The algorithm deciding $H \preccurlyeq_i G$ assumes that all nodes z in $NTD(G)$ is associated with a set containing all elements of $\Gamma_{f,\mathrm{ch}}(P, z)$, for all frequent patterns $P \in \mathcal{F}_H \setminus \{H\}$. It visits the nodes of $NTD(G)$ in postorder traversal and calculates first $\Gamma_f(P, z)$ for all nodes z visited; this can be done in time polynomial in the size of H by (iv) of Theorem 6. It then computes $\Gamma_{f,\mathrm{ch}}(P, z)$ by testing for all $\xi = (S, \mathcal{D}, K, \psi) \in \Gamma_f(P, z)$ whether ξ is a characteristic. Depending on the type of z, this test can be performed by checking the condition given in the corresponding case below:

LEAF: By case LEAF of Lemma 2, ξ is a characteristic iff $\mathcal{D} = \emptyset$.

SEPARATOR: Let z' be the child of z in $NTD(G)$ and let $\mathfrak{S}(\xi)$ be the set of all iso-quadruples $\xi' \in \Gamma(H, z')$ that satisfy conditions (S.a)–(S.c) of Lemma 2. Using similar arguments as in the proof of Lemma 16 in [5], one can show that (i) ξ is a characteristic iff $\Gamma_{f,\mathrm{ch}}(H, z') \cap \mathfrak{S}(\xi) \neq \emptyset$ and (ii) $\mathfrak{S}(\xi) \subseteq \Gamma_f(H, z')$ and thus, it can be computed in time polynomial in the size of H.

JOIN: Let z_1 and z_2 be the two children of z in $NTD(G)$. To give the condition for this case, we need a definition. Let $\xi_i = (S_i, \mathcal{D}_i, K_i, \psi_i) \in \Gamma(P_i, z_i)$ for some $P_i \in \mathcal{F}_H \setminus \{H\}$ $(i = 1, 2)$. We assume without loss of generality that K, K_1, and K_2 are pairwise vertex disjoint. The *join* of ξ_1 and ξ_2 with respect to ξ, denoted $\oplus_\xi(\xi_1, \xi_2)$, is an iso-quadruple $(S', \mathcal{D}_1 \cup \mathcal{D}_2, K', \psi')$ relative to z obtained from $(S_1 \cup S_2, \mathcal{D}_1 \cup \mathcal{D}_2, K_1 \cup K_2, \psi_1 \cup \psi_2)$ by (i) replacing u_1 and u_2 in $S_1 \cup S_2$, $K_1 \cup K_2$, and $\psi_1 \cup \psi_2$ with a new vertex u for all vertex pairs $u_1 \in S_1$ and $u_2 \in S_2$ with $\psi_1(u_1) = \psi_2(u_2)$ and by (ii) connecting in K' all original vertices $u, v \in S'$ with $u \in S_1$ and $v \in S_2$ by an edge labeled by ℓ if the vertices $u', v' \in S$ with $\psi(u') = \psi_1(u)$ and $\psi(v') = \psi_2(v)$ are connected in K by an edge labeled with ℓ. One can check that this definition is in fact an adaptation of conditions (J.a)–(J.c) of Lemma 2. In a way similar to the proof of Lemma 17 in [5], one can show that (i) ξ is a characteristic iff there are $\xi_i = (S_i, \mathcal{D}_i, K_i, \psi_i) \in \bigcup_{P \in \mathcal{F}_H} \Gamma_{\mathrm{f,ch}}(P, z_i)$ for $i = 1, 2$ with $\xi \equiv \oplus_\xi(\xi_1, \xi_2)$ and that (ii) $\oplus_\xi(\xi_1, \xi_2)$ can be computed in time polynomial in the size of ξ, ξ_1, and ξ_2 for any ξ_1, ξ_2, implying that it can be decided in time polynomial in the size of $\mathcal{F}_H$, i.e., in *incremental polynomial time*, whether ξ is a characteristic.

Combining the arguments above with Lemma 1, we get Theorem 7 below for condition (v) of Theorem 2. Together with Theorems 4 and 5, this completes the proof of our main result stated in Theorem 3.

Theorem 7. *Let $\mathcal{G}$ be the class of bounded tree-width graphs. For every $H, G \in \mathcal{G}$ such that H is a strong candidate pattern generated by Algorithm 1, it can be decided in time polynomial in the combined size of the input DB and the set of frequent patterns listed before H whether $H \preccurlyeq_i G$.*

5   Concluding Remarks

By the main result of this paper, the FCISM problem can be solved in incremental polynomial time for bounded tree-width graphs. The positive results on the FCSM problem in [5] and on the FCISM problem in this work suggest the investigation of further, computationally hard pattern matching operators for bounded tree-width graphs, such as, for example (induced) homeomorphism or (induced) minor embedding. We suspect that the systematic study of these and other pattern matching operators will result in an efficient *parameterized* frequent pattern mining algorithm for graphs of bounded tree-width, with the pattern matching operator as the parameter. Designing such a *generic* pattern mining algorithm is a very challenging project because, as the results in [5] and in this paper show, different pattern matching operators may require entirely different pattern refinement operators and entirely different combinatorial characterizations of feasible iso-quadruples.

The results of this paper raise some interesting open problems. For example, it is an open question whether the positive result formulated in Theorem 3 can further be strengthened. In particular, can the FCISM problem be solved with *polynomial delay* for bounded tree-width graphs? By setting the frequency threshold t to 1, our main result implies that one can efficiently generate all *distinct* connected induced subgraphs of a bounded tree-width graph. Does this positive result hold for arbitrary graphs as well? Or does the negative result given in Theorem 1 apply even to the special case that the database contains a single (arbitrary) graph and the frequency threshold is set to 1?

Finally we note that we are going to design and implement a practically fast algorithm listing frequent connected induced subgraphs for graphs of tree-width at most 3. For this graph class, motivated practically e.g. by pharmacological molecules (see the statistics with the ZINC dataset in Section 4), there are linear time recognition algorithms [1]. Though the arguments used for join nodes in Section 4.3 might suggest that we need time quadratic in the size of $\mathcal{F}_H$, one can show that this test can be carried out actually in time only *linear* in it.

Acknowledgments. Part of this work was supported by the German Science Foundation (DFG) under the reference number "GA 1615/1-1". Keisuke Otaki is supported by the "Scholarship for Japanese Graduate Students Learning Abroad" of the KDDI Foundation, Japan. This research was conducted during his stay at the University of Bonn and Fraunhofer IAIS in Sankt Augustin, Germany. Jan Ramon is supported by ERC Starting Grant 240186 "MiGraNT".

References

1. Arnborg, S., Proskurowski, A.: Characterization and recognition of partial 3-trees. SIAM Journal Algebraic Discrete Methods 7(2), 305–314 (1986)
2. Arnborg, S., Proskurowski, A.: Linear time algorithms for NP-hard problems on graphs embedded in k-trees. Discrete Applied Mathematics 23, 11–24 (1989)
3. Bodlaender, H.L.: A partial k-arboretum of graphs with bounded treewidth. Theoretical Computer Science 209(1-2), 1–45 (1998)
4. Hajiaghayi, M., Nishimura, N.: Subgraph isomorphism, log-bounded fragmentation, and graphs of (locally) bounded treewidth. Journal of Computer and System Sciences 73(5), 755–768 (2007)
5. Horváth, T., Ramon, J.: Efficient frequent connected subgraph mining in graphs of bounded tree-width. Theoretical Computer Science 411(31-33), 2784–2797 (2010)
6. Johnson, D.S., Papadimitriou, C.H., Yannakakis, M.: On generating all maximal independent sets. Information Processing Letters 27(3), 119–123 (1988)
7. Mannila, H., Toivonen, H.: Levelwise search and borders of theories in knowledge discovery. Data Mining and Knowledge Discovery 1(3), 241–258 (1997)
8. Matousek, J., Thomas, R.: On the complexity of finding iso- and other morphisms for partial k-trees. Discrete Mathematics 108(1-3), 343–364 (1992)
9. Nienhuys-Cheng, S.-H., de Wolf, R. (eds.): Foundations of Inductive Logic Programming. LNCS, vol. 1228. Springer, Heidelberg (1997)
10. Robertson, N., Seymour, P.D.: Graph minors. II. Algorithmic Aspects of Tree-Width. Journal of Algorithms 7(3), 309–322 (1986)
11. Syslo, M.M.: The subgraph isomorphism problem for outerplanar graphs. Theoretical Computer Science 17, 91–97 (1982)

Continuous Similarity Computation over Streaming Graphs

Elena Valari and Apostolos N. Papadopoulos

Data Engineering Lab., Department of Informatics, Aristotle University,
54124 Thessaloniki, Greece
{evalari,papadopo}@csd.auth.gr

Abstract. Large network analysis is a very important topic in data mining. A significant body of work in the area studies the problem of node similarity. One way to express node similarity is to associate with each node the set of 1-hop neighbors and compute the Jaccard similarity between these sets. This information can be used subsequently for more complex operations like link prediction, clustering or dense subgraph discovery. In this work, we study algorithms to monitor the result of a similarity join between nodes continuously, assuming a sliding window accommodating graph edges. Since the arrival of a new edge or the expiration of an existing one may change the similarity between several node pairs, the challenge is to maintain the similarity join result as efficiently as possible. Our theoretical study is validated by a thorough experimental evaluation, based on real-world as well as synthetically generated graphs, demonstrating the superiority of the proposed technique in comparison to baseline approaches.

Keywords: mining streaming graphs, continuous similarity processing.

1 Introduction

Graphs play an important role in modern world [1], due to their widespread use for modeling, representing and organizing linked data. Taking into consideration that most of the "killer" applications require a graph-based representation (e.g., the Web, social network management, protein interaction networks), efficient query processing and analysis techniques are required, not only because these graphs are massive but also because the operations that must be supported are complex, requiring significant computational resources.

A graph $G(V, E)$, in its simplest form, is composed of a node-set V, representing the entities (objects), and an edge-set E, representing the relationship among the entities. Each edge $e_{u,v} \in E$ connects a pair of nodes u, v, denoting that these nodes are directly related in a meaningful manner. For example, if nodes represent authors, then an edge between two authors may denote that they have collaborated in at least one paper. As another example, in a social network application (e.g., Facebook), an edge may denote that two users are connected by a friendship relationship.

H. Blockeel et al. (Eds.): ECML PKDD 2013, Part I, LNAI 8188, pp. 638–653, 2013.
© Springer-Verlag Berlin Heidelberg 2013

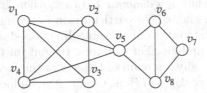

Fig. 1. Graph example

Motivation. A significant operation in a graph is the computation of the *similarity* between nodes. The similarity between nodes u and v may be expressed in several ways, depending on application or user requirements. For example, we may express similarity by means of shortest paths, maximum flow, random walks or a combination of measures. In general, similarity is expressed by a function $V \times V \to [0, 1]$, where a value close to 0 means low similarity and a value close to 1 denotes a high similarity between a node pair. In this work, we express similarity by means of the *Jaccard similarity coefficient*, which enjoys a widespread use in diverse areas such as link prediction and recommendation [15], data cleaning [3], near duplicate detection [19], diversity analysis [9], whereas it is one of the most important measures for set similarity. We associate with each node u the set of its immediate neighbors $N(u)$ (u inclusive). Then, the similarity between nodes u and v is computed as the fraction of their common neighborhood size over the cardinality of their neighborhood union, i.e.:

$$S_J(u, v) = \frac{|N(u) \cap N(v)|}{|N(u) \cup N(v)|} \tag{1}$$

Example 1. Figure 1 depicts a small graph where $|V| = 8$ and $|E| = 14$. Based on our similarity definition, it can be verified easily that: $S_J(v_1, v_2) = 5/5 = 1$, $S_J(v_2, v_6) = 1/8$, $S_J(v_6, v_8) = 4/4 = 1$ and $S_J(v_1, v_7) = 0$. We observe that node pairs that share the same set of immediate neighbors (e.g., v_1 and v_2) have a similarity of 1, whereas node pairs without common neighbors have a similarity of 0 (e.g., v_1 and v_7). ∎

An important operation which is based on pair-wise node similarities is the *similarity join*. More specifically, given a set of objects, a similarity function and a threshold ϑ, the similarity join operator reports all object pairs with a similarity at least ϑ. The output of this operator may be used subsequently for more complex mining tasks like clustering, dense subgraph discovery, association and link prediction. Regarding our setting, the similarity join result set R between graph nodes is defined as the set of node pairs $< u, v >$ such that the Jaccard similarity between their neighborhoods is at least ϑ. More formally:

$$R = \{(u, v) : u \in V, v \in V, S_J(u, v) \geq \vartheta\} \tag{2}$$

Our Contributions. Although similarity joins have been studied before (see for example [14]), to the best of our knowledge, there is a lack of research in maintaining the join result in a *dynamic network*, where insertions and deletions of nodes and edges are allowed. In particular, in many modern applications, sequential access to the data is the only feasible direction, due to huge data volumes or because of frequent updates. For example, the output of a network router, in its simplest form, is usually a stream of triplets of the form $< IP_1, IP_2, t >$, denoting that IP_1 sent a packet to IP_2 at time t. Any online processing performed on the router output must be based on sequential access, since the order of the output is completely random, whereas the frequency of the stream prevents the use of expensive data structures to organize the data on-the-fly. Based on this, we assume that the graph is available in the form of a *data stream* [17], where edges should be processed as they are presented to the algorithm.

More specifically, we study two different alternatives of the *streaming graph* model. In the *turnstile model*, the graph is presented as a sequence of edge insertions and edge deletions. For example, the sequence $+e_{u,v}, +e_{u,x}, +e_{x,y}, -e_{u,x}$ represents a streaming graph which is constructed by inserting edges $e_{u,v}$, $e_{u,x}$ and $e_{x,y}$ (we use a plus sign in front of an insertion) and deleting the edge $e_{u,x}$ (we use the minus sign in front of the edge). A special case of the turnstile model is the *sliding window model*, where the last w elements are maintained in a first-in first-out fashion. In this setting, the arrival of a new edge $e_{u,v}$ is followed by the expiration of an existing edge $e_{u',v'}$. In fact, the expired edge is the one with the oldest timestamp. Based on this model, at any given time, the *active* set of edges forms a subgraph of the streaming graph, representing the last m interactions among the graph nodes. This simple model may be generalized in several directions. For example, in some cases there is a whole set of newly arrived edges, meaning that an equal number of edges must expire. Another option is to have a *time-based sliding window*, where the window maintains the interactions that took place in the last h hours. To keep the presentation and the algorithms simple, we base our work on *count-based sliding windows*, where at any given time, exactly w edges are maintained in memory, whereas arrivals and expirations refer to single edges.

The main goal of this paper is to study efficient algorithms for continuous similarity monitoring of the nodes of an evolving graph, which is presented in the form of a stream of edges. In particular, our contributions are as follows:

- To the best of our knowledge, this is the first work that studies continuous similarity computation over streaming graphs using sliding windows. Taking into consideration that node similarity is the base for more complex tasks, the results of our study can be used for clustering or community discovery over streaming graphs.
- We propose efficient algorithms to maintain the similarity join result both when insertions and deletions of edges are arbitrary and when they follow a sliding window scenario, thus, enabling the use of our techniques in any dynamic network. The proposed algorithm uses effective pruning techniques to avoid the recomputation of Jaccard similarity wherever this is possible.

- We offer a thorough experimental evaluation based on large real-world as well as synthetically generated graphs, showing that the proposed algorithms offer significant performance improvement in comparison to baseline approaches.

Roadmap. The rest of the article is organized as follows. Section 2 presents some research contributions that are highly related to our work. Algorithms for continuous similarity computation over streaming graphs are given in Section 3. Performance evaluation results based on real-world as well as synthetic networks are offered in Section 4. Finally, Section 5 concludes our work.

2   Related Work

The issues studied in this paper, lie in the intersection between graph mining [1] and data streams [11,17]. Mining streaming graphs is challenging, mainly due to the data massiveness and also because of the inherent difficulty in solving complex graph problems in the streaming model of computation [8,20].

Node similarity in graphs plays an important role in graph mining because it is often the base for supporting more complex operations such as clustering and community detection [10]. To express the similarity between graph nodes, a meaningful similarity measure is required. One such measure is the Jaccard similarity, which has been applied successfully in areas such as duplicate detection [6,19], link prediction [15], similarity evaluation in wikipedia [4], triangle counting in massive graphs [5] and diversity analysis in documents [9].

Based on the importance of the Jaccard similarity, in this work we focus on the application of this measure to detect node pairs of a dynamic network, with a high degree of similarity. In particular, network dynamics are controlled by a sliding window of a fixed size w, which maintains the most recent edges of the streaming graph. Our work is inspired by previous research approaches to process complex queries over sliding window data streams. The work in [16] studies the problem of top-k query processing over a multidimensional data stream for any monotone ranking function. In a similar manner, [13] proposes efficient algorithms for top-k dominating queries whereas [12] focuses on outlier mining over general metric streams. Those works focus on multidimensional or metric streams.

Although there is a significant body of work dealing with processing over streaming graphs [8,20,2], none of the existing works handles similarity computation over a streaming graph using sliding windows. A research topic that is closely related to similarity computation is *triangle counting*. Algorithms for counting triangles in streaming graphs have been reported in [5] where the semi-streaming model is used, in [7] where sampling is used. Those works aim at reducing the space requirements and thus the solutions they provide are approximate. Moreover, since those techniques are based on either minhashing or sampling, they cannot support deletions efficiently.

An important challenge is that apart from the fact that, in contrast to relational join processing, the insertion/deletion of an edge affects the similarity of other node pairs in the graph, the result set is composed of two types of node pairs: i) node pairs joined by an edge and ii) non-adjacent node pairs. This feature is unique in graphs and requires attention because edge pruning cannot be performed easily.

3   Continuous Similarity Computation

3.1   Preliminaries

In this section, we present some fundamental concepts related to continuous Jaccard similarity computation in a streaming environment. Formally, the problem we attack is the following:

PROBLEM DEFINITION. *Given a streaming graph G and a count-based sliding window of size w, monitor all node pairs v_i, v_j such that $S_J(v_i, v_j) \geq \vartheta$, where $\vartheta \in [0, 1]$ is a user-defined similarity threshold.*

To facilitate efficient processing, the graph is organized by an adjacency list representation, where each node points to its immediate neighbors. Since in a streaming environment insertions and deletions of edges are very frequent, node information is stored in a hashmap for fast lookups. This allows us to locate each node in $O(1)$ expected time. Likewise, the result set R which contains the node pairs having similarity larger than the threshold ϑ, is also organized by a hashtable. This way, checking if a node pair is in the result set involves a lookup in the hashtable using as key a combination of the node identifiers. The indexing schemes used by our techniques are shown in Figure 2. Notice that, R may contain node pairs that either are not joined by an edge or are direct neighbors. This means that some node pairs in R correspond to adjacency list entries and some do not. For example, the entries of R shown shaded in Figure 2 correspond to disjoint node pairs, whereas the rest correspond to node pairs connected by an edge of G.

An important issue in the data organization is the way the set of neighbors $N(v)$ of a node v is arranged, since this has a direct impact on the efficiency of the Jaccard similarity computation. To provide the best possible solution we have to take into account that: i) insertions and deletions in $N(v)$ must be handled efficiently and ii) the Jaccard similarity computation between two nodes must be also computed efficiently. We distinguish among the following cases, assuming that currently, $|N(v)| = k$:

Unordered List (UL). The set of neighbors $N(v)$ is organized as a simple unordered list. This offers $O(1)$ worst case time for inserting a new neighbor in $N(v)$, but requires linear cost to find or delete a neighbor. For Jaccard computation, if both nodes have k neighbors, then in $O(k)$ expected time we can compute the intersection and the union of the neighborhoods using hashing.

Ordered List (OL). If both lists are ordered, then the computation of the intersection and the union can be completed in $O(k)$ worst case. Likewise, insertions and deletions also require linear cost in the worst case.

Binary Search Tree (BST). With a BST, insertions and deletions are handled in $O(\log n)$ worst case, whereas intersections and union operations are executed in linear $O(k)$ time worst case.

Hash Table (HT). In this scheme, instead of having an unordered list, the set of neighbors is organized in a hashtable. This provides $O(1)$ expected cost for insertions and deletions, and also $O(k)$ expected cost for computing the intersection and the union.

Based on the previous discussion, HT is the most promising technique for Jaccard similarity computation, and this is also validated by the experimental results we report in Section 4.

3.2   Algorithmics

In this section, we study algorithmic techniques toward continuous Jaccard similarity computation over a streaming graph. Initially, we provide a baseline approach to solve the problem, followed by an efficient algorithm that can handle insertions and deletions of edges. Finally, we propose a more sophisticated algorithm which is more appropriate for the sliding-window case.

Definition 1. *The set of affected pairs of an edge $e_{u,v}$, denoted as $SAP(e_{u,v})$ or simply $SAP(u,v)$, is the set of node pairs whose Jaccard similarity is affected by the arrival or the expiration of the edge $e_{u,v}$.*

Based on the previous definition, the similarity of a node pair contained in $SAP(e_{u,v})$ may be increased or decreased, according to the structure of the graph. The following lemma explains which pairs are contained in $SAP(e_{u,v})$ and how their similarity is affected.

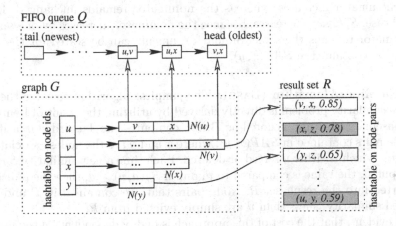

Fig. 2. Indexing techniques employed

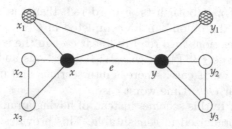

Fig. 3. Example graph used in the proof of Lemma 1

Lemma 1. *Let $e_{u,v}$ be an inserted/deleted edge and u, v the associated nodes. The similarities that are affected by this insertion/deletion are those defined by: i) the pair (u,v), ii) u and all neighbors of u, iii) u and all neighbors of v, iv) v and all neighbors of v and v) v and all neighbors of u.*

Proof. We focus on the case where the edge e is inserted, because the deletion is handled symmetrically. Therefore, let e be a new edge joining the nodes x and y as it is indicated in Figure 3. Based on the definition of the Jaccard similarity, $SAP(x, y)$ contains only the pairs mentioned above, since by using contradiction, it is impossible that the similarity of a node pair not belonging to one of these cases will change due to the insertion of e. Next we show how the similarities of the node pairs contained in $SAP(x, y)$ are modified.

The similarity between x and y is increased, since now y becomes a direct neighbor of x and x becomes a direct neighbor of y. Consequently, the set $N(x) \cap N(y)$ gets two new members, x and y, resulting in an increase of the value of $S_J(x, y)$. Next, we examine what is the impact of inserting e to the similarity between x and each of its direct neighbors, denoted as x_i. There are two cases to examine here: in the first case, y is not a neighbor of x_i (this is the case for x_2 and x_3), whereas in the second case, y is a neighbor of x_i (e.g., when x_i is x_1). In the first case, the value of $S_J(x, x_i)$ decreases, because only the denominator increases, whereas the nominator remains unchanged. In the second case, $S_J(x, x_i)$ increases, because the nominator increases by one and the denominator remains the same. Similar arguments can be stated for the other node pairs contained in $SAP(x, y)$.   □

The Baseline Algorithm (BASE). The simplest algorithm to solve the continuous similarity problem is directly derived by utilizing the result of Lemma 1. This baseline algorithm, denoted as BASE, computes the Jaccard similarity for all node pairs contained in $SAP(x, y)$, where x and y are the nodes associated to an edge e which is either inserted or deleted. Each time a new similarity $S_J(u, v)$ is computed, the value is compared to ϑ, and if $S_J(u, v) \geq \vartheta$, then the pair (u, v) is inserted into the result set R. Node pairs that are contained in R and their updated similarity is less than ϑ are simply evicted from R.

It is evident, that the cost of this approach is highly dependent on the number of Jaccard similarity computations executed. To reduce this number, we first

Algorithm 1. BASE

Input: G: the graph, e: the new or expiring edge between x and y, R: result set
Output: updated result set R

```
 1: determine the set SAP(x, y);
 2: for each node pair (u, v) ∈ SAP(x, y)
 3:     compute UB_J(u, v)
 4:     if (UB_J(u, v) < ϑ)
 5:         if ((u, v) ∈ R)  R ← R − {(u, v)};
 6:     else
 7:         recompute S_J(u, v);
 8:         if (S_J(u, v) ≥ ϑ)
 9:             if ((u, v) ∉ R)  R ← R + {(u, v)}; /* insert (u, v) into R */
10:         else
11:             if ((u, v) ∈ R)  R ← R − {(u, v)}; /* remove (u, v) from R */
12: return;
```

enforce an upper bound, and if the node pair still survives the test, only then the Jaccard similarity is computed. In particular, given two nodes u and v, their Jaccard similarity satisfies the following inequality:

$$S_J(u, v) \leq UB_J(u, v) = \frac{min(|N(u)|, |N(v)|)}{max(|N(u)|, |N(v)|)} \tag{3}$$

The outline of BASE is given in Algorithm 1. The upper bound test is performed at Line 4, and the Jaccard computation is executed at Line 7. Although the use of the upper bound reduces the number of Jaccard similarity computations, more sophisticated techniques are required to decrease the cost further.

The Counter-Based Algorithm (COUNTER). The main drawback of BASE is that there is a significant number of Jaccard similarity computations, leading to performance deterioration. To overcome this limitation, the next algorithm (COUNTER) is based on keeping separate counters for the cardinality of the intersection (nominator) and the cardinality of the union (denominator), thus reducing the cost of computing Jaccard similarities significantly.

The key idea of the COUNTER algorithm is that when a new edge e joining x and y is inserted, we compute the value $S_J(x, y)$ and we maintain two separate counters $C_\cap$ and $C_\cup$ for the cardinality of the intersection and the union of the neighborhoods respectively, i.e. $C_\cap(x, y) = |N(x) \cap N(y)|$ and $C_\cup(x, y) = |N(x) \cup N(y)|$. Thus, whenever there is a need to recompute the value of $S_J(x, y)$, we need only adjust the values of $C_\cap(x, y)$ and $C_\cup(x, y)$ and just perform the division $C_\cap(x, y)/C_\cup(x, y)$. In addition, intersection and union counters are maintained for node pairs that are not connected by an edge but are included in the result set R. Subsequent recomputations of the Jaccard similarity are executed fast, avoiding unnecessary set-oriented operations among the neighborhoods. In the

---

Algorithm 2. COUNTER

Input: G: the graph, e: the new edge between x and y, R: result set
Output: updated result set R

```
 1: determine the set SAP(x, y);
 2: for each node pair (u, v) ∈ SAP(x, y)
 3:     if ((u, v) ∈ R)
 4:         update counters C∩(u, v) and C∪(u, v) for (u, v);
 5:         if (C∩(u, v)/C∪(u, v) < ϑ)
 6:             R ← R − {(u, v)};
 7:     else if ((u, v) ∈ E) /* edge (u, v) exists */
 8:         update counters C∩(u, v) and C∪(u, v) for (u, v);
 9:         if (C∩(u, v)/C∪(u, v) ≥ ϑ)
10:             R ← R + {(u, v)};
11:     else
12:         compute UBJ(u, v)
13:         if (UBJ(u, v) ≥ ϑ)
14:             compute SJ(u, v);
15:             if (SJ(u, v) ≥ ϑ)   R ← R + {(u, v)};
16: return;
```

---

sequel, we examine only the insertion case, since deletions are handled in a similar manner.

Let e be a new edge that is inserted in G, linking nodes x and y. COUNTER first checks if the pair (x, y) is already in the result set R. If yes, then definitely there exist counters for the intersection and the union that have been set previously. Therefore, the new value of $S_J(x, y)$ is computed easily. The outline of COUNTER is given in Algorithm 2. Notice that, before the computation of the Jaccard similarity in Line 12, the algorithm first checks if the node pair is in R or the corresponding edge exists in E. Then, the intersection and union counters are updated based on the cases reported in Lemma 1. To avoid confusion, we use the term *set-based Jaccard computation* to refer to the Jaccard computation when there are no precomputed counters, and use the term *counter-based Jaccard computation* otherwise.

The Slide-Oriented Algorithm (SLIDE). Although COUNTER is more efficient than BASE, it is designed to support insertion and deletion of arbitrary edges. However, our goal is to support continuous evaluation over a sliding window of size w. In this case, we know exactly the expiration time of an edge, since edges arrive and depart in a FIFO fashion. This means that additional optimizations can be applied toward the design of an algorithm which is more appropriate for the sliding-window case.

In this section, we provide the details of the SLIDE algorithm, which has been designed for the sliding-window scenario. The key idea of SLIDE is that if we could determine the time instance when a node pair (u, v) will enter the result set R, then we could decide if (u,v) is promising or not. It turns out that such a prediction is possible, resulting in an effective mechanism to determine node pairs that can be eliminated safely. More specifically, when a new edge is inserted, we make an optimistic prediction, determining the closer time instance that the nodes associated with the edge can be included in the result set R. The prediction is optimistic, in the sense that the estimated time instance is computed assuming the best possible scenario for this edge. In addition, as we show in the sequel, this estimation produces only false positives and never false dismissals.

Lemma 2. *Let e be a newly arrived edge joining nodes x and y. Let $t^*(x,y)$ denote the closer time instance into the future in order for (x,y) to enter the result set R. Then, it holds that:*

$$t^*(x,y) = t_{now} + \min\{C_\cap^*(e) - C_\cap(e), C_\cup(e) - C_\cup^*(e)\} \tag{4}$$

where t_{now} is the current time, $C_\cap(e)$ and $C_\cup(e)$ are the values of intersection and union counters computed upon examination of e, and $C_\cap^*(e)$ and $C_\cup^*(e)$ are the values of intersection and union counters when e is expected to be inserted into R.

Proof. Assume that e joins the nodes x and y and it is checked at the current time t_{now}. Let also $C_\cap(e)$ and $C_\cup(e)$ denote the values of the intersection and union counters for e at time t_{now}. We assume that the node pair (x,y) will enter the result set R at some time in the future, and let $t^*(x,y)$ denote this particular time instance. We are looking for the smallest possible value of $t^*(x,y)$. If $C_\cap^*(e)$ and $C_\cup^*(e)$ are the values of the intersection and union counters when (x,y) enters R, then clearly we have that: $C_\cap^*(e)/C_\cup^*(e) \geq \vartheta$. Based on the previous discussion, every time $S_J(x,y)$ is affected, exactly one of the following five cases is true: i) only $C_\cap(e)$ increases, ii) only $C_\cup(e)$ increases, iii) only $C_\cap(e)$ decreases, iv) only $C_\cup(e)$ decreases, iv) both $C_\cap(e)$ and $C_\cup(e)$ increase or v) both $C_\cap(e)$ and $C_\cup(e)$ decrease. Among the previous cases, the ones that may lead faster to the inclusion of (x,y) into R are the first two. Indeed, the fraction $C_\cap(e)/C_\cup(e)$ increases faster if either the nominator increases (keeping the denominator fixed) or the denominator decreases (keeping the nominator fixed). Note, that these two events cannot happen at the same time. Consequently, to gain the additional $\Delta\vartheta$ similarity score required to enter R, it suffixes to wait for $\min(C_\cap^*(e) - C_\cap(e), C_\cup(e) - C_\cup^*(e))$ time instances at best, assuming the most optimistic scenario. □

Lemma 3. *If for an edge e it holds that $t_{exp}(e) < t^*(e)$, then it is safe to skip the Jaccard similarity computation for this edge.*

Proof. Recall that $t^*(e)$ is the closest time instance when e will enter R, considering the most favorable scenario for e, i.e., by increasing the nominator and decreasing the denominator as much as possible. Consequently, if the expiration time of e is less than $t^*(e)$, then it is impossible for e to enter the result set R. Therefore, the Jaccard computation between the nodes joined by e may be skipped safely. □

Algorithm 3. SLIDE

Input: G: the graph, e: the new edge between x and y, R: result set
Output: updated result set R

```
 1: determine the set SAP(x, y);
 2: for each node pair (u, v) ∈ SAP(x, y)
 3:    if ((u, v) ∈ R)
 4:        update counters C∩(u, v) and C∪(u, v) for (u, v);
 5:        if (C∩(u, v)/C∪(u, v) < ϑ)
 6:            R ← R − {(u', v')};
 7:    else if ((u, v) ∈ E) /* edge (u, v) exists */
 8:        if (texp(u, v) < t*(u, v))
 9:            reject (u, v) from further consideration;
10:        else
11:            update counters C∩(u, v) and C∪(u, v) for (u, v);
12:            if (C∩(u, v)/C∪(u, v) ≥ ϑ)
13:                R ← R + {(u, v)};
14:    else
15:        compute UBJ(u, v)
16:        if (UBJ(u, v) ≥ ϑ)
17:            compute SJ(u, v);
18:            if (SJ(u, v) ≥ ϑ)   R ← R + {(u, v)};
19: return;
```

If an edge e joining nodes x and y satisfies the inequality of Lemma 3, then there is no need to test the pair (x, y) again, and consequently there is no need to maintain intersection and union counters, since it is guaranteed that e will never enter R for the rest of its lifespan. The outline of SLIDE is given in Algorithm 3. The expiration time pruning is applied in Lines 8 and 9.

4   Performance Evaluation

In this section, we report some representative performance results showing the efficiency and scalability of the proposed approach. All algorithms have been implemented in JAVA and the experiments have been conducted on an Intel Core i5@2.7GHz machine. We study the performance of the algorithms in terms of their runtime and their pruning capabilities, by varying the most important parameters, such as the window size (w) and the value of the similarity threshold (ϑ). The default values for the parameters, if not otherwise specified, are: $w = 1,000,000$ and $\vartheta = 0.8$. The computational cost of the algorithms is given in terms of the expected time required by an update (an insertion followed by a deletion). This value determines the processing capabilities of the algorithms, since it is inversely proportional to the number of updates that can be served per time unit, which is an important measure in applications managing data streams.

4.1   Data Description

To study the performance of the algorithms we have used both real-world and synthetic data sets. The real-world data sets are described briefly in Table 1 and are freely available for download at http://snap.stanford.edu/data/index.html.

Table 1. Real-world data sets (*source http://snap.stanford.edu/data/index.html*)

Data	#Nodes	#Edges	Description
Wiki-Talk	2,394,385	5,021,410	pages editing between wikipedia users
Web-BerkStan	685,230	7,600,595	web from berkeley.edu and stanford.edu
Soc-LiveJournal1	4,847,571	68,993,773	users' connections in LiveJournal social network

The synthetic graphs have been generated by using the GenGraph tool [18]. This generator produces graphs obeying power-law degree distributions. In particular, GenGraph generates a set of n integers in the interval $[d_{min}, d_{max}]$ obeying a power-law distribution with exponent a. Therefore, according to the degree distribution produced, a random power-law graph is generated. The default values for the parameters of the generator are: $d_{min} = 0.1\%$ of the number of vertices, $d_{max} = 0.8\%$ of the number of vertices and $a \in \{1.8, 2, 2.2, 2.5\}$. The maximum number of vertices has been set to 10,000.

4.2   Experimental Results for Real-Life Data

The first result involves the way Jaccard computations are computed, which is highly related to the way adjacency lists are maintained, as it has been described in Section 3.1. In particular, Figure 4 depicts the performance of the three studied algorithms for the Soc-LiveJournal1 data set. As expected, the HT organization, which relies on hashing, shows the best performance. Therefore, we apply the HT technique in the performance evaluation discussed in the sequel.

Figures 5 and 6 demonstrate the scalability of the algorithms by varying the windows size w. Figure 5 shows the runtime per update. All algorithms are affected negatively when the number of active edges increases. However, we observe

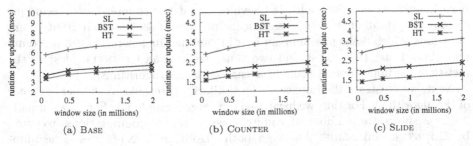

(a) BASE   (b) COUNTER   (c) SLIDE

Fig. 4. Comparison of adjacency list maintenance using Soc-LiveJournal1

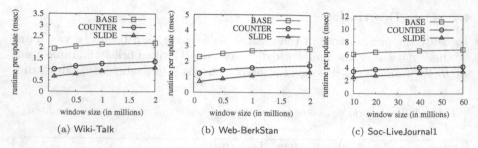

Fig. 5. Runtime vs window size

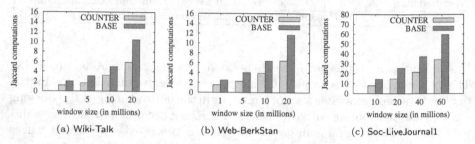

Fig. 6. Jaccard similarity computations vs window size

Table 2. Number of counter-based Jaccard computations (Soc-LiveJournal1)

threshold ϑ	result set R	SLIDE executed	saved	COUNTER executed
0.1	294,117	308,011,791	8,913,518	316,925,309
0.3	179,213	321,129,674	16,828,460	337,958,134
0.5	98,173	333,849,425	27,991,030	361,840,455
0.7	63,171	341,328,740	36,990,891	378,319,631
0.9	304	350,187,993	45,882,785	396,070,778

that COUNTER and SLIDE are consistently more efficient than BASE. This is explained by studying the number of Jaccard computations performed by each algorithm. Figure 6 compares BASE and COUNTER with respect to the number of similarity computations. It is evident, that the counter-based technique employed by COUNTER saves a significant number of set-based similarity computations, which is the predominant cost in runtime. In general, SLIDE is around four times faster than BASE and two times faster than COUNTER, despite the fact that the upper bound pruning is enabled for all algorithms.

Next, we illustrate the impact of the similarity threshold to the performance of the algorithms. For this, we have used our largest graph, i.e., Soc-LiveJournal1. Table 2 shows the number of counter-based Jaccard computations executed by COUNTER and SLIDE. Although both algorithms execute the same number of set-based Jaccard computations, SLIDE manages to reduce the number of

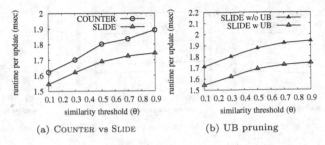

(a) COUNTER vs SLIDE

(b) UB pruning

Fig. 7. Performance vs threshold (Soc-LiveJournal1)

counter-based Jaccard computations due to the expiration time pruning technique employed. This leads to a significant performance improvement.

Figure 7(a) shows the runtime comparison between COUNTER and SLIDE algorithms. We observe that the performance gap between the algorithms increases by increasing the similarity threshold. Note that, as it is also shown in Table 2, the larger the similarity threshold the fewer node pairs manage to enter the result set. This means that we are going to have less precomputed information in R and therefore SLIDE benefits more by this situation since it can skip more counter-based Jaccard computations.

Finally, in Figure 7(b) we report on the pruning power of the upper bound given in Equation 3, when applied to the SLIDE algorithm. We clearly see that there is a performance gain ranging between 12% and 20%, which is very important, since the runtime per update defines the throughput (edges per time unit) that can be processed by the algorithm.

4.3 Experimental Results for Synthetic Data

In the sequel, we report some evaluation results showing the efficiency of the proposed approach over synthetic streaming graphs. These graphs in which the experiments were performed are denser than the real-life graphs explored previously. In particular, as the value of parameter a (power-law exponent) decreases, the graph generated by GenGraph [18] contains more nodes with large degree, resulting in a graph with larger density. This means that the density of a graph with $a = 1.8$ is larger than that of a graph with $a = 2.2$.

Figure 8 shows the performance of the algorithms, for different values of the window size w and the power-law exponent a. Again, as in the case of real-world data, we observe that SLIDE shows the best performance in terms of runtime (Figure 8(a)) and this is also true for different values of the power-law exponent (Figure 8(b)). The small performance difference of all algorithms when the power-law exponent increases, is due to the impact of a on the graph density, because the cardinality of the SAP of a node pair is highly dependent on the density of the graph. The number of set-based Jaccard computations are given in Figure 8(c). Again, the precomputed counters save a significant number of set-based Jaccard computations, resulting in performance improvement.

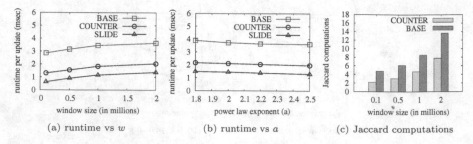

(a) runtime vs w (b) runtime vs a (c) Jaccard computations

Fig. 8. Performance for synthetic data sets

5 Conclusions

Node similarity in graphs is an important operation, because it allows the execution of more complex analysis tasks such as clustering and community discovery. In this work, we have studied algorithms for continuous evaluation of pair-wise similarities, where the graph is accessed as a random sequence of edges in a sliding window scenario. More specifically, given a similarity threshold ϑ, we are interested in determining all node pairs with Jaccard similarity at least ϑ. This problem arises frequently in data streams, and especially in streaming graphs, where a sliding window retains the last w entity interactions.

Three algorithms have been studied and evaluated, namely BASE, which is the baseline approach, COUNTER an algorithm that supports insertion and deletion of any edge and it is based on precomputed counters and finally SLIDE which is designed for a streaming scenario and uses a pruning technique to ignore node pairs that will never make it to the result set. Experimental results based on real-world and synthetic data sets have demonstrated that SLIDE is consistently more efficient that the other algorithms.

There are several interesting directions for future work, such as: i) the design of algorithm for top-k most similar pairs, ii) the generalization of our techniques to consider h-hop neighbors for similarity computation and iii) the use of sketch-based techniques to enable performance boost by penalizing the accuracy of the result. With respect to the last direction, graph-specific sketches, like the gSketch [21] or cascading summaries [8], may be applied to allow for low-space similarity computation.

References

1. Aggarwal, C., Wang, H.: Managing and Mining Graph Data. Springer (2010)
2. Aggarwal, C., Zhao, Y., Yu, P.S.: On Clustering Graph Streams. In: Proceedings of SIAM SDM, pp. 478–489 (2010)
3. Arasu, A., Ganti, V., Kaushik, R.: Efficient Exact Set-Similarity Joins. In: Proceedings of VLDB, pp. 918–929 (2006)
4. Bank, J., Cole, B.: Calculating the Jaccard Similarity Coefficient with Map Reduce for Entity Pairs in Wikipedia (2008),
   http://weblab.infosci.cornell.edu/weblab/papers/Bank2008.pdf

5. Becchetti, L., Boldi, P., Castillo, C., Gionis, A.: Efficient Semi-Streaming Algorithms for Local Triangle Counting in Massive Graphs. In: Proceedings of ACM
SIGKDD, pp. 16–24 (2008)
6. Broder, A.Z., Glassman, S.C., Manasse, M.S., Zweig, G.: Syntactic clustering of
the web. In: Proceedings of WWW, pp. 1157–1166 (1997)
7. Buriol, L.S., Frahling, G., Leonardi, S., Marchetti-Spaccamela, A., Sohler, C.:
Counting triangles in data streams. In: Proceedings of ACM PODS, pp. 253–262
(2006)
8. Cormode, G., Muthukrishnan, S.: Space Efficient Mining of Multigraph Streams.
In: Proceedings of ACM PODS, pp. 271–282 (2005)
9. Deng, F., Siersdorfer, S., Zerr, S.: Efficient Jaccard-based Diversity Analysis of
Large Document Collections. In: Proceedings of ACM CIKM, pp. 1402–1411 (2012)
10. Fortunato, S.: Community Detection in Graphs, arXiv:0906.0612v2 [physics.soc-
ph] (2009)
11. Garofalakis, M., Gehrke, J., Rastogi, R.: Querying and Mining Data Streams: You
Only Get One Look. In: Proceedings of ACM SIGMOD, Tutorial (2002)
12. Kontaki, M., Gounaris, A., Papadopoulos, A.N., Tsichlas, K., Manolopoulos, Y.:
Continuous Monitoring of Distance-Based Outliers over Data Streams. In: Proceedings of IEEE ICDE, pp. 135–146 (2011)
13. Kontaki, M., Papadopoulos, A.N., Manolopoulos, Y.: Continuous Top-k Dominating Queries. IEEE Transactions on Knowledge and Data Engineering 24(5),
840–853 (2012)
14. Lian, X., Chen, L.: Efficient Join Processing on Uncertain Data Streams. In: Proceedings of ACM CIKM, pp. 857–866 (2009)
15. Liben-Nowell, D., Kleinberg, J.: The Link Prediction Problem for Social Networks.
In: Proceedings of ACM CIKM, pp. 556–559 (2003)
16. Mouratidis, K., Bakiras, S., Papadias, D.: Continuous Monitoring of Top-k Queries
over Sliding Windows. In: Proceedings of ACM SIGMOD, pp. 635–646 (2006)
17. Muthukrishnan, S.: Data Streams: Algorithms and Applications. Foundations and
Trends in Theoretical Computer Science 1(2), 117–236 (2005)
18. Viger, F., Latapy, M.: Efficient and Simple Generation of Random Simple Connected Graphs with Prescribed Degree Sequence. In: Wang, L. (ed.) COCOON
2005. LNCS, vol. 3595, pp. 440–449. Springer, Heidelberg (2005)
19. Xiao, C., Wang, W., Lin, X., Yu, J.X., Wang, G.: Efficient Similarity Joins for
Near Duplicate Detection. ACM Transactions on Database Systems 15, 15:1–15:41
(2011)
20. Zelke, M.: Algorithms for Streaming Graphs. PhD Dissertation, Humboldt University of Berlin (2009)
21. Zhao, P., Aggarwal, C.C., Wang, M.: gSketch: On Query Estimation in Graph
Streams. In: Proceedings of VLDB, pp. 193–204 (2011)

Trend Mining in Dynamic Attributed Graphs

Elise Desmier[1], Marc Plantevit[2], Céline Robardet[1],
and Jean-François Boulicaut[1]

[1] Université de Lyon, CNRS, INSA-Lyon, LIRIS UMR5205,
F-69621 Villeurbanne, France
elise.desmier@liris.cnrs.fr,
{celine.robardet,Jean-Francois.Boulicaut}@insa-lyon.fr
[2] Université de Lyon, CNRS, Univ. Lyon1, LIRIS UMR5205,
F-69622 Villeurbanne, France
marc.plantevit@liris.cnrs.fr

Abstract. Many applications see huge demands of discovering important patterns in dynamic attributed graph. In this paper, we introduce the problem of discovering trend sub-graphs in dynamic attributed graphs. This new kind of pattern relies on the graph structure and the temporal evolution of the attribute values. Several interestingness measures are introduced to focus on the most relevant patterns with regard to the graph structure, the vertex attributes, and the time. We design an efficient algorithm that benefits from various constraint properties and provide an extensive empirical study from several real-world dynamic attributed graphs.

1 Introduction

Data mining techniques are now sufficiently mature to investigate complex data such as graph, whose vertices stand for entities and edges represent their relationships or interactions. With the rapid development of social media, sensor technologies and bioinformatic assay tools, real-world graph data has become ubiquitous and new dedicated data mining techniques have been developed. Whereas dynamic graphs [2,4,13,15] and attributed graphs [12,14,16] have been separately considered so-far, we focus on the extraction of valuable information from dynamic attributed graphs. The simultaneous consideration of the graph structure, the vertex attributes and their evolution through time makes possible to tackle a wide variety of mining problems. A timely challenge is to provide tools and methods to describe the evolution of the whole graph but also the specific evolution of some particular sub-graphs.

The second problem was recently tackled in [6], where an algorithm that mines cohesive co-evolution patterns is proposed. These patterns identify sets of vertices that are similar from the point of view of their attribute values and of the vertices in their neighborhood. However, as this method under-utilizes the topological structure of the vertex sets (i.e., only similarity measure are computed from two vertex adjacency lists), it tends to fragment some reliable patterns.

H. Blockeel et al. (Eds.): ECML PKDD 2013, Part I, LNAI 8188, pp. 654–669, 2013.

In this paper, we propose to mine maximal dynamic attributed sub-graphs that satisfy some constraints on the graph topology and on the attribute values. To be more robust towards intrinsic inter-individual variability, we do not compare raw numerical values, but their trends, that is, their derivative at time stamp t. The connectivity of the dynamic sub-graphs is constrained by a maximum diameter value that limits the length of the longest shortest path between two vertices. Additional interestingness measures are used to assess the interest of the trend dynamic sub-graphs and guide their search by user-parameterized constraints. These constraints aim at answering the following questions:

- How similar are the vertices outside the trend dynamic sub-graph to the ones inside it?
- Are trends specific to the vertices of the pattern?
- What about the dynamic of the pattern? Does it appear suddenly or continuously?

The algorithm designed to compute these patterns traverses the lattice of dynamic attributed sub-graphs in a depth-first manner. It prunes and propagates constraints that are fully or partially monotonic or anti-monotonic [5], and thus takes advantage of a large variety of constraints that are usually not exploited by standard lattice-based approaches. To summarize, the main contributions of this paper are:

- The introduction of a novel problem: the discovery of trend dynamic sub-graphs in dynamic attributed graph. We define the trend dynamic sub-graph as a suitable mathematical notion for the study of dynamic attributed graphs and introduce the notions of vertex specificity, temporal dynamic, and trend relevancy characterizations.
- The design of an efficient algorithm that exploits the constraints, even those that are neither monotonic nor anti-monotonic.
- A quantitative and qualitative empirical study. We report on the evaluation of the efficiency and the effectiveness of the algorithm on several real-world dynamic attributed graphs.

The remainder of the paper is organized as follows. Section 2 defines the trend dynamic sub-graphs and their related interestingness measures. It also formalizes a new data mining task. Section 3 presents the algorithm that computes trend dynamic sub-graphs. An empirical evaluation on real-world attributed dynamic graphs is reported in Section 4. Section 5 discusses the related work. A conclusion ends the paper in Section 6.

2   Trend Dynamic Sub-graphs and Their Related Constraints

2.1   Trend Dynamic Sub-graphs

The input of our mining task is a dynamic graph $\mathcal{G} = \{G_t \mid t = 1\ldots t_{\max}\}$ over a discrete time span $T = [\![1, t_{\max}]\!]$. Each static graph is a non-directed attributed graph $G_t = (V, E_t, A)$ where V is a set of n vertices $\{v_1, \ldots, v_n\}$ that is fixed throughout the time, $\{E_t \mid t \in T\}$ is a sequence of sets of edges that connect vertices of V at time t ($E_t \subseteq V \times V$), and A is a set of p ordinal attributes $\{a_1, \ldots, a_p\}$ whose values are defined for each vertex at each time step ($a_i : V \times T \to \mathbb{D}_i$, where $\mathbb{D}_i$ is the domain of a_i).

Intuitively, a *trend dynamic sub-graph* is an *induced* dynamic graph of $\mathcal{G}(V, T)$ whose vertices follow the same trend over a subset of attributes of A. Formally, given a subset of vertices $U \subseteq V$ and a subsequence $S = \langle t_1, \cdots, t_s \rangle$ of time stamps of T, the dynamic sub-graph of $\mathcal{G}$ induced by (U, S) is $\mathcal{G}(U, S) = \{G_t(U) \mid t \in S\}$ and $G_t(U)$ contains all the edges in E_t that have both ends in U. The induced dynamic graphs that are apt to convey a useful piece of information are those whose vertices follow a similar trend for a set of attributes, that is to say whose attribute value derivative at a time stamp t has the same sign over all the vertices and the time stamps of the dynamic sub-graph. We say that an attribute a shows an increasing trend over $\mathcal{G}(U, S)$, denoted a^+, if $\forall u \in U$ and $\forall t \in S$, $a(u, t) < a(u, t+1)$. In a similar way, we also consider decreasing trend, a^-. Many trend dynamic sub-graph can be observed over a dynamic attributed graph, but those that are particularly important occur in nodes that are closely related through the induced sub-graph topology. To that end, we are looking for trend dynamic sub-graphs whose static induce sub-graphs have a small diameter. To summarize, a trend dynamic sub-graph is defined as follows:

Definition 1 (Trend Dynamic Sub-graph). *A trend dynamic sub-graph of an attributed dynamic graph $\Big(\mathcal{G}(V, T), A \times \{+, -\}\Big)$ is composed by (1) the induced dynamic sub-graph $\mathcal{G}(U, S) = \{G_t(U) \mid t \in S\}$ where $U \subseteq V$ and $S = \langle t_1, \ldots, t_s \rangle$ is a subsequence of T, and (2) a subset of signed attributes Ω, with $\Omega \subseteq A \times \{+, -\}$. It is denoted $\big(\mathcal{G}(U, S), \Omega\big)$ and satisfies the following properties :*

1. *At each time stamp $t \in S$, the sub-graph induced by U is $G_t(U) = (U, F_t)$ with $F_t = E_t \cap (U \times U)$.*
2. *At each time stamp $t \in S$, the diameter of the graph $G_t(U)$ is less than or equal to k, where k is a user-defined threshold. I.e., for any two vertices $v, w \in U$, there exists a path connecting them whose length is smaller than or equal to k. Formally, let $d_{G_t(U)}(v, w)$ be the shortest path length between the vertices v and w in $G_t(U)$. The diameter of G is thus defined by*

$$diam_{G_t(U)} \equiv \max_{v, w \in U} d_{G_t(U)}(v, w)$$

and the diameter constraint, that is $diam_{G_t(U)} \leq k$, $\forall t \in S$, is denoted $diameter\big(\mathcal{G}(U, S), \Omega\big)$.
3. *Each signed attribute $(a, m) \in \Omega$ defined a trend that has to be satisfied by any vertex $u \in U$ at any timestamp $t \in S$:*

$$\begin{cases} a^+(u, t) \equiv a(u, t) < a(u, t+1), \text{ if } m = + \\ a^-(u, t) \equiv a(u, t) > a(u, t+1), \text{ if } m = - \end{cases}$$

This constraint is denoted $trend\big(\mathcal{G}(U, S), \Omega\big)$.

4. *If $(\mathcal{G}(U,S), \Omega)$ is maximal, then the sets U and Ω, as well as the sequence S cannot be enlarged without invalidating one or more of the above properties. This constraint is denoted* $maximal(\mathcal{G}(U,S), \Omega)$.

2.2 Interestingness Measures

To further guide the extraction of trend dynamic sub-graphs toward most relevant ones, we propose several interstingness measures that offer the possibility to the end-users to express their needs. An interestingness measure is a function which assigns a value to a pattern according to its quality. Such a measure can easily be used as a constraint by specifying a user-defined threshold that makes possible the selection of patterns having a high or a low value on these measures.

Size measures: As most simple interestingness measures are often the most useful ones, we first consider size measures that characterize a pattern by the number of elements it contains: $sz\_vertices(\mathcal{G}(U,S), \Omega) = |U|$, $sz\_times(\mathcal{G}(U,S), \Omega) = |S|$ and $sz\_attributes(\mathcal{G}(U,S), \Omega) = |\Omega|$. These measures are generally used to constrain patterns to a minimal size.

Volume measure: In some contexts, it can also be useful to combine the three size measures in a single value: $volume(\mathcal{G}(U,S), \Omega) = \frac{|U|}{|V|} \times \frac{|S|}{|T|} \times \frac{|\Omega|}{|A|}$. This measure is also generally used to constrain patterns to a minimal volume.

Measure of vertex specificity: The question that aims to answer this measure is: How similar are the vertices outside the trend dynamic sub-graph to the ones inside it? We want to quantify the average proportion of trends that are satisfied by outside pattern vertices:

$$vertex\_specificity(\mathcal{G}(U,S), \Omega) = \frac{\sum_{w \in V \setminus U} \sum_{(a,m) \in \Omega} \sum_{t \in S} \delta_{a^m(w,t)}}{|V \setminus U| \times |\Omega| \times |S|}$$

where $\delta_{condition}$ is the Kronecker function that is equal to 1 if *condition* is satisfied, or 0 otherwise. The more the trend dynamic sub-graph is made of specific vertices with respect to attribute trends, the lower this measure.

Measure of trend relevancy: The question that aims to answer this measure is: Does the attributes that do not belong to Ω have an homogeneous trend on $\mathcal{G}(U,S)$? To that end, we evaluate the entropy of the attribute trends and consider the one that has the smallest entropy. Let

$$P_1(b^m, \mathcal{G}(U,S)) = \frac{\sum_{u \in U} \sum_{t \in S} \delta_{b^m(u,t)}}{\sum_{u \in U} \sum_{t \in S} \left(\delta_{b^-(u,t)} + \delta_{b^+(u,t)} \right)}$$

be the proportion of the trend m of attribute b on the vertices and time stamps of $\mathcal{G}(U,S)$. Then the trend relevancy interestingness measure is:

$$trend\_relevancy(\mathcal{G}(U,S), \Omega) = \min_{b \in A \setminus \Omega} \sum_{m \in \{-,+\}} -P_1(b^m, \mathcal{G}(U,S)) \log P_1(b^m, \mathcal{G}(U,S))$$

The more a trend dynamic sub-graph is trend relevant, the higher this measure.

Measure of temporal dynamic: The question that aims to answer this measure is: How does a pattern appear in the time? Does it burst? To that end, we evaluate the dynamic of the proportion of vertices and attributes that satisfy the pattern before and after the time stamps of S: $P_2(t, (\mathcal{G}(U,S), \Omega)) =$

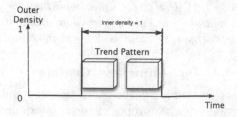

Fig. 1. Does the pattern burst ?

$\frac{\sum_{u \in U} \sum_{(a,m) \in \Omega} \delta_a{}^m{}_{(u,t)}}{|U|.|\Omega|}$. If a trend dynamic sub-graph bursts, then the proportion P_2 is below a threshold at every time stamps not in S:

$$temporal\_dynamic(\mathcal{G}(U,S), \Omega) = \max_{t \in T \setminus S} P_2(t, (\mathcal{G}(U,S), \Omega))$$

3   Trend Sub-graph Enumeration

To compute all the trend attributed sub-graphs that satisfy the interestingness measures, we design MINTAG algorithm (for MINing Trend Attributed Graph) that enumerates induced dynamic sub-graphs based on the next partial order.

Definition 2 (Partial Order on Attributed Induced Dynamic Sub-graphs). *Let $Q_1 = (\mathcal{G}(U_1, S_1), \Omega_1)$ and $Q_2 = (\mathcal{G}(U_2, S_2), \Omega_2)$ be two attributed induced dynamic sub-graphs. We say that Q_1 is more specific than Q_2, $Q_1 \preceq Q_2$, iff $U_1 \subseteq U_2$ and $S_1 \subseteq S_2$ and $\Omega_1 \subseteq \Omega_2$.*

This partial order forms a lattice: for any nonempty finite subset of attributed induced dynamic sub-graphs $\mathcal{F} = \{Q_i \mid i = 1 \ldots k\}$, $\mathcal{F}^\vee = (\mathcal{G}(\bigcup U_i, \bigcup S_i), \bigcup \Omega_i)$ and $\mathcal{F}^\wedge = (\mathcal{G}(\bigcap U_i, \bigcap S_i), \bigcap \Omega_i)$ are respectively the join and meet elements. The bounds of the lattice are $Q_\top = (\mathcal{G}(V,T), A \times \{+,-\})$ and $Q_\bot = (\mathcal{G}(\emptyset, \emptyset), \emptyset)$. The enumeration strategy used by MINTAG is a binary partition [17]. In order to enumerate all the trend attributed sub-graphs $\mathcal{R}$ induced from $(\mathcal{G}(V,T), A \times \{+,-\})$, a binary partition algorithm consists in choosing an element $e \in E = V \cup T \cup A \times \{+,-\}$ and divides $\mathcal{R}$ into two sets $\mathcal{R}_{+e}$ and $\mathcal{R}_{-e}$ so that $\mathcal{R}_{+e}$ consists of all the elements of $\mathcal{R}$ including e, and $\mathcal{R}_{-e}$ consists of those that do not include e. Therefore, e belongs to $\mathcal{R}_{+e}^\wedge$ and e does not belong to $\mathcal{R}_{-e}^\vee$. If $\mathcal{R}_{+e}$ (resp. $\mathcal{R}_{-e}$) is not empty and $\mathcal{R}_{+e}^\vee \neq \mathcal{R}_{+e}^\wedge$ (resp. $\mathcal{R}_{-e}^\vee \neq \mathcal{R}_{-e}^\wedge$), it is recursively divided by choosing another element in $\mathcal{R}_{+e}^\vee \setminus \mathcal{R}_{+e}^\wedge$ (resp. $\mathcal{R}_{-e}^\vee \setminus \mathcal{R}_{-e}^\wedge$). The number of iterations of a binary partition algorithm is linear in $|\mathcal{R}|$, which is the output size, if it is possible to check whether either $\mathcal{R}_{+e}$ or $\mathcal{R}_{-e}$ are empty. In the following, we explain how this test is performed.

3.1   Constraint Checking and Propagation Mechanisms

Let I and O be two subsets of E. We denote by $\mathcal{R}_{IO}$ a search space such that I is the set of elements that are included in all the patterns of $\mathcal{R}_{IO}$ and O is the set of elements that cannot be included in any pattern of $\mathcal{R}_{IO}$. $\mathcal{R}_{IO}^{\vee}$ and $\mathcal{R}_{IO}^{\wedge}$ are respectively the join and meet elements of this search space and $I \subseteq \mathcal{R}_{IO}^{\wedge}$ and $O \cap \mathcal{R}_{IO}^{\vee} = \emptyset$. Checking whether the search space is empty can be done by evaluating the constraints on the join or the meet elements. Indeed, if a monotonic constraint is not satisfied by the join element, then $\mathcal{R}_{IO}$ is empty. Similarly, if an anti-monotonic constraint is not satisfied by the meet element, then $\mathcal{R}_{IO}$ is also empty. Constraints that are partially monotonic or anti-monotonic can also be pushed [5], as it is explained below.

Trend Sub-graph Constraints

Trend constraint: This constraint is anti-monotonic with respect to $\preceq$. That is, if Q_1 and Q_2 are two attributed induced dynamic sub-graphs such that $Q_1 \preceq Q_2$, then, $trend(Q_2) \Rightarrow trend(Q_1)$. The anti-monotonic property of the *trend* constraint implies that if $trend(\mathcal{R}_{IO}^{\wedge})$ is not satisfied, then $\mathcal{R}_{IO}$ is empty. In MINTAG algorithm, this constraint is propagated using the following procedure: if there exists e in $\mathcal{R}_{IO}^{\vee} \setminus \mathcal{R}_{IO}^{\wedge}$ such that $trend(\mathcal{R}_{IO}^{\wedge} \cup e)$ is not satisfied, then e is removed from $\mathcal{R}_{IO}^{\vee}$.

Diameter constraint: This constraint is neither monotonic nor anti-monotonic with respect to $\preceq$. However, noting that this constraint is monotonic or anti-monotonic in each of its parameters, we can derive a propagation mechanism of this constraint. That is, for all vertex v and all time stamp t in the trend sub-graph, we should have $\max_{w \in U_1} d_{G_t(U_2)}(v, w) \leq k$. This constraint is anti-monotonic on U_1 and monotonic on U_2, that is (a) if the constraint is satisfied on U_1, it is also satisfied for any of its subsets; (b) if the constraint is satisfied on a graph $G_t(U_2)$, then, adding some vertices and edges to $G_t(U_2)$ will not increase its value. Therefore, in MINTAG algorithm, this constraint is propagated using the following mechanisms: (1) if there exists $v \in \mathcal{R}_{IO}^{\vee} \setminus \mathcal{R}_{IO}^{\wedge}$, $w \in \mathcal{R}_{IO}^{\wedge}$ and $t \in \mathcal{R}_{IO}^{\wedge}$ such that $d_{G_t(\mathcal{R}_{IO}^{\vee} \cap V)}(v, w) > k$ then v is removed from $\mathcal{R}_{IO}^{\vee}$; (2) if there exists $t \in \mathcal{R}_{IO}^{\vee} \setminus \mathcal{R}_{IO}^{\wedge}$, $v \in \mathcal{R}_{IO}^{\wedge}$ and $w \in \mathcal{R}_{IO}^{\wedge}$ such that $d_{G_t(\mathcal{R}_{IO}^{\vee} \cap V)}(v, w) > k$ then t is removed from $\mathcal{R}_{IO}^{\vee}$.

Other Interestingness Constraints

Minimal size constraints: As these constraints are monotonic, if $sz\_vertices$ $(\mathcal{R}_{IO}^{\vee} \cap V) < \min\_sz\_vertices$ or $sz\_attributes(\mathcal{R}_{IO}^{\vee} \cap A \times \{+, -\}) < \min\_sz\_attributes$ or $sz\_times(\mathcal{R}_{IO}^{\vee} \cap T) < \min\_sz\_times$, then $\mathcal{R}_{IO}$ is empty.

Minimal volume constraint: Similarly, this constraint is monotonic and if $volume(\mathcal{R}_{IO}^{\vee}) < \min\_volume$, then $\mathcal{R}_{IO}$ is empty.

Maximal vertex_specificity constraint: As the diameter constraint, this constraint is monotonic or anti-monotonic on each of its parameters. Considering the equation $\frac{\sum_{w \in V \setminus U_1} \sum_{(a,m) \in \Omega_1} \sum_{t \in S_1} \delta_{a^m}(w,t)}{|V \setminus U_2| \times |\Omega_2| \times |S_2|} \leq$ max_vertex_spec, we can observe that it is monotonic on U_1, S_2 and Ω_2 and anti-monotonic on U_2, S_1 and Ω_1. Thus, $\mathcal{R}_{IO}$ is empty if

$$\frac{\sum_{w \in (\mathcal{R}_{IO}^{\vee} \cap V)} \sum_{(a,m) \in (\mathcal{R}_{IO}^{\wedge} \cap (A \times \{+,-\}))} \sum_{t \in (\mathcal{R}_{IO}^{\wedge} \cap T)} \delta_{a^m}(w,t)}{|\mathcal{R}_{IO}^{\wedge} \cap V| \times |\mathcal{R}_{IO}^{\vee} \cap (A \times \{+,-\})| \times |\mathcal{R}_{IO}^{\vee} \cap T|} > \text{max\_vertex\_spec}$$

Minimal trend_relevancy constraint: Handling this constraint is a little more tricky. Let us first consider the entropy function with two probability values: $f(x) = -x \log(x) - (1-x)\log(1-x)$. This function increases on $[0, \frac{1}{2}]$ and decreases on $[\frac{1}{2}, 1]$. Using this notation, the minimal trend_relevancy can be rewritten as $\min_{b \in A \setminus \Omega} f(P_1(b^+, \mathcal{G}(U,S)) \geq \text{min\_trend\_rel}$.[1] Second, we can derive the following upper bound on $P_1(b^m, \mathcal{G}(U,S))$:

$$P_1(b^m, \mathcal{G}(U,S)) \leq \frac{\sum_{u \in (\mathcal{R}_{IO}^{\vee} \cap U)} \sum_{t \in (\mathcal{R}_{IO}^{\vee} \cap S)} \delta_{b^m}(u,t)}{\sum_{u \in (\mathcal{R}_{IO}^{\wedge} \cap U)} \sum_{t \in (\mathcal{R}_{IO}^{\wedge} \cap S)} \left(\delta_{b^-}(u,t) + \delta_{b^+}(u,t)\right)} = UB(b^m)$$

as P_1 is monotonic on its numerator parameters, and anti-monotonic on its denominator ones. Similarly, we can derive a lower bound[2] $LB(b^m) \leq P_1(b^m, \mathcal{G}(U,S))$. Thus, if $UB(b^m) \leq \frac{1}{2}$, then f is increasing and $f(P_1(b^m, \mathcal{G}(U,S))) \leq f(UB(b^m))$. Similarly, if $LB(b^m) \geq \frac{1}{2}$, then f is decreasing and $f(P_1(b^m, \mathcal{G}(U,S))) \leq f(LB(b^m))$.

Therefore, if there exists $b \in A \setminus \mathcal{R}_{IO}^{\vee}$ and $m \in \{+,-\}$ such that either (1) $UB(b^m) \leq \frac{1}{2}$ and $f(UB(b^m)) < \text{min\_trend\_rel}$, or (2) $LB(b^m) \geq \frac{1}{2}$ and $f(LB(b^m)) < \text{min\_trend\_rel}$ then $f(P_1(b^m, \mathcal{G}(U,S))) < \text{min\_trend\_rel}$ and we can conclude that $\mathcal{R}_{IO}$ is empty.

Maximal temporal_dynamic constraint: This constraint is anti-monotonic on its parameters on the numerator and monotonic on the ones on the denominator:

$$\max_{t \in T \setminus S} \frac{\sum_{u \in U} \sum_{(a,m) \in \Omega} \delta_{a^m}(u,t)}{|U|.|\Omega|} \leq \text{max\_temp\_dyn}$$

Therefore, if there exists $t \in T \setminus \mathcal{R}_{IO}^{\vee}$ such that $\frac{\sum_{u \in \mathcal{R}_{IO}^{\wedge} \cap U} \sum_{(a,m) \in \mathcal{R}_{IO}^{\wedge} \cap \Omega} \delta_{a^m}(u,t)}{|\mathcal{R}_{IO}^{\vee} \cap U|.|\mathcal{R}_{IO}^{\vee} \cap \Omega|} >$ max_temp_dyn, then we can conclude that $\mathcal{R}_{IO}$ is empty.

3.2   MINTAG Algorithm

Algorithm 1 presents the main steps of MINTAG. Lines 1 and 2 initialize I and O to the emptyset. Line 3 and 4 initialize the sub-space join value to the lattice top and the meet value to the lattice bottom. Line 5 is the first call to MINTAG_Enum

---

[1] This is equivalent to $\min_{b \in A \setminus \Omega} f(P_1(b^-, \mathcal{G}(U,S)) \geq \text{min\_trend\_rel}$ as $P_1(b^+, \mathcal{G}(U,S)) = 1 - P_1(b^-, \mathcal{G}(U,S))$.

[2] $LB(b^m) = \frac{\sum_{u \in (\mathcal{R}_{IO}^{\wedge} \cap U)} \sum_{t \in (\mathcal{R}_{IO}^{\wedge} \cap S)} \delta_{b^m}(u,t)}{\sum_{u \in (\mathcal{R}_{IO}^{\vee} \cap U)} \sum_{t \in (\mathcal{R}_{IO}^{\vee} \cap S)} \left(\delta_{b^-}(u,t) + \delta_{b^+}(u,t)\right)} \leq P_1(b^m, \mathcal{G}(U,S))$.

function which enumerates once and only once each trend dynamic sub-graph. The first line of the function tests if the search space contains a single trend dynamic sub-graph. If so, it is output. Line 4 reduces the search space join by removing elements whose enumeration will emptied the search space due to the trend or the diameter constraints. Line 5 checks if the search space is empty by considering the maximality, minimal size, minimal volume, maximal vertex specificity, minimal trend relevancy and maximal temporal dynamic constraints. If one of these constraints is not relevant for the end-user, she can set the corresponding threshold to 0, for the minimal constraints, or to 1 for the other ones. In that case, these constraints do not coerce the result. If the search space is not empty, a new element, that belongs to the join but not to the meet, is enumerated. This element is first added to the search space meet before the recursive call (lines 7 and 8), and then it is removed from the search space join before the recursive call (lines 10 and 11).

Algorithm 1. MINTAG

Require: An attributed dynamic graph $\mathcal{G} = \{G_t = (V, E_t, A) \mid t \in T\}$ with $A\{a_1, \ldots, a_p\}$, $a_i : V \times T \to \mathbb{D}_i$ and the parameters.

Ensure: All trend dynamic sub-graph that satisfy the constraints.

1: $I \leftarrow \emptyset$
2: $O \leftarrow \emptyset$
3: $\mathcal{R}_{IO}^{\vee} \leftarrow \left(\mathcal{G}(V, T), A \times \{+, -\} \right)$
4: $\mathcal{R}_{IO}^{\wedge} \leftarrow \left(\mathcal{G}(\emptyset, \emptyset), \emptyset \right)$
5: MINTAG_Enum($\mathcal{R}_{IO}^{\vee}, \mathcal{R}_{IO}^{\wedge}$)

Function MINTAG_Enum($\mathcal{R}_{IO}^{\vee}, \mathcal{R}_{IO}^{\wedge}$)

1: **if** $\mathcal{R}_{IO}^{\vee} = \mathcal{R}_{IO}^{\wedge}$ **then**
2: $\quad$ Ouput($\mathcal{R}_{IO}^{\vee}$)
3: **else**
4: $\quad \mathcal{R}_{IO}^{\vee} \leftarrow$ Constraint_Propagation($\mathcal{R}_{IO}^{\vee}, \mathcal{R}_{IO}^{\wedge}$)
5: $\quad$ **if not** Empty_Search_Space($\mathcal{R}_{IO}^{\vee}, \mathcal{R}_{IO}^{\wedge}$) **then**
6: $\quad\quad$ **for all** $e \in \mathcal{R}_{IO}^{\vee} \setminus \mathcal{R}_{IO}^{\wedge}$ **do**
7: $\quad\quad\quad I \leftarrow I \cup \{e\}$
8: $\quad\quad\quad$ MINTAG_Enum($\mathcal{R}_{IO}^{\vee}, \mathcal{R}_{IO}^{\wedge} \cup \{e\}$)
9: $\quad\quad\quad I \leftarrow I \setminus \{e\}$
10: $\quad\quad\quad O \leftarrow O \cup \{e\}$
11: $\quad\quad\quad$ MINTAG_Enum($\mathcal{R}_{IO}^{\vee} \setminus \{e\}, \mathcal{R}_{IO}^{\wedge}$)
12: $\quad\quad\quad O \leftarrow O \setminus \{e\}$
13: $\quad\quad$ **end for**
14: $\quad$ **end if**
15: **end if**

---

4 Experimental Study

In this section, we report on experimental results to illustrate the interest of the proposed approach. We start by describing the different real-world dynamic attributed graphs we use, as well as the questions we aim to answer. Then, we provide a performance study and give some qualitative results. All experiments were performed on a cluster. Nodes are equipped with 2 processors at 2.5GHz and 16GB of RAM under Linux operating systems. MINTAG algorithm is implemented in standard C++.

4.1   Real-World Dynamic Attributed Graphs Description

We considered 3 real-world dynamic attributed graphs whose characteristics are given in Figure 4.1.

DBLP: This co-authorship graph is built from the DBLP digital library [3]. Each vertex represents an author who published at least ten papers in one of the major conferences and journals of the Data Mining and Database com-

Dynamic attributed graph		$|V|$	$|T|$	$|A|$	density
DBLP		2145	10	43	1.3×10^{-3}
US Flights	Last 20 years	361	20	8	3.2×10^{-2}
	September 2001	220	30	6	5.7×10^{-2}
	Two years around 9/11	234	25	8	5.7×10^{-2}
	Katrina	280	8	8	5×10^{-2}
Brazil landslides		394885	2	11	5.7×10^{-4}

Fig. 2. Main characteristics of the dynamic attributed graphs

munities between January 1990 and December 2012. This time period is divided in 10 timestamps. Each timestamp describes the co-authorship relations and the publication records of the authors over 5 consecutive years. For sake of consistency in the data, two consecutive periods have a 3 year overlap[4]. Each edge at a time stamp t links two authors who co-authored at least one paper in this time interval. The vertex properties are the number of publications in each of the 43 journals or conferences.

US Flights: RITA "On-Time Performance" database[5] contains on-time arrival data for non-stop US domestic flights by major air carriers. From this database, we generated 4 dynamic attributed graphs that aggregate data over different period of time. Graph vertices stand for US airports and are connected by an edge if there is at least a flight connecting them during the time period. We consider 8 vertex attributes that are the number of departures/arrivals, the number of canceled flights, the number of flights whose destination airport has been diverted, the mean delay of departure/arrival and the ground waiting time departure/arrival. The four dynamic graphs are:

- Last 20 years: Data are aggregated over each year.
- September 2001: Data are aggregated over each day of September 2001.
- Two years around 9/11: Data are aggregated over each month between September 2000 and September 2002.
- Katrina: To study the consequences of hurricane Katrina on US airports, data are aggregated over each week between 01/08/2005 and 25/09/2005.

Brazil landslides: This dynamic attributed graph is derived from two satellite images taken before and after huge landslides in Brazil. It is composed of 394885 vertices that stand for image shapes (segmented areas), two time stamps and 11 attributes that are the spectral response in infra-red, red, blue green and indices computed from these values. There is an edge between two vertices if the corresponding shapes are contiguous.

---

[3] http://dblp.uni-trier.de/

[4] [1990-1994][1992-1996][1994-1998]...[2008-2012].

[5] http://www.transtats.bts.gov

The ensuing experimental study aims at answering the following questions: *What is the efficiency of* MINTAG *with regard to the graph characteristics that may affect its execution time? How effective are* MINTAG's *pruning properties? Does* MINTAG *scale? What about* MINTAG's *trend dynamic sub-graph relevancy?*

4.2 Quantitative Results

We conduct intensive experiments to evaluate the performance of MINTAG in terms of computational cost and number of trend dynamic sub-graphs on different dynamic attributed graphs. Figure 3 shows the number of extracted patterns

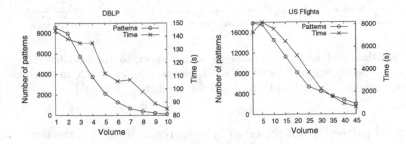

Fig. 3. Number of patterns and runtime for DBLP (left) and US flights (right) with respect to volume: max_vertex_spec = 0.5, min_trend_rel = 0.05 and max_temp_dyn = 0.8. The diameter is set to 2 on (left) and to 1 on (right).

and the execution times of MINTAG on DBLP and US Flights with respect to the volume threshold. When the minimum volume threshold decreases, more execution time is required since more trend dynamic sub-graph are obtained. Yet, MINTAG is able to extract trend dynamic sub-graphs when the minimum volume threshold is minimal, that is to say equals 1, since we report absolute volume values. MINTAG does not exhibit a similar monotonic behavior when varying the diameter constraint: the time computation is no more proportional to the number of extracted patterns. Actually, pushing this constraint needs to compute shortest paths in the graph, that is costly.

Figure 4 reports the execution times and the number of patterns with respect to the other interestingness measures: vertex specificity, trend relevancy and temporal dynamic. We can observe that for the graphs DBLP and US Flights, the less stringent the constraints, the higher the execution times and the number of patterns are. In most of the cases, the number of patterns increases dramatically. This behavior shows that our approach push efficiently these constraints that are neither monotonic nor anti-monotonic. It is noteworthy that in Figure 4, the execution time of MINTAG on DBLP for min_trend_rel = 0 is not available because the process was killed after several hours.

Figure 5 reports on the scability of MINTAG. We used DBLP and replicated alternatively the number of vertices, time stamps and attributes. As the number

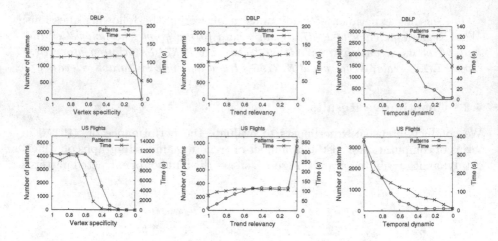

Fig. 4. Runtime and number of patterns with respect to the specificity measures (max_vertex_spec = 0.3, min_trend_rel = 0.1, max_temp_dyn = 0.5, min_volume = 5 and max_diameter = 2 for DBLP (top) or 1 for US flights (bottom))

of extracted patterns is not preserved by these replications (i.e., the vertex replication adds connected components while the time replication introduces new variations involving the last time stamp) we report the runtime per pattern. It appears that MINTAG is more robust to the increase of the number of attributes and to the number of vertices than to the number of time stamps. This is a good point since, in practice, the number of vertices is often large while the numbers of attributes and mainly the number of time stamps are rather small.

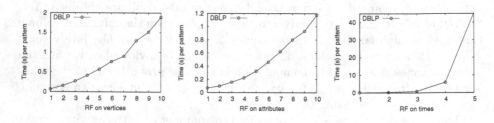

Fig. 5. Runtime per pattern with respect to replication factors on vertices, attributes and time stamps (max_vertex_spec = 0.3, min_trend_rel = 0.1, max_temp_dyn = 0.5, min_volume = 5 and max_diameter = 2

We study the effectiveness of each constraint on both DBLP and US Flights, when varying the different thresholds (volume, vertex specificity, temporal dynamic and trend relevancy). To this end, we count the number of pruned unpromising candidates by each constraint. The results are shown in Figure 6 for DBLP (top) and US Flights (bottom). It is noteworthy that all the constraints enable to prune unpromising candidates and they have different impact on both graphs. We can observe that the trend relevancy constraint is effective on the two graphs and prunes almost 50% of the unpromising candidates on DBLP in most of the cases. Even if this constraint has no anti-monotonic property, it is efficiently pushed in MINTAG. The volume constraint, more effective on DBLP than US Flights, makes possible to prune large part of the search space. This behavior is much more expected since this constraint is anti-monotonic. The pruning impact of the temporal_dynamic constraint is not negligible, since it prunes nearly 20% of the candidates on DBLP and up to 60% on US flights. This important difference is mainly due to the temporal regularity of US Flights. This can also explain the fact that the vertex specificity constraint plays a prominent role on the US Flights while having a limited impact of the DBLP dynamic graph.

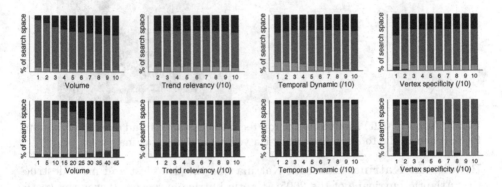

Fig. 6. Constraint efficiency on DBLP (top) and US Flights (bottom) w.r.t. specificity measures. From top to bottom: volume (black), trend_relevancy (red), temporal_dynamic (green) and vertex_specificity (blue). Same parameters as in Fig. 3 and 4.

4.3 Qualitative Results

Results on DBLP: We perform an extraction on DBLP dynamic attributed graph with max_diameter set to infinity (vertices belong to the same connected component) and min_volume = 5. Other constraints threshold are set so as not to constrain the result. We obtained 112 trend dynamic sub-graphs in less than 4 seconds. The top 2 largest patterns depict the same well-known phenomenon, explained below. The first pattern involves 171 authors having an increasing number of publications in PVLDB between 2004 and 2012. The second one involves 164 authors that have a decreasing number of publications in VLDB during the same period. These patterns reflect the new policy of the VLDB endowment. Indeed, PVLDB appeared in 2008 and, in 2010, the review process of the VLDB conference series was done in collaboration with, and entirely through PVLDB in 2011. Then, we carry out a new extraction taking into account all the constraints (max_diameter = 2, max_vertex_spec = 0.3, max_temp_dyn = 0.5) except min_trend_rel that was set to 0. We obtained 41 patterns in 8 seconds.

We first consider the pattern that has the longest duration and involves the most recent period, that is [2008-2012]. It implies the vertices related to Jimeng Sun and Christos Faloutsos, who have an increasing number of publications in KDD and SDM, while having a decreasing number of publications in VLDB. We consider another pattern which has the best *temporal_dynamic* value among the patterns having their *trend_relevancy* greater than 0.1. It involves two authors, Rong Zhou and Eric A. Hansen, and the time stamps between 1998 and 2008. On this period, the authors have an increasing number of publications in AAAI conference series. This pattern has good values on *vertex_specificity* (0.12), *temporal_dynamic* (0) and *trend_relevancy* (0.81). This publication trend is rare with regard to the whole graph.

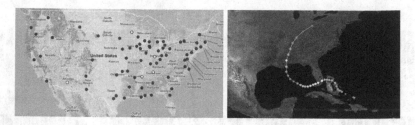

Fig. 7. Airports (left) involved in the top temporal_dynamic trend dynamic sub-graph (in red) and in the top trend_relevancy (in yellow) and the Katrina's track (right)

Results on Katrina: Hurricane Katrina was the deadliest and most destructive Atlantic hurricane of the 2005 Atlantic hurricane season. It was the costliest natural disaster, as well as one of the five deadliest hurricanes, in the history of the United States. Among recorded Atlantic hurricanes, it was the sixth strongest overall. In this experiment, we aim to characterize the impact of this hurricane on the US domestic flights. To this end, we set constraints as follows: min_volume = 10, max_vertex_spec = 0.6, min_trend_rel = 0.1, max_temp_dyn = 0.2 and max_diameter = ∞. We extract 37 patterns in 14 seconds. We look for two patterns: (i) the trend dynamic sub-graph with largest *temporal_dynamic* value, and (ii) the pattern with the highest *trend_relevancy* value. These patterns and Katrina's track[6] are shown in Figure 7. Pattern (i) involves 71 airports (in red on Figure 7 (left)) whose arrival delays increase over 3 weeks. One week is not related to the hurricane but the two others are the two weeks after Katrina caused severe destruction along the Gulf coast. This pattern has a *temporal_dynamic* = 0, which means that arrival delays never increased in these airports during another week. The hurricane strongly influenced the domestic flight organization. Pattern (ii) has a *trend_relevancy* value equal to 0.81 and includes 5 airports (in yellow on Figure 7 (left)) whose number of departures and arrivals increased over the three weeks following Katrina hurricane. Three out of the 5 airports are in the Katrina's trajectory while the two other ones were

---

[6] Map from ©2013 Google, INEGI, Inav/Geosistemas SRL, MapLink
http://commons.wikimedia.org/wiki/File:Katrina_2005_track

Table 1. Trend dynamic sub-graphs extracted by `MINTAG` on September 2001 graph

Pattern	$|V|$	Days	A	vertex_spec.	temp._dyn.	trend_rel.
P1	179	10, 11	#Cancel.$^+$	0.5	0.41	0.94
P2	111	13, 15	#Cancel.$^-$	0.52	0.83	·0.9
P3	102	13, 14, 15	#Cancel.$^-$	0.6	0.84	0.81

impacted because of their connections to airports from damaged areas. Substitutions flights were provided from these airports during this period. The values on the other interestingness measures show that this behavior is rather rare in the rest of the graph ($vertex\_specificity = 0.29$, $temporal\_dynamic = 0.2$).

Results on September 2001: To characterize the impact of September 11 attacks, we look for patterns involving many airports (at least 100) whose trends are relevant ($trend\_relevancy = 0.8$). Given this setting, `MINTAG` returns 3 trend dynamic sub-graphs in 8 seconds. These patterns are reported in Table 1. They depict a large number of airports, whose number of canceled flights increased on September 11 and 12 compared to the previous days, and then decreased two days after the terrorist attacks (between the 13th and 16th September). These patterns identify the time required for a return to normal domestic traffic.

Results on Two years around 9/11: Considering longer periods before and after the September attacks, with more restrictive threshold values ($temporal\_dynamic = 1$, $vertex\_specificity = 0.5$ and $trend\_relevancy = 0.8$), we obtain 87 patterns in 67 seconds. The top $trend\_relevancy$ pattern involves 159 airports that have an increasing number of canceled flights in September 2001 and December 2000. Obviously, the number of canceled flights in September 2001 is related to terrorist attack. It is noteworthy that December 2000 snow storm had a similar impact on the cancellation of flights, because we do not quantify the strength of the trends. Actually, the number of canceled flights in September 2001 is four times bigger than the one in December 2000.

Results on Brazil landslides: In this series of 2 satellite images, the goal is to identify regions in which a landslide appears in the second image. Generally, the main consequence of a landslide if the disappearance of the vegetation. Therefore, we focus on the patterns that involve $NDVI^-$, since $NDVI$ is a computed index that quantifies the level of vegetation. `MINTAG` returns 4821 patterns in 2 hours that involve 34275 regions that are reported on Figure 8. These results were evaluated by an expert who testified that 69% of the true landslide regions appear in the computed patterns. These regions represent 46% of the extracted regions. The 54% remaining regions belong to one of the 4 following categories:(1) regions nearby true

Fig. 8. Regions involved in the patterns: true landslides (red) and other phenomena (white).

landslides which have not been interpreted as landslides by the expert (border effect), (2) deforested area not due to landslides (e.g., human activity), (3) regions found due to misalignment of the segmentation technique and (4) regions that represent cities and human activity footprints.

5   Related Work

Many proposals intend to characterize graph evolution by means of patterns or rules. Borgwardt et al. [4] introduce the problem of mining frequent sub-graphs in dynamic graphs. Lahiri and Berger-Wolf [10] extract frequent sub-graphs that appear periodically. Inokuchi and Washio [7] define frequent induced sub-graph subsequence whose isomorphic occurrences appear frequently in graph sequences. Ahmed et al. [1] propose to mine time-persistent edges and captures all maximal non-redundant evolution paths among them. You and Cook [18] compute graph rewriting rules that describe the evolution of consecutive graphs. Berlingerio et al. [2] extract patterns based on frequency and derive graph evolution rules. Descriptive n-ary association rules are defined in [13]. More recently, dynamic attributed graphs have received a particular interest. Boden et al. [3] propose to extract clusters in each static attributed graph and associate time consecutive clusters that are similar. Jin et al. [8] consider dynamic graph whose vertices are weighted. They extract groups of connected vertices whose vertex weights follow a similar evolution, increasing or decreasing, on consecutive time stamps. Desmier et al. [6] discover neighborhood similar set of vertices whose attributes follow the same trends. All the above works only assess the interest of the patterns by means of frequency-based constraints. They do not specify additional interestingness measures to guide the search toward relevant patterns. However, such constraints have been extensively studied in itemset mining, but not yet in dynamic attributed graph settings. To name a few, Morishita et al. [11] define a theoretical framework to compute significant association rules according to statistical measures and Kuznetsov [9] defines the stability of a formal concept.

6   Conclusion

In this paper, we propose to extract dynamic sub-graphs that have a small diameter. These dynamic sub-graphs are characterized by the attributes that have the same trend over the pattern vertices at each pattern time stamps. To only compute the most significant trend dynamic sub-graphs, we define three interestingness measures. We design an algorithm that actively uses all the constraints, even those that are neither monotonic nor anti-monotonic. It reduces the search space while preserving the completeness of the extraction. We provide experiments that prove that MINTAG computes the trend dynamic sub-graph in a feasible time. Moreover, experiments on real-world dynamic attributed graphs show that our method allows to extract truly relevant patterns.

Acknowledgements. The authors thank ANR for supporting this work through the FOSTER project (ANR-2010-COSI-012-02). They also acknowledge support from the CNRS/IN2P3 Computing Center and the ICube laboratory for providing and preprocessing the Brazil landslide data.

References

1. Ahmed, R., Karypis, G.: Algorithms for Mining the Evolution of Conserved Relational States in Dynamic Networks. In: ICDM, pp. 1–10. IEEE (2011)
2. Berlingerio, M., Bonchi, F., Bringmann, B., Gionis, A.: Mining Graph Evolution Rules. In: Buntine, W., Grobelnik, M., Mladenić, D., Shawe-Taylor, J. (eds.) ECML PKDD 2009, Part I. LNCS, vol. 5781, pp. 115–130. Springer, Heidelberg (2009)
3. Boden, B., Günnemann, S., Seidl, T.: Tracing clusters in evolving graphs with node attributes. In: CIKM, pp. 2331–2334 (2012)
4. Borgwardt, K.M., Kriegel, H.P., Wackersreuther, P.: Pattern mining in frequent dynamic subgraphs. In: ICDM, pp. 818–822. IEEE (2006)
5. Cerf, L., Besson, J., Robardet, C., Boulicaut, J.F.: Closed patterns meet n-ary relations. TKDD 3(1), 3:1–3:36 (2009)
6. Desmier, E., Plantevit, M., Robardet, C., Boulicaut, J.-F.: Cohesive co-evolution patterns in dynamic attributed graphs. In: Ganascia, J.-G., Lenca, P., Petit, J.-M. (eds.) DS 2012. LNCS, vol. 7569, pp. 110–124. Springer, Heidelberg (2012)
7. Inokuchi, A., Washio, T.: Mining frequent graph sequence patterns induced by vertices. In: SDM, pp. 466–477. SIAM (2010)
8. Jin, R., McCallen, S., Almaas, E.: Trend Motif: A Graph Mining Approach for Analysis of Dynamic Complex Networks. In: ICDM, pp. 541–546. IEEE (2007)
9. Kuznetsov, S.O.: On stability of a formal concept. Ann. Math. Artif. Intell. 49(1-4), 101–115 (2007)
10. Lahiri, M., Berger-Wolf, T.Y.: Mining periodic behavior in dynamic social networks. In: ICDM, pp. 373–382. IEEE (2008)
11. Morishita, S., Sese, J.: Traversing itemset lattice with statistical metric pruning. In: PODS, pp. 226–236 (2000)
12. Moser, F., Colak, R., Rafiey, A., Ester, M.: Mining cohesive patterns from graphs with feature vectors. In: SDM, pp. 593–604. SIAM (2009)
13. Nguyen, K.N., Cerf, L., Plantevit, M., Boulicaut, J.F.: Discovering descriptive rules in relational dynamic graphs. Intell. Data Anal. 17(1), 49–69 (2013)
14. Prado, A., Plantevit, M., Robardet, C., Boulicaut, J.F.: Mining graph topological patterns. In: IEEE TKDE, pp. 1–14 (2013)
15. Robardet, C.: Constraint-Based Pattern Mining in Dynamic Graphs. In: ICDM, pp. 950–955. IEEE (2009)
16. Silva, A., Meira Jr., W., Zaki, M.J.: Mining attribute-structure correlated patterns in large attributed graphs. PVLDB 5(5), 466–477 (2012)
17. Uno, T.: An efficient algorithm for solving pseudo clique enumeration problem. Algorithmica 56(1), 3–16 (2010)
18. You, C.H., Holder, L.B., Cook, D.J.: Learning Patterns in the Dynamics of Biological Networks. In: KDD, pp. 977–985. ACM (2009)

Sparse Relational Topic Models
for Document Networks

Aonan Zhang, Jun Zhu, and Bo Zhang

Department of Computer Science and Technology, Tsinghua University
{zan12,dcszj,dcszb}@mail.tsinghua.edu.cn

Abstract. Learning latent representations is playing a pivotal role in machine learning and many application areas. Previous work on relational topic models (RTM) has shown promise on learning latent topical representations for describing relational document networks and predicting pairwise links. However under a probabilistic formulation with normalization constraints, RTM could be ineffective in controlling the sparsity of the topical representations, and may often need to make strict mean-field assumptions for approximate inference. This paper presents sparse relational topic models (SRTM) under a non-probabilistic formulation that can effectively control the sparsity via a sparsity-inducing regularizer. Our model can also handle imbalance issues in real networks via introducing various cost parameters for positive and negative links. The deterministic optimization problem of SRTM admits efficient coordinate descent algorithms. We also present a generalization to consider all pairwise topic interactions. Our empirical results on several real network datasets demonstrate better performance on link prediction, sparser latent representations, and faster running time than the competitors under a probabilistic formulation.

1 Introduction

Given the fast growth of the Internet and data collection technologies, statistical network data analysis is playing an increasingly important role in both scientific and engineering areas, such as biology, social science, data mining, etc. A network is normally represented by a set of vertices (i.e., entities) and a set of edges (i.e., links) between these entities. Link prediction is a fundamental task in network analysis [1], and building link prediction models can provide solutions like suggesting friends for social network users or recommending products.

Many approaches have been developed for link prediction, including both parametric [2–4] and nonparametric [5, 6] Bayesian models as well as matrix factorization methods [7]. Most of these approaches focus on modeling the network structure. One work that accounts for both network structure and entity contents is the relational topic model (RTM) [8], an extension of latent Dirichlet allocation (LDA) [9] to model document networks. Because of its probabilistic formulation, RTM has some restrictions on modeling real networks, which can be highly complex and imbalanced. For example, real networks normally have very

H. Blockeel et al. (Eds.): ECML PKDD 2013, Part I, LNAI 8188, pp. 670–685, 2013.

few positive links while most are negative; but the standard maximum likelihood estimation (MLE) or Bayesian inference of RTM cannot handle this imbalance issue. Furthermore, sparsity is an important property in learning latent representations that are semantically meaningful and interpretable [10], especially in large-scale applications; but RTM cannot effectively control the sparsity of latent representations due to its probabilistic formulation with normalization constraints.

To deal with the above issues, we present an alternative formulation of relational topic models that discover nonnegative latent representations of words and documents and make predictions on unseen links. With a non-probabilistic formulation [11] and no normalization constraints, we can effectively control the sparsity of the latent representations by using a sparsity-inducing ℓ_1-norm regularizer; by using different regularization parameters on the positive link likelihood and negative link likelihood respectively, the sparse relational topic model (SRTM) can effectively deal with the imbalance issue of common real networks. Furthermore, SRTM can be generalized to capture all pairwise topic interactions in a link likelihood model and is applicable to both symmetric and asymmetric networks. Finally, SRTM admits efficient and simple coordinate descent algorithms. Empirical results on several real network datasets demonstrate better link prediction performance, sparser latent representations, as well as faster running time than the competitors under a probabilistic formulation.

The paper is structured as follows. Section 2 discusses related works. Section 3 introduces our sparse relational topic model as a cost-sensitive Maximum-a-Posteriori (MAP) estimate, as well as a coordinate descent optimization algorithm. In Section 4 we show empirical results and Section 5 concludes.

2   Related Work

Link prediction [1] has been considered as an important task in statistical network analysis. One promising branch for predicting links is to build latent variable models. Hoff et al. [3] proposed a Bayesian parametric latent variable model in which the relationship between two entities is measured by the distance between them in a latent "social space". Hoff [4] then extended the model by exploiting the low rank structure in the network link matrix. Airoldi et al. [2] built hierarchical Bayesian mixed membership block models where each entity pair has a local membership assignment and all the entity pairs are also governed by a global block matrix. To infer the dimension of the latent representations for entities from data, Miller et al. [5] developed non-parametric Bayesian models for link prediction and their max-margin variants under the regularized Bayesian framework were proposed by Zhu [6].

One drawback of the above models is that they do not account for contents of entities. This issue is even more important when we analyze document networks, where the semantic meaning of documents can be very useful for predicting links among them. Chang et al. [8] proposed probabilistic relational topic models (RTMs) built on latent Dirichlet allocation to consider both the network structure and the contents of each entity when predicting links, and their performance exceeds several baseline methods that do not consider contents. Liu et al. [12] further considered the author communities behind the document networks in their models when predicting links among documents. Our SRTM model is a non-probabilistic variant of RTM.

SRTM is based on a non-probabilistic topic model named sparse topical coding (STC) [11], which is essentially a hierarchical non-negative matrix factorization method [10]. STC builds a two-level hierarchy by assigning codes for documents and each word in them. By relaxing normalization constraints and enforcing codes to be non-negative, STC can put an ℓ_1-norm regularizer on the word level and this makes STC a flexible model to control word code sparsity [11], which is a good property for learning topical representations especially in large-scale applications. The effectiveness of STC has been demonstrated on several domains including text [11], images and videos [13–15].

SRTM presents an extension of STC to address the challenging problem of link prediction, as we stated above. While sharing the merit of STC to learn sparse codes, SRTM can handle the imbalance issues among networks.

3   Sparse Relational Topic Models

In this section, we present the sparse relational topic model that solves a deterministic optimization problem. By relaxing the normalization constraints as in probabilistic models, SRTM can learn sparse word codes with an ℓ_1-norm regularizer and admits an efficient coordinate descent algorithm. In contrast, the probabilistic RTM often makes mean-field assumptions for approximate inference. Though SRTM can be defined from a regularized loss minimization perspective, for the ease of understanding we first introduce a probabilistic generative process and then cast SRTM as solving a MAP estimate with cost-sensitive regularization parameters to deal with imbalance issues of real networks.

3.1   A Generative Process for SRTM

Let $V = \{1, 2, \cdots, N\}$ be a vocabulary containing N terms and $\mathcal{D} = \{\mathbf{W}, \mathbf{Y}\}$ be a training dataset, where $\mathbf{W} = \{\mathbf{w}_d\}_{d=1}^D$ represents a corpus of D documents and $\mathbf{Y}$ denotes the set of pairwise links between documents. We will use $\mathcal{I}$ to denote the set of document pairs whose links are in the training set, i.e., $\mathcal{I} = \{(d, d') : y_{d,d'} \in \mathbf{Y}\}$. We adopt the conventional bag-of-words model, i.e., each document is represented as a set $\mathbf{w_d} = \{w_{dn}, n \in I_d\}$, where w_{dn} is the *word count* for the nth term in the dictionary and I_d is the set of terms in document d. Let $y_{d,d'}$ denote the label of the link between document d and d'. Though SRTM can be easily extended to do multi-type link prediction, for clarity we consider binary links, that is $y_{d,d'} = 1$ if there is a link between document[1] d and d', and $y_{d,d'} = -1$ otherwise.

---

[1] For asymmetric networks, $y_{d,d'}$ denotes the link from document d to document d'.

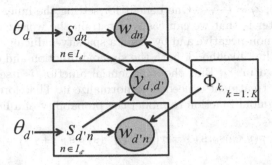

Fig. 1. Graphical Model for SRTM considering only one document pair as an illustration

As a relational topic model, SRTM models words $\mathbf{W}$ and links $\mathbf{Y}$ with two closely connected components. The first component is a hierarchical sparse topical coding (STC) [11] to describe words by using a topical dictionary $\boldsymbol{\Phi} \in \mathbb{R}^{K \times N}$ with K topical bases, that is, each row $\boldsymbol{\Phi}_{k.}$ is a normalized distributional vector over the given vocabulary. We use $\boldsymbol{\Phi}_{.n}$ to denote the nth column of $\boldsymbol{\Phi}$. Each document d has a topical representation $\boldsymbol{\theta}_d \in \mathbb{R}^K$ (i.e., *document code*) and each words in the document has an individual *word code* $\mathbf{s}_{dn} \in \mathbb{R}^K$ ($n \in I_d$). Note that here we do not put normalization constraints on document codes or word codes. This relaxation enables us to build a more flexible topic model. In fact, we can achieve sparse word codes by imposing non-negative constraints and a sparsity-inducing regularizer [10, 11]. SRTM also assumes that word codes in one document are independent given the document code and the word count w_{dn} follows a distribution whose mean parameter is $\mathbf{s}_{dn}^\top \boldsymbol{\Phi}_{.n}$ [11]. The second component of SRTM defines a likelihood model of the links between documents. Formally, the generative procedure of SRTM on document words and links can be described as:

1. for each document d
   (1) draw a document code $\boldsymbol{\theta}_d$ from $p(\boldsymbol{\theta}_d)$.
   (2) for each observed word $n \in I_d$
       (a) draw the word code $\mathbf{s}_{dn}$ from $p(\mathbf{s}_{dn}|\boldsymbol{\theta}_d)$
       (b) draw the observed word count w_{dn} from $p(w_{dn}|\mathbf{s}_{dn}^\top \boldsymbol{\Phi}_{.n})$.
2. for each document pair (d, d'), draw a link from $p(y_{d,d'}|\bar{\mathbf{s}}_d, \bar{\mathbf{s}}_{d'})$.

where $\bar{\mathbf{s}}_d = \frac{1}{|I_d|} \sum_{n \in I_d} \mathbf{s}_{dn}$ is the average word code of document d, a representation of document d in the latent topic space. For the clarity of presentation, we show a graphical model of SRTM considering only one document pair in Fig. 1, and it can be easily extended to model a large network of documents. To fully specify the model, we need to define the word likelihood model $p(w_{dn}|\mathbf{s}_{dn}, \boldsymbol{\Phi})$ and the link likelihood model $p(y_{d,d'}|\bar{\mathbf{s}}_d, \bar{\mathbf{s}}_{d'})$. For word counts, since w_{dn} is a positive integer, we choose the commonly used Poisson distribution and set $\mathbf{s}_{dn}^\top \boldsymbol{\Phi}_{.n}$ as the mean parameter:

$$p(w_{dn}|\mathbf{s}_{dn}, \boldsymbol{\Phi}) = Poisson(w_{dn}, \mathbf{s}_{dn}^\top \boldsymbol{\Phi}_{.n}), \tag{1}$$

where $Poisson(x, \nu) = \frac{\nu^x e^{-\nu}}{x!}$. One benefit for setting the inner product $\mathbf{s}_{dn}^\top \mathbf{\Phi}_{\cdot n}$ as mean parameter is that we can easily constrain the word code space by enforcing $\mathbf{s}_{dn}$ to be non-negative and by using a sparsity-inducing ℓ_1-norm regularizer [10]. For the link likelihood, both the sigmoid function and exponential link function were used in [8]. But, the exponential function is itself unnormalized and some special treatment is needed to normalize it. Therefore, we choose the more common sigmoid function to model the probability of a link:

$$p(y_{d,d'} | \bar{\mathbf{s}}_d, \bar{\mathbf{s}}_{d'}) = \sigma\Big(y_{d,d'} (\boldsymbol{\eta}^\top (\bar{\mathbf{s}}_d \circ \bar{\mathbf{s}}_{d'}) + \nu) \Big), \qquad (2)$$

where $\sigma(x) = \frac{1}{1+e^{-x}}$; $\boldsymbol{\eta} = (\eta_1, \eta_2, \cdots, \eta_K)^\top$ are the parameters describing how likely there is a link between two documents when they share a specific topic; and ν denotes the offset for the link probability. The symbol $\circ$ denotes the element-wise product.

3.2   Cost-Sensitive MAP Estimate

Let $\mathbf{\Theta} = \{\boldsymbol{\theta}_d\}$ and $\mathbf{S} = \{\mathbf{s}_d\}$ denote the latent representations of documents and words respectively. Then joint distribution of SRTM can be written as:

$$p(\mathbf{W}, \mathbf{Y}, \mathbf{\Theta}, \mathbf{S} | \mathbf{\Phi}) = \prod_d \Big(p(\boldsymbol{\theta}_d) \prod_{n \in I_d} p(\mathbf{s}_{dn} | \boldsymbol{\theta}_d) p(w_{dn} | \mathbf{s}_{dn}, \mathbf{\Phi}) \Big) \prod_{(d,d') \in \mathcal{I}} p(y_{d,d'} | \bar{\mathbf{s}}_d, \bar{\mathbf{s}}_{d'})$$
$$(3)$$

We naturally impose a normal prior on $\boldsymbol{\theta}_d$ so that $p(\boldsymbol{\theta}_d) \propto \exp(-\lambda \|\boldsymbol{\theta}_d\|_2^2)$. For the word code $\mathbf{s}_{dn}$ we use a Laplace prior to achieve sparsity [16]. Furthermore, we restrict the word codes not too far away from the document code by a normal regularizer. This results in a composite prior $p(\mathbf{s}_{dn} | \boldsymbol{\theta}_d) \propto \exp(-\gamma \|\boldsymbol{\theta}_d - \mathbf{s}_{dn}\|_2^2 - \rho \|\mathbf{s}_{dn}\|_1)$, which is super-Gaussian [17] and the ℓ_1-term drives our estimates to be sparse. The hyper-parameters (λ, γ, ρ) can be predefined or selected using cross-validation. We will provide sensitivity analysis to these parameters in experiments.

With the above joint distribution, a standard MAP estimate with dictionary learning can be formulated as solving the problem:

$$\min_{\mathbf{\Theta}, \mathbf{S}, \mathbf{\Phi}} \quad \ell(\mathbf{S}, \mathbf{\Phi}; \mathbf{W}) + \ell(\mathbf{S}, \boldsymbol{\eta}; \mathbf{Y}) + \Omega(\mathbf{\Theta}, \mathbf{S})$$
$$\text{s.t.:} \quad \boldsymbol{\theta}_d \geq 0, \forall d; \mathbf{s}_{dn} \geq 0, \forall d, n \in I_d; \mathbf{\Phi}_{k \cdot} \in \mathcal{P}, \forall k, \qquad (4)$$

where $\ell(\mathbf{S}, \mathbf{\Phi}; \mathbf{W}) = \sum_{d,n \in I_d} \ell(\mathbf{s}_{dn}, \mathbf{\Phi}) = -\sum_{d,n \in I_d} \log Poisson(w_{dn}, \mathbf{s}_{dn}^\top \mathbf{\Phi}_{\cdot n})$ is the negative log-likelihood of word counts; $\ell(\mathbf{S}, \boldsymbol{\eta}; \mathbf{Y}) = \sum_{(d,d') \in \mathcal{I}} \ell(\mathbf{s}_d, \mathbf{s}_{d'}; y_{d,d'}) = -\sum_{(d,d') \in \mathcal{I}} \log p(y_{d,d'} | \bar{\mathbf{s}}_d, \bar{\mathbf{s}}_{d'})$ is the negative log-likelihood of links; $\Omega(\mathbf{\Theta}, \mathbf{S}) = \lambda \sum_d \|\boldsymbol{\theta}_d\|_2^2 + \sum_{d,n \in I_d} (\gamma \|\mathbf{s}_{dn} - \boldsymbol{\theta}_d\|_2^2 + \rho \|\mathbf{s}_{dn}\|_1)$ is the regularization term; $\mathcal{P}$ is the $(N-1)$-dimensional simplex. The negative log-likelihood is usually called a log-loss. We have imposed non-negative constraints on the latent representations in order to obtain good interpretability, as a non-negative code can be interpreted as the importance of a topic. Moreover, non-negative constraints are good for our objective of a sparse estimate.

It is worth noting that there could be two imbalance issues with the standard MAP estimate. Firstly, for each pair of documents there is only one link variable while there could be hundreds of words. This difference would lead to an imbalanced combination of word likelihood and link likelihood in problem (4). Secondly, in common real networks only a few links are positive while most are negative, e.g., the widely used Cora citation network [8] has about 0.1% positive links. This difference would lead to an imbalanced combination of positive link likelihood and negative link likelihood. To address these imbalance issues, we can easily extend the regularized log-loss minimization problem to a cost-sensitive MAP estimate by introducing different regularization parameters for the positive and negative links respectively. Specifically, we replace the standard log-loss of links with the following cost-sensitive log-loss:

$$\ell(\mathbf{S}, \boldsymbol{\eta}; \mathbf{Y}) = C_+ \sum_{(d,d') \in \mathcal{I}_+} \ell(\mathbf{s}_d, \mathbf{s}_{d'}; y_{d,d'}) + C_- \sum_{(d,d') \in \mathcal{I}_-} \ell(\mathbf{s}_d, \mathbf{s}_{d'}; y_{d,d'}), \quad (5)$$

where $\mathcal{I}_+ = \{(d, d') \in \mathcal{I} : y_{d,d'} = 1\}$ and $\mathcal{I}_- = \mathcal{I} \backslash \mathcal{I}_+$. Then, by setting C_+ and C_- at a value larger than 1, we can improve the influence of links and overcome the imbalance issue between words and links; and by setting C_+ at a value larger than C_-, we can better balance the influence of positive links and negative links. We will provide more insights in the experiment section.

If we look back at the generative formulation, which is easy to understand, an intuitive understanding of the regularization parameters C_+ and C_- is that they are *pseudo-counts* of the links, and the likelihood of the links are correspondingly:

$$p(y_{d,d'} = 1 | \bar{\mathbf{s}}_d, \bar{\mathbf{s}}_{d'}) = \sigma(\boldsymbol{\eta}^\top (\bar{\mathbf{s}}_d \circ \bar{\mathbf{s}}_{d'}) + \nu)^{C_+}$$
$$p(y_{d,d'} = -1 | \bar{\mathbf{s}}_d, \bar{\mathbf{s}}_{d'}) = \sigma(-\boldsymbol{\eta}^\top (\bar{\mathbf{s}}_d \circ \bar{\mathbf{s}}_{d'}) - \nu)^{C_-}.$$

Note that these likelihood functions are unnormalized if the pseudo-counts are not 1. But the un-normalization does not affect our estimates in the cost-sensitive log-loss minimization framework.

3.3 Optimization Algorithms

We first present our learning algorithm for solving problem (4). Since the optimization problem is bi-convex, i.e. convex over $\boldsymbol{\Theta}$ and $\mathbf{S}$ given the dictionary $\boldsymbol{\Phi}$ and the networks parameters $\boldsymbol{\eta}$ and ν; and convex over $\boldsymbol{\Phi}$, $\boldsymbol{\eta}$, and ν given the document codes $\boldsymbol{\Theta}$ and the word codes $\mathbf{S}$, we use a coordinate descent algorithm to iteratively optimize the objective function. As outlined in Algorithm 1, the algorithm iteratively solves three subproblems:

1. *Hierarchical Sparse Coding*: learns document codes and sparse word codes for the documents;
2. *Dictionary Learning*: learns the topical dictionary with document codes and word codes given;

Algorithm 1. Sparse Relational Topic Models

1: Initialize $\Phi, \Theta, \mathbf{S}, \boldsymbol{\eta}, \nu$
2: read corpus $\mathcal{D}$
3: **while** not converge **do**
4:     $(\Theta, \mathbf{S}) = HierarchicalSparseCoding(\Phi, \boldsymbol{\eta}, \nu)$;
5:     $\Phi = DictionaryLearning(\mathbf{S})$;
6:     $(\boldsymbol{\eta}, \nu) = LinkModelLearning(\mathbf{S})$;
7: **end while**

3. *Link Model Learning*: learns the link likelihood model with the codes and
   topical dictionary given.

Below, we discuss each step in detail. For notation simplicity, we will set $C_+ = C_- = C$.

Hierarchical Sparse Coding: This step involves solving for the word codes
and document codes. Since the subproblem is convex, we can apply a generic
algorithm to solve it. Here, we use the similar coordinate descent method as
in [11]. For *document codes*, since the documents are independent, we can solve
for each $\boldsymbol{\theta}_d$ separately and this results in a convex subproblem:

$$\min_{\boldsymbol{\theta}_d} \lambda \|\boldsymbol{\theta}_d\|_2^2 + \gamma \sum_{n \in I_d} \|\mathbf{s}_{dn} - \boldsymbol{\theta}_d\|_2^2, \text{ s.t.: } \boldsymbol{\theta}_d \geq 0. \tag{6}$$

It can be shown that the optimum solution is $\boldsymbol{\theta}_d = \frac{\gamma \sum_{n \in I_d} \mathbf{s}_{dn}}{\lambda + \gamma |I_d|}$, that is, the
document code is the average (with some re-scaling) of word codes.

For *word codes*, again we can treat each document separately. Formally, the
optimization problem for word codes of document d is:

$$\min_{\mathbf{s}_d} \sum_{n \in I_d} \ell(\mathbf{s}_{dn}, \beta) + \sum_{n \in I_d} (\gamma \|\mathbf{s}_{dn} - \boldsymbol{\theta}_d\|_2^2 + \rho \|\mathbf{s}_{dn}\|_1) + C \sum_{d' \in \mathcal{N}_d} \ell(\mathbf{s}_d, \mathbf{s}_{d'}; y_{d,d'})$$
$$\text{s.t.: } \mathbf{s}_{dn} \geq 0, \forall n \in I_d, \tag{7}$$

where $\mathcal{N}_d = \{d' : (d, d') \in \mathcal{I}\}$ is the neighborhood of document d in the training
network. For the sigmoid link function, the log-loss of links is

$$\ell(\mathbf{s}_d, \mathbf{s}_{d'}; y_{d,d'}) \doteq \log \left(1 + \exp(-y_{d,d'}(\boldsymbol{\eta}^\top (\bar{\mathbf{s}}_d \circ \bar{\mathbf{s}}_{d'}) + \nu)) \right). \tag{8}$$

Since the objective function w.r.t. a single word code is convex given other word
codes, we can iteratively optimize each word code $\mathbf{s}_{dn}$ by solving:

$$\min_{\mathbf{s}_{dn}} \ell(\mathbf{s}_{dn}, \Phi) + \gamma \|\mathbf{s}_{dn} - \boldsymbol{\theta}_d\|_2^2 + \rho \|\mathbf{s}_{dn}\|_1 + C \sum_{d' \in \mathcal{N}_d} \ell(\mathbf{s}_d, \mathbf{s}_{d'}; y_{d,d'})$$
$$\text{s.t.: } \mathbf{s}_{dn} \geq 0. \tag{9}$$

This subproblem does not have a closed-form solution because of the nonlinearity
of the sigmoid likelihood. Therefore, we resort to numerical methods using projected gradient descent [18] to take care of the constraints. Precisely, we take a

gradient descent step with a stepsize selected with line search, and then perform projection onto the convex feasible domain. Formally, the projected gradient descent is to update:

$$\mathbf{s}_{dn}^{new} = \Pi_P(\mathbf{s}_{dn}^{old} - t\nabla_{\mathbf{s}_{dn}}\mathcal{L})$$

where t is the step size; Π_P is a projection operator; and $\Pi_P(x) = \arg\min_{x' \in P} d(x, x')$. Here P is the positive half space of $\mathbb{R}^K$ and $d(\cdot, \cdot)$ stands for the Euclidian distance. Let $\mathcal{L}$ be the objective function of the subproblem (9). We can verify that $s_{dnk}^{new} = 0$ if $s_{dnk}^{old} - t\nabla_{s_{dnk}}\mathcal{L} < 0$ and $s_{dnk}^{new} = s_{dnk}^{old} - t\nabla_{s_{dnk}}\mathcal{L}$ otherwise. To simplify notation, we first calculate the derivative of the sigmoid link function in Eq. (8) w.r.t. to $\mathbf{s}_{dn}$

$$\nabla_{\mathbf{s}_{dn}}\ell(\mathbf{s}_d, \mathbf{s}_{d'}; y_{d,d'}) = \frac{\partial\ell}{\partial z_{d,d'}} \cdot \frac{\partial z_{d,d'}}{\partial\mathbf{s}_{dn}} = \frac{-y_{d,d'}\exp(z_{d,d'})}{1 + \exp(z_{d,d'})} \cdot \frac{\eta_k\bar{\mathbf{s}}_{d'}}{|I_d|}, \qquad (10)$$

where $z_{d,d'} = -y_{d,d'}(\boldsymbol{\eta}^\top(\bar{\mathbf{s}}_d \circ \bar{\mathbf{s}}_{d'}) + \nu)$. Then, the gradient w.r.t. $\mathbf{s}_{dn}$ is

$$\nabla_{\mathbf{s}_{dn}}\mathcal{L} = (1 - \frac{w_{dn}}{\mathbf{s}_{dn}^\top \mathbf{\Phi}_{\cdot n}})\mathbf{\Phi}_{\cdot n} + 2\gamma(\mathbf{s}_{dn} - \boldsymbol{\theta}_d) + \rho + C\sum_{d' \in \mathcal{N}_d} \nabla_{\mathbf{s}_{dn}}\ell(\mathbf{s}_d, \mathbf{s}_{d'}; y_{d,d'}). \quad (11)$$

Dictionary Learning: This step involves solving for the topical dictionary $\mathbf{\Phi}$. Since $\mathbf{\Phi}$ is constrained on a probabilistic simplex, we can use projected gradient descent to update $\mathbf{\Phi}$ and then project each row onto an ℓ_1-simplex [11]. Efficient linear time projection methods are available to make this step fast [19].

Link Likelihood Learning: This step involves solving for the parameters $\boldsymbol{\eta}$ and ν of the link likelihood model. In this step we only need to account for the link part $\sum_{(d,d') \in \mathcal{I}} \ell(\mathbf{s}_d, \mathbf{s}_{d'}, y_{d,d'})$. The objective for each link is convex so the summation is also convex for $\boldsymbol{\eta}$ and ν. Simply taking gradient we get

$$\nabla_{\eta_k}\mathcal{L} = C\sum_{(d,d') \in \mathcal{I}} \frac{-y_{d,d'}\bar{s}_{dk}\bar{s}_{d'k}\exp(z_{d,d'})}{1 + \exp(z_{d,d'})}$$

$$\nabla_{\nu}\mathcal{L} = C\sum_{(d,d') \in \mathcal{I}} \frac{-y_{d,d'}\exp(z_{d,d'})}{1 + \exp(z_{d,d'})}$$

and we can use gradient descent with line search to solve the problem.

3.4 A Generalized Sparse Relational Topic Model

It is worth noticing that in SRTM we define the strength of a link between two documents by $\boldsymbol{\eta}^\top(\bar{\mathbf{s}}_d \circ \bar{\mathbf{s}}_{d'}) + \nu = \bar{\mathbf{s}}_d\top\text{diag}(\boldsymbol{\eta})\bar{\mathbf{s}}_{d'} + \nu$, where $\text{diag}(\boldsymbol{\eta})$ is a diagonal matrix with the diagonal elements being those of $\boldsymbol{\eta}$. Therefore, SRTM can only capture the *same-topic-interactions* (i.e., only when two documents have the same topic, there is a nonzero contribution to the link likelihood); and it could be unsuitable for modeling asymmetric networks because of the symmetric nature of diagonal matrices. To relax these constraints and capture *all-pairwise-topic-interactions*, one straightforward extension is to use a full weight matrix $H^{K \times K}$ and define the link likelihood model as:

$$p(y_{d,d'}|\bar{\mathbf{s}}_d, \bar{\mathbf{s}}_{d'}) = \sigma(y_{d,d'}(\bar{\mathbf{s}}_d^\top H\bar{\mathbf{s}}_{d'} + \nu)). \qquad (12)$$

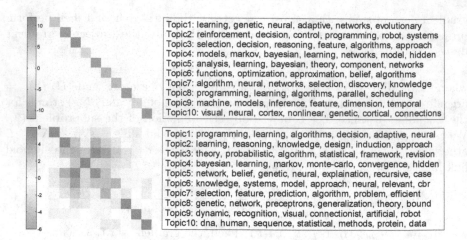

Fig. 2. Weight matrix and according representative words for each topic learned by SRTM (first row) and gSRTM (second row) on the Cora citation network data

where H_{ij} represents the strength of two documents being connected when they have topic i and topic j respectively. We denote this generalized SRTM by gSRTM. Formally, using the sigmoid likelihood function we have a similar optimization problem, and a similar coordinate descent algorithm can be applied with few changes for learning word codes and link likelihood models when taking the gradient descent steps.

Before presenting all the details of our experiments, we first illustrate the latent semantic structures learned by the sparse relational topic models and compare the diagonal SRTM and the generalized SRTM with a full weight matrix. Specifically, Fig. 2 shows the weight matrices learned by SRTM and gSRTM on the Cora citation network data (details are in the next section), as well as the top words of each of the 10 topics, respectively. For the diagonal SRTM, since the latent features $\bar{s}_d$ in the link likelihood are nonnegative, the learned weight matrix must have some negative diagonal entries although most diagonal entries are positive in order to fit the training data with binary links. The negative diagonal entries somehow conflict our intuition that papers with the same topic should be more likely to have a citation link. In contrast, the full weight matrix learned by gSRTM has only positive diagonal entries, which are consistent with our intuition; and many off-diagonal entries are negative, again consistent with our intuition that papers with different topics are less likely to have a citation relation. We also note that some topics are generic, and papers with these topics are likely to get cited by or cite the papers with other closely related topics. For example, Topic3 in gSRTM is a generic topic about *theory, probabilistic, algorithm* and *statistical*; and the papers with Topic3 are likely to have a citation relationship with the papers with the related topics, such as Topic4 (*Bayesian, learning, Markov,* etc.), Topic5 (*network, belief, genetic,* etc.), and Topic6 (*knowledge, systems, model,* etc.).

Table 1. Statistics of the datasets used in our experiments

Dataset	# Entities	# Terms (N)	# Links	Link sparsity ratio
Cora [20]	2,708	1,433	5,429	0.07%
WebKB [21]	877	1,608	1,703	0.2%
CiteSeer	3,312	3,703	4,714	0.04%

4 Experiments

In this section, we present more experimental results and compare with several models on link prediction tasks. We further present a sensitivity analysis over some built-in hyper-parameters to verify that SRTM can handle the imbalance issues in real networks while effectively learning sparse word codes.

4.1 Datasets and Models

Our experiments are conducted on three publicly available datasets. All the datasets contain very sparse positive links, as detailed below:

- The *Cora* dataset [20] consists of 2,708 research papers with a vocabulary of 1,433 terms in total. Among the papers there are 5,429 positive links, each representing a citation from one paper to the other. So on average each paper has about 2 citations and the ratio of positive links is roughly 0.07%;

- The *WebKB* dataset [21] consists of 877 webpages collected from computer science departments of four universities, with 1,608 hyper-links among pages. In total, there are 1,703 terms in the dictionary. Again, this network is sparse and about 0.2% of the pairs have links;

- The *CiteSeer* dataset is another sparse document network consisting of 3,312 papers and 3,703 citations among those papers (i.e., the link sparsity ratio is about 0.04%). Its dictionary consists of 4,712 individual words.

Since RTM has been shown to outperform several baseline models on link prediction [8], our empirical studies are concentrated on analyzing the effectiveness of sparse learning in relational topic models. We use RTM as our competitive baseline method, and compare all the methods on the above three real network datasets. In summary, the methods we compare are the followings:

- **RTM** [8]: the probabilistic relational topic model built on LDA using variational methods with mean-field assumptions to approximately infer the posterior distribution. We consider the case where the logistic link function is used to model links with a diagonal weight matrix;

- **STC+Regression**: a two-step model in which we first train an unsupervised sparse topic coding (STC) [11] to discover the latent representations of all documents and then learn a logistic regression model on training links to predict the links of testing document pairs. Note that the link information does not affect the latent representations in this method;

– **SRTM**: the proposed sparse relational topic model that uses a diagonal weight matrix in the logistic link likelihood function;

– **gSRTM**: the generalized SRTM with a full symmetric weight matrix in the logistic link likelihood model.

4.2   Results on Link Prediction

We follow the same approach as in [8] to predict links for unseen documents. Namely, for each testing document, we predict its links to the training documents. For SRTM models (i.e., SRTM and gSRTM), this can be done by first inferring the latent representation of the testing document through solving a hierarchical sparse coding step, and then applying the logistic link likelihood function to compute the probability of existing a link. Given a link's probability, we can make binary decision, that is, if the probability is larger than 0.5, there is a link exists; otherwise, no link exists. Here, we use *link rank*[2] as the performance measure, the same as in [8]. We also compare the training time to analyze the efficiency of various methods. Since all the methods are very efficient in testing, we omit the comparison on testing time.

To partly address the serious imbalance issues of the real networks and improve time efficiency, we randomly sample 0.2% of the negative links[3] and form the training data together with all the positive links to learn the sparse topic models, including SRTM, gSRTM and the de-coupled approach of STC+Regression. For the probabilistic RTM, since there is no effective mechanism on balancing positive and negative links, we found that using the same down-sampling strategy would produce worse results on both link prediction and time efficiency than the "regularization" strategy suggested in [8]. Thus, we choose to use only positive links and put a regularizer over η and ν to make sure that they will not diverge.

Fig. 3 shows the results on link prediction and training time. We tune hyper-parameters for all the models to their best settings for link prediction. For RTM, we tune the Dirichlet hyper-parameters α and for the SRTM models we fix $\lambda = \gamma$ and tune the ratio ρ/γ. Those hyper-parameters will affect link prediction results and the sparsity of word codes, and we will provide a sensitivity analysis on them in Section 4.3. But, in general, SRTM models have a wide range of these parameters to get good link prediction and sparsity of word codes. Overall, we

---

[2] For a document, its link rank is defined an average over the ranks of positive links in the list of all testing pairs. Then, the overall link rank is an average of the link rank over all testing documents.

[3] Other sampling ratios (e.g., 1%, 0.5%, 0.1%, etc.) do not affect the link prediction results of the SRTM models much, due to the effective balancing strategy by tuning regularization parameters.

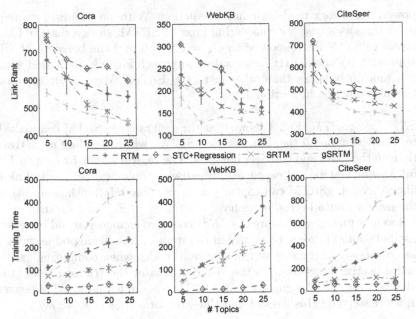

Fig. 3. First row: Link rank on three datasets using different models when changing the number of topics. Second row: Training time (in seconds) on three datasets using different models when changing the number of topics.

can see that the sparse relational topic models obtain significantly better results on all datasets. A closer examination can be done by comparing the following model pairs:

– *RTM vs. SRTM*: On all the datasets, SRTM makes more accurate link prediction (e.g., SRTM improves the average link rank by about 100 on the Cora dataset) and uses less (about 2 times when the number of topics is relatively large) time than RTM. These improvements are attributed to several factors. First, SRTM accounts for the imbalance issues in the network, which can affect the link prediction performance, while RTM cannot handle that within its Bayesian framework. Second, RTM makes mean-field assumptions, which can be too strict [22], while SRTM avoids making this assumption by solving a deterministic optimization problem. Finally, SRTM uses coordinate descent methods to optimize the objective function, where each step breaks down to very quick projected gradient methods. All these factors make SRTM perform better in link prediction while still faster than RTM, even though RTM does not use negative links;

– *STC+Regression vs. SRTM*: Since SRTM takes link information into account during the hierarchical sparse coding step, its latent representations could be more discriminative for link prediction and thus SRTM obtains a huge gain in link prediction as shown in Fig. 3. With moderate values of C_+ and C_-, SRTM accounts for both links and words to produce a much powerful network model for link prediction. With no surprise, we require more time as a cost for considering links in SRTM. Notice that SRTM collapses to STC+Regression when $C_+ = C_- = 0$ and the behavior of SRTM approximates the matrix factorization approach for link prediction when C is significantly larger than other factors in Eq. (4). We will further analyze this phenomenon in Section 4.3;

– *SRTM vs. gSRTM*: Fig. 3 shows that the generalized gSRTM can make better prediction on all the datasets than SRTM, while spending more training time on the Cora and CiteSeer datasets. The reason is that by using a $K \times K$ full weight matrix and capturing all pairwise topic interactions in link likelihood model, gSRTM can capture valuable topic relationships and thus fit the network data better as we have illustrated in Fig. 2. Of course, using a full weight matrix with more (i.e., K^2) non-zero elements would increase the computational burden, obviously in the steps of link likelihood learning and less obviously in the step of learning word codes when computing gradients and objective functions. On the WebKB dataset the training time of both SRTM and gSRTM seems comparable. The reason is that gSRTM converges in fewer steps on this dataset and thus the total time cost is low.

4.3   Sensitivity Analysis

Word Code Sparsity. The strength of SRTM partly lies on its flexibility to learn sparse word codes by adjusting the hyper-parameters (λ, γ, ρ). Following [11] we fix $\lambda = \gamma$ and only tune the ratio ρ/γ. By checking problem (4) we can clearly see that when setting ρ/γ to a relative large value, SRTM is encouraged to learn sparse word codes. But this can cause a high divergence between word codes and the corresponding document code. From our experiments we verify that balancing the two factors can let the model generalize well to unseen data while effectively learning sparse word codes. For the RTM model, the Dirichlet hyper-parameters α control the sparsity level[4]. As it will be shown in the experiments, RTM cannot learn sparse word codes while maintaining good link prediction performance by tuning α.

In Fig. 4(a) and Fig. 4(b) we compare the sparsity ratio of word codes[5] between RTM and SRTM with different numbers of topics when tuning their hyperparameters. For RTM, we tune the Dirichlet parameter α and for SRTM we fix γ to a constant and tune[6] ρ. This results in a change of ρ/γ. Fig. 4(a) shows a sharp drop of sparsity ratio when α grows to a certain level in RTM[7]. This is due to the property of the Dirichlet prior, where a little shift can cause the

---

[4] We use the common symmetric Dirichlet prior for the topic mixing proportions in RTMs.

[5] The sparsity ratio is defined as the average ratio of zero elements in word codes.

[6] Changing both γ and ρ will lead to even better link prediction results.

[7] In theory, RTM does not have sparse word codes if $\alpha > 0$. Here, we treat a small value ϵ (e.g., $\epsilon < 0.001$) as zero.

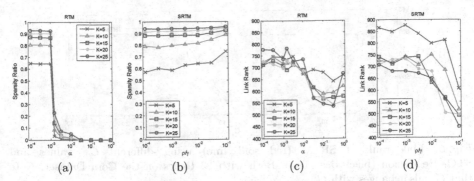

Fig. 4. Sparsity ratio (a) and link rank (c) for RTM with different number of topics when tuning hyper-parameter α on the Cora dataset; Sparsity ratio (b) and link rank (d) for SRTM with different number of topics when tuning the ratio of hyper-parameters ρ/γ on the Cora dataset.

"sharpness" of the prior changes significantly. For SRTM, Fig. 4(b) demonstrates that the sparsity ratio stays at a relative high level. When the number of topics is relatively small, changing ρ can gradually affect the sparsity ratio. There is a trend that SRTM does not learn a dense word code, which is probably due to a clear meaning of words in the dataset that each word only has a few topical meanings.

We also analyze how the hyper-parameters affect link prediction accuracy. Fig. 4(c) shows that the best link prediction results of RTM can be reached when α is around 0.1. At this point, the sparsity ratio is zero. So on the Cora dataset, RTM tends to perform better when learning dense codes. This is not a coincidence because a small α can produce a very "sharp" Dirichlet prior, which can dramatically bias the model and result in an inefficient control of sparsity ratio. In contrast, from Fig. 4(d) we can see that for SRTM there is a gradual change in link rank when ρ grows. Finally, the model reaches its best link rank result at a high sparsity ratio when ρ/γ is around 0.1. The reason is that SRTM relaxes the probability constraints of codes and thus effectively learn sparse codes by introducing ℓ_1-norm constraints at the word code level. SRTM achieves a built-in sparsity control mechanism by constructing a two-level hierarchical topic model.

The Hyper-Parameter C. As we have discussed, a relational topic model might have two imbalance issues, i.e., the imbalance between modeling words and links, and the imbalance between positive links and negative links. To address both issues, SRTM introduces the hyper-parameter C_+ for positive links and C_- for negative links. For the first issue, we can fix a reasonable value of C_+ and C_- to balance words and links. For the second issue, since negative links usually dominate positive links, we can either tune the C_+/C_- ratio or sub-sample the negative links. In our experiments, we use both strategies and find that sub-sampling a few negative links while tuning C_+/C_- can make very good prediction results.

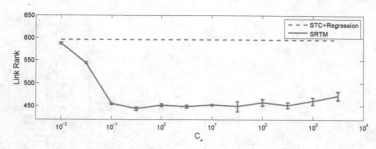

Fig. 5. Link rank of SRTM (red sold line) using different C_+ values and STC+Regression (black dash line). Both with 25 topics on the Cora Dataset. Note that C_- also changes with C_+.

To analyze the sensitivity, we fix a reasonable ratio $C_+ = 10C_-$ to balance the links[8] and tune C_+ for training SRTM with 25 topics on the Cora dataset. The link ranks for different C_+ values are shown in Fig. 5. We can see that SRTM performs best when C_+ is not too large nor too small, e.g., in the wide range between 0.1 and 100. When C_+ approaches zero SRTM collapses to a sparse topical coding followed by regression. On the other end, when C_+ grows large, the link part dominates the whole objective function. Thus, the behavior of SRTM approximates the matrix factorization approach for link prediction. SRTM does better link prediction, both utilizing words and links with a moderate C_+ than merely using any one of them. This verifies that SRTM successfully combines the knowledge of each part to get an overall better model.

5   Conclusions and Discussions

We present sparse relational topic models (SRTM), a non-probabilistic formulation of relational topic models to understand document networks and predict missing links. By relaxing the normalization constraints of probabilistic models and introducing appropriate regularization terms, SRTM can handle the common imbalance issues in real networks and efficiently learn sparse latent representations. SRTM admits a simple coordinate descent algorithm, and it can be naturally extended to capture all pairwise topic interactions for predicting links among document networks. Empirical results show that our models perform significantly better than probabilistic relational topic models in link prediction, training time, and discovering sparse representations.

The current batch algorithm to learn the topical dictionary and link likelihood model may cause limitations on applying SRTM to large-scale applications. Therefore, it is worth investigating stochastic gradient descent methods [23] in the future. Furthermore, though a restricted grid search works well as we have done in the experiments, in general it is hard to search for the optimal hyper-parameters for SRTM, and developing more efficient methods for hyper-parameter estimation is an interesting topic.

---

[8] As in the link prediction experiments, we sub-sample 0.2% of negative links as our training data.

References

1. Liben-Nowell, D., Kleinberg, J.: The link prediction problem for social networks. In: CIKM (2003)
2. Airoldi, E.M., Blei, D.M., Fienberg, S.E., Xing, E.P.: Mixed membership stochastic blockmodels. Journal of Machine Learning Research 9, 1981–2014 (2008)
3. Hoff, P., Raftery, A., Handcock, M.: Latent space approches to social network analysis. Journal of American Statistical Association 97, 1090–1098 (2002)
4. Hoff, P.: Modeling homophily and stochastic equivalence in symmetric relational data. In: NIPS (2007)
5. Miller, K., Griffiths, T., Jordan, M.: Nonparametric latent feature models for link prediction. In: NIPS (2009)
6. Zhu, J.: Max-margin nonparametric latent feature models for link prediction. In: ICML (2012)
7. Menon, A.K., Elkan, C.: Link prediction via matrix factorization. In: Gunopulos, D., Hofmann, T., Malerba, D., Vazirgiannis, M. (eds.) ECML PKDD 2011, Part II. LNCS, vol. 6912, pp. 437–452. Springer, Heidelberg (2011)
8. Chang, J., Blei, D.: Relational topic models for document networks. In: AISTATS (2009)
9. Blei, D., Ng, A., Jordan, M.: Latent Dirichlet allocation. Journal of Machine Learning Research 3, 993–1022 (2003)
10. Lee, D., Seung, H.: Learning the parts of objects by non-negative matrix factorization. Nature 401, 788–791 (1999)
11. Zhu, J., Xing, E.: Sparse topical coding. In: UAI (2011)
12. Liu, Y., Niculescu-Mizil, A., Gryc, W.: Topic-link lda: Joint models of topic and author community. In: ICML (2009)
13. Fu, W., Wang, J., Li, Z., Lu, H., Ma, S.: Learning semantic motion patterns for dynamic scenes by improved sparse topical coding. In: ICME (2012)
14. Ji, R., Duan, L., Chen, J., Gao, W.: Towards compact topical descriptors. In: CVPR (2012)
15. Li, L.-J., Zhu, J., Su, H., Xing, E.P., Fei-Fei, L.: Multi-level structured image coding on high-dimensional image representation. In: Lee, K.M., Matsushita, Y., Rehg, J.M., Hu, Z. (eds.) ACCV 2012, Part II. LNCS, vol. 7725, pp. 147–161. Springer, Heidelberg (2013)
16. Tibshirani, R.: Regression shrinkage and selection via the lasso. J. Royal. Statist. Soc. B58, 267–288 (1996)
17. Hyvarinen, A.: Sparse code shrinkage: Denoising of nongaussian data by maximum likelihood estimation. Neural Computation 11, 1739–1768 (1999)
18. Boyd, S., Vandenberghe, L.: Convex Optimization. Cambridge University Press (2004)
19. Duchi, J., Shalev-Shwartz, S., Singer, Y., Chandra, T.: Efficient projections onto the ℓ_1-ball for learning in high dimensions. In: ICML (2008)
20. McCallum, A., Nigam, K., Rennie, J., Seymore, K.: Automating the construction of internet portals with machine learning. Information Retrieval (2000)
21. Craven, M., Dipasquo, D., Freitag, D., McCallum, A.: Learning to extract symbolic knowledge from the world wide web. In: AAAI (1998)
22. Jordan, M.I., Ghahramani, Z., Jaakkola, T., Saul, L.K.: An introduction to variational methods for graphical models. In: Jordan, M.I. (ed.) Learning in Graphical Models. MIT Press, Cambridge (1999)
23. Zhang, A., Zhu, J., Zhang, B.: Sparse online topic models. In: WWW (2013)

Author Index

Abbasian, Houman   III-33
Abbasnejad, M. Ehsan   II-515
Abraham, Zubin   II-320
Acharya, Ayan   II-194, II-369
Adam, Antoine   III-681
Adiga, Abhijin   I-541
Aertsen, Abram   III-681
Agarwal, Pankaj   III-579
Agosta, John Mark   I-162
Agrawal, Priyanka   II-564
Akbarinia, Reza   I-493
Almeida, Ezilda   I-480
Alpaydın, Ethem   II-675
Ammar, Haitham Bou   II-449
Anagnostopoulos, Georgios C.   III-224
Antoniuk, Kostiantyn   III-96
Archambeau, Cédric   II-80
Assent, Ira   III-304
Auger, David   I-194
Azar, Mohammad Gheshlaghi   I-97
Azimi, Javad   I-210

Babagholami-Mohamadabadi,
    Behnam   III-192
Bachman, Philip   I-241
Baghshah, Mahdieh Soleymani   III-192
Bailey, James   II-483
Balasubramanyan, Ramnath   II-628
Barbieri, Nicola   II-48
Bar-Noy, Amotz   I-525
Barros, Rodrigo C.   II-385
Basu, Prithwish   I-525
Batista, Gustavo E.A.P.A.   III-160
Baumer, Ben   I-525
Becker, Martin   III-288
Beling, Peter A.   I-33
Bento, José   I-257
Berlingerio, Michele   III-663
Berthold, Michael R.   III-645
Bhaduri, Kanishka   III-321
Bhattacharya, Indrajit   II-465, II-564
Bian, Yatao   III-81
Bifet, Albert   I-465
Biggio, Battista   III-387

Bischl, Bernd   III-645
Bischoff, Bastian   I-49
Blanco, Lorenzo   I-417
Blaschko, Matthew B.   II-304
Blasiak, Sam   I-401
Blockeel, Hendrik   III-681
B.N., Ranganath   II-465
Boella, Guido   II-64
Bogdanov, Petko   I-525
Boley, Mario   III-370
Bonchi, Francesco   II-48
Bonilla, Edwin V.   II-515
Bouchard, Guillaume   II-80
Boulicaut, Jean-François   I-654
Bourguignon, Pierre-Yves   I-321
Bousmalis, Konstantinos   II-531
Boyd, Kendrick   III-451
Brunskill, Emma   I-97
Buffoni, David   I-225
Burnside, Elizabeth S.   III-595

Calabrese, Francesco   III-663
Callan, Jamie   III-128
Canini, Kevin   III-17
Canu, Stéphane   II-145
Cao, Longbing   III-563
Cao, Mingqi   III-81
Carreira-Perpiñán, Miguel Á.   III-256
Cerri, Ricardo   II-385
Carvalho, André C.P.L.F. de   II-385
Carvalho Chanel, Caroline P.   I-145
Chan, Jeffrey   II-483
Chang, Kai-Wei   II-401, III-176
Chawla, Nitesh V.   II-16
Chawla, Sanjay   III-337
Chen, Da   II-161
Chen, Fang   II-483
Chen, Zheng   II-336
Cohen, William W.   II-628, III-128
Collautti, Marco   III-435
Contal, Emile   I-225
Corona, Igino   III-387
Couëtoux, Adrien   I-194
Cule, Boris   I-353

Dalvi, Bhavana   II-628, III-128
Dang,· Xuan Hong   III-304
Dao, Thi-Bich-Hanh   III-419
Darling, William   II-80
Das, Kamalika   III-321
De Bie, Tijl   II-256, III-612
Delporte, Julien   II-145
Desmier, Elise   I-654
Destercke, Sebastien   II-112
Dexter, Philip   I-113
Di Caro, Luigi   II-64
Di Lorenzo, Giusy   III-663
Ding, Chris   II-177
Dojchinovski, Milan   III-654
Domeniconi, Carlotta   I-574
Dong, Yuxiao   II-16
Driessens, Kurt   II-449
Dror, Gideon   III-499
Drummond, Chris   III-33
Duong, Khanh-Chuong   III-419

Eggeling, Ralf   I-321
Eng, Kevin H.   III-451

Fan, Wei   III-483
Fang, Meng   I-273
Feng, Ziming   III-483
Fern, Alan   I-178
Fern, Xiaoli   I-210
Ferreira, Carlos   I-480
Fischer, Simon   III-645
Franc, Vojtěch   III-96
Freitas, Alex A.   II-385
Frongillo, Rafael   III-17
Fukuchi, Kazuto   II-499

Galland, Stéphane   III-641
Gallinari, Patrick   II-161
Gama, João   I-480
Gao, Bo   III-645
Gao, Jing   I-557, III-483
Gao, Jun   II-1
Gao, Sheng   II-161
Gärtner, Thomas   III-672
Gaud, Nicolas   III-641
Gaur, Neeraj   II-194
Geiger, Bernhard C.   II-612
Geist, Matthieu   I-1, I-17
Geng, Guanggang   II-660
Georgiopoulos, Michael   III-224

Ghahramani, Zoubin   II-531
Ghosh, Joydeep   II-194
Giacinto, Giorgio   III-387
Gionis, Aristides   II-32
Gkoulalas-Divanis, Aris   III-353
Goethals, Bart   I-353
Gohr, André   I-321
Goldszmidt, Moises   II-95
Govers, Sander   III-681
Greene, Derek   III-677
Gretton, Arthur   II-304
Grosse, Ivo   I-321
Grosskreutz, Henrik   I-369
Gunasekar, Suriya   II-194
Guo, Hongyu   I-433
Guo, Jun   II-161
Guo, Yuhong   II-417
Gupta, Manish   I-557
Gutmann, Michael U.   II-596

Habrard, Amaury   II-433
Han, Jiawei   I-557
Hayes, Conor   III-677
He, Qing   II-353, III-208
Henelius, Andreas   I-337
Herrera-Joancomartí, Jordi   I-590
Hilaire, Vincent   III-641
Hlaváč, Václav   III-96
Hollmén, Jaakko   I-449
Holmes, Geoff   I-465
Horváth, Tamás   I-622
Hruschka, Eduardo R.   II-369
Huang, Kaizhu   II-660
Huang, Sheng   III-467
Huang, Yinjie   III-224
Hulpuş, Ioana   III-677

Ioannidis, Stratis   I-257

Jabbour, Said   III-403
Jaeger, Manfred   III-112
Jalali, Ali   I-210
Japkowicz, Nathalie   III-33
Ji, Tengfei   II-1
Jin, Xin   II-353
Joachims, Thorsten   II-128
Joshi, Saket   I-178
Jozwowicz, Marek   III-677

Kamishima, Toshihiro   II-499
Kamp, Michael   III-370

Karatzoglou, Alexandros   II-145
Karnstedt, Marcel   III-677
Keerthi, S. Sathiya   III-176
Khan, Omar Zia   I-162
Khardon, Roni   I-178
Klein, Edouard   I-1
Kliegr, Tomáš   III-654
Knoll, Alois   I-49
Kober, Jens   III-627
Koller, Torsten   I-49
Kong, Deguang   II-177
Kong, Shu   III-240
Kontonasios, Kleanthis-Nikolaos   II-256
Kopp, Christine   III-370
Korpela, Jussi   I-337
Krömer, Oliver   III-627
Kunapuli, Gautam   III-1
Kuusisto, Finn   III-595

Lang, Bastian   I-369
Laskey, Kathryn B.   I-401
Laskov, Pavel   III-387
Lauri, Fabrice   III-641
Lavrač, Nada   III-650
Lazaric, Alessandro   I-97
Leckie, Christopher   II-483
Lee, Tai Sing   III-49
van Leeuwen, Matthijs   III-272
Lemmerich, Florian   III-288
Li, Chengtao   II-336
Li, Cong   III-224
Li, Shantao   II-161
Li, Xiong   III-49, III-81
Liao, Qing   II-242
Lijffijt, Jefrey   I-385
Lin, Zhouchen   II-226
Liszewska, Malgorzata   II-320
Liu, Cheng-Lin   II-660
Liu, Dehua   II-210
Liu, Song   II-596
Liu, Wei   II-483
Liu, Yuncai   III-49, III-81
Loscalzo, Steven   I-113
Lou, Tiancheng   II-16
Loukides, Grigorios   III-353
Lu, Haiping   II-288
Lu, Yiqi   III-467
Luo, Hao   II-161
Luo, Ping   III-208
Luttinen, Jaakko   I-305

Maarek, Yoelle   III-499
Maiorca, Davide   III-387
Maldjian, Joseph A.   III-1
Malitsky, Yuri   III-435
Markert, Heiner   I-49
Masseglia, Florent   I-493
Matuszczyk, Tomasz   II-145
Matwin, Stan   III-33
May, Michael   III-370
Mehmood, Yasir   II-48
Mehta, Deepak   III-435
Metzen, Jan Hendrik   I-81
Micenková, Barbora   III-304
Miettinen, Pauli   I-509
Mirkin, Shachar   II-80
Mirylenka, Daniil   III-667
Mitra, Adway   II-465
Mladenić, Dunja   II-643, III-637
Mocanu, Decebal Constantin   II-449
Mock, Michael   III-370
Mooney, Raymond J.   II-369
Morency, Louis–Philippe   II-531
Morik, Katharina   III-321
Mukherjee, Indraneel   III-17
Mülling, Katharina   III-627
Muthukrishnan, S.   I-257

Nair, Rahul   III-663
Najork, Marc   II-95
Nassif, Houssam   III-595
Natarajan, Sriraam   II-580, III-1
Nelson, Blaine   III-387
Neumann, Gerhard   III-627
Ng, Raymond T.   III-304
Nguyen-Tuong, Duy   I-49
Nickel, Maximilian   II-272, III-617
Ntoutsi, Eirini   III-622
Ntrigkogias, Christos   III-659

Obradovic, Zoran   III-579
O'Sullivan, Barry   III-435
Otaki, Keisuke   I-622

Page, C. David   III-451, III-547, III-595
Pajarinen, Joni   I-129
Panagiotakopoulos, Constantinos   III-65
Pantic, Maja   II-531
Papadopoulos, Apostolos N.   I-638
Paparizos, Stelios   II-95
Papotti, Paolo   I-417

Passerini, Andrea   III-667
Paurat, Daniel   III-672
Peharz, Robert   II-612
Pelekis, Nikos   III-659
Peltonen, Jaakko   I-129
Perdinan   II-320
Pérez-Solà, Cristina   I-590
Pernkopf, Franz   II-612
Peters, Jan   III-627
Peyrache, Jean-Philippe   II-433
Pfahringer, Bernhard   I-465
Pietquin, Olivier   I-1, I-17
Pinelli, Fabio   III-663
Pinto, Fábio   III-531
Piot, Bilal   I-1, I-17
Plantevit, Marc   I-654
Poulis, Giorgos   III-353
Poupart, Pascal   I-162
Precup, Doina   I-241
Puolamäki, Kai   I-337
Puppe, Frank   III-288

Qian, Hui   II-210
Qiao, Qifeng   I-33
Qiu, Disheng   I-417
Quattoni, Ariadna   I-289
Quinn, John A.   II-596

Raghavan, Aswin   I-178
Ramamohanarao, Kotagiri   II-483
Raman, Karthik   II-128
Ramon, Jan   I-509, I-622
Rangwala, Huzefa   I-401, I-574
Rawal, Aditya   II-369
Ray, Nilanjan   III-1
Read, Jesse   I-465
Recasens, Adrià   I-289
Rezende, Solange O.   III-160
Rijn, Jan N. van   III-645
Rish, Irina   III-632
Robardet, Céline   I-654
Robicquet, Alexandre   I-225
Roli, Fabio   III-387
Roth, Dan   II-401
Rudin, Cynthia   III-515

Saha, Baidya Nath   III-1
Sais, Lakhdar   III-403
Sakuma, Jun   II-499
Salhi, Yakoub   III-403

Sangeux, Morgan   III-563
Sanner, Scott   II-515
Santos Costa, Vítor   III-595
Sbodio, Marco Luca   III-663
Schult, Rene   III-622
Schuurmans, Dale   II-417
Sebban, Marc   II-433
Semerci, Murat   II-675
Sevieri, Rich   III-515
Shavlik, Jude   III-595
Shevade, Shirish   III-144
Shi, Zhongzhi   II-353, III-208
Sideridis, Stylianos   III-659
Singer, Yoram   III-17
Singh, Ambuj K.   I-525
Skiadopoulos, Spiros   III-353
Sluban, Borut   III-650
Šmídl, Václav   II-548
Soares, Carlos   III-531
Song, Yin   III-563
Sousa, Celso André R. de   III-160
Spiliopoulou, Myra   III-622
Spyropoulou, Eirini   III-612
Srijith, P.K.   III-144
Srikumar, Vivek   II-401
Šrndić, Nedim   III-387
Stolpe, Marco   III-321
Stone, Peter   I-65
Sugiyama, Masashi   II-596
Sundararajan, S.   III-144, III-176
Surian, Didi   III-337
Szpektor, Idan   III-499

Tadepalli, Prasad   I-178
Tampakis, Panagiotis   III-659
Tan, Pang-Ning   II-320
Tang, Jie   II-16
Tatti, Nikolaj   II-32
Taylor, Matthew E.   II-449
Teichteil-Königsbuch, Florent   I-145
Tekumalla, Lavanya Sita   II-564
Teytaud, Olivier   I-194
Theodoridis, Yannis   III-622, III-659
Tichý, Ondřej   II-548
Tomašev, Nenad   II-643, III-637
Torgo, Luis   III-645
Trabold, Daniel   I-369
Tresp, Volker   II-272, III-617
Tsampouka, Petroula   III-65
Tuyls, Karl   II-449

Ukkonen, Antti   II-48, III-272
Umaashankar, Venkatesh   III-645
Urieli, Daniel   I-65

Valari, Elena   I-638
Vanschoren, Joaquin   III-645
Vayatis, Nicolas   I-225
Vladymyrov, Max   III-256
Vrain, Christel   III-419
Vreeken, Jilles   I-509, II-256
Vries, Gerben K.D. de   I-606
Vullikanti, Anil Kumar S.   I-541

Wagner, Daniel   III-515
Wang, Bin   III-49
Wang, Donghui   III-240
Wang, Shuhui   II-353
Wang, Tong   III-515
Wang, Wei   III-467
Weiss, Gerhard   II-449
Weiss, Jeremy C.   III-547
Wingender, Edgar   I-321
Winkler, Julie   II-320
Winter, Patrick   III-645
Wiswedel, Bernd   III-645
Wright, Robert   I-113
Wu, Bin   II-16
Wu, Xian   III-483

Xiao, Yanghua   III-467
Xu, Congfu   II-210
Xu, Xiaomin   III-467

Yan, Jinyun   I-257
Yang, Dongqing   II-1

Yang, Shuo   II-580
Yin, Jie   I-273
Yu, Guoxian   I-574
Yu, Lei   I-113
Yu, Wenchao   III-208
Yu, Yong   III-483

Zafeiriou, Stefanos   II-531
Zaremba, Wojciech   II-304
Zarghami, Ali   III-192
Zeng, Guangxiang   III-208
Zhang, Aonan   I-670
Zhang, Bo   I-670
Zhang, Chao   II-226
Zhang, Guoji   I-574
Zhang, Hongyang   II-226
Zhang, Jian   III-563
Zhang, Jianwen   II-336
Zhang, Miao   II-177
Zhang, Ping   III-579
Zhang, Qian   II-242
Zhang, Ruofei   I-210
Zhang, Yan-Ming   II-660
Zhang, Zhihua   II-210
Zhong, Shiyuan   II-320
Zhou, Cheng   I-353
Zhou, Tengfei   II-210
Zhu, Jun   I-670
Zhu, Xingquan   I-273
Zhuang, Fuzhen   II-353, III-208
Žliobaitė, Indrė   I-449, I-465
Zolfaghari, Mohammadreza   III-192